Geophysical Monograph Series

Including

IUGG Volumes
Maurice Ewing Volumes
Mineral Physics Volumes

Geophysical Monograph Series

106 **Faulting and Magmatism at Mid-Ocean Ridges** *W. Roger Buck, Paul T. Delaney, Jeffrey A. Karson, and Yves Lagabrielle (Eds.)*

107 **Rivers Over Rock: Fluvial Processes in Bedrock Channels** *Keith J. Tinkler and Ellen E. Wohl (Eds.)*

108 **Assessment of Non-Point Source Pollution in the Vadose Zone** *Dennis L. Corwin, Keith Loague, and Timothy R. Ellsworth (Eds.)*

109 **Sun-Earth Plasma Interactions** *J. L. Burch, R. L. Carovillano, and S. K. Antiochos (Eds.)*

110 **The Controlled Flood in Grand Canyon** *Robert H. Webb, John C. Schmidt, G. Richard Marzolf, and Richard A. Valdez (Eds.)*

111 **Magnetic Helicity in Space and Laboratory Plasmas** *Michael R. Brown, Richard C. Canfield, and Alexei A. Pevtsov (Eds.)*

112 **Mechanisms of Global Climate Change at Millennial Time Scales** *Peter U. Clark, Robert S. Webb, and Lloyd D. Keigwin (Eds.)*

113 **Faults and Subsurface Fluid Flow in the Shallow Crust** *William C. Haneberg, Peter S. Mozley, J. Casey Moore, and Laurel B. Goodwin (Eds.)*

114 **Inverse Methods in Global Biogeochemical Cycles** *Prasad Kasibhatla, Martin Heimann, Peter Rayner, Natalie Mahowald, Ronald G. Prinn, and Dana E. Hartley (Eds.)*

115 **Atlantic Rifts and Continental Margins** *Webster Mohriak and Manik Talwani (Eds.)*

116 **Remote Sensing of Active Volcanism** *Peter J. Mouginis-Mark, Joy A. Crisp, and Jonathan H. Fink (Eds.)*

117 **Earth's Deep Interior: Mineral Physics and Tomography From the Atomic to the Global Scale** *Shun-ichiro Karato, Alessandro Forte, Robert Liebermann, Guy Masters, and Lars Stixrude (Eds.)*

118 **Magnetospheric Current Systems** *Shin-ichi Ohtani, Ryoichi Fujii, Michael Hesse, and Robert L. Lysak (Eds.)*

119 **Radio Astronomy at Long Wavelengths** *Robert G. Stone, Kurt W. Weiler, Melvyn L. Goldstein, and Jean-Louis Bougeret (Eds.)*

120 **GeoComplexity and the Physics of Earthquakes** *John B. Rundle, Donald L. Turcotte, and William Klein (Eds.)*

121 **The History and Dynamics of Global Plate Motions** *Mark A. Richards, Richard G. Gordon, and Rob D. van der Hilst (Eds.)*

122 **Dynamics of Fluids in Fractured Rock** *Boris Faybishenko, Paul A. Witherspoon, and Sally M. Benson (Eds.)*

123 **Atmospheric Science Across the Stratopause** *David E. Siskind, Stephen D. Eckerman, and Michael E. Summers (Eds.)*

124 **Natural Gas Hydrates: Occurrence, Distribution, and Detection** *Charles K. Paull and Willam P. Dillon (Eds.)*

125 **Space Weather** *Paul Song, Howard J. Singer, and George L. Siscoe (Eds.)*

126 **The Oceans and Rapid Climate Change: Past, Present, and Future** *Dan Seidov, Bernd J. Haupt, and Mark Maslin (Eds.)*

127 **Gas Transfer at Water Surfaces** *M. A. Donelan, W. M. Drennan, E. S. Saltzman, and R. Wanninkhof (Eds.)*

128 **Hawaiian Volcanoes: Deep Underwater Perspectives** *Eiichi Takahashi, Peter W. Lipman, Michael O. Garcia, Jiro Naka, and Shigeo Aramaki (Eds.)*

129 **Environmental Mechanics: Water, Mass and Energy Transfer in the Biosphere** *Peter A.C. Raats, David Smiles, and Arthur W. Warrick (Eds.)*

130 **Atmospheres in the Solar System: Comparative Aeronomy** *Michael Mendillo, Andrew Nagy, and J. H. Waite (Eds.)*

131 **The Ostracoda: Applications in Quaternary Research** *Jonathan A. Holmes and Allan R. Chivas (Eds.)*

132 **Mountain Building in the Uralides Pangea to the Present** *Dennis Brown, Christopher Juhlin, and Victor Puchkov (Eds.)*

133 **Earth's Low-Latitude Boundary Layer** *Patrick T. Newell and Terry Onsage (Eds.)*

134 **The North Atlantic Oscillation: Climatic Significance and Environmental Impact** *James W. Hurrell, Yochanan Kushnir, Geir Ottersen, and Martin Visbeck (Eds.)*

135 **Prediction in Geomorphology** *Peter R. Wilcock and Richard M. Iverson (Eds.)*

136 **The Central Atlantic Magmatic Province: Insights from Fragments of Pangea** *W. Hames, J. G. McHone, P. Renne, and C. Ruppel (Eds.)*

137 **Earth's Climate and Orbital Eccentricity: The Marine Isotope Stage 11 Question** *André W. Droxler, Richard Z. Poore, and Lloyd H. Burckle (Eds.)*

138 **Inside the Subduction Factory** *John Eiler (Ed.)*

139 **Volcanism and the Earth's Atmosphere** *Alan Robock and Clive Oppenheimer (Eds.)*

Geophysical Monograph 140

Explosive Subaqueous Volcanism

James D. L. White
John L. Smellie
David A. Clague
Editors

American Geophysical Union
Washington, DC

Published under the aegis of the AGU Books Board

Jean-Louis Bougeret, Chair; Gray E. Bebout, Carl T. Friedrichs, James L. Horwitz, Lisa A. Levin, W. Berry Lyons, Kenneth R. Minschwaner, Darrell Strobel, and William R. Young, members.

Library of Congress Cataloging-in-Publication Data

Explosive subaqueous volcanism / James D. L. White, John L. Smellie, and David A. Clague, editors.
       p. cm.-- (Geophysical monograph ; 140)
   Includes bibliographical references.
   ISBN 0-87590-999-X
       1. Sea-floor spreading. 2. Explosive volcanic eruptions. 3. Hydrothermal vents. 4. Marine sediments. I. White, James D. L. II. Smellie, J. L. III. Clague, D. A. IV. Series.

QE511.7.E97 2003
551.1'36--dc22
                                          2003058351

ISSN 0065-8448
ISBN 0-87590-999-X

Copyright 2003 by the American Geophysical Union
2000 Florida Avenue, N.W.
Washington, DC 20009

Figures, tables, and short excerpts may be reprinted in scientific books and journals if the source is properly cited.

Authorization to photocopy items for internal or personal use, or the internal or personal use of specific clients, is granted by the American Geophysical Union for libraries and other users registered with the Copyright Clearance Center (CCC) Transactional Reporting Service, provided that the base fee of $1.50 per copy plus $0.35 per page is paid directly to CCC, 222 Rosewood Dr., Danvers, MA 01923. 0065-8448/03/$01.50+0.35.

This consent does not extend to other kinds of copying, such as copying for creating new collective works or for resale. The reproduction of multiple copies and the use of full articles or the use of extracts, including figures and tables, for commercial purposes requires permission from the American Geophysical Union.

Printed in the United States of America.

CONTENTS

Preface
James White ix

Introduction
James D. L. White, John L. Smellie, and David A. Clague 1

Section I: Subaqueous Eruption Dynamics

Water/Magma Interaction: Physical Considerations for the Deep Submarine Environment
Kenneth H. Wohletz 25

Phreatomagmatic Explosions in Subaqueous Volcanism
Bernd Zimanowski and Ralf Büttner 51

Melting of Ice by Magma-Ice-Water Interactions During Subglacial Eruptions as an Indicator of Heat Transfer in Subaqueous Eruptions
Magnús T. Gudmundsson 61

Pyroclastic and Hydroclastic Deposits on Loihi Seamount, Hawaii
David A. Clague, R. Batiza, James W. Head III, and Alicé S. Davis 73

Large-Scale Interaction of Lake Water and Rhyolitic Magma During the 1.8 ka Taupo Eruption, New Zealand
B. F. Houghton, B. J. Hobden, K. V. Cashman, C. J. N. Wilson, and R. T. Smith 97

Section II: Explosive Eruptions in the Modern Deep Sea

Submarine Strombolian Eruptions on the Gorda Mid-Ocean Ridge
David A. Clague, Alicé S. Davis, and Jacqueline E. Dixon 111

Hyaloclastite from Miocene Seamounts Offshore Central California: Compositions, Eruption Styles, and Depositional Processes
Alicé S. Davis and David A. Clague 129

Recent MORB Volcaniclastic Explosive Deposits Formed Between 500 and 1750 m.b.s.l. on the Axis of the Mid-Atlantic Ridge, South of the Azores
Jean-Philippe Eissen, Yves Fouquet, Delphine Hardy, Hélène Ondréas 143

Section III: Explosive Shallow-Marine (Surteyan) Eruptions and Their Kin

A Cluster Of Surtseyan Volcanoes at Lookout Bluff, North Otago, New Zealand: Aspects of Edifice Spacing and Time
Doris Maicher 167

CONTENTS

Eruptive and Depositional Mechanisms of an Eocene Shallow Submarine Volcano, Moeraki Peninsula, New Zealand
Benjamin Andrews 179

A Subaqueous Eruption Model For Shallow-Water, Small Volume Eruptions: Evidence From Two Precambrian Examples
Wulf U. Mueller 189

Basaltic Lava Balloons Produced During the 1998-2001 Serreta Submarine Ridge Eruption (Azores)
João L. Gaspar, Gabriela Queiroz, José M. Pacheco, Teresa Ferreira, Nicolau Wallenstein, Maria H. Almeida, and Rui Coutinho 205

Section III: Pumiceous Subsea Silicic Eruptions From the Modern Seafloor

Subaqueous Pumice Eruptions and Their Products: A Review
K. Kano 213

Submarine Silicic Calderas on the Northern Shichito-Iwojima Ridge, Izu-Ogasawara (Bonin) Arc, Western Pacific
Makoto Yuasa and Kazuhiko Kano 231

Section IV: Subaqueous Pumiceous Deposits and Their Interpretation

Submarine, Silicic, Syn-Eruptive Pyroclastic Units in the Mount Read Volcanics, Western Tasmania: Influence of Vent Setting and Proximity on Lithofacies Characteristics
Jocelyn McPhie and Rodney L. Allen 245

Vesiculation and Eruption Processes of Submarine Effusive and Explosive Rocks from the Middle Miocene Ogi Basalt, Sado Island, Japan
Norie Fujibayashi and Umio Sakai 259

The Submarine Record of a Large-Scale Explosive Eruption in the Vanuatu Arc: ~1 Ma Efaté Pumice Formation
Alison M. Raos and Jocelyn McPhie 273

Products of Explosive Subaqueous Felsic Eruptions Based on Examples From the Hellenic Island Arc, Greece
S.R. Allen and A.L. Stewart 285

Miocene Submarine Fire Fountain Deposits, Ryugazaki Headland, Oshoro Peninsula, Hokkaido, Japan: Implications for Submarine Fountain Dynamics and Fragmentation Processes
R. A. F. Cas, H. Yamagishi, L. Moore and C. Scutter 299

An Archean Submarine Pyroclastic Flow Due to Submarine Dome Collapse: The Hurd Deposit, Harker Township, Ontario, Canada
C. R. Scott, D. Richard, and A. D. Fowler 317

CONTENTS

Section V: Economic Significance of Explosive Submarine Eruptions

Deep Marine Pumice from the Woodlark and Manus Basins, Papua New Guinea
Raymond A. Binns 329

Morphology, Distribution, and Estimated Eruption Volumes for Intracaldera Tuffs Associated With Volcanic-Hosted Massive Sulfide Deposits in the Archean Sturgeon Lake Caldera Complex, Northwestern Ontario
George J. Hudak, Ronald L. Morton, James M. Franklin, and Dean. M. Peterson 345

Analysis of VHMS-Hosting Ignimbrites Erupted at Bathyal Water Depths (Ordovician Bald Mountain Sequence, Northern Maine)
Lowell G. Kessel and Cathy J. Busby 361

PREFACE

Subaqueous explosive eruptions are common, and in earth's early history were ubiquitous. Although they are unlike eruptions we find on land, they operate with the same fundamental processes. Deep-sea eruptions modify important seafloor hydrothermal systems and their coupled habitats for extremophile organisms, and large eruptions on the continental shelf presumably have as yet unknown effects on a wide range of marine organisms. Shallow eruptions that can affect shipping lanes and threaten coastal environments, either directly or by generation of tsunami, also appear to produce deposits and conditions closely linked with formation of significant chunks of the world's mineral resources.

For volcanologists, marine geologists, and marine biologists, such eruptions have prompted intense interest. Economic geologists, mining engineers, biochemists and geneticists working with extremophiles have found the products of such eruptions vital sources of study. The broader earth-science community recognizes seafloor volcanism as a part of the earth-control system feeding heat and elemental fertilizer to the ocean, with large seafloor eruptions perhaps forcing oceanographic and climatic changes. For civil engineers of coastal settings and civil-defense planners, knowing what sort of eruption potential lurks offshore certainly merits concern. Results from recent submersible dive programs have provided new and compelling evidence for explosive caldera-forming eruptions, and for explosive basaltic eruptions in the unexpected setting of the Mid-Atlantic Ridge. These results are focusing increased attention on submarine volcanoes in all settings. Other recent results have demonstrated strong links between ore deposits, seafloor calderas, and fragmental deposits (rather than lavas).

Through research cruises, submersible dives, geochemical analysis, physical modeling, and laboratory experiments, the editors of this volume have investigated marine eruptions through their deposits for more than half a century in aggregate, and the volume's contributors amplify this depth of experience manyfold. The work of the contributors has spanned the globe, while that of the editors has been particularly informed by studies of active eruptions in Hawaii, Iceland, and New Zealand, with further fieldwork on subaqueous deposits, both young and old, formed in environments ranging from the deep-sea floor to sublacustrine and subglacial settings.

Most importantly, the contributions in this book offer a counterpoint to what has long been a commonplace inference: that there is no significant explosivity during eruptions taking place at substantial water depths. Studies of rocks from the ancient and modern seafloor by marine geologists and geophysicists amply demonstrate that lava flows are the major product of subsea eruptions. Under some circumstances, however, explosive eruptions take place in water depths exceeding a kilometer, and some of these eruptions produce substantial deposits of volcanic ash. The most important of these settings, particularly in the past but also today for places such as Japan and New Zealand, are continental shelves in which evolved magmas such as rhyolite erupt. Other settings include the deep sea where, in back-arc basins and on seamounts, gas-rich magmas are occasionally produced which are able to expand strongly even under high confining pressures.

Processes forming ash deposits from subaqueous eruptions have received little mainstream volcanological attention, and the presence of such deposits is scarcely addressed in volcanological textbooks. Nevertheless, given the preservation bias in favor of sub-wavebase marine deposits in the geological record, it is likely that deposits of subaqueous explosive eruptions exceed in volume and economic significance those of subaerial ones. The role of explosivity in subaqueous eruptions, particularly in the sea and at large scales, is a topic of both high interest and acknowledged disagreement.

All subaqueous eruptions have as necessary participants both magma and water, but the controls on explosivity driven by magmatic gases are quite different from those on explosive magma-water interactions.

For explosive submarine eruptions driven by magmatic gases, there is a comparatively well-understood process not involving external water upon which to base our inferences. Simple pressure-volume-temperature and gas-volume calculations suggest that explosive eruptions of rhyolitic magmas with typical volatile contents are likely in water depths extending to almost a kilometer, and hence including most continental shelf areas and the submerged parts of island arcs. Detailed information is only now being systematically collected for volcanic deposits of submerged shelves, despite their known association with mineralization and the relative abundance of deposits from such settings known from the ancient record.

The ability of basaltic magmas to produce explosive "Hawaiian" style lava-fountain eruptions at seafloor depths has been under consideration for over two decades, but despite repeated study of the beds of mm-scale glass fragments, including tiny bubble-wall fragments, that are inferred to be their products, uncertainty remains. Basic eruption modeling published last year suggests that conditions for such eruptions would be very difficult to meet with plausible magma compositions, though it is now suggested that gas accumulation prior to eruption might produce lava-fountain eruptions involving only small volumes of magma. Alternatively, a mechanism involving modest expansion of entrapped seawater in fast-moving, highly fluid, magma has been shown capable of producing such deposits at depths shallower than the critical depth for water, and an untested hypothesis is that supercritical fluids at greater depths may also do so.

Explosive magma-water interactions are generally considered to result from fuel coolant interaction (FCI), a type of explosive interaction first identified as having potential to produce megaton-scale explosions while plans were being made to dynamite the front of lava flows entering Heimay harbor in 1973. Although the dynamiting never took place (depriving volcanology of a potentially exciting field test) the lava flow stopped of its own accord short of blocking the harbor. Thirty years of analysis and experiments later, some workers infer that such interactions are unlikely to occur at water depths exceeding 100 m. This strict pressure control is challenged by others, however, who have broadly explored the governing thermodynamics of magma-water explosions and re-examined experimental results to conclude that explosions are possible even at pressures exceeding those of typical abyssal plains.

A final question, not yet amenable to rigorous physical analysis, is how explosivity resulting from magma degassing interacts with that resulting from magma-water interaction. For any substantial eruption, magma degassing is needed to propel the magma from the vent. It is commonly considered that any fragmentation resulting from this volatile expansion would be broadly additive to any subsequent fragmentation from magma-water interaction.

Laboratory results, however, suggest that the presence of vapor-filled bubbles within a melt inhibits, rather than enhances, explosivity of magma-water interactions. Similarly, the highly variable gas-bubble content typical of fragments from eruptions involving magma-water interaction argues against such simple addition, because the varying bubbliness reflects suppression of gas-bubble growth by chilling. The variety of gas in the bubbles may matter as well; gases such as carbon dioxide, which is important for eruptions under high confining pressures, have been argued to be particularly strong suppressants of explosive interactions. On the other hand, there must be an important role for gas release from magmas in the "driving" of seafloor eruptions, and it is increasingly recognized that the rate and style at which an erupting magma encounters water is a critical control on magma-water explosions. Rather than adding two fragmentation events, the focus now is on understanding how acceleration and mixing affect explosivity and particle transport from the eruptions.

Although not a focus of contributions to this volume, a better understanding of the abundance, size and style of subaqueous, particularly submarine, eruptions has practical implications, for minerals exploration and for our understanding of hazardous natural processes. Many important mineral deposits are hosted in the fragmental deposits of subaqueous volcanoes, and similar deposits are ongoing exploration targets. Offshore of Japan, ships are excluded from waters above Myojin-sho volcano, an eruption of which in 1952 sunk a research vessel with the loss of all onboard. A similar exclusion zone attends Kick-em Jenny volcano offshore in the West Indies.

Acknowledgments. The editors express appreciation to the publications staff at AGU for flexible and helpful handling of this volume, which derives from an AGU Chapman Conference of the same name convened by White and B.F. Houghton and held in Dunedin, New Zealand in January 2002.

Dedication. This volume is dedicated to the memory of Richard Virgil Fisher, 1928 - 2002. "RV" was an inspiration to many in the investigation of what happens when magma meets water, and made contributions addressing pressure controls on underwater eruptions, how erupted particles travel and are deposited, how magma and water interact, and even the eruption aftermath of aqueous reentrainment and resedimentation of tephra. He was also endlessly inquisitive and open to new ideas and information, qualities ideally suited for work on the information-challenged study of eruptions that take place out of sight and out of reach.

James White

Introduction: A Deductive Outline and Topical Overview of Subaqueous Explosive Volcanism

James D.L. White

Geology Department, University of Otago, Dunedin, New Zealand

John L. Smellie

British Antarctic Survey, Cambridge, United Kingdom

David A. Clague

Monterey Bay Aquarium Research Institute, Moss Landing, California

Subaqueous eruptions are the most abundant on earth, and the proportion of such eruptions that are explosive is larger than generally appreciated though still poorly constrained. Subaqueous eruptions are fundamentally affected by water's ability to vaporize upon contact with magma, its high density (in comparison with air) and the accompanying increase in confining pressure with depth of eruption, its greater viscosity than air, and its high heat capacity and thermal conductivity. The effects are both on source dynamics of the eruptions (exit conditions, fragmentation) and the transport and deposition of eruptive products. Interpretation of ancient subaqueous deposits is important both practically, because they host significant mineral deposits, and more broadly in order to understand how volcanoes work on the ¾ of our planet beneath water. Successful interpretation requires an understanding of the full range of water's effects on eruptions in order to sail backwards from deposit characteristics through deposition and transport processes and back into the vents. Investigation of modern seafloor volcanoes demonstrates a range of volcano and eruption styles, and is also providing insight into mineralization sites and processes within still-active magmatic systems.

INTRODUCTION

The role of explosive subaqueous eruptions has long been considered unimportant, although the significance of subaqueous volcanism globally is unequivocal [e.g. *Crisp*, 1984]. Work presented in this volume (Figure 1) suggests, by its range of topics, the contents of individual studies, and the variety of new insights, that this consensus view requires reassessment. Why? It overlooks the economic significance, the sheer abundance (Figure 2), the concomitant significance in terms of the global geochemicothermal budget (Table 1) and its tempo of change, and the insight into fundamental eruptive and depositional processes available from study of these eruptions and their products. If submarine eruptions yield on average even a sixteenth the explosivity of subaerial ones (Table 1), it implies that about

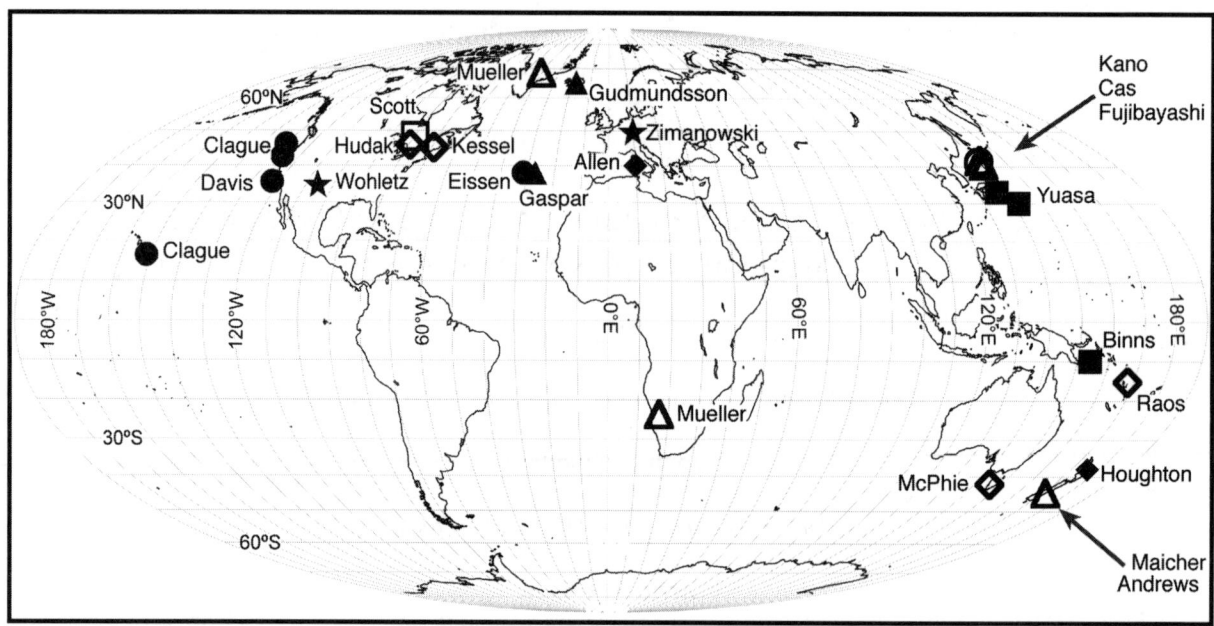

Figure 1. Locations and types of eruptions or laboratory investigations for papers in this volume, labelled by first author surnames for contributions. Stars = sites of experimental work; Filled circles = deep-marine modern; Open circles = deep-marine ancient; Filled triangles = shallow-water basaltic modern; Open triangles = shallow-water basaltic ancient; Filled diamonds = subaqueous silicic calderas, modern; Open diamonds = subaqueous silicic calderas, ancient; Filled squares = subaqueous silicic domes, modern; Open squares = subaqueous silicic domes, ancient. (Note that most domes associated with calderas - see text.)

a quarter of earth's explosive activity occurs under water. If we take a more traditional view and estimate that only 2% of subaqueous volcanism is explosive, it still yields an explosive eruptive output about a tenth that of subaerial eruptions. The uncertainty is whether explosive subaqueous volcanism is merely significant, or whether it is of subequal importance to the subaerial style we know so well.

Explosive subaqueous eruptions are unlikely to yield the richness of eruption observation that is provided by subaerial eruptions (Table 2). Not only are there logistical difficulties, but the limited optical transparency of water means that there can be no "distant" visual observation of subaqueous eruptions and their plumes; the need to have ships on-site for observation imposes an additional transport impediment, because ships are quite significantly slower than jet aircraft. Development of a better understanding of these eruptions, then, must rely on interpretation of their deposits, informed by physical models and relevant experimental data.

Understanding subaqueous eruptions from their deposits is challenging. Some aspects of eruption processes can be deciphered from particle properties combined with geochemical analysis of fragments, phenocrysts, and inclusions. Other aspects, addressed using deposit data for subaerial eruptions, can be addressed for few, if any, subaqueous eruptions. The data lacking for subaqueous eruptions can be roughly subdivided into (a) whole-deposit information from isopach, isopleth, and isomass mapping, and (b) local information on bedding thickness, grainsize and sorting, grading, and internal stratification.

Uplifted and eroded deposits of ancient subaqueous eruptions can provide elements of the latter, but only a tiny proportion of seafloor rocks are preserved and uplifted, and these, though abundant, are typically tectonized and incompletely exposed. Bedding thickness, grainsize, sorting and grading characteristics can be determined from drill-core or piston-core sampling of modern seafloor deposits, but coarse deposits are difficult to recover, cores of any sort tend to be rather widely spaced, and, critically, cores do not allow determination of internal stratification characteristics at the critical meters to tens of meters scale available from outcrop observations.

Whole-deposit data are needed to infer eruption volumes and dispersal characteristics of eruption products, and from their combination, the dynamics of eruptions. On land, whole-deposit data are collected from scores of samples from sites at kilometer to tens of kilometers spacings, over areas that may exceed a million km^2 (Table 2).

Meaningful intepretation of whole-deposit data, as well as integrated interpretation of subaqueous eruptions, cannot be accomplished without good local information. The rea-

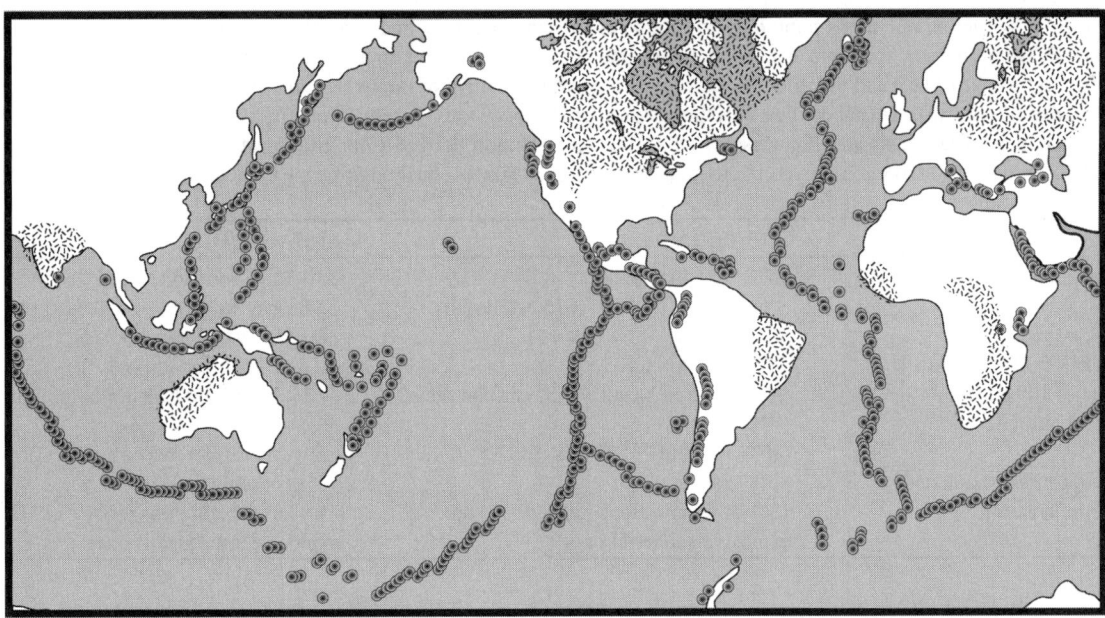

Figure 2. Most volcanism on earth takes place under water. Map shows zones of subaqueous volcanism, including spreading ridges and island arcs, as schematic alignments of volcanoes rather than the usual tectonic lines. Not all back-arc spreading ridges are shown, nor are all intra-plate oceanic volcanoes. Subaqueously formed volcanic rocks are predominant in the rock record as well, with major ar-eas indicated being the seafloor (shaded), and ancient sea-floors of Precambrian greenstone and greenstone-granite terranes (patterned). Locations of intraplate volcanoes and volcanoes on land taken from Fisher and Schmincke [1984].

son for this is that local information holds the signature of depositional process, and reflects the final transport processes. In cases where this final transport results from erosion and re-entrainment of debris from an eruption, no further direct inferences of eruption process are possible. Dispersion characteristics of such secondary deposits will reflect wholly non-volcanic processes. Readers of this monograph will also become familiar with the primary deposits of subaqueous eruptions. These may be termed "eruption-fed" deposits [e.g. *White*, 2000; *White and Houghton*, 2000], because the particles in the deposits have been *injected* into water directly by the eruption [*Head and Wilson*, 2003], and the dispersion and local depositional characteristics reflect the dynamics of emergent to subaqueous plumes or jets, as mediated by a stage of particle dispersion in which volcanic energy played a direct role. Eruption plumes and eruption-fed density currents produce the deposits that accumulate to form subaqueous clastic mounds and cones; no edifice having positive depositional relief can be formed entirely of reworked deposits.

SUBAQUEOUS ERUPTION DYNAMICS

Subaqueous eruptions bring magma toward the lake- or seafloor, where arriving lava or a mixture of gas and hot pyroclasts encounters water. Fundamental questions for any analysis of subaqueous eruptions are two. What effect has the water of the enclosing and overlying ocean or lake had on the eruption? At what point or points was the eruption affected? For some subaqueous eruptions, the effects are fundamental and pervasive through every aspect of eruption behavior. For other eruptions, the effects are significant, but less pervasive, and allow some eruption processes that are fundamentally magmatic to operate much as they would subaerially. High-intensity eruptions from the shallow seafloor, or from lakes, can produce substantial subaerial eruption plumes [e.g. *Houghton et al.*, this volume] and gas-particle supported density currents [e.g. *Wilson*, 2000] similar to those from wholly subaerial eruptions.

Water differs from air in many ways (Table 3), and we recognize four major roles for water in affecting the characteristics of subaqueous eruptions. (1) The role of steam. Water will boil and expand dramatically in shallow water, where explosive eruptions are well known. This process is damped with increased depth, and there is no phase change during expansion below roughly 3 kilometers' depth in the oceans. (2) The role of pressure. Increased ambient pressures characterize eruption sites beneath water, particularly deep water, and affect the solubility, expansion, and release of magmatic volatiles, as well as the development of steam.

Table 1. Subaqueous eruptions are more commonly basaltic than are subaerial ones, and constitute a much larger share of total global volcanism. Although the proportion of all subaqueous eruptions that is explosive may be small, such eruptions are nevertheless likely to play important roles in transfer of heat and volatiles (sulfur provided as representative of non-water volatiles) on a global basis. Volume values derived from Crisp (1984), with 1/6 subduction volcanism considered subaqueous; basalt vs. non-basalt proportions estimated. Sulfur and CO_2 contents estimated after Palais and Sigurdsson (1989); heat contribution derived from eruption temperatures of 1200C (basalt) and 800C (average non-basalt). Proportions of eruptions that are explosive (i.e. produce significant kinetic energy release and fragmented magma) are rough estimates.

	SUBAERIAL	SUBAQUEOUS
volume % of global volcanism	15% (.615 km^3/yr)	85% (3.485 km^3/yr)
basalt (estimated)	.185 km^3/yr (30% of .615 est. basalt)	2.788 km^3/yr (80% of 3.485 est. basalt)
non-basalt	.430 km^3/yr (70% of .615)	.697 km^3/yr (20% of 3.485)
sulfur (basalt: ~ 5x10^6 tons/km^3)	~2.35 million tons/yr	~ 14.62 million tons/yr
(other: ~ 10^6 tons/km^3)	(~ .93 million basalt + 1.35 million other)	(~ 13.94 million basalt +.68 million other)
heat (% of global total)	14%	86%
	(30% at 1200C + 70% at 800 C)	(80% at 1200C + 20% at 800 C)
explosive proportions, estimated	80% (generous)	~5% (pessimistic)
explosive abundance	13% of all eruptions (80% of 17) are explosive subaerial ones	~4 % of all eruptions (~5% of 83) are explosive subaqueous ones

(3) The role of heat capacity and conductivity. Water has very high heat capacity and conductivity (compared with that of air; Table 3), which causes rapid cooling of magma and erupted gases or steam. (4) The role of water rheology. Compared to air, water is dense and viscous, which strongly affects development of eruption plumes and the dispersal of ejecta to depositional sites (Table 4).

The Role of Steam

Fragmentation. Water vaporizes in contact with magma above the boiling point, and its expansion can enable or effect fragmentation. If water is enclosed in magma when expansion takes place, it can tear or break the molten lava apart. This process has been observed to form limu o Pele and other "aerodynamic" achnelithic fragments in littoral environments [e.g. *Mattox and Mangan, 1997*], and has been proposed to produce similar but much smaller scale sideromelane fragments in the shape of curved and folded sheets at depth [*Batiza et al., 1996; Maicher, 1999; Maicher et al., 2000; Clague et al., 2000; Maicher and White, 2001*].

Kokelaar [1983; 1986a] inferred that a more dynamic version of steam expansion, termed "bulk interaction", is an important process in shallow explosions of surtseyan eruptions. Vaporization of water is also a key feature of explosive {molten} fuel-coolant interactions, or {M}FCI, which represent the most effective mode of heat-transfer possible in magma-water interactions [*Colgate and Sigurgeirsson, 1973; Sheridan and Wohletz, 1981; Zimanowski and Buettner, this volume*], referred to by Kokelaar (1986) as "contact-surface explosivity". In an MFCI, magma momentum and fluid dynamics result in premixing of magma with water [*Zimanowski and Buettner, this volume*]. This leads to superheating of water, thermohydraulic fragmentation of melt and fragment dispersion within superheated water, followed by sudden (explosive) expansion of the water as it flashes to steam [*Zimanowski et al., 1997*]. Energy from expansion drives further mixing leading to further brittle fracturing of the magma and hence more fragmentation. Although not the sole means of achieving rapid vaporization of water, (M)FCI is both the only experimentally verified means of achieving explosive magma-water interactions, and by far the most thermo-kinetically efficient means of converting magma's heat energy into fragmentation energy and violent steam-driven expansion.

Dispersal. Steam produced in subaqueous eruptions affects dispersal by having low density, high buoyancy, *and a limited lifetime.* Effects of these properties on plume and density current properties have been addressed to varying degrees by many authors [e.g. *Fiske and Matsuda, 1964; Cashman and Fiske, 1991; Kano et al., 1994; Mueller and White, 1992; Head and Wilson, 2003*] and White [2000] used simple buoyancy calculations to infer minimum clast densities and concentrations capable of sustaining subaqueous gas-supported density currents. Steam also strongly modifies the thermal environment of clasts in transit, allowing some to attain achnelithic forms [e.g. spatter; *Fujibayashi, and Sakai, this volume*]. In some cases steam provides sufficient insulation for clasts to reach subaqueous depositional sites while still hot [*Sparks et al., 1980; Howells et al., 1986; Kokelaar, 1986; Mueller and White, 1992; Kano et al., 1994*].

Table 2. Comparison of different reasons to study subaerial versus submarine explosive eruptions, and of different aspects of research methodologies.

SUBAERIAL ERUPTIONS	SUBAQUEOUS (MOSTLY MARINE) ERUPTIONS
incentives to study	
• major hazards to people and infrastructure	• minor hazards to people and infrastructure
• some economic significance	• great economic significance
• hone understanding of eruption processes	• begin to understand eruption processes
eruption observation and data collection	
• eruptions generally observed	• eruptions rarely observed
• eruptions commonly filmed, locally instrumented	• eruptions not filmed, rarely locally instrumented
• visibility 10s of thousands of metres	• visibility tens of metres
• samples commonly collected during an eruption	• samples rarely collected during an eruption
data collection & costs	
• deposits inexpensively and non-destructively sampled from widespread sites	• deposits expensively sampled, often with partial destruction of fabric and/or layering
• soil, algae, lichens may obscure deposits	• manganese encrustations may obscure deposits
• variable disturbance by plants and plant roots, but easily avoided in sampling of young deposits	• ubiquitous bioturbation of thin deposits in most ocean waters
• supporting data from satellites, aerial photographs, ground photographs, topographic maps, handheld GPS	• supporting data from ?satellites (not yet done for fully subaqueous eruption, but may be possible), bathymetry, GPS
• fieldwork: tens of dollars per day	• fieldwork: thousands of dollars per day
• per-sample collection time: seconds to minutes	• per-sample collection time: minutes to tens of minutes
• *in situ* examination by hand lens	• no in situ examination

The Role of Pressure

Fragmentation. Increasing pressure decreases fragmentation since volatile volume expansion is decreased [e.g. *Fisher*, 1984; *Fisher and Schmincke*, 1984; *Staudigel and Schmincke*, 1984]. At increasing pressures, the magmatic volatile content must be larger to attain the same amount of fragmentation; because volatiles are retained in the melt, lower viscosities may result [*Yamagishi and Dimroth*, 1985], further affecting fragmentation style [e.g. *Mueller and White*, 1992]. For explosive magma-water interactions, the situation is less clear, although there is a well-established general tendency for all types of explosivity to decrease with depth [e.g. *Fisher*, 1984; *Fisher and Schmincke*, 1984; *Staudigel and Schmincke*, 1984]. This general tendency is not fully understood with respect to magma-water explosivity; Wohletz [this volume] presents experimental evidence that, at a fixed magma-water ratio, fuel-coolant interactions tend to *increase* in violence as confining pressure is increased, and suggests that explosive magma-water interaction is possible to depths exceeding those of most of the deep ocean floor. Uncertainties in this relationship concern the rheological role of the experimental containment vessel, which may reflect shock waves in a way that deep water does not, and the thermal role of effectively infinite amounts of enclosing water on the sites of explosive interaction.

Dispersal. To the extent that particle dispersal is initiated by gas-driven eruptions, whether that gas is magmatic gas or vaporized water, the effect of significant hydrostatic pressure upon an existing gas is to allow considerable adiabatic expansion with rise to lesser depths. The effects of this phenomenon have yet to be investigated in detail, in part because condensation (see role of heat capacity and thermal conductivity) is a more limiting factor. On the other hand, increased confining pressure commonly limits explosive expansion, and aspects of particle dispersal relying on such expansion, whether in gas-thrust eruption columns or discreet explosions projecting particles outward from an explosion site, will be less effective overall as pressure increases.

The Role of Heat Capacity and Thermal Conductivity

Fragmentation. The very high heat capacity and thermal conductivity of water, particularly as compared to air, allows it to very rapidly quench erupting melt that comes into direct contact with it. Such rapid chilling causes the melt to solidify and shrink. If its encounter with water is sufficiently energetic [*Thorarinsson*, 1967; *Moore et al.*, 1973; *Kokelaar*, 1986], or if melt domains are sufficiently small [*Carlisle*, 1963], quench granulation results. Rapid solidification to form an external glassy "shell" on flowing lava has also been considered to make the surficial glass susceptible to "dynamothermal spalling" [*Kokelaar*, 1986]

Table 3. Comparison of some important properties of water versus air, and their effects on eruptions. Note the similar values for steam's viscosity and heat capacity to those of air. Heat capacity per volume for both air and steam is much lower than that of water, because the values are per kilogram. Water's thermal conductivity is about 20 times that of air, but steam, surprisingly has a thermal conductivity almost 50 times that of water. Source for steam viscosity: http://pump.net/otherdata/viscsteamwater.htm; source for other physical data: http://hypertextbook.com/physics/

AIR	WATER (* STEAM)
Density	Density
1.239 kg/m^3 (cold dry air at sea level)	1000 kg/m^3 (fresh water, standard conditions)
decreases with altitude	1025 kg/m^3 (typical surface seawater)
Viscosity	Viscosity
0.0179 mPa s (millipascal) at 15 degrees C, STP	1.00 mPa s (millipascal) at 20 degrees C, std conditions
	* 0.01 mPa s (millipascal) saturated steam, std conditions
Specific Heat Capacity	Specific Heat Capacity
1158 J/kg K (at 300 degrees K)	4148.8 J/kg K (liquid water 20 degrees C)
	* 1039.2 J/kg K (water vapor at 100 degrees C)
Thermal Conductivity	Thermal Conductivity
0.025 W/m K (air at sea level)	0.56 W/m K (liquid water at 273 degrees K)
	* 27.0 (water vapor at 400 degrees K)
	** 2.8 (ice at 223 degrees K)

in response to stresses exerted by the lava. This process seems to best produce fragmentation with *intrusion* into seafloor sediments, however [e.g. *Skilling et al.*, 2002], whereas observations on the Juan de Fuca Ridge and elsewhere show that there is little spalled glass present among pillows of very young and fresh lava flows. Neither Moore et al. [1973] nor subsequent workers have noted significant fragmentation associated with pillow inflation. Moore et al. [1973] also describe a viscous lobe from a'a lava that moved across sand without releasing significant fragments, even though it had a deeply cracked surface attributed to flow inflation; a puzzling contradiction.

For eruptions that fragment magma before or upon contact with water, the high heat capacity and thermal conductivity of water leads both to rapid heat transfer into enclosing water [*Gudmundsson*, this volume] and potentially to an additional stage of fragmentation. This can occur either when hot particles come into direct contact with water and become involved in fuel-coolant interactions [e.g. *Kokelaar*, 1986], or if particles of appropriate size cool more slowly and are quenched to granules [*Carlisle*, 1963].

Magma clots or bombs are common subaerial products where a fluid magma is fragmented during eruption, as in a magma fountain or some strombolian bursts. If these fluidal fragments land while still plastic, spatter deposits or clastogenic lava flows are formed. In a subaqueous setting, fluidal clasts do not form if there is direct contact of magma with water during fragmentation, because rapid chilling causes the magma to become brittle and fragment. If water is not present continuously at the fragmentation site, however, fluidal clasts may form. Absence of water at a subaqueous fragmentation site may take place where magmatic volatiles effectively exclude water, or where steam generated by the contact of some magma with water is sufficiently abundant to similarly exclude water from part of the zone of fragmentation. Fluidal clasts formed during subaqueous eruptions will inevitably come into contact with water at some point. If this occurs while the clasts are still very hot, and with sufficient vigor, fuel-coolant interactions may lead to additional brittle fragmentation. Fluidal clasts can be preserved where they cool more slowly in the presence of steam, either because of heat in the transporting current [e.g. *Mueller and White*, 1992; *Doucet et al.*, 1994; *Cousineau and Bedard*, 2000], in the accumulating deposit [probably the case for *Cas et al.*, this volume; *Fujibayashi and Sakai*, this volume; *Mueller*, this volume; *Mueller et al.*, 2000; *Simpson and McPhie*, 2001], or because they are relatively small [*Lonsdale and Batiza*, 1980; *Batiza et al.*, 1984; *Smith and Batiza*, 1989; *Clague et al.*, 2000; *Maicher and White*, 2001; *Clague et al.*, this volume a; *Clague et al.*, this volume b]. Steam is able to "insulate" clasts in this context because it has little capacity to absorb or take up the heat itself (heat capacity; Table 3). So despite steam's surprisingly good heat conductivity (Table 3), extraction of heat from individual clasts is limited by the absorption of heat by enclosing water across the steam-water interface.

Dispersal. To the extent that particle dispersal is initiated by gas-driven eruptions, whether that gas be magmatic gas or vaporized external water, the effect of water's high heat capacity compared to that of steam is to rapidly cool the vapor to or toward its condensation point. For steam bub-

Table 4. Comparison of some important environmental factors for subaqueous and subaerial eruptions.

PHENOMENON	SUBAQUEOUS	EFFECT (+/- TREND)	SUBAERIAL	EFFECT
steam from interaction with magma, hot particles, and/or as magmatic volatile	ubiquitously formed above critical depths by interaction of magma with ambient water; films on hot clasts; from magma at shallower than critical depths	expansion (may be violent), high buoyancy, low heat capacity compared to water; steam formation suppressed with depth; disappears at ~3 km in seawater	steam from interaction with magma only in "wet" sites; steam in eruption plume also from heating of air entrained, and from magma	expansion (may be violent), buoyant when hot, condensing water alters particle transport properties (e.g. adhesion) heat capacity similar to air
pressure	hydrostatic pressure	damps expansion of steam from boiling and of magmatic gases; in combination with cooling, condenses gas in eruption plumes to produce aqueous plumes or currents; effect increases strongly with depth	atmospheric pressure	allows expansion of gases; eruption plumes are at maximum pressure near vent exit, and pressure decreases gradually with height in atmosphere
thermal behavior	high heat capacity	rapid cooling of magma, hot rock (but see "steam" above) can cause fragmentation by granulation	low heat capacity	slow cooling of magma, hot rock; granulation not effective, but dynamothermal spalling for some lavas
rheology	high density, high viscosity	low clast settling velocities, slower movement or expansion of plumes, currents; hot particles may be temporarily buoyant, and some pumice persistently buoyant; gas-supported currents require very high particle concentrations to remain negatively buoyant	low density, low viscosity	high clast settling velocities, granular collisions more important in transport; all clasts more dense than atmosphere at all times; gas-supported currents negatively buoyant even at low to moderate particle concentrations

bles rising from the seafloor, the water's great heat capacity is what keeps the bubbles from growing as they otherwise would during rise to progressively lower confining pressures. Few steam bubbles breach the surface until shoaling volcanoes, such Surtsey in 1963 and its satellites in succeeding years, grow to within wavebase or shallower [*Thorarinsson et al.*, 1964; *Thorarinsson*, 1967], only meters to perhaps a couple of tens of meters below the surface. Though the temperature of erupting magma and ambient pressures on the Icelandic shelf seafloor ensure that steam was created from the outset of the eruption, steam bubbles condensed before they could reach the surface until just before shoaling. Gaspar and others [this volume] report bubbles appearing at the surface above deeper vents, but they are associated with unusual "lava balloons", which may have released vapor bubbles after rising upward from the vent; the same sort of explanation would apply to the similar observations of Siebe et al. [1995].

The Role of Water Rheology

Fragmentation. Any aggregate or porous structure, whether fixed or in motion, is less permeable to fluids of higher viscosity, such as water, than to those of very low viscosity such as air. Explosive fragmentation during volcanic eruptions typically involves expansion rates that greatly exceed static infiltration rates for water. The result is that at the fragmentation front and in the gas-thrust region, conditions during subaqueous eruptions are much as they would be subaerially except for lessened degrees of expansion due to increased ambient pressure. For magmatic eruptions, exsolution and expansion during eruption results in an expanded mixture of gas (or supercritical fluid) and particles, isolated from enclosing water by their own coupled momentum and by a vapor barrier if the water is above boiling point per depth at the margin of the dispersion. For *phreatomagmatic* eruptions, including for instance phreatoplinian ones [*Self*, 1983; *Wilson*, 2000; *Houghton et al.*, this volume], there seems abundant evidence for a combination of high eruption rates, intense fragmentation, and water involvement, but the dynamic mechanism allowing pervasive contact of water with magma across a high-velocity jet of large cross-sectional area remains unclear.

Dispersal. The comparatively high viscosity and density of water compared to air have tremendous implications for dispersal of particles from subaqueous eruptions. For individually dense and massive particles, such as fragments of seafloor or clots of dense magma, inertia-driven (ballistic) transport that would be important subaerially becomes almost wholly ineffective; beyond limited water-exclusion zones above or immediately enclosing a vent [cf. *Kokelaar*, 1986; *White*, 1996], particle inertia is insufficient to carry clasts away from the vent, and rapid settling to the seafloor results. In contrast, particles that are temporarily buoyant can be carried passively to the surface and carried from the eruption site by currents. Basaltic clasts may float in this manner while still hot, but typically sink as they cool [e.g. *Tribble*, 1991; *Siebe et al.*, 1995; *Gaspar et al.*, this volume]. Subaqueously erupted pumice clasts, in contrast, may either buoy up and float temporarily before sinking in the same way as they cool and draw water into their pores [*Whitham and Sparks*, 1986; *Manville et al.*, 1998; *White et al.*, 2001], or may float for long periods [e.g. *Coombs and Landis*, 1966], particularly if their pore structure is sufficiently constricted to retard saturation, as for subaqueously erupted pumiceous domes [e.g. *Wilson and Walker*, 1985; *Kano*, this volume].

For smaller particles, delivered *en masse* to water from eruption columns of subaqueous eruptions, the transition to a buoyant *aqueous* eruption column precludes passage of the tephra across the water-air interface, but can allow significant lateral transport in plumes of warm water [e.g. *Cashman and Fiske*, 1991; *Fiske et al.*, 1998]. Subsequent settling by suspension from drifting aqueous plumes of tephra can produce deposits with a distinctive size relationship between dense and pumiceous grains that reflects a difference in relative settling velocities of the tephra particles in water as compared to air [*Cashman and Fiske*, 1991].

The *en masse* settling of suspended particles in vertical density currents, rather than indivually, has long been of sedimentological interest [e.g. *Allen*, 1982], and is increasingly regarded as an important control on aqueous sedimentation of tephra delivered to the water surface [*Wiesner et al.*, 1995; *Carey*, 1997] or injected by subaqueous eruptions into the water column [*White*, 1996; *White*, 2000]. In water, the lessened effective density of the particles, and the higher fluid viscosity, significantly increase the tendency toward formation of such currents over that in subaerial eruption plumes. When such vertical currents arrive at the depositional surface, they will expand laterally and/or move away downslope, leaving a record of seafloor current transport in their deposits.

Density currents consisting of water loaded by an erupting volcano with tephra are not merely "sub" aqueous, which also includes gas-supported currents flowing beneath water, but "aqueous". These eruption-fed aqueous density currents have gone under a variety of names, [*Fiske*, 1963; *Fiske and Matsuda*, 1964; *McPhie et al.*, 1993; *White*, 1996; *Fiske et al.*, 1998; *White*, 2000], but the growing appreciation of their importance in primary dispersal of tephra from subaqueous eruptions is clearly indicated by the contributions here [e.g. *Andrews*, this volume; *Maicher*, this volume; *McPhie and Allen*, this volume; *Mueller*, this volume; *Raos and McPhie*, this volume]. The behavior of such currents is still under study; understanding them (and being able to distinguish their products from those of density currents generated by remobilization of tephra from primary deposits) is the key to being able to interpret processes of subaqueous eruption from their primary deposits.

Gas-supported subaqueous density currents, considered true subaqueous pyroclastic flows, must have high particle concentrations and a limited pumice content to remain negatively buoyant and flow along the seafloor [*White*, 2000]; the low excess density relative to water, and water's high density relative to air, ensure that they will travel at low velocity. Such gas-supported subaqueous currents can be expected to form deposits similar to those of many subaerial pyroclastic flows of relatively low velocity; welding is a distinct possibility given both the chemical effects of seawater [*Sparks et al.*, 1980], and the likelihood of high depositional rates in proximal settings that would accompany the slow advance speed of the currents. Unwelded deposits of such gas-supported subaqueous pyroclastic flows have been identified using paleomagnetic evidence of deposition above the Curie point [e.g. *Yamazaki et al.*, 1973; *Kano et al.*, 1994]. As a current cools, gas condenses to water and current volume shrinks; only larger clasts, which have high heat contents that allow them to remain enclosed in a vapor film even when transported in aqueous currents, will show evidence for high-temperature deposition [e.g. *Mueller and White*, 1992].

Subaqueous "surges", which are turbulent gaseous density currents having low overall particle concentration [a few percent solids: *Fisher and Schmincke*, 1984], are of necessity buoyant in subaqueous settings, and cannot flow along the seafloor.

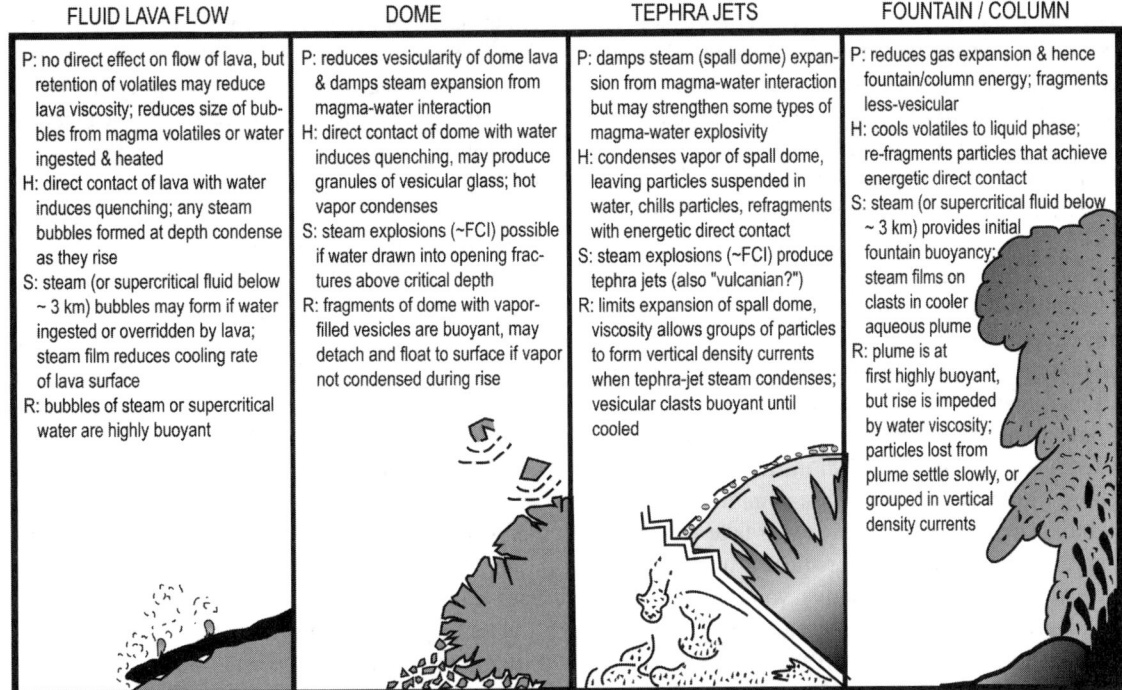

Figure 3. Summary of some major influences of pressure (P), high heat capacity and conductivity of water (H), steam (S) and water rheology (R) on different eruptions. See text for further discussion.

Fuel-Coolant Interaction in Subaqueous Eruptions

The factors addressed above affect all subaqueous eruptions to one degree or another. One critical aspect of eruptions into water has been addressed only in passing; the explosive and dynamic interaction of magma with water that is commonly described as a fuel-coolant interaction. This fundamental aspect of eruptions involving water is well addressed in different ways by Zimanowski and Buettner [this volume], and Wohletz [this volume]. It is worth emphasizing, however, that whether or not water interacts explosively with magma during eruption, there can still be explosive (magmatic) eruptions in water, and moreover that both dispersal and deposition from such eruptions will be strongly influenced by the water in which they occur.

VOLCANIC ERUPTIONS IN OPEN WATER

Having addressed deductively the effects of water on subaqueous eruptions, we now turn to a brief consideration of the characteristics of known subaqueous eruptions and their products that are the focus of this book.

Perhaps surprisingly, the huge advances made in understanding tectonic and petrogenetic aspects of mid-ocean ridge volcanism as a result of five decades of post-war ocean exploration have only recently begun to be accompanied by increased understanding of subaqueous eruption processes. In part, this is because the bread and butter observations of volcanology take place at a far smaller scale than those of plate tectonicists, or of petrologists interested in magma origins and evolution. Physical volcanologists need to know the specific context of rocks to relate them to a sequence of eruption, and volcanologists interested in explosive eruptions have an even more stringent requirement; they require sampling of particulate debris, preferably in systematic ways.

Submarines and remote vehicles, core drills and sampling by manipulator meet physical volcanological requirements, but have only recently become available in any abundance, and remain extremely costly. The result is that the depth of physical volcanological understanding of marine volcanism lags significantly behind that of subaerial equivalents.

Explosive Shallow Water (Surtseyan) Eruptions

Explosive subaqueous eruptions of basaltic Surtseyan type occur from vents in water less than a few hundred metres deep (i.e. shelfal depths or less). An important distinction from eruptions forming maars or tuff rings is that the latter are essentially subaerial and caused by interaction between magma and (limited) groundwater (i.e. within an aquifer), whereas at Surtseyan centres magma interacts with abundant surface water and the vents are flooded. There is also an important distinction between basaltic and silicic magmas: in basaltic eruptions, the juvenile water content is low (though CO_2 content may be quite high), and fragmentation of the magma by decompression and volatile exsolution is less efficient than in (water-rich) silicic magmas. Surtseyan volcanoes occur in every shallow subaqueous setting (submarine, lacustrine, glacial). They are explosive by virtue of the ambient hydrostatic pressure being less than that at which explosive gas expansion is suppressed. For basaltic magmas, explosive eruptions generally occur at water depths no greater than 100–200 m but in some examples (e.g. water rich, alkaline magmas) much greater fragmentation depths are inferred [> 700 m; *Kokelaar*, 1986, and see below]. Surtseyan eruptions are commonly monogenetic and build pyroclastic cones to sea- or lake-level and above. Erupted volumes are characteristically small for the individual centres, although cumulative volumes for geographically coherent monogenetic volcanic fields may be much greater [*Moore*, 1985; *White*, 1991; *Smellie*, 1999; *Maicher*, this volume]. Single centre activity normally lasts from a few days to a few years.

Surtseyan cones are typically steep-sided and constructed of fallout- and density current-deposited phreatomagmatic tephra. Many centres have an initial pillow-lava-dominated effusive phase [e.g. *Jones*, 1969; *Staudigel and Schmincke*, 1984; *Skilling*, 1994; *McPhie*, 1995; *Andrews*, this volume], and a few are dominated by pillow lava [e.g. *Smellie and Hole*, 1997]. Additional common and volumetrically important coeval lithofacies are mainly mass-flow deposits formed from post-depositional gravitational collapses of the volcano flanks on a range of scales (surface-sediment slumps to sector collapses). The larger collapses typically leave behind prominent curved slump-scar surfaces draped by younger beds [*Smellie and Hole*, 1997; *Smellie*, 2001]. The basal pillowed section is commonly called a pillow volcano [*Schmincke and Bednarz*, 1990], whilst the overlying tephra-dominated part is a subaqueous to subaerial tuff cone which, for subglacially constructed volcanoes, is called a tindar [*Jones*, 1969; *Smellie*, 2000].

The dynamics of eruptions in shallow water follow a broadly similar pattern to those of subaerial volcanoes, but with several important differences distinctive of subaqueous vents [*Cashman and Fiske*, 1991; *Kokelaar and Busby*, 1992; *Koyaguchi and Woods*, 1996]. During decompression of magma rising to the sea- or lake-floor, volatiles are exsolved, causing expansion of the magma—bubble mixture to form a foam. In most cases, magma-water interaction occurs at a high level in the vent conduit, probably in the crater itself, as evidenced by the paucity of accidental lithic fragments eroded from the vent walls. Much of the water occurs in a water-saturated tephra slurry [*Kokelaar*, 1983]. The timing of the interaction relative to the extent of volatile exsolution in the magma determines the vesicularity of juvenile clasts in the erupted deposits [*Houghton and Wilson*, 1989]. The foam or slurry is predicted to exit the vent with a pressure in excess of ambient, the value of which is determined by a) the elevation difference between the volatile fragmentation depth and the sea or lake floor [e.g. *Wilson and Head*, 1981], and b) the conduit radius – e.g. if it is essentially fixed (solid parallel rock walls), the erupting mixture may be very overpressured [*Woods*, 1998]. Conversely, for low mass flux eruptions or higher ambient pressures, the jet may be so relatively underpressured that a slow dense fountain of ash and gas/water may be developed that simply "boils over" the crater rim [cf. *Woods*, 1998].

As a subaqueous eruption gets underway, a cloud of gas/steam, water and ash spreads out radially from the vent. It is accompanied by a series of hemispherical shock waves (alternating compression and rarefaction waves), which advance ahead of the erupted material and are seen as spectacular expanding spall domes [*White*, 1996]. The size of a spall dome and the energy of associated tephra jets are dependent on explosion magnitude and water depth [*White*, 1996]. Shock waves are probably linked to subaqueous incandescent fire flashes observed in these eruptions, which are interpreted to signify explosive hydrovolcanic fragmentation events. With efficient fragmentation, magma particles are small (mm to sub-mm) and thermal equilibration occurs over times shorter than 1 s [*Woods and Bursik*, 1991; *Gudmundsson*, this volume]. Close to the vent, a gas thrust region gives way upward to a column that may be buoyant and convecting due to the increased gaseous (steam) component of the mixture. Thereafter, the particular thermodynamic path followed by the magma-water mixture depends on the detailed dynamics of mixing [*Koyaguchi and Woods*, 1996].

As mentioned previously, submarine gas-thrust regions experience more rapid deceleration than subaerial ones because of the greater density and viscosity of the overlying water, and more rapid cooling caused by water entrainment and endothermic conversion of that water to steam [*Cashman and Fiske*, 1991; *Kokelaar and Busby*, 1992]. Conversely, the conversion from gas thrust to a buoyant column may be more easily achieved in water because of the smaller density contrast between pyroclasts and water, and even the relatively slow upward velocities in much less energetic columns may be sufficient to entrain particles [*Cashman and Fiske*, 1991]. A buoyant convective column is likely to rise to the water surface, but the interface with air is probably an important density barrier that can only be penetrated during very shallow eruptions [metres to few tens of metres: *Kokelaar*, 1983; *McInnes*, 2000], by highly energetic jets or when vigorous columns have extremely high inertia. During the relatively deep water eruptive phase, only white steam clouds are evident above the water and most of the tephra-bearing column spreads out laterally as a large mushroom "cloud" just below the surface, from which tephra particles rain out onto the sea floor. Once ash particles penetrate the water surface, the column becomes grey in colour. The rapid flux of ash falling out of a subaerial column leads to saturation of particles in the upper layer of the water. When the bulk density of the particle—water mixture becomes high enough to promote convective instability, plumes begin to grow and detach downward, transforming into vertically descending sediment gravity currents [*Carey*, 1997]. Where these currents reach the sea floor, they are likely to continue as sediment gravity flows down the steep cone flanks.

All recent descriptions of Surtseyan volcanoes have emphasized the overwhelming dominance of sediment gravity flow deposits among the lithofacies [e.g. *Skilling*, 1994; *McPhie*, 1995; *White*, 1996; *Smellie and Hole*, 1997; *Smellie*, 2001; *Maicher*, this volume; *McPhie and Allen*, this volume; *Mueller*, this volume]. White [2000] grouped these observations in a unified model of eruption-fed deposits linked to column dynamics [see also *Mueller and White*, 1992; *Mueller*, this volume]. The behaviour of a subaqueous eruption column is dominated by the dynamics of mixing of the two principal components: mass flux of magma versus mass fraction of (external) water [cf. *Koyaguchi and Woods*, 1996]. The influence of these two components is antithetic and they have different implications for generating eruption-fed deposits (Figure 4):

1) With a small mass fraction of water (including external water) in the column, for example as a result of high magma flux, the column is dense and insufficient water, converted to steam, is available for the column to become buoyant. It is thus predisposed to form a low fountain and to collapse prematurely, thus generating dense pyroclastic gravity flows directly from the crater. In a steadily sustained high-concentration particulate flow with enough hot basalt clasts, water may be excluded during transport and the resulting deposit has the potential to be welded [cf. *Kokelaar and Busby*, 1992].

2) With increasing mass fraction of water (including external water), the column density decreases, it decelerates more slowly, more steam is created and it becomes buoyant. Pyroclastic density currents will be

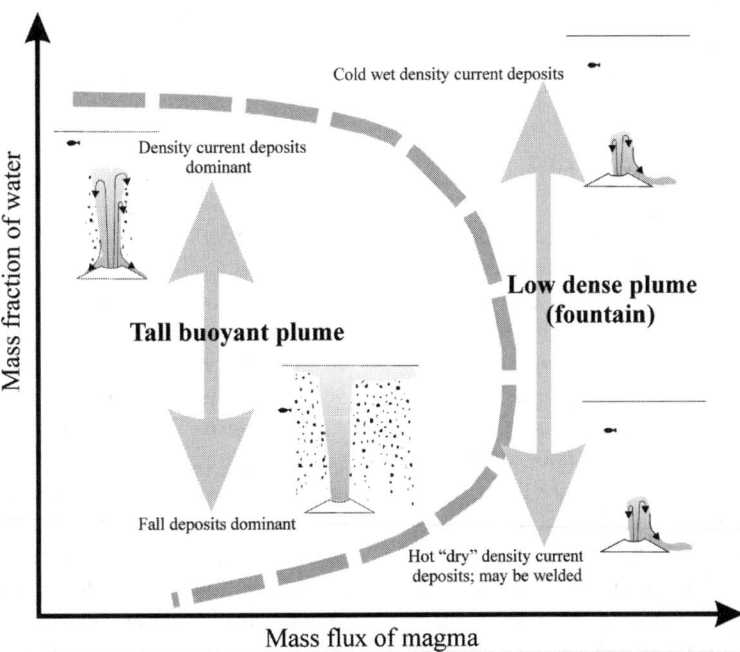

Figure 4. Schematic illustration of the effects on the dynamics of subaqueous eruption plumes of varying the water mass fraction and mass flux of basaltic magma, and predicted types of eruption-fed deposits formed as a result. Modified after Koyaguchi and Woods (1996).

created at the point where the column velocity falls to zero and collapses as column margin-fed density currents. If the column reaches the water surface and intrudes laterally as a neutral-density mushroom "cloud", tephra released from the cloud will form widespread graded fallout layers [*Cashman and Fiske*, 1991]. With sufficient inertia, the upper column may become subaerial, and vertical gravity currents will be generated secondarily (after subaerial tephra fall) from the ash-charged water surface [sensu *Carey*, 1997].

3) Finally, the resulting decrease in plume temperature due to ingestion of even higher mass fractions of cold water suppresses column buoyancy, and it becomes unstable, collapsing to generating wet dense pyroclastic gravity flows. Unlike deposits generated from dense magma fountains with low water mass fractions, these gravity flow deposits will always be too cold to weld.

The columns described so far are characteristic of the quasi steady-state Surtseyan behaviour known as continuous-uprush activity, which is dominated by high magma discharge and rapid aggradation of the volcanic pile [*Kokelaar*, 1983; *Smellie*, 2001]. Such vigorous sustained activity may produce local water-exclusion zones at the vent margin in which clasts are transported in part through steam [*Kokelaar*, 1983; *White*, 1996]. The existence of these zones appears to be a prerequisite to explain hitherto enigmatic features of many Surtseyan piles, such as fluidal-surfaced bombs, impact structures and armoured lapilli, which are often considered characteristic of "dry" eruptions [*Kokelaar and Durant*, 1983; *Mueller and White*, 1992; *White*, 1996; *White and Houghton*, 2000].

Surtseyan eruptions are also characterised by tephra jetting behaviour, caused by intermittent explosions associated with much lower magma discharge rates. The transport, deposition and sedimentary characteristics of deposits linked to individual jets are very poorly known and have been treated theoretically. White [1996] postulated that steam-rich jets charged with quenched hyaloclasts would follow ballistic trajectories behind expanding spall domes. As their steam condensed, the jets were envisaged stalling, collapsing and releasing their particles, which would fall to the sea/lake floor, entraining water and flowing laterally as sediment gravity flows down the inclined flanks of edifices. Intuitively, the sedimentary characteristics of deposits formed by these jets are likely to be well-defined. The beds should be sharply bounded and internally well structured, in contrast to the much cruder stratification (gradational bed surfaces) and faint sedimentary structures shown by beds correlated with continuous-uprush eruptions [*Smellie*, 2000; *Smellie*, 2001].

Explosive Deep-Marine Eruptions of Basalt and Other Mafic Magmas, Modern and Ancient

The Surtseyan basaltic eruptions outlined above are limited to water depths less than a few hundred meters, but other styles of mafic explosive eruptions occur in deeper water. Many such eruptions are fundamentally pyroclastic, being driven by release of magmatic volatiles. A recent overview of subaqueous pyroclastic eruptions [*Head and Wilson*, 2003] summarises the conditions required for Hawaiian-style fountaining and for strombolian eruptions. Head and Wilson conclude that the high concentrations of magmatic volatiles required for fragmentation of mafic magmas at elevated pressure exceed the estimated primary volatile contents of most magmas (except perhaps the most extremely alkaline). They suggest, however, that bubbles filled with exsolved magmatic volatiles may accumulate in some magmas prior to eruption. The dominant volatile constituent in mafic magmas is generally CO_2 [*Dixon et al.*, 1997; *Clague et al.*, this volume a; *Clague et al.*, this volume b; *Eissen et al.*, this volume] except in arc and back-arc settings, where it is H_2O [*Gill et al.*, 1990]. Other volatile constituents, such as S, play a subsidiary role in bubble growth. Bubbles of exsolved volatiles may accumulate at the top of magma reservoirs at shallow sub-seafloor depths, thus allowing magma fragmentation to occur within or just below the seafloor vent.

Young, submarine, pyroclastic basaltic deposits, erupted at known depths, now include nephelinite, basanite, and alkalic basalt from 4300 m in the North Arch volcanic field [*Clague et al.*, 1990; *Clague and Moore*, 2002; *Davis and Clague*, this volume], Hawaiian alkalic basalt, transitional basalt, and tholeiitic basalt from at least 1000 m on Loihi Seamount [*Clague et al.*, this volume a], Hawaiian tholeiitic basalt from as deep as several km on Kilauea's submarine east rift zone [*Clague et al.*, 2000], enriched mid-ocean ridge basalt from 400 to 1700 m on the Mid-Atlantic Ridge [*Fouquet et al.*, 1998; *Eissen et al.*, this volume], and depleted and enriched mid-ocean ridge basalt from as deep as 3800 m on the Gorda Ridge [*Clague et al.*, this volume b].

The deposits formed range from extensive, thick, layered volcaniclastic units, such as on Loihi Seamount [*Clague et al.*, this volume a] and the Mid-Atlantic Ridge [*Eissen et al.*, this volume] to dispersed sand-sized pyroclasts deposited in otherwise hemipelagic sedimentary deposits, such as along the Gorda mid-ocean ridge [*Clague et al.*, this volume b]. Other deposits are small, local drifts of volcanic sand, as observed along Kilauea's submarine east rift zone at about 2100 m. These deposits reflect the dispersal and depositional characteristics of the eruption plumes and the volume of pyroclasts produced. In general, the subaqueous eruption plumes are of limited height [*Head and Wilson*, 2003], so only the finer and more hydrodynamically efficient particles, such as limu o Pele, are dispersed more than a few tens of meters from the vents. The pyroclasts formed also display a range of morphologies including angular vesicular to dense glass fragments, fluidal clasts or spatter, limu o Pele, and Pele's hair [*Clague et al.*, this volume a; *Clague et al.*, this volume b; *Eissen et al.*, this volume]. Grainsize analysis of the Loihi deposits shows a range from sandy gravel, gravelly sand, sand, muddy sand, to sandy mud [*Clague et al.*, this volume a], with most samples moderately to well sorted. Similar results were obtained for the deposits from the Mid-Atlanic Ridge [*Eissen et al.*, this volume], reflecting elutriation of fines in the water column and perhaps even post-deposition.

The evidence from all these deposits suggests that exsolved magmatic volatiles, dominated by CO_2, accumulated during upward transport or storage in crustal magma reservoirs, leading to enhanced volatile contents sufficient to cause magma fragmentation and drive explosive eruptions, even at depths as great as 4300 m [*Clague et al.*, 2002; *Davis and Clague*, this volume]. Such accumulation of volatiles takes place in volatile-rich magmas including nephelinite, basanite, alkalic basalt, and E-MORB [*Clague et al.*, 2002; *Davis and Clague*, this volume; *Eissen et al.*, this volume], the eruptions of which range from Strombolian to Hawaiian, and in volatile-poor magmas like Hawaiian tholeiite [*Clague et al.*, this volume a] and N-MORB [*Clague et al.*, this volume b], the eruptions of which are most similar to Strombolian eruptions.

Other submarine volcaniclastic deposits are known from near-ridge seamounts, near the East Pacific Rise [*Batiza et al.*, 1984; *Batiza and White*, 2000] and the Gorda Ridge [*Davis and Clague*, 2000]. Such deposits, as well as some from Loihi Seamount [*Clague et al.*, 2000] were originally interpreted to have formed during hydroclastic eruptions, perhaps enhanced by quench granulation, although Clague et al. [this volume b] suggest that many may actually be pyroclastic deposits. Other bedded volcaniclastic deposits, mainly of hawaiite composition, observed and sampled on Miocene Seamounts offshore central California [*Davis and Clague*, this volume], are also inferred to have formed during pyroclastic eruptions, although their depth of eruption is unknown. The submarine deposits are difficult and expensive to map and sample thoroughly, so many of the characteristics and distribution of facies remain largely undocumented. These shortcomings of the submarine studies can be remedied somewhat by detailed study of similar deposits now uplifted and exposed on land. For example, Fujibayashi and Sakai [this volume] propose that strombolian fire fountaining in moderate water depths produced fluidal clasts in deposits from the Miocene Ogi Basalt in Japan.

Several authors have emphasized the complexity of such submarine pyroclastic eruptions [*Eissen et al.*, this volume; *Fujibayashi and Sakai*, this volume; *Gaspar et al.*, this volume] and called upon fragmentation that is initially pyroclastic, but which rapidly evolves into hydroclastic fragmentation due to the rapid increase in surface area of hot or molten clasts in contact with the surrounding water. At water depth shallower than the critical depth of seawater (about 3 km; ~2940 m in the Pacific Ocean), production of steam is possible, and steam explosivity could combine with initial pyroclastic fragmentation to produce more violent activity. At greater depths, sea water expands upon heating, but no phase change occurs and many workers infer that explosive eruptions are necessarily pyroclastic [see discussion in *Clague et al. b*, this volume]. Expansion of both water and carbon dioxide at such depths occurs in the supercritical state, so it is clear that formation of vapor (gas) is not absolutely necessary to produce either vesicular clasts or bubble-burst fragments (limu). Uncertainty also remains regarding the ability of magma-water interaction to fuel explosive interactions at greater than critical depths. Wohletz [this volume] provides evidence of hydromagmatic interactions under high confining pressures, and an analysis emphasizing both the physical plausibility of such interactions in the deep sea and the weakness of extant evidence against them.

Subaqueous hydromagmatic (thermohydraulic) explosive mafic eruptions can also occur, although their likelihood and the energy released apparently decreases with increasing depth below about 100 m [*Zimanowski and Buettner*, this volume]. The deepest documented case of such activity is on Loihi Seamount, where volcaniclastic deposits include layers that contain fragments of hydrothermal stockwork and altered basalt that were fragmented and ejected during phreatomagmatic and, perhaps, during phreatic eruptive events [*Clague et al.*, this volume a]. These authors infer that the eruptions accompanied collapse of pit craters and calderas, as observed at Kilauea's summit in 1924, when water came in contact with large amounts of hot rock exposed during the collapse events. The Loihi phreatomagmatic eruptions are inferred to have occurred at a minimum depth of 1356 m, the depth of the bottom of the collapse pits.

Pumiceous Submarine Silicic Eruptions

Production of pumice, a glassy and highly vesicular rock, typically of silicic composition and capable of floating in water when dry, has been considered a characteristic indicator of explosive eruptions, perhaps in part because magma is often considered to fragment when pumice-like vesicularities are achieved [*Sparks*, 1978]. Pumice also forms in parts of lava flows of evolved compositions both subaerially [e.g. *Fink and Manley*, 1987] and subaqueously [*Wilson and Walker*, 1985; *Binns*, this volume; *Yuasa and Kano*, this volume]. Non-explosive production of pumice in subaqueous settings introduces particular interpretive uncertainties, because unlike in subaerial settings, significant dispersal of subaqueously effused pumice is possible because of its buoyancy and hence ability to float away in currents or driven by wind. In marine or lacustrine environments, identification of dispersed pumice is not a necessary indicator of any sort of explosive eruption.

For a very thorough review of pumice-forming submarine eruptions, both explosive and non explosive, see Kano [this volume]. He divides subaqueous eruptions into four groups: subaqueous plinian-type ones; those generating subaqueous (eruption-fed aqueous) flows; those involving explosive bulk interaction of vesicular magma and water, and; the sort of non-explosive release of pumice clasts mentioned above. Fountaining eruptions involving rhyolitic magma tend not to produce true pumice [e.g. *Mueller and White*, 1992], but rather less-vesicular clasts more akin in texture to the scoria of basaltic fountains. Raos and McPhie [this volume] present results from study of the Efaté Pumice in the Vanuatu arc, which comprises thick beds of coarse trachytic pumice overlain by tens of meters of shard-rich sand and silt in thin beds with ubiquitous foraminera. They interpret the succession to record a subaqueous plinian-type eruption that waned to a more phreatomagmatic stage. Deposition of the lower beds was from eruption-fed aqueous density currents, but the upper beds may in part result from redeposition.

McPhie and Allen [this volume] similarly provide a careful interpretation of a pumiceus succession associated with Tasmania's mining district. Primary syn-eruptive deposition from eruption-fed density currents is inferred to have produced thick pumice-dominated beds, with a range of more crystal-rich associated facies, some of them welded. Welding of subaqueously erupted and deposited pumice has been repeatedly inferred [*Sparks et al.*, 1980; *Kokelaar et al.*, 1985; *Howells et al.*, 1986; *Kokelaar and Busby*, 1992; *Schneider et al.*, 1992; *White and McPhie*, 1997], but there is disagreement over the conditions under which it can be deposited [*Cas and Wright*, 1991], and an apparently limited range of particle assemblages for which subaqueous density flow of gas-supported currents is plausible [*White*, 2000].

Another area of uncertainty concerns subaqueous eruptions sufficiently powerful to produce subaerial plumes. Allen and Stewart [this volume] assess eruption styles for different pumiceous deposits of subaqueous eruptions in the Hellenic island arc, emphasizing the roles of eruption magnitude and water depth in determining whether eruptions breach the sea surface or not, and on styles of fragmentation and particle dispersion from such eruptions. Houghton and others [this volume] give a detailed analysis of one phreatoplinian suite produced from sub-lacustrine intracaldera vents during the Taupo 1.8 ka eruption.

EXPLOSIVE SUBMARINE ERUPTIONS AND ECONOMICALLY IMPORTANT DEPOSITS

Identification of submarine pyroclastic deposits in modern [*Wright and Gamble*, 1999; *Fiske et al.*, 2001] and ancient subaqueous, felsic-dominated caldera structures [*Lichtblau*, 1989; *Gibson*, 1990; *Hudak et al.*, this volume; *Kessel and Busby*, this volume] is significant for mining exploration because these deposits are particularly favourable hosts for hydrothermal massive sulfide formation [*Gibson and Watkinson*, 1990; *Stix et al.*, 2003]. The association of pyroclastic rocks with volcanogenic massive sulfide (VMS) mineralization is well-known in widespread Archean greenstone belts (Figure 2). More specifically, mineralization is often coupled with subaqueous caldera development, and in Canada's Abitibi greenstone belt alone this link has been documented for the 2728 Ma Joutel volcanic complex [*Lafrance et al.*, 2000], the ca. 2735 Ma Sturgeon Lake caldera complex [*Morton et al.*, 1991; *Hudak et al.*, 2002], the 2734-2728 Ma Hunter Mine volcanic complex [*Mueller and Mortensen*, 2002], and the 2703-2698 Ma Central Noranda volcanic complex [*De Rosen-Spence et al.*, 1980; *Chartrand and Cattalani*, 1990; *Gibson*, 1990].

Distinguishing in the ancient rock record between subaqueously deposited tephra originating directly from eruptions, versus deposits formed of clasts reworked from initial deposits, has been a contentious issue [*Cas and Wright*, 1991; *Mueller*, 2001], especially for deposits of what Fiske and Matsuda [1964] termed "subaqueous pyroclastic flows" (p. 102). These flows were described as having transported tepha away from an eruption in aqueous density currents, and White [2000] suggested the unambiguous term "eruption-fed aqueous density currents" for them. Both terms emphasize the primary nature of transport and direct link with eruption processes [*Mueller et al.*, 2000], but we prefer to reserve "pyroclastic flow" for transport with an interstitial gas-phase. In any case, the deposits from such currents are considered primary pyroclastic rocks, because there is no preceding episode of deposition, and hence no "re"-working or redeposition involved. These deposits are also more than simply "syn-eruptive" [*McPhie et al.*, 1993], which is a broader term including both primary deposits and those formed of particles redeposited during the period of eruption. Fiske et al. [2001] have recently interpreted deposits of Myojin Knoll, a relatively deep-water caldera that is actively producing a gold-rich VMS deposit, as similarly formed by direct deposition from a subaqueous eruption. Busby-Spera [1984; 1986] described primary subaqueous pyroclastic deposits of eruption-fed aqueous density currents from the ancient submarine Mineral King caldera collapse structure.

Subaqueous primary pyroclastic deposits are difficult to identify unless there is a combination of heat retention features, such as columnar joints, eutaxitic texture, rootless segregation pipes, vapour phase crystallization, or fiamme [*Fisher and Schmincke*, 1984; *Stix*, 1991; *McPhie et al.*, 1993], or thermoremanent magnetization studies indicate elevated emplacement temperatures ≥350 °C [*Yamazaki et al.*, 1973; *Tamura et al.*, 1991; *Mandeville et al.*, 1994].

The distinction between water-saturated primary deposits of subaqueous eruptions and deposits resulting from redeposition or reworking is critically important because 1) reworking is often associated with redistribution to more distal sites, and 2) VMS deposits are favoured in vent-proximal sites. The characteristics of subaqueous fire-fountains are of particular interest because their deposits known from Archean caldera floor settings and composite edifices associated with VMS deposits [*Gibson et al.*, 1989; *Mueller et al.*, 1994; *Gibson et al.*, 1997; *Mueller and Mortensen*, 2002]. These edifices, commonly < 1 km in diameter, occupy small segments of a caldera's floor.

Subaqueous Fountaining Eruptions

Mafic and felsic fountaining eruptions typically produce 1-10's of m-thick units characterized by fining-upward sequences that form small-scale stratified volcanic edifices [*Mueller and White*, 1992; *Doucet et al.*, 1994]. A fining-upward sequence represents a single eruptive event, and in many cases records a shift in final fragmentation process from magmatic to hydroclastic. Such deposits may accumulate to hundreds of meters' thickness [*Simpson and McPhie*, 2001; *Fujibayashi and Sakai*, this volume]

Explosive Dome-Collapse Breccias

Explosive dome collapse is a process known from subaerial dome-flow complexes whereby a mass of dome-building lava fails and progressively and "explosively" shatters to form an avalanching mass of hot pyroclastic debris. In a subaqueous setting, this process is unlikely to be effective, because the density and viscosity of water are such that the clast-clast collisions that drive fragmentation in the subaerial setting would be very strongly damped. This analysis is challenged, however, in the paper presented by Scott and others [this volume], who infer a dome-collapse origin for columnar-jointed pyroclastic rocks of Archean age in Ontario. In contrast, hydroclastic fragmentation, resulting from thermal contraction granulation and dynamothermal shattering of a flow or dome carapace, may be very effective in forming breccia-grade fragments from growing subaqueous domes. Water drawn into opening fracture systems in the hot lava can cause phreatomagmatic explosions [*Colgate and Sigurgeirsson*, 1973; *Wohletz*, 1983; *Brooks*, 1995] that disrupt the dome or lava. Subaqueously extruded domes have been identified in several ancient sequences [*Gibson et al.*, 1989; *McPhie and Allen*, 1992; *Doyle and McPhie*, 2000; *De Rita et al.*, 2001], and where within central calderas are highly favourable hosts for massive sulfides. Domes are three dimensional structures composed of stubby lava flows and 3-D structures that cannot be easily distinguished from thick felsic flows unless km-scale mapping shows a circular distribution pattern and cross-cutting relationships [*Lafrance et al.*, 2000]. Many of the documented ancient domes are now recognized as intrusive [e.g. *Goto and McPhie*, 1998], and therefore show the importance of detailed mapping in establishing clearly a stratigraphic context before interpreting them in terms of subaqueous eruption processes.

Subaqueous Calderas and Massive Sulfides

Subaqueous calderas are of economic importance because of their association with VMS deposits [*Gibson*, 1990; *Iizasa et al.*, 1998]. All calderas are collapse structures, and known modes of origin include (1) rapid evacuation of magma by large explosive eruptions that can produce hundreds of km^3 of pyroclastic debris in single eruption episodes, (2) outpouring of magma via extensive fountaining eruptions and lava flows that may also arise from associated rift systems, (3) draining of magma into satellite chambers along rift zones linked with a major, shallow-level (1–5km depth) magma chamber below the volcano summit, or (4) numerous low-volume pyroclastic events coupled with abundant effusive volcanism and tectonic extension. Felsic calderas related to large ash flows [*Smith and Bailey*, 1968; *Wilson*, 1993; *Lipman*, 1997] and mafic calderas associated with well-developed rift zones [*Tilling and Dvorak*, 1993] are the best known.

Incremental collapse may follow repeated low-volume eruptions, and has been advocated for the Hunter Mine caldera [*Mueller and Mortensen*, 2002]. Gudmundsson [1998] argues that caldera eruptions typically vent along their bounding faults [see also *Wilson and Hildreth*, 1997].

The greatest displacements typically occur along the outermost set of encircling ring faults [*Walker*, 1984]. Mapping of calderas on the modern seafloor is based primarily on bathymetry, with a few sites mapped and sampled by submersibles; much of our understanding of seafloor calderas arises from mapping and analysis of ancient, dissected caldera sequences. Fracture systems and faults act as hydrothermal fluid pathways, and intense alteration and mineralization take place while the magma system is active.

Proximal Pyroclastic Deposits in Calderas

Calderas range to tens of km in diameter, and explosive eruptions and vents occur at various sites on the caldera floor. Pyroclastic or volcaniclastic deposits from these intra-caldera and caldera-edge vents may be proximal in the sense of lying within the caldera, yet distal with respect to their vent sites and to pathways for mineralizing fluids. It is hence particularly important to identify primary pyroclastic debris formed by smaller fountaining, surtseyan, or dome eruptions, because these vents need not lie along the major caldera-bounding faults. Mineralized zones are directly related to synvolcanic fault systems, along which these small volcanic structures or subcentres are typically localized on the caldera floor as has been shown for the Hunter Mine caldera [*Mueller and Donaldson*, 2002].

In terms of caldera settings, it is a misconception to consider relatively fine-grained tuff as a clear indicator of deposition distal to the volcanic source. Within a subaqueous caldera, both eruption-fed turbidites and those arising from redeposition can extend for many kilometres yet remain within the caldera and above a shallow magma reservoir. Numerous such tuffs in Archean complexes have been altered into carbonate and oxide iron-formations by percolating Fe-rich caldera-floor hydrothermal fluids [*Chown et al.*, 2000]. Such "distal" tuffs were deposited onto fracture or fault systems linked with shallow magma, and these fine tuffs are favourable prospects for VMS exploration.

Most VMS deposits occur at depth [> 1000 m – *Ishibashi and Urabe*, 1995], but formation in relatively shallow-water environments is also possible, as is inferred for the Cambro-Ordovician, Mount Windsor deposits in Queensland [*Doyle*, 2000] and Tasmanian VMS deposits [*McPhie and Allen*, 1992; *White and McPhie*, 1997]. Small volcanic edifices generated by fire-fountaining eruptions, and explosive dome collapse structures that form parts of dome-flow complexes may occur at depth, whereas subaqueous Surtseyan eruptions forming tuff cones of mafic or felsic composition [*Mueller et al.*, 2002] may be confined to water depths <200 m.

Caldera-bounding faults are particularly favourable for the precipitation of ores, but are rarely observed in ancient sequences. Caldera margin settings are inferred from the

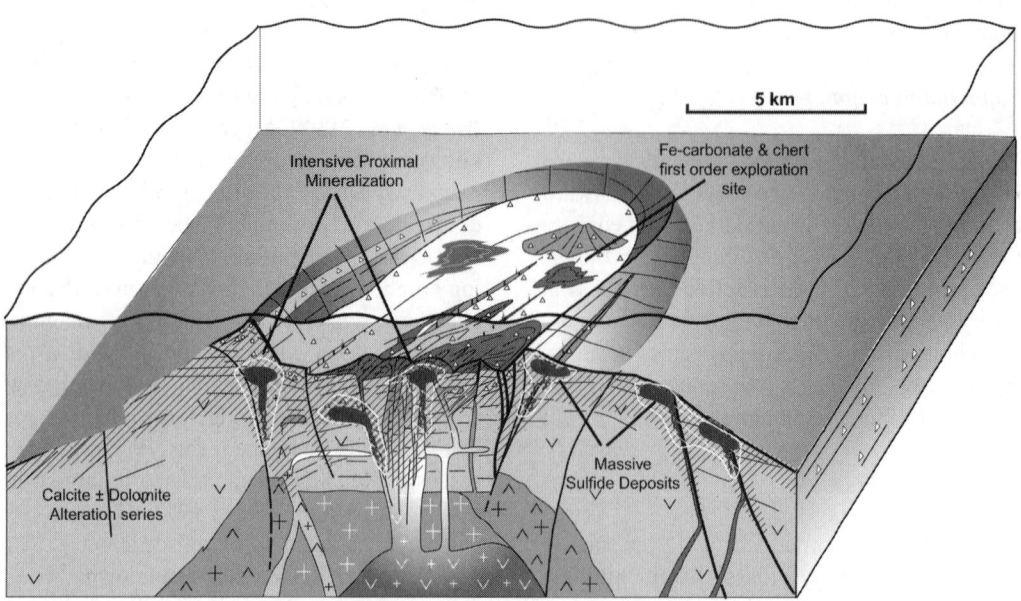

Figure 5. Mineral deposits associated with products of subaqueous eruptions in a typical subaqueous caldera; the Hunter Mine Caldera. Small volcanoes on caldera floor are sites of intra-caldera fountaining or phreatomagmatic eruptions assoiated with feeder dikes and associated faults and fissures that are of high exploration potential for carbonate, chert and oxide dominated ore deposits. (Drawing simplified after Mueller & White, unpublished manuscript.)

presence of chaotic breccia assemblages containing rotated megablocks and slumped composite blocks of the caldera wall [e.g. *Lipman*, 1976]. Abundant fractures filled with silica and/or iron-formation are associated with these faults, as are mafic and felsic dykes [*Mueller and Mortensen*, 2002]. The gold-rich VMS Sunrise deposit of the Myojin Knoll volcano, located at the foot of the caldera wall at 1210-1360 m depth [*Iizasa et al.*, 1999; *Fiske et al.*, 2001], exemplifies the importance of such bounding faults.

Caldera-floor faults are also common, and graben-style structures may be occupied by dome-flow complexes and explosive dome collapse breccias. The dome complexes, as well as their endogenous counterparts, are prevalent in the central segment of the caldera floor and represent prime exploration targets because domes, lavas and fractured hypabyssal intrusions represent both heat sources and suitable roof rocks for hydrothermal precipitation. An extensive felsic sheeted dyke complex in the Hunter Mine Group marks significant draining of the magma reservoir onto the overlying caldera floor [*Mueller and Donaldson*, 1992; *Dostal and Mueller*, 1996], and also served as a primary circulation path for hydrothermal fluids. The domes' feeder dykes typically intrude along synvolcanic faults of the broken caldera floor, and the fractured, leaky caldera floor probably also plays a role in development of the devolatilized and viscous dome magmas by allowing volatiles to stream away from underlying magma. Subaqueous fountaining and surtseyan eruptions producing less viscous magma than the domes may be favored nearer the walls of the caldera. Primary pyroclastic debris, such as the fining-upward deposits of fountaining, eruption-fed density currents and explosive collapse, as well as their reworked counterparts, provide the necessary porous and permeable host rock for subsurface VMS precipitation.

Hydrothermal alteration patterns are first order indicators of mineralization. There are extensive and distinct patterns of alteration around the Hunter Mine VMS deposits [*Chown et al.*, 2000] and similar sites [*Franklin*, 1990; *Offler and Withford*, 1992; *Galley*, 1993; *Large et al.*, 2001], and alteration zones are especially well developed in volcaniclastic deposits because of their high porosity and permeability. Semi-conformable carbonate alteration halos with calcite-dolomite-ankerite-siderite-magnesite [e.g. *Galley*, 1993] represent the most pervasive hydrothermal pattern in Archean caldera complexes [Mattabi-type; *Morton and Franklin*, 1987], and may extend for 30–50 km. Intense alteration is focused along syneruptive fault systems.

Acknowledgments. The section on economically important deposits, and accompanying figure, are modified from part of a review manuscript (Mueller and White, in prep) to be submitted. Many thanks are owed to the production staff at AGU for forbearance with the lead editor's slow delivery of this paper. Review comments from Wulf Mueller, Magnus Gudmundsson, and Pierre-Simon Ross helped improve the presentation.

REFERENCES

Allen, J.R.L., *Sedimentary Structures. Their Character and Physical Basis*, 593 pp., Elsevier, Amsterdam, 1982.

Andrews, B., Eruptive and Depositional Mechanisms of an Eocene Shallow Submarine Volcano, Moeraki Peninsula, New Zealand, in *Explosive Subaqueous Volcanism*, edited by J.D.L. White, J.L. Smellie, and D.A. Clague, American Geophysical Union, Washington DC, this volume.

Batiza, R., N. Becker, D. Bercovici, T. Coleman, T. Gorman, J. Head III, L. Holloway, J. Karsten, A. Kelly, L. Keszthelyi, D. Maicher, W. Mueller, J. Muller, L. Norby, J. Paduan, G. Parker, L. Prockter, D. Stakes, and J. White, New evidence from Alvin for the origin of deep-sea eruptive hyaloclastite on Seamount 6, Cocos Plate, 12°43' N, in *1996 Fall Meeting Abstracts*, *EOS*, *77*, F319, 1996.

Batiza, R., D.J. Fornari, D.A. Vanko, and P. Lonsdale, Craters, caldereas and hyaloclastites on young Pacific seamounts, *Journal of Geophysical Research*, *89*, 8371-8390, 1984.

Batiza, R., and J.D.L. White, Submarine lava and hyaloclastite, in *Encyclopedia of Volcanoes*, edited by H. Sigurdsson, B. Houghton, S. McNutt, H. Rymer, and J. Stix, pp. 361-382, Academic Press, New York, 2000.

Binns, R., Deep Marine Pumice from the Woodlark and Manus Basins, Papua New Guinea, in *Explosive Subaqueous Volcanism*, edited by J.D.L. White, J.L. Smellie, and D.A. Clague, American Geophysical Union, Washington DC, this volume.

Brooks, E.R., Paleozoic fluidization, folding and peperite formation, Northern Sierra Neveda, California, *Canadian Journal of Earth Sciences*, *32*, 314-324, 1995.

Busby-Spera, C.J., Large volume rhyolite ash-flow eruptions and submarine caldera collapse in the Lower Mesozoic Sierra Nevada, California, *Journal of Geophysical Research*, *89*, 8417-8427, 1984.

Busby-Spera, C.J., Depositional features of rhyolitic and andesitic volcaniclastic rocks of the Mineral King submarine caldera complex, Sierra Nevada, California, *Journal of Volcanology and Geothermal Research*, *27*, 43-76, 1986.

Carey, S., Influence of convective sedimentation on the formation of widespread tephra fall layers in the deep sea, *Geology*, *25*, 839-842, 1997.

Carlisle, D., Pillow breccias and their aquagene tuffs, Quadra Island, British Columbia, *Journal of Geology*, *71*, 48-71, 1963.

Cas, R.A.F., and J.V. Wright, Subaqueous pyroclastic flows and ignimbrites: an assessment, *Bulletin of Volcanology*, *53*, 357-380, 1991.

Cas, R.A.F., H. Yamagishi, L. Moore, and C.R. Scutter, Miocene Submarine Fire Fountain Deposits, Ryugazaki Headland,

Oshoro Peninsula, Hokkaido, Japan: Implications for Submarine Foun-tain Dynamics and Fragmentation Processes, in *Explosive Subaqueous Volcanism*, edited by J.D.L. White, J.L. Smellie, and D.A. Clague, American Geophysical Union, Washington DC, this volume.

Cashman, K.V., and R.S. Fiske, Fallout of pyroclastic debris from submarine volcanic eruptions, *Science, 253*, 275-280, 1991.

Chartrand, F., and S. Cattalani, Massive sulfide deposits in northwestern Quebec, in *The northwestern Quebec polymetallic belt; a summary of 60 years of mining exploration; proceedings of the Rouyn-Noranda 1990 symposium, Canadian Institute of Mining and Metallurgy Special Volume, 43*, pp. 77-91, Canadian Institute of Mining and Metallurgy, Montreal, PQ, Canada, 1990.

Chown, E.H., E. Ndah, and W.U. Mueller, The relation between iron-formation and low temperature hydrothermal alteration in an Archean volcanic environment, *Precambrian Research, 101*, 263-275, 2000.

Clague, D., R. Batiza, J.W. Head, and A. Davis, Pyroclastic and Hydroclastic Deposits on Loihi Seamount, Hawaii, in *Explosive Subaqueous Volcanism*, edited by J.D.L. White, J.L. Smellie, and D.A. Clague, American Geophysical Union, Washington D.C., this volume a.

Clague, D.A., A.S. Davis, J.L. Bischoff, J.E. Dixon, and R. Geyer, Lava bubble-wall fragments formed by submarine hydrovolcanic explosions on Loihi Seamount and Kilauea Volcano, *Bulletin of Volcanology, 61*, 437-449, 2000.

Clague, D.A., A.S. Davis, and J.E. Dixon, Submarine strombolian eruptions on the Gorda mid-ocean ridge, in *Explosive Subaqueous Volcanism*, edited by J.D.L. White, J.L. Smellie, and D.A. Clague, American Geophysical Union, Washington DC, this volume b.

Clague, D.A., R.T. Holcomb, J.M. Sinton, R.S. Detrick, and M.E. Torresan, Pliocene and Pleistocene alkalic flood basalts on the seafloor north of the Hawaiian Islands, *Earth and Planetary Science Letters, 98*, 175-191, 1990.

Clague, D.A., and J.G. Moore, The proximal part of the giant submarine Wailau Landslide, Molokai, Hawaii, *Journal of Volcanology and Geothermal Research, 113*, 259-287, 2002.

Clague, D.A., J.G. Moore, and J.R. Reynolds, Formation of submarine flat-topped volcanic cones in Hawai'i, *Bulletin of Volcanology, 62*, 214-233, 2000b.

Colgate, S.A., and T. Sigurgeirsson, Dynamic mixing of water and lava, *Nature, 244*, 552-555, 1973.

Coombs, D.S., and C.A. Landis, Pumice from the South Sandwich eruption of March 1962 reaches New Zealand, *Nature (London), 209*, 289-290, 1966.

Cousineau, P.A., and J.H. Bedard, Sedimentation in a subaqueous arc/back-arc setting: the Bobby Cove Formation, Snooks Arms Group, Newfoundland, *Precambrian Research, 101*, 111-134, 2000.

Crisp, J.A., Rates of magma emplacement and volcanic output, *Journal of Volcanology and Geothermal Research, 20*, 177-211, 1984.

Davis, A., and D.A. Clague, Hyaloclastite From Miocene Seamounts Offshore Central California: Compositions, Eruption Styles, and Depositional Processes, in *Explosive Subaqueous Volcanism*, edited by J.D.L. White, J.L. Smellie, and D.A. Clague, American Geophysical Union, Washington DC, this volume.

Davis, A.S., and D.A. Clague, President Jackson Seamounts, northern Gorda Ridge: Tectonomagmatic relationship between on- and off-axis volcanism, *Journal of Geophysical Research-Solid Earth, 105*, 27939-27956, 2000.

De Rita, D., G. Giordano, and A. Cecili, A model for submarine rhyolite dome growth; Ponza Island (central Italy), *Journal of Volcanology and Geothermal Research, 107*, 221-239, 2001.

De Rosen-Spence, A.F., G. Provost, E. Dimroth, K. Gochnauer, and V. Owen, Archean subaqueous felsic flows, Rouyn-Noranda, Quebec, Canada, and their Quarternary equivalents, *Precambrian Research, 12*, 43-77, 1980.

Dixon, J.E., D.A. Clague, P. Wallace, and R. Poreda, Volatiles In Alkalic Basalts From the North Arch Volcanic Field, Hawaii - Extensive Degassing Of Deep Submarine-Erupted Alkalic Series Lavas, *Journal of Petrology, 38*, 911-939, 1997.

Dostal, J., and W. Mueller, An Archean oceanic felsic dyke swarm in a nascent arc; the Hunter Mine Group, Abitibi greenstone belt, Canada, *Journal of Volcanology and Geothermal Research, 72*, 37-57, 1996.

Doucet, P., W. Mueller, and F. Chartrand, Archean, deep-marine, volcanic eruptive products associated with the Coniagas massive sulfide deposit, Quebec, Canada, *Canadian Journal of Earth Sciences, 31*, 1569-1584, 1994.

Doyle, M.G., Clast shape and textural associations in peperite as a guide to hydromagmatic interactions: Upper Permian basaltic and basaltic andesite examples from Kiama, Australia, *Australian Journal of Earth Sciences, 47*, 167-177, 2000.

Doyle, M.G., and J. McPhie, Facies architecture of a silicic intrusion-dominated volcanic centre at Highway-Reward, Queensland, Australia, *Journal of Volcanology and Geothermal Research, 99*, 79-96, 2000.

Eissen, J.-P., Y. Fouquet, D. Hardy, and H. Ondréas, Recent MORB Volcaniclastic Explosive Deposits Formed Between 500 and 1750 m.b.s.l. on the Axis of the Mid-Atlantic Ridge, South of the Azores, in *Explosive Subaqueous Volcanism*, edited by J.D.L. White, J.L. Smellie, and D.A. Clague, American Geophysical Union, Washington DC, this volume.

Fink, J.H., and C.R. Manley, Origin of pumiceous and glassy textures in rhyolite flows and domes, in *The emplacement of silicic domes and lava flows, Geological Society of America Special Paper, 212*, 77-88, 1987.

Fisher, R.V., Submarine volcaniclastic rocks, in *Marginal Basin Geology: Volcanic and Associated Sedimentary and Tectonic Processes in modern and Ancient Marginal Basins*, edited by B.P. Kokelaar, and M.F. Howells, *Geological Society of London Special Publication, 16*, 5-27, 1984.

Fisher, R.V., and H.-U. Schmincke, *Pyroclastic Rocks*, 472 pp., Springer-Verlag, Berlin, 1984.

Fiske, R.S., Subaqueous pyroclastic flows in the Ohanapecosh Formation, Washington, *Geological Society of America Bulletin, 74*, 391-406, 1963.

Fiske, R.S., K.V. Cashman, A. Shibata, and K. Watanabe, Tephra dispersal from Myojinsho, Japan, during its shallow submarine eruption of 1952-1953, *Bulletin of Volcanology*, 59, 262-275, 1998.

Fiske, R.S., and T. Matsuda, Submarine equivalents of ash flows in the Tokiwa Formation, Japan, *American Journal of Science*, 262, 76-106, 1964.

Fiske, R.S., J. Naka, K. Iizasa, M. Yuasa, and A. Klaus, Submarine silicic caldera at the front of the Izu-Bonin Arc, Japan; voluminous seafloor eruptions of rhyolite pumice, *Geological Society of America Bulletin*, 113, 813-824, 2001.

Fouquet, Y., J.P. Eissen, H. Ondreas, F. Barriga, R. Batiza, and L. Danyushevsky, Extensive volcaniclastic deposits at the Mid-Atlantic Ridge axis: results of deep-water basaltic explosive volcanic activity?, *Terra Nova*, 10, 280- 286, 1998.

Franklin, J.M., Volcanic-associated massive sulphide deposits, in *Gold and base-metal mineralization in the Abitibi subprovince, Canada, with emphasis on the Quebec segment: short-course notes*, edited by S.E. Ho, F. Roberts, and D.I. Groves, *The University of Western Australia Publication, 24*, pp. 211-241, 1990.

Fujibayashi, N., and U. Sakai, Vesicularity and vesicle size distribution in glassy clasts of subaqueous lava fountain deposits from the Middle Miocene Ogi Basalts, Sado Island, Japan., in *Explosive Subaqueous Volcanism*, edited by J.D.L. White, J.L. Smellie, and D.A. Clague, American Geophysical Union, Washington DC, this volume.

Galley, A.G., Characteristics of semi-conformable alteration zones associated with volcanogenic massive sulphide districts, *Journal of Geochemical Exploration*, 48, 175-200, 1993.

Gaspar, J.L., G. Queiroz, J.M. Pacheco, T. Ferreira, N. Wallenstein, M.H. Almeida, and R. Coutinho, Basaltic Lava Balloons Produced During the 1998-2001 Serreta Submarine Ridge Eruption (Azores), in *Explosive Subaqueous Volcanism*, edited by J.D.L. White, J.L. Smellie, and D.A. Clague, American Geophysical Union, Washington D.C., this volume.

Gibson, H.L., The Mine Sequence of the Central Noranda volcanic complex: Geology, alteration, massive sulfide deposits and volcanological reconstruction, PhD thesis, Carleton University, Ottawa, 1990.

Gibson, H.L., R. Morton, and G. Hudak, Subaqueous volcanism: environments and controls on VMS mineralization, in *Volcanic-Associated Massive Sulfide Deposits: Processes and Examples in Modern and Ancient Settings*, edited by T.C. Barrie, and M.D. Hannington, pp. 13-51, 1997.

Gibson, H.L., and D.H. Watkinson, Volcanogenic massive sulphide deposits of the Noranda cauldron and shield volcano, Quebec, in *The northwestern Quebec polymetallic belt; a summary of 60 years of mining exploration; proceedings of the Rouyn-Noranda 1990 symposium, Canadian Institute of Mining and Metallurgy Special Volume*, 43, 119-132, 1990.

Gibson, H.L., D.H. Watkinson, and C.D.A. Comba, Subaqueous phreatomagmatic explosion breccias at Buttercup Hill, Noranda, Quebec, *Canadian Journal of Earth Sciences*, 26, 1428-1439, 1989.

Gill, J., P. Torssandu, H. Lapierre, R. Taylor, K. Kaiho, M. Koyama, M. Kusakabe, J. Aitchison, C. Cisonski, K. Dadey, K. Fujioka, A. Klaus, M. Lovell, K. Marsaglia, P. Pesard, B. Taylor, and K. Tazaki, Explosive deep water basalt in the Sumisu backarc rift, *Science*, 248, 1214-1217, 1990.

Goto, Y., and J. McPhie, Endogenous growth of a Miocene submarine dacite cryptodome, Rebun Island, Hokkaido, Japan, *Journal of Volcanology and Geothermal Research*, 84, 273-286, 1998.

Gudmundsson, A., Magma Chambers Modeled As Cavities Explain the Formation Of Rift Zone Central Volcanoes and Their Eruption and Intrusion Statistics, *Journal Of Geophysical Research-Solid Earth*, 103, 7401-7412, 1998.

Gudmundsson, M.T., Melting of ice by magma-ice-water interactions during subglacial eruptions as an indicator of heat transfer in subaqueous eruptions, in *Explosive Subaqueous Volcanism*, edited by J.D.L. White, J.L. Smellie, and D.A. Clague, American Geophysical Union, Washington D.C., this volume.

Head, J.W., and L. Wilson, Deep submarine pyroclastic eruptions: theory and predicted landforms and deposits, *Journal of Volcanology and Geothermal Research*, 151, 155-193, 2003.

Houghton, B.F., B.J. Hobden, K.V. Cashman, C.J.N. Wilson, and R.T. Smith, Large-scale Interaction of Lake Water and Rhyolitic Magma During the 1.8 ka Taupo Eruption, New Zealand., in *Explosive Subaqueous Volcanism*, edited by J.D.L. White, J.L. Smellie, and D.A. Clague, American Geophysical Union, Washington D.C., this volume.

Houghton, B.F., and C.J.N. Wilson, A vesicularity index for pyroclastic deposits, *Bulletin of Volcanology*, 51, 451-462, 1989.

Howells, M.F., A.J. Reedman, and S.D.G. Campbell, The submarine eruption and emplacement of the Lower Rhyolitic Tuff Formation (Ordovician), N Wales, *Journal of the Geological Society of London*, 143, 411-423, 1986.

Hudak, G.J., R.L. Morton, J.M. Franklin, and D.M. Peterson, Morphology, distribution, and estimated eruption volumes for intracaldera felsic tuff deposits in the Archean Sturgeon Lake subaqueous caldera complex, Northwestern Ontario, in *AGU Chapman Conference on Explosive Subaqueous Volcanism*, edited by J.D.L. White, and B.F. Houghton, pp. 30, Dunedin, New Zealand, 2002.

Hudak, G.J., R.L. Morton, J.M. Franklin, and D.M. Peterson, Morphology, distribution, and estimated eruption volumes for intracaldera tuffs associated with volcanic-hosted massive sulfide deposits in the Archean Sturgeon Caldera Complex, northwestern Ontario, in *Explosive Subaqueous Volcanism*, edited by J.D.L. White, J.L. Smellie, and D.A. Clague, American Geophysical Union, Washington DC, this volume.

Ishibashi, J.-I., and T. Urabe, Hydrothermal activity related to arc-backarc magmatism in the western Pacific, in *Backarc Basins: Tectonics and Magmatism*, edited by B. Taylor, pp. 451-495, Plenum Press, New York, 1995.

Iizasa, K., R.S. Fiske, O. Ishizuka, M. Yuasa, J. Hashimoto, J. Ishibashi, J. Naka, Y. Horii, Y. Fujiwara, A. Imai, and S. Koyama, A Kuroko-type polymetallic sulfide deposit in a submarine silicic caldera, *Science*, 283, 975-977, 1999.

Iizasa, K., K. Kawasaki, K. Maeda, T. Matsumoto, N. Saito, and K. Hirai, Hydrothermal Sulfide-Bearing Fe-Si Oxyhydroxide Deposits From the Coriolis Troughs, Vanuatu Backarc, Southwestern Pacific, *Marine Geology, 145*, 1-21, 1998.

Jones, J.G., Intraglacial volcanoes of the Laugarvatn region, southwest Iceland, *Journal of the Geological Society of London, 124*, 197-211, 1969.

Kano, K., Subaqueous pumice eruptions and their products, in *Explosive Subaqueous Volcanism*, edited by J.D.L. White, J.L. Smellie, and D.A. Clague, American Geophysical Union, Washington DC, this volume.

Kano, K., G.J. Orton, and T. Kano, A hot Miocene subaqueous scoria-flow deposit in the Shimane Peninsula, SW Japan, *Journal of Volcanology and Geothermal Research, 60*, 1-14, 1994.

Kessel, L.G., and C.J. Busby, Analysis of VHMS-hosting ignimbrites erupted at bathyal water depths (Ordovician Bald Mountain Sequence, Northern Maine), in *Explosive Subaqueous Volcanism*, edited by J.D.L. White, J.L. Smellie, and D.A. Clague, American Geophysical Union, Washington DC, this volume.

Kokelaar, B.P., The mechanism of Surtseyan volcanism, *Journal of the Geological Society of London, 140*, 939-944, 1983.

Kokelaar, B.P., Magma-water interactions in subaqueous and emergent basaltic volcanism, *Bulletin of Volcanology, 48*, 275-289, 1986.

Kokelaar, B.P., R.E. Bevins, and R.A. Roach, Submarine silicic volcanism and associated sedimentary and tectonic processes, Ramsey Island, SW Wales, *Journal of the Geological Society of London, 142*, 591-613, 1985.

Kokelaar, B.P., and C.J. Busby, Subaqueous explosive eruption and welding of pyroclastic deposits, *Science, 257*, 196-201, 1992.

Kokelaar, B.P., and G.P. Durant, The submarine eruption and erosion of Surtla (Surtsey), Iceland, *Journal of Volcanology and Geothermal Research, 19*, 239-246, 1983.

Koyaguchi, T., and A.W. Woods, On the formation of eruption columns following explosive mixing of magma and surface water, *Journal of Geophysical Research, 101*, 5561-5574, 1996.

Lafrance, B., W.U. Mueller, R. Daigneault, and N. Dupras, Evolution of a submerged composite arc volcano: volcanology and geochemistry of the Normetal volcanic complex, Abitibi greenstone belt, Quebec, Canada, *Precambrian Research, 101*, 277-311, 2000.

Large, R.R., J. McPhie, J.B. Gemmell, W. Herrmann, and G.J. Davidson, The spectrum of ore deposit types, volcanic environments, alteration halos, and related exploration vectors in submarine volcanic successions: Some examples from Australia, *Economic Geology, 96*, 913-938, 2001.

Lichtblau, A., Stratigraphy and facies at the south margin of the Archean Noranda Caldera, MSc thesis, Université du Québec à Chicoutimi, Chicoutimi, 1989.

Lipman, P.W., Caldera-collapse breccias in the western San Juan Mountains, Colorado, *Geological Society of America Bulletin, 87*, 1397-1410, 1976.

Lipman, P.W., Subsidence of ash-flow calderas: relation to caldera size and magma-chamber geometry, *Bulletin of Volcanology, 59*, 198-218, 1997.

Lonsdale, P., and R. Batiza, Hyaloclastite and lava flows on young seamounts examined with a submersible, *Geological Society of America Bulletin, 91*, 545-554, 1980.

Maicher, D., Hyaloclastite beds of shelf and seamount: roles of exsolution, entrapment and entrainment, unpublished Phd dissertation, University of Otago, Dunedin, 1999.

Maicher, D., A cluster of surtseyan volcanoes at Lookout Bluff, North Otago, New Zealand: Aspects of edifice spacing and time, in *Explosive Subaqueous Volcanism*, edited by J.D.L. White, J.L. Smellie, and D.A. Clague, American Geophysical Union, Washington DC, this volume.

Maicher, D., and J.D.L. White, The formation of deep-sea Limu o Pele, *Bulletin of Volcanology, 63*, 482-496, 2001.

Maicher, D., J.D.L. White, and R. Batiza, Sheet hyaloclastite: density-current deposits of quench and bubble-burst fragments from thin, glassy sheet lava flows, Seamount Six, Eastern Pacific Ocean, *Marine Geology, 171*, 75-94, 2000.

Mandeville, C.W., S.N. Carey, H. Sigurdsson, and J. King, Paleomagnetic evidence for high-temperature emplacement of the 1883 subaqueous pyroclastic flows from Krakatau volcano, Indonesia, *Journal of Geophysical Research, 99*, 9487-9504, 1994.

Manville, V., J.D.L. White, B.F. Houghton, and C.J.N. Wilson, The saturation behaviour of pumice and some sedimentological implications, *Sedimentary Geology, 119*, 5-16, 1998.

Mattox, T.N., and M.T. Mangan, Littoral hydrovolcanic explosions: a case study of lava-seawater interaction at Kilauea Volcano, *Journal of Volcanology and Geothermal Research, 75*, 1-17, 1997.

McInnes, B.I.A., Kavachi eruption (video), CSIRO, Australia, 2000.

McPhie, J., A Pliocene Shoaling Basaltic Seamount - Ba Volcanic Group At Rakiraki, Fiji, *Journal of Volcanology and Geothermal Research, 64*, 193-210, 1995.

McPhie, J., and R.L. Allen, Facies architecture of mineralized submarine volcanic sequences: Cambrian Mount read volcanics, Western Tasmania, *Economic Geology, 87*, 587-596, 1992.

McPhie, J., and R.L. Allen, Submarine, Silicic, Syn-eruptive Pyroclastic Units in the Mount Read Volcanics, Western Tasmania: Influence of Vent Setting and Proximity on Lithofacies Characteristics, in *Explosive Subaqueous Volcanism*, edited by J.D.L. White, J.L. Smellie, and D.A. Clague, American Geophysical Union, Washington DC, this volume.

McPhie, J., M. Doyle, and R. Allen, *Volcanic Textures: a guide to the interpretation of textures in volcanic rocks*, 198 pp., CODES Key Centre, University of Tasmania, Hobart, 1993b.

Moore, J.G., Structure and eruptive mechanisms at Surtsey Volcano, Iceland, *Geological Magazine, 122*, 649-661, 1985.

Moore, J.G., R.L. Phillips, R.W. Grigg, D.W. Peterson, and D.A. Swanson, Flow of lava into the sea 1969-1971, Kilauea Volcano, Hawaii, *Geological Society of America Bulletin, 84*, 537-546, 1973.

Morton, R.L., and J.M. Franklin, Two-fold classification of Archean volcanic-associated massive sulfide deposits, *Economic Geology and the Bulletin of the Society of Economic Geologists, 82*, 1057-1063, 1987.

Morton, R.L., J.S. Walker, G.J. Hudak, and J.M. Franklin, The early development of an Archean submarine caldera complex with emphasis on the Mattabi ash-flow tuff and its relationship to the Mattabi massive sulfide deposit, *Economic Geology, 86*, 1002-1011, 1991.

Mueller, W.U., E.H. Chown, and R. Potvin, Substorm wave base felsic hydroclastic deposits in the Archean Lac des Vents volcanic complex, Abitibi Belt, Canada, *Journal of Volcanology and Geothermal Research, 60*, 273-300, 1994.

Mueller, W.U., and J.A. Donaldson, A felsic dyke swarm formed under the sea: the Archean Hunter Mine Group, south-central Abitibi Belt, Quebec, Canada, *Bulletin of Volcanology, 54*, 602-610, 1992.

Mueller, W.U., and J.D.L. White, Felsic fire-fountaining bneath Archean seas: Pyroclastic deposits of the 2730Ma Hunter Mine Group, Quebec, Canada, *Journal of Volcanology and Geothermal Research, 54*, 117-134, 1992.

Mueller, W.U., Subaqueous eruption-fed density currents from small volume mafic eruptions: the crossover from volcanology to sedimentology, *Commission on volcanogenic sediments (CVS) Newsletter, 20*, 2-9, 2001.

Mueller, W.U., A subaqueous eruption model for shallow-water, small volume eruptions: Evidence from two Precambrian examples, in *Explosive Subaqueous Volcanism*, edited by J.D.L. White, J.L. Smellie, and D.A. Clague, American Geophysical Union, Washington DC, this volume.

Mueller, W.U., A. Aubin, H. van der Berg, R. van der Walt, and D. Pretorius, Recognition of Precambrian eruption-fed density currents: the response to shallow-water, small volume eruptions, in *AGU Chapman Conference on Explosive Subaqueous Volcanism*, edited by J.D.L. White, and B.F. Houghton, pp. 28, Dunedin, New Zealand, 2002.

Mueller, W.U., A.A. Garde, and H. Stendal, Shallow-water, eruption-fed, mafic pyroclastic deposits along a Paleoproterozoic coastline; Kangerluluk volcano-sedimentary sequence, Southeast Greenland, in *Processes in physical volcanology and volcaniclastic sedimentation; modern and ancient*, edited by W.U. Mueller, E.H. Chown, and P.C. Thurston, *Precambrian Research, 101*, 163-192, 2000.

Mueller, W.U., and J. Mortensen, Age constraints and characteristics of subaqueous volcanic construction, the Archean Hunter Mine Group, Abitibi greenstone belt, *Precambrian Research, in press*, 2002.

Offler, R., and D.J. Withford, Wall-rock alteration and metamorphism of a volcanic-hosted massive sulfide deposit at Que river, Tasmania: Petrology and mineralogy, *Economic Geology, 87*, 686-705, 1992.

Palais, J.M., and H. Sigurdsson, Petrologic evidence of volatile emissions from major historic and pre-historic volcanic eruptions, in *Understanding Climate Change*, edited by A. Berger, R.E. Dickinson, and J.W. Kidson, *Geophysical Monograph, 7*, 31-53, 1989.

Raos, A.M., and J. McPhie, The submarine record of a large-scale explosive eruption in the Vanuatu arc: ~1 Ma Efaté Pumice Formation, in *Explosive Subaqueous Volcanism*, edited by J.D.L. White, J.L. Smellie, and D.A. Clague, American Geophysical Union, Washington DC, this volume.

Schmincke, H.-U., and U. Bednarz, Pillow, sheet flow and breccia flow volcanoes and volcano-tectonic hydrothermal cycles in the Extrusive Series of the northeastern Troodos ophiolite (Cyprus), in *Ophiolites, Oceanic Crustal Analogues: proceedings of the Symposium "Troodos 1987"*, edited by J. Malpas, E.M. Moores, A. Panayiotou, and C. Xenophontos, pp. 185-206, Geological Survey Dept, Ministry of Agriculture and Natural Resources, Nicosia, Cyprus, 1990.

Schneider, J.-L., C. Fourquin, and J.-L. Paicheler, Two examples of subaqueously welded ash-flow tuffs: the Visean of southern Vosges (France) and the upper Createcous of northern Anatolia (Turkey), *Journal of Volcanology and Geothermal Research, 49*, 365-383, 1992.

Self, S., Large-scale phreatomagmatic silicic volcanism: a case study from New Zealand, *Journal of Volcanology and Geothermal Research, 17*, 433-469, 1983.

Sheridan, M.F., and K.H. Wohletz, Hydrovolcanic explosions: the systematics of water-pyroclast equilibration, *Science, 212*, 1387-1389, 1981.

Siebe, C., J.-C. Komorowski, C. Navarro, J. McHone, H. Delgado, and A. Cortes, Submarine eruption near Socorro Island, Mexico: Geochemistry and scanning electron microscopy studies of floating scoria and reticulite, *Journal of Volcanology and Geothermal Research, 68*, 239-271, 1995.

Simpson, K., and J. McPhie, Fluidal-clast breccia generated by submarine fire fountaining, Trooper Creek Formation, Queensland, Australia, *Journal of Volcanology and Geothermal Research, 109*, 339-355, 2001.

Skilling, I., J.D.L. White, and J. McPhie, Peperite: a review of magma-sediment mingling, in *Peperites: processes and products of magma-sediment mingling*, edited by I. Skilling, J.D.L. White, and J. McPhie, *Journal of Volcanology and Geothermal Research, 114*, 1-17, 2002.

Skilling, I.P., Evolution of an englacial volcano: Brown Bluff, Antarctica, *Bulletin of Volcanology, 56*, 573-591, 1994.

Smellie, J.L., Lithostratigraphy of Miocene-Recent, alkaline volcanic fields in the Antarctic Peninsula and eastern Ellsworth Land, *Antarctic Science, 11*, 362-378, 1999.

Smellie, J.L., Subglacial Eruptions, in *Encyclopedia of Volcanoes*, edited by H. Sigurdsson, B.F. Houghton, S.R. McNutt, H. Rymer, and J. Stix, pp. 403-418, Academic Press, San Diego, 2000.

Smellie, J.L., Lithofacies architecture and construction of volcanoes erupted in englacial lakes; icefall nunatak, Mount Murphy, eastern Marie Byrd Land, Antarctica, in *Volcaniclastic sedimentation in lacustrine settings*, edited by J.D.L. White, and N.R. Riggs, pp. 9-34, Blackwell, Oxford, International, 2001.

Smellie, J.L., and M.J. Hole, Products and processes in Pliocene-Recent, subaqueous to emergent volcanism in the Antarctic Peninsula: examples of englacial Surtseyan volcano construction, *Bulletin of Volcanology, 58*, 628-646, 1997.

Smith, R.L., and R.A. Bailey, Resurgent cauldrons, in *Studies in Volcanology (Howell Williams volume)*, edited by R.R. Coats,

R.L. Hay, and C.A. Anderson, pp. 153-210, Geological Society of America, 1968.

Smith, T.L., and R. Batiza, New field and laboratory evidence for the origin of hyaloclastite flows on seamount summits, *Bulletin of Volcanology, 51*, 96-114, 1989.

Sparks, R.S.J., The dynamics of bubble formation and growth in magmas: a review and analysis, *Journal og Volcanology and Geothermal Research, 3*, 1-37, 1978.

Sparks, R.S.J., H. Sigurdsson, and S.N. Carey, The entrance of pyroclastic flows into the sea, II. Theoretical considerations on subaqueous emplacement and welding, *Journal of Volcanology and Geothermal Research, 7*, 97-105, 1980.

Staudigel, H., and H.-U. Schmincke, The Pliocene seamount series of La Palma/Canary Islands, *Journal of Geophysical Research, 89*, 11195-11215, 1984.

Stix, J., Subaqueous, intermediate to silicic-composition explosive volcanism; a review, *Earth-Science Reviews, 31*, 21-53, 1991.

Stix, J., B. Kennedy, M. Hannington, H. Gibson, R. Fiske, W. Mueller, and J. Franklin, Caldera-forming processes and the origin of submarine volcanogenic massive sulfide deposits, *Geology (Boulder), 31*, 375-378, 2003.

Tamura, Y., M. Koyama, and R.S. Fiske, Paleomagnetic evidence for hot pyroclastic debris flow in the shallow submarine Shirahama Group (upper Miocene-Pliocene), Japan, *Journal of Geophysical Research, B, Solid Earth and Planets, 96*, 21,779-21,787, 1991.

Thorarinsson, S., *Surtsey. The New Island in the North Atlantic*, 47 pp., The Viking Press, New York, 1967.

Thorarinsson, S., T. Einarsson, S.G. E., and G. Elisson, The submarine eruption off the Vestmann Islands, 1963-64, *Bulletin of Volcanology, 27*, 434-445, 1964.

Tilling, R.I., and J.J. Dvorak, Anatomy of a basaltic volcano, *Nature, London, 363*, 125-133, 1993.

Tribble, G.W., Underwater observations of active lava flows from Kilauea volcano, Hawaii, *Geology, 19*, 633-636, 1991.

Walker, G.P.L., Downsag calderas, ring faults, caldera sizes, and incremental caldera growth, *journal of Geophysical Research, 89*, 8407-8416, 1984.

White, J.D.L., The depositional record of small, monogenetic volcanoes within terrestrial basins, in *Sedimentation in Volcanic Settings*, edited by R.V. Fisher, and G.A. Smith, *Society of Economic Paleontologists and Mineralogists Special Publication, 45*, 155-171, 1991.

White, J.D.L., Pre-emergent construction of a lacustrine basaltic volcano, Pahvant Butte, Utah (USA), *Bulletin of Volcanology, 58*, 249-262, 1996.

White, J.D.L., Subaqueous eruption-fed density currents and their deposits, in *Processes in physical volcanology and volcaniclastic sedimentation: modern and ancient*, edited by W. Mueller, E.H. Chown, and P.C. Thurston, *Precambrian Research, 101*, 87-109, 2000.

White, J.D.L., and B.F. Houghton, Surtseyan and related eruptions, in *Encyclopedia of Volcanoes*, edited by H. Sigurdsson, B. Houghton, S. McNutt, H. Rymer, and J. Stix, pp. 1417, Academic Press, New York, 2000.

White, J.D.L., V. Manville, C.J.N. Wilson, B.F. Houghton, N.R. Riggs, and M. Ort, Settling and deposition of 181 A.D. Taupo pumice in lacustrine and associated environments, in *Volcaniclastic Sedimentation in Lacustrine Settings*, edited by J.D.L. White, and N.R. Riggs, *International Association of Sedimentologists Special Publication, 30*, 141-150, 2001.

White, M.J., and J. McPhie, A submarine welded ignimbrite - crystal-rich sandstone facies association in the Cambrian Tyndall Group, western Tasmania, Australia, *Journal of Volcanology and Geothermal Research, 76*, 277-295, 1997.

Whitham, A.G., and R.S.J. Sparks, Pumice, *Bulletin of Volcanology, 48*, 209-223, 1986.

Wiesner, M.G., Y. Wang, and L. Zheng, Fallout of volcanic ash to the deep South China Sea induced by the 1991 eruption of Mount Pinatubo (Phillipines), *Geology*, 885-888, 1995.

Wilson, C.J.N., Stratigraphy, chronology, styles and dynamics of late Quaternary eruptions from Taupo volcano, New Zealand, *Philosophical transactions of the Royal Society of London, 343*, 205-306, 1993.

Wilson, C.J.N., The 26.5 ka Oruanui eruption, New Zealand: an introduction and overview, *Journal of Volcanology and Geothermal Research, 112*, 133-174, 2000.

Wilson, C.J.N., and W. Hildreth, The Bishop Tuff - New Insights From Eruptive Stratigraphy, *Journal Of Geology, 105*, 407-439, 1997.

Wilson, C.J.N., and G.P.L. Walker, The Taupo eruption, New Zealand I. General aspects, *Philosophical Transactions of the Royal Society of London Series A, 314*, 199-228, 1985.

Wilson, L., and J.W. Head, Ascent and eruption of basaltic magma on the Earth and Moon, *Journal of Geophysical Research, 86*, 2971-3001, 1981.

Wohletz, K.H., Mechanisms of hydrovolcanic pyroclast formation, grain-size, scanning electron microscopy, and experimental studies, *Journal of Volcanology and Geothermal Research, 17*, 31-63, 1983.

Wohletz, K.H., Water/magma interaction: physical considerations for the deep submarine environment, in *Explosive Subaqueous Volcanism*, edited by J.D.L. White, J.L. Smellie, and D.A. Clague, American Geophysical Union, Washington DC, this volume.

Woods, A.W., Observations and models of volcanic eruption columns, in *The physics of explosive volcanic eruptions*, pp. 91-114, Geological Society of London, London, United Kingdom, 1998.

Woods, A.W., and M.I. Bursik, Particle fallout, thermal disequilibrium and volcanic plumes, *Bulletin of Volcanology, 53*, 559-570, 1991.

Wright, I.C., and J.A. Gamble, Southern Kermadec submarine caldera arc volcanoes (SW Pacific); caldera formation by effusive and pyroclastic eruption, *Marine Geology, 161*, 209-229, 1999.

Yamagishi, H., and E. Dimroth, A comparison of Miocene and Archean rhyolite hyaloclastites: evidence for a hot and fluid rhyolite lava, *Journal of Volcanology and Geothermal Research, 23*, 337-355, 1985.

Yamazaki, T., I. Kato, I. Muroi, and M. Abe, Textural analysis and flow mechanism of the Donzurubo subaqueous pyroclastic flow deposits., *Bulletin of Volcanology*, *37*, 231-244, 1973.

Yuasa, M., and K. Kano, Submarine silicic calderas on the northern Shichito-Iwojima Ridge, Izu-Ogasawara (Bonin) Arc, western Pacific, in *Explosive Subaqueous Volcanism*, edited by J.D.L. White, J.L. Smellie, and D.A. Clague, American Geophysical Union, Washington DC, this volume.

Zimanowski, B., and R. Buettner, Phreatomagmatic explosions in subaqueous volcanism, in *Explosive Subaqueous Volcanism*, edited by J.D.L. White, J.L. Smellie, and D.A. Clague, American Geophysical Union, Washington DC, this volume.

Zimanowski, B., R. Buettner, V. Lorenz, and H.-G. Haefele, Fragmentation of basaltic melt in the course of explosive volcanism, *J. Geophys. Res.*, *102B*, 803-814, 1997.

James D. L. White, Geology Department, University of Otago, P.O. Box 56, Dunedin, New Zealand 9015.

John L. Smellie, British Antarctic Survey, High Cross, Madingley Road, Cambridge CB3 0ET, United Kingdom.

David A. Clague, Monterey Bay Aquarium Research Institute, 7700 Sandholdt Road, Moss Landing, CA 95039-9644.

Water/Magma Interaction: Physical Considerations for the Deep Submarine Environment

Kenneth H. Wohletz

Earth and Environmental Sciences, Los Alamos National Laboratory, Los Alamos, New Mexico, USA

One might conclude that in deep submarine environments, where hydrostatic pressure is in excess of water's critical pressure, water/magma interaction does not produce expanding vapor and explosive behavior cannot occur. This conclusion is supported by the apparent paucity of hydroclastic material in samples recovered from deep submarine environments. Analog molten fuel-coolant interaction (MFCI) experiments, however, demonstrate explosive dynamics for conditions where water is pressurized above its critical pressure before interaction; MFCI theory further indicates this explosive potential. Thermodynamic predictions show that the conversion of thermal to mechanical energy is only high enough to support explosive behavior for a narrow range of water/magma mass ratios. In submarine environments, apparent mass ratios are too high for explosive behavior, but effective mass ratios (those determined from the water and magma directly involved during interaction) depend upon characteristic times, determined by the sound speed of the water and interface geometry. At high pressure, a supercritical fluid film grows at the water/magma contact surface and can become unstable. With instability the film oscillates, rapidly expanding and collapsing, with a periodicity of milliseconds or less. Each film collapse imparts kinetic energy into the magma, causing magma fragmentation, especially where quench contraction has weakened the magma. With fragmentation more magma surface area is exposed to water, and the film growth/collapse process escalates. When perturbed by some external pressure wave, the unstable film is prone to a detonation-like phenomenon that causes rapid, localized vapor expansion even at high ambient pressure.

1. INTRODUCTION

The earth's oceans cover over 60% of its surface; hence, much of earth's volcanism and volcanic products are hidden from direct view by deep water. Deep seafloor observations by submersible vessels have provided visual and sample documentation of only a miniscule portion of submarine volcanic terrain, and likewise, seafloor drilling samples represent only an insignificant portion of the volcanic products thought to underlie much of the deep ocean basins. Submarine volcanism is significant not only because of volume considerations but also because magma is known to interact dynamically with water. Called *hydrovolcanism* in general, volcanism involving contact of magma with external water occurs in a wide variety of environments and exhibits a range of eruptive phenomena from passive lava quenching to development of extensive hydrothermal systems to enormous explosive eruptions, characterized by production of steam in vapor explosions. In all cases of hydrovolcanism that have been observed, steam production is a key phenomenon that signifies the interaction.

Submarine hydrovolcanism [*Bonatti*, 1967] occurs within deep (>200 m) saline water [*Honorez and Kirst*, 1975] as opposed to shallower (<200 m) *epeiric* and *littoral* hydrovolcanism [*Wentworth*, 1938; and *Mattox and Mangan*, 1997]. This type of volcanism is thought to be most common at oceanic spreading centers and on large submarine volcanoes that form flat-topped *guyots* [*Cotton*, 1969], consisting of pillow basalts and *hyaloclastites*, which are the angular, glassy shards formed by rapid quenching of magma during its interaction with external water in a subaqueous environment.

Whereas shallow submarine to littoral eruptions are known to be capable of explosive behavior (e.g., Myojin Reef in 1952; Capelinhos in 1958; Surtsey in 1963) explosive hydrovolcanic eruptions in the deep submarine environment (>2000 m) have not been directly observed but only surmised from observations of deep seafloor hyaloclastites and related phenomena [*Lonsdale and Batiza*, 1980; *Batiza et al.*, 1984; *Haymon et al.*, 1993; *Clague et al.*, 2000; and *Clague et al.*, this volume]. *Smith and Batiza* [1989] found that hyaloclastite deposits occur commonly around summits of seamounts near the East Pacific Rise, showing evidence of hydrovolcanic but not necessarily explosive origins. In some cases these deposits of hyaloclastite resemble deposits of pyroclastic density currents that have spread out from unidentified vents; these have been termed *sheet hyaloclastite*, and their origin is interpreted as disruption of lava flows by water-saturated sediments trapped beneath them [*Maicher et al.*, 2000; and *Maicher and White*, 2001].

Smith and Batiza [1989] incorrectly concluded (as will be shown later) that water's reduced volume expansion at depths >2300 m prevents steam explosivity; however they did support the idea of explosive mixing of magma and seawater, presumably by cooling-contraction granulation causing high rates of heat exchange.

Haymon et al. [1993] made observations of an eruption on the seafloor that included phenomena such as bottom-water murkiness caused by suspended particulates, near-critical temperature and low salinity vent fluids, and venting of white vapor that transformed to gray smoke above the vent. Those authors discussed these observations as possible evidence for explosive activity, but they misinterpreted the lack of cuspate shapes among the blocky, angular shards they collected as not diagnostic of *phreatomagmatic* eruption (hydrovolcanism within the zone of saturation), when in fact they are. However, they did mention the shards' similarity to explosive hydrovolcanic ashes, which seems contradictory if not just a terminology problem.

Kokelaar [1986] defined four classes of subaqueous clastic volcanism, including: *magmatic explosivity*, *contact-surface steam explosivity*, *bulk interaction steam explosivity*, and *cooling-contraction granulation*. These definitions are strictly qualitative and thus somewhat ambiguous in physical application, but *Kokelaar* [1986] provided a guide for the depth limitation of these classes. However, those depth limitations appear to be simplistic and not fully representative of magmatic processes [e.g., *Dudás*, 1983] or hydrovolcanism (as will be discussed). For example, *Kokelaar* [1986] limited bulk interaction steam explosivity to depths at which hydrostatic pressure is less than water's critical pressure, based on his interpretations of experiments and observations, but he did not provide hydrodynamic justification for this limit.

Because observed explosive hydrovolcanism is always associated with production of vast quantities of steam and the explosive energy is thought to be derived from the thermodynamic work of steam expansion, an important question arises about the effects of ambient pressure in the deep submarine environment below a depth of 2200 to 3000 m (depending on salinity), where hydrostatic pressure is greater than the critical pressure of seawater. Over the last 40 years many workers have followed stipulations posed by *McBirney* [1963] and assumed that supercritical ambient pressure precludes the formation of steam, and thus explosive dynamics are not possible. Because the hydroclastic products (hyaloclastite) that result from magma fragmentation associated with explosive interaction are rarely observed in deep sea cores (although on seamounts they may be common) and pillow lavas are typically observed, one may conclude that high hydrostatic pressure in the deep submarine environment does in fact make explosive interactions unlikely or impossible.

One must also recognize that not all tephra produced by deep submarine eruption are hydroclastic. The magmatic mechanism of tephra production may also be important. *Burnham* [1981 and 1983] points out that volatile-constituent exsolution that occurs because of crystallization (second boiling) or rapid pressure decrease (e.g., by failure of vent rocks surrounding a submarine conduit) can produce volume increases up to 50% or more even at ambient pressures of 50 MPa (~5 km depth). The thermodynamic work caused by this volume increase has the potential for explosive release, and it may account for extensive magma fragmentation such as that indicated by the volcanic products associated with Kuroko ore deposits [*Tanimura et al.*, 1983], which formed in deep water (up to 3500 m). Such findings are also emphasized by *Dudás* [1983].

In a treatment complimentary to the broad overview and discussion of deep submarine pyroclastic eruptions recently published by *Head and Wilson* [2003], this paper focuses on experimental results and theoretical considerations that bear on whether or not hydrovolcanism in the deep submarine environment can produce explosive magma fragmentation. A working definition of explosion is the sudden and rapid production of gas, heat, noise, pressure, and in many cases, a shock wave. Two basic types of explosion are: (1) *detonation*, which is a supersonic propagation of a combustion wave that causes nearly instantaneous vapor release and expansion; and (2) *deflagration*, which is a rapid but subsonic propagation of a combustion wave. For certain circumstances, the word *combustion* can be replaced by *vaporization* in these definitions of explosion.

In the following discussions of explosive submarine hydrovolcanism, I first discuss aspects of the deep submarine environment that play an important role, namely the mode of magma extrusion, the compositional effects of seawater, and water's thermodynamic behavior and variability near the critical point (or critical curve for seawater, see below). Although many molten fuel-coolant interaction (MFCI or FCI) experiments [e.g., *Zimanowski et al.*, 1997a] are not directly linked to the high water abundance and hydrostatic pressure conditions in deep submarine environments, those discussed in this paper do include these important links. Pertinent MFCI experiments help establish the potential for explosive behavior that can then be considered in light of water/magma interaction physics. The physics are complex and in no way can be completely addressed in this paper, but following on previous studies of hydrovolcanism [e.g., *Sheridan and Wohletz*, 1983] the important parameters of water/magma mass ratio and confining pressure will be specifically addressed. Other theoretical aspects to be considered are the pressure- and temperature-dependent thermal equilibrium between magma and water, the hydrodynamic behavior of supercritical water in response to pressure fluctuations, and the role of *detonation* physics in explosive vaporization.

Overall, the reader should consider this topic an "open book" and realize the limitations of geological observations as well as those of theory and experimentation. In doing so, I hope that the discussions presented help set a basis for future observational and diagnostic studies of submarine hydrovolcanism, stimulate open mindedness to the realm of possibilities, and promote the idea that the problem is not easily constrained.

2. THE DEEP SUBMARINE HYDROVOLCANIC ENVIRONMENT

Two fundamental aspects of the deep submarine hydrovolcanic environment are the magma and the seawater. For the magma its composition and extrusion rate determine how it is introduced to the seawater. For the seawater its phase relationships and thermodynamic properties are important. An additional consideration is the quantities of magma and water that are directly involved during a submarine eruption.

2.1. Magma

Because of the mechanics of seafloor spreading and hot spot volcanism, basaltic compositions are found to dominate the deep submarine environment; however, silicic compositions also exist, especially in arc settings but also along spreading ridges as well [*Stoffers et al.*, 2002]. Four generalized modes of magma extrusion are schematically portrayed in Figure 1, but these are by no means comprehensive. Depending upon the flow rate of magma within a conduit below the seafloor, the violence of the eruption, and the magnitude and rapidity of magma and fluid volume changes, seismic disturbances may accompany extrusion. All of these factors are important when considering the dynamics of water/magma interaction.

For mafic (basaltic) magmas, one can consider two end-member extrusion types: (1) slow extrusion rates that tend to form lava flows whose thickness is largely controlled by rheological properties (Figure 1a); and (2) fast extrusion rates that tend to produce a fountain-like structure (Figure

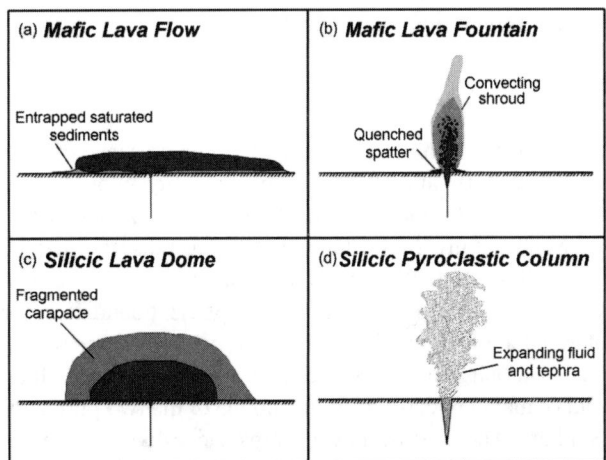

Figure 1. Schematic illustration of four hypothetical modes of deep submarine extrusion.

1b). For fast extrusion rates, magma's volume change by crystallization and volatile exsolution may play a role, and the motion of the extruding magma is typically vertical, causing it to rise like a fountain above the seafloor before gravitation forces cause it to settle downward. During its relatively rapid rise from the seafloor, shear stresses on the surface of the magma fountain may exceed surface tension resulting in tearing and separation of magma globules and smaller fragments [cf. *Head and Wilson*, 2003]. The larger globules then may cascade to the seafloor, producing a kind of spatter rampart around the vent, whereas smaller fragments may be entrained in convectively rising seawater that is heated by the magma. These smaller fragments produce a shroud around the fountain, perhaps resembling the fluid issuing from black smokers at spreading-center hydrothermal vents.

Silicic magmas can contain significant portions of dissolved volatiles. During extrusion, magma volume may change significantly by crystallization and volatile exsolution. Crystallization caused by magma cooling and extrusive pressure release promotes volatile oversaturation and exsolution. This process has been well described for the submarine environment by *Burnham* [1983]. The mechanical work involved with magma volume changes results in magma fragmentation by both brittle and viscous processes. In this fashion an extruding dome of viscous silicic magma may develop a carapace of hyaloclastite (Figure 1c). If the volatile oversaturation is high and extrusion dynamics (such as abrupt failure of confining conduit walls) promote a catastrophic decompression of magma, a vent may erupt a column of supercritical fluid and magma fragments (Figure 1d). As the column rises, the fluid expands and produces momentum that allows further decompression within the conduit, prolonging the eruption.

2.2. Seawater

Seawater is a solution dominated by the presence of salts (mostly NaCl), and its thermodynamic behavior can be approximated by the two-component system of pure water and NaCl. Figure 2a illustrates a P-T phase diagram for the system NaCl-H_2O that shows phase boundaries of the pure components and projections of the phase boundaries for intermediate compositions. The salinity of seawater results in critical behavior not occurring at a single point but along a curve that connects the critical points of the two pure endmembers. These phase relationships show that at any temperature two fluid phases can coexist and a single critical point does not exist if solid NaCl is present. Depending upon local salinity, critical behavior occurs at pressures and

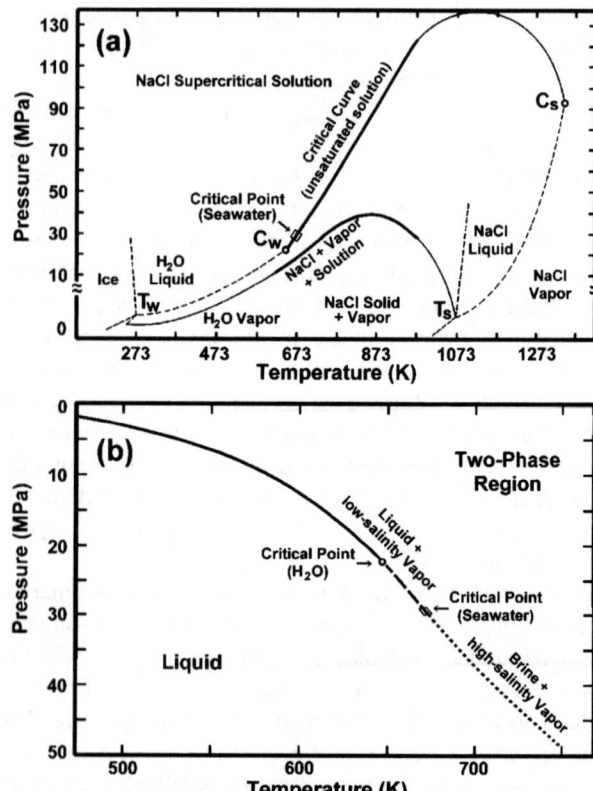

Figure 2. (a) Pressure-temperature diagram for the system NaCl-H_2O adapted from *Krauskopf* [1967] from experimental data from *Sourirajan and Kennedy* [1962]. This diagram is an overlay for the pure H_2O and NaCl endmembers (dashed lines). The solid lines (bold are experimental data) schematically represent the phase boundaries connecting the pure endmembers. T_w and C_w denote the triple point and critical points of H_2O respectively; T_s and C_s those points for NaCl. (b) The two-phase curve for standard seawater (3.2% NaCl) as a function of pressure and temperature, based on data from Bischoff and Rosenbauer [1984]. Note that in this plot, pressure increases downward. The solid curve designates the boundary where pure water and seawater boundary are nearly coincidental. The boundary for pure water terminates at its critical point, whereas the boundary for seawater extends (dashed curve) to its respective critical point, along which it separates the stability regions of liquid and a mixture of low-salinity vapor and liquid. The phase boundary extends (dotted curve) from seawater's critical point to higher temperature and pressures, separating the liquid region from that of a mixture of high-salinity vapor and brine.

temperatures elevated from those of pure water (22 MPa and 647 K) to values approaching ~30 MPa and ~680 K for seawater with a salinity of 3.2 wt% NaCl [*Bischoff and Rosenbauer*, 1988], and critical pressure is expected at a depth of ~3 km in the submarine environment. Also, Figure 2a shows that as seawater is heated, solid NaCl is precipitated, which may greatly affect vapor nucleation [cf. *White*, 1996].

The two-phase boundary of seawater [*Bischoff and Rosenbauer*, 1984] is similar to that of pure water for subcritical conditions in pressure-temperature space (Figure 2b), but unlike pure water, it does not end at the critical point, but projects nearly linearly to higher pressures and temperatures (~680 K at 30 MPa to ~750 K at 50 MPa). Below the critical point the two-phase region of seawater consists of liquid and low-salinity vapor, and above the critical point it consists of brine and high-salinity vapor. These phase relationships indicate that a vapor phase can exist in seawater at supercritical pressures.

Another effect of the dissolved solids in seawater is a decreased heat capacity compared to that of pure water. This effect can be approximated [*Buntebarth and Schopper*, 1998] as:

$$C_{sw} = x_w C_w + (1 - x_w) C_s, \quad (1)$$

where C_w and C_s denote the constant volume heat capacities of pure water and dissolved constituents respectively and x_w is the pure water mass fraction. Dissolved solids in seawater range from about 0.7 to 4.5% by mass with the standard average being 3.5% [*Turekian*, 1968]. Using Eq. (1) the heat capacity of seawater (C_{sw}) is several percent lower than that of pure water; however, as discussed below, since the heat capacity of pure water varies by approximately a factor of 2 over most of the range of pressure and temperatures typical of hydrovolcanism, the bulk heat-capacity effect of dissolved solids in seawater is relatively small. The following discussions assume that heat capacity and phase transition effects of seawater have offsetting effects for situations of rapid heating such that pure water provides a workable proxy.

Because water/magma interaction can result in a relatively high-pressure and high-temperature water phase, it is important to consider the variability of water near critical point (curve) conditions. Figure 3 illustrates the variation of heat capacity, viscosity, and expansion coefficients at 30 and 60 MPa, analogous to deep submarine environments at about 3000 and 6000 m depth, respectively. Note that heat capacity, viscosity, and the isobaric expansion coefficient vary rapidly in the range 600–800 K, near water's critical-point temperature (647 K). This means that small changes in temperature produce large changes in properties that determine how water behaves thermodynamically and hydrodynamically. If large thermal gradients exist near the contact of water with magma, which is to be expected in hydrovolcanism, then high pressure and velocity gradients will also

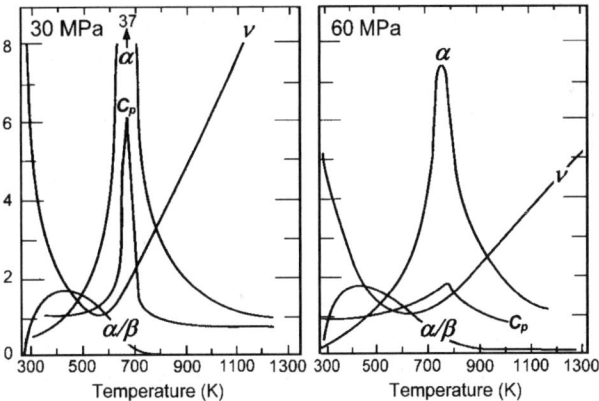

Figure 3. Variation of physical properties of water at 30 MPa (~3000 m depth) and 60 MPa (~6000 m depth) as a function of temperature. The symbols and units for the curves are: α—isobaric expansion coefficient (10^3 K^{-1}); α/β—pressure coefficient [$(dp/dt)_v$; 10^{-2} MPa K^{-1}] where β is the isothermal expansion coefficient; v—kinematic viscosity (10^{-7} m^2 s^{-1}); and C_p—the constant pressure heat capacity (4.184 x kJ kg^{-1} K^{-1}). Note the sharp inflections and discontinuities apparent near the critical temperature.

exist, and perturbations in water movement result in the likelihood of hydrodynamic instability. For example, a supercritical fluid subjected to small pressure perturbations may tend to oscillate [*Greer and Moldover*, 1981] in density between liquid and vapor states (e.g., growth and collapse of vapor bubbles). The speed at which water flows from higher to lower pressure regimes depends not only on the magnitude of the pressure gradient but also on its viscosity; rapid viscosity fluctuation may enhance or dampen convective currents. It is a chaotic *thermal-hydraulic* system that has received considerable attention for over two decades from nuclear engineers concerned with coolant flow stability in nuclear reactors [e.g., *Ruggles et al.*, 1989 and 1997].

2.3. Water/Magma Interaction

Interaction of water with magma involves heat transfer from the magma to the water and chemical species migration driven by solution and precipitation. In submarine environments, the mass of water present typically exceeds that of magma by orders of magnitude, such that heat transfer produces only localized water temperature and compositional gradients that rapidly dissipate. However, the amount of water that dynamically interacts with the magma can be quite variable. Where water becomes entrapped by magma either within the vent conduit, beneath lava flows (as wet sediment), or by engulfment during rapid extrusion, the mass of magma may exceed that of water during interaction.

On the other hand, at lava flow top surfaces, convective currents might involve a much larger mass of water than that of magma, promoting passive interaction. The chemical dynamics of water/magma interaction are very important to the evolution of seawater and hydroclast compositions; however the physical dynamics are the focus of this paper.

3. MOLTEN FUEL-COOLANT INTERACTION EXPERIMENTS

Peckover et al. [1973] gave the first published accounts of molten fuel-coolant interaction in submarine volcanic explosions. Their work closely followed *Colgate and Sigurgeirsson's* [1973] study of the MFCI explosive hazard of lava-flow diversion by quenching it with water. In 1976 after discussions with Stirling Colgate, I began MFCI experiments as analogs for water/magma interaction [e.g., *Wohletz*, 1980; *Wohletz and McQueen*, 1984; and *Wohletz et al.* 1995], drawing on the expertise rapidly developing in the fields of nuclear and mechanical engineering in application to nuclear reactor safety [e.g., *Buxton and Benedict*, 1979]. Bernd Zimanowski and his colleagues began similar experimentation in the mid 1980s, and they continue a very fruitful MFCI experimental program to the present time [e.g., *Zimanowski et al.*, 1986, 1991, and 1997a]. These experiments demonstrate a wide range of dynamic phenomena caused by the interaction of a melt (fuel) with water (coolant). Experimentation has also shown that confining pressure likely plays an important role in controlling the nature of the phenomena. One must consider the limitations of such experiments in reproducing the submarine environment, and such limitation concern the experimental design (geometry) and method by which the melt is introduced to the water. In this light, the MFCI experiments conducted by *Wohletz et al.* [1995] and *Zimanowski et al.* [1991 and 1997a] have notable differences, and thus experimental results have distinct contrasts, but general similarities persist nonetheless. For the following discussions, I focus on the *Wohletz et al.* [1995] results, noting that *Zimanowski* [this volume] has different conclusions, based on his experimental results.

Previous papers [*Wohletz et al.*, 1995; and *Wohletz*, 2002] have addressed the adequacy of using a thermite melt as a basalt analog and for study for the initial seconds of water/melt interaction. Melt temperature, density, viscosity, enthalpy, and surface tension are important factors, and the thermite analog is similar to basalt except its enthalpy is about 3 times that of basalt. A second consideration for application of these experimental analogs is the manner by which high pressure is attained prior to water/melt interaction; the experiments are self-pressurized, which will be discussed later. Finally, experimental results have shown that the bulk water/melt mass ratio appears to be a primary factor controlling the dynamics of the interaction. For submarine volcanism, this ratio is apparently very high, but the actual amounts or water and magma contributing to an interaction may be strongly dependent upon the control volume considered. The following experimental results for high water/melt ratios (assumed to be analogous to submarine conditions) have important bearing on understanding factors that control interactions in an environment where the volume of water is much greater than that of magma.

The *water-box* experiments [*Wohletz et al.*, 1995] utilized Plexiglas box of ~1-m width, length, and height (Plate 1). Thermite was placed within the box in a steel canister stack, ~0.3 m in diameter, ~1 m in height, and containing ~100 kg of thermite. The box was then filled with about 900 kg of water, giving a high water/melt ratio. After the thermite was ignited and became fully molten, it eventually melted through the canisters and contacted the water. The ensuing water/melt interactions lasted a number a minutes, dominated by flow of molten globs of melt at the base of the box and ballistic, Strombolian-like activity near the top of the box (Plate 2). These experiments varied in their violence with time, and one that displayed mild Strombolian play for several minutes ended abruptly in a violent Surtseyan-like blast. Investigation of the products from these experiments showed that less than half of the thermite melt formed pillow-lava-like globs that remained within the box and the rest was fragmented to scoriaceous, centimeter-sized fragments mostly ejected out of the box. These experiments not only demonstrate the quenching ability of water at high interaction ratios but also the variability of interaction violence, depending on the location of the interaction at the base of the box (passive) or at the surface of the water (mildly explosive). However, these experiments did not address the effects of hydrostatic pressure, which will be covered in the following results for high pressure experiments.

In a review of experiments [*Wohletz et al.*, 1995] that best simulate high hydrostatic (confining) pressure, strong evidence is given for the potential of supercritical explosion in MFCI phenomena. Those experiments involved the use of a confinement vessel (Plate 3) that allowed development of supercritical pressure prior to burst. Pressure vessels of several sizes were used, and they contained from 3 to 90 kg of molten thermite in which quartzo-feldspathic sand was mixed to more closely simulate basaltic compositions. Water (typically one-quarter to one-third the mass of thermite) filled the base of the vessels, separated from the ther-

Plate 1. Photograph of the *water box* experiment, consisting of a Plexiglas box about 1 m on a side, filled with water, and enclosing a cylindrical container of thermite, prior to ignition.

Plate 2. Photograph of a *water box* experiment in action. The interaction ejected centimeter-size fragments of molten thermite is ballistic trajectories like a Strombolian eruption. At the same time, not visible in this picture, molten globs of thermite spread like pillow lava over the floor of the box.

mite above by an aluminum plate. The melting process initiated at the top of the vessel and proceeded downward until the molten thermite perforated an aluminum plate and then directly contacted water. This design simulated the rise and injection of magma onto the seafloor by utilizing gravitational force of the melt. Although this design is inverted compared to natural systems, a vent pipe was used in some experiment designs; it extended from the vessel top into the water compartment and insured that the release of water pressure was not impeded by the melt. A burst valve at the bottom of the vent pipe allowed pressure to rise to a specified level prior to onset of vapor expansion.

Plate 4 shows a supersonic jet of melt fragments and superheated steam rising >30 m above the experimental vessel and a plume of micrometer-sized dust (quenched melt fragments) convectively rising above the jet. This example illustrates the typical burst phenomena for high-pressure experiments. Example pressure records from high-pressure experiments are displayed in Figure 4. In order to interpret these pressure records, several aspects of this experimental design must be mentioned, and these considerations clarify the analogy to submarine hydrovolcanism. The zero-time for the pressure records is arbitrary, since the time required for the thermite to become fully molten and contact the water varied among the experiments. Important to this experimental design is that pore-gas expansion during thermite melting contributed much of the vessel pressurization prior to the contact of the melt with water; it also led to premature vent failure in some experiments for which the ejected debris showed lumps and clots typical of incomplete interaction. The time of burst (vapor explosion) is marked by the rapid pressure decline, as confirmed by cinematography. Typically, the thermite gradually pressurized the system in ~1 s prior to contacting the water, after which the pressure rose precipitously in a few milliseconds before bursting.

Difficulties in monitoring pressure in these experiments include gauge damage by the violence of the interaction (Figure 4a) and gauges recording different pressure histories at different positions within the vessel (Figure 4b). Figure 4a shows results for an experiment where one pressure gauge showed little or no response (damage confirmed by post-experiment examination) while another gauge recorded a pressure of approximately 35 MPa prior to burst. Figure 4b shows an initial burst (dashed curve) occurred at the designed bursting pressure (6.8 MPa) and a secondary burst occurred ~0.5 s later at ~9 MPa. This double-burst was recorded by another transducer as a single event that reached a pressure of 23 MPa, demonstrating that interaction pressure is not the same for all locations in this dynamic system, possibly a manifestation of multiphase effects and multiple shock-wave domains. Figure 4c shows the interaction pressure rapidly rising to >50 MPa approximately 0.5 s before the major burst. A small pressure spike ~0.1 s before the main pressurization likely reflects an initial vapor-film growth and collapse event before wholesale interaction occurred, while the third pressure spike was an event caused by residual water and melt in the vessel after the main burst. Figure 4d records burst at 23 MPa just above critical pressure. Figure 4e shows bursting from pressure exceeding 60 MPa, whereas Figure 4f records bursting at a pressure just below critical. It is important to note that the time scale for these pressure plots is too large to show detail of the pressure history caused by water/magma interaction just prior to burst. Our interpretations of many pressure records is that once full water/melt interaction begins, pressure builds to its maximum within a few milliseconds or less before burst. This means that for these experiments full

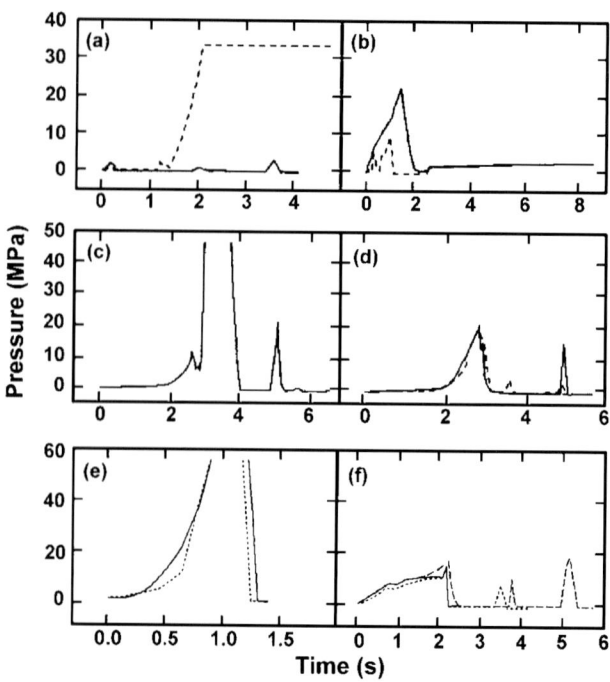

Figure 4. Example pressure records from MFCI experiments designed to study interaction at high confining pressure. Dashed and dotted curves are records from multiple pressure transducers. The designed bursting pressure for these examples are (a) 16.3 MPa, (b) 6.8 MPa, (c) 16.3 MPa, (d) 35.7 MPa, (e) 35.7 MPa, and (f) 16.3 MPa. For the example shown in d, the pressure records do not show the experiment reached designed burst pressure even though burst did occur as shown by the rapid pressure fall off; this behavior was later found to be caused by failure of the burst diaphragm thermal insulation.

Plate 3. Confinement vessel being pressure-tested. The vessel is about 1 m high and 0.4 m in diameter. The central vent tube (~0.1m diameter) visible at the vessel top extends to the base of the cylinder into a compartment holding water with the thermite place above. The base of the vent tube is sealed with a burst diaphragm welded in place. Burst diaphragms were constructed from aluminum plates and machined with crossing grooves, the depth of which determined the burst strength. These diaphragms (also called *petal valves* because their failure resembled the opening flower petals) were calibrated by the above pressure test, and the maximum confinement design was 35.7 MPa.

Plate 4. A supersonic jet of molten thermite dust and superheated steam vents to heights of >30 m from an experimental vessel about 1 m high (just visible in the far right photograph). These images display the violence of high-pressure interaction of water with melt.

interaction and bursting are nearly simultaneous events and most of the interaction occurred near the peaks of the pressure curves shown in Figure 4.

General results indicate that experiments designed to pressurize above critical pressure did in fact show explosive interaction. A smaller fraction of the high-pressure experiments attained explosive interaction than did low-pressure ones; however, those high-pressure experiments that exploded, did so with markedly increased energy [*Wohletz et al.*, 1995]. These experiments also showed maximum pressure well in excess of designed burst pressure, indicating pressure grew faster than it could be relieved, a shockwave phenomena. Two other general results are noted but not well understood. When designed confining pressure was >35 MPa, experiments showed exponential increases to supercritical pressures in less than a second prior to vapor explosion (commencement of rapid expansion), whereas those confined at pressures near critical showed slower rises and lower maximum pressures. Overall these experiments demonstrate that not only can water/melt interaction produce supercritical burst pressures, they can also occur where confining pressure is supercritical prior to burst.

4. THERMODYNAMICS AND WATER/MAGMA MASS RATIOS

The water/melt mass ratio, R, was experimentally identified by *Wohletz and McQueen* [1984] as an important control of interaction dynamics, determining the violence of the interaction and whether it is explosive or not. Although R is difficult to quantify in nature [cf. *White*, 1996], thermodynamic predictions at atmospheric pressure and numerous semi-quantitative and qualitative observations for near-surface environments support this theory [e.g., *Wohletz*, 1986]. In order to apply this theory to the deep submarine environment, one can adapt thermodynamic calculations for high ambient pressures.

Where careful and detailed observations constrain magma and water flux, such as in a terrestrial setting of a known aquifer and a witnessed eruption [e.g., *Ort et al.*, 2000], inferred values of R for various phreatomagmatic eruption types fit quite closely with thermodynamic predictions. The abundance of pillow lava in submarine settings also qualitatively supports experimental results. However, where water is abundant, it is difficult to establish constraints on just how much of it actually interacts with erupting magma. Thus for the submarine environment, only *apparent* mass ratios (those ratios unconstrained by a defined control volume) can be determined ($R \gg 10$). From the aforementioned experimental evidence and by the theory to be presented, such mass ratios are much too high to explain explosive interactions in the submarine environment. As discussed earlier, the actual involvement of seawater with magma may be influenced by partial or full enclosure of a volume of water by magma, in which cases the *effective* mass ratio (a mass ratio that measures the actual amounts of water and magma involved in heat exchange) may be <1.

4.1. Hydrostatic Pressure Effect

The approach to calculating the thermodynamic work produced by water/magma interaction involves three primary assumptions, based on the *Hick-Menzies* [1965] approach as discussed by *Wohletz* [1986] and *Wohletz et al.* [1995]. These assumptions are: (1) during interaction water can be treated as a single-component (pure) substance; (2) water and magma reach an initial equilibrium temperature at nearly constant volume prior to expansion; and (3) the expansion phase can be approximated by two thermodynamic cases. These thermodynamic cases are: (1) *isentropic fluid* in which the water expands at constant entropy as an idealized "frictionless" adiabatic process; and (2) *isentropic mixture* in which the water expands while being continuously heated by magma fragments enclosed by the water. For brevity I will refer to these two cases as *fluid* and *mixture*, respectively. Whereas *Kieffer and Delany* [1979] critically assess *isentropic fluid* decompression within a geological context, there is little precedence for evaluation of *isentropic mixture* decompression in this context. *Self et al.* [1979] suggested using the term *isothermal* for the nearly constant-temperature expansion of vapor in contact with tiny magma fragments. This terminology, which *Wohletz* [1986] and *Wohletz et al.* [1995] attempted to fit to MFCI, implies the assumption that the mass of fragments greatly exceeds that of the vapor and heat transfer over the large surface area of the fragments is fast enough to keep the vapor at nearly a constant temperature during expansion. For water/magma interaction where R and fragment size can vary over an order of magnitude, the *isothermal* terminology does not strictly apply.

The initial equilibrium and expansion assumptions are idealizations that allow one to constrain the maximum pressure, temperature, and volume that water attains during interaction. Certainly many factors limit the validity of these assumptions, and these factors include most notably the time scales on which heat transfer and phase separation occur. If water expands and separates from the magma heat source prior to reaching initial thermal equilibrium with the magma, then the interaction thermodynamic work will be minimized. Although this possibility is best modeled by the *fluid* expansion case, that model still assumes an initial thermal equilibrium. On the other hand, the *mixture* expansion case assumes no separation of water from the magma during expansion, and thus, it is difficult to constrain for a situation where the initial thermal equilibrium does not occur.

Heat transfer between the magma and water is dominated by conduction and convection; the thermal absorption coefficient of water is too low for effective radiative heat transfer within a control volume, except in certain cases of film boiling [*Dinh et al.*, 1998]. Experiments show that water/melt interaction can involve a commingling phenomenon that occurs prior to thermodynamic expansion and flow. This commingling involves the rapid breakdown of the melt into a mixture of fine particles with water that has been called an *explosive premixture*. The melt fragmentation in this premixture produces the high surface areas needed for heat transfer rates fast enough for thermal equilibrium to occur before expansion. The premixture also promotes thermodynamic expansion of the water in intimate contact with fine melt particles, as modeled by the *mixture* case, which predicts an upper limit for thermodynamic work production. If during expansion the water separates from the melt particles, then the *fluid* case applies and provides a measure of the lower limit for thermodynamic work production. As shown later, for deep submarine conditions where ambient pressure is high, *mixture* expansion may be a necessary condition for explosive behavior.

Based on the assumptions above, some thermodynamic predictions (Appendix A) highlight the potential effects of water/magma mass ratio and hydrostatic pressure on interaction dynamics. Figure 5 shows details about calculated, pre-expansion, equilibrium conditions and the thermodynamic paths for expansion for the *fluid* and *mixture* cases. For the following examples, water is at 277 K (near its densest state) and the magma is basaltic (~1500 K) because it is volumetrically dominant in the deep submarine environment. Similar results are expected for much less common silicic magmas. Silicic magmas have lower extrusion temperatures than the basalt considered here, and about 75% of its heat content [*Wohletz*, 2002], with the differences having a corresponding effect on results presented here.

Figure 5a shows that calculated initial equilibrium states are a strong function of R and exceed critical pressure for R values <2.0. The initial equilibrium states also show specific volume increasing with decreasing R; but that increase amounts to a <2% increase in the mixture volume at $R = 0.01$. Although increasing ambient pressure decreases the equilibrium entropy, this effect is too small to portray in Figure 5a. Water expansion follows isentropes (Figure 5a) to a pressure of 0.1 MPa (atmospheric pressure); for expansion to higher ambient pressures typical of deep submarine conditions, the final expansion state in Figure 5a is simply the intersection of an isentrope with a horizontal line (isobar) at the ambient pressure. Clearly greater expansion is achieved for interactions at lower R values, which by Appendix A, indicates greater thermodynamic work per unit water mass. For most R values, expansion to 0.1 MPa ends in the steam dome (two-phase region of liquid plus vapor). For cases of expansion to 10 MPa, if $R > 1$ then there is lit-

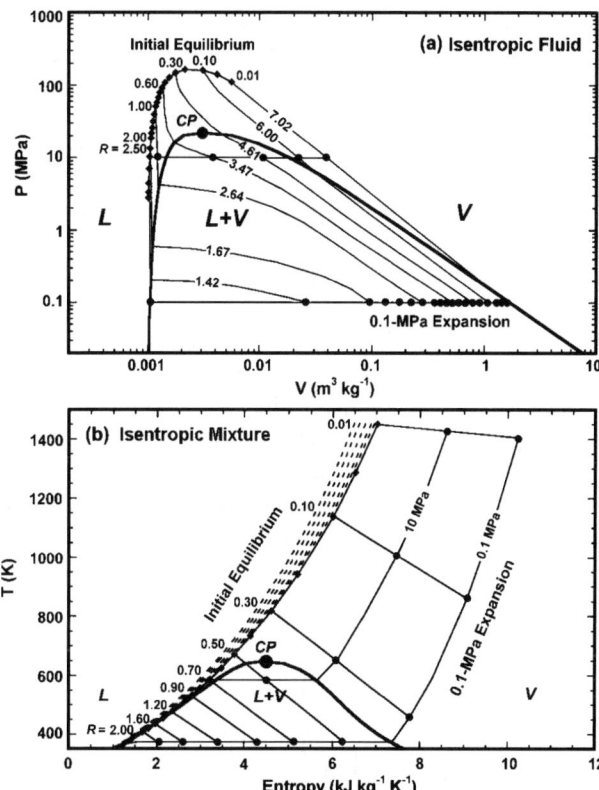

Figure 5. Thermodynamic phase diagrams (L is liquid, V is vapor) for calculated water/magma interactions at different water/magma mass ratios, R. Initial equilibrium states are diamond-shaped points; filled circles represent expanded states; and CP is the critical point. (a) A pressure-volume phase diagram illustrates *isentropic fluid* expansion with example expansion isentropes labeled. Note that all 0.1-MPa expanded states are within the steam dome, but those at 10.0 MPa are not. (b) A temperature-entropy phase diagram shows calculated effects of ambient pressure on the initial equilibrium state as a function of R. The solid line with diamond-shaped points is are equilibrium states at 0.1-MPa ambient pressure, dashed lines are equilibrium states with increasing ambient pressures (10, 20, 40, 60, and 80 MPa) plotting increasingly to the left. This plot also shows *isentropic mixture* expansion final states (0.1 and 10.0 MPa) with schematically drawn lines connecting initial and final states for example R values.

tle or no vapor production and thermodynamic work; in contrast, if $R < 0.1$ then expansion produces superheated vapor and high amounts of thermodynamic work per unit mass water. For cases where the ambient pressure is supercritical, if $R > 0.5$ then isentropes are so steep that little volume change occurs along them and the resulting thermodynamic work is small. However, with increasing supercritical ambient pressure, the initial equilibrium pressure increases so that the thermodynamic work also increases.

Figure 5b illustrates the *mixture* case for which water entropy increases during expansion because the water is continuously heated by hydroclasts entrained in it. The entropy increase with expansion is shown by straight lines in Figure 5b for simplicity, but these lines may be slightly curved because the temperature and pressure dependence of conductivities, especially in the steam dome where conduction to liquid and vapor components does not occur at the same rate. Because *mixture* expansion takes water to higher entropies, it produces more thermodynamic work than does expansion in the *fluid* case. The dashed lines in Figure 5b show the small but finite effect of hydrostatic pressure on the initial equilibrium at various R values. With increasing ambient pressure, the initial equilibrium entropy decreases. For the expanded states, example isobars drawn at 0.1 and 10 MPa show that increasing hydrostatic pressure limits expansion to decreasing final entropy (lower volume) states. Accordingly, the pressure-volume work is limited by hydrostatic pressure. However, as will be discussed later, it is not correct to assume that expansion stops when ambient pressure is reached for dynamic interactions where shock waves are formed and sound speeds vary considerably over short distances.

What do these thermodynamic calculations show other than the fact that they predict hypothetical initial equilibrium states that are above the critical pressure for all R values <2 and above 100 MPa (~10 km depth) for R values <0.6? First of all one should note that in the supercritical region, isentropes become steeper with increasing pressure such that pressure change produces less and less volume change, especially for R values >0.5. This observation suggests that as supercritical ambient pressure increases, *fluid* expansion produces less and less thermodynamic work. In contrast, the *mixture* case shows that for expansion at R values <0.4, water entropy increases (while the mixture is entropic) from 4 kJ kg^{-1} K^{-1} to values that reflect higher volumes; thus, in the supercritical region, the *mixture* case produces more thermodynamic work than does the *fluid* case.

To evaluate the thermodynamic work potential for interactions at different R values, Figure 6 shows plots of thermodynamic conversion ratios, which represent the percentage of the magma's thermal energy converted to thermodynamic work, approximately half of which might be manifested as melt-fragment kinetic energy [*Wohletz et al.*, 1995]. For the cases of *fluid* and *mixture* expansion, two end-member final states are shown, one for expansion to hydrostatic (ambient) pressure and one for expansion to 0.1 MPa. Whereas expan-

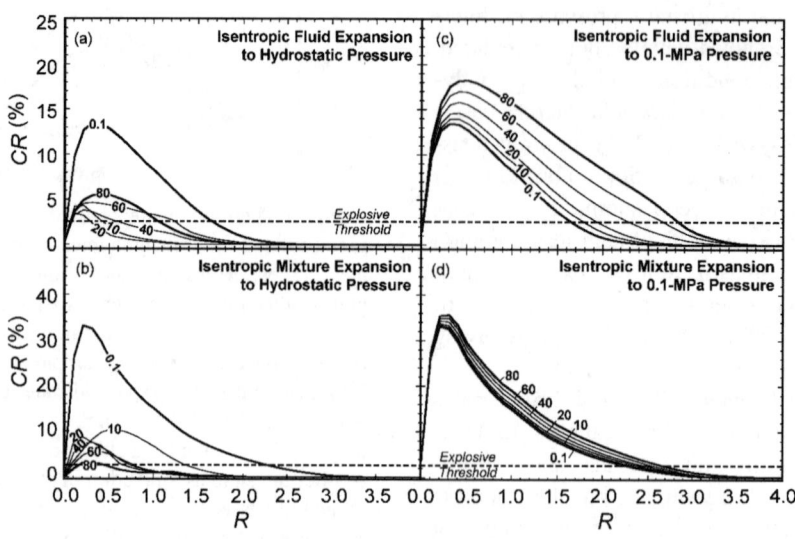

Figure 6. Thermodynamic conversion ratios for *isentropic fluid* plotted as a function of R (water/magma mass ratio). Curves are numbered according to ambient hydrostatic pressure (MPa), and the division between in explosive and effusive is arbitrarily fit to observations of terrestrial (0.1 MPa) eruptions. The four plots illustrate two expansion cases (*isentropic fluid* and *isentropic mixture*) each with two endmember expansion endmembers (hydrostatic and 0.1 MPa).

sion to hydrostatic pressure is intuitive, the other end-member represents full expansion to atmospheric pressure, which is approachable under certain conditions of shock-wave propagation and other factors to be discussed. First of all, the *mixture* cases produce higher conversion ratios than do the *fluid* cases, as also do full expansions to 0.1 MPa compared with those limited to expansion to hydrostatic pressure. For the *fluid* hydrostatic end-member (Figure 6a), the effects of ambient pressure are not intuitive (because of the effect of R on the initial equilibrium state). Below the critical pressure, conversion ratios decrease with increasing ambient pressure (i.e., 0.1, 10, and 20 MPa), but above the critical pressure conversion ratios rise with increasing ambient pressure (i.e., 40, 60, and 80 MPa). The trend for the *mixture* hydrostatic end-member (Figure 6c) shows decreasing conversion ratios with increasing ambient pressure. In contrast for the 0.1-MPa end-members (Figure 6b and 6d), both the *fluid* and *mixture* cases show increasing conversion ratios with increasing ambient pressure.

In order to further describe these conversion ratio calculations, a horizontal line is drawn in Figure 6 at a conversion ratio of 2.5%, designating an explosive threshold. This value represents the minimum conversion ratios calculated for experiments [*Wohletz et al.*, 1995] that produced demonstrably explosive behavior and complete melt fragmentation. In deference to the limitations of drawing analogies from these experimental results, this explosive threshold is considered to be arbitrary. The result of this consideration is that water/magma interactions are capable of being explosive up to hydrostatic pressures of 80 MPa, which represents a depth of 8000 m. Higher hydrostatic pressures, which have not been calculated, are also expected to be capable of explosive work production with the notable exception of the *mixture* hydrostatic end-member (Figure 6c), which shows conversion ratios declining below the explosive threshold as ambient pressure exceeds 80 MPa. Another aspect of the results shown in Figure 6 is the range of R values over which explosive behavior might be attained for subsurface interactions (hydrostatic pressure >0.1 MPa). This range extends from $R \approx 0.1$-1.3 (hydrostatic end-members) and from $R \approx 0.1$-3.0 (0.1-MPa end-members). Whereas increasing pressure always increases this range for both 0.1-MPa end-members, it decreases the range to $R \approx 0.1$-0.7 for hydrostatic end-members at ambient pressures found at depths from 1000 to 4000 m. The range of explosive R values for the mixture hydrostatic end-member strongly decreases with increasing pressure, becoming negligible at pressures above 80 MPa.

In summary of hydrostatic pressure considerations, both 0.1-MPa end-members show the greatest likelihood (with respect to R values and pressure) for conditions necessary for deep submarine explosive interactions. For the hydrostatic end-members, increasing pressure generally decreases the likelihood of explosive interaction. Overall, predicted explosive interaction requires R values that seem very small, considering the abundance of water in submarine environments; thus, the question arises as to whether such conditions are really applicable, which is the subject of the next section.

4.2. Effective Water/Magma Mass Ratios

Effective mass ratios depend upon characteristic times and lengths, determined by the propagation speed of thermal and pressure waves and by the interface geometry. Such parameters control just how much water actually is involved in the interaction heat exchange. For magma, conductive heat transfer times and lengths are important. For seawater, conductive and convective transport dominate, and radiative transport plays a role only where supercritical water loses its transparency *Dinh et al.* [1998]. On the other hand, a multitude of geometric possibilities can be imagined, from a simple planar interface to entrapment and engulfing configurations to rapid mixing of water and magma fragments in an erupting lava fountain [e.g., *Batiza et al.*, 1984; *Smith and Batiza*, 1989; *Head and Wilson*, this volume; and *Clague*, this volume]. For each possibility the linear dimensions determined by characteristic lengths and times can be constrained. The example given in Figure 7 is just one possible configuration that serves as an example how an effective mass ratio might be evaluated.

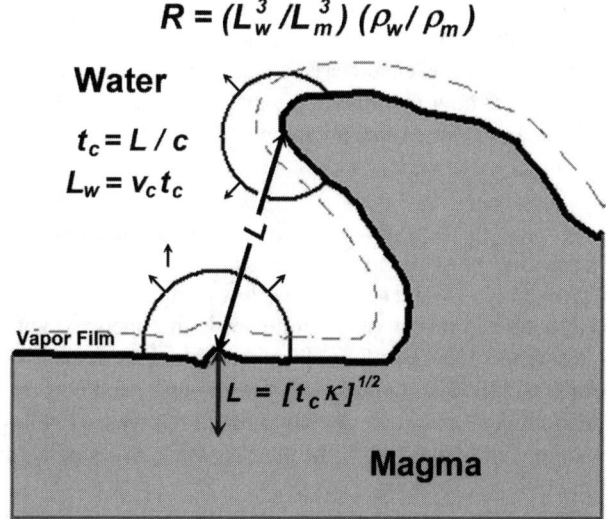

Figure 7. Schematic illustration of determination of characteristic length scales that determine effective water/magma mass ratios, R.

Figure 7 depicts a hypothetical contact of seawater over a rugose lava flow surface, such as might be caused by rapid extrusion rates or development of pressure ridges [cf. *Maicher et al.*, 2000]. For this case one can simply estimate the effective interaction ratio by volume ratios, defined by characteristic lengths. In order to define characteristic lengths (L; subscripts w and m denote water and lava respectively), a characteristic time, t_c, must be defined that takes into account both the thermal and fluid dynamics involved. From the experiments discussed earlier, this characteristic time is linked to the periodicity of water pressurization and expansion.

Consider the growth of a supercritical-fluid film at the contact interface shown in Figure 7. Expansion of this film continues while its pressure is greater than the surrounding seawater. It also creates a pressure wave that propagates into the seawater at the speed of sound and eventually reflects off an impedance (the product of density and sound speed) discontinuity, which for this example is a spine of lava at a distance, L, from the film surface. Such an impedance boundary might also be the substrate for a situation where water is trapped below a lava flow or the other side of a cavity in which water is surrounded by lava. When the reflected wave strikes the surface of the film, the film is partially or fully destabilized, setting a characteristic time:

$$t_c = 2L/c, \qquad (2)$$

where c is the sound speed of water (~1500 m s⁻¹ for seawater). This reasoning is very simplistic: a multitude of pressure perturbations likely exist with a spectrum of travel times for which some period or range of periods is dominant. For this argument, let Eq. (2) define an average characteristic time. The characteristic length for seawater, L_w, is a measure of how far a thermal wave moves into the water in the characteristic time; it can be simply stated as:

$$L_w = v_c t_c, \qquad (3)$$

where v_c is growth speed of the supercritical fluid layer. *Dinh et al.* [1998] quantify *film boiling* heat transfer for FCI conditions and report vapor film speeds that increase with water temperature. For fully developed film boiling at temperatures <1000 K, these speeds are <5 m s⁻¹, and because film boiling takes time to develop fully, film speeds are likely to be well below 1 m s⁻¹ in the first few milliseconds of film boiling [*Corradini*, 1981].

Because the heat flux in the seawater is limited by the heat conducted from the lava, the amount of heat transferred from the lava depends on the characteristic depth (length), L_m, of a thermal wave penetration into the lava over the period of time, t_c:

$$L_m \approx \sqrt{t_c \kappa}, \qquad (4)$$

where κ is the thermal diffusivity of the lava. From this greatly simplified estimation of L_w and L_m and values for the densities of seawater (r_w) and lava (r_m), the *effective* water/magma mass ratio R is:

$$R = \left(L_w^3 / L_m^3\right)(\rho_w / \rho_m). \qquad (5)$$

Although the above estimation of R is quite hypothetical, the following example shows how it might be evaluated. Consider a submarine basaltic lava flow with a rugose surface [cf. *Maicher et al.*, 2000]. For a characteristic dimension L of 1.5 m (Figure 7) T_c would be ~2 ms. Using typical k values for basalt (~1x10⁻⁶ m² s⁻¹), $L_m \approx 45$ μm from Eq. (4). Because the thermal diffusivity of water at pressures above 22 MPa averages about one-half that of basaltic lava, L_w should be ~0.7 L_m for a conduction-dominated system. However, Eq. (3) also takes into account convective and radiative heat transfer; thus, L_w ranges from a conductive minimum of ~20 μm to an incipient film boiling value of ~200 μm. In order to evaluate Eq. (5) densities must be factored in. Basaltic lava density varies with composition and crystallinity, and here I will use a value of 2500 kg/m³. The density of the supercritical fluid at a temperature of ~1000 K and pressure of 22 to 50 MPa is between 50 and 110 kg/m³. From these values Eq. (5) predicts and *effective R* in the range of 0.01 to 3.8; the lower extent of this range is compatible with R values thought to be typical of terrestrial explosive hydrovolcanism.

Certainly, a more rigorous approach to estimating effective values of R can be developed, and the one shown above is only to illustrate some of the parametric constraints that might be considered. But if characteristic lengths and times are truly important in determining an *effective R*, then one might conclude that explosive eruptions are indeed possible in submarine situations, especially for cases where smaller characteristic lengths and shorter times are involved, such as might be associated with high extrusive rates. On the other hand, as characteristic lengths and times increase, film boiling has more time to mature such that L_w increases more rapidly than does L_m, leading to much higher *effective R* values.

With thermodynamic and geometrical considerations that suggest an explosive potential for deep submarine water/magma interactions, the details of how heat transfer from the magma to water can proceed at a rate required to produce dynamic effects needs attention. The dynamics at the water/magma interface involve not only heat transfer phenomena but also hydrodynamic phenomena, those kinetics that describe how water and magma behave in the presence of high thermal and pressure gradients. The following discussion covers considerations that demonstrate that contact interface dynamics might in fact be relatively insensitive to water/magma mass ratios, making the preceding discussions of secondary importance.

5. CONTACT INTERFACE DYNAMICS

Kokelaar [1986] identified two classes of steam explosivity in subaqueous basaltic volcanism: *contact-surface* and *bulk interaction*. The former class concerns the dynamics along an interface between a free body of water and magma, and the latter case applies to the dynamics of a volume of magma that confines water or water-rich clastic materials either within the magma or at its margins. In either case the dynamics begin with an initial contact of magma with water. During this initial exposure, a very thin film of water is nearly instantaneously heated. From considerations of characteristic heat diffusion times for the water and magma, an estimate of this instantaneous contact temperature, T_i, can be made:

$$T_i = \frac{T_m[\kappa/\sqrt{\alpha}]_m + T_w[\kappa/\sqrt{\alpha}]_w}{[\kappa/\sqrt{\alpha}]_m + [\kappa/\sqrt{\alpha}]_w}, \qquad (6)$$

in which subscripts denote properties of the magma (*m*) and water (*w*), κ is the thermal conductivity, α is the thermal diffusivity ($\alpha = \kappa/\rho C$; C = specific heat capacity; ρ = density), and T is the initial temperature. For submarine conditions, the values for water are: $T_w \approx 277$ K, $C \approx 4.2$ kJ kg^{-1} K^{-1}, and $\kappa \approx 0.6$ W m^{-1} K^{-1} with C falling and κ increasing about 1% for every 10 MPa increase in pressure. Magma composition generally dictates T_m with basalts erupting at 1473 to 1523 K; more silicic compositions range from 900 to 1200 K. For magma the specific heat capacity is typically 1.0 to 1.2 kJ/kg-K, and κ ranges from 1.1 to 4.8 W/m-K. These values set T_i in the range of 800-1000 K, which for neatly constant-volume equilibrium requires an instantaneous supercritical pressure.

5.1. Film Instability

Instantaneous heating of a film of water produces a small shock wave, which moves away from the contact at or above the speed of sound. The film expands in its wake and eventually develops into a region of film boiling, which is not necessarily a stable state. Stability of this vapor film requires that the rate heat is supplied to the film from the magma equals that transferred to the surrounding water. Instability arises where the film expands so fast that it exceeds the volume where it is in thermodynamic equilibrium. It then abruptly condenses, collapsing back on itself. The collapse of the film causes it to impact the magma surface and produce a finite strain in the magma. For certain contact surface geometries, the film collapse can be axisymmetric, and then produces tiny jets of water that penetrate the magma surface. After collapse, the film is recreated, repeatedly growing and collapsing in a cyclic fashion at a characteristic frequency of several kilohertz or more. The vapor-film oscillation gradually heats the water in the vicinity of the magma surface and causes strain to accumulate in the magma by repeated film impacts and jetting as well as by the volumetric changes caused by rapid magma cooling (quench contraction); the accumulated strain generally produces magma fragmentation. In contrast to this scenario of film instability leading to magma fragmentation, growth of a stable film interface effectively insulates the magma such that fragmentation and quench granulation do not develop, perhaps a reason why pillow lava stays intact.

Figure 8 illustrates heat transfer associated with film growth and collapse in an idealized spherical system. The conductive factor is the differential change in heat transfer rate with the film thickness (radius) for a constant thermal differential [*Wohletz*, 1983]. As vapor forms and expands around the melt sphere, it cools, its pressure decreases, and the area over which it conducts heat to the surrounding liquid increases, leading to a decreasing conductive factor. The momentum of the film growth may cause its over-expansion

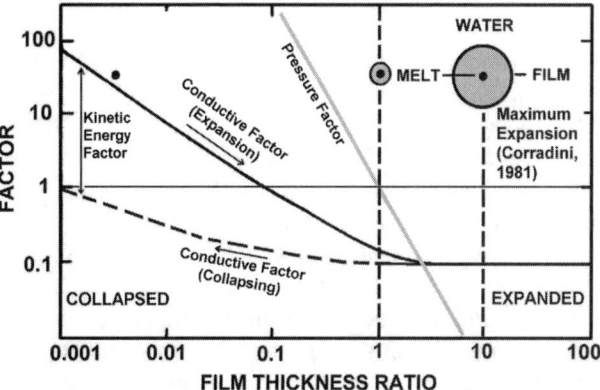

Figure 8. Schematic illustration of vapor film expansion and collapse, diagrammatically plotting the conductive factor as a function of film thickness and the pressure factor (a ratio of film pressure to ambient), both dimensionless.

to a thickness where the heat supplied is less than the heat lost and the film pressure is less than ambient. In this case growth is halted when the film spontaneously condenses and collapses. Because of viscous effects, this expansion and collapse cycle is irreversible. During collapse the conductive factor increases at a slower rate than it decreases during expansion, and heat is converted to kinetic energy by an amount that is proportional to the hysteresis of the system. In contrast, a stable film might oscillate around values of unity for conductive factor and film thickness, but it would not grow out of equilibrium nor collapse completely. Stable films might exist where convective heat transport around the film balances conduction from the magma, a situation where passive melt quenching occurs such as in the case of pillow lava formation.

One fate of such contact interface dynamics is production of a coarse mixture of magma fragments, liquid, and vapor. With increasing magma surface area for conductivity, heat transport grows exponentially as an escalating system. Convective currents that develop in response to these dynamics tend to dampen the system so that catastrophic mixing of fragmented magma and water does not occur. If, on the other hand, catastrophic mixing does occur, then the system is not only thermodynamically unstable (because of the film expansion and collapse process), it is also hydrodynamically unstable because of large pressure, density, sound speed, and conductivity gradients produced by the film. Such instability is prone to a kind of detonation, termed *thermal detonation*, especially if perturbed by some external pressure wave, such as that produced by volcanic seismicity.

5.2. Thermal Detonation

The term *thermal detonation* originated during early studies of fuel-coolant interactions for nuclear reactor safety analysis [*Board et al.*, 1975; *Bankoff and Jo*, 1976; *Fauske*, 1977; *Sharon and Bankoff*, 1981; and *Yuen and Theofanous*, 1995]. Stimulated by the proposal of *Fowles* [1979] that vapor explosions constituted a kind of elementary detonation, a very radical idea, considering that detonation is classically tied to chemical reactions, *Rabie et al.* [1979] did a rigorous study of rapid phase-change dynamics and concluded that certain materials could display the phenomenon of *polymorphic detonation*. *Harlow and Ruppel* [1981] used numerical multiphase simulations to demonstrate the plausibility of explosion wave propagation along the contact of two liquids, one above the boiling-point temperature of the other.

The concept of thermal detonation as originally conceived has many shortcomings when applied to real situations of MFCI, several of which are: (1) it requires an unrealistically high trigger pressure; (2) it involves hydrodynamic fragmentation that may not be fast enough to support detonation; (3) it is based on the classical, single-phase, Chapman-Jouguet detonation theory that is difficult to prove for heterogeneous mixtures; and (4) the pressure-wave attenuation caused by the mixture tends to prevent sustained shock-wave propagation. Because detonation is such a specific concept many have argued that it just does not apply to MFCIs, especially from standpoints of required fragmentation rates and premixture ratios [*Condiff*, 1982; *Fletcher and Theofanous*, 1995; and *Yuen and Theofanous*, 1999].

Yuen and Theofanous [1999] show calculations that illustrate why the now classical theory of multiphase thermal detonation of *Board et al.* [1975] is not physically possible. Their calculations focus on fuel-coolant premixture ratios (volumetric ratio of fuel to water-plus-vapor). For lean ratios, detonation is only possible where the void (vapor) fraction is nearly zero (i.e., $R > 1.5$), a situation where thermodynamic conversions ratios are very low and necessary film boiling is unrealistically precluded. For rich ratios with the physically required film boiling (i.e., $R < 0.05$) only weak detonation is possible. Only for a rather limited range of intermediate premixture ratios (the case examined by *Board et al.* [1975]) for which the volume fractions of fuel, water, and vapor are equal (i.e., $R \approx 0.5$), did *Yuen and Theofanous* [1999] calculate a stable detonation with a pressure of ~150 MPa. However, experimental [*Angelini et al.*, 1992; 1995] and analytical studies [*Fletcher and Thyagaraja*, 1991] show that such a premixture is not physically possible for MFCI. Overall, the main argument of these calculations is that for all premixtures (other than those of unreasonably high melt concentration), the shock wave sweeps in additional coolant such that thermal equilibration of the fragmented melt does not produce the amount of water expansion needed to sustain the wave.

Yuen and Theofanous, [1994] recognized this fundamental problem with the classical thermal detonation theory but also acknowledged experimental evidence of MFCI explosion phenomena that produce strong shock waves. Those authors developed the *microinteractions* model of thermal detonation, which hypothesized that the rate of water mixing with fragmented melt is proportional to the melt fragmentation rate. *Chen et al.* [1995] experimentally verified this model and showed detonation dynamics limited to only what is termed the *m-fluid* (a mixture of fragmented debris and entrained coolant). With those results and successful simulation of MFCI detonation dynamics, *Yuen and Theofanous* [1999] emphasize that the microinteractions model of thermal detonation is viable; it occurs under much less restrictive premixture conditions and avoids other problems of the classical theory.

For application to the dynamics of water/magma interaction, let us just assume that thermal detonation simply entails shock-wave dynamics that lead to catastrophic fragmentation and expansion of a mixture of magma fragments and water. This generalized view is intended to include the phenomena such as *thermohydraulic fracturing* and *brittle reaction* that *Büttner and Zimanowski* [1998] and *Zimanowski et al.* [1997b] describe. *Zimanowski et al.* [1997a] describe MFCI experiments that show development of intense shock waves in less than a millisecond with extreme cooling (>10^6 K s^{-1}) and stress rates (>3 GPa m^{-2}). These phenomena constitute a brittle reaction that occurs on a very fine scale; the brittle reaction is quite different in concept from the detonation idea presented above. However, one can argue that the thermodynamic and hydrodynamic conditions of this brittle reaction are also those that pertain to detonation. For this reason, I will further consider the broad, generalized concept of thermal detonation and how it might apply to the deep submarine environment.

Figure 9 schematically illustrates the basic idea of thermal detonation. Consider the vapor-film dynamics discussed above and imagine that the contact between a submarine extrusion and seawater forms a selvage zone of unstable vapor and magma fragments. The process leading to explosion of this mixture involves the propagation of a pressure disturbance in the seawater caused by a dynamic event such as volcanic seismicity, vent collapse, or energetic film collapse. As this wave moves through the mixture of hydroclasts, vapor, and water, if its overpressure is great enough, it can compress the vapor into liquid, causing intimate contact with the magma. This contact induces an abrupt increase in overall heat transfer to the water, leading to a pressure jump behind the wave that can drive the wave as a shock. As the shock moves through the mixture, its steep pressure gradient accelerates the water and hydroclasts proportional to their density. This differential acceleration produces a slip velocity between the water and hydroclasts high enough to tear the hydroclasts into micrometer-sized particles, increasing the surface area for heat transfer by orders of magnitude. The increased heat transport caused by the shock wave and the fine fragmentation and expansion in its wake tend to sustain and even enhance the shock. An idealized shock has a *N-wave* profile, and falling pressure in its wake allows expansion and release of thermodynamic work even at high ambient pressure.

In order for this phenomenon to be considered a detonation, the acceleration of the mixture by the shock wave must

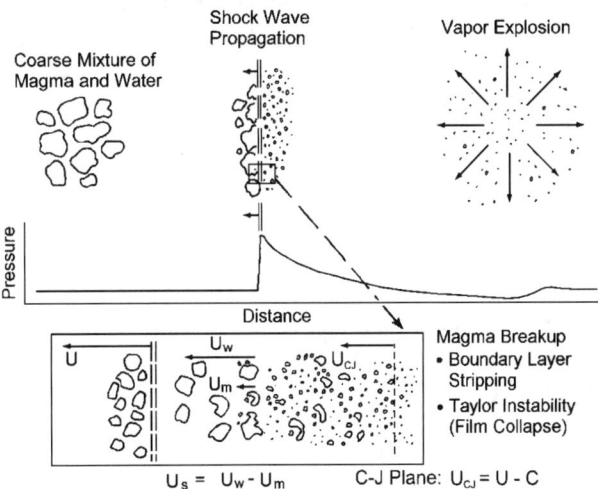

Figure 9. Schematic illustration of the concept of thermal detonation [adapted from *Wohletz*, 1986; and *Board et al.*, 1975], showing the propagation of a shock wave through a coarse mixture of magma fragments and water (vapor and liquid). The shock wave moves at a velocity u and differentially accelerates the water and magma to velocities of u_w and u_m, respectively, resulting in a slip velocity u_s, which decays behind the shock. The slip velocity must be of sufficient amplitude to cause fine fragmentation of the magma fragments by mechanisms such as boundary layer stripping and Taylor instability before the arrival of the C-J plane. At the C-J plane the average mixture velocity is just sonic (c) with respect to the shock wave. The fine fragmentation causes an exponential rise in heat transfer from the magma fragments to the water and catastrophic vapor expansion.

produce a relative velocity high enough to satisfy the Chapman-Jouguet (C-J) condition: the relative velocity, u_r, is the speed of the shocked material relative to the shock front, and u_r must equal the sonic velocity of the shocked material [*Courant and Friedrichs*, 1948; and *Zel'dovich and Raizer*, 1966]. The C-J condition can be evaluated on a pressure-volume diagram that shows the shock adiabat (termed the *shock Hugoniot* and defined as the locus of points representing pressure-volume states achievable by shocking a material from an initial state) and the release adiabat (called the *detonation curve* or *detonation Hugoniot*). These adiabats are concave upward and the detonation curve exists at higher volume states than the shock Hugoniot. Two points on the shock Hugoniot, one at the initial pressure and the other at the pressure of the shock front (the *von Neumann spike*), define a line called the *Rayleigh line*. A C-J condition only exists if the Rayleigh line intersects the detonation curve at a single point of tangency. The points behind a propagating shock at which the C-J condition exists define a surface known as the *C-J plane* or the *detonation front* (not to be confused with the shock front).

Board et al. [1975] and *Fauske* [1977] suggested that the C-J condition for a water/melt mixture can be met if a propagating shock wave causes melt fragmentation in a time shorter than that required for water-melt velocity equilibration (zero slip velocity). Those authors describe how melt breakup and velocity equilibration times can be assessed by a Bond number function. The Bond number is 3/8 the product of the coefficient of drag and the Weber number (a ratio of inertial forces to surface tension forces); it is used in calculations of momentum transfer in general, especially for assessing atomization and motion of bubbles and droplets. The Bond number function as envisioned by *Bankoff and Jo* [1976] includes the effects of phase densities and volume fractions, initial premixture fragment size, the water/melt slip velocity, and the pressure at the shock front an at the C-J plane. With the microinteractions model of *Yuen and Theofanous* [1999] this function predicts that thermal detonation can occur in MFCIs.

In application of detonation theory there are mitigating factors in hydrovolcanic systems that one should consider. One major factor is that water/magma interaction systems likely involve spatially varying mixture densities and thermodynamic states. Such variations predict nonuniform C-J conditions that tend to destabilize a detonation wave. A second factor is geometry. It is likely that the mixture zone is discontinuous, thin in some places, thick in others. This situation leads to 3-D effects that cause large lateral slip velocities along the shock front, and such slip velocities tend to degrade the sonic conditions behind the front necessary for detonation.

Because the microinteractions model of *Yuen and Theofanous* [1999] addresses many of the mitigating circumstances in MFCI detonation theory and allows successful prediction of experimental explosions, there is some justification for applying general aspects of that theory to assess the effects of ambient pressure on hydrovolcanism. By assuming that a C-J condition is satisfied by interaction dynamics and that the slip velocity between the shocked melt fragments and water is at least as large as the C-J plane relative velocity [*Board et al.*, 1975], then the Bond number function can be calculated. *Board et al.* [1975] use the Rankine-Hugoniot jump condition [*Landau and Lifshitz*, 1959; and *Zel'dovich and Raizer*, 1966] of the propagating shock wave to determine its velocity as ~300 m s⁻¹. The relative velocity, u_r, of the shocked mixture leaving the front is given by a function of the mixture's pressure, p, and specific volume, $V = V(R)$, at ambient (subscript i) and C-J (subscript cj) conditions:

$$u_r = \sqrt{(p_{cj} - p_i)(V_i - V_{cj})} \quad . \qquad (7)$$

For an idealized thermal detonation in which $p_{cj} \approx 100$ MPa, *Board et al.* [1975] calculated u_r at ~100 m s⁻¹. For MFCI volcano analogs, *Wohletz* [1986] used the approach suggested by *Corradini* [1981] to estimate a minimum u_r at 60 m s⁻¹. *Drumheller* [1979] combined the requirements for relative velocity and melt breakup time into what can be called a critical Bond number. By assuming a constant p_{cj}, *Wohletz* [1986] evaluated the critical Bond number with respect to FCI experimental data [*Wohletz and McQueen*, 1984] to predict the effects of R and ambient pressure on the development of relative velocities and magma particle sizes. These results are plotted in Figure 10, which shows optimal conditions for thermal detonation at $0.5 < R < 2.0$ for ambient pressures at or below 40 MPa. With increasing ambient pressure, the predicted relative velocities fall, eventually going below 60 m s⁻¹, which *Wohletz* [1986] considered as the lower limit for sustaining a detonation. With increasing ambient pressure, particle fragmentation is also decreased, meaning less thermal energy is released in the wake of the shock wave. If these results have any bearing on shock-wave dynamics for water/magma interactions in the submarine environment, then they do suggest that a thermal detonation is not likely at water depths greater than about 4000 m.

6. DISCUSSION

Because observational evidence of deep submarine eruptions is sparse, much of the information I have presented is conceptual and highly theoretical with factual basis going

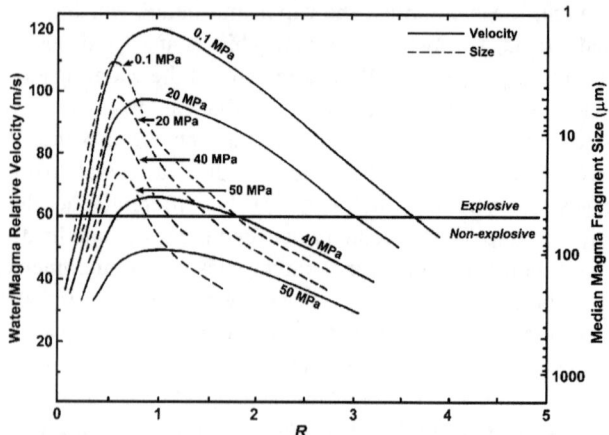

Figure 10. Calculated effect of ambient pressure on thermal detonation relative velocities and fragment size as a function of R (water/magma mass ratio) and ambient pressure [from *Wohletz*, 1986].

only so far as what analog MFCI experiments and thermodynamic constraints permit. Overall I have described some general factors governing submarine hydrovolcanism and specifically addressed issues concerning the possibility of explosive water/magma interaction at high hydrostatic pressure. I have little or no proof of the predictions presented here, which is a major weakness of this contribution. I have relied upon results of analog experiments designed to quantify the controls of water/melt interaction, attempted thermodynamic analysis of the somewhat complex heat exchange between magma and water, and presented a short review of dominant physical processes that govern the character of interaction.

Chief concerns about application of experimental results to deep submarine eruptions are the experimental water/magma ratios used in high-pressure MFCI studies and the method by which high ambient pressure is experimentally produced. Even though the *water-box* experiments approached R values of 10, these are in no way even close to the *apparent R* values in submarine conditions, which are potentially so large that they are practically impossible to quantify. As I pointed out, from the heat flow mechanisms involved, characteristic times and lengths of thermal diffusion, pressure-wave propagation, and film boiling do limit the amount of water that can really be involved in interaction over short periods of time. It is because of these limitations, that the *water-box* experiments produced both pillow-lava-like debris, presumably from more passive quenching processes, as well as explosive fragmental debris. These results suggest the upper range of R that is practical for consideration is no larger than 10.

The *water-box* experiments did not employ any kind of confinement to simulate the hydrostatic pressure typical of submarine conditions. For this reason, the high-pressure experiments were examined to test if explosive interaction could occur at ambient pressures above critical pressure. The main concern here is if the ambient pressure generated at the instant of water/melt contact but prior to burst is a plausible analog for hydrostatic pressure. From the standpoint of thermodynamics and physical properties of water, I conclude that this concern is negligible. Because initiation of water/melt interaction causes pressures to rise to near-maximum values in $<<1$ s, it is safe to assume that most of the dynamic heat exchange occurred at the burst pressure. One might argue that in the deep submarine environment this pressure exists before any heat exchange occurs, in contrast to the experiments where the pressure rose in milliseconds prior to interaction. However, the thermodynamic models show that for the initial equilibrium conditions, ambient pressure plays only a small role.

The thermodynamic models presented are limited in applicability because they assume pure water. Whereas this assumption is justifiable for subcritical thermodynamic calculations, because seawater's heat capacity is within a few percent of that of pure water, such may not be the case for supercritical seawater. The effects of phase separation and the extension of the two-phase boundary to supercritical pressures and temperatures place additional energy constraints on isentropic expansion from high temperatures and pressures; further work is needed address the magnitude of these effects for MFCIs. The model discussed in this paper was designed to be applicable to experimental results; its validity is how well it predicts experimental results. For FCI experiments that constrained conversion ratios by kinetic energy measurements, the thermodynamic model proved to be reliable. Unlike controlled experiments, it is not really known how much of water and melt are involved at any given time during submarine hydrovolcanism; the thermodynamic model applied to *apparent R* values tends to give minimum estimates of explosive energy.

From my experience with MFCI experiments and theoretical modeling of water/magma interaction I have focused primarily on R as a controlling parameter. Water/magma mass ratio is really difficult to evaluate, but not impossible as shown by *Wohletz* [2002]. No matter what approach ones uses to constraining R, the *effective R* is controlled by a system's characteristic length and time scales for heat transfer and hydrodynamics. The example discussed earlier emphasizes that the length scales are a function of diffusivities and sound speed with characteristic times about of about a millisecond per meter of characteristic length for pressure waves propagation. Considering thermal diffusion rates, the volume of water heated is within an order of magnitude of the volume of magma cooled. Thus even though the volume of water available is practically infinite for submarine hydrovolcanism, I suggest that *effective R* values are less than ~10 for submarine volcanism.

As a function of R, calculated thermodynamic conversion ratios are a measure of thermodynamic work done by water expansion. One cannot assume that all of this work is manifested as melt-fragment kinetic energy. *Wohletz et al.* [1995] measured FCI ejecta kinetic energy and found it to be typically one-third to one-half the thermodynamic work predicted. Much of the work is spent on melt fragmentation and deformation, seismic and acoustic waves, and viscous losses. This consideration suggests that measures of hydrovolcanic ejecta dispersal (a typical approach to studying terrestrial eruptions) do not fully constrain the eruption energetics. On the other hand, energy calculations based upon crater dimensions do account for most energy expenditures because these calculations are empirical [*Wohletz and Heiken*, 1992]. For the submarine environment, hydrovolcanic energies may be difficult to constrain because measurement of crater excavation and fragment dispersal are limited by seafloor imaging techniques. That is not to say estimates cannot be done. I recommend making volume estimates of fragmental deposits as a measure of how much magma might have been involved in a submarine hydrovolcanic eruption. The seawater volume involved in such an eruption is more interpretive and requires deduction of the mode of interaction and application of logic similar to that presented in Equations (2-4). Then evaluation of Eq. (5) provides a measure of the *effective R* for the eruption from which a thermodynamic conversion ratio can be calculated.

In modeling the expansion work caused by the depressurization of heated water, I presented two bounding thermodynamic cases, *isentropic fluid* and *isentropic mixture* for calculating the conversion of thermal to mechanical energy. If expansion only extends to local hydrostatic pressure, then calculated conversion ratios at high hydrostatic pressures (>200 MPa) exceed the experimentally constrained explosive threshold for a rather limited range of R values, which suggests that explosive hydrovolcanism may not be common in deep environments and the *fluid* case of calculation is appropriate. This point of view might be supported by the much more common observations of lava than hydroclasts from deep ocean cores. However, if explosive interaction is triggered, the *mixture* expansion case may be more appropriate, since magma fragmentation adds the potential for continued heat transfer during water expansion leading to higher conversion ratios.

Thermodynamic modeling also includes two end-member expansion states, a conservative one for which expansion ends at hydrostatic pressure and the other allowing full expansion to atmospheric pressure. If there is real potential for shock-wave dynamics, as in the case of the generalized concept of thermal detonation, localized expansion to pressure much lower than ambient might occur because of spatially varying sonic conditions. Recalling that local sound speeds can vary over an order of magnitude in multiphase systems [e.g., *Kieffer*, 1977], the rapidly expanding mixture will not "know" when it has reached hydrostatic pressure until a finite time after it has expanded beyond that point to much lower pressures. This aspect is a fundamental of supersonic flow and shock waves, which brings up another aspect of ambient pressure above the critical pressure. Whereas arguments can be made that explosion will not occur at these pressures because the expansion is supercritical and does not involve a large volume change, I emphasize that the dynamics are mostly controlled by how fast the expansion occurs, whether or not a distinct vapor phase is present. I also note that most chemical explosives operate far above their critical pressures [e.g., *Fordham*, 1966], and this has been a criterion that FCI studies [e.g., *Yuen and Theofanous*, 1999] have used to differentiate detonation phenomena. Based on these considerations, if explosive interaction involves *isentropic mixture* expansion, then the calculations shown in Figure 6 indicate that mechanical energy release increases with ambient pressure, and the details of how high interaction pressure and heat transfer might occur come into question.

In discussion of the effects of ambient pressure on propagation of a hypothetical thermal detonation within a water/melt mixture, I presented some theory and calculations supported by experimental evidence. These calculations showed that with increasing ambient pressure, the range of R over which relative velocities support detonation narrows and becomes nonexistent at ambient pressures above 40 MPa. Considering the C-J conditions for a thermal detonation, isothermally or adiabatically increasing ambient pressure decreases water's specific volume and, assuming a constant detonation curve, increases the slope of the Rayleigh line. If in this case the Rayleigh line were to intersect the detonation curve at two points, a situation of overdriven detonation would occur, which would not be stable. On the other hand, if the Rayleigh line were still tangent to the detonation curve, then its point of tangency would stipulate a higher C-J pressure. A higher C-J pressure causes a higher pressure to exist in the water/melt mixture. If the sound speed of the mixture behaves like the liquid-gas mixtures studied by *Kieffer* [1977] and increases with pressure, then the relative velocity of the C-J plane should also increase. This consideration (not recognized by *Wohletz* [1986]) suggests that curves shown in Figure 10 might plot at higher values of relative velocity for ambient pressures above atmospheric; hence, ambient pressure may not necessarily suppress MFCI detonation.

At this point in time the work of *Zimanowski et al.* [1997a] is perhaps the most plausible description of a mechanism for extremely high energy transfer rates in MFCIs. The tremendously dynamic, brittle reaction discovered by these workers has been photographically documented [*Zimanowski et al.*, 1997b], and it provides a new way for understanding high interaction pressures and heat transfer rates. The hydrodynamics of the brittle reaction have not been linked to detonation, and to do so requires documentation of propagation speeds, sound speeds, and the shock Hugoniot of a water/melt mixture. If a link to detonation were to be established, then a robust predictive capability could be established by scaling experimental work. One such prediction, mentioned earlier, is the hypothetical effect of ambient pressure on the potential for detonation. Even though increasing pressure (depth) might increase the mechanical energy released in a detonation wave, the slip velocity (between the water and melt in a shocked mixture) and the degree of melt fragmentation behind the shock may decrease with increasing pressure, leading to conditions not capable of sustaining a detonation wave. This line of reasoning seems to fit results of high-pressure MFCI experiments discussed above in which the highest energy bursts occurred at the highest confining pressures but the likelihood of non-burst was also higher.

Some aspects of how magma is erupted on the seafloor have not been fully considered. Passive extrusion of flows and domes are suspected to be capable of entrapping water beneath them (producing low *effective R* values) and forming surface selvages of unstable, supercritical fluids and hydroclasts, situations that potentially can be explosive. However, the situation of magma fountaining with magmatic volatile exsolution is another situation seemingly prone to explosive interaction. The fountaining pre-fragments the magma, resulting in a much greater surface area for heat transfer, and fountaining drives mixing of water with hydroclasts; both processes are important for explosive interaction. If, however, exsolved volatiles are rich in noncondensible gases, such as CO_2, explosive interaction might be damped. Noncondensible gases tend to limit intimate contact of water with the magma by forming a stable insulating film at the magma surface.

7. CONCLUSIONS

By definition, submarine volcanism is hydrovolcanism, but for deep submarine environments it is apparently overwhelmingly nonexplosive hydrovolcanism, with products of explosive events rarely observed in deep seafloor samples. Do these observations indicate that high hydrostatic pressure prevents explosive interaction, or do they more simply point to the fact that deep seafloor observations are too sparse relative to the vast expanses of the oceans to adequately assess the frequency of explosive products? At this point I conclude that this issue is unresolved.

I also conclude that experimental evidence and theoretical considerations indicate that explosive hydrovolcanism is certainly possible for depths extending to greater than 4000 m. Furthermore, theory points to the potential that if explosions do occur at great depth, then they may release more mechanical energy per unit mass of magma than they would at or near the sea surface. What remains to be studied is how a violent explosion might affect the deep submarine environment (e.g., hydroclast characteristics and dispersal, deposit morphology, vent shape, and seawater currents), taking into consideration the much greater viscous and thermal dissipations existing in the aqueous environment compared to those in subaerial settings.

I further conclude that high ambient pressure may have the potential of decreasing the probability of thermal detonation as a mechanism for explosive interaction. But I do emphasize that this conclusion is largely conceptual. The detonation curve for MFCI is yet to be well established, and higher C-J pressures might result for systems at higher ambient pressure. As a final note about thermal detonation, I suggest that MFCI explosions may be better represented by detonations other than the Chapman-Jouguet type [e.g., *Wood and Kirkwood*, 1954; and *Rabie et al.*, 1979].

At the time of writing this paper, over 30 years have passed since focused research on water/magma interaction (hydrovolcanism) began, then in an effort to understand the origins of maar craters and how they might be differentiated from those formed by bolide impacts. The early studies [e.g., *Fisher*, 1968; *Lorenz*, 1970; *Waters and Fisher*, 1971; and *Heiken*, 1971] were perceptive and led to quantitative field techniques, theoretical considerations, and experimental studies [e.g., *Sheridan and Wohletz*, 1981; *Zimanowski et al.*, 1986; and *Kokelaar*, 1986]. The identification of phenomenological commonality with industrial vapor explosions, especially those of concern to nuclear reactor safety [e.g., *Marshall*, 1986], has greatly enhanced the appreciation of the physical controls of hydrovolcanism. However, I conclude that research has a long way to go in developing a fuller understanding of hydrovolcanism, not only of its physical controls but also of its geochemical significance, especially in the submarine environment.

Acknowledgments. This work was done under the auspices of the U.S. Department of Energy. The author thanks James White for enlightening information about submarine volcanism, stimulating discussion of various points of view, and editorial help in adapting this work to suit this volume. Larry Mastin and Magnús Gudmundsson provided detailed and insightful reviews that highlighted the strengths and weaknesses of this work, leading to its improvement. I am indebted to Bernd Zimanowski and Ralf Büttner for sharing their unique insights into experimental results. Although our conclusions may differ, our common grounds of experience will no doubt stimulate future advances on understanding molten fuel-coolant interactions.

APPENDIX A. CALCULATION OF INTERACTION THERMODYNAMIC WORK

The ratio of thermodynamic work to the magma's heat energy (in excess of ambient) is termed the *conversion ratio*, which is a measure of how dynamic a water/magma interaction is. Passive interactions result in little pressure-volume work so that the conversion ratio is a few percent or less while explosive interactions may show conversion ratios reaching 20% or more. Explosive interactions that tend to keep the expanding water and magma fragments in constant contact and thermal equilibrium show higher conversion ratios than do those where the water separates from the magma during expansion.

Calculation of thermodynamic conversion ratios as a function of R (water/magma mass ratio) provides a theoretical basis for predicting the effects of hydrostatic pressure. These calculations begin with the following assumptions: (1) water is initially saturated (no vapor present); (2) all heat transferred from the magma during interaction is to the water (adiabatic boundary); (3) liquid water is incompressible; and (4) heat exchange is sufficiently rapid that water and magma reach an equilibrium temperature, T_e, before the water expands:

$$T_e = \frac{RC_v T_w + C_m T_m}{RC_v + C_m} , \qquad (A\text{-}1)$$

where T is temperature (subscripts e for equilibrium, w for water, and m for magma), and C is specific heat, assumed to be constant (subscripts v for water at constant volume and m for magma). Because most of the water stays saturated during attainment of initial thermal equilibrium with magma, its specific heat is nearly constant between T_w and T_e and Eq. (A-1) also assumes this constancy.

The *Hicks-Menzies* [1965] assumption of rapid heat exchange includes the idea that water does not experience much volume change in reaching equilibrium with the melt so that volume terms can be ignored in the derivative of entropy:

$$dS = \frac{C_v}{T} dT + \left(\frac{\partial p}{\partial T}\right)_v dV , \qquad (A\text{-}2)$$

for which S is entropy, C_v is the constant volume heat capacity, T is temperature, p is pressure and V is volume. Solving Eq. (A-2) for constant volume yields:

$$S_e = S_i + C_v \ln(T_e / T_i) , \qquad (A\text{-}3)$$

where the subscripts i and e denote the initial (T ambient) and equilibrium states, respectively.

T_e and S_e alone are not sufficient to predict other thermodynamic properties without further considerations. The rapid heating to initial equilibrium temperature and pressure can be idealized as *isochoric* (constant volume), but that idealization is only approached for shock compression of water at low pressure, and it does not allow for the creation of a vapor film at water/melt interfaces, which is documented in MFCI experiments. Accordingly, for the calculated T_e and S_e values, equilibrium states of pressure, volume, and other thermodynamic parameters are determined by fitting polynomial functions to steam-table data [e.g., *Haar et al.*, 1984]. Figure 5a shows water's initial equilibrium specific volume increasing with decreasing R; however, water's total volume fraction in the mixture decreases with decreasing R such that the effect on the mixture volume is always less than 10%.

With the equilibrium state defined as a function of R, calculation of the final expanded thermodynamic state depends upon whether an *isentropic fluid* or *isentropic mixture* expansion path is followed during water expansion and the final pressure. For calculations, the *fluid* case just requires finding thermodynamic parameters for the desired final pressure at the equilibrium entropy (isentropic expansion). The *mixture* case is a bit more complex [*Wohletz*, 1986], requiring calculation of the slope of the expansion curve in temperature-entropy space and finding thermodynamic parameters for points (if any) where this curve intersects the saturation curve. The total work for the *mixture* calculation then becomes the sum of the work for each leg of the expansion path.

The First Law of Thermodynamics provides a starting point for calculation of thermodynamic work:

$$dU = dQ + dW , \qquad (A\text{-}4)$$

for which U is the energy of the system, Q is heat, W is thermodynamic work, p is pressure, and V is volume such that $dW = -pdV$. For a system at constant pressure, $U = H - pV$, $dQ = C_p dT$, and $-W = U_f - U_i$, where H is enthalpy, C_p is the constant pressure heat capacity, and the subscripts f and i denote the final and initial states of the system, respectively. Furthermore at constant pressure, $dU = dQ - pdV$ and $C_p = dU/dT + p(dV/dT) = dQ/dT$. Alternatively, for a system at constant volume, $dU = dQ$, and $C_v = dU/dT = dQ/dT$.

Using the above thermodynamic relationships and the definition of an isentropic adiabatic process as one where $dQ = 0$, $dW = dU$, the mechanical work involved is $dW = (U_i - U_f)$ so that

$$W = (H - pV)_i - (H - pV)_f . \qquad (A\text{-}5)$$

For an *isentropic mixture* process involving magma and water (where m and w are their respective masses), thermodynamic work is the sum of heat transfer and internal energy change, $-dW = dQ - dU$, where heat transfer is $dQ = (mC_m + wC_v)dT + pdV$:

$$-dW = [(mC_m + wC_v)dT + pdV] - dU \quad , \quad \text{(A-6)}$$

integrating:

$$\begin{aligned}-W =\ & [(mC_m + wC_v)(T_f - T_i)] \\ & + [H_f + X_f H_f^* - p(V_f + X_f V_f^*)] \\ & - [H_i + X_i H_i^* - p(V_i + X_i V_i^*)] \\ & - [(H_f - pV_f) - (H_i - pV_i)] \quad , \end{aligned} \quad \text{(A-7)}$$

and combining terms:

$$\begin{aligned}W =\ & [(mC_m + wC_v)(T_f - T_i)] \\ & + [X_i(H_i^* - pV_i^*) - X_f(H_f^* - pV_f^*)] \quad . \end{aligned} \quad \text{(A-8)}$$

Expansion calculations for the steam dome involves specification of pressure-dependent phase-change enthalpy, H^*, and volume V^*, and calculation of steam fractions, X, by the ratio of entropy S over the phase-change entropy. Whereas calculation of the *isentropic fluid* case by adiabatic expansion is relatively simple, application of Eq. (A-8) requires careful consideration of thermodynamic path. If expansion does not go through the steam dome, phase-change terms do not apply so that one should use:

$$\begin{aligned}W =\ & [(mC_m + wC_v)(T_f - T_i)] \\ & - [(H_f - pV_f) - (H_i - pV_i)] \quad . \end{aligned} \quad \text{(A-9)}$$

Numerical implementation of this calculation involves steam-table lookups for which the author has written a computer program with built-in thermodynamic property calculations that are accurate within 1% of values listed in steam tables [*Lemmon et al.*, 2001]. This program may be downloaded from the Internet by contacting the author.

REFERENCES

Angelini, S., Takara, E., Yuen, W. W., and T. G. Theofanous, Multiphase transients in the premixing of steam explosions, *Nucl. Eng. Des.*, 146, 83-95, 1994.

Angelini, S., Theofanous, T. G. and W. W. Yuen, The mixing of particle clouds plunging into water, *Nucl. Eng. Des.*, 177, 285-301, 1994.

Bankoff, S. G., and J. H. Jo, Existence of steady-state fuel-coolant thermal detonation waves, ERDA Rep. NU-2512-6 (CONF-760328-8), National Technical Information Service, Springfield, VA, 1976.

Batiza, R., D. J. Fornari, D. A. Vanko, and P. Lonsdale, Craters, calderas, and hyaloclastites on young Pacific seamounts, *J. Geophys. Res.*, 89, 8371-8390, 1984.

Bischoff, J. L., and R. J. Rosenbauer, The critical point and two-phase boundary of seawater, 200-500°C, *Earth Planet. Sci. Let.*, 68, 172-180, 1984.

Bischoff, J. L., and R. J. Rosenbauer, Liquid-vapor relations in the critical region of the system NaCl-H2O from 380 to 415°C: a refined determination of the critical point and two-phase boundary of seawater, *Geochim. Cosmochim. Acta*, 52, 2121-2126, 1988.

Board, S. J., R. W. Hall, and R. S. Hall, Detonation of fuel coolant explosions, *Nature*, 254, 319-321, 1975.

Bonatti, E., Mechanisms of deep sea volcanism in the South Pacific, in *Researches in Geochemistry*, 2, edited by P. H. Ableson, pp. 453-491, John Wiley & Sons, New York, 1967.

Buntebarth, G., and J. R Schopper, Experimental and theoretical investigations on the influence of fluids, solids and interactions between them on thermal properties of porous rocks, *Phys. Chem. Earth, (23)9-10*: 1141-1146, 1998.

Burnham, C. W., Energy of explosive volcanic eruptions, *Sci.*, 213, 69-70, 1981.

Burnham, C., Deep submarine pyroclastic eruptions, *Econ. Geol., Monograph 5*, 142-148, 1983.

Büttner, R., and B. Zimanowski, Physics of thermohydraulic explosions, *Phys. Review E*, 57-5, 1-4, 1998.

Buxton, L. D.; and W. B. Benedict, Steam explosion efficiency studies, Sandia National Laboratories Report, SAND79-1399, NUREG/CR-0947, 62 pp., NTIS, Springfield, Virginia, 1979.

Chen, X., Yuen, W. W., and T. G. Theofanous, On the constitutive description of the microinteractions concept in steam explosions, *Nucl. Eng. Des.*, 177, 303-319.

Clague, D. A., Davis, A. S., Bischoff, J. L., Dixon, H. E., and R. Geyer, Lava bubble-wall fragments formed by submarine hydrovolcanic explosions on Loihi Seamount and Kilauea Volcano, *Bull. Volcanol.*, 61, 437-449, 2000.

Clague, D. A., Davis, A. S., and H. E. Dixon, Submarine Strombolian eruptions on the Gorda mid-ocean ridge, (this volume).

Colgate, S. A., and T. Sigurgeirsson, Dynamic mixing of water and lava, *Nature*, 244, 552-555., 1973..

Condiff, D., Contributions concerning quasi-steady propagation of thermal detonations, *Int. J. Heat Mass Transfer*, 25, 87-98, 1982.

Corradini, M. L., Analysis and modeling of steam explosion experiments, Sandia National Laboratories SAND80-2131, NUREG/CR-2072, 114 pp, NTIS, Springfield, Virginia, 1981.

Cotton, C. A., The pedestals of oceanic volcanic islands, *Geol. Soc. Amer. Bull.*, 80, 749-760, 1969.

Courant, R., and K. O. Friedrichs, *Supersonic Flow and Shock Waves*, 464 pp., Springer, New York, 1948.

Dinh, T.N., Dinh, A. T., Nourgaliev, R. R., and B. R. Sehgal, Investigation of film boiling thermal hydraulics under FCI conditions, in *Proc. of OECD/CSNI Specialists Meeting on Fuel*

Coolant Interactions, Tokai, Japan, May 19-21, 1997, pp. 674-695, Nuclear Energy Agency (France) CSNI R(97)26, 1998.

Dudás, F. O., The effect of volatile content on the vesiculation of submarine basalt. *Econ. Geol. Mono. 5*, 134-141, 1983.

Drumheller, D. S., The initiation of melt fragmentation in fuel-coolant interactions, *Nucl. Sci. Eng.*, 72, 347-356, 1979.

Fauske, H. K., Some comments on shock-induced fragmentation and detonating thermal explosions, *Am. Nucl. Soc. Trans.*, 27, 666-667, 1977.

Fisher, R. V., Puu Hou littoral cones, Hawaii, *Geol. Rundschau*, *57*, 837-864, 1968.

Fletcher, D. F., and T. G. Theofanous, Heat transfer and fluid dynamic aspects of explosive melt-water interactions, *Adv. Heat Trans.: Heat Trans. Nucl. React. Safety*, 29, 129-213, 1997.

Fletcher, D. F., and A. Thyagaraja, The CHYMES coarse mixing model, *Prog. Nucl. Ener.*, 26, 31-61, 1991.

Fordham, S., *High Explosives and Propellants*, Pergamon, New York, 1966.

Fowles, G. R., Vapor phase explosions: elementary detonations? *Sci.*, 204, 168-169, 1979.

Haar, L., Gallagher, J. S., and G. S. Kell, *NBS/NRC Steam Tables*, McGraw-Hill International Book Company, London, 320 pp, 1984.

Harlow, F. H., and H. M. Ruppel, Propagation of a liquid-liquid explosion, Los Alamos National Laboratory Report, LA-8971-MS, 11 pp., NTIS, Springfield, Virginia, 1981.

Haymon, R. M., Fornari D. J., Von Damm, K. L., Lilley, M. D., Perfit, M. R., Edmond, J. M., Shanks, III, W. C., Lutz, R. A., Grebmeier, J. M., Carbotte, S., Wright, D., McLaughlin, E., Smith, M., Beedle, N., and E. Olsen, Volcanic eruption of the mid-ocean ridge along the East Pacific Rise crest at 9°45-52'N: Direct submersible observations of seafloor phenomena associated with an eruption event in April, 1991, *Earth Planet. Sci. Let.*, 119, 85-101, 1993.

Head, J.W. III, and L. Wilson, Deep submarine pyroclastic eruptions: theory and predicted landforms and deposits, *J. Volcanol. Geotherm. Res.*, 121, 155-193, 2003.

Heiken, G., Tuff rings: examples from the Fort Rock-Christmas Lake Valley, south-central Oregon, *J. Geophys. Res.*, 76, 5615-5626, 1971.

Hicks, E. P., and D. C. Menzies, Theoretical studies on the fast reactor maximum accident, *in Proceedings of the Conference on Safety, Fuels, and Core Design in Large Fast Power Reactors*, USAEC Rep. ANL-7120, pp. 654-670, Nat. Tech. Inform. Serv., Springfield, Virginia, 1965.

Honnorez, H., and P. Kirst, Submarine basaltic volcanism: morphometric parameters for discriminating hyaloclastites from hyalotuffs, *Bull. Volcanol.*, 39, 441-465, 1975.

Kieffer, S. W., Sound speeds in liquid-gas mixtures: water-air and water-steam, *J. Geophys. Res.*, 82, 2895-2904, 1977.

Kieffer, S. W., and J. M. Delany, Isentropic decompression of fluids from crustal and mantle pressures, *J. Geophys. Res.*, 84, 1611-1620, 1979.

Kokelaar, P., Magma-water interactions in subaqueous and emergent basaltic volcanism, *Bull. Volcanol.* 48, 275-290, 1986.

Krauskopf, K. B., *Introduction to Geochemistry*, McGraw-Hill, New York, 721 pp., 1967.

Landau, L. D., and B. M. Lifshitz, *Fluid Mechanics: Vol 6, Course of Theoretical Physics*, Pergamon, New York, 1959

Lemmon, W., M.O. McLinden, and D.G. Friend, Thermophysical properties of fluid systems: *in NIST Chemistry WebBook, NIST Standard Reference Database Number 69*, edited by P. J. Linstrom and W.G. Mallard, National Institute of Standards and Technology, Gaithersburg MD, 2001.

Lonsdale, P., and R. Batiza, Submersible study of hyaloclastite and lava flows on young seamounts at the mantle of the Gulf of California, *Bull. Geol. Soc. Amer.*, 91, 545-554, 1980.

Lorenz, V., Some aspects of the eruption mechanism of the Big Hole Maar, central Oregon, *Geol. Soc. Amer. Bull.*, 81, 1823-1830, 1970.

Maicher, D., White, J. D. L., and R. Batiza, Sheet hyaloclastite: density-current deposits of quench and bubble-burst fragments from thin, glassy sheet lava flows, Seamount Six, Eastern Pacific Ocean, *Marine Geol.*, 171, 75-94, 2000.

Maicher D, and J. D. L. White, The formation of deep-sea limu o' Pelee, *Bull. Volcanol.*, 63, 482-496, 2001.

Marshall, E., The lessons of Chernobyl, *Sci.*, 233, 1375-1376, 1986.

Mattox, T. N., and M. T. Mangan, Littoral hydrovolcanic explosions: a case study of lava-seawater interaction at Kilauea Volcano, *J. Volcanol, Geotherm. Res.*, 75, 1-17, 1997

McBirney, A. R., Factors governing the nature of submarine volcanism, *Bull. Volcanol.*, 26, 455-469, 1963.

Ort M.H., Wohletz, K., Hooten, J. A., Neal, C. A., and V. S. McConnel, The Ukinrek maars eruption, Alaska, 1977: a natural laboratory for the study of phreatomagmatic processes at maars, *Terra Nostra, 2000/6*, 396-400, 2000.

Peckover, R.S., Buchanan, D. J., and D. E. Ashby, Fuel-coolant interactions in submarine volcanism, *Nature*, 245, 308-308, 1973.

Rabie, R. L., Fowles, G. R., and W. Fickett, The polymorphic detonation, *Phys. Fluids*, 22, 422-435, 1979.

Ruggles, A E., Drew, D. A., Lahey, R. T., Jr., and H. A. Scarton, The relationship between standing waves, pressure pulse propagation and the critical flow rate in two-phase mixtures, *J. Heat Trans.*, .111(2), 467-473, 1989.

Ruggles, A. E., Vasiliev, A. D., Brown, N. W., and M. W. Wendel, role of heater thermal response in reactor thermal limits during oscillatory two-phase flows, *Nucl. Sci. Engin.*, 125, 75-83, 1997.

Self, S., Wilson, L., and I. A. Nairn, Vulcanian eruption mechanisms, *Nature*, 277, 440-443, 1979.

Sharon, A., and S. G. Bankoff, On the existence of steady supercritical plane thermal detonations, *Int. J. Mass Heat Trans.*, 24, 1561-1572, 1981.

Sheridan, M. F.; and K. H. Wohletz, Hydrovolcanic eruptions I. The systematics of water-pyroclast equilibration. *Sci.*, 212, 1387-1389, 1981.

Sheridan, M. F., and K. H. Wohletz, Hydrovolcanism: basic considerations and review, *J. Volcanol. Geotherm. Res.*, 17, 1-29, 1983.

Smith, T. L, and R. Batiza, New field and laboratory evidence for the origin of hyaloclastite flows on seamount summits, *Bull. Volcanol., 51*, 96-114, 1989.

Sourirajan, S., and G. C. Kennedy, The system H_2O-NaCl at elevated temperatures and pressures, *Am. J. Sci., 260*, 115-141, 1962.

Stoffers, P., Worthington, T., Hekinian, R., Petersen, S., Hannington, M., Rurkay, M., and the SO 157 Shipboard Scientific Party, Silicic volcanism and hydrothermal activity documented at Pacific-Antarctic ridge, *EOS Trans. Amer. Geophys. Un., 83(28)*, 301-304, 2002.

Tanimura, S., Date, J. Takahashi, T., and H. Ohmoto, Geologic setting of the Kuroko deposits in the Hokuroku district: Part II. Stratigraphy and structure of the Hokoroku district, *Econ. Geol. Mono., 5*, 24-38, 1983.

Turekian, K. K., *Oceans*, Prentice-Hall, New Jersey, 150 pp, 1968.

Waters, A. C., and R. V Fisher, Base surges and their deposits: Capelinhos and Taal volcanoes, *J. Geophys. Res., 76*, 5596-5614, 1971.

Wentworth, C. K., Ash formations of the island of Hawaii. Hawaii Volcano Observatory, 3rd Spec. Rep., 173 pp., Hawaiian Volcano Research Society, Honolulu, 1938.

White, J. D. L., Impure coolants and interaction dynamics of phreatomagmatic eruptions, *Jour. Volcanol. Geotherm. Res., 74*, 155-170, 1996.

Wohletz, K. H., Mechanisms of hydrovolcanic pyroclast formation: grain-size, scanning electron microscopy, and experimental results, *Jour. Volcanol. Geotherm. Res., 17*, 31-63, 1983.

Wohletz, K. H., Explosive magma-water interactions: Thermodynamics, explosion mechanisms, and field studies, *Bull. Volcanol., 48*, 245-264, 1986.

Wohletz, K. H., Water/magma interaction: some theory and experiments on peperite formation. *J. Volcanol. Geotherm. Res., 114*, 19-35, 2002.

Wohletz, K. H., and G. Heiken, *Volcanology and geothermal energy*, University of California Press, Berkeley, California, 1992.

Wohletz, K., and R. G. McQueen, Experimental studies of hydromagmatic volcanism. *in Explosive Volcanism: Inception, Evolution, and Hazards*, pp. 158-169, Studies in Geophysics, National Academy Press, Washington, 1984.

Wohletz, K. H., McQueen, R. G, and M. Morrissey, Analysis of fuel-coolant interaction experimental analogs of hydrovolcanism, *in Intense Multiphase Interactions*, edited by T. G. Theofanous, and M. Akiyama, pp. 287-317, Proceedings of US (NSF) Japan (JSPS) Joint Seminar, Santa Barbara, CA, June 8-13, 1995, 1995.

Wood, W. W.; and J. G. Kirkwood, Diameter effect in condensed explosives, *J. Chem. Phys., 22*, 1920-1924, 1954.

Yuen, W.W., and T.G. Theofanous, The prediction of 2D thermal detonations and resulting damage potential, *Nucl. Eng. Design, 155*, 289-309, 1994.

Yuen, W.W., and T.G. Theofanous, On the existence of multiphase thermal detonations, *Proc. Int. J. Multiphase Flow, 25*, 1505-1519, 1999.

Zel'dovich, Ya. B., and Yu. P. Raizer, *Physics of Shock Waves and High-Temperature Hydrodynamic Phenomena*, Volume I and II, Academic Press, New York, 1966.

Zimanowski, B., and R. Büttner, Phreatomagmatic explosions in subaqueous volcanism? (this volume).

Zimanowski, B., Büttner, R., Lorenz, V., and H.-G. Häfele, Fragmentation of basaltic melt in the course of explosive volcanism, *J. Geophys. Res., 107*, 803-814, 1997a.

Zimanowski, B., Büttner, R., and J. Nestler, , Brittle reaction of a high-temperature ion melt, *Europhys. Let., 38(4)*, 285-289, 1997b.

Zimanowski, B., Lorenz V., and G. Fröhlich, Experiments on phreatomagmatic explosions with silicate and carbonatitic melts, *J. Volcanol. Geotherm. Res., 30*, 149-153, 1986.

Zimanowski, B., Fröhlich, G., and V. Lorenz, Quantitative experiments on phreatomagmatic explosions, *J. Volcanol. Geotherm. Res., 48*, 341-358, 1991.

Kenneth H. Wohletz, Earth and Environmental Sciences, Los Alamos National Laboratory, Los Alamos, New Mexico 87545 USA

Phreatomagmatic Explosions in Subaqueous Volcanism

Bernd Zimanowski and Ralf Büttner

Physikalisch Vulkanologisches Labor, Institut für Geologie, Universität Würzburg, Würzburg, Germany

Pyroclastic deposits, produced during subaqueous volcanic eruptions, point to the existence of explosive processes. Magma/water interaction is a possible source of these explosions. Under atmospheric pressure a thermohydraulic explosion mechanism was identified that can explain the high kinetic energy release of phreatomagmatic explosion and the formation of typical subaerial phreatomagmatic pyroclastic deposits. The applicability of this mechanism under subaqueous physical conditions is discussed. Whereas the efficacy of heat energy transfer from magma to water in general increases with rising hydrostatic pressure, the conditions for the formation of a critical magma-water premix volume increasingly decline. Our analysis indicates that subaqueous volcanic thermohydraulic explosions should become increasingly improbable at water depths exceeding 100 m and practically impossible at water depths in excess of 1 km. Pyroclastic deposits found at greater depths, bearing signatures of phreatomagmatic origin therefore should be the result of comparably low energy magma-water interaction.

1. INTRODUCTION

Phreatomagmatic explosions, perhaps the most intensive and hazardous single events in volcanism, can be observed during the contact of magma and water [*Colgate and Sigurgeirsson*, 1973; *Wohletz*, 1983, *Sheridan and Wohletz*, 1983; *Fisher and Schmicke*, 1984; *Lorenz*, 1986, 1987; *Zimanowski*, 1998]. As the earth is a "water planet", contact between magma that rises to the surface and water (ocean, lake, and/or groundwater) is the normal case. Phreatomagmatic explosions, however, nearly exclusively have been observed when water entered an emergent volcanic vent. The growing database on subaqueous volcanism (this volume) points to the existence of explosive mechanisms, documented by the presence of pyroclastic deposits on the ocean floor. Experiments on thermohydraulic magma-water interaction using remelted volcanic rocks as magma equivalent have been performed at atmospheric pressure. Scaling was achieved by the reproduction of natural phreatomagmatic volcanic ash [*Büttner et al.*, 1999, 2002; *Zimanowski et al.*, 2002]. In this article we summarize the results of these experimental studies and discuss their applicability to a subaqueous environment.

2. DEFINITIONS

As some terms that occur in the following sections may be used with different meaning in the multidisciplinary community of earth sciences, some definitions follow.

Explosion is a single non-steady state event that causes a rapid expansion within a surrounding medium [e.g., atmosphere, hydrosphere, rock). It is commonly accepted that an explosion has to be accompanied by a sudden sound, which occurs once the expansion has supersonic qualities.

Detonation is a special case of a pressure building process that is acting significantly faster than the speed of sound within the respective volume of the pressurized substance.

In this way a shock wave is sustained and amplified to produce a so-called detonation wave [Board and Hall, 1975]. This process always causes a macroscopic explosion in the surrounding medium.

The *homogeneous nucleation temperature (HNT)* marks the transition from the boiling of a liquid to the volumetric transition of a liquid into vapor. In the case of water, the boiling regime under atmospheric conditions starts at 373 K and HNT is reached at about 583 K.

Film boiling occurs if a liquid is heated at the contact to a hot surface where the contact temperature exceeds the HNT of the liquid. Depending on the frame conditions, stable or meta stable film boiling can be observed. A well-known example is water vapor film boiling, the so-called "Leidenfrost Phenomenon".

Molten fuel coolant interaction (MFCI), a term used in engineering science, and is equivalent to phreatomagmatic explosion in earth sciences. It characterizes an explosion caused by the interaction between water and a hot melt (e.g. metal or magma) at temperatures above HNT.

Viscosity characterizes the inertial forces acting in a fluid (liquid or gas) against flow. The commonly used model for the determination of viscosity is the so-called "Newtonian flow regime", i.e. the referenced viscosity of a fluid without a shear modulus. It is important to understand, that real liquids are more or less "non-Newtonian" fluids. Magmas, in fact, are very non-Newtonian [Bottinga, 1994]. Thus, the values given for the viscosity of magmas are restricted to specified flow regimes and scales.

3. PHYSICS OF PHREATOMAGMATIC EXPLOSIONS

During phreatomagmatic explosion (PE) a complex multiphase and multicomponent system (melt, crystals, gas bubbles, vapor, water) interacts under non-equilibrium thermodynamic conditions and geometrical restrictions [Henry and Fauske, 1981; Theofanous, 1995; Büttner and Zimanowski, 1998], meaning that this system cannot be described by the well established means of thermodynamics. The empirical and phenomenological description of the physics of PE given here resulted from experimental studies using melts with magma-like physical properties under atmospheric conditions [Wohletz and McQueen, 1984; Zimanowski et al., 1991, 1997a; Fröhlich et al., 1993]. The mechanism of PE can be divided in four phases (Figure 1).

3.1. Phase 1

Low energetic hydrodynamic mingling of water and magma under stable film boiling conditions results in

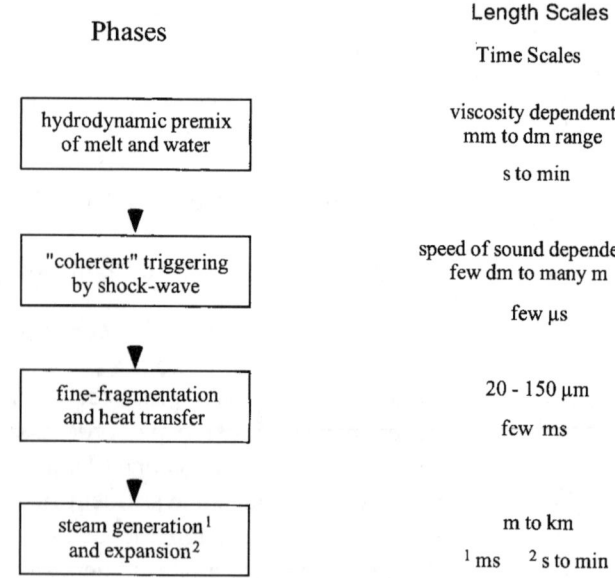

Figure 1. Phases of phreatomagmatic explosion (from Zimanowski et al. 1997a).

domains of water entrapped in a larger volume of magma. The mingling energy to create this pre-explosive mixture (premix) may be drawn from the ascending magma or from intruding water. As the heat transfer is limited by the stable vapor films that surround the water domains, the premix phase can take place on a time scale of several minutes.

3.2. Phase 2

Induction and propagation of rapid break-down of all insulating vapor films in the premix volume is triggered by a sharp pressure pulse that causes a sudden condensation of the vapor films. Such pulses in nature can be produced by stoss condensation of vapor or by brittle failure of wall rocks. Within less than a second a transition of the premix from stable film boiling to direct contact of water and magma takes place.

3.3. Phase 3

Escalative evolution of heat transfer by pressurization of the rapidly heated entrapped water and fragmentation of the rapidly cooled surrounding magma constitute the main phase of the PE, with thermal energy converted into mechanical energy on a time scale of some milliseconds. This process has some qualities of a detonation. A consequence is the heating of the water to a supercritical state, probably well above the NHT.

3.4. Phase 4

Quantitative vaporization of the supercritical entrapped water and consequent expansion of the magma-vapor-water-system to ambient pressure constitute the expansion phase. This phase represents the phreatomagmatic eruption, and delivers an additional, but minor, part of the kinetic energy.

3.5. Phenomenological Description of PE

If water contacts magma the temperature contrast is always well above the NHT of water, even in the case of the coldest known magma. Therefore, at hydrostatic pressures well below the critical pressure of water, explosive mixtures of water in magma can form in the stable film-boiling regime. Depending on the rheology of magma (viscosity >0.1 Pa s and <10^{10} Pa s), the premix formation is controlled by the supplied hydrodynamic mingling energy [*Zimanowski et al.*, 1997b; *Zimanowski and Büttner*, 2002]: the higher the viscosity of the magma, the higher the mechanical mingling energy needed. However, each magma-water configuration has an optimum with respect to explosivity. At high relative mingling energy the film boiling will become metastable, due to increasing influence of hydrodynamic interface instabilities [*Chandrasekhar*, 1961; *Sharp*, 1984]. If the amplitudes of the interfacial waves are in the same range as the thickness of the vapor films, stable film boiling is no longer possible. In addition, the lifespan of smaller water domains that result from increased mingling energy will be shortened by the increased heat flux.

In contrast, at low relative mingling energy, the prolonged mingling time needed to produce an explosive premix will cause significant vaporization of water and subcooling (i.e. glass transition) of the melt. From the experimental results, an optimum pre-explosive mixture is characterized by centimeter to decimeter sized water domains that are entrapped in excess magma [*Zimanowski et al.*, 1997b; *Büttner and Zimanowski*, 1998]. Owing to the low degree of superheat of magma (i.e. temperature above solidus), the optimum water-magma volume ratio is significantly smaller than 1 [*Wohletz et al.*, 1995; *Morrisey et al.*, 2000]. The energy of the resulting explosions was found to be proportional to the pre-explosive water-magma interface area [*Zimanowski et al.*, 1997a, 1997b]. Thus, the better the premix quality and the larger its volume, the stronger will be the explosion. The thermohydraulic explosion mechanism is triggered by the quasi-coherent collapse of all vapor films of the premix, synchronized by a pressure pulse [*Fletcher*, 1995]. The optimum trigger signals are shock waves; because of their supersonic qualities they can affect a large premix volume quasi-contemporaneously and are well known to induce rapid condensation [*Kobayashi et al.*, 1996]. Such pressure signals are not exotic in volcanic environments, in fact they are very common. The cracking of host rock, caused by thermal stress, is a typical source for such shock waves [*Zimanowski*, 1998].

The trigger phase leads to direct contact between water and magma [*Zimanowski et al.*, 1997c], i.e. the transition of the premix from a liquid water, steam, and liquid magma system (i.e. 2-phase-system) into a liquid water and liquid magma system (i.e. single-phase-system). The consequence of this transition is a strong thermal and mechanical coupling of both liquids and a significant increase of the heat flux and the speed of sound. Due to system inertia and the large difference between the thermal expansion coefficients of water and magma, heating of the water and cooling of the magma under quasi-isochoric conditions causes a local pressure increase within a time interval of the order of milliseconds. If the load pressure on the magma exceeds a critical value (depending on composition and temperature) the magma reacts as a stressed solid body in which cracks form on a mm to cm scale. Pressurized water will quickly intrude into these cracks, increasing the contact area and thus the heat flux. Extremely high cooling rates of the magma in the vicinity of the magma/water interface lead to high thermally induced stresses during the fast approach of the glass transition temperature. Triggered by the seismic energy released during the formation of the cracks described above, a second brittle-type fragmentation cycle is started on a micrometer scale [*Chaudri and Liangyi*, 1986]. A fragmentation front thus forms which expands into the surrounding magma following the propagation of the leading cracks at a considerably slower speed [*Zimanowski et al.*, 1997c]. Both fragmentation mechanisms lead to an escalative behavior of heat transfer thus generating a positive feedback mechanism. Relaxation of the stored structural energy within the magma in a short time period by these brittle processes causes an explosion shock wave that propagates into the surrounding and is capable of intensive fragmentation of the host rocks.

This thermohydraulic mechanism [*Büttner and Zimanowski*, 1998] is terminated once the system starts to expand. Vaporization of superheated water occurs and the system is thermally and mechanically decoupled. During the following expansion phase the superheated water vaporizes and the generated steam expands to ambient pressure, driving a phreatomagmatic eruption. (Figure 2) The expansion phase in principle can be described as a "steam engine" and the energy release depends nearly exclusively on the degree of superheat of the water.

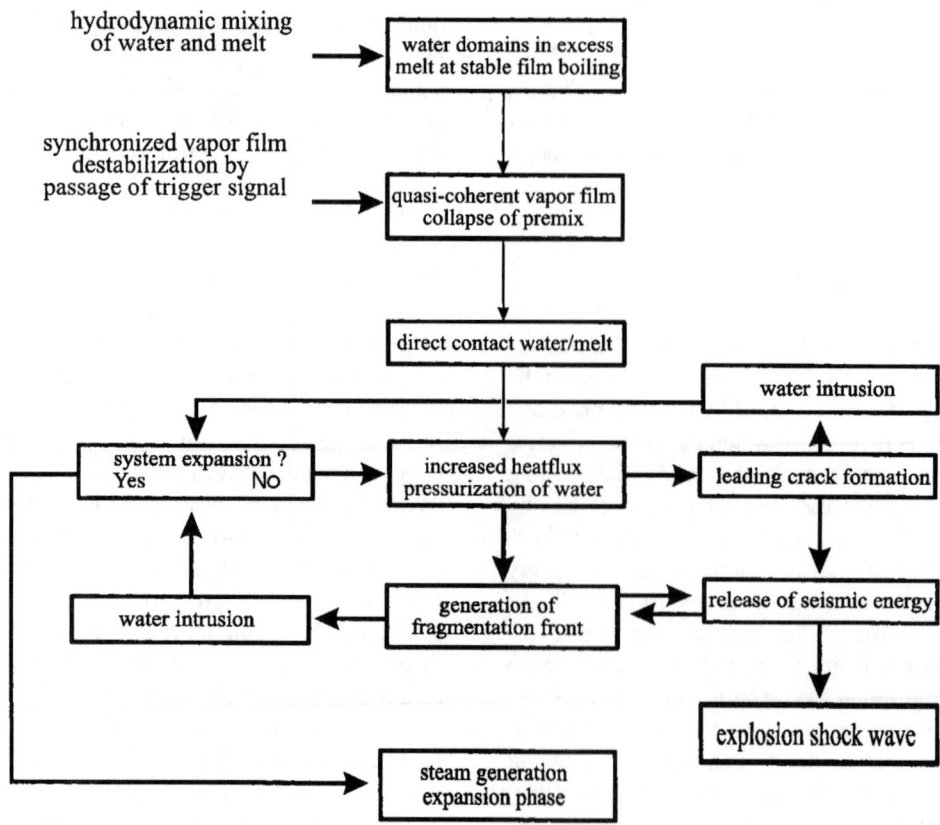

Figure 2. Phenomenological model of phreatomagmatic explosion (from *Büttner and Zimanowski* 1998).

4. DISCUSSION

Experimental research on PE so far was restricted to atmospheric conditions. However, the behavior of the water-steam system is experimentally well described up to temperatures > 1300 K and pressures > 100 MPa. Thus it is possible to expand the experimental data on PE to higher ambient pressures and to submarine volcanic conditions (max. 10 km water column).

In the premix phase a hydrodynamic mingling of magma and water (Figure 3) is required, that is limited by the heat flux: The formation of water/magma mixes is complicated by the extreme temperature contrast. As the superheat of magmas normally does not exceed 150 K, the time window for hydrodynamic mingling processes is strongly limited by the cooling of the magma. Once the magma approaches its glass transition temperature due to heat loss, hydrodynamic mingling will be hindered by rising viscosity and will finally be stopped when the magma reaches solid state behaviour [*Zimanowski and Büttner*, 2002]. In the case of magma dispersed in water this cooling is a critical feature of the system because of its limited heat content (Figure 4).

Hydrodynamic mingling occurs exclusively between two immiscible liquids, i.e. with liquids between which an interfacial tension exists [*Ottino*, 1990]. The size distribution of the dispersed liquid reflects the physical parameters of both liquids and the mingling energy, i.e. the shear rate or differential flow speed. Under otherwise constant physical conditions the mean size of the dispersed liquid domains decreases with increasing differential flow rates (Figure 5). The most important physical properties of liquids with respect to hydrodynamic mingling are the viscosity and the interfacial tension. Furthermore, it is a crucial question whether the hydrodynamics can be assumed to act in a laminar flow regime or if turbulence may occur. In the latter case only crude experimental modeling (experimental physics or simplified empirically calibrated numerical analog systems) will be possible. Occurrence of turbulence can be considered using the critical Reynolds Number (Re_{crit}) that depends on flow diameter, density of flowing liquids, flow speed, viscosities, and the roughness of the geometrical flow boundaries. Under smooth conduit conditions Re_{crit} for Newtonian liquids is about 20,000, whereas under rough conditions (e.g. sharp edges or conduit roughness in the

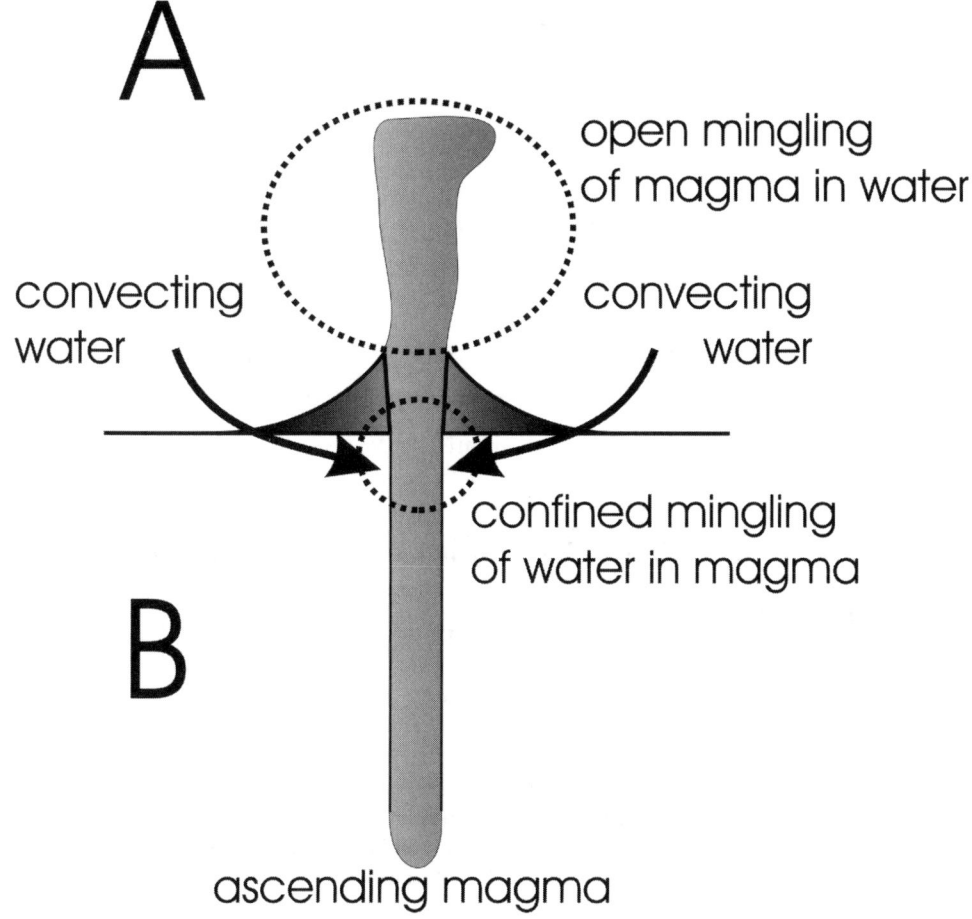

Figure 3. Schematic drawing of water/magma contact modes in subaqueous volcanism. Regime A: non confined water body (ocean or lake). Regime B: confined water transport by porous or gap flow.

range of 10 % of the conduit diameter) it is in the range of 1000 [*Grossmann*, 2000]. To mingle a basaltic magma and water in a turbulent mode under "smooth conditions" (i.e. not confined), a differential flow speed of about 1000 m/s is needed. Therefore, turbulent mingling of water and magma on a larger scale can practically be excluded from volcanism on earth.

The hydrodynamic properties of the system magma/water do not change drastically with increasing pressure. Little is known about the effect of pressure on interfacial tension. However, it can be expected that this influence is small in the range of pressures reflecting the submarine volcanic conditions. The viscosity of water increases slightly with increasing pressure, but decreases more with increasing temperature. A salt content of some 3 to 5 wt.% also will have no significant influence on viscosity or interfacial tension. The effect of increased pressure on the viscosity of magma depends mostly on the behavior of dissolved volatiles and, at least in the case of basaltic compositions, has not yet been measured. However, it can be expected that the viscosity will only slightly decrease with increasing hydrostatic pressure. Despite the fact that the viscosity contrast is the governing factor for hydrodynamic mingling, little if any effect is to be expected on the hydrodynamics at increasing pressure.

In contrast, the thermal behavior of the magma/water contact changes dramatically with increasing hydrostatic pressure (Figure 6). In general the effects of mineralized or salt water on the thermal properties are small. Saturated aqueous solutions, as they occur in hydrothermal systems, however, may show a different behavior. The transport of heat from the magma to the water takes place in three regimes (Figure 7), with increasing heat flux from (i) to (iii):

(i) Stable film boiling: at temperature contrasts well above NHT of water a macroscopic vapor film forms at the interface between water and magma and restricts the heat flux. The vapor film roughly consists of two regions: a thicker "layer" of superheated unsaturated steam at the hot

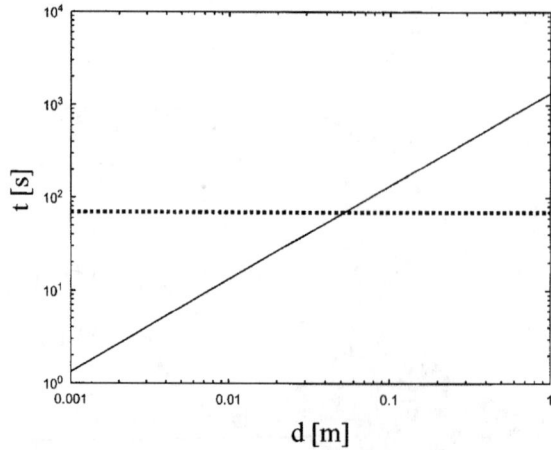

cooling model at stable film boiling conditions

$\alpha = 2.1 \; [10^3 \; W/(m^2 \; K)]$

$\rho = 2800 \; [kg/m^3]$

$C_p = 1500 \; [J/kg \; K]$

Figure 4. Cooling model for domains of basaltic melt (with diameters d) dispersed in water matrix. Starting conditions: melt temperature 1370 K, constant water temperature (sink temperature) 290 K The time axis represents the maximum mingling time, i.e. the time window between mingling onset and solidification of the melt. The dotted line represents the minimum time estimated for production of a premix volume that can drive an explosive eruption.

side with "magmatic" temperature and a thinner "layer" at the cold side, within which an "equilibrium" of condensation and vaporization exists. The temperature of the latter layer, the so-called "thermal boundary layer", can be defined only microscopically. The thickness of this film depends on the temperature contrast and the ambient pressure: the higher the temperature contrast and the lower the ambient pressure, the thicker is the vapor film.

(ii) Metastable film boiling: at temperature contrast above the boiling point, but below NHT of water a macroscopic vapor film does not form, however, the heat flux is modified by local formation of steam at the interface. At temperature contrasts well above NHT metastable film boiling takes place in two cases. Case 1: the ambient pressure causes a reduction of the thickness of the superheated steam layer of the film in such a way that the thickness reaches the same range as the thermal boundary layer. Case 2: hydrodynamic disturbances occur that cause interfacial instability waves with amplitudes of the same scale as the total film thickness.

(iii) Direct contact: at temperatures below the boiling point of water and at ambient pressures above the critical pressure of water the heat flux from the magma to the water can be described using equilibrium thermodynamics and depends on the thermal properties of the liquids and the hydrodynamics at the interface. The thermal properties of magma are not strongly affected by increasing ambient pressure, but the properties of water can change dramatically. Direct contact as a short time effect during PE is characterized by a very high heat flow (> 20 MW/m²) that results from the high-velocity hydrodynamic conditions and formation of a new hot surface at a very high rate (fragmentation of magma) during the thermohydraulic explosion phase of the PE.

Considering the effect of increasing hydrostatic pressure on the heat transfer from magma to water during subaqueous volcanism, the stable film boiling regime is restricted to ambient pressures below about 1 MPa, as the vapor film thickness at magmatic temperatures of the "hot layer" becomes critical, reaching the scale of the thermally and hydrodynamically induced surface instability (Figure 6). Above ambient pressures in excess of 10 MPa, the moderating effect of metastable film boiling can practically be excluded. At thermal power plants, where an optimum heat flux is required, the water pressure of the heat exchange system usually is adjusted to 15 to 20 MPa at temperature contrasts comparable to magmatic ones. The heat transfer from magma to water during subaqueous volcanism at water depths exceeding 1 km therefore can be described in good approximation in the direct contact regime.

5. CONSEQUENCES, PREDICTIONS, AND SPECULATIONS

By definition, every kind of extrusive subaqueous volcanism on earth is phreatomagmatic, as a magma/water interaction must take place. Compared to subaerial volcanism a major difference is the stronger thermal and mechanical coupling to the medium within which the magma erupts. The amount of heat transferred into the surrounding per unit interfacial area will be at least one order of magnitude greater. On the other hand the inertial forces counteracting an expansion are more than one order of magnitude higher. Therefore, the macroscopic expansion rates of a subaqueous eruption should be much smaller compared to a subaerial eruption at identical magma production rates. However, as the intensity of an eruption should be described by the release of mechanical energy per unit time and as the expansion energy is only part of the total mechanical energy released, the methods used for classification of the explosivity of subaerial eruptions are not applicable.

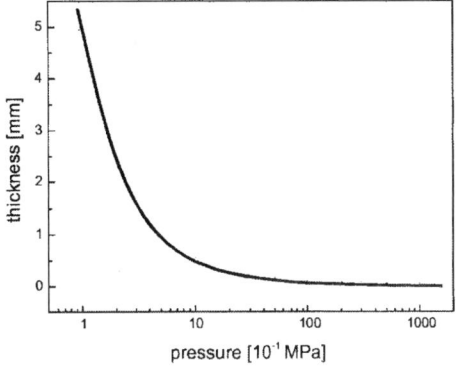

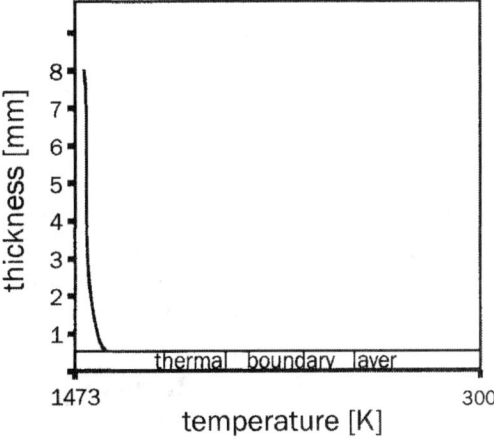

Figure 6. Vapor film thickness against hydrostatic pressure in standing water at constant temperature (upper graph). In the lower graph a temperature profile of a vapor film is given in example. In the region of the thermal boundary layer a local equilibrium of vaporization and condensation exists, thus no general temperature can be assigned to this layer. Under differential flow conditions the vapor film stability will decrease.

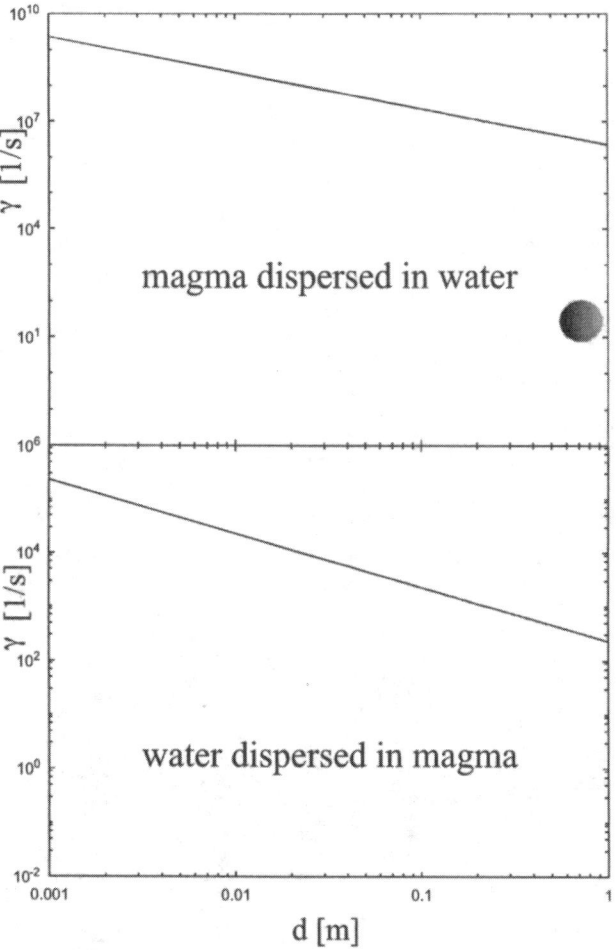

Figure 5. Mingling model for the 2 liquid system water/magma. Viscosity of basaltic melt is at 1.5 Pa s, viscosity of water is 0.001 Pa s, density of melt is 2800 kg/m^3, interfacial tension is at 0.25 N/m. The mingling energy is expressed by the shear rate (γ) needed to produce domain sizes with diameters d. The shear rate needed to disperse magma in water is about 10000 times larger than in the case of water dispersed in magma. The big dot marks the position of a lava fountain of about 1 m in diameter and an exit velocity of 100 to 200 m/s.

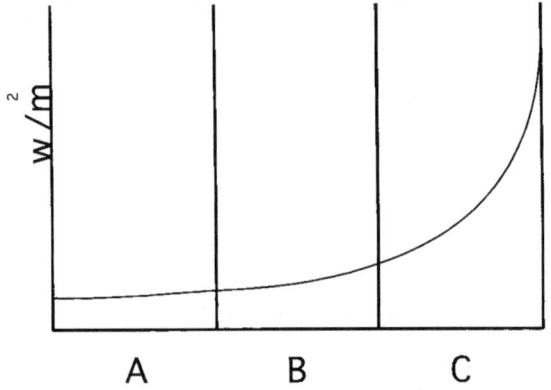

Figure 7. Qualitative graph of the heat flux at the water/magma interface at constant temperature. Starting form region A at stable film boiling conditions, through region B at meta stable film boiling to region C at direct contact the heat flux increases by about 10000.

Understanding the processes of volcanic eruption in the past was strongly coupled to direct observations in the field. Owing to the operational restrictions of submarine research, it cannot be expected in the near future, that the observational data for subaqueous eruptions will be as satisfying for understanding important mechanisms. However, if volcanologists understand the physics of subaerial volcanism, good approximations and predictions can be made for the physics of subaqueous volcanism, which is by far the most prominent volcanism on this planet.

Concerning the importance of PE in subaqueous volcanism, the step by step analysis of the physical processes of the current model leads to the following prediction. As PE requires the formation of a water/magma premix under stable film boiling conditions, volcanic thermohydraulic explosions should become increasingly improbable at water depths exceeding 100 m and practically impossible at water depths in excess of 1 km. Furthermore, the source regions of phreatomagmatic eruptions, being the consequence of quantitative vaporization of water and the formation of superheated steam, from the same line of arguments should be restricted to water depths of less than 1 km.

Observation made during the eruption of a small sea mount near Izu Peninsula (Japan) in 1989 by *Yamamoto et al.* [1991] are in good agreement with these predictions. Observations of *Clague et al.* [2003, this volume], however, show that effective production of pyroclasts can be observed at depths exceeding 1.35 km water column.

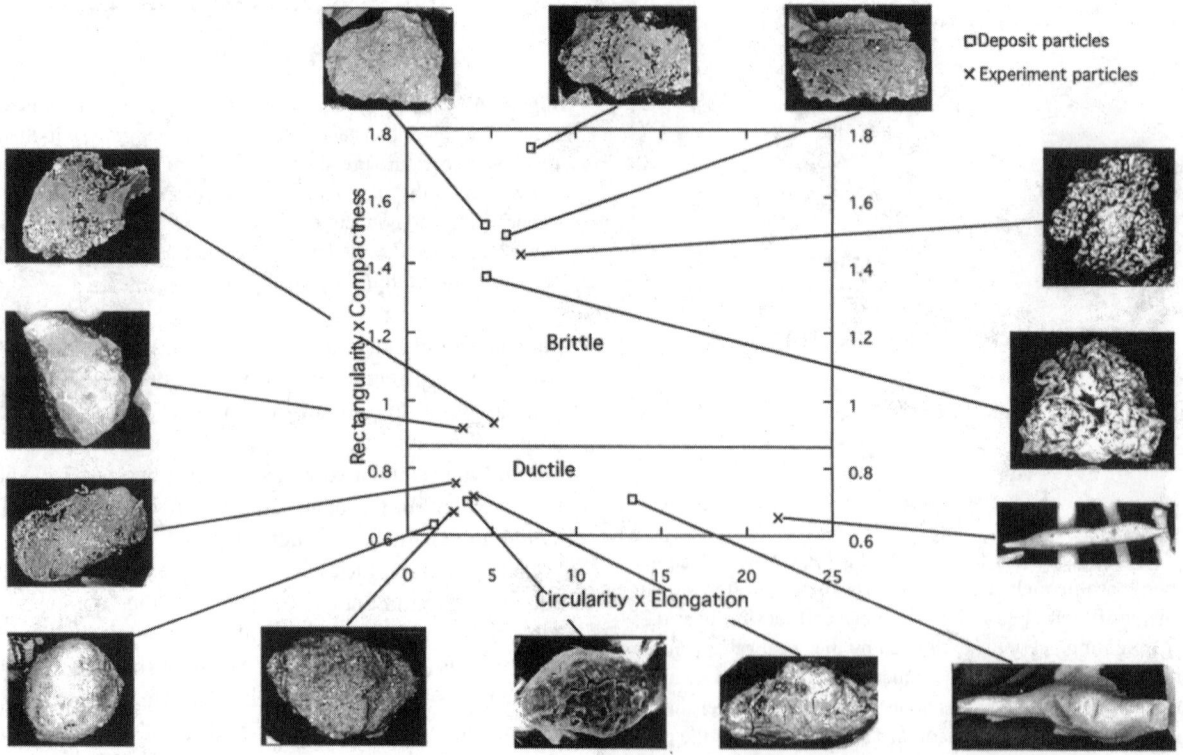

Figure 8. Example of classification of particle families and scaling. Particles with known history (experiment particles) are compared to their natural analogs (deposit particles). This way laboratory data can be assigned to natural processes (from *Büttner et al.* 2002).

Despite the fact that thermohydraulic explosions, and thus PE, seem to be very improbable in the course of deep seated marine and lacustrine volcanism, it is not yet possible to exclude all types of explosive magma/water interaction. The only way to provoke the escalation of the endothermal process into explosive behavior is to increase the heat flux from the magma to the surrounding water on a short time scale. At direct contact, representing the magma/water thermal coupling at greater water depths, the heat flow is relatively high. It seems, that at least on a micro scale, a thermal coupling of a fragmentation mechanism is possible [*Wohletz*, 2003, this volume].

6. OUTLOOK

As any experimental approach to explosive subaqueous volcanism will be a very costly and time consuming enterprise it appears more useful to concentrate on the analysis of samples from submarine pyroclast deposits. From past research a data base exists that was successfully used to recognize typical pyroclast sizes and shape and to identify the respective processes of formation [*Heiken and Wohletz*, 1985].

The relevance of this method has recently been confirmed by experimental studies [*Büttner et al.*, 1999]. In these studies particles were produced under controlled laboratory conditions using standardized fragmentation experiments. These particles, resp. particle families, therefore can be directly linked to specific physical processes acting during their formation (Figure 8). This technique was applied to comparative studies on natural deposits. By using identical melt composition and adjusting the experimental parameters, the natural pyroclasts were reproduced experimentally (Figure 9). Now it was possible to assign mechanisms and energy budgets of the experiments to the natural scale [*Büttner et al.*, 2002].

From this experience we recommend studies of the respective characteristics of subaqueous pyroclasts, using similar techniques of classification. As only one physical parameter is changed from near surface magma/water interaction to deep-seated magma/water interaction (i.e. hydrostatic pressure), a comparison to pyroclast families of known origin should be possible.

Acknowledgments. The manuscript was improved by helpful and insightful reviews by A. McBirney and T.G. Theofanous. We

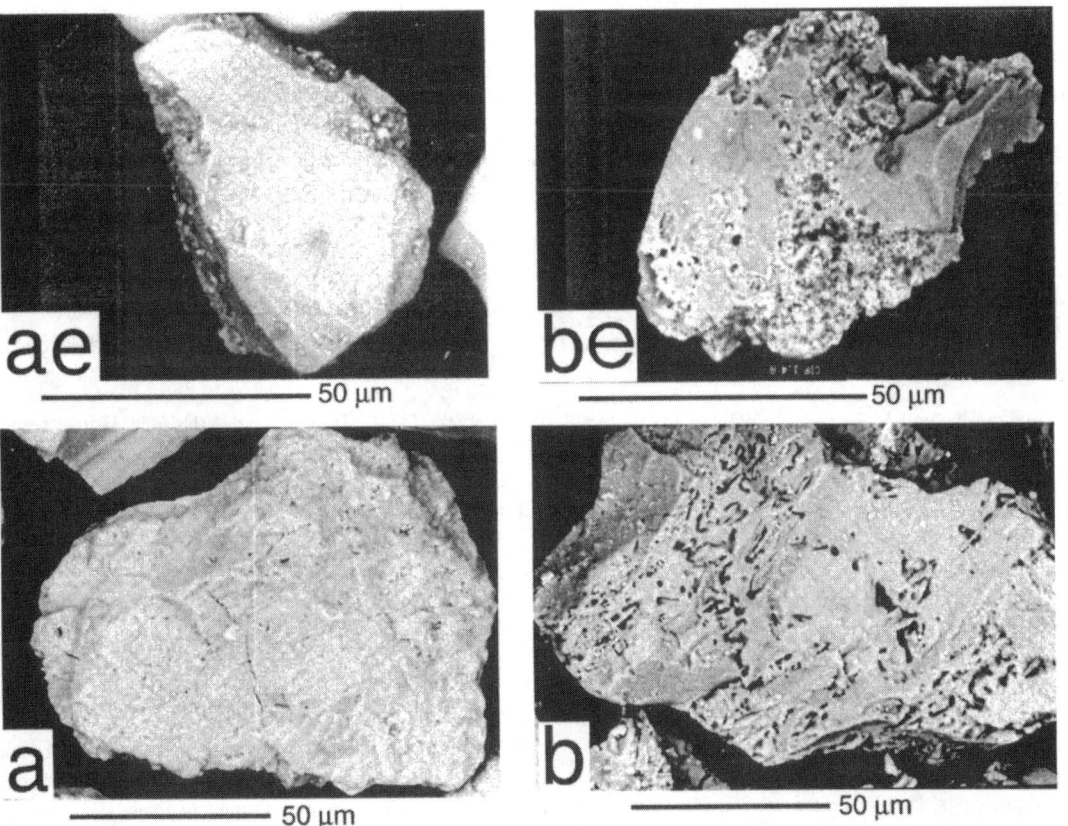

Figure 9. Example of particles exclusively formed during PE. Particles from experiments (ae, be) and their natural analogs (a, b).

like to express our special thanks to the editors James White and Dave Clague, who not only provided important scientific hints, but also helped to improve the formal quality of the manuscript.

REFERENCES

Board, S. J., and R. W. Hall, Detonation of fuel-coolant explosions, *Nature, 245*, 319-321, 1975.

Bottinga, Y., Configurational entropy and the non-Newtonian rheology of homogenous silicate liquids, *Phys. Rev. B, 49*, 95-99, 1994.

Büttner, R., and B. Zimanowski, Physics of thermohydraulic explosions, *Phys. Rev. E, 57, 5*, 5726-5729, 1998.

Büttner, R., P. Dellino, and B. Zimanowski, Identifying modes of magma/water interaction from the surface features of ash particles, *Nature, 401*, 688-690, 1999.

Büttner, R., P. Dellino, L. LaVolpe, V. Lorenz, and B. Zimanowski, Thermohydraulic explosions in phreatomagmatic eruptions as evidenced by the comparison between pyroclasts and products from Molten Fuel Coolant Interaction experiments, *J. Geophys. Res.*, 107, doi: 10.1029/2001JB000792, 2002.

Chandrasekhar, S., *Hydrodynamic and hydromagnetic stability*, 187 p., Oxford, 1961.

Chaudhri, M. M., and C. Liangyi, The catastrophic failure of thermally tempered glass caused by small-particle impact, *Nature, 320*, 48-50, 1986.

Clague, D.A., R. Batiza, J.W. Head III, and A.S. Davis, Pyroclastic and Hydroclastic Deposits on Loihi Seamount, Hawaii, this volume, 2003.

Colgate, S. A., and T. Sigurgeirsson, Dynamic mixing of water and lava, *Nature, 244*, 552-555, 1973.

Grossmann, S., The onset of shear flow turbulence, *Rev. Modern Physics, 72*, 603-618, 2000.

Fisher, R. V., and H.-U. Schmincke, *Pyroclastic Rocks*, 472 p., Springer-Verlag Berlin, 1984.

Fletcher, D. F., Steam explosion triggering: a review of theoretical and experimental investigations, *Nucl. Eng. Des. 155*, 27-36, 1995.

Fröhlich, G., B. Zimanowski, and V. Lorenz, Explosive thermal interactions with molten lava and water, in *Proc. 3rd World Conf. on Fluid Mechanics and Thermodynamics, Honolulu, USA*, 1459-1468, Elsevier, Amsterdam, 1993.

Heiken G, and K. H. Wohletz, *Volcanic ash*, 246 p., University of California Press, Berkeley, 1985.

Henry, R. E., and H. K. Fauske, Required initial conditions for energetic steam explosions, in *Fuel-Coolant Interactions. Am. Soc. of Mech. Eng., New York, Rep. HTD-V19*, 42 p., 1981.

Kobayashi, Y., T. Watanabe, and N. Nagai, Vapor condensation behind a shock wave propagating through a large molecular-mass medium, *Shock Waves, 5*, 287-292, 1996.

Lorenz, V., On the growth of maars and diatremes and its relevance to the formation of tuff-rings, *Bull. Volcanol. 48*, 265-274, 1986.

Lorenz, V., Phreatomagmatism and its relevance, *Chem. Geol. 62*, 149-156, 1987.

Morrissey, M., B. Zimanowski, K.H. Wohletz, and R. Büttner, Phreatomagmatic Fragmentation, in *Encyclopedia of Volcanism*, edited by H. Sigurdsson, Academic Press, London, 2000.

Ottino, J. M., Mixing, chaotic advection, and turbulence, *Annu. Rev. Fluid Mech. 22*, 207-253, 1990.

Sharp, D. H., An overview of Rayleigh-Taylor instability, *Physica, 12D, 3*, 3-18, 1984.

Sheridan, M. F., and K. H. Wohletz, Hydrovolcanism: Basic considerations and review, *J. Volcanol. Geotherm. Res., 17*, 1-29, 1983.

Theofanous, T. G., The study of steam explosion in nuclear systems, *Nucl. Engrg. Des., 155*, 1-26, 1995.

Wohletz, K.H., Mechanisms of hydrovolcanic pyroclast formation: grain-size, scanning electron microscopy, and experimental studies, *J. Volcanol. Geotherm. Res., 17*, 31-64, 1983.

Wohletz, K. H., and R.G. McQueen, Experimental studies of hydrovolcanic volcanism, in *Studies in Geophysics*, 158-169, Natl. Acad. Press., Washington, 1984.

Wohletz, K. H., R. G. McQueen, and M. Morrissey, Experimental study of hydrovolcanism by fuel-coolant interaction analogues, in *Proc. NSF/JSPS AMIGO-IMI Seminar, Santa Barbara, CA, June 8-13*, 287-317, 1995.

Yamamoto, T., T. Soya, S. Suto, K. Uto, A. Takada, K. Sakaguchi, and K. Ono, The 1989 submarine eruption off eastern Izu Peninsula, Japan: ejecta and eruption mechanisms, *Bull. Volcanol., 53*, 301-308, 1991.

Zimanowski, B., Phreatomagmatic explosions, in *From magma to tephra, Developments in volcanology 4,* edited by A. Freundt and M. Rosi, 25-54, Elsevier, Amsterdam, 1998.

Zimanowski, B., G. Fröhlich, and V. Lorenz, Quantitative experiments on phreatomagmatic explosions, *J. Volcanol. Geotherm. Res., 48*, 341-358, 1991.

Zimanowski, B., R. Büttner, V. Lorenz, and H. G. Häfele, Fragmentation of basaltic melt in the course of explosion volcanism, *J. Geophys. Res., 102*, 803-814, 1997a.

Zimanowski, B., R. Büttner, and V. Lorenz, Premixing of magma and water in MFCI experiments, *Bull. Volc., 58*, 491-495, 1997b.

Zimanowski, B., R. Büttner, and J. Nestler, Brittle reaction of a high temperature ion melt, *Europhys. Lett., 38*, 285-289, 1997c.

Zimanowski, B., and R. Büttner, Dynamic mingling of magma and liquefied sediments, *J. Volcanol. Geotherm. Res., 114*, 37-44, 2002.

Zimanowski, B., K.H. Wohletz, P. Dellino, and R. Büttner, The volcanic ash problem, *J. Volcanol. Geotherm. Res.*, doi: 10.1016/S0377-0273(02)00471-7, 2002.

B. Zimanowski and R.Büttner, Physikalisch Vulkanologisches Labor, Institut für Geologie der Universität Würzburg, Pleicherwall 1, D-97070 Würzburg, Germany. (zimano@geologie.uni-wuerzburg.de)

Melting of Ice by Magma-Ice-Water Interactions During Subglacial Eruptions as an Indicator of Heat Transfer in Subaqueous Eruptions

Magnús T. Gudmundsson

Science Institute, University of Iceland, Reykjavík, Iceland

Eruptions within glaciers are characterized by fast cooling of volcanic deposits, rapid melting of ice and heating of meltwater. Heat transfer rates in subglacial eruptions may be monitored through melting rates of ice and simple calorimetric calculations used to infer heat fluxes and estimate the efficiency of heat transfer from magma. Cooling models of effusive basaltic eruptions forming pillow lava indicate that thermal efficiency of such eruptions is 10–45%, and highest when the eruption rates are low and pillows are exposed to surrounding meltwater for a comparatively long time. When magma fragmentation occurs by non-explosive granulation or explosive activity the glass particles formed have diffusion times mainly in the range 10^{-3} s to 10^2 s depending on grain size, the mean being of the order of 1 s. Limited observational data on ice-melting rates and models of cooling times suggest that the efficiency of heat transfer from fragments may commonly be 70–80%. Correspondingly, total heat transfer rates associated with fragmentation are several times higher than for pillow lava at the same eruption rate. The contrasts in efficiency imply that variation in heat transfer rates during fragmentation may closely correlate with variations in magma eruption rate, whereas for pillow lava eruptions changes in heat transfer lag well behind changes in eruption rate. Though pillows may still have molten cores when buried in a growing volcanic pile, the temperature of volcanic glass created during subaqueous fragmentation should be no greater than 250–300°C at the time of deposition.

INTRODUCTION

Interaction of magma and water in subaqueous eruptions can cause rapid quenching of magma into glass, pillow lava formation with partial to almost complete crystallization, yet in some cases sheet lavas still form [e.g. *Moore*, 1975; *Wohletz*, 1983; *Griffiths* and *Fink*, 1992]. The rapid cooling reflected in the formation of fine-grained glass particles, quenched, glassy, pillow rinds and, in some cases almost fully quenched and glassy sheet lavas, implies fast heat transfer from magma to the surroundings, i.e. ocean, lake or groundwater. The quantification of these processes (heat transfer rates etc.) is difficult, especially so in the ocean due to a paucity of well-instrumented observations of submarine eruptions.

In most cases eruptions under glaciers are essentially subaqueous, with the same cooling processes occurring as in eruptions under open water. Volcanic areas characterized by hyaloclastite formations and pillow lava were formed during the Pleistocene in Canada, Iceland and Antarctica [*Hickson*, 2000; *Kjartansson*, 1959; *Smellie*, 1999]. High stratovolcanoes throughout the world have partly ice-covered slopes, and ice-(water-)magma interaction influences their eruption behavior [*Major* and *Newhall*, 1989]. In Iceland, eruptions within glaciers are common and hyaloclastite mountains are still being formed under ice caps [*Gudmundsson et al.* 1997, 2002; *Larsen*, 2002]. For eruptions of stratovolcanoes where magma-ice interactions take place, the ice is often thin, and most of the energy is usually lost to the atmosphere through the eruption plume; only a minor part is dissipated through melting of the shallow ice covering the slopes. In contrast, when eruptions occur within large glaciers, ice caps or large ice-filled calderas a large part of the magmatic heat is used to melt ice (Figure 1) and calculations of melting rates and heat transfer can be made through measurements of the volume of ice melted. This provides important insight into rates of cooling processes in subaqueous eruptions in lakes and in the ocean.

SUBGLACIAL ERUPTIONS

Most eruptions under glaciers lead to jökulhlaups or lahars when meltwater is released. Such flooding is one of the main hazards associated with subglacial eruptions [*e.g. Björnsson*, 1975]. Several observations of ice-magma interaction have been reported from stratovolcanoes [*e.g. Major and Newhall*, 1989]. Direct observations of eruptions in which ice melting takes up a large part of the thermal energy of the eruption are not as widely reported. Recognition of exclusively subglacial eruptions is rare. Most observed eruptions within glaciers in Iceland have an early phase during which the activity is confined under the ice (Figures 2 and 3). Usually, the eruption melts its way through the ice cover and then becomes explosive, typically producing surtseyan, phreatomagmatic, activity (Figure 4), and sustaining eruption plumes that carry tephra into the atmosphere [*Larsen et al.* 1998, *Larsen* 2002]. Confirmed exclusively subglacial eruptions include the eruption in Gjálp, Vatnajökull in 1938 [*Björnsson* 1988; *Gudmundsson and Björnsson*, 1991]. The subaerial phase of the more recent Gjálp eruption, in 1996, constituted only a minor part of the eruption in terms of magma and energy transport [*Gudmundsson et al.* 1997, 2003]. In both eruptions >95% of the total available thermal energy was used for ice melting, with the great majority of heat for melting transferred through water. Three eruptions have been directly observed at Grímsvötn, Iceland, in 1934, 1983 and 1998 [*Áskelsson*, 1936; *Tryggvason*, 1960; *Grönvold* and *Jóhannesson*, 1984; *Sigmundsson et al.*, 1999]. These eruptions rapidly opened a pathway for magma through 100–150 m of ice and were subsequently characterized by surtseyan activity (Figure 4). Some melting of ice occurred in these eruptions. However, the volume of tephra deposited from the eruption plumes was such a large fraction of the total magma erupted that the heat dissipated to the atmosphere from the eruption plumes was probably greater than that used to melt ice.

The best-documented example of an eruption within a thick ice cap is the Gjálp eruption in 1996 (Figure 3). It lasted from September 30 until October 13. Ice thickness at the site was 550–750 m before the eruption, during which a 6 km-long and up to 450 m-high hyaloclastite ridge was formed by 0.8 km^3 of mostly volcanic glass, equivalent to 0.45 km^3 of magma [*Gudmundsson et al.* 2002] of basaltic icelandite composition [*Steinthorsson et al.* 2000]. During the eruption about 3 km^3 of ice were melted. About 2 km^3 of ice was melted in the first four days, indicating a heat

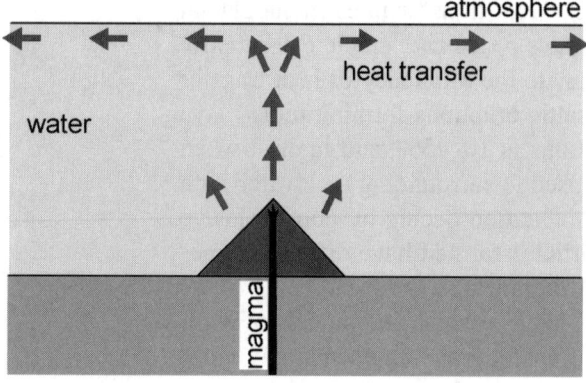

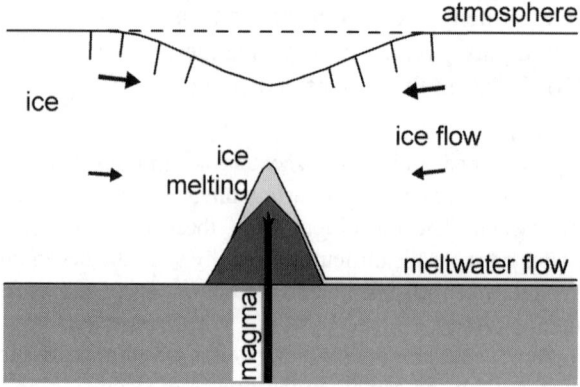

Figure 1. Schematic settings for, a) a subaqueous eruption, and, b) a fully subglacial eruption. In both cases the volcano and the water reservoir form a closed system. However, in a) heat transfer rates are very difficult to measure, whereas in the subglacial case fairly accurate estimates can be done by estimating meltwater production and/or ice loss, based on a combination of visual observations at the eruption site and possibly stream gauging downstream.

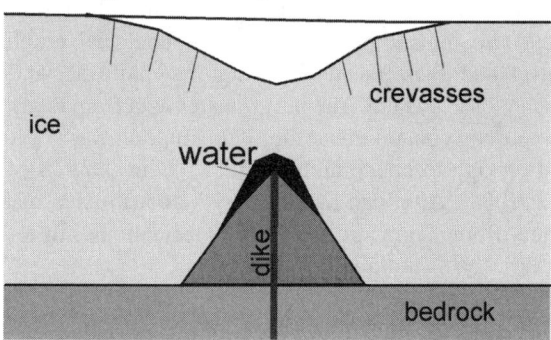

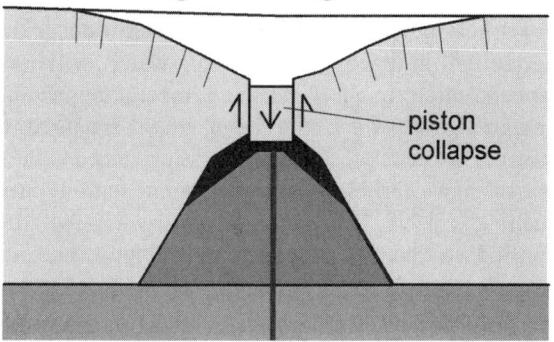

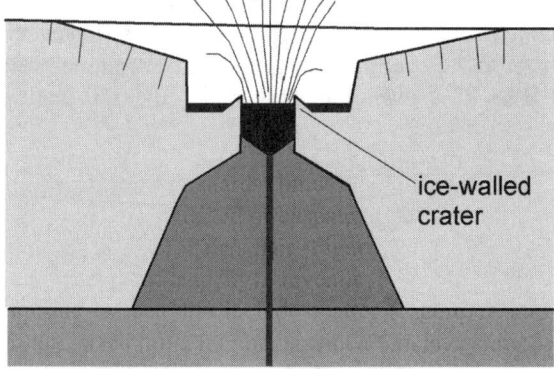

Figure 2. Schematic sections showing how volcanic eruptions may penetrate thick ice, based on observations from the Gjálp eruption in Iceland in 1996. In (1), heat from the eruption rapidly melted the overlying ice, and as water drained away subglacially, depressions formed in the ice surface. (2) As the eruption advanced brittle failure of the ice overlying the vent caused collapse of a piston of ice, which was rapidly melted. In (3) a subaerial vent has opened through the hole left by the melted ice piston. [Modified from *Gudmundsson et al., 1997*].

Figure 3. The surface expression of a fully subglacial eruption in Gjálp on October 1 1996, about 15 hours after its start. Depressions were forming in the ice surface over subglacial vents.

transfer rate of about 2×10^{12} W [*Gudmundsson et al.* 1997]. The corresponding heat flux, found by dividing the heat transfer rate by the area of the volcanic edifice, was $5–6 \times 10^5$ W m^{-2}; about an order of magnitude too high to be explained by cooling and solidification of pillow lavas, and hence implying turbulent mixing of quenched ash fragments and meltwater [*Gudmundsson et al.* 1997; 2003].

The morphology of erupted products from past subglacial eruptions provides indications of heat transfer rates during the formation of these deposits, which are found in several volcanic regions around the world [e.g. *Mathews*, 1947, *Bemmelen* and *Rutten*, 1955; *Kjartansson*, 1959; *Jones*, 1969; *Furnes et al.*, 1980; *Smellie*, 1999; *Hickson*, 2000;

Figure 4. The Grímsvötn eruption in December 1998 was englacial, but the vents ejected material subaerially from the start because ice thickness at the eruption site was only 50–150 m. The photo shows the eruption in its fifth day. A tuff cone was forming from surtseyan explosive activity. A large part of the erupted tephra banked up against ice walls, demonstrating that it had lost most of its heat as it was deposited. The cliff is about 300 m high.

Tuffen et al. 2001]. For basalts, the complete stratigraphic sequence formed in such eruptions can be divided into four main units:

Unit 1: Pillow lavas usually make up the basal part of the sequence. The pillow lavas are erupted in the initial effusive phase when water pressure in the subglacial vault is high. The pillows are commonly 0.5–1.0 m in diameter [e.g. *Jones*, 1969] and have a glassy rind and crystalline core.

Unit 2: On top of the basal pillow lavas, fragmented volcanic glass piles up when granulation and explosive activity become the dominant styles of activity as the volcanic edifice rises and confining pressure is reduced. In hyaloclastite mountains formed during glacial periods the volcanic glass has to a large degree altered to palagonite [e.g. *Jones*, 1969; *Jakobsson*, 1979; *Smellie* and *Skilling*, 1994]. The term hyaloclastite is used here for this type of fragmented volcanic glass; it is commonly used for all clastic volcanic material produced both in explosive and non-explosive magma-water interaction [e.g. *Fisher and Schmincke*, 1984; *Werner and Schmincke*, 1999]. As the hyaloclastite pile grows the eruption may melt its way through the ice, leading to surtseyan explosive activity within an englacial lake.

Unit 3: After the crater of the growing hyaloclastite edifice has emerged from the water, the vent may eventually isolate itself from the surrounding englacial lake. The style of the eruption will then change, and effusive activity lead to subaerial flow of lava.

Unit 4: As the lavas of unit 3 enter the water a delta forms, which is composed of dipping layers of pillows, pillow fragments and hyaloclastite. *Jones* [1969] and *Jones and Nelson* [1970] introduced the term flow-foot breccia for these deltas.

The term tuya is used for mountains having lava caps and formed by eruptions within glaciers. The similar mountain with no lava cap commonly indicates that the eruption did not last long enough to become effusive during its subaerial stage, or that a relatively thin cap was removed by erosion. This applies to most of the hyaloclastite ridges formed in subglacial fissure eruptions in Iceland [*Kjartansson*, 1959; *Jones*, 1969; *Allen*, 1980], since usually only units (1) and (2) are found in the ridges. Some ridges and tuyas may lack the basal pillows, especially ridges of modest height. Both pillow lavas and hyaloclastites are diagnostic of rapid heat transfer from magma to the surroundings.

In the following discussion on heat transfer in eruption of basaltic to mildly intermediate magma, only the formation of the two units that occur in truly subglacial eruptions (units 1 and 2) is considered, when magma, water and ice form a closed system. The heat transfer data from Gjálp are an important constraint but the Gjálp eruption was not long-lived enough to create units 3 and 4 [*Gudmundsson et al.*, 1997; 2003]. Although models may be constructed of heat transfer from magma to water during formation of flow-foot breccias, this is not attempted here.

CALORIMETRY

The processes that occur during rapid cooling and solidification of magma, especially fragmentation due to magma-water interaction, are very complex. In addition to high rates of heat transfer from magma to meltwater and ice, vaporization and condensation of water will occur. Expansion of steam accelerates and imparts momentum to the magma-steam-water mix, the magma is fractured, seismic waves are generated, and energy is expended in the creation of new particle surfaces (fragmentation energy) [*Zimanowski*, 1998; *Buettner and Zimanowski*, 1998]. In the simplified treatment of calorimetry and eruptive heat fluxes provided here, only the heat transfer is considered and most of the secondary processes are ignored. This simplification can be easily justified for the truly subglacial case, in which the volcano, ice and meltwater may form a closed system where the heat of the magma is used to melt ice (Figure 5) and no heat is lost to the atmosphere through a plume. In some cases the meltwater may have a temperature above the melting point when it drains from the glacier. However, heat transfer to the overlying ice from fast-flowing meltwater at the base of a glacier is very rapid; the 8°C meltwater

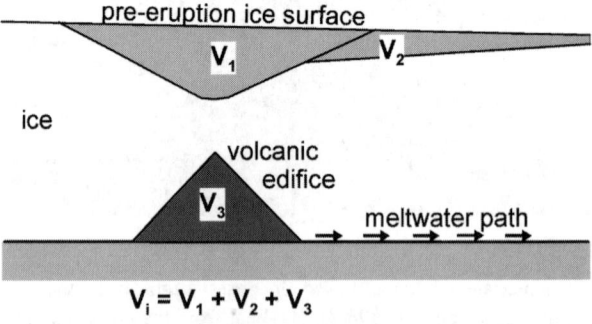

Figure 5. Schematic illustration of how large ice volumes can be melted in subglacial eruptions. V_i, can be estimated from volume of depressions in the ice surface (V_1 and V_2) and the edifice (V_3). A small but significant contribution to V_1 is the volume of void created by crevasses. The volume V_2 is created as the meltwater loses its heat during passage along the glacier bed.

released from Grímsvötn in Vatnajökull on November 5 1996 lost its excess heat over a path of only 6 km [Björnsson et al., 2001; Jóhannesson, 2002].

In the case where magma at the liquidus crystallizes and cools down to ambient temperatures, the energy contained in a volume V_m of magma with density ρ_m is given by

$$E = \rho_m V_m \left[L_m + C_m (T_i - T_f) \right] \quad (1)$$

Here L_m is the latent heat of crystallization of the magma, C_m is the specific heat capacity of the crystalline lava, T_i is the initial magma temperature and T_f is the temperature of the surroundings. If phenocrysts make up a significant fraction of the volume of the magma, this can be taken into account by only using the latent heat of the liquid fraction. When magma is fragmented into glass and no crystallization takes place, the heat content is calculated from

$$E = \rho_m V_m C_g (T_i - T_f) \quad (2)$$

No latent heat is released, but the specific heat capacity of glass is somewhat higher than that of crystalline material of the same composition [Bacon, 1977]. The heat capacity decreases with decreasing temperature [Bacon, 1977, Spera, 2000] but for the large temperature changes characteristic for ice-volcano interaction, the use of a single mean value, C_g, should be sufficiently accurate in most cases.

The magmatic heat is dissipated by ice heating, ice melting, meltwater heating and vaporization of a fraction of the meltwater. While the system remains closed vaporisation can be ignored since steam will condense as it melts ice. Thus, the loss of heat during melting of a volume V_i of ice is given by

$$Q = \rho_i V_i \left[C_i (T_0 - T_1) + L_i + C_w (T_2 - T_0) \right] \quad (3)$$

where C_i and C_w are respectively the heat capacities of ice and liquid water, L_i is the latent heat of fusion for water, T_1 is the initial temperature of the ice, T_0 the melting temperature of the ice and T_2 is the final temperature of the meltwater. In most cases the meltwater will lose its heat as it flows along the glacier bed, and in the case of temperate glaciers $T_1 = T_0$. Thus, for a temperate glacier the heat dissipated may simply be $Q = \rho_i V_i L_i$.

Although heat transfer can be extremely fast, not all the magmatic heat is given off instantly. The efficiency of heat transfer [Höskuldsson and Sparks, 1997, Gudmundsson et al. 2003] can be defined as the ratio of the rate of heat transferred through the vent by magma to the rate of heat transferred to the surroundings by melting of ice and heating of the meltwater. The rate at which heat is brought through the vent, dE/dt, depends on the magma flow rate $\rho_m dV_m/dt$, so the rate of heat transferred to the surroundings is given by dQ/dt. The efficiency of heat transfer F is then given by

$$f = \frac{dQ/dt}{dE/dt} \quad (4)$$

This efficiency is difficult to estimate; here data on melting rates in actual eruptions are used and inferences made on the basis of models of heat flux. Melting rates were measured in the Gjálp eruption in 1996 [Gudmundsson et al. 1997; 2003]. Time-average values of f were obtained from the total mass of ice melted and total mass of erupted material [Gudmundsson et al. 2003]. Two different definitions of f were used for the Gjálp eruption. Firstly, there is the efficiency of heat transfer for ice melting, f_i. Secondly, there is the efficiency of heat transfer from magma f_m. The first value only takes into account ice melted at the eruption site itself while in the second, heating of the meltwater is also included. The values obtained for the two efficiencies were f_i = 50–60% and f_m = 63–77% [Gudmundsson et al. 2003]. The difference between the two reflects the residual heat of the meltwater, which had a temperature of about 20°C as it flowed from the eruption site. This heat was released along the path of the meltwater, downslope of the eruption site itself [Gudmundsson et al. 1997; 2003].

MECHANISMS OF HEAT TRANSFER

Effusive Eruptions

Heat transfer in effusive subglacial eruptions (those forming pillow lava) has been studied by Allen [1980], Höskuldsson and Sparks [1997], Wilson and Head [2002] and Tuffen et al. [2002]. Allen [1980] estimated the heat flow from pillows and concluded that it was rapid enough to melt a volume of ice larger than the volume of effused lava, thereby refuting an argument against tuya formation in subglacial eruptions put forward by Einarsson [1966]. Höskuldsson and Sparks [1997] and Wilson and Head [2002] used solutions to the Stefan problem [Carslaw and Jaeger, 1959; Turcotte and Schubert, 1982] to study heat fluxes from a layer of pillows. Stefan's problem deals with the heat flux from a solidifying infinitely-wide and thick layer (a half-space) where the temperature at the layer boundary remains fixed at the ambient temperature (T_f). Heat is lost by solidification of the melt and cooling of the solidified crust that separates the melt region from the surroundings. To simplify the problem it is assumed that the magma has a single well-defined melting point T_i. The depth to the solidification front y_m, is found from

$$y_m = 2\lambda\sqrt{\kappa t} \qquad (5)$$

where κ is the thermal diffusivity of the lava and λ is a dimensionless parameter obtained from

$$\frac{L_m\sqrt{\pi}}{C_m(T_i-T_f)} = \frac{e^{-\lambda^2}}{\lambda \, erf \, \lambda} \qquad (6)$$

Previously defined parameters are as before and erf is the error function. A temperature profile can be obtained for the crust (the region $0<y<y_m$) as well as an equation for the average heat flux over time t since the emplacement of the molten layer

$$q_{av} = \frac{\rho y_m}{t}\left[L_m + C_m\left(T_i - \frac{1}{y_m}\int_0^{y_m} T(y)dy\right)\right] \qquad (7)$$

The thickness y_m is obtained from (5) and (6). To a good approximation (7) can be simplified by assuming a linear temperature gradient in the crust, yielding

$$q_{av} \approx \frac{\rho y_m}{t}\left[L_m + \frac{C_m(T_i-T_f)}{2}\right] \qquad (8)$$

Using a slightly different equation for q, *Höskuldsson and Sparks* [1997] calculated heat fluxes for various extrusion rates, assuming that one layer of pillows is emplaced on top of another, with the layers having a fixed thickness. The time t in (7) and (8) is the exposure time of the pillow lava layer. This simple model probably underestimates the heat flux from a layer of cooling pillows, at least the early phase of the cooling. Pillows have a partly glassy outer surface that will lose heat more rapidly than assumed in the model. Cooling cracks also form, commonly with a spacing of ~0.1 m, allowing access of water to the interior of the pillow. Figure 6 shows the heat flux from a 0.5 m thick pillow lava layer assumed to cover an area of 2 km². The layer is an idealization, but 0.5 m is the typical diameter of basaltic pillows [*Jones*, 1969]. The heat flux increases with increasing extrusion rate, but the heat exchange efficiency falls due to shorter exposure time. In reality pillows buried within a pile of pillow lava continue to give off heat but this must happen at a rate much reduced compared to pillows that are directly exposed to liquid water, even though a film of steam may partly insulate pillows at the seafloor from water [*Moore*, 1975]. To acknowledge these shortfalls in the model, the heat flux from the pillow layer in Figure 6 is shown as a region, with the lower bound being according to (8) and the upper bound double that value. However, for the lowest flow rates (<100 m³s⁻¹) only the results for (8) are shown, because when the exposure time approaches the same order

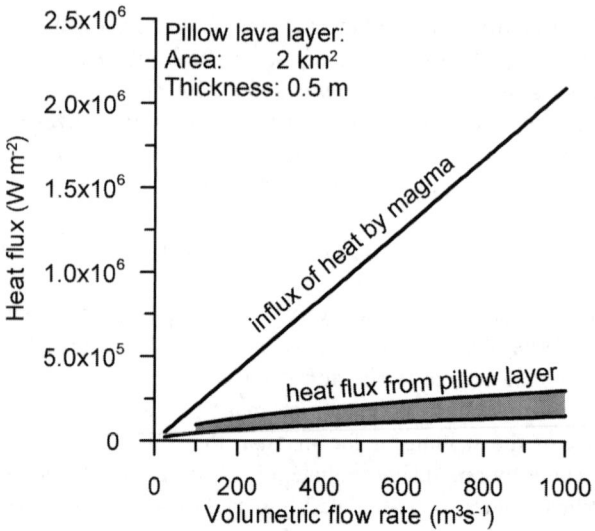

Figure 6. Estimates of heat fluxes in a subglacial basaltic effusive eruption. The influx is estimated from eq. (1) assuming magma density $\rho=2600$ kg m⁻³, latent heat of fusion for magma $L_m=4\times10^5$ J kg⁻¹ [*Spera*, 2000], specific heat capacity of magma $C_m=1100$ J kg⁻¹°C⁻¹, $T_i=1200°C$ and $T_f=100°C$. The heat flux from the pillow lava layer is obtained from eq. (8) assuming a thermal diffusivity of $\kappa=10^{-6}$ m²s⁻¹.

as the time it would take to release most of the thermal energy in the layer, the average heat flux cannot be significantly higher than predicted by (8). The rather ad-hoc estimate of the upper bound in heat flux demonstrates the uncertainty in the model; the results, especially for the higher flow rates, should be regarded as order-of-magnitude estimates.

Magma Fragmentation

The abundance of hyaloclastites in subaqueously and subglacially-erupted sequences testifies to the importance of magma fragmentation in such eruptions. Clearly, heat transfer from magma/volcanic glass to the surroundings cannot be described by the solution to Stefan´s problem (equations 7 and 8). Grain size of glass particles formed in "hydrovolcanic" fragmentation is commonly in the millimeter to sub-millimeter range (Figure 7) [*Wohletz*, 1983, *Werner and Schmincke*, 1999]. To describe the heat flux from the magma/glass particles to the surroundings it is convenient to use the concept of diffusion time [*Colgate and Sigurgeirsson*, 1973; *Turcotte and Schubert*, 1982; *Wohletz*, 1983]. The thermal diffusion time (or thermal equilibration time) of a particle with thermal diffusivity κ and diameter/thickness d is

$$\tau = \frac{d^2}{4\kappa} \qquad (9)$$

It is assumed that heat flow within the particle occurs by conduction, and that convection in the surrounding fluid keeps the surface of the particle at or close to the fluid temperature. This fluid may be liquid water or steam. For a particle cooled from all sides, about 90% of its excess heat relative to the surroundings is lost during its diffusion time. In the case of subglacial eruptions the surroundings are the meltwater; for other subaqueous eruptions it is a pre-existing water body.

The intense heat transfer from magma to water should lead to steam generation but the role of steam in heat transfer in the magma-water-ice system is poorly known. It is clear from observations [e.g. *Thorarinsson*, 1967], theoretical considerations [e.g. *Wohletz*, 1983] and experiments [*Zimanowski*, 1998] that steam generation and expansion is a major factor in shallow-water explosive activity and the formation of plumes in such settings. This situation will, however, not be considered further here. The following discussion is limitied to the closed magma-water-ice system and the effect steam may have on the rates of heat transfer from magma to the overlying ice.

Steam films may cover some and perhaps most particles [e.g. *Wohletz*, 1983] for a part of their exposure time. Steam would influence the heat transfer in two ways. Firstly, the temperature of saturated steam may be 100–200°C higher

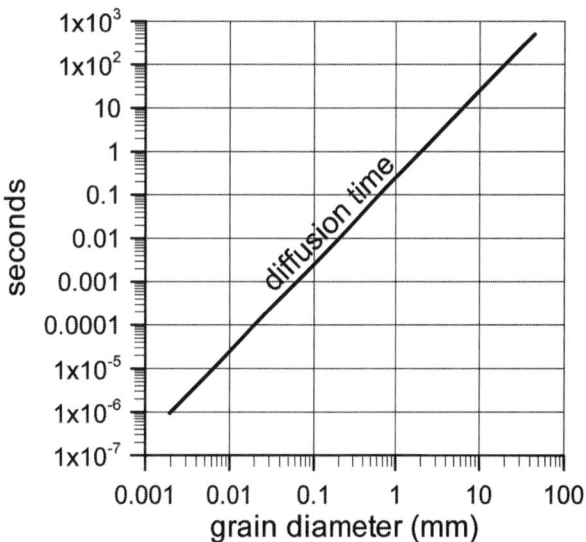

Figure 8. Dependence of diffusion time (thermal equilibrium time) on particle diameter (thermal diffusivity $\kappa = 10^{-6}$ m^{2}s^{-1}).

than that of the meltwater; secondly, thermal conductivity of steam is much lower than for liquid water. The first effect may provide "ambient" temperature that is much higher than that of the meltwater. The second effect reduces the rate of heat loss from the particles to the surroundings although it may be partly offset by steam convection and radiation.

Since the heat loss is related to the square of the diameter, the heat flux is strongly dependent on the size of grains into which the magma fragments (Figure 8). Samples were taken from the upper slopes of the Gjálp edifice in June 1997 (Figure 7). It is possible that some of this material was produced in surtseyan explosive activity during the latter stages of the eruption. Thus, the top part may not be fully representative of the bulk of the edifice. However, gravity modeling of the edifice indicates that it is mainly made of hyaloclastite [*Gudmundsson et al.*, 2002] and the calculated heat fluxes [*Gudmundsson et al.* 1997; 2003] suggest that fragmentation was a major process in the eruption. The grain sizes also conform to reported analyses of subglacially-erupted hyaloclastites [*Werner and Schmincke*, 1999]. The Gjálp data are therefore used here to demonstrate cooling rates during magma fragmentation. For the Gjálp data, the mean diameter of the particles is just under 2 mm; such a particle has a heat diffusion time of 1 s. The diffusion time for the largest particles sampled for Gjalp is >10^2 s, but it is <10^{-3} s for the finest material (Figure 9).

Heat Loss During Magma Fragmentation

The absence of good constraints on meltwater temperature, the importance of heat transfer by steam, and the exact

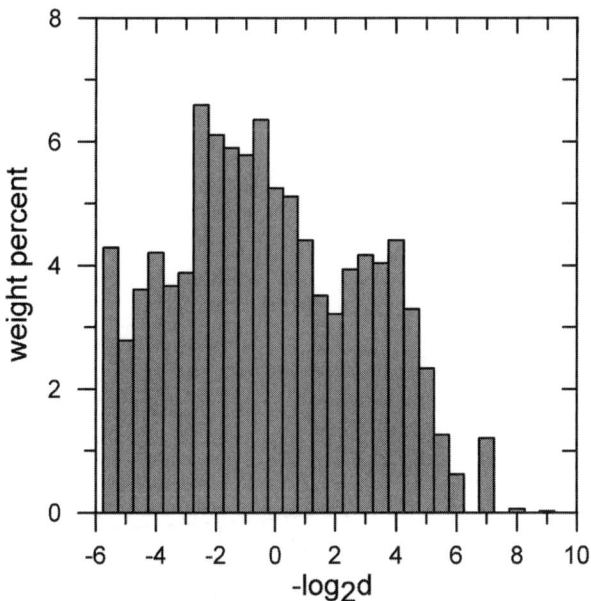

Figure 7. Grain size distribution of samples from Gjálp, collected on June 10 1997 when ice melting after the eruption had exposed the uppermost 40 m of the subglacially-formed edifice. The sites where the samples were collected correspond to the upper slopes of the edifice of stage 3 on Figure 2. The grain size d is in mm. The largest particles had diameters of 45 mm, the mean was just under 2 mm while the smallest grain size was 2x10^{-3} mm.

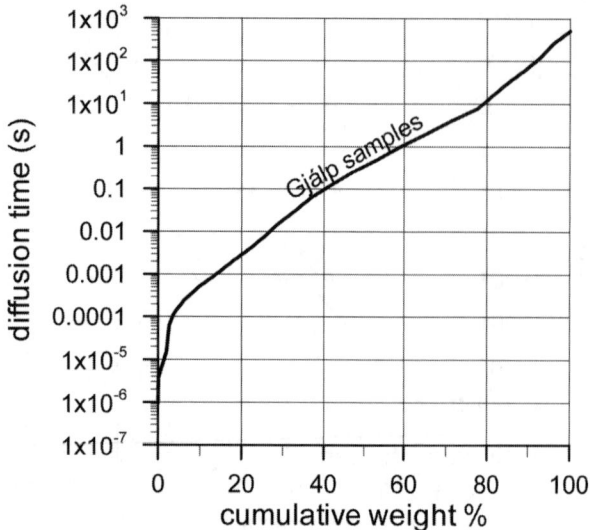

Figure 9. Diffusion time of the Gjálp samples; the curve shows the weight percent of the samples that have diffusion times equal to or smaller than the value on the vertical axis.

nature of the fragmentation process makes the following discussion somewhat speculative. However, it may be of use to consider likely processes and their possible consequences for heat transfer and temperatures in a growing subglacial volcano. In order to get some idea of the temperature of the glass particles as they settle and get buried in the volcanic pile, we need to consider the length of time that the particles are exposed to meltwater-temperatures and also the temperature of the fluid (meltwater/meltwater+steam?) in the subglacial vault above the vent. The exposure time of the particles can be divided into two stages. Firstly, there is the settling time, i.e. the time it takes a particle to rise out of the vent and then sink through the water body onto the slopes of the edifice, possibly impacting the overlying ice roof in the process. Secondly, there is the burial time, the time it takes a particle to get buried in the growing volcanic pile once it has settled. Burial will not stop heat transfer but should slow it down, provided all the heat has not been lost to the overlying fluid.

If fragmentation occurs mostly through non-explosive granulation, the rise of particles above the vent may be only a few meters; their transport may mainly be in a turbulent to laminar gravity current down the sides of the edifice onto which they settle [*Smith and Batiza*, 1989]. For high-concentration flows of hot particles, the temperature of the fluid part of the current could be close to that of saturated steam. In that case the particles may never be exposed to temperatures less than 200–250°C (boiling-point corresponding to water depths of 200–500 m, e.g. *Ingebritsen and Sanford* [1998]). Burial time (Figure 10) would be of order 10 s, suggesting that particles comprising the coarsest 20% of the total mass would retain a significant fraction of their heat as they get buried (Figure 9).

If fragmentation is explosive, the particles formed will be ejected out of the vent by rapid expansion of steam [e.g. *Zimanowski*, 1998]. The settling velocity [*Cashman and Fiske*, 1991] would be 0.1–1 m s^{-1}, with the largest grains having the highest velocity. The settling time would be 10^1–10^2 s for a water layer with a thickness of order 10 m. Only the very largest particles (~10% of mass) would have settling times short compared to their diffusion time (Figure 9); hence, only about 10% of the mass would retain a significant part of its heat by the time it is added to the top of the pile.

The above considerations indicate that probably 85–90% of the heat of the particles is lost to the fluid in the subglacial vault prior to burial in the volcanic pile. The heat released to the fluid is subsequently used to melt ice. If possible complications due to pulsating activity are ignored, some sort of equilibrium should exist between eruption rate, vault-fluid temperature and ice melting rate. Ultimately, the ambient temperature is that of the ice (~0°C) but the fluid (meltwater/meltwater+steam) in which the heat of the magma is released is hotter. Although the mean temperature of the meltwater as it flowed from the eruption site at Gjálp

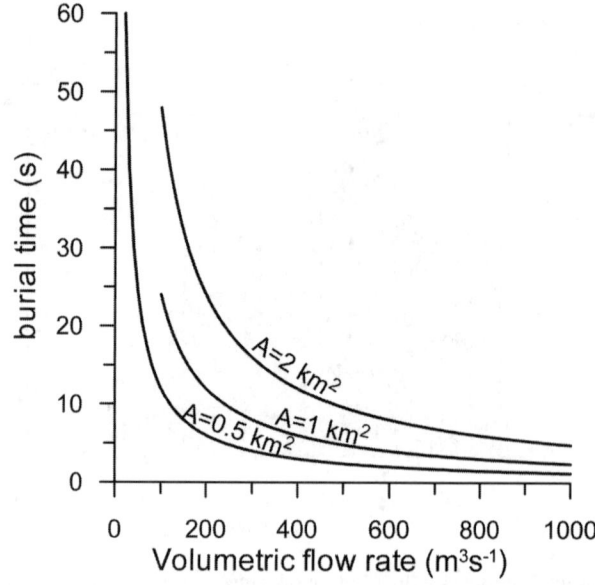

Figure 10. Time of deposition of a 4 mm thick layer as a function of volumetric magma flow rate dV_m/dt. The time is taken as equal to $Ah(1-\varphi)/(dV_m/dt)$ where A is the area over which the material is dispersed, h is layer thickness, φ is porosity of the volcanic pile (taken as 0.4).

was 20°C, its temperature was probably considerably higher during the height of the eruption [*Gudmundsson et al.*, 1997; 2003]. *Höskuldsson and Sparks* [1997] calculated temperatures as high as 100°C in a convecting water body for heat fluxes that were lower than observed at Gjálp. Meltwater temperatures of 50–100°C or even higher in vigorous eruptions do, therefore, seem plausible, and gravity currents down the slope of the edifice may be considerably warmer. Thus, although a particle may have reached thermal equilibrium with the vault fluid, it may still retain some 10% of its heat relative to the ice. The heat loss from the fragmented material to the surroundings prior to burial could therefore amount to 75–80% of the available magmatic heat. Considering also that the energy used to fragment the magma, amounting to a few percent of the total thermal energy [*Zimanowski*, 1998; *Buttner and Zimanowski*, 1998], is not available for melting, 70-80% may be a realistic estimate of the instantaneous efficiency of heat transfer (equation 4) for a grain size distribution like that obtained at Gjálp (Figure 11). A similar estimate of 63–77% was obtained as an average for the Gjálp eruption by calorimetric considerations based on ice melting [*Gudmundsson et al.* 2003].

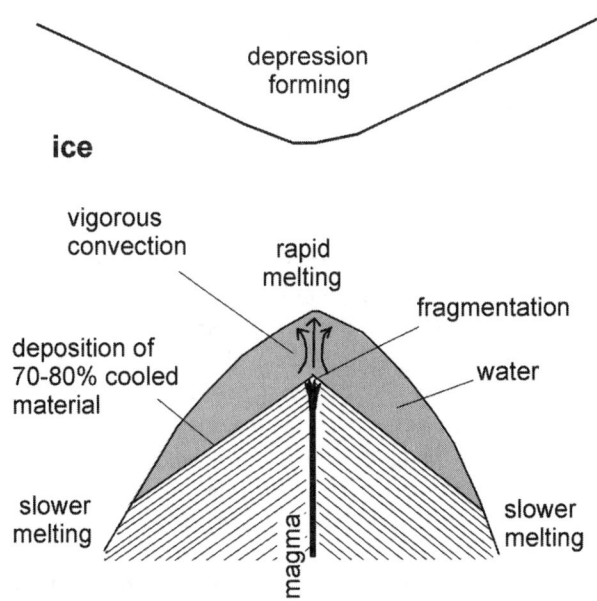

Figure 11. A schematic illustration of how heat transfer, ice melting, magma fragmentation and material deposition within a water-filled vault may occur during a subglacial volcanic eruption. The layering indicated represents material deposited at the same time; it need not reflect pulsating activity. The dip of layers is uncertain but the volcanic material may bank up against ice walls at the vault margins.

Flow-Foot Breccias

The treatment of heat transfer has here focused on the formation of pillow lava (stratigraphic unit 1) and magma fragmentation prior to emergence of a volcano in an englacial lake (unit 2). The basic mechanisms of heat transfer during formation of flow-foot breccias (unit 4) should be the same. Pillows and pillow fragments are abundant and heat fluxes to meltwater from breccias may therefore be intermediate between those suggested for pillow lavas and fragmentation.

Rhyolitic Eruptions

Studied eruptions have been basaltic, or the magma has been of mildly intermediate composition. No rhyolitic subglacial eruptions have been observed, but subglacially-erupted rhyolites are found in Iceland, notably in the Torfajökull region [*Furnes et al.*, 1980; *Tuffen et al.* 2001, 2002]. Magma composition and its effect on viscosity and magma temperature may affect fragmentation mechanisms [*Zimanowski*, 1998]. Due to lower magma temperature, and lower latent heat of fusion and heat capacity of acidic rock, the heat content of rhyolitic magma is only 60-70% of that of basaltic magma [e.g. *Höskuldsson and Sparks* 1997; *Spera*, 2000]. Thus, rhyolitic magma has less ice-melting potential. Moreover, rhyolitic pillows are considerably larger than basaltic ones [*Furnes et al.*, 1980; *Tuffen et al.* 2001, 2002], and their cooling times are therefore considerably longer and heat fluxes correspondingly lower [*Höskuldsson and Sparks*, 1997]. However, melting and heat transfer rates for magma fragmentation should be somewhat reduced, but of the same order as in a basaltic eruption, given the same magma eruption rate.

MECHANISM OF MAGMA FRAGMENTATION

Experimental work on phreatomagmatic explosions has shown that fragmentation during magma-water interaction results from processes ranging from non-explosive granulation through moderately energetic steam explosions to very high-energy thermohydraulic explosions caused by molten fuel-coolant interactions (MFCI) [*Zimanowski et al.* 1997; *Zimanowski*, 1998]. The MFCI explosions result in very intense fragmentation, and the fines in the Gjálp samples (Figure 7) may have been formed in such explosions. The abundance of coarser material indicates that less-energetic fragmentation was also important. It should be kept in mind that the samples at Gjálp represent the top part of the edifice, where the overlying water column amounted to only a few tens of meters. At least for basaltic melts it is likely that at higher confining pressures, during early phases of subglacial eruptions where fragmentation takes place, that non-explosive granulation is a more important style of activity.

THERMAL EFFICIENCY AND EDIFICE TEMPERATURE

The difference in thermal efficiency as defined by (4) for the two processes of effusive activity (eruption of pillow-lava) and magma fragmentation is quite marked (Figure 12). The efficiency during pillow lava formation is 10–45%, the precise value being dependent mainly on exposure time of the pillow lavas. The efficiency of fragmentation may be more variable than shown, although the range for magma fragmentation used on Figure 12 is based on the extreme bounds obtained for the Gjálp data.

The contrasts in heat transfer rates during magma effusion and magma fragmentation lead to considerable differences in residual heat left in the pile after deposition in the growing volcanic edifice. A rapidly growing pile of pillow lava may have pillows with hot and partly molten cores. The heat in this pile will diffuse out of the pillows and be transported out of the edifice by hydrothermal convection. The heat transfer rate of a subglacial or subaqueous effusive eruption will be smoothly varying, and may continue at levels comparable to those observed during an eruption, for a time considerably longer than the duration of the eruption. The peak in heat transfer from magma to water or ice may be much lower than the peak in influx of heat with the magma. In contrast, for an eruption where efficient magma fragmentation is a dominant process, heat transfer rates to water or ice at any time during the eruption will be of similar magnitude as the influx of magmatic heat. Heat flux will fall by orders of magnitude as soon as eruption stops.

An interesting consequence of the 70–80% efficiency for eruptions that produce grain-size populations similar to that of Gjálp's deposits is that temperature in the volcanic pile will initially be highly variable over distances of centimeters to decimeters. The largest grains may have temperatures of a few hundred degrees while the surrounding mass of finer material may have temperatures of 50–100°C. The efficiency suggests that a mean temperature of the pile may be 250–300°C, some 150–200°C higher than that of the surrounding meltwater. It is likely that meltwater will quickly percolate into the edifice and lead to mean temperatures of tephra and pore water of perhaps 120–180°C. If magma is intruded into the edifice during the eruption it will increase the temperature of the pile. Heating caused by intrusions during the final stages of activity in the 1963–1967 Surtsey eruption is considered to have led to the formation of a geothermal area on the island [*Axelsson et al.* 1982]. After an eruption ends, heat flux out of the pile will be relatively slow, a few orders of magnitude lower than during eruption and initial cooling. This residual heat in the pile will eventually be removed by conduction, hydrothermal convection and groundwater advection.

In terms of heat transfer mechanisms operating, the difference between subglacial and other subaqueous eruptions should be small. As a consequence, similar heat fluxes and heat transfer rates are to be expected in both cases. Submarine eruptions will be exposed to the very large thermal reservoir of the ocean; the eruption may not alter the temperature of the surroundings to any degree [e.g. *Thorarinsson*, 1967]. Within lakes, eruptions may lead to considerable heating of the lake water. For example, in a lake with a volume of 1 km³ the release of 50% of the thermal energy of 0.1 km³ of basaltic magma erupted at its bottom would heat the body of water by about 40°C.

CONCLUSIONS

Heat transfer rates and heat fluxes during magma-water interaction may be studied by applying calorimetry to ice-melting rates in subglacial eruptions. For fully subaqueous/subglacial eruptions the form of eruption may be either effusive, usually leading to formation of pillow lavas, or the magma may fragment, leading to the formation of a pile of volcanic glass. Cooling models indicate heat fluxes several times lower for pillow lava formation, and heat exchange efficiencies of 10–45%. In contrast, both observations from

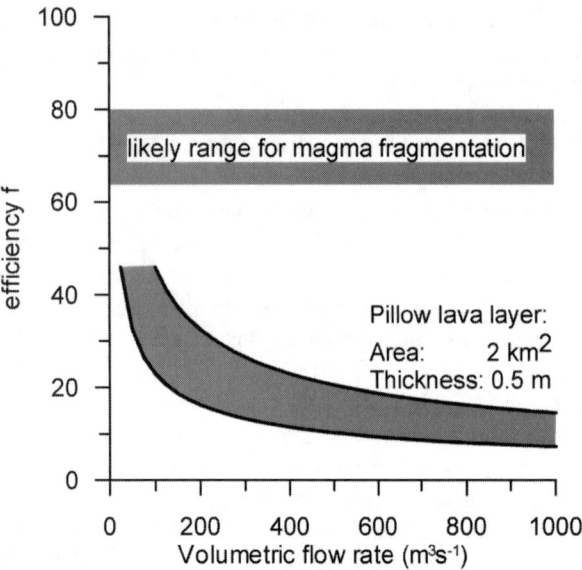

Figure 12. Bounds on efficiency of heat exchange for magma fragmentation (63–80%) and deposition of a layer of pillow lava (10–45%).

the Gjálp eruption, and results from a simple cooling model indicate that fragmentation of magma into glass particles leads to a short-term efficiency of heat transfer of 70–80% and the formation of a "cold" volcanic pile that may have an average temperature of 150–300°C immediately after deposition. The remaining heat in pillow lavas (55–90% of the total magmatic heat) and a pile of fragmented glass (20–30% of the total) is lost to the surroundings by hydrothermal convection at a much slower rate than during the fast initial cooling.

Although simple heat transfer models have been applied to subglacial and subaqueous eruptions, and recent eruptions have provided important data, more work is required to gain further insight into heat transfer processes and the thermal histories of subaqueously-formed volcanic edifices. Grain size analysis and analysis of grain morphology [Buettner et al., 1999] in hyaloclastites at different levels in hyaloclastite mountains may provide insight into how mechanisms of fragmentation vary with ambient pressure. Cooling models of pillow lavas need refinement, taking into account the cooling rates of the glassy outer margins and the effect of cooling cracks in pillows. In this way further understanding of ice-volcano interaction may be achieved by work on deposits of subaqueous and subglacial eruptions, field observations of eruptions, and both experimental and theoretical work.

Acknowledgments. Fruitful discussions with Sveinn P. Jakobsson and John Smellie on fragmentation, hyaloclastites and other aspects of subglacial volcanism are acknowledged. Gudrún Larsen made volcanic samples of Gjálp available for grain size analysis. The sieving of the Gjálp samples was done by Sæunn Halldórsdóttir. Fieldwork that provided the background for this work was done with the assistance of the Iceland Glaciological Society, Iceland Aviation Authority and the National Power Company in Iceland. Helpful comments by Hugh Tuffen, Leo Kristjansson, James White and the reviewers Bernd Zimanowski and Ian Skilling improved the quality of this paper. The Public Road Administration of Iceland and the University of Iceland Research Fund provided financial support.

REFERENCES

Allen, C.C., Icelandic subglacial volcanism: thermal and physical studies, *J. Geol.*, 88, 108-117, 1980.

Áskelsson, J., On the last eruptions in Vatnajökull, *Societas Scientiarum Islandica*, 18, 1-68, 1936.

Axelsson, G., V. Stefánsson, G. Gudmundsson, and B. Steingrímsson, Thermal condition of Surtsey, *Surtsey Research Progress Report*, IX, 102-110, 1982.

Bacon, C.R., High temperature heat content and heat capacity of silicate glasses: Experimental determination and a model for calculation, *American Journal of Science,* 277, 109-135, 1977.

Bemmelen, R.W. van, and M.G. Rutten, *Tablemountains of northern Iceland*, E.J. Brill, Leiden, 217 pp., 1955.

Björnsson, H., Subglacial water reservoirs, jökulhlaups and volcanic eruptions, *Jökull*, 25, 1-11, 1975.

Björnsson, H., Hydrology of ice caps in volcanic regions, *Societas Scientarium Islandica*, 45, Reykjavík, pp. 1-139, 1988.

Björnsson, H., F. Pálsson, G.E. Flowers, and M.T. Gudmundsson, The extraordinary 1996 jökulhlaup from Grímsvötn, Vatnajökull, Iceland, *Eos Trans. AGU*, 82, 47, 2001.

Buttner, R., and B. Zimanowski, Physics of thermohydraulic explosions, *Phys. Rev, E*, 57, 5726-5729, 1998.

Buttner, R., P. Dellino, and B. Zimanowski, Identifying magma-water interaction from surface features of ash particles, *Nature*, 401, 688-690, 1999.

Carslaw, H.S., and J.C. Jaeger, *Conduction of heat in solids*, 2nd edition, Oxford University Press, Oxford, 510 pp., 1959.

Cashman, K.V., and R.S. Fiske, Fallout of pyroclastic debris from submarine volcanic eruptions, *Science*, 253, 275-280, 1991.

Colgate, S.A., and Th. Sigurgeirsson, Dynamic mixing of water and lava, *Nature*, 244, 552-555, 1973.

Einarsson, T., Physical aspects of sub-glacial eruptions, *Jökull*, 16, 167-174, 1966.

Fisher, R.V., and H.U. Schmincke, *Pyroclastic rocks*, Springer, Berlin. 472 pp., 1984.

Furnes, H., I.B. Fridleifsson, and F.B. Atkins, Subglacial volcanics —on the formation of acid hyaloclastites, *J. Volc. Geoth. Res.*, 8, 95-110, 1980.

Griffiths, R.W., and J.H. Fink, Solidification and morphology of submarine lavas: a dependence on exstrusion rate, *J. Geophys. Res.*, 97, 19729-19737, 1992.

Grönvold, K., and H. Jóhannesson, Eruption in Grímsvötn: Course of events and chemical studies of the tephra, *Jökull*, 34, 1-11, 1984.

Gudmundsson, M.T., and H. Björnsson, Eruptions in Grímsvötn 1934-1991, *Jökull*, 41, 21-46, 1991.

Gudmundsson, M.T., F. Sigmundsson, and H. Björnsson, Ice-volcano interaction of the 1996 Gjálp subglacial eruption, Vatnajökull, Iceland, *Nature*, 389, 954-957, 1997.

Gudmundsson, M.T., F. Pálsson, H. Björnsson, and Th. Högnadóttir, The hyaloclastite ridge formed in the subglacial 1996 eruption in Gjálp, Vatnajökull, Iceland: present day shape and future preservation, in *Volcano-Ice interaction on Earth and Mars, Geological Society of London Spec. Publ.*, 202 edited by J.L. Smellie and M. Chapman, pp. 319-335, 2002.

Gudmundsson, M.T., F. Sigmundsson, H. Björnsson, and Th. Högnadóttir, The 1996 eruption at Gjálp, Vatnajökull ice cap, Iceland: Efficiency of heat transfer, ice deformation and subglacial water pressure, *Bull. Volc. DOI:10.1007/s00445-003-0295-9*, 2003.

Hickson, C.J., Physical controls and resulting morphologic forms of Quaternary ice-contact volcanoes in western Canada, *Geomorphology*, 32, 239-261, 2000.

Höskuldsson, Á., and R.S.J. Sparks, Thermodynamics and fluid dynamics of effusive subglacial eruptions, *Bull. Volc.*, 59, 219-230, 1997.

Ingebritsen, S.E., and W.E. Sanford, *Groundwater in geologic processes,* Cambridge University Press, 341 pp., 1998.

Jakobsson, S.P., Outline of the petrology of Iceland, *Jökull,* 29, 57-73, 1979.

Jóhannesson, T., Propagation of a subglacial flood wave during the initiation of a jökulhlaup, *Hydrological Sciences,* 47, 417-434, 2002.

Jones, J.G., Intraglacial volcanoes of the Laugarvatn region, southwest Iceland - I., *Quarterly Journal of the Geol. Soc. Lond.,* 124, 197-211, 1969.

Jones, J.G., and Nelson, P.H.H., The flow of basalt lava from air into water—its structural expression and stratigraphic significance, *Geol. Mag.,* 107, 13-19, 1970.

Kjartansson, G., The moberg formation, *Geografiska Annaler,* 41, 139-143, 1959.

Larsen, G., A brief overview of eruptions from ice-covered and ice-capped volcanic systems in Iceland during the past 11 centuries: frequency, periodicity and implications, in *Volcano-Ice Interactions on Earth and Mars, Geological Society Of London Spec. Publ.,* 202, edited by J.L. Smellie, and M.G. Chapman, pp. 81-90, 2002.

Larsen, G., M.T. Gudmundsson, and H. Björnsson, Eight centuries of periodic volcanism at the center of the Iceland Hot Spot revealed by glacier tephrostratigraphy, *Geology,* 26, 943-946, 1998.

Major, J.J., and C.H. Newhall, Snow and ice perturbation during historical volcanic eruptions and the formation of lahars and floods, *Bull. Volc.,* 51, 1-27, 1989.

Mathews, W.H., "Tuyas". Flat topped volcanoes in northern British Columbia, *American Journal of Science,* 245, 560-570, 1947.

Moore, J.G., Mechanism of formation of pillow lava, *Am. Sci.,* 63, 269-277, 1975.

Sigmundsson, F., M.T. Gudmundsson, G. Sverrisdóttir, N. Óskarsson, and P. Einarsson, Style of basaltic eruptions in shallow ice/water depths: the 1998 Grímsvötn eruption, Vatnajökull ice cap, Iceland, *Eos Trans. AGU,* 46, 1084, 1999.

Smellie, J.L., Lithostratigraphy of Miocene-Recent, alkaline volcanic fields in the Antarctic Peninsula and eastern Ellsworth Land, *Antarctic Science,* 11, 347-363, 1999.

Smellie, J.L., and I. P. Skilling, Products of volcanic eruptions under different ice thicknesses: two examples from Antarctica, *Sed. Geol.,* 91, 115-129, 1994.

Smith, T.L., and R. Batiza, New field and laboratory evidence for the origin of hyaloclastite flow on seamount summits, *Bull. Volc.,* 51, 96-114, 1989.

Spera, F.J., Physical properties of magmas, in *Encyclopaedia of Volcanoes, Academic Press,* edited by H. Sigurdsson, pp. 171-190, 2000.

Steinthorsson, S., B.S. Hardarson, R.M. Ellam, and G. Larsen, Petrochemistry of the Gjálp 1996 subglacial eruption, Vatnajökull, SE Iceland, *J. Volc. Geoth. Res.,* 98, 79-90, 2000.

Thorarinsson, S., *Surtsey, the new Island in the North Atlantic,* Viking Press, New York, 100 pp, 1967.

Tryggvason, E., Earthquakes, jökulhlaups and subglacial eruptions, *Jökull,* 10, 18-22, 1960.

Tuffen, H., J.S. Gilbert, and D.W. McGarvie, Products of an effusive subglacial rhyolite eruption Bláhnúkur, Torfajökull, Iceland, *Bull. Volc.,* 63, 179-190, 2001.

Tuffen, H., H. Pinkerton., D.W. McGarvie, and J.S. Gilbert, Melting at the glacier base during a small-volume subglacial rhyolite eruption: evidence from Bláhnúkur, Iceland. *Sed. Geol.,* 149, 183-198, 2002.

Turcotte, D.L., and G. Schubert, *Geodynamics,* John Wiley & Sons, New York, 450 pp., 1982.

Werner, R., and H.-U. Schmincke, Englacial vs lacustrine origin of volcanic table mountains: evidence from Iceland. *Bull. Volc.,* 60, 335-354, 1999.

Wilson, L., and J.W. Head, Heat transfer and melting in subglacial basaltic volcanic eruptions: implications for volcanic deposit morphology and meltwater volumes, in *Volcano-Ice Interactions on Earth and Mars, Geological Society Of London Spec. Publ.,* 202, edited by J. L. Smellie, and M. G. Chapman, pp. 5-26, 2002.

Wohletz, K.H., Mechanisms of hydrovolcanic pyroclast formation: grain size, scanning electron microscopy, and experimental studies, *J. Volc. Geoth. Res.,* 17, 31-64, 1983.

Zimanowski, B., Phreatomagmatic explosions, in *From magma to tephra, Elsevier, Amsterdam,* edited by A. Freundt and M. Rosi, pp. 25-54, 1998.

Zimanowski, B., R. Buttner, and V. Lorenz, Premixing of magma and water in MFCI experiments, *Bull. Volc.,* 58, 491-495, 1997.

Magnús T. Gudmundsson, Science Institute, University of Iceland, Dunhaga 3, 107 Reykjavík, Iceland. (e-mail: mtg@raunvis.hi.is)

Pyroclastic and Hydroclastic Deposits on Loihi Seamount, Hawaii

David A. Clague[1], R. Batiza[2], James W. Head III[3], and Alicé S. Davis[1]

Layered volcaniclastic deposits up to 11-m thick crop out along caldera-bounding faults on the summit of Loihi Seamount. The layers include unconsolidated volcanic sands and gravels and volcanic silt-to-mudstone. Fragments in volcaniclastic units include fluidal clasts, bubble-wall fragments (limu-o Pele), highly vesicular to scoriaceous fragments, and Pele's hair formed during pyroclastic eruptions, and lithic fragments coated in lava, coarse-grained basalt fragments, hydrothermally altered basalt and glass fragments, and hydrothermal stockwork fragments of pyrite and barite formed during hydromagmatic (phreatic and phreatomagmatic) eruptions. Pyroclasts of tholeiitic and transitional compositions tend to be dense and probably formed during strombolian activity whereas those of alkalic compositions are highly vesicular to scoriaceous and could have formed during strombolian or hawaiian eruptions. Scoria fragments from subaerial eruptions of Kilauea, intercalated with locally derived volcaniclastic deposits, are most likely from the ca. 1790 A.D. Keanakakoi eruption, and suggest that the volcaniclastic units on Loihi are younger than a few thousand years. Exposed sections decrease in thickness to the north on the summit, suggesting sources located mainly in the southern part of the summit platform. Based on analogy with Kilauea, we infer that hydrovolcanic eruptions are linked to pit crater formation, as occurred in 1996, and to earlier caldera formation. The bottom of the deepest pit crater is at 1356 m depth, so explosive hydrovolcanic activity can occur at least this deep. Loihi Seamount is an unparalleled natural laboratory to study the eruptive style of submarine basaltic explosive eruptions, deposition of the ejecta, and its redistribution and winnowing by currents.

1. INTRODUCTION

Explosive pyroclastic and hydrovolcanic eruptions are common at subaerial basaltic volcanoes, but also occur in the subaqueous environment. On land, specific eruption styles are documented to produce specific deposit and clast types that can be used to identify eruption styles in ancient deposits. Pyroclastic eruptions, for example, produce bubbly lava clots ranging from Pele's hair and tears to scoria and reticulite. The ejecta accumulate close to the vent and commonly form spatter cones during hawaiian eruptions and cinder cones during strombolian eruptions [*Vergniolle and Mangan*, 2000].

Subaqueous explosive activity has been inferred from theoretical and experimental analysis [e.g., *Dullforce et al.*, 1976; *Froelich*, 1997; *Zimanowski et al.*, 1991, 1997, this volume; *Head and Wilson*, 2003], mapping in ancient submarine pyroclastic rocks exposed on land [e.g., *Staudigel*

---

[1]Monterey Bay Aquarium Research Institute, Moss Landing, California
[2]National Science Foundation, Washington, D.C.
[3]Brown University, Providence, Rhode Island

Explosive Subaqueous Volcanism
Geophysical Monograph 140
Copyright 2003 by the American Geophysical Union
10.1029/140GM05

and Schmincke, 1984; Fisher and Schmincke, 1984; Cas and Wright, 1987], and observations of recent volcaniclastic deposits exposed on the seafloor [e.g., Lonsdale and Batiza, 1982; Batiza et al., 1984, 1996; Smith and Batiza, 1989; Binard et al., 1992; Clague et al., 1990, 2000; Maicher et al., 2000; Maicher and White, 2001]. The limiting conditions for such eruptions have profound implications for understanding the hazards they may pose on Earth [Mastin and Witter, 2000], as well as understanding planetary volcanic processes [Wilson and Head, 1983].

Of the three types of study to evaluate submarine eruption style, direct observations of recent deposits in the oceans provide the least equivocal information about eruption conditions and style. Available evidence suggests that submarine explosive eruptions are possible as deep as several kilometers [Head and Wilson, 2002], with depth critically depending on whether the explosions are pyroclastic and caused by exsolution of magmatic volatiles or hydroclastic and caused by formation of steam during magma-water interaction [Wohletz, 1986, this volume; Heiken and Wohletz, 1986; Kokelaar, 1986]. The maximum depth for significantly energetic pyroclastic explosions is estimated to be between ~500 and ~1800 m [McBirney, 1963; Staudigel and Schmincke, 1984; Gill et al., 1990; Fouquet et al., 1998]. Pyroclastic deposits of highly vesicular fragments of alkaline composition, however, have formed at least as deep as 4100 m [Clague et al., 1990]. Submarine hydrovolcanic explosions can be extremely violent in shallow water (0-300 m) but become decreasingly violent with increasing depth [Zimanowski and Büttner, this volume]. This change is due to the decrease in the expansivity of seawater as pressures approach the critical point of seawater at 407°C and 298 bars (equivalent to roughly 2940 m depth in the oceans) [see Clague et al., this volume].

There are few active volcanoes whose summits are located within the depth interval 500-1700 m, but Loihi Seamount offshore Hawaii has a summit platform at about 1150 m depth. The depth, combined with the wide range of basalt compositions erupted at Loihi [Moore et al., 1982], make it a natural laboratory to study submarine explosive eruptions and fragmentation processes [e.g., Heiken and Wohletz, 1991; Cashman et al., 2000]. Volcaniclastic deposits were first discovered on Loihi during submersible dives in 1996 and 1997 [Clague et al., 2000]. The deposits consisted of rippled unconsolidated basaltic sediment, commonly thin enough that pillow basalt lobes protruded through the deposits. Examination of collected samples showed that a significant portion of the basaltic glass (sideromelane) fragments were limu o Pele fragments, which Clague et al. [2000] proposed were formed during mild steam explosions. Based on the similarity of these fragments to others formed at pressures above the critical point of seawater, Clague et al. [this volume] have reconsidered and now propose that these fragments form during strombolian eruptions. Additional evidence that pyroclastic deposits might be more widespread on Loihi included an observation of a volcaniclastic unit several m thick in the steep slopes on the east side of the summit platform [Garcia et al., 1995].

Previous dives to assess the possible presence and nature of pyroclastic deposits at great depths were reported in the exploration of Seamount 6, at depths of 1400-2000 m [Batiza et al., 1984, 1996]. The limited amount of pyroclastics and hydroclasitics observed at these depths [Batiza et al., 1996; Maicher et al., 2000; Maicher and White, 2001] and the combination of factors outlined above led us to propose a small dive program specifically to search for thicker, and perhaps more varied, pyroclastic units on Loihi. Our goal was to determine the range of eruptive styles that occur on Loihi's summit at roughly 100 bars pressure. On Kilauea Volcano, fault scarps expose volcanic ash produced during explosive eruptions. The faults there include caldera-bounding faults and those associated with seaward slumping of Kiluaea's south flank. Using Kilauea as a model, we successfully focused our dives to explore several caldera-bounding faults on the eastern part of the summit platform and along the headwall scarp of a large slump on the eastern side of the volcano.

2. GEOLOGIC SETTING

Loihi Seamount is the southeasternmost, and the youngest, volcano in the Hawaiian-Emperor volcanic chain [Moore et al., 1982]. The volcano is entirely submarine and is located about 30 km off the southeastern coast of Hawaii (Figure 1). The volcano is elongate in form and has two prominent rift zones extending north and then northeast and south-southeast from a summit platform (Figure 2) that is punctuated by three collapse pits; the southwestern one, Pele's pit, formed in 1996 [Loihi Science Team, 1997]. The south-southeastern rift is about 17 km long and the one to the north and then northeast is about 8 km long. A third, poorly developed, rift zone extends north for about 2.5 km from the northeastern corner of the summit platform and parallels the upper part of the north rift. The summit platform averages about 1150 m deep and is approximately 3 by 4.5 km, elongate in the north-south direction. The three pit craters, North, East, and Pele's, are located in the southern part of the summit platform and their floors are at 1302, 1356, and 1330 m, with depths below their rims averaging about 300 m, 200 m, and 180 m, respectively (Figure 2).

North Pit is nested within a low shield, much like Halema'uma'u pit in Kilauea's summit caldera. It is also bounded on its west side by the shallowest area on Loihi, at 980 m depth. Prior to the collapse event in 1996, the shallowest point on Loihi was 969 m deep, but this region collapsed into the new pit crater. The second shallowest area on the summit, at 985 m depth, is located on the northeast rim of the summit platform. A series of inward-facing, roughly circumferential, nested fault scarps indicate that the nearly flat summit region is a caldera complex, as originally proposed by *Malahoff* [1987], but subsequently argued against by *Fornari et al.* [1988]. The three pit craters are nested within the summit caldera complex. The formation of the pit craters and the larger caldera complex are integral to our interpretation of eruption style on Loihi Seamount.

3. PREVIOUS WORK ON LOIHI SEAMOUNT

Unconsolidated black volcanic sand was observed and sampled on the summit and uppermost south rift zone of Loihi Seamount during *Pisces V* submersible dives in 1996 and its distribution along the upper south rift zone was explored in 1997 [*Clague et al.*, 2000]. The sands contained abundant fragments of limu o Pele, thought at that time to have formed during mild steam explosions as lava entrapped seawater, in a manner analogous to that taking place where lava from Kiluaea's current eruption enters the sea [*Hon et al.*, 1988, *Mattox and Mangan*, 1997]. The basaltic sands were at first thought to be derived from an

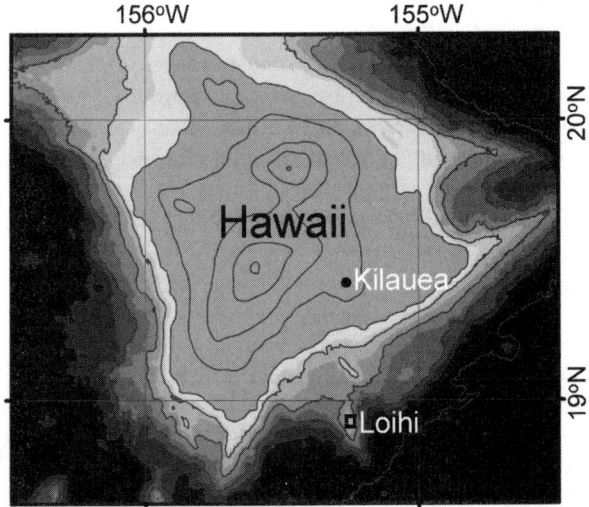

Figure 1. Location of Loihi Seamount on the southeastern side of Hawaii Island. Contours are 1000 m above and below sea level. The box outlines the detailed bathymetric map shown in Figure 2.

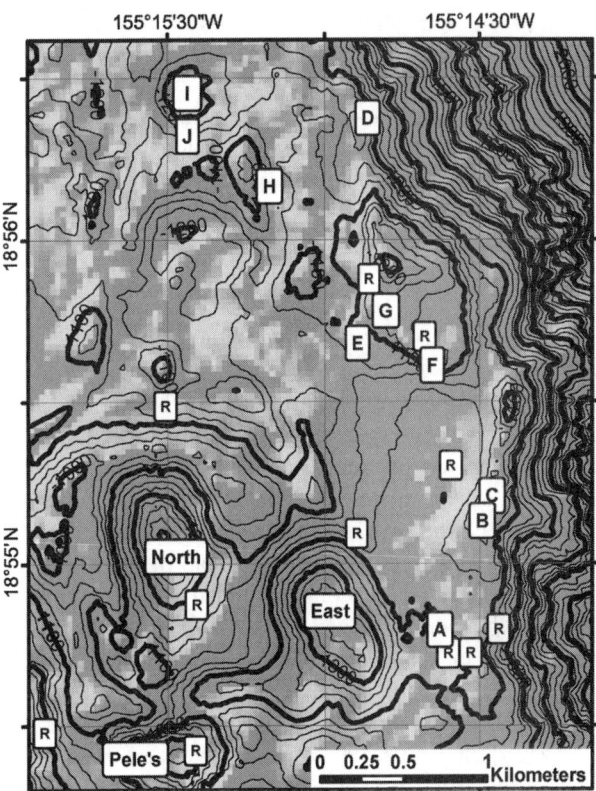

Figure 2. Illuminated bathymetric map of the summit region of Loihi Seamount. Loihi has been mapped repeatedly with swath sonar mapping systems [*Malahoff*, 1987; *Fornari et al.*, 1988; *Chadwick et al.*, 1993; *MBARI Mapping Team*, 2000; *Caress and Chayes.*, 2000; *Smith et al.*, 2002]. This map uses all available data, weighted by resolution. The grid has 5-m resolution and the contour interval shown is 25 m. Three pit craters discussed in the text are labeled. The letters A to J indicate the locations of volcaniclastic deposits described in the text. Locations of sampled rippled basaltic sediment deposits of unknown thickness (many from *Clague et al.*, [2000]) indicated by the symbol R.

eruption in 1996 that predated or accompanied the collapse of the summit and formation of the new pit crater (Pele's pit, Figure 2). The wide range of glass compositions, ranging from alkalic to tholeiitic with MgO from 4.1 to 8.7 wt%, analyzed from the 40 sediment samples studied, however, indicated that the glass fragments formed during numerous eruptions on Loihi [*Clague et al.*, 2000].

The observations in 1996 and 1997 suggested that the volcanic glass fragments formed a thin, current-transported and rippled sedimentary veneer over much of the upper south rift zone and south-central part of the summit platform of Loihi. The observed fragment shapes included angular dense fragments typical of thermal shock granulation [*Kokalaar*,

1986], limu o Pele bubble-wall fragments thought to form during mild steam explosions [*Clague et al.*, 2000; *Maicher and White*, 2001], Pele's hair, and vesicular to scoriacious fragments similar to those formed in Hawaiian lava fountains. However, the locations of the source vents for the fragments from these sedimentary deposits were unknown, as was their distribution away from those vents.

4. RESULTS

Descriptions of Volcaniclastic Deposits

The volcaniclastic deposits observed from the *Pisces V* submersible during dives 399, 401, and 404 in 1998 range from thin deposits of rippled sediment, like those observed on *Pisces V* dives in 1996 and 1997, to an 11-m thick bedded volcaniclastic section. The specific deposits described below are located on a bathymetric map of the summit in Figure 2, as are the locations of rippled sediments, mostly described previously [*Clague et al.*, 2000]. The following descriptions are grouped by their locations on the summit.

4.1.1. Southeast summit. Deposit A (Figure 3a) was observed during dive 399 to the southeastern part of the summit platform. It consists of roughly a meter thick sequence of interbedded black volcanic sands and silts, grey-to-white sediments, and orange-brown layers of hydrothermal nontronite. There are 7 main units with the top (1) being the rippled gray surface sediment. Downsection, the units are (2) a tan to tan-brown nontronite layer, (3) a thick black layer of volcanic sand and silt, (4) a layer of grey over white over gray sediments with finer gray laminae, (5) a 10-cm thick yellow-brown-tan unit with distinctive fine black laminations, (6) a brown mottled unlaminated layer with coarse mottling, over a basal coarse layer of black fragments. The white units contain abundant foraminiferal tests and the grey ones are mixtures of silt-to-clay sized volcanic fragments and foraminiferal tests. The basal unit forms a small break-in-slope and is partly covered by material slumped from the overlying layers. The thin laminations in unit 5 are continuous over a 1-3 m scale along strike. Three sediment scoop samples were collected, with one (399S1) a composite of the upper units and the other two from the basal unit about 1 m apart (399S4 and 399S5).

Deposit B was also observed during dive 399. It consists of a low mound draped by coarse black sediment, partly covered by nontronite that has deposited along a polygonal set of cracks. There is no obvious layering at this site, as seen in holes excavated using the manipulator. The larger fragments are angular fragments of vesicular basalt ranging up to a few cm across. Two sediment scoop samples were collected here with 399S3 from the side of the mound and 399S9 from near the base.

Deposit C, also observed on dive 399, is a black sand deposit draping a low elongate ridge of pillow lava. This ridge also had orange-brown hydrothermal nontronite on the surface. There is a thin laminated layer of grey-white sediment below the nontronite but above the black sands.

4.1.2. Northeastern and east central summit. Deposit D (Figure 3b), observed on dive 401, consists of basaltic sand, lapilli, and pillow fragments in a crudely layered unit exposed at 1160 m depth in the headwall of a large slump on the northeast side of summit platform. The volcaniclastic deposit here was observed and previously sampled (sample 13) during Pisces V dive 186 [*Garcia et al.*, 1995] during an upslope transect to collect a stratigraphic suite of samples. We returned and observed the deposit along strike. The unit varies in thickness along strike, has crude bedding (Figure 3b), contains pillow fragments up to about 30 cm across, and is cut by a near-vertical dike. The lapilli are spatter and scoria clasts and the sand is coarsely vesicular,

Figure 3. Video framegrabs of the deposits described in the text. a) Deposit A showing the thinly laminated zone overlying a more massive black sand unit. The surface is coated with hydrothermal nontronite. Outcrop is approximately 1 m high. b) Deposit D, a crudely bedded deposit of scoria and coarsely vesicular basalt clasts and sand, draping a pillow basalt outcrop to the left. Section illustrated is approximately 3 m thick. c) The uppermost layers at deposit G. The buff-colored layers are fine grained, semi-lithified volcanic mudstone, and overhang the less consolidated black sand and gravel layers. Image shows approximately the top 2 m of an 11-m thick section exposed along an inward facing caldera bounding fault. d) The lower-middle part of deposit G showing thin buff-colored semi-lithified volcanic mudstone layers, but mainly less consolidated black sand to gravel layers. The image shows approximately 5–6 m of section. e) Close-up of the lower buff-colored layers shown in (d), with the 30-cm pushcore for scale. f) Close-up of the base of the section at deposit G showing that many of the layers vary in thickness along strike, pinch out, and drape the underlying pillow lavas. The upper part of the photo shows a thick unconsolidated unit of black sand and gravel, with some clasts that have slid down from higher layers. Photo shows approximately 1 m of section. g) Deposit H. The top and bottom light-colored layers are bright yellow-orange nontronite whereas the middle layer is a fine-grained volcaniclastic unit. The black layers are basaltic sand and gravel. Image shows approximately 2 m of section. h) Deposit I showing thin, yellow-colored, fine-grained volcanic mudstone layers interbedded with black sand layers. Section shown is approximately 2 m thick.

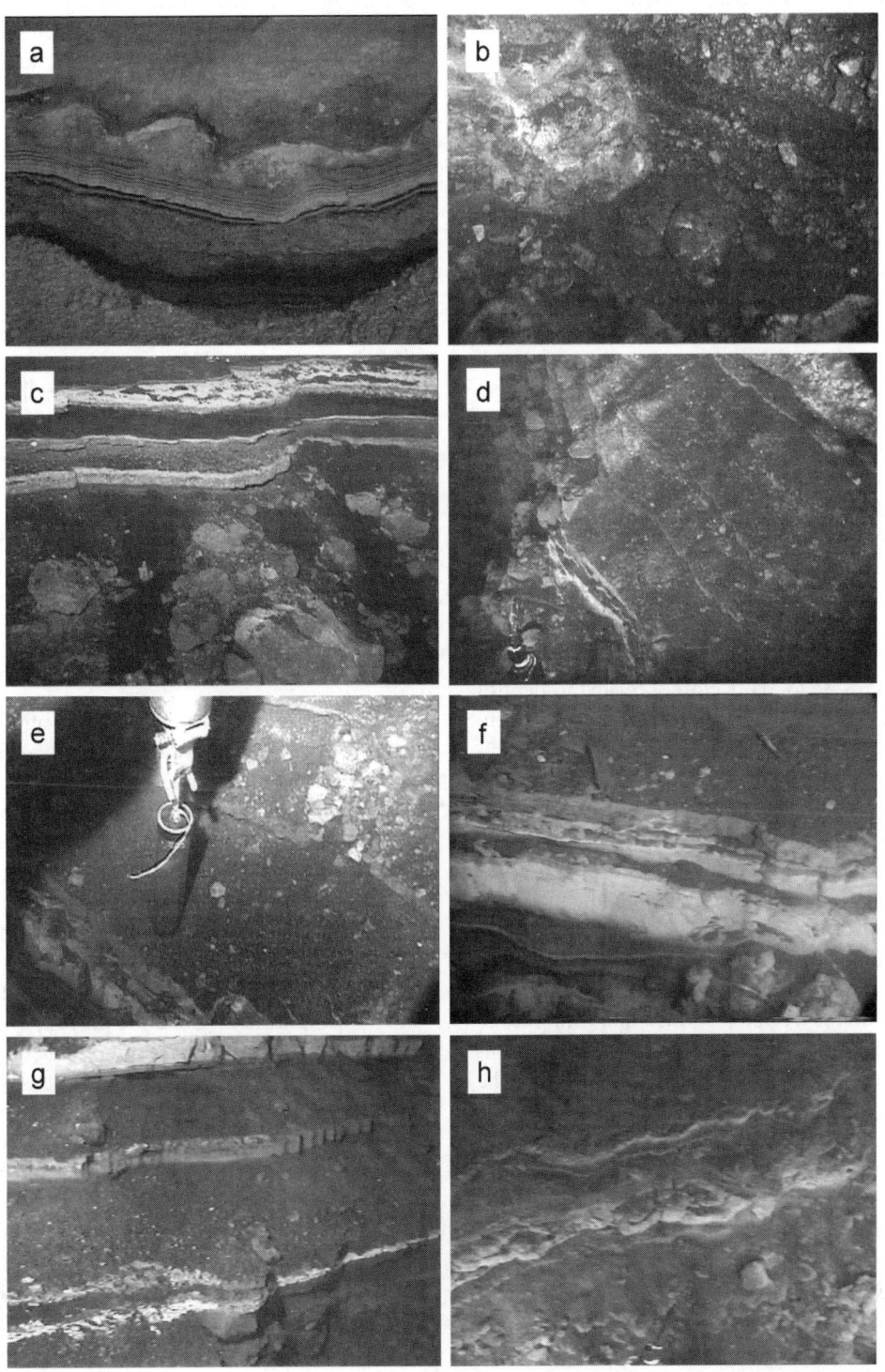

so that most fragments appear dense. One observation made after samples were recovered is that most of these olivine-rich samples have been hydrothermally altered, as seen in white amorphous SiO_2 coatings on the olivine phenocrysts. The alteration is significantly less pronounced in the talus unit than in the finer parts of the deposit. Five samples were collected from this unit (401S6 from top, 401S8 from middle, 401S2 from the near base, 401S3 from the contact with underlying pillow lava, and 401S1 from talus).

Deposit E, at 1083 m depth, was observed during a southerly transect from deposit D during dive 401. This deposit was not sampled but consisted of horizontal layers of volcanic sandstone exposed on the northeast side of a large block. It is most likely that deposits E, F, and G are different outcrops of essentially the same section, since all are located within about 500 m of each other.

Deposit F, at 1058 m depth, was observed near the southern end of a transect on dive 401. The deposit occurs as a sequence of volcanic sandstone layers above truncated pillow lava. The fault scarp offsets the deposit, although some secondary slumping has modified the scarp. A single sample, 401S9 was collected from the layered sequence. A second sample (401S5) was collected from the surficial rippled basaltic sediment at the top of this unit several hundred m farther south.

Deposit G (Figures 3c to 3f) is closely related to deposit F, although it is exposed on a different nearby fault. It consists of an 11-m thick exposure of layered volcanic sediments overlying pillow lava and was sampled extensively on dive 404. This section is the thickest observed during the 3 dives and is exposed on a northwest-southeast trending fault scarp with the southwest side down. It consists of thick units of coarse black silt, sand, and gravel interbedded with hard fine-grained pale tan-to-brown layers of volcanic ash. The lower layers drape the underlying pillow basalt, but the layering is increasingly uniform and horizontal upsection. The top of the section is a rippled surface that continues to the east. As at deposit G, the fault scarp offsets the volcaniclastic unit and therefore faulting postdates the deposit. The section shown in Figure 4 is constructed as a composite from several traverses of the section. We collected 2 pushcores (404PC1 and PC2), 4 sediment scoop samples (404S5, S2, S1 and S4), and several rocks from the hard fine-grained layers. The positions of these samples are shown in Figure 4. Several of these samples retained the layering and were subdivided into multiple units for later analysis (9 layers in 404PC1, 6 in 404SC4, and 3 in 404SC5 and 404SC1). The section is apparently repeated in a roughly parallel scarp about 50 m to the west, but the western fault has a larger throw and exposes a thick basalt unit with columnar joints below the volcaniclastic section.

4.1.3. North-central summit. Deposit H (Figure 3g) was found during a transect to the north during dive 404. It occurs along a north-south trending fault along the eastern side of a hill. The outcrop is several meters thick and consists mainly of orange-brown hydrothermal nontronite with layers of black volcanic sand and semi-lithified, buff-colored volcanic siltstone to claystone. We collected sediment scoop 404S8 above 404S9 and two rock samples of the hard layers.

Deposit I (Figure 3h) is layered volcanic sand on the floor of a crater in the north central part of the summit. The sand drapes drained pillow lava and is best exposed in the walls of small collapse structures where the sediment filtered through holes and cracks in the underlying hollow pillows. We collected box core 404BC and sediment scoop 404S3 at this site and subdivided the structureless black sand in the boxcore into upper, middle, and lower thirds.

Deposit J is a coarse fragmental deposit exposed on the inner wall of a collapse structure. The inner wall is generally coated with nontronite, but slumps have removed the nontronite coating in places and exposed deposits of black lapilli-sized fragments. The lapilli are highly vesicular and the deposit appears to be part of a small scoria cone exposed on the south rim of the collapse pit.

4.2. Grain-Size Analyses of Deposits

Simple grain size analyses of 54 samples determined by wet sieving are presented in Table 1. The clay particles were cohesive in many samples, so the fine fractions are commonly underdetermined and the coarse fractions are slightly enhanced. The data are plotted on a ternary diagram of gravel, sand, and mud (silt + clay) sized fractions in Figure 5a. The samples range from sandy gravel with up to 71% >2mm through gravelly sand, sand, muddy sand, to sandy mud with as much as 88% <74 μ. Most samples plot near the mud-sand or the sand-gravel edges of the diagram, with only 6 samples falling into the gravelly muddy sand classification. The same samples are plotted on a second ternary diagram showing the relative proportions of coarse, medium and fine sand in Figure 5b. There are a number of samples whose sand component is dominantly coarse sand, but most samples contain either mostly fine and medium sand or mostly medium and coarse sand.

There is a large range in grain size from individual deposits. Deposit A samples are dominated by sand but samples are muddy sand, gravelly sand, and sand, using nomenclature from *Folk* [1966]. Deposit B samples are either sandy gravel (399S3) or muddy sand (399S9). The two samples 399S6) from Deposit C are muddy sand whereas the 4 samples (401S1, 401S3, 401S6, and 401S8) from Deposit D are all sandy gravel and the single sample (401S9) from

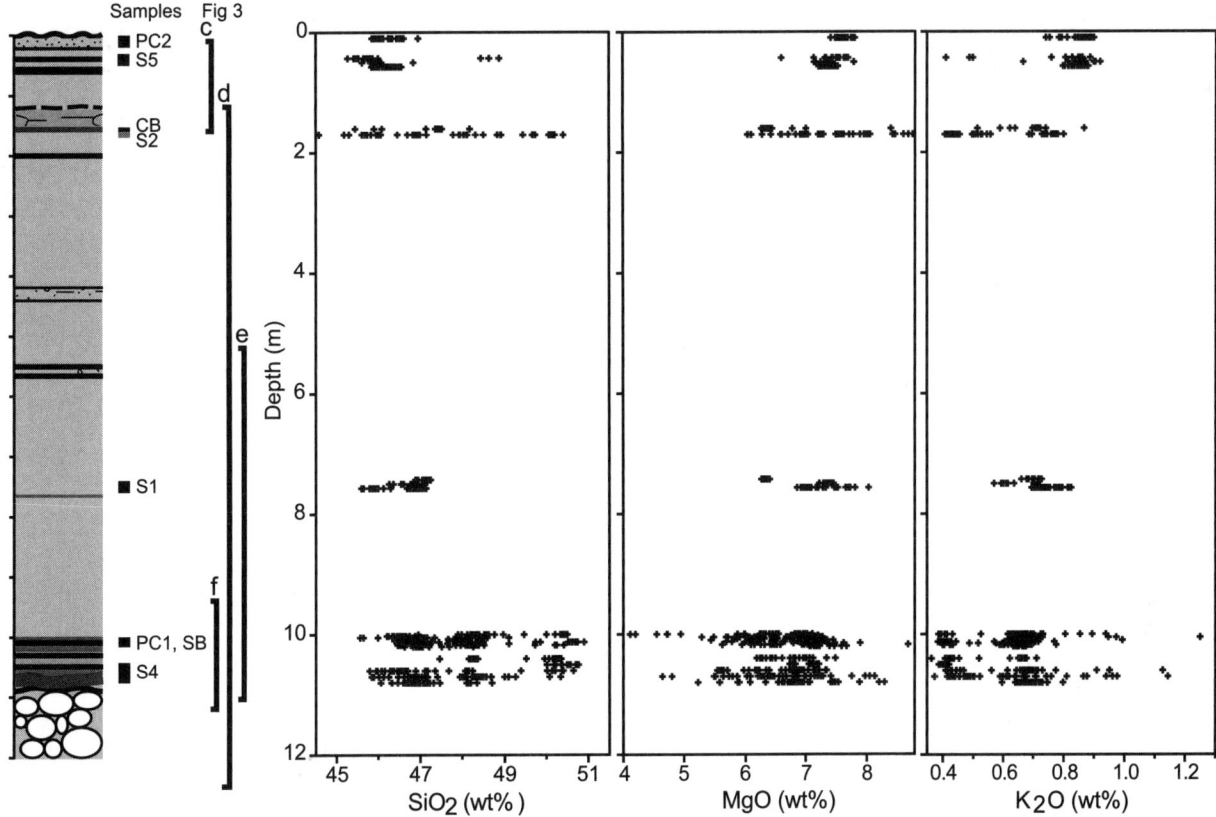

Figure 4. Diagram showing the general stratigraphy of deposit G, the approximate locations of the samples recovered, the approximate intervals shown in the photographs in Figure 3, and plots SiO_2, MgO, and K_2O contents as a function of depth in the section.

Deposit F is gravelly sand. The 22 samples from Deposit G cover a large range from sandy mud (almost mud) to muddy sand, gravelly muddy sand, gravelly sand, and sandy gravel, although none of the samples contains as much gravel as samples from Deposits D, B, J or sample 341S2 of rippled sediment. Many of the most mud-rich samples analyzed are from this deposit. These samples are the buff-colored layers, commonly collected using the manipulator rather than sediment scoops or push-cores due to their semi-lithified nature. Three samples (404S8 and 404S9) from Deposit H are muddy sand or gravelly muddy sand whereas 4 samples from Deposit I are sandy mud or muddy sand. The single sample from Deposit J is a sandy gravel, similar to some samples from deposits B and D. The surface rippled sediment deposits range from sandy gravel to gravelly muddy sand, muddy sand, and sand.

We have also examined the data on plots (not shown) of cumulative percent coarser as a function of grain size to determine standard statistical measures of grain size distribution [*Inman*, 1952]. However, 9 samples have median grain size ($M_d=\Phi_{50}$) greater than our coarsest sieve size (2 mm, $\Phi=-2$) and 7 others have more than 50% smaller than our finest sieve size (74 μ, $\Phi=3.75$). Therefore the median grain size of these 16 samples cannot be determined. The remaining samples have M_d ranging from $\Phi=-0.7$ (about 1.6 mm) to $\Phi=3.9$ (about 0.07 mm) with a large group clustered around $\Phi=1.4$ (about 0.37 mm). The mean $M=(\Phi_{16}+\Phi_{84})/2$ and sorting coefficient $\sigma=(\Phi_{84}-\Phi_{16})/2$, can be determined on only 16 samples because either Φ_{16} or Φ_{84} are contained within our largest or smallest size fraction. The median and mean grain sizes and sorting coefficients of the samples are listed in Table 1.

4.3. Componentry of the Deposits

The silt-, sand-, and gravel-sized particles in these deposits are highly variable in shape and composition, and provide significant clues to the styles of eruptive activity. The components that make up these samples are summarized in Table 2. Most juvenile components are composed of glass and include dense angular fragments (Figure 6a), vesicular angular fragments (Figure 6f), slightly curved

Table 1. Grain Size Data

Sample	Location	>2 mm	600 μm	250 μm	125 μm	74 μm	<74 μm	Φ_{50}	M	σ
399S1	Deposit A	1.38	5.29	29.89	34.08	2.90	26.46	2.4	–	–
399S4 Top	Deposit A	6.95	24.66	32.93	17.99	7.64	9.84	1.45	1.45	1.75
399S4 Bottom	Deposit A	7.76	45.06	27.68	10.31	1.12	8.07	0.65	0.83	1.52
399S5 Top	Deposit A	1.06	19.90	34.47	29.33	1.79	13.45	1.8	1.68	1.32
399S5 Bottom	Deposit A	0.66	37.90	26.01	9.88	1.93	23.62	1.3	–	–
399S3 Top 1/3	Deposit B	67.58	24.57	4.90	0.92	0.12	1.91	–	–	–
399S3 Middle 1/3	Deposit B	68.71	26.43	2.71	0.51	0.16	1.49	–	–	–
399S3 Bottom 1/3	Deposit B	58.59	34.42	4.79	0.62	0.10	1.48	–	–	–
399S9 Top	Deposit B	1.14	24.37	47.84	10.92	1.71	14.02	1.4	1.55	1.45
399S9 Middle	Deposit B	0.53	23.78	46.24	12.92	2.07	14.46	1.45	1.65	1.50
399S9 Bottom	Deposit B	0.42	22.61	49.60	10.30	1.96	15.10	1.4	1.83	1.58
399S6 Top	Deposit C	1.97	26.93	46.87	4.58	2.79	16.86	1.3	–	–
399S6 Bottom	Deposit C	2.22	21.40	41.49	18.88	1.46	14.54	1.55	1.65	1.45
401S1	Deposit D	68.67	25.35	3.77	0.77	0.07	1.37	–	–	–
401S3	Deposit D	43.05	39.40	12.56	2.49	0.35	2.14	–0.7	–	–
401S6	Deposit D	70.86	21.82	5.20	0.74	0.08	1.31	–	–	–
401S8	Deposit D	68.14	25.39	4.38	0.56	0.11	1.42	–	–	–
401S9	Deposit F	13.38	39.66	33.37	4.87	1.69	7.04	0.6	0.5	1.4
404PC1-1(top)	Deposit G	3.52	34.72	17.34	15.88	2.56	25.98	1.6	–	–
404PC1-2	Deposit G	3.67	5.83	9.91	13.05	14.28	53.26	–	–	–
404PC1-3	Deposit G	1.22	4.74	17.64	13.61	7.88	54.90	–	–	–
404PC1-4	Deposit G	1.04	7.72	22.52	13.95	3.05	51.72	–	–	–
404PC1-5 (mud)	Deposit G	0.00	0.07	1.85	4.82	5.04	88.21	–	–	–
404PC1-6	Deposit G	4.50	8.54	8.37	9.82	7.01	61.75	–	–	–
404PC1-7	Deposit G	5.90	7.66	9.41	11.64	14.35	51.04	3.9	–	–
404PC1-8H (hard)	Deposit G	0.55	8.73	21.31	20.63	4.76	44.02	2.95	–	–
404PC1-8 (bottom)	Deposit G	3.49	13.59	21.62	17.16	5.60	38.54	2.65	–	–
404PC2	Deposit G	0.80	37.16	33.52	14.26	2.04	12.22	1.2	1.3	1.6
404S1 top	Deposit G	11.43	19.64	31.43	16.53	2.41	18.57	1.5	–	–
404S1 bottom	Deposit G	23.55	7.85	55.22	5.20	2.03	6.15	1.2	–	–
404S2	Deposit G	43.28	39.98	12.54	2.53	0.43	1.25	–0.7	–	–
404S4-1 (top)	Deposit G	0.12	12.47	41.13	8.85	5.64	31.80	1.9	–	–
404S4-2	Deposit G	0.00	2.47	15.93	28.11	7.41	46.08	3.35	–	–
404S4-3	Deposit G	14.40	15.66	18.54	5.61	3.42	42.37	2.2	–	–
404S4-3 (mud)	Deposit G	3.90	6.36	9.62	3.26	4.69	72.18	–	–	–
404S4-4	Deposit G	2.36	9.16	16.00	11.66	6.43	54.39	–	–	–
404S4-5 (bottom)	Deposit G	3.73	17.29	23.02	9.77	6.04	40.15	2.6	–	–
404S5 top	Deposit G	2.59	26.86	41.95	10.16	2.38	16.05	1.4	1.9	2.0
404S5 middle	Deposit G	3.74	68.75	13.63	4.34	0.45	9.09	0.2	0.52	1.27
404S5 bottom	Deposit G	8.64	56.93	20.08	5.76	0.50	8.10	0.25	0.55	1.35
404S8 nontronite	Deposit H	15.62	22.38	18.48	5.65	3.00	34.87	1.6	–	–
404S8 main	Deposit H	0.77	7.08	34.50	16.51	2.99	38.14	2.5	–	–
404S9	Deposit H	10.32	23.44	32.39	9.73	1.55	22.58	1.4	–	–
404S3	Deposit I	0.00	0.96	11.69	12.27	3.09	71.99	–	–	–
404BC top	Deposit I	0.36	7.53	14.34	47.01	3.99	26.77	2.6	–	–
404BC middle	Deposit I	0.28	1.29	15.76	61.37	2.59	18.72	2.5	–	–
404BC bottom	Deposit I	0.05	12.49	63.94	9.01	2.48	12.02	1.5	1.8	1.0
404S6	Deposit J	60.06	30.57	6.57	0.80	0.11	1.89	–	–	–
341S2	sediment	59.97	30.40	6.68	1.37	0.23	1.34	–	–	–
393 @Marker 10	sediment	7.39	4.02	4.42	49.13	0.70	34.34	2.7	–	–
399S2 Top	sediment	1.67	34.31	40.71	9.95	2.74	10.62	1.2	1.25	1.5
399S2 Bottom	sediment	6.37	34.23	29.12	6.05	1.99	22.24	1.15	–	–
401S4	sediment	0.64	55.39	33.83	5.05	0.25	4.84	0.55	0.62	1.12
401S5	sediment	0.95	68.04	20.52	4.84	0.46	5.19	0.6	0.55	1.15

$\Phi_{50}=Md_\Phi$, $M=(\Phi_{84}+\Phi_{16})/2$, and $\sigma=(\Phi_{84}-\Phi_{16})/2$

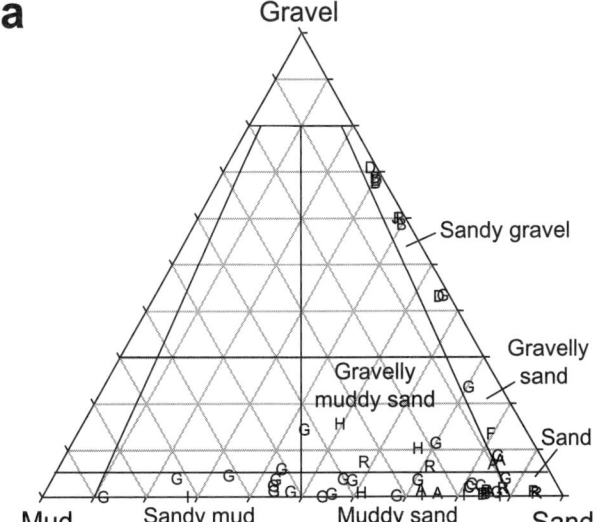

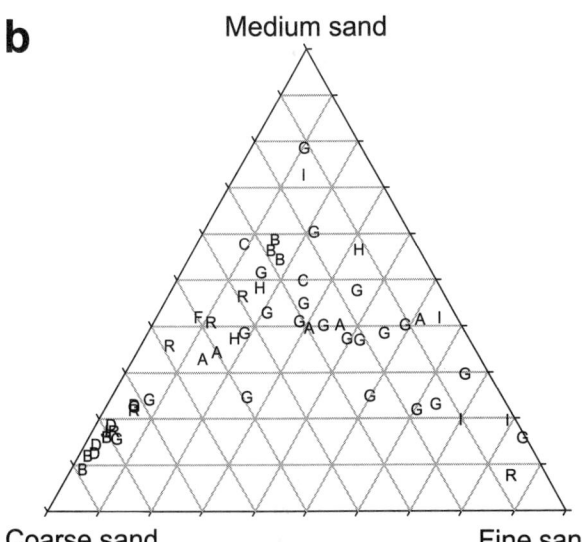

Figure 5. Plots of grain-size data for 54 samples derived by wet sieving. a), gravel comprises the >2 mm fraction, sand is 74 μm to 2 mm, and mud is the <74μm fraction (combining silt and clay). Fields for different textural classes from *Folk* [1966]. b), coarse sand is 500 μm to 2 mm, medium sand is 125 μm to 500 μm, and fine sand is 74 μm to 125 μm. We used 74 μm instead of the usual 62.5 μm to divide very fine sand from coarse silt, simply because of the sieve set used. Samples are identified by deposit, as shown in Figure 2.

bubble-wall or platy shards (limu o Pele) fragments (Figure 6d), folded and contorted or cuspate bubble-wall fragments (Figure 6c and 6e), Pele's hair (Figure 6h), and scoria fragments (Figure 6f and 6g). Juvenile components that are composed of tachylite include spatter (Figure 7a and 7b), and sand-sized lithic fragments with a thin lava skin (Figure 7c); these resemble miniature cored bombs. Fragment vesicularity covers a wide range from dense (<2% vesicles) to scoriaceous (70–85% vesicles). We have subdivided vesicular fragments into weakly vesicular (2–20% vesicles), moderately vesicular (20–40% vesicles, and highly vesicular (40–70% vesicles).

Non-juvenile components include a variety of basaltic fragments: 1) flat to elongate, usually dense, fragments of sideromelane to tachylite with strong conchoidal fractures (note thin edges in Figure 6b), which are most likely glass rinds from pillow lava fragments; 2) microcrystalline basalt fragments and 3) diabasic-textured basalt fragments (Figure 7d and 7e), both most likely from flow interiors; and 4) altered crystalline basalt and altered glass fragments (Figure 7f). Other non-juvenile fragments are hydrothermal barite (Figure 7g) and pyrite (Figure 7h). Broken or euhedral crystal fragments (olivine, plagioclase, and clinopyroxene) that could be either juvenile or non-juvenile comprise an additional component. Other components are biogenic and consist mainly of foraminifers (planktonic and benthic) and sponge spicules. In most samples, the silt and finer particles mainly consist of glass in one form or another. The particle types and compositions record information about eruptive style, as will be discussed below.

4.4. Chemical Compositions of Glass Fragments

Compositional data for glass fragments can help to define whether the particles in a layer are cogenetic and likely deposited during a single magmatic eruption, or if they are variable in composition and could have formed during hydrovolcanic eruptions that produce explosions below the seafloor that eject and mix glass fragments formed partly during previous eruptions. Mixed lithologies could also result from sedimentary processes that transport and deposit fragments formed over either a large area or at various times. In addition, the dissolved volatile contents of the glass fragments can be used to confirm that the fragments actually formed under deep submarine conditions. In particular, we have used S analyses, easily determined by electron microprobe, to characterize the submarine glass fragments and to distinguish a group of fragments formed under subaerial conditions on nearby Kilauea Volcano.

The wide range of glass compositions on Loihi is well documented [*Moore et al.*, 1982; *Garcia et al.*, 1989, 1993, 1995, 1998; *Clague et al.*, 2000; *Dixon and Clague*, 2001]. The vast majority of previous glass analyses from Loihi are of grains in volcanic sands recovered in 1996 and 1997 [*Clague et al.*, 2000]. That data set, mainly from the southern part of the summit and the upper south rift zone, did not include samples of basanite composition, found previously

Table 2. Componentry of Sand Fraction of Volcaniclastic Samples

Sample	Deposit	Fine Ash	Pillow glass	Dense glass	Ves. glass	Scoria	Limu o Pele	Pele's hair	Spatter	Crystalline Lithics	Crystals	Altered basalt	Hydro. frags.	Forams
399S1	A	N	A	C	C	-	C	C	-	-	-	R	-	R
399S4-B,T	A	-	A	C	-	-	R	-	-	A	C-ol+pc	R	-	-
399S5-B,T	A	-	A	C	R	-	R	-	-	A	C-pc/ol	-	-	-
399S3-B,M,T	B	N	-	C	C	C	R	-	C	A	C-ol	C	C	-
399S9-B	B	N	-	C	A	-	R	R	-	C	R-ol	-	-	-
399S9-M	B	N	-	R	A	-	C	R	-	R	R-ol	-	-	-
399S9-T	B	N	-	C	A	-	C	C	-	-	-	-	-	-
399S6-B,T	C	-	-	-	A	-	A	C	-	R	R-ol	-	-	R
401S1	D	-	-	C	A	-	-	-	-	-	C-ol	-	-	-
401S2	D	-	-	-	A	-	-	-	-	-	C-ol	-	-	-
401S3	D	-	-	A	C	-	-	-	-	-	C-ol	-	-	-
401S6	D	-	-	C	A	-	R	-	-	-	C-ol	-	-	-
401S8	D	-	-	-	A	-	-	-	-	-	C-ol	-	-	-
401S9	F	-	-	C	A	-	A	-	-	A	C-ol	C	R	-
404CB	G	-	-	-	-	-	C	C	-	A	C-ol/pc	-	-	-
404PC1-1	G	A	-	C	C	-	C	-	-	A	C-ol/pc	-	R	-
404PC1-2H	G	C	-	-	C	-	C	R	-	C	C-pc	-	-	-
404PC1-3	G	C	-	-	-	-	C	C	R	A	A	-	-	R
404PC1-4H	G	A	-	-	C	-	R	-	-	C	R-pc	-	-	-
404PC1-5	G	A	-	C	C	-	C	-	-	C	C-pc/ol	-	-	-
404PC1-6	G	A	-	A	C	-	C	C	-	C	R-ol	C	-	-
404PC1-7H	G	C	-	-	A	-	-	-	-	C	A-pc	R	-	-
404PC1-8	G	A	-	C	C	R	C	C	-	C	C-pc/ol/px	-	-	R
404PC1-8H	G	A	-	-	C	-	C	-	-	-	-	-	-	-
404PC2	G	C	-	-	A	-	A	C	-	C	C-ol	-	-	C
404S1-B	G	-	-	R	A	-	A	A	R	R	-	-	-	-
404S1-M	G	-	-	C	A	-	R	R	-	R	R-ol	-	-	-
404S1-T	G	-	-	C	A	-	A	C	-	A	C-ol	-	-	-
404S2	G	-	-	C	C	-	-	R	C	A	A-ol	R	-	-
404S3H	G	A	-	-	C	-	-	-	-	C	C-ol/pc	R	-	-
404S4 mix	G	-	-	-	C	-	A	C	-	R	-	-	C	-
404S4-1	G	A	-	C	A	-	A	R	-	C	R-ol	-	-	R
404S4-2	G	A	-	-	C	-	C	-	-	R	-	-	-	C
404S4-3	G	A	-	C	C	-	-	R	R	A	C-ol	-	-	R
404S4-3H	G	A	-	-	C	R	C	-	-	A	c-pc/ol	-	-	-
404S4-4	G	A	-	R	A	-	C	C	R	C	C-ol	R	R	-
404S4-5	G	C	-	C	A	-	A	A	-	R	R-ol	-	R	-
404S5-T	G	-	-	-	A	-	C	R	-	C	R-ol	R	-	R
404S5-B	G	A	-	R	A	-	A	R	-	A	R-ol	-	-	R
404S5-M	G	C	-	-	A	-	A	C	-	C	R-ol	-	-	R
404SB	G	A	-	-	-	C	C	-	-	C	C-ol/pc	R	-	R
404PA	H	C	-	-	A	R	R	-	-	A	C-ol/pc	R	-	-
404S8	H	-	-	A	-	-	R	-	-	A	C-ol/pc	-	-	-
404S8H	H	N	-	-	R	-	R	-	-	C	C-pc	-	-	-
404S9	H	N	-	A	R	R	C	-	R	A	C-ol/pc	R	-	-
404S9H	H	C	-	-	-	-	C	-	-	A	C-ol/pc	-	-	-
404B1-B	I	N	-	C	C	-	A	C	-	R	R-ol/pc	-	-	R
404B1-M	I	C	-	-	-	R	A	C	-	-	C-ol/pc	-	-	C
404B1-T	I	-	-	-	C	R	A	C	-	-	-	-	-	C
404S3	J	A	-	C	A	-	A	R	-	C	R-ol	-	-	R
404S6	J	N	-	-	A	C	-	-	-	-	-	-	-	R
399S2-B	R	-	A	A	-	-	R	-	-	C	R-ol	-	-	-
399S2-T	R	-	A	-	-	-	R	C	-	-	-	-	-	-
401S4	R	-	A	A	C	-	A	R	-	C	R-pc	R	-	R
401S5	R	-	A	C	A	-	C	R	-	R	R-pc	-	-	R

Ves.=vesicular, Hydro.=hydrothermal A=abundant, C=common, R=rare. N=nontronite. pc=plagioclase, ol=olivine, px=pyroxene

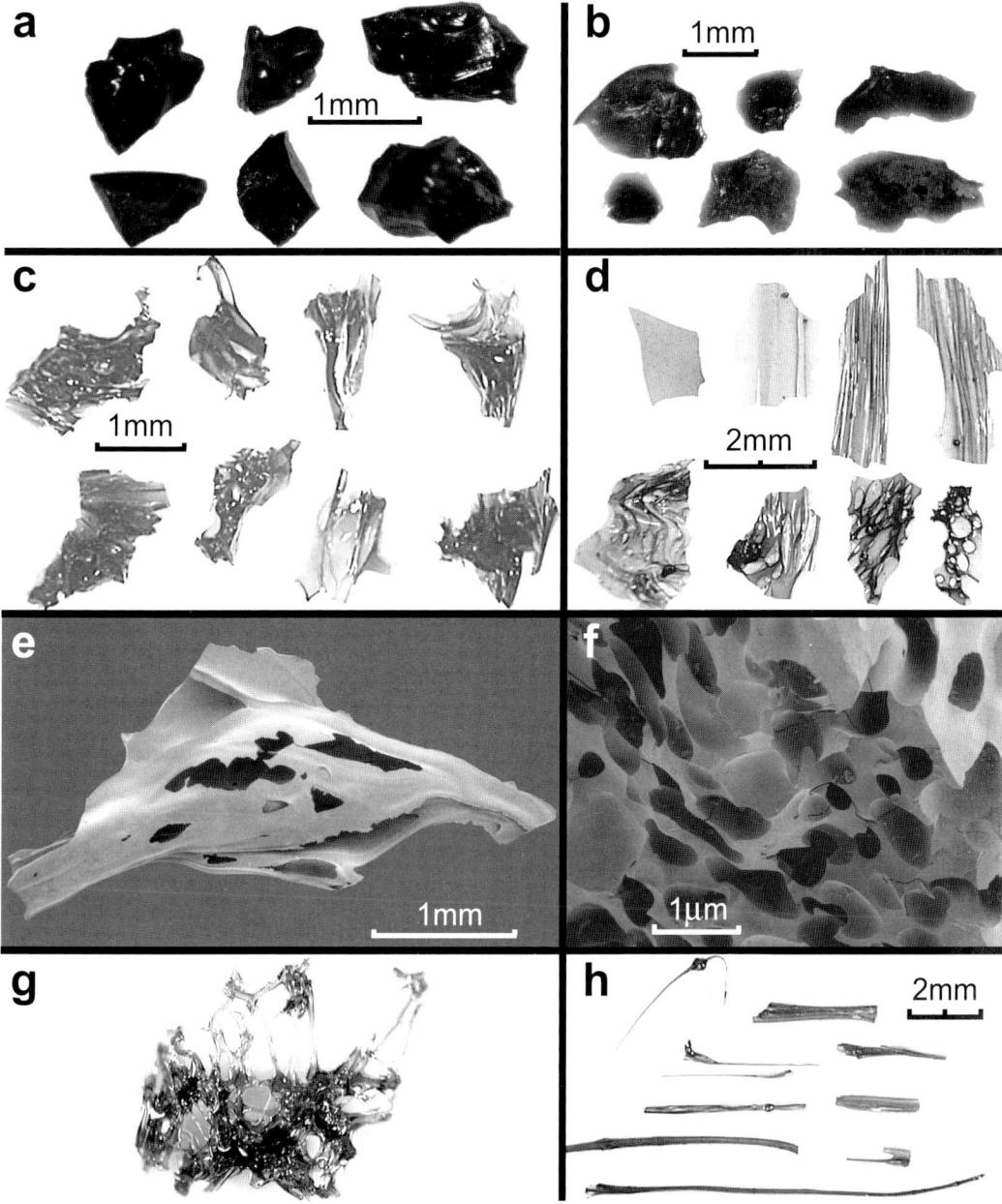

Figure 6. Photographs of various types of juvenile (a, c–h) and non-juvenile (b) fragments that comprise the black sand layers. a) Dense angular glass fragments from sample 399S2-T, rippled sediment. Most have original surface on one side. b) Angular to flat glass to tachylite fragments with conchoidal fractures from sample 399S2-T, rippled sediment. These are most likely fragments of pillow rinds and are found in volcanic sediments only near pillow basalt outcrops. c) Fragments of highly contorted vesicular glass from sample 399S9, deposit C. d) Glass fragments from sample 401S9, deposit F, ranging from lightly curved limu o Pele (upper left) to stretched and contorted limu (upper right). Lower fragments are stretched vesicular glass fragments, with the amount of stretching decreasing from left to right. (e) Scanning electron photomicrograph from sample 404P2, deposit G, showing stretched hollow glass fragment shaped like a spindle bomb. (f) Scanning electron photomicrograph of sample 404SB (hard layer within 404P1 sample), deposit G, showing the highly vesicular character of fragments that form near-vent scoria-like deposits. (g) Glass "taffy" from sample 399S1, deposit A, demonstrating the submarine formation of Pele's hair. (h) Pele's hair fragments from sample 401S9, deposit F.

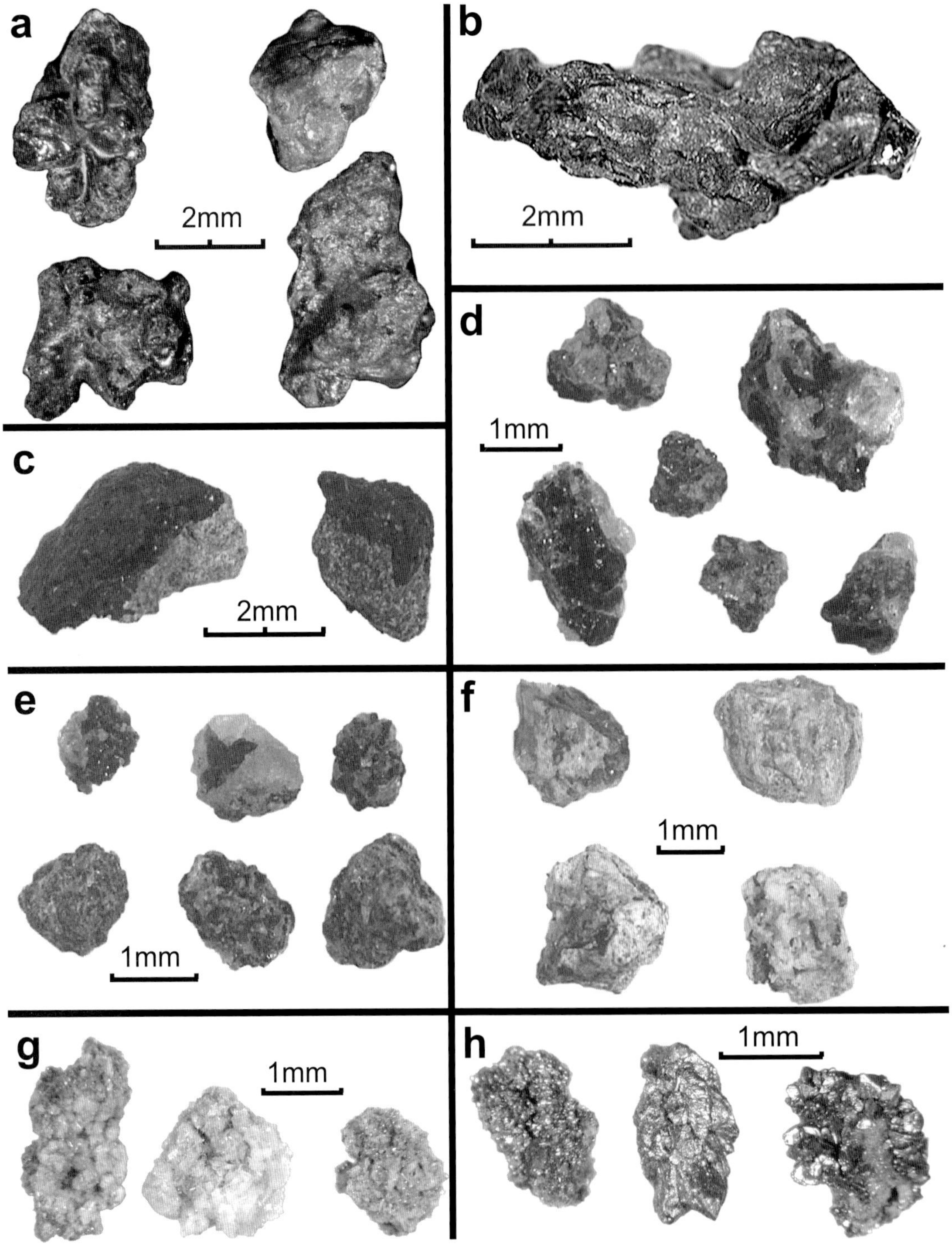

Figure 7. Photographs of various types of juvenile (a–c) and non-juvenile (d–h) fragments recovered from volcaniclastic deposits. a) and b) Small volcanic spatter fragments of tachylite with dull surfaces from sample 404S2, deposit G. (c) Lithic fragments with thin lava rind, similar to miniature cored bombs. Sample 399S3-B, deposit B. d) and e) Coarse-grained basalt fragments from flow interiors; these approach diabase textures. Samples 399S3 and 404S2, deposits C and G, respectively. f) Hydrothermally altered basaltic glass and basalt fragments from sample 399S3-M, deposit C. g) Hydrothermal barite fragments from sample 404S4-B, deposit G. h) Hydrothermal pyrite (some with barite) fragments from sample 404S4-B, deposit G.

as pillow rinds [*Moore et al.*, 1982; *Garcia et al.*, 1995], nor hawaiite [*Garcia et al.*, 1995]. We have analyzed an additional 1036 glass fragments from the 1998 volcaniclastic sediment samples (Table 3) and have combined the results with the previous 659 grain analyses presented in *Clague et al.* [2000].

The new analyses fall within the ranges reported previously except for a group of mugearite analyses from deposit I (analysis 404S3 in Table 3). The range of compositions from each deposit is plotted on alkali-silica diagrams in Plate 1. Deposit A (Plate 1a) contains alkalic basalt limu o Pele and vesicular fragments, and some dense tholeiitic limu o Pele. Deposit B (Plate 1a) includes vesicular alkalic basalt fragments, limu o Pele of tholeiitic, transitional, and alkalic basalt, and scoria and vesicular fragments of hawaiite. Deposit C (Plate 1a) contains vesicular fragments of tholeiitic and alkalic basalt. The rippled sediment collected mainly east of deposit A (Plate 1a) contains mostly tholeiitic limu o Pele and vesicular alkalic basalt fragments. Deposit D (Plate 1b) contains only angular dense tholeiitic fragments and only alkalic limu o Pele was analyzed from deposit F (Plate 1b). Surface sediment collected near deposit F (Plate 1b) includes alkalic and tholeiitic limu o pele fragments.

Deposit G (Plate 1c), the 11-m thick section, has a wide range of glass compositions. Tholeiitic glasses are mainly limu o Pele or dense fragments. Transitional basalt and alkalic basalt are present as limu o Pele and weakly vesicular fragments, whereas hawaiite and basanite fragments are mostly highly vesicular. Weakly vesicular spatter fragments are tholeiitic and alkalic basalt.

Deposit H (Plate 1d) contains scoria erupted subaerially at Kilauea Volcano, vesicular fragments mostly of alkalic composition but with rare tholeiitic ones, and limu o Pele of tholeiitic, transitional, and alkalic composition. Deposit I (Plate 1e) also contains scoria fragments erupted subaerially at Kilauea Volcano (although with different compositions than those at deposit H), limu o Pele of tholeiitic, transitional, and alkalic basalt, as well as of mugearite and vesicular alkalic basalt fragments. Deposit J fragments (Plate 1e) are vesicular transitional basalt.

Only rare sand grains have compositions similar to that of the breccia fragments erupted in early 1996 [*Garcia et al.*, 1998], and little, if any, of the deposits appear to be related to that eruption. Deposit D was sampled by *Garcia et al.* [1995] who determined a low-K_2O tholeiitic composition for the glass (SiO_2=49.45, MgO=7.65, K_2O=0.43, analysis 196-13 in their Table 1) compared with a high-K_2O tholeiite composition from our analyses (analysis 401S6 in Table 3; an average of 19 analyses is SiO_2=48.81, MgO=7.53, and K_2O=0.74). Our sample consists of lapilli–sized glass fragments collected with a sediment scoop from within the deposit whereas the *Garcia et al.* [1995] sample is a rind on a pillow fragment from the volcaniclastic bed that may represent either the underlying pillow flow or a substrate-derived block in the volcaniclastic deposit.

We have also plotted the glass analyses from the 11-m thick section at deposit G as depth plots (Figure 4). These plots demonstrate that tholeiitic, transitional, and alkalic basalt are deposited in random sequence, although there is an overall tendency for more of the shallower younger glass fragments to be alkalic and more of the deeper older glass fragments to be tholeiitic. This observation is contrary to the proposal by *Moore et al.* [1982], based on the thickness of palagonite alteration on dredged samples, that Loihi evolved from predominantly alkalic to predominantly tholeiitic eruptions, or of the concurring assessment by *Garcia et al.* [1993, 1995] based on stratigraphic sequences of lava flows. The volcanic history recorded in the volcaniclastic sections may, however, be biased by inclusion of more alkalic basalt samples due to their greater dispersal, as will be discussed below.

One of the main objectives of the geochemical analyses, to determine if individual layers were compositionally homogeneous, proved extremely difficult to accomplish in practice due to the unconsolidated nature of most of the sampled sections. Particles from overlying layers slump along the face of the outcrop and the sampling was generally not precise enough to allow sampling of discrete layers. The main exceptions to this generalization are the semi-lithified, buff-colored fine-grained layers. In general, these layers contain only minor quantities of glass fragments coarse enough to analyze by microprobe, but they appear, on the basis of low contents of substrate-derived lithic fragments or biogenic components, to be deposited during single magmatic eruptions. Several other subsamples contain only closely related, though not identical, usually moderately vesicular glass fragments that also probably formed during single magmatic eruptions.

Subaerially erupted Kilauea ash falls into the sea above Loihi and is found in the volcaniclastic deposits on the seamount. These Kilauea glass fragments are identified by their low S (<200 ppm) and Cl (<240 ppm) contents (Table 3), consistent with subaerial eruption and with lack of assimilation of seawater Cl [*Kent et al.*, 1999; *Clague and Dixon*, 2000; *Dixon and Clague*, 2001]. *Clague et al.* [2000] observed some fragments with Kilauea compositions and we have found more during this study, although they are restricted to specific samples and are mainly highly vesicular fragments. Kilauea fragments are relatively common in the upper 2/3 of the box-core sample from deposit I, for example. They were also found in deposit H, where they

Table 3. Representative Electron Microprobe Glass Analyses

Sample	399S3-B	399S3-B	399S3-B	399S3-T	399S9-M	401S6	404S8	404S9	404S9H
Fragment	Scoria 12	Scoria 7	Vesicular 7	Vesicular 7	Limu 61	Dense 2	Vesicular 7	Limu 5	Limu 29
Lithology	Hawaiite	Transitional	Transitional	Hawaiite	Alk.Basalt	Tholeiite	Kilauea	Tholeiite	Alk. Basalt
SiO_2	50.58	48.46	48.01	51.19	46.92	48.82	50.56	50.57	46.72
TiO_2	3.51	2.93	2.64	3.38	2.47	2.78	2.63	2.60	2.43
Al_2O_3	14.26	14.36	13.44	14.31	13.29	12.73	13.52	13.58	12.51
FeO	12.42	12.20	11.44	11.85	12.35	11.94	11.44	11.58	12.20
MnO	0.16	0.19	0.18	0.19	0.18	0.17	0.15	0.17	0.15
MgO	3.95	6.05	7.62	3.74	8.20	7.53	7.84	6.82	9.26
CaO	8.45	11.39	12.17	7.88	12.35	11.77	11.14	11.08	12.18
Na_2O	3.75	2.68	2.57	3.76	2.69	2.30	2.23	2.43	2.54
K_2O	1.40	0.63	0.65	1.52	0.61	0.75	0.41	0.38	0.63
P_2O_5	0.68	0.31	0.33	0.72	0.29	0.29	0.28	0.23	0.27
Cl	0.208	0.066	0.048	0.223	0.062	0.046	0.013	0.074	0.084
S	0.092	0.082	0.072	0.072	0.081	0.061	0.014	0.161	0.142
Total	99.46	99.35	99.17	98.83	99.49	99.19	100.23	99.68	99.12
Sample	404PA	404B1-T	404B1-M	404B1-M	404S3	404P2	404S5-1	404S2	404S2
Fragment	Vesicular 6	Scoria 5	Scoria 6	Vesicular 9	Limu 11	Vesicular11	Vesicular15	Vesicular14	Spatter 1
Lithology	Alk.Basalt	Kilauea	Kilauea	Alk. Basalt	Mugearite	Alk.Basalt	Alk.Basalt	Alk.Basalt	Transitional
SiO_2	46.26	49.73	50.94	47.99	53.82	46.28	45.82	45.61	48.12
TiO_2	3.61	2.41	3.13	3.51	2.88	2.81	2.88	2.30	2.70
Al_2O_3	14.73	12.73	13.61	14.94	15.01	13.20	13.29	13.34	13.58
FeO	13.30	11.78	12.05	12.32	10.33	12.10	12.07	12.13	11.46
MnO	0.14	0.14	0.16	0.20	0.17	0.13	0.12	0.15	0.16
MgO	4.71	10.10	6.04	4.85	3.46	7.56	7.68	8.06	7.50
CaO	10.44	10.34	10.67	10.12	7.04	12.98	13.12	13.57	12.22
Na_2O	3.71	2.05	2.37	3.55	4.16	2.81	2.84	2.70	2.46
K_2O	1.12	0.42	0.49	1.03	1.46	0.86	0.83	0.76	0.51
P_2O_5	0.58	0.22	0.30	0.47	0.73	0.35	0.34	0.30	0.27
Cl	0.116	0.022	0.013	0.117	0.143	0.086	0.084	0.062	0.060
S	0.079	0.044	0.020	0.133	0.060	0.114	0.098	0.070	0.188
Total	98.80	99.99	99.79	99.23	99.26	99.28	99.17	99.05	99.23
Sample	404S2	404S1-M	404P1-1	404P1-3	404P1-5	404P1-8	404P1-SB	404S4-2	404S4-4
Fragment	Spatter 12	Vesicular11	Scoria 4	Vesicular15	Vesicular 4	Vesicular13	Scoria 11	Limu 36	Limu 11
Lithology	Alk. Basalt	Alk. Basalt	Hawaiite	Basanitoid	Alk. Basalt	Alk. Basalt	Alk. Basalt	Tholeiite	Alk. Basalt
SiO_2	46.46	46.96	49.46	45.56	46.74	48.21	47.98	50.77	47.11
TiO_2	2.66	2.85	3.13	3.65	2.75	3.37	3.22	2.52	2.56
Al_2O_3	13.03	13.50	15.53	14.14	13.80	13.67	13.68	13.32	13.67
FeO	11.62	11.97	12.48	12.27	11.82	12.02	11.83	11.55	11.97
MnO	0.15	0.13	0.14	0.17	0.12	0.14	0.16	0.14	0.17
MgO	8.61	7.40	4.11	5.29	7.17	6.41	6.49	7.09	6.79
CaO	12.73	12.34	8.93	11.27	13.06	11.69	11.80	10.84	13.11
Na_2O	2.62	2.80	3.55	3.59	2.65	2.80	2.74	2.34	2.57
K_2O	0.54	0.60	0.94	1.25	0.63	0.63	0.66	0.41	0.71
P_2O_5	0.32	0.34	0.40	0.61	0.27	0.33	0.40	0.27	0.33
Cl	0.062	0.075	0.133	0.122	0.079	0.107	0.102	0.057	0.061
S	0.226	0.092	0.086	0.111	0.127	0.093	0.062	0.172	0.091
Total	99.03	99.06	98.89	98.03	99.22	99.47	99.12	99.48	99.14

Under sample: the first 3 digits are the dive number, the next digit is S for sediment scoop, P for pushcore, B for boxcore, or rock designation (PA), next digit is number of sediment scoop, etc, followed by a hyphen and T for top, M for middle, or B for bottom, or a number for position in core with 1 for top increasing downwards. Under fragment: description of the fragment type and the analysis number. Under lithology: Alk. Basalt is alkalic basalt and includes compositions transitional to hawaiite. Samples labeled Kilauea are fall deposits of tholeiite from Kilauea Volcano.

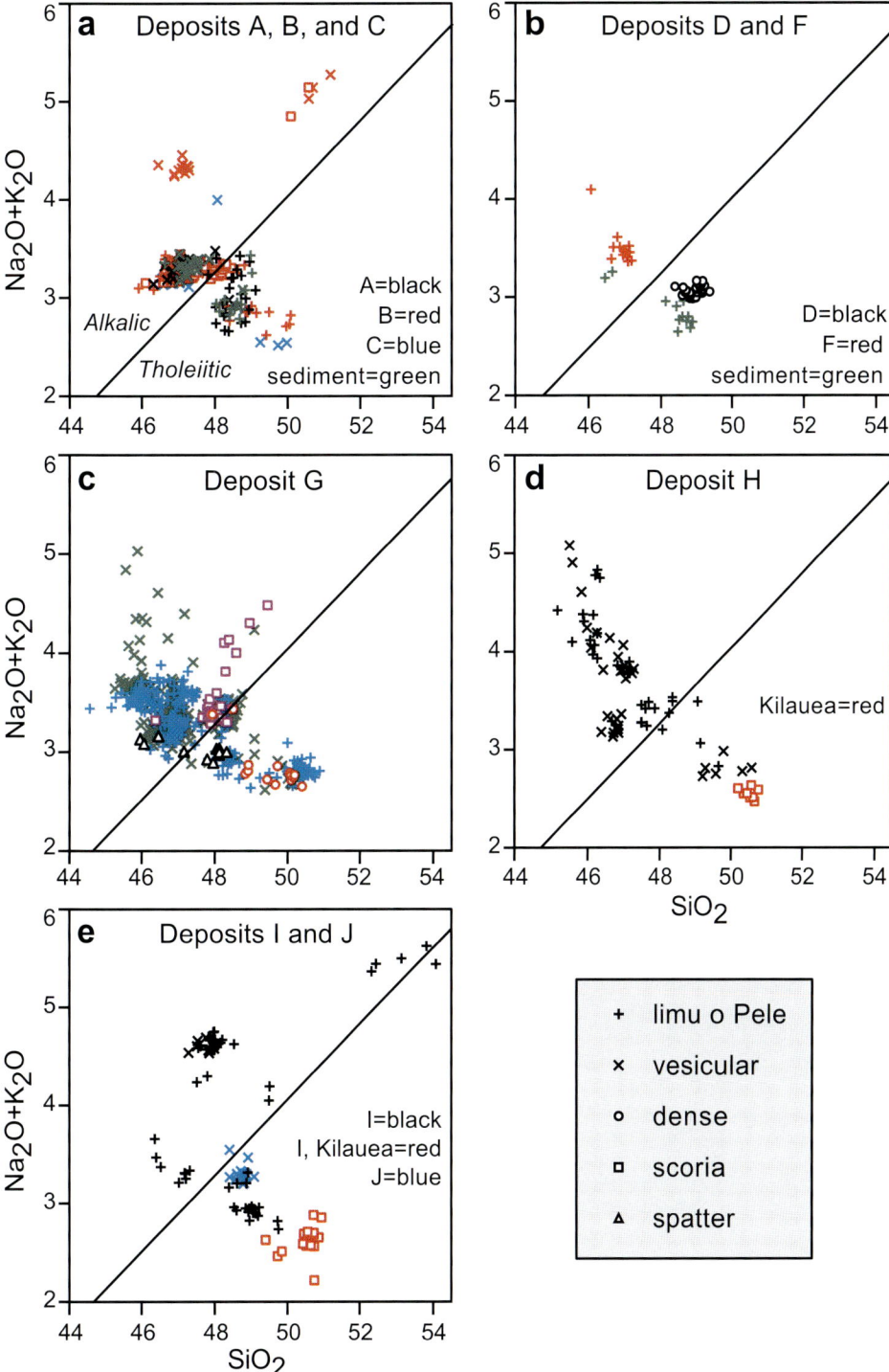

Plate 1. Alkali-silica plots for glass from deposits A, B, C (in a), D, F (in b) G (in c), H (in d) I, and J (in e) as shown in Figure 2, subdivided by deposit and fragment type. The same symbol is used in all plots for the same fragment type. The Macdonald-Katsura boundary separating Hawaiian tholeiitic and alkalic basalt is shown and the fields labeled in part a). The different colors for fragments from deposit G in part c) simply emphasize the different symbols. Compositions of fragments from deposit G in part c) are also shown as a function of stratigraphic position in Figure 4.

occur in sample 404SC8, from near the top of the section, and in semi-lithified sample 404PA of fine-grained volcanic siltstone to mudstone collected near the base of the section by the manipulator. The dispersal of these mm-sized Kilauea fragments to the summit of Loihi was almost certainly through the atmosphere. Loihi's summit is, however, nearly 60 km south of Kilauea's summit and in line with neither the direction of the tradewinds (WSW) nor of the higher-elevation counter-tradewinds (ENE). The ca. 1790 A.D. Keanakakoi Ash on Kilauea is only mm thick near the top of the Hilina Pali only 10 km south-southwest of the summit [*McPhie et al.*, 1990]. Other older ash units are considerably thicker where exposed in this fault scarp, although their thickness mainly reflects surge deposition [*Clague et al.*, 1995, in prep].

Clague et al. [2000] showed that vesicular glass fragments from Loihi commonly had intermediate S contents (commonly 600-900 ppm), consistent with partial loss of S during vesiculation [*Davis et al.*, 1991]. Our new data (glass from deposit G shown in Figure 8) confirm this observation and suggest that some vesicular glass has lost an even greater proportion of its initial S, resulting in S contents as low as 300 ppm, despite eruption at depths >1000 m. The S data therefore documents extensive loss of magmatic volatiles prior to and during eruptions that produced most of the highly vesicular to scoriaceous pyroclasts on Loihi Seamount.

These S contents are, for the most part, still significantly higher than those found in glass fragments from littoral deposits. We analyzed glass fragments from 27 littoral cones on Hawaii and found that the glass averaged only 70 ppm S, with a range from undetectable to 160 ppm. Thus, the concentrations of dissolved volatiles in glass fragments are a key characteristic in assessing the physical setting of basaltic volcaniclastic deposits.

4.5. Constraints on the Age of the Deposits

We do not have reliable control on the age of the deposits on Loihi Seamount. However, the compositions of the scoria fragments derived from Kilauea Volcano provide information that helps constrain their age. At site I, some of the Kilauea glass fragments recovered in the top third of a box core have compositions with up to 10.15 wt% MgO (Table 3). Analyses of glass from all the large ash deposits exposed on Kilauea [*Clague et al.*, 1995, in prep], show that such MgO-rich compositions only occur in the ca.1790 AD Keanakakoi Ash and in the Pahala Ash, erupted between 22.4 [*Easton*, 1987] and 28.8 ka [*Clague et al.*, 1995]. It seems most likely that the glass scoria fragments from near the top of the box core date from ca. 1790 A.D. The same box core contains additional lower-MgO Kilauea scoria fragments in the middle third of the sample. On Kilauea, the next large phreatomagmatic unit below the Keanakakoi Ash Member is the Uwekahuna Ash Member, which erupted between 2,100 and 2,800 B.P. [*Dzurisin et al.*, 1995]. The glass compositions and stratigraphic position below the probable 1790 A.D. fragments are consistent with these scoria fragments being from the Uwekahuna Ash Member. This interpretation of the stratigraphy would suggest that Loihi pyroclastic deposits no more than about 15 cm below the surface could be as old as several thousand years, at least at the northerly site of deposit H. The thick section exposed as deposit G would then be likely to represent several tens to perhaps a hundred thousand years of activity on Loihi or to be located much closer to the vents. A more likely scenario is that both occurrences of Kilauea scoria in the box core are from the ca. 1790 Keanakakoi Ash, which erupted in three main phases [*McPhie et al.*, 1990], perhaps over several hundreds of years [*Swanson et al.*, 1998]. This interpretation suggests that the entire section of volcaniclastic deposits on Loihi is much younger, perhaps as little as several thousands of years for the 11-m section at deposit G, and probably <1 ka for the 1-m section at site I. The Kilauea scoria fragments in deposit H do not have distinctive compositions and only indicate that the deposit is older than 1790 A.D.

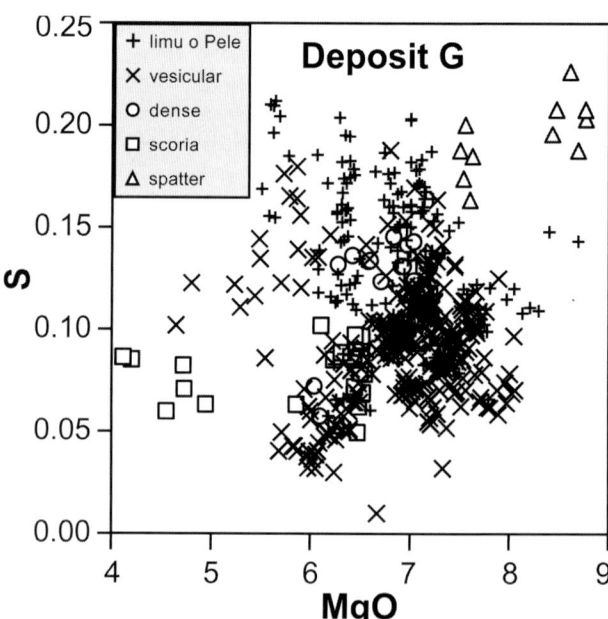

Figure 8. S versus MgO showing that the vesicular and scoriaceous glass (those from deposit G are shown) have lost significant S compared to limu o Pele, spatter, or angular dense glass fragments.

5. DISCUSSION

5.1. Distribution and Dispersal of Volcaniclastic Deposits

The volcaniclastic sediments on Loihi Seamount reflect variations in eruptive process, initial deposition, and subsequent reworking by currents. Secondary sedimentary reworking is widespread, as seen in the common occurrence of ripple marks on the surface of the deposits, and obscures information about eruptive process and initial deposition. *Head and Wilson* [2003] evaluated the production of submarine eruption plumes formed from the interaction of a steadily erupting gas-pyroclast jet and the overlying water. Their analysis suggests that any magmatic H_2O in the jet condenses and that magmatic CO_2 in the jet is dissolved in the entrained water. The consequence of this is that the density of the mixed plume is always greater than that of the surrounding water and the plume of pyroclasts cannot rise more than a few hundred meters above the vent. For conditions similar to those at the summit of Loihi, the plume of pyroclasts ejected from a vent at 1250 m depth will rise only 200 m for a volatile rich magma containing 5.4 wt% CO_2 and 1.9 wt% H_2O. For the case of a volatile foam accumulated under the roof of a magma chamber, the maximum height of the eruption plume is about 500 meters for 75% gas contents in the foam, a vent depth of 1500 m, and the reservoir roof 1000 m below the vent. Thus, the dispersal of the pyroclasts during any submarine magmatic eruption is likely to be limited and the deposits should be localized around the vent. However, depending on the current velocities in the water column and the grain size and density of the pyroclasts, even a plume only a few hundred m tall could disperse pyroclasts over an area larger than the roughly 11 km^2 summit of Loihi. The deposits should thicken and coarsen closer to vents, assuming that sedimentary redistribution has not transported particles long distances.

In several cases on Loihi the deposits are stabilized by precipitated hydrothermal clays that form surface crusts. Sediment like that present at the site of the HUGO junction box, consist almost entirely of clay. Such sediments are not observed in the exposed stratigraphic sections, suggesting that currents may be too strong for settling of fine-gained sediment at these sites or that subsequent to deposition, currents winnow the deposits and remove the fine-grained portion over time. The fine mud would presumably be redeposited on the deep flanks of the volcano, leaving behind the winnowed sandy and gravelly units.

Deposits E, F, and G are almost certainly sections of the same unit, exposed along different faults or parts of faults. The much thinner deposits observed at H and I are likely to be more distal outcrops of the same deposit observed at E, F, and G. Likewise, deposits A, B, and C may be the upper, younger, parts of the thicker deposits observed at E, F, and G. We suspect that this volcaniclastic unit covers a large part of the summit, particularly the east-central part, as no outcrops of lava flows were seen during more than a kilometer of transect on either dive 404 or 401. However, correlation of individual layers from location to location is not possible with the sparse sampling we were able to accomplish during three dives. The outcrops of volcaniclastic units are also unlikely to be similar in different locations since local lava flows could subdivide the section in near-vent locations into separate units and other local pyroclastic units may be intercalated in some locations.

The observations to date show that volcaniclastic units are widespread on Loihi and that they are well exposed along caldera-bounding faults, which offset the deposits. The analyses of glass fragments from these units demonstrate that numerous geochemical compositions are represented in the deposits, which we presume represent different eruptions. However, despite our ability to recognize that many eruptions of different types contributed to the deposits observed, we still have little information about the locations of the eruptive vents, with the exception of two scoria-spatter deposits (deposits D and J) that are clearly near-vent deposits. Glass grains with the same compositions were not encountered in the other deposits, however, suggesting limited dispersal from these vents. Kilauea scoria fragments were found deeper in the section at deposit H than at the more northerly deposit I location. Assuming that the scoria at both sites represents the same Kilauea eruption, most probably the 1790 A.D. eruption, then the Loihi deposits increase in thickness to the south.

5.2. Evidence for Pyroclastic Eruptions

Pyroclastic magmatic eruptions of basaltic magma are typically strombolian, hawaiian, or gradational between the two. Strombolian eruptions are characterized by discharge of discrete bursts of magmatic volatiles which disrupt the top of the magma column and eject incandescent pyroclasts [*Blackburn et al.*, 1976; *Ripepe et al.*, 1993; *Vergniolle and Mangan*, 2000]. Hawaiian eruptions are similar to strombolian eruptions in being of relatively low energy, but the bursts of released magmatic volatiles sustain continuous lava fountains [*Head and Wilson*, 1987, 1989; *Parfitt and Wilson*, 1995; *Vergniolle and Mangan*, 2000]. In subaerial eruptions of both types, as well as in explosive littoral activity, lava fragments into bubbly molten clots, most commonly forming spatter, Pele's tears, Pele's hair, and scoria, that accumulate near the vent to form spatter or cinder cones. In general, strombolian eruptions produce significant pyroclastic deposits whereas hawaiian fountains produce mainly rootless flows of coalesced spatter. The pyroclastic eruptive products of hawaiian and strombolian eruptions overlap, however, so determination of the eruptive style from limited sampling of the deposits is difficult.

On Loihi Seamount, fragments indicative of pyroclastic activity include glassy spatter fragments and small botryoidal tachylitic spatter grains with dull surfaces (Figure 5a and 5b). In two cases, we have identified deposits of spatter-like fragments of angular broken vesicular lava (deposits D, Figure 3b, and J) as having been formed during submarine pyroclastic activity. Although the deposits appear similar to subaerial cinder deposits with spatter-shaped clasts, many of the fragments are broken glassy spatter with lower vesicularities than typical subaerial scoria (Figure 5f). Spatter fragments in these two deposits have transitional to tholeiitic compositions (Figure 8 and Plate 1).

Most of the fine-grained, light-colored layers in the various deposits on Loihi have a few percent sub-millimeter-sized highly vesicular to scoriaceous glass fragments in a matrix of silt and clay-sized glass and altered glass fragments. These layers are interpreted to be distal aqueous fall deposits [*Cashman and Fiske*, 1991] from submarine pyroclastic eruptions on Loihi Seamount. Other layers consist almost entirely of well-sorted sand-sized particles of moderately to highly vesicular glass fragments of uniform composition that are interpreted to be proximal aqueous fall deposits of submarine pyroclastic eruptions on Loihi Seamount. These layers are impoverished in fines, which were either transported away from the vent in suspension in the water column or were winnowed by currents following deposition. Other particle types that form during pyroclastic eruptions are Pele's hair (Figure 6h), taffy-like fragments (Figure 6g), and vesicular contorted limu-like fragments (Figure 6c and lower right of Figure 6d). These fragment types consist predominantly of alkalic basalt, which has a higher initial volatile inventory [*Dixon and Clague*, 2001] than tholeiitic and transitional basalt. Loihi volcaniclastic deposits formed on nearly horizontal surfaces and there is no evidence for deposition by eruption-fed density currents.

Nearly all the layered volcaniclastic and rippled basaltic sediments from Loihi Seamount contain common to abundant slightly curved limu o Pele fragments (Figure 6d, upper left), dominantly of tholeiitic composition [*Clague et al.*, 2000]. Such fragments have been proposed to form during mild steam explosions when molten lava trapped seawater [*Clague et al.*, 2000; *Maicher and White*, 2002] in a process analogous to that described by *Hon et al.* [1988] and *Mattox and Mangan* [1997] from observations of lava flowing into the sea on Kilauea Volcano. *Clague et al.* [this volume] reinterpreted submarine limu o Pele produced deeper than the critical point of seawater (>2936 m), as forming during mild strombolian eruptions. The supercritical limu o Pele from the Gorda Ridge [*Clague et al.*, this volume] and that from Loihi Seamount [*Clague et al.*, 2000] are physically identical, which leads us to propose that the limu o Pele on Loihi Seamount also forms during strombolian eruptions. Limu o Pele may be the most common pyroclast formed during submarine strombolian eruptions [*Clague et al.*, this volume]. Spatter is the other type of pyroclast that probably forms during strombolian activity, as discussed below.

Strombolian activity is generally viewed as arising from a gas phase separated from the melt into large bubbles prior to arrival of the melt/large-bubble mixture in the conduit. One way to accomplish this is by coalescence of exsolved and growing gas bubbles in a slowly rising magma column [*Parfitt and Wilson*, 1995]. Another way to produce large coalesced bubbles is if bubbles nucleate, grow, and coalesce during storage within the volcanic edifice [*Bottinga and Javoy*, 1989; *Vergniolle and Jaupart*, 1990], in which case the narrow dike widths and slow rise speeds calculated [*Head and Wilson*, 2003] to allow for bubble coalescence may not be required for strombolian activity. During subaerial strombolian eruptions, additional bubbles filled with H_2O and S form in the shallow conduit so pyroclasts are highly vesicular. However, in the submarine environment, ambient pressure prevents degassing of H_2O and S in the shallow conduit and pyroclasts can be dense glass fragments ejected by centimeter to decimeter CO_2-filled bubbles that rise through the magma in the vent and burst through the surface of low-level fountains. In the submarine environment, such processes seem the only magmatic method to produce dense or slightly vesicular pyroclastic fragments, such as the majority of limu o Pele and spatter fragments from Loihi. Strombolian activity that produces dense pyroclasts, therefore, appears to characterize eruptions of tholeiitic and transitional basalt with low initial volatile contents. More vesicular pyroclasts could be produced during hawaiian or strombolian activity from magmas with higher initial volatile contents, such as alkalic magmas.

Hawaiian-style eruptions require exsolution of large amounts of magmatic gas, particularly at high eruption pressures, to allow for explosive disruption of the magma. Such degassing is well documented for subaerial tholeiitic eruptions at Kilauea [*Gerlach*, 1986; *Vergniolle and Jaupart*, 1986, 1990; *Hilton et al.*, 1997], and evidence of this subaerial degassing of volatiles is also recorded in submarine hybrid lavas on Kilauea [*Dixon et al.*, 1991]. Hawaiian alkalic magmas initially contain more magmatic volatiles and also degas large amounts of magmatic gas [*Dixon*, 1997; *Dixon et al.*, 1997; *Hilton et al.*, 1998; *Dixon and Clague*, 2001]. However, the calculated initial volatile contents of alkalic basalts from Loihi are roughly a factor of 2–3 below the gas contents required to sustain hawaiian fountains at pressures of 10 Mpa, roughly the pressure at Loihi's summit, according to calculations by *Head and Wilson* [2003]. Their calculations suggest minimum gas contents for hawaiian eruptions at 10 Mpa of 2.3% pure water or 4.2% pure CO_2, or some combination of the two. *Dixon* [1997] showed that most of the exsolved gas phase consists of CO_2 in all but the most strongly alkalic basalts. If hawaiian eruptions do occur on Loihi, the additional magmatic volatiles to drive such eruptions must accumulate in the magmas prior to eruption, most likely near the top of the magma conduit as a foam during shallow storage as proposed previously by *Vergniolle and Jaupart* [1990]. Hawaiian activity will produce vesicular to scoriaceous pyroclasts, like many of the Loihi pyroclasts, but as noted above, so could strombolian activity if the magma is alkalic and contains a large inventory of volatiles.

5.3. Evidence for Hydroclastic Eruptions

Phreatic eruptions are steam-driven and result from interaction of water and hot rock, in the absence of magma or molten lava [e.g., *Barberi et al.*, 1992; *Germanovich and Lowell*, 1995; *Mastin*, 1995]. Eruption products are lithic and crystal fragments that gradationally decrease in size and thickness away from the vent. The deposits on Loihi contain a surprising array of such lithic fragments including the crystalline interiors of lava flows (Figure 7d and e), altered lava and glass fragments (Figure 7f), and, most unexpectedly, fragments of coarse-grained pyrite and barite from hydrothermal stockwork (Figures 7g and 7h). The hydrothermal fragments occur at several horizons in the volcaniclastic deposits, suggesting that phreatic explosions repeatedly disrupted hydrothermal circulation systems that overlie the top of magma conduits, as seen at Kilauea [*Almendros et al.*, 2001]. The fine grain sizes (mostly sand- and silt-sized) of these particles suggest these are distal facies.

Phreatomagmatic eruptions result from the interaction of water and magma or molten lava [e.g., *Colgate and Sigurgeirsson*, 1973; *Wohletz*, 1986, *Morrissey et al.*, 2000]. The only clear evidence that phreatomagmatic eruptions occurred on Loihi is the presence of a few tiny lithic fragments coated by tachylite (Figure 7c). Deposits from phreatomagmtic eruptions could constitute a large fraction of the observed sections on Loihi Seamount. Our sampling techniques and post-depositional reworking by currents, however, mix fragments from multiple layers, making it impossible for us to determine if the layers rich in lithics also contain juvenile glass fragments.

The lithic fragments might have been produced during vulcanian eruptions [*Morrissey and Mastin*, 2000], although most vulcanian eruptions occur at volcanoes characterized by viscous magmas unlike the fluid basalt erupted at Loihi Seamount. Phreatic and phreatomagmatic eruptions also seem more likely, given the ready availability of seawater. These eruptions took place no shallower than 1000 m depth and probably closer to 1300-1400 m, making them the deepest high-energy eruptions of this type known.

Kilauea Volcano is the only volcano in Hawaii with historic explosive eruptions that might serve as a model for similar eruptions on Loihi. At Kilauea, phreatic and phreatomagmatic eruptions are thought to be closely linked to infrequent collapses forming either pit craters or calderas. The last caldera-forming collapse occurred between 1470 A.D and 1790 A.D [*Clague et al.*, 1999], and the previous one predated about 1290 A.D. However, smaller collapses that form pit craters, the most recent in 1924 A.D., occur more frequently, perhaps on the order of every few hundred years.

Kilauea and Loihi both have summit calderas, of roughly the same size, and both have several large pit craters punctuating their summit regions. At Kilauea, the Kilauea Iki and Halema'uma'u pit craters are similar in size to the East Pit, West Pit, and recently formed Pele's Pit on Loihi. Not only are the summit structures of the two volcanoes similar, but the timeframe for explosive eruptions also appears to be on the order of hundreds to a few thousands of years. These similarities lead us to hypothesize that phreatic and phreatomagmatic eruptions on Loihi Seamount are probably also closely linked to caldera or pit crater formation.

The most recent collapse event in Hawaii took place in the summer of 1996, when a large swarm of earthquakes occurred on Loihi Seamount [*Loihi Science Team*, 1997]. A rapid response cruise was organized that made a new bathymetric map of Loihi and supported dives with the *Pisces V* submersible. The new map, when compared to previous maps, revealed that the southernmost part of the summit had collapsed into a new pit crater. The crater was about 600 m across and the floor of the crater was about 300 m below the original surface. The collapse downdropped roughly 0.17 km^3 of rock [*Davis and Clague*, 1998]. Scoria-like blocks of tholeiitic basalt found in the vicinity of the new pit erupted several months prior to the collapse [*Garcia et al.*, 1998]. Thus, the collapse apparently followed magma withdrawal from the conduit system.

Sediment samples recovered from the walls and floor of Pele's Pit include coarse-grained pyrite, pyrrhotite, wurtzite, and barite crystals inferred to be from the stockwork beneath the pre-existing low-temperature hydrothermal vents at Pele's vents [*Davis and Clague*, 1998]. The fragments are similar to those found in deposits B, F, and G. The 1996 events also injected light-gray clay and microbial flock into the water, dramatically decreasing visibility in and near the new pit. The sediment deposited in the pit contained abundant glass fragments with a range of compositions as well as lithic fragments. The observations suggest that phreatic and possibly phreatomagmatic eruption(s) accompanied the collapse of Pele's Pit and that these eruptions disrupted the subsurface hydrothermal stockwork. The depth of the hydrothermal fragments in the stratigraphic sections, particularly at deposit G, suggests that they were not emplaced during the 1996 events, but during previous phreatic and phreatomagmatic eruptions. We suspect that these earlier explosive eruptions also accompanied collapse that formed pit craters on the summit of Loihi.

Davis and Clague [1998] compared the Pele's Pit collapse event to the 1924 collapse in Halema'uma'u Crater at Kilauea's summit, which had a series of phreatic eruptions following magma withdrawal and the collapse of the crater floor. The 1924 eruption lasted 18 days and was characterized by 0 to 13 discrete explosions per day that lasted from a few minutes to 7 hours [*Decker and Christiansen*, 1984]. The most violent explosions occurred on the ninth day of the eruption. The ca. 1790 A.D. phreatomagmatic eruption on Kilauea apparently followed magma withdrawal and collapse of the caldera floor, as the ash drapes the caldera walls [*McPhie et al.*, 1990]. Availability of water to drive such eruptions is a limiting factor for Kilauea explosive eruptions and has led *Mastin* [1997] to propose that water stored in a crater lake drained into the hot conduit, leading to explosive eruptions driven mainly by steam.

In the submarine environment, as at Loihi, availability of water is not a limiting factor. Caldera or pit crater collapse following magma withdrawal led to phreatic and phreatomagmatic, eruptions at the summit of Loihi, most recently in 1996. A likely scenario is that the top of the magma conduit system, probably located about 1 km below the surface by analogy with Halema'uma'u crater on Kilauea [*Ohminato et al.*, 1998], drained during down-rift eruptions or intrusions. The magma withdrawal led to collapse of the overlying lid of lava and hydrothermal stockwork, with ingress of seawater that resulted in phreatic and phreatomagmatic eruptions from within and below the collapsing pits, even at or below 1350 m water depth.

6. CONCLUSIONS

1) Sections of layered volcaniclastic deposits up to 11 m thick are exposed on inward-facing caldera bounding faults on the eastern part of Loihi Seamount's summit platform.

2) Layers in the deposits include well-sorted sands as well as clay and silt-rich ones. The sandy units were probably winnowed by currents and the fines removed by currents and deposited on the deep flanks of Loihi.

3) The volcaniclastic sands consist of a variety of glass fragments including dense angular fragments, dense limu o Pele fragments, folded vesicular limu o Pele fragments, highly vesicular fragments, spatter fragments, and small irregular bombs. Lithic fragments, common in many layers, include altered basalt and altered glass fragments, coarse crystalline basalt fragments, tiny lava-coated fragments, and hydrothermal pyrite and barite fragments. Broken crystals of olivine, plagioclase, and rare pyroxene are also present, though rarely abundant.

4) The different fragment types suggest a variety of pyroclastic and hydroclastic eruption styles on Loihi's summit, including strombolian (limu o Pele and spatter fragments) and probably hawaiian (highly vesicular fragments and Pele's hair) to phreatic (range of lithic fragments) and phreatomagmatic (lava coated lithic fragments).

5) We infer that eruptions of tholeiitic basalt are most commonly strombolian whereas those of alkalic basalt appear to be hawaiian, reflecting the higher initial volatile contents of the latter. The hawaiian eruptions are only possible if magmatic volatiles are accumulated as a separated volatile phase (i.e., bubbles) in a magma storage zone prior to eruption.

6) The phreatic and phreatomagmatic eruptions are related to summit collapse events that form pit craters or perhaps to larger caldera collapses that formed the flat summit platform. Magma withdrawal due to eruptions or intrusions downrift led to collapse of the pits, ingress of seawater, and susequent phreatic and phreatomagmatic explosions that deposited lithic-rich sand and gravel, apparently as far as 2–3 km from the vent. These violent steam explosions occur at depths at least as great as 1350 m, the depth of the bottom of the deepest pit crater, and probably somewhat deeper within the collapsed conduits beneath the pits.

7) Loihi Seamount is a natural laboratory to study submarine eruptive processes and deposition and reworking of volcaniclastic deposits. Future work may determine the ages of some of the volcaniclastic deposits using AMS radiocarbon dating on the rare foraminifers in some of the layers. Deposits nearer to the rims of the pit craters should be explored and sampled to determine the distribution of grain-size and glass compositions

that could constrain the emplacement style of the material ejected during the explosive events. More detailed vertical coring of the volcaniclastic sediments on Loihi may define the frequency and magnitude of such eruptions, particularly if the sampling were closer to the three pit craters, which are the most likely source vents.

8) The 1996 seismic crisis at Loihi Seamount resulted not only in formation of a new pit crater, but in phreatic and perhaps phreatomagmatic explosions that may still have been ongoing during the rapid response cruise. Rapid response using manned submersibles to future seismic/volcanic crises on relatively shallow submarine volcanoes with summit collapse structures should be carefully evaluated to minimize risk.

Acknowledgments. The authors thank the ship crew of the R/V Ka`imikai-o-Kanoloa and the *Pisces V* submersible team, particularly chief pilot Terry Kerby, for their support during the dive program in 1998. Jenny Paduan skillfully assisted with the production of the graphics used in the paper. The manuscript was improved by helpful and insightful reviews by Kathy Cashman, Peter Kokelaar, and editor James White. The project was supported by NOAA's Hawaii Undersea Research Program. Support from the David and Lucile Packard Foundation to D. Clague, from NSF to R. Batiza, and from NASA Grant NAG5-10690 to J. Head III are gratefully acknowledged.

REFERENCES

Almendros, J., B. Chouet, and P. Dawson, Spatial extent of a hydrothermal system at Kilauea Volcano, Hawaii, determined from array analyses of shallow long-period seismicity 2. Results, *J. Geophys. Res., 106*, 13,581-13,597, 2001.

Barberi, F., A. Bertagnini, P. Landi, and C. Principe, A review of phreatic eruptions and their precursors, *J. Volcanol. Geotherm. Res., 52*, 231-246, 1992.

Batiza, R., N. Becker, D. Bercovici, T. Coleman, T. Gorman, J.W. Head III, L. Holloway, J. Karsten, A. Kelly, L.P. Keszthelyi, D. maicher, W. Mueller, J. Muller, L. Norby, J. Paduan, G. Parker, L. Proctor, D.Stakes, and J.D.L. White, New evidence from Alvin for the origin of deep-sea eruptive hyaloclastite on Seamount 6: Cocos plate, 12°43'N, *EOS Trans. Am. Geophys. Union, 77*, 319, 1996.

Batiza, R., D.J. Fornari, D.A. Vanko, and P. Lonsdale, Craters, calderas, and hyaloclastites on young Pacific seamounts, *J. Geophys. Res., 89*, 8371-8390, 1984.

Binard, N., R. Hekinian, J.L. Cheminee, and P. Stoffers, Styles of eruptive activity on intraplate volcanoes in the Society and Austral hot spots regions: bathymetry, petrology and submersible observations, *J. Geophys. Res., 97*, 13,999-14,015, 1992.

Blackburn, E.A., L. Wilson, and R.S.J. Sparks, Mechanisms and dynamics of strombolian activity, *J. Geol. Soc. London, 132*, 429-440, 1976.

Bottinga, Y., and M. Javoy, Mid-ocean ridge basalt degassing: bubble nucleation, *J. Geophys. Res., 95*, 5125-5131, 1989.

Caress, D.W., and D.N. Chayes, Optimal navigation adjustment for poorly navigated swath bathymetry surveys, *EOS Trans. Am. Geophys. Union, 81*, F1096, 2000.

Cas, R.A.F., and J.V. Wright, Volcanic successions: modern and ancient, Allen and Unwin, London, 528 p, 1987.

Cashman, K.V., and R.S. Fiske, Fallout of pyroclastic debris from submarine volcanic eruptions, *Science, 253*, 275-280, 1991.

Cashman, K.V., B. Sturtevat, P. Papale, and O. Navon, Magmatic fragmentation, in *Encyclopedia of Volcanoes*, ed. by H. Sigurdsson, Academic Press, San Diego, 421-430, 2000.

Chadwick, W.W. Jr., J. Smith, J.G. Moore, D.A. Clague, M.O. Garcia, and C. Fox, Bathymetry of the south flank of Kilauea Volcano, Hawaii, *U.S. Geol. Surv. Misc. Field Map, MF-2231*, 1993.

Clague, D.A., Beeson, M.H., Denlinger, R.P., and Mastin, L.G., Ancient ash deposits and calderas at Kilauea Volcano, *EOS Trans. Am. Geophys. Union., 76*, F666, 1995.

Clague, D.A., A.S. Davis, J.L. Bischoff, J.E. Dixon, and R. Geyer, Lava bubble-wall fragments formed by submarine hydrovolcanic explosions on Loihi Seamount and Kilauea Volcano, *Bull. Volcanol., 61*, 437-449, 2000.

Clague, D.A., and J.E. Dixon, Extrinsic controls on the evolution of Hawaiian ocean island volcanoes, *Geochem. Geophys. Geosystems, 1*, Paper number 1999GC000023, 12 p., 2000.

Clague, D.A., J.T. Hagstrum, D.E. Champion, and M.H. Beeson, Kilauea summit overflows: their ages and distribution in the Puna District, Hawaii, *Bull. Volcanol., 61,* 363-381, 1999.

Clague, D.A., R.T. Holcomb, J.M. Sinton, R.S. Detrick, and M.E. Torresan, Pliocene and Pleistocene alkalic flood basalts on the seafloor north of the Hawaiian Islands, *Earth Planet. Sci. Lett., 98*, 175-191, 1990.

Colgate, S.A., and T. Sigurgeirsson, Dynamic mixing of water and lava, *Nature, 244*, 552-555, 1973.

Davis, A.S., and D.A. Clague, Changes in the hydrothermal system at Loihi Seamount after the formation of Pele's pit in 1996, *Geology, 26*, 399-402, 1998.

Davis, A.S., D.A. Clague, M.S. Schulz, and J.R. Hein, Low sulfur content in submarine lavas: an unreliable indicator of subaerial eruption, *Geology, 19*, 750-753, 1991.

Decker, R.W., and R.L. Christiansen, Explosive eruptions at Kilauea Volcano, Hawaii, *in* Explosive volcanism: inception, evolution, and hazards, Natl. Res. Council, Nat. Acad. Sci., Nat. Acad. Press, Washington, D.C., 82, 122-132, 1984.

Dixon, J.E., Degassing of alkalic basalts, *Am. Mineral., 82*, 368-378, 1997.

Dixon, J.E., and D.A. Clague, Volatiles in basaltic glasses from Loihi Seamount, Hawaii: evidence for a relatively dry plume component, *J. Petrol., 42*, 627-654, 2001.

Dixon, J.E., D.A. Clague, and E.M. Stolper, Degassing history of water, sulfur, and carbon in submarine lavas from Kilauea Volcano, Hawaii, *J. Geo. 99*, 371-394, 1991.

Dixon, J.E., D.A. Clague, P. Wallace, and R. Poreda, Volatiles in alkalic basalts from the North Arch Volcanic Field, Hawaii:

extensive degassing of deep submarine-erupted alkalic series lavas, *J. Petrol.*, 38, 911-938, 1997.

Dullforce, T.A., D.J. Buchanan, and R.S. Peckover, Self-triggering of small-scale fuel-coolant interactions: I. Experiments, *J. Physics D: Applied Physics*, 9, 1295-1303, 1976.

Dzurisin, D., J.P. Lockwood, T.C. Casadevall, and M. Rubin, The Uwekahuna Ash member of the Puna Basalt: product of violent phreatomagmatic eruptions at Kilauea volcano, Hawaii, between 2800 and 2100 14C years ago, *J. Volcanol. Geotherm. Res.*, 66, 163-184, 1995.

Easton, R.M., Stratigraphy of Kilauea Volcano, *U.S. Geol. Surv. Prof. Paper*, 1350, 243-260, 1987.

Fisher, R.V., and H.-U. Schmincke, Pyroclastic rocks, Springer-Verlag, Berlin, 472 p., 1984.

Folk, R.L., A review of grain size parameters, *Sedimentology*, 6, 73-93, 1966.

Fornari, D.J., M.O. Garcia, R.C. Tyce, and D.G. Gallo, Morphology and structure of Loihi seamount based on multibeam sonar mapping, *J. Geophys. Res.*, 93, 15,277-15,238, 1988.

Fouquet, Y., J.-P. Eissen, H. Ondreas, R. Batiza, and L. Danyushevsky, Deep-water (~1700m) explosive basalt volcanism at the Mid-Atlantic Ridge?, *Terra Nova*, 10, 280-286, 1998.

Froehlich, G., Interaction experiments between water and hot melts in entrapment and stratification configurations, *Chem. Geol.*, 62, 137-147, 1997.

Garcia, M.O., D.J.P. Foss, H.B. West, and J.J. Mahoney, Geochemical and isotopic evolution of Loihi Volcano, Hawaii, *J. Petrol.*, 36, 1647-1674, 1995.

Garcia, M.O., B.A. Jorgenson, and J.J. Mahoney, An evaluation of temporal geochemical evolution of Loihi summit lavas: results from *Alvin* submersible dives, *J. Geophys. Res.*, 98, 537-550, 1993.

Garcia, M.O., D.W. Muenow, K.E. Aggrey, and J.R. O'Neil, Major element, volatile, and stable isotope geochemistry of Hawaiian submarine tholeiitic glasses, *J. Geophys. Res.*, 94, 10,525-10,538, 1989.

Garcia, M.O., K.H. Rubin, M.D. Norman, J.M. Rhodes, D.W. Graham, D.W. Muenow, and K. Spencer, Petrology and geochemistry of basalt breccia from the 1996 earthquake swarm of Loihi seamount, Hawaii; magmatic histroy of its 1996 eruption, *Bull. Volcanol.*, 59, 577-592, 1998.

Gerlach, T.M., Exsolution of H_2O, CO_2, and S during eruptive episodes at Kilauea Volcano, Hawaii, *J. Geophys. Res.*, 91, 12,177-12185, 1986.

Germanovich, L.N., and R.P. Lowell, The mechanism of phreatic eruptions, *J. Geophys. Res.*, 100, 8417-8434, 1995.

Gill, J., P. Torssander, H. LaPierre, R. Taylor, K. Kaiho, M. Koyama, M. Kusakabe, J. Aitchison, S. Cisowski, K. Dadey, K. Fujioka, A. Klaus, M. Lovell, K. Marsaglia, P. Pezard, B. Taylor, and K. Tazaki, Explosive deep water basalt in the Sumisu backarc rift, *Science*, 248, 1214-1217, 1990.

Head, J.W., III, and L. Wilson, Lava fountain heights at Pu'u O'o, Kilauea, Hawaii: Indicators of amount and variations of exsolved magma volatiles, *J. Geophys. Res.*, 92, 13,715-13,719, 1987.

Head, J.W., III, and L. Wilson, Basaltic pyroclastic eruptions: Influence of gas-release patterns and volume fluxes on fountain structure and the formation of cinder cones, spatter cones, rootless cones, lava ponds and lava flows, *J. Volcanol. Geotherm. Res.*, 37, 261-271, 1989.

Head, J.W., III, and L. Wilson, Deep submarine pyroclastic eruptions: Theory and predicted landforms and deposits, *J. Volcano. Geotherm. Res.*, 121, 155-193, 2002.

Heiken, G.H., and K.H. Wohletz, *Volcanic ash*, University of California Press, Berkeley, 246 p., 1986.

Heiken, G.H., and K.H. Wohletz, Fragmentation processes in explosive volcanic eruptions, in Sedimentation in volcanic settings, R.V. Fisher and G.A. Smith, eds., *SEPM Spec. Pub.*, 45, 19-26, 1991.

Hilton, D.R., G.M. McMurtry, and R. Kreulen, Evidence for extensive degassing of the Hawaiian mantle plume from helium-carbon relationships at Kilauea Volcano, *Geophys. Res. Lett.*, 24, 3065-3068, 1997.

Hilton, D.R., G.M. McMurtry, and F. Goff, Large variations in vent fluid $CO_2/^3He$ ratios signal rapid changes in magma chemistry at Loihi Seamount, Hawaii, *Nature*, 396, 359-362, 1998.

Hon, K, C. Heiker, and J.I. Kjargaard, Limu o Pele: a new kind of hydroclastic tephra from Kilauea Volcano, Hawaii, *Geol. Soc. Am. Abst. Prog.*, 20(7), 112-113, 1988.

Inman, D..L., Measures of describing the size distribution of sediments. *J. Sed. Petrol.*, 22, 125-145, 1952.

Kent, A.J., D.A. Clague, M. Honda, E.M. Stolper, I.D. Hutcheon, and M.D. Norman, Widespread assimilation of a sea-water-derived component at Loihi Seamount, Hawaii, *Geochim. Cosmochim. Acta*, 63, 2749-2761, 1999.

Kokelaar, B.P., Magma-water interactions in subaqueous and emergent basaltic volcanism, *Bull. Volcanol.*, 48, 275-289, 1986.

Loihi Science Team, Rapid response to submarine activity at Loihi Volcano, Hawaii, *EOS Trans. Am. Geophys. Union*, 78, 229-233, 1997.

Lonsdale, P., and R. Batiza, Submersible study of hyaloclastite and lava flows on young seamounts at the mouth of the Gulf of California, *Bull. Geol. Soc. Am.*, 91, 545-554, 1980.

Maicher, D., and J.D.L. White, The formation of deep-sea limu o Pele, *Bull. Volcanol.*, 63, 482-496, 2001.

Maicher, D., J.D.L. White, and R. Batiza, Sheet hyaloclastite: density current deposits of quench and bubble-burst fragments from thin glassy sheet lava flows, Seamount 6, Eastern Pacific Ocean, *Mar. Geol.*, 171, 75-94, 2000.

Malahoff, A., Geology of the summit of Loihi submarine volcano, *U.S. Geol. Surv. Prof. Paper*, 1350, 133-144, 1987.

Mastin, L.G., Thermodynamics of gas and steam-blast eruptions, *B. Volcanol.*, 57, 85-98, 1995.

Mastin, L.G., Evidence for water influx from a caldera lake during the explosive hydromagmatic eruption of 1790, Kilauea volcano, Hawaii, *J. Geophys. Res.*, 102, 20,093-20,109, 1997.

Mastin, L.G., and J.B. Witter, The hazards of eruptions through lakes and seawater, *J. Volcanol. Geotherm. Res.*, 97, 195-214, 2000.

Mattox, T.N., and M.T. Mangan, Littoral hydrovolcanic explosions: a case study of lava-seawater interaction at Kilauea Volcano, *J. Volcanol. Geotherm. Res.*, 75, 1-17, 1997.

MBARI Mapping Team, 2000, MBARI Hawaii multibeam survey, Monterey Bay Aquarium Research Institute, Digital data Series No. 2.

McBirney, A.R., Factors governing the nature of submarine volcanism, *Bull. Volcanol., 26*, 455-469, 1963.

McPhie, J., G.P.L. Walker, and R.L. Christiansen, Phreatomagmatic and phreatic fall and surge deposits from explosions at Kilauea Volcano, Hawaii, 1790 A.D.: Keanakakoi Ash Member, *Bull. Volcanol., 52*, 334-354, 1990.

Moore, J.G., D.A. Clague, and W.R. Normark, Diverse basalt types from Loihi seamount, Hawaii, *Geology, 10*, 88-92, 1982.

Morrissey, M.M., and L.G. Mastin, Vulcanian eruptions, in *Encyclopedia of Volcanoes*, ed. by H. Sigurdsson, Academic Press, San Diego, 463-475, 2000.

Morrissey, M.M., B. Zimanowski, K. Wohletz, and R. Buettner, Phreatomagmatic eruptions, in *Encyclopedia of Volcanoes*, ed. by H. Sigurdsson, Academic Press, San Diego, 431-445, 2000.

Ohminato, T., B.A. Chouet, P. Dawson, and S. Kedar, Waveform inversion of very long period impulsive signals associated with magmatic injection beneath Kilauea Volcano, Hawaii, *J. Geophys. Res. 103*, 23,839-23,862, 1998.

Parfitt, E.A., and L. Wilson, Explosive volcanic eruptions-IX. The transition between Hawaiian-style lava fountaining and Strombolian explosive activity, *Geophys. J. Internat., 121*, 215-232, 1995.

Ripepe, M., M. Rossi, and G. Saccorotti, Image processing of explosive activity at Stromboli, *J. Volcanol. Geotherm. Res., 54*, 335-351, 1993.

Smith, J.R., K. Satake, and K. Suyehiro, Deepwater multibeam sonar surveys along the southeastern Hawaiian Ridge: Guide to the CD-ROM, *in Hawaiian volcanoes: deep underwater perspectives*, ed. by E. Takahashi, et al., *Am. Geophys. Union Monograph 128*, 3-10, 2002.

Smith, T., and R. Batiza, New field and laboratory evidence for the origin of hyaloclastite flows on seamount summits, *Bull. Volcanol., 51*, 96-114, 1989.

Staudigel, H., and H.-U. Schmincke, The Pliocene seamount series of La Palma (Canary Islands), *J. Geophys. Res., 89*, 11,195-11,215, 1984.

Swanson, D.A., R.S. Fiske, T.R. Rose, and C.L. Kenedi, Prolonged deposition of the Keanakakoi Ash Member, Kilauea, *EOS Trans. Am. Geophys. Union, 79*, F937, 1998.

Vergniolle, S., and C. Jaupart, Separated two-phase flow and basaltic eruptions, *J. Geophys. Res., 91*, 12842-12860, 1986.

Vergniolle, S., and C. Jaupart, Dynamics of degassing at Kilauea Volcano, Hawaii, *J. Geophys. Res., 95*, 2793-2809, 1990.

Vergniolle, S. and M. Mangan, Hawaiian and strombolian eruptions, in *Encyclopedia of Volcanoes*, ed. by H. Sigurdsson, Academic Press, San Diego, 447-461, 2000.

Wilson, L., and J.W. Head III, A comparison of volcanic eruption processes on Earth, Moon, Mars, Io, and Venus, *Nature, 302*, 663-669, 1983.

Wohletz, K.H., Explosive magma-water interactions: thermodynamics, explosion mechanisms, and field studies, *Bull. Volcanol., 48*, 245-264, 1986.

Zimanowski, B., G. Froehlich, and V. Lorenz, Quantitative experiments on phreatomagmatic eruptions, *J. Volcanol. Geotherm. Res., 48*, 341-358, 1991.

Zimanowski, B., R. Buettner, V. Lorenz, and H.-G. Hafele, Fragmentation of basaltic melt in the course of explosive volcanism, *J. Geophys. Res.,102*, 803-814, 1997.

David A. Clague and Alicé S. Davis, Monterey Bay Aquarium Research Institute, 7700 Sandholdt Road, Moss Landing, CA 95039-9644.

Rodey Batiza, National Science Foundation, 4201 Wilson Boulevard, Arlington, Virginia 22230.

James W. Head III, Department of Geological Sciences, Brown University, Providence, RI 02912.

Large-Scale Interaction of Lake Water and Rhyolitic Magma During the 1.8 ka Taupo Eruption, New Zealand.

B. F. Houghton[1], B. J. Hobden[2], K. V. Cashman[3], C. J. N. Wilson[4], R. T. Smith[2]

The 1.8 ka Taupo eruption was one of the largest and most powerful of eruptions world wide in the last 5000 years, and notable for its uniform chemical composition, yet diversity of eruptive styles. Three phreatomagmatic phases (phases 1,3,4) involved extensive interaction with a pre-existing caldera lake. The eruption shifted abruptly between "wet" and "dry" eruptive conditions. These shifts accompanied changes in vent position that led to discharge of melt that had undergone quite different ascent and degassing histories. Vesicle size distributions and morphologies in the 1.8 ka juvenile ejecta supply constraints on these processes. Pumice clasts in units 1 and 3 have microtextures reflecting vesicle growth and coalescence during steady ascent of magma, with slight contrasts probably due to contrasting faster (phase 3) and slower (phase 1) rates of rise. There is no unambiguous evidence for magma:water interaction triggering, or contributing to fragmentation during these phases and we suspect that the principal role of external water was *after* fragmentation. Unit 4 clasts have microtextures reflecting bubble collapse. It is clear that this melt was disrupted after its peak of vesiculation; a scenario most compatible with a staged ascent of the melt. Low vesicularities and vesicle number densities and ample evidence for bubble collapse suggest that magma:water interaction played the dominant role in fragmentation for this phase. It is clear from the deposits of the Taupo eruption that interaction between surface water and silicic magma took place under widely varying conditions and that the physical state of the melt (in turn reflecting the ascent and degassing histories) was the determinant variable. At Taupo, extensive interaction of the rising magma with the caldera lake only occurred when slow final magma ascent permitted egress of lake water to the shallow conduit.

[1]Department of Geology & Geophysics, University of Hawaii, Hawaii, USA
[2]Department of Earth Sciences, University of Waikato, Hamilton, New Zealand
[3]Department of Geological Sciences, University of Oregon, Oregon, Eugene, USA
[4]Institute of Geological & Nuclear Sciences, Wairakei Research Centre, Taupo, New Zealand

Explosive Subaqueous Volcanism
Geophysical Monograph 140
Copyright 2003 by the American Geophysical Union
10.1029/140GM06

1. INTRODUCTION

1.1. Phreatoplinian Volcanism

Phreatoplinian eruptions have been only rarely studied since *Self and Sparks* [1978] introduced the term. Phreatoplinian refers to eruptions producing widely dispersed fall deposits with a grain size distribution rich in fine ash and reflecting interaction of magma with large volumes of external water (Figure 1). *Self and Sparks* [1978] described 4 type-examples of such widely dispersed, fine-grained pyroclastic fall deposits, all of which formed when

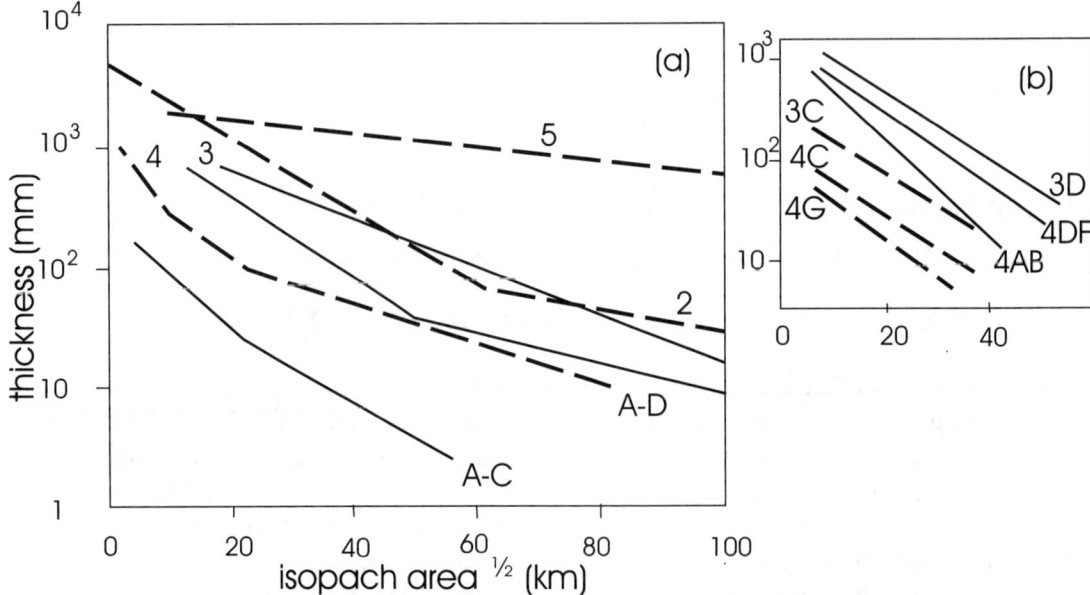

Figure 1. (a) A comparison of the dispersal characteristics of phreatoplinian (solid lines labeled 3 and 4) and associated plinian units (dashed lines labeled 2 and 5) at Taupo. Data from *Smith and Houghton* [1995a], and *Walker* [1980, 1981b]. For comparison data is included for the phreatoplinian (A-C) and plinian (A-D) phases of the 1875 Askja eruption, after *Sparks et al.* [1981]. (b) Comparable data for contrasting wetter (solid lines labeled 3D, 4DF and 4AB) and drier (dashed lines labeled 3C, 4C and 4G) subunits of the Taupo phreatoplinian units, note that wetter phases have slighter lower b_t values (i.e., steeper slopes on the diagram).

silicic magma interacted with large quantities of lake water. Three of these examples were from Taupo, two of which are the focus of this paper. Phreatoplinian volcanism remains poorly understood today, largely because of the lack of witnessed historic examples: there were no observed 20th century examples of such large eruptions with the high water/magma ratios inferred to typify the phreatoplinian style.

In many prehistoric phreatoplinian eruptions, the eruptive regime apparently shifted abruptly between "wet" (phreatoplinian) and "dry" (plinian) conditions; during the Taupo 1.8 ka eruption this happened three times. An unanswered question is what drives such sudden shifts in magma:water ratio:—changed availability of water, or changes in some aspect of magma ascent and behavior in the conduit? In this paper we present evidence in favor of the latter.

1.2. Taupo Volcano

Taupo volcano is unusual on a world scale in having experienced two young rhyolitic caldera-forming eruptions at 26.5 and 1.8 ka (Figure 2). Between these events, 26 other explosive and dome-building eruptions have occurred, making Taupo the world's most frequently active rhyolitic volcano [*Wilson*, 1993]. Almost all of these eruptions took place from vents within a substantial lake and many show indications of involvement of lake water in the eruptive phases. The 1.8 ka Taupo eruption was the largest eruption at Taupo in the last 26,500 years, and the most powerful world wide in the last 5000 years [*Wilson and Walker*, 1985]. It is equally notable for its diversity of eruptive style and intensity yet uniform chemical composition of the magma. The ~35 km³ (DRE) eruption ejected phenocryst-poor rhyolite, in effusive and "wet" and "dry" explosive eruptive styles and over a wide range of intensities. Existing work [*Walker* 1980, 1981a, 1981b; *Wilson and Walker* 1985; *Wilson* 1985; and *Smith and Houghton* 1995a, 1995b] has modeled each phase of this complex eruption but not fully explained the abrupt shifts in style/intensity.

1.3. 1.8 ka Eruption

The 1.8 ka eruption had seven discrete phases (Table 1, Figure 3) from at least three vents (Figure 2); this paper focuses on the three phases with clear phreatomagmatic signatures (phases 1, 3, and 4). Phase 1 of the eruption was phreatomagmatic, creating a locally dispersed fall deposit (Unit 1) that is absent beyond 15 km from vent. Phase 2 of the eruption was a sustained plinian outburst of moderate intensity that produced a uniform, non-graded 6-km³ lapilli fall deposit, Unit 2 [*Walker*, 1981a]. The two type phreatoplinian phases (3 and 4) followed, depositing 1.75 and 0.8 km³ of fine-grained fall and minor pyroclastic density current deposits, respectively, separated by a syn-eruptive erosional unconformity [*Walker*, 1981b; *Smith and Houghton*, 1995a; *Houghton et al.* 2000]. Unit 3 is dm-scale bedded, rich in fine ash, and dominated by highly vesicular pumice. Unit 4 has similar dispersal and grain size characteristics to Unit 3 (Figure 1) but contains a predominance of grey dense juvenile clasts.

Phase 5 marked a reversion to dry eruptive conditions and the establishment of a sustained high eruptive plume. It generated Unit 5a, the exceptionally powerfully dispersed Taupo plinian deposit, described by *Walker* [1980] as "ultraplinian", together with at least 15 pyroclastic flow units (Unit 5b) deposited by weakly energetic, concentrated pyroclastic density currents associated with partial collapses of the plume. Explosive volcanism culminated in phase 6 with generation of the violently emplaced Taupo ignimbrite (Unit 6). This deposit shows many different ignimbrite facies, reflecting a highly energetic nature of the parent density current [*Wilson*, 1985].

During phase 7, rhyolitic lava domes, the remnants of which are thought to be represented at the present-day by the Horomatangi Reefs and Waitahanui Bank (location 3, Figure 2), were extruded onto the floor of the re-formed Lake Taupo, probably following an interval of several years to a few decades. Large pumiceous blocks (forming Unit 7) detached from the dome carapaces, floated ashore [*Wilson and Walker*, 1985] and were incorporated in lacustrine shoreline sediments around the eastern edge of the re-formed lake.

2. PHREATOMAGMATIC DEPOSITS OF THE 1.8 KA ERUPTION

2.1. Unit 1

Unit 1 is of minor volume (0.015 km³) when compared with the following units, and of very limited dispersal [*Wilson and Walker*, 1985]. Thickness data and isopachs presented by *Wilson and Walker* [1985] can be reinterpreted to suggest that Unit 1 came from a vent (location 1, Figure 2) situated in the same area as proposed for Unit 3 by *Smith and Houghton* [1995a]. In all but the thickest exposures, Unit 1 is uniformly fine-grained and rich in pumiceous ash, and is interpreted, on the basis of the grain size data, as the product of mild phreatomagmatic activity [*Wilson and Walker*, 1985]. Vesicularities of pumice lapilli are similar on average to those in Unit 3 but show a broader range) [*Houghton and Wilson*, 1989, and below].

2.2. Unit 3

Unit 3 covers an area of >15 000 km² and has b_t values (the thickness half-distance of *Pyle*, 1989) of 4.4 to 5.5 km. In medial outcrops (8–60 km from source) it can be divided into 4 subunits. Subunits 3A, 3B and 3D show relatively symmetrical isopach patterns; isopachs for subunit 3C are elongated to the east, suggesting the plume was more strongly influenced by the westerly wind flow. Each subunit contains multiple beds. Subunit 3A is very poorly sorted (σ_ϕ= 2.6–3.9, Table 1) with a massive lower half of fines-bearing coarse ash to medium lapilli and a bedded upper half consisting of planar cm-scale alternations of fines-free and fines-bearing lapilli fall. Subunits 3B and 3D consist of aggregate-bearing ash

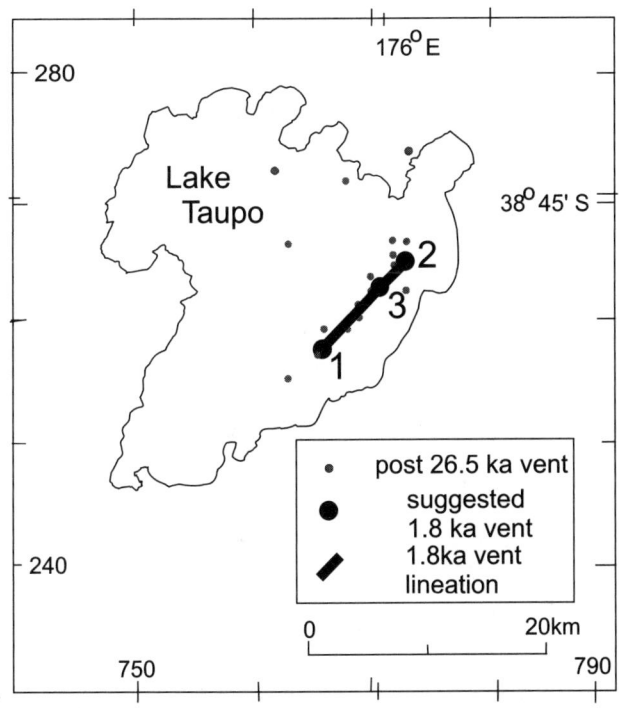

Figure 2. Location map showing position of post 26.5 ka vents at Taupo and the distribution of the 1.8 ka vents along a 10-km-long vent system, that is coincident with the main locus of all post 26.5 ka volcanism. After *Smith and Houghton* [1995a] and *Wilson* [1993]. Vent 1 was postulated as the source of Units 1 and 3 and vent 2 as the source of Unit 4 by *Smith and Houghton* [1995a], Vent 3 is the location of the single vent proposed for the eruption by *Walker* [180, 1981b] and *Wilson and Walker* [1985].

Table 1. Summary of the deposits and eruptive styles during the 1.8 ka Taupo eruption, largely taken from *Wilson and Walker* [1985].

Unit	Stratigraphic name	Wilson 1993	Type of Deposit	Notes	Rate (kg-s^{-1})
1	Initial ash	Y-1	Fall	Weak phreatomagmatism	10^5
2	Hatepe plinian pumice	Y-2	Fall	Moderate plinian activity	10^7
3	Hatepe ash	Y-3	fall (+pdc)	Phreatoplinian, pumiceous juvenile clasts	10^6
4	Rotongaio ash	Y-4	fall (+pdc)	Phreatoplinian, dense juvenile clasts	10^5
5	Taupoplinian/early ignimbrite flow units	Y-5	fall+pdc	Powerful plinian and local pyroclastic density currents	10^8
6	Taupo ignimbrite	Y-6	Pdc	Devastating, widespread pyroclastic density current	10^{11}
7	floating blocks	Z	(lava dome)	Subaqueous dome growth	?

falls (Md$_\phi$ = 2.4–4.5) separated by dm-spaced, laterally continuous thin partings of coarse ash and pumice lapilli fall (Md$_\phi$ = 0.4–1.0). Subunit 3C resembles in appearance a plinian fall deposit, with coarse pumice lapilli up to 100 mm across (Md$_\phi$ = –2.2–0.6), and smaller wall rock lithic and dense juvenile clasts up to 30 mm in diameter. In the most proximal exposures (4–8 km from vent), Unit 3 contains thick sequences of massive and wavy-bedded ash containing discontinuous laminae of pumice lapilli that pinch-and-swell- or vanish over short lateral distances [*Smith*, 1998]. These deposits are interpreted as the products of dilute, turbulent pyroclastic density currents (PDCs).

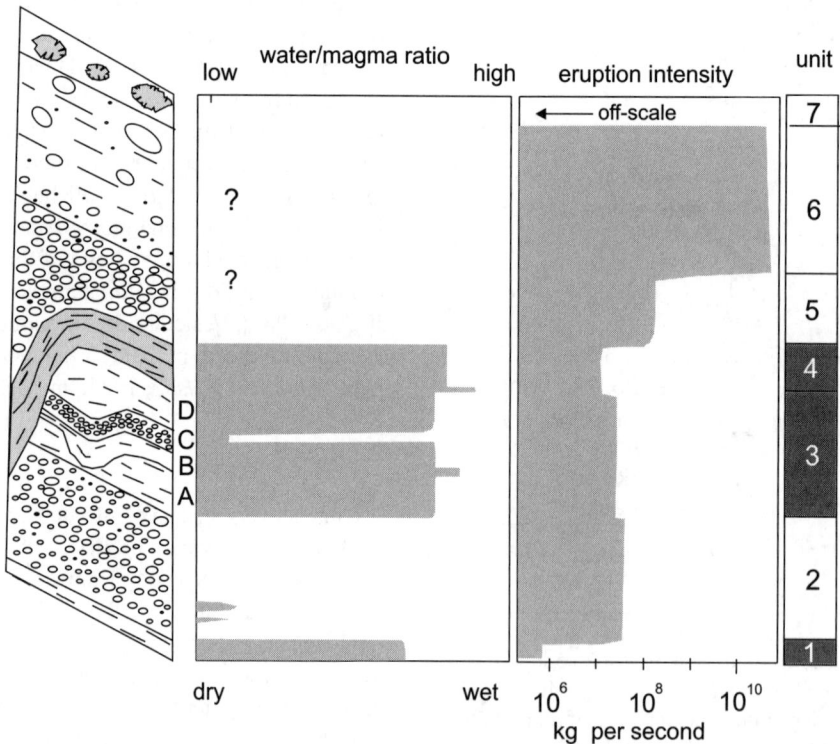

Figure 3. Stratigraphy and inferred eruptive conditions of the 1.81 ka Taupo eruption, modified after *Wilson and Walker* [1985]. The second and third columns are inferred relative water/magma ratios and discharge rates used by *Walker and Wilson*. In our current model these magma:water ratios would apply more to the eruption plume than to the fragmentation process. In the fourth column phreatomagmatic units are shown as gray.

In contrast to the wide variation in grain size amongst Unit 3 beds, component proportions and juvenile clast vesicularities are very uniform. All subunits, ash-rich and ash-poor alike, are uniformly low in wall rock clasts. Lithic clast contents (dominantly flow-banded rhyolite lava) are up to 5–8 wt% in fractions coarser than 2 mm, and typically 2–5 wt% in fractions between 2 mm and 63 μm. Juvenile lapilli are typically sub-rounded to sub-angular pumice. Ash-grade juvenile clasts are all highly vesicular, with bubble-wall textures.

2.3. Unit 4

Unit 4 is the most distinctive part of the 1.8 ka sequence because of its dark steel-gray color due to a predominance of dark, poorly to moderately vesicular juvenile clasts. Its source was a vent at the northeastern end of the 1.81 ka vent system (location 3, Figure 2). The medial Unit 4 can be subdivided into 8 subunits, each of which is a package of numerous mm-thick beds. Subunits have b_t values of 2.9 to 5 km. Two relatively coarse-grained subunits have isopachs that are markedly elongate downwind to the east; the remainder show more symmetrical isopach patterns. In the proximal area, the Unit 4 deposits are exceptionally fine grained, even relative to Unit 3, and contain equally finely laminated beds to those seen in medial to distal areas. However, many of these beds cannot be traced into the medial environment and are interpreted as the products of less intense discrete explosions. Like Unit 3, Unit 4 contains interbedded massive to weakly cross-bedded PDC deposits in some proximal areas.

3. VESICULARITY OF JUVENILE CLASTS

We have examined the vesicularity of juvenile clasts from the phreatomagmatic deposits to establish if they were characterized by different vesiculation histories than the "dry" or plinian phases. Samples of at least 100 lapilli were collected over narrow vertical intervals (1–3 clasts high) from the phreatomagmatic units and their enclosing plinian fall deposits. For Unit 4 only one subunit was sufficiently coarse to sample, but microscopic investigation has confirmed that the vesicularity of its clasts are typical of Unit 4 as a whole. The densities of single clasts were measured following *Houghton and Wilson* [1989] and the data used as a "filter" to select between 3 and 7 clasts from each sample that represented the mean and extreme values of density and hence vesicularity. Vesicle numbers, shapes and abundances in these clasts were quantified using a combination of images captured on petrographic and electron microscopes.

Typically 14 images, at a range of magnifications from 12.5x to 250x, were transformed to binary images (glass and vesicles), and analysed to obtain areas, perimeter lengths and elongation ratios. Typical images are shown in Figure 5 including pumices from Unit 2 as representative of the plinian phases of the eruption. Size distributions were converted to volume distributions using the stereological techniques of *Sahagian and Proussevitch* [1998].

3.1. Clast Densities and Bulk Vesicularities

The new density data (Figure 4) largely confirm the conclusions of *Houghton and Wilson* [1989]. Two of the three phreatomagmatic deposits (units 1 and 3) are dominated by highly vesicular pumices similar to the two plinian units (2 and 5a) but have a "tail" of clasts of density 800 to 1100 kg m^{-3}, equivalent to vesicularity values of 53 to 66%. In contrast juvenile clasts in Unit 4 are distinctly different from those of Unit 3 and the plinian phases with a density mean of 1200 kg m^{-3} and a range from 500 to 2100 kg m^{-3}. Unit 4 has densest (least vesicular) clasts and widest density ranges seen in the entire eruption including the late-stage dome.

3.2. Microscopic Textures of the Juvenile Clasts

All clasts from Unit 1 are characterised by a population of relatively coarse vesicles (0.5 to 1.5 mm) with shapes that imply a major role for bubble coalescence (Figure 5 m–o); they contrast markedly with the population in the overlying plinian Unit 2 (Figure 5 j–l) in the large size and irregular form of many of the vesicles. In two pumices, typical of the mean and highest densities, these large vesicles occur in clusters or weakly defined bands. In low-density pumices, abundant large vesicles have a more uniform distribution. Smaller vesicles in all three pumices show a wide range of shapes from spherical to complexly contorted. The simpler forms typify the smallest vesicles (<50 μm), a feature typical of all Taupo pumices.

We have studied vesicle size distributions (VSDs) in two subunits from Unit 3; subunit 3C (the lapilli bed) that is similar in grain size to the plinian falls and from an ash-rich interval within 3D. In this way we hoped to contrast 'drier' and 'wetter' extremes of this phreatoplinian deposit. The six Unit 3 pumices show perhaps the most diverse range of textures that we studied. Two clasts show marked stretching of vesicles of all sizes leading to a pronounced anisotropic 'long-tube' fabric (Figure 5 f, g). Two others have local domains of intense bubble deformation amongst the moderate-to-large sized vesicles (Figure 5 e, i), and two are characterised by densely packed, equant vesicles with complex

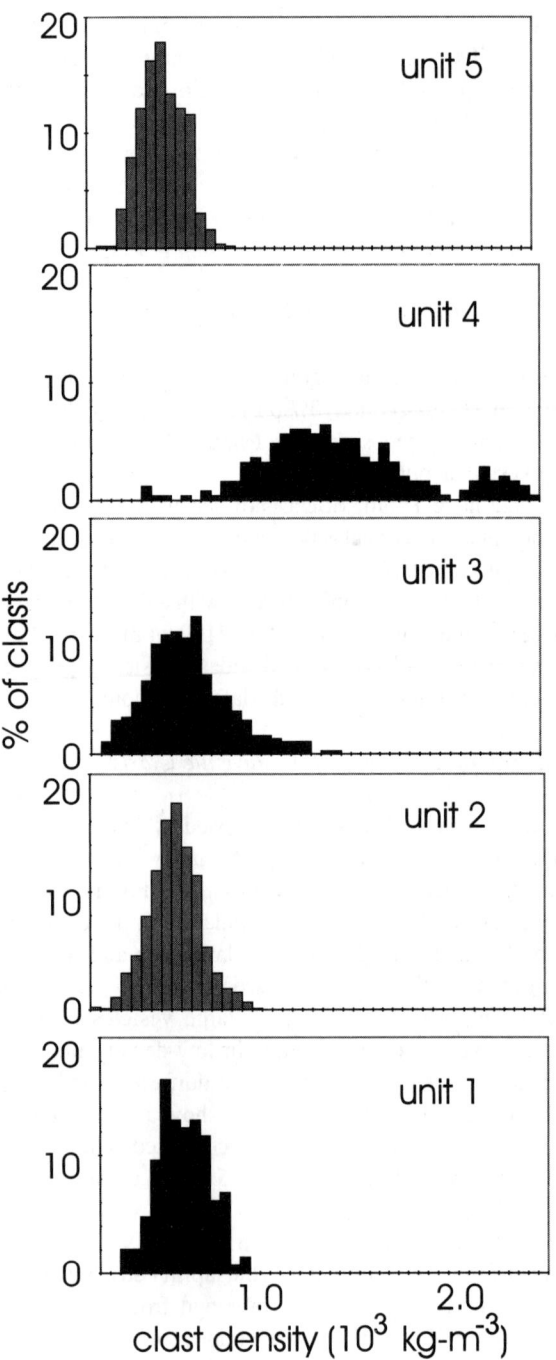

Figure 4. Bulk density of representative pumices from samples for the phreatomagmatic and plinian phases of the Taupo eruption. Samples consist of up to 1500 clasts. Plinian units are shown in gray and phreatomagmatic units in black. Note the marked contrast between the poorly to moderately vesicular clasts in Unit 4 versus the pumices that characterize all the remaining units.

shapes that reflect interaction and coalescence between bubbles (Figure 5 d, h). There is, however, no correlation between the degree of stretching and clast vesicularity; for example, the two long-tube pumices represent the extremes of Unit 3 vesicularity at 58% and 93%.

We analysed four clasts from Unit 4 in recognition of the broad distribution of clast densities and hence vesicularities. These clasts have highly distinctive textures, with few features in common with the pumices of the immediately over- and under-lying units. There is a striking similarity to textures preserved in subaerial obsidian flows (e.g. *Fink, 1983; Fink and Manley, 1987*). The Unit 4 clasts have thick glass walls and wall thicknesses predictably increase with increasing density (Figure 5a–c). A striking feature is the occurrence of mm-long zones of larger (200–500 µm) bubbles with complex indented walls surrounded by zones characterised by 10–30 µm flattened but isolated vesicles with thickened walls (Fig 5c). Even widely spaced vesicles no longer in contact with their neighbors have mostly complex curvilinear and fiamme-like shapes which we can only explain as due to an earlier history of bubble interaction and coalescence followed by total collapse of all except the larger bubbles (Figure 5c). All Unit 4 clasts contain small quantities of microlites a feature lacking in the pumices from the other eruptive units.

3.3. Quantitative Measurements of Vesicle Size

Vesicle size distributions. Vesicle size distributions are depicted by plotting the population number density n (the slope of the cumulative number density, dN/dL, where N is the vesicle density per unit volume and L is the vesicle size). A linear trend on a $\ln(n)$ versus L plot would be consistent with constant rates of nucleation and growth of bubbles (Klug et al. 2002). All the Taupo size distributions are strikingly non-linear (Figure 6), almost certainly due to nonsteady state bubble growth [*Blower et al.* 2002] and mass balance constraints [*Klug. et al.* 2002].

Vesicle volume distributions. Vesicle volume distributions for the median pumices are essentially unimodal although most display a subordinate subpopulation of 1–4 mm vesicles. The Taupo samples are less obviously polymodal than for many silicic tephras [*Cashman and Mangan*, 1994; *Klug et al.*, 2002]. The data for pumices of mean density/vesicularity (Figure 7a) reflect changes with time through the alternating 'wet' and 'dry' eruptive phases. There is a predominant size mode at approximately 30–40 µm for the analysed mean-density pumices (Table 2) except that of Unit 1 (160 µm).

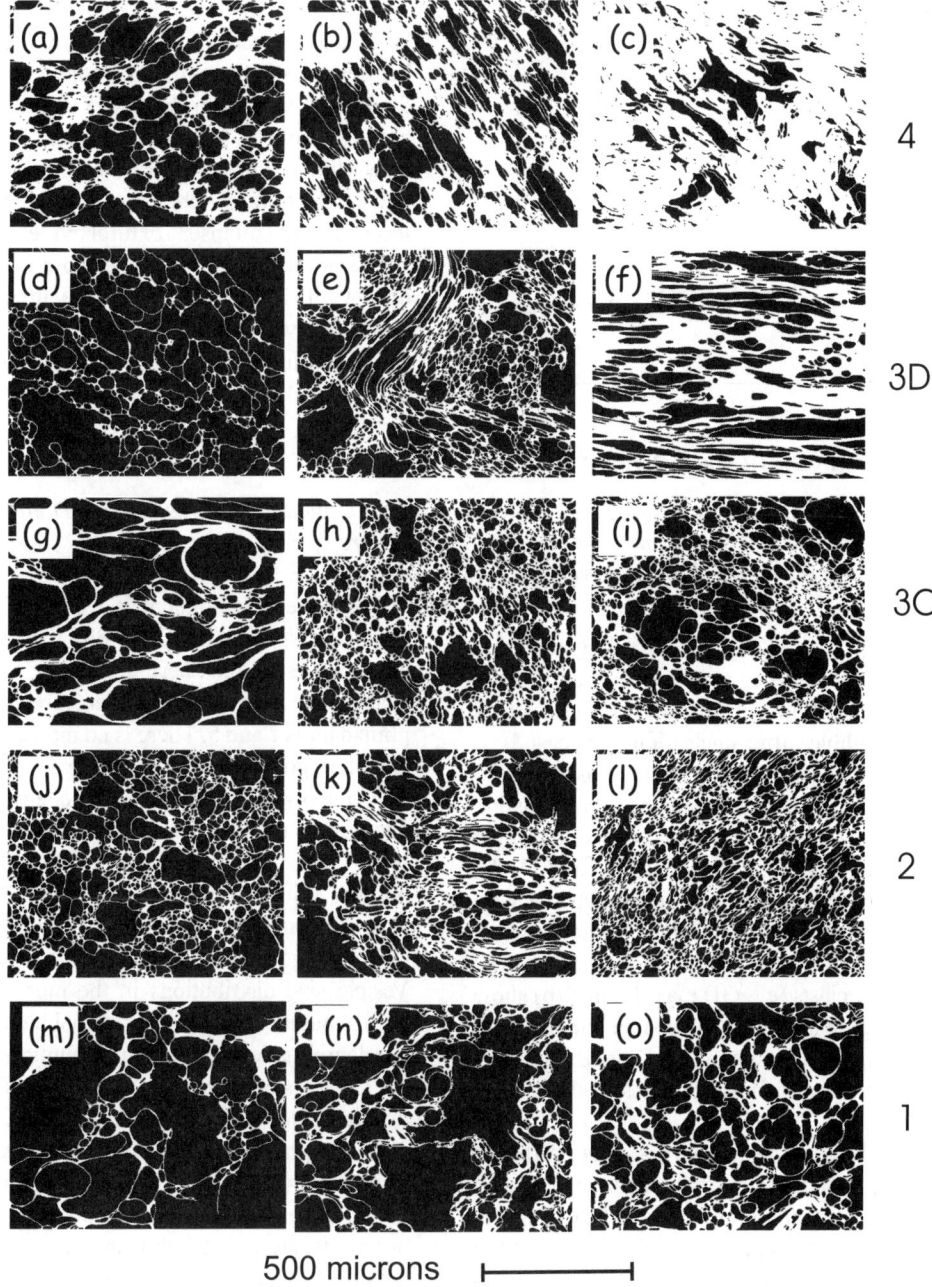

Figure 5. Representative 100x scanning electron microscope images from Taupo clasts. Vesicles are shown in black and groundmass in white. For each sample we have included one image from clasts chosen to represent the maximum (left column), mean (center column) and minimum (right column) vesicularity.

This is also apparent from the plot of cumulative volume distributions for the median clasts (Figure 8). The form of the curves are very similar again for all except Unit 1, and dominated by a single segment that reflects the essentially unimodal nature of the Taupo vesicle populations. This predominance of a single population of relatively small bubbles suggests that most of the vesiculation took place over a limited time period. However, the form of the curves departs from a normal distribution at large vesicle sizes. The plot shows very clearly how the size distribution for the

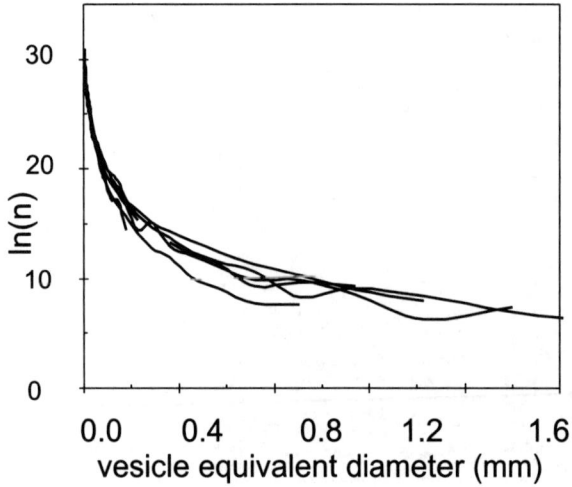

Figure 6. Vesicle size distributions for the Taupo clasts. All data are strongly non-linear, probably reflecting mass balance constraints and non-steady-state and disequilibrium degassing.

mean pumice for Unit 1 is skewed toward larger bubble sizes. This curve shows a near-normal distribution of bubbles that would be consistent with regular expansion of a population of small bubbles after nucleation had ceased.

We can also examine the diversity present across the fragmentation surface at a single instant in time by contrasting clasts of mean, maximum and minimum density for single samples (Figure 7b,c; Figure 9). For subunits 3C and 3D the mean and least vesicular pumices form a tight cluster with volume-weighted median diameters (the diameters at a cumulative volume fraction of 0.5) around 40 µm whereas the two most vesicular clasts (e.g. 3D max, Figure 7b) show a shift in the vesicle population toward coarse bubbles with medians of 87 and 170 µm respectively (Figure 9a). This pattern is identical to that shown by samples of the plinian phases of the eruption.

As seen qualitatively, clasts from Unit 4 are very different from every other Taupo sample. Vesicularities are markedly lower, as remarked before, and there are few or no bubbles larger than 200 µm in every clast (Figure 9b). There is little overlap in vesicularity between clasts in Units 3 and 4 (Figure 4) but comparison between the least vesicular clast in Unit 3 (3 min 58% vesicles) and the most vesicular in Unit 4 (4 max 63% vesicles) shows this contrast in volume-weighted median diameter clearly (Figure 9b).

Vesicle number density and cumulative number density. The Taupo clasts cluster on a log-log plot of $N_v(>L)$ versus L to define a power law relationship with an exponential decay value of approximately −3.2 (Figure 10). Vesicle number densities (N_t) summed for all bubbles larger than 7 µm and calculated on a vesicle-free (i.e. melt-referenced) basis are between 10^8 and 10^{10} cm^{-3} (Table 2). Vesicle number densities in the pumices from the two samples from Unit 3 are very similar to the data for plinian Units 2 and 5. In contrast number densities are lower for Unit 4 and particularly Unit 1. These lower numbers could reflect either limited nucleation of bubbles or loss of bubbles through coalescence and collapse. Textural evidence summarized above suggests that the latter is the more likely explanation. For only Unit 1 there is a decrease in number density with increasing vesicularity suggesting bubble growth by coalescence continued for a period after nucleation had ceased.

4. INTERPRETATION AND DISCUSSION

4.1. Implications of Vesicularity Data for Timing and Nature of Fragmentation

The three phreatomagmatic units of the 1.8 ka eruption contain clasts with markedly different vesicularities. We discuss the vesicle data here in order of increasing complexity. Clasts from phreatoplinian Unit 3 are very similar to those in plinian units 2 and 5. There is no indication that vesicle nucleation and growth during phase 3 was terminated prematurely by magma:water interaction and quenching, as vesicularities and vesicle number densities are similar for units 2, 3 and 5. In fact, vesiculation was sufficiently advanced in parts of the magma ejected in both subunit 3C (relatively low water/magma ratio) and 3D (relatively high water/magma ratio) to generate pumices with vesicularities of >90%. Vesicle size distributions in the most vesicular pumices in both subunits, reflect extensive bubble coalescence to a similar degree as seen in the plinian pumices. The very high number densities in Unit 3 pumices suggest that the reduced eruption rate with respect to plinian phase 2 (*Wilson* and *Walker* 1985) was not accompanied by a decrease in ascent velocity. The combination of reduced discharge rate, rapid ascent and a marked increase in the abundance of stretched or long-tube pumice is interpreted below in terms of a role for reduction in the effective radius of the conduit.

Unit 1 is distinctive in that all pumices show a marked shift toward relatively coarse median vesicle diameters (100–260 µm). In addition, the vesicle number densities are in general 0.5 to 1.0 orders of magnitude lower in the Unit 1 clasts, particularly the most vesicular clast, than in corresponding clasts from the other units. *Mangan and Sisson* [2000] have suggested that the bubble size decreases and number density increases with increasing decompression (and hence ascent) rate, and that coalescence increases in importance at lower decompression rates. These observa-

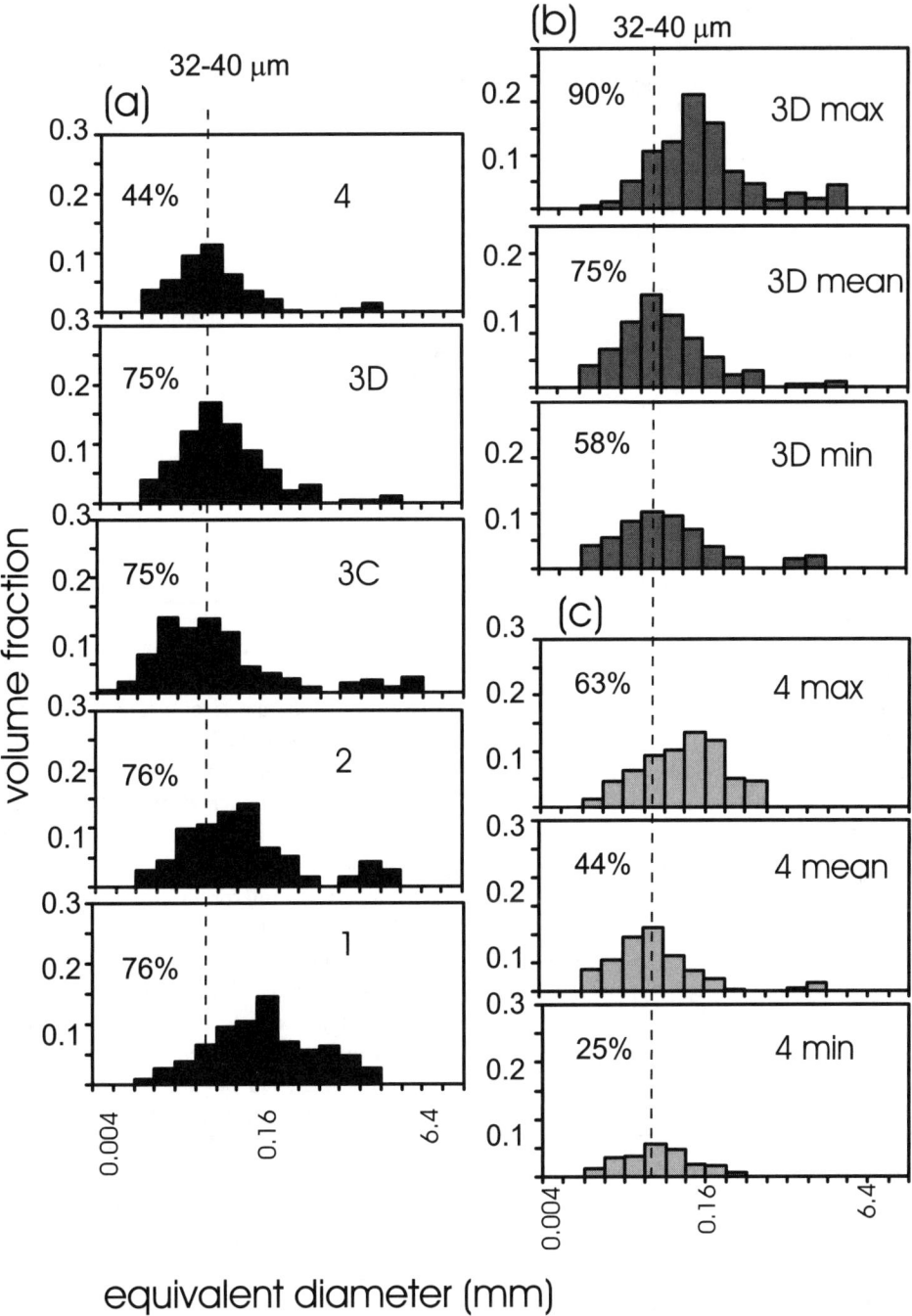

Figure 7. Vesicle volume distributions for Taupo clasts, plotted for the geometric bins used in the statistical analysis. Dashed line in each case is the 32 to 40 μm bin. Percentage values are the determined bulk vesicularity of each clast. (a) Clasts of mean density/vesicularity for each sample. Note the displacement of the vesicularity mode for Unit 1 relative to the other samples. (b) and (c) are the clasts of maximum, mean and minimum vesicularity from subunit 3D and Unit 4 respectively. Note in both cases the vesicle population for the highest vesicularity clast is displaced towards higher values.

Table 2. Vesicularity data for selected pumices from the Taupo 1.8 ka eruption. Primary modes of vesicle distribution are calculated from histograms (e.g. Fig. 7) and volume-based medians from cumulative plots (Figs. 8 9).

Unit	Clast #	Vesicularity	Mode of vesicle distribution	Vol. based median	N_t (log no cm^{-3})	N_t^m (log no cm^{-3})
5	MF01103	79%	40 μm	38 μm	8.3	9.0
4	MF07115	63%	100 μm	71 μm	8.1	8.5
4	MF07101	52%	160 μm	110 μm	8.1	8.4
4	MF07111	44%	40 μm	28 μm	8.3	8.5
4	MF07121	25%	40 μm	30 μm	8.0	8.1
3D	MF09102	90%	100 μm	87 μm	7.8	8.8
3D	MF09105	75%	40 μm	40 μm	8.4	9.0
3D	MF09106	58%	40 μm	42 μm	8.1	8.7
3C	MF09204	93%	250 μm	170 μm	8.3	9.4
3C	MF09206	75%	40 μm	36 μm	8.9	9.5
3C	MF09253	65%	40 μm	43 μm	8.2	8.8
2	MF02241	76%	100 μm	67 μm	8.2	8.8
1	MF05106	86%	600 μm	260 μm	7.0	7.8
1	MF05124	76%	160 μm	120 μm	4.8	8.4
1	MF05138	65%	100 μm	100 μm	8.1	8.6

tions are consistent with an interpretation that the magma for Unit 1 ascended more slowly than that for Unit 3 and that nucleation had ceased prior to fragmentation and quenching of the pumice.

The dominant texture in the Unit 4 clasts is one of bubble collapse, unlike units 1 and 3 where vesicle textures reflect strong influences of bubble growth and coalescence (the latter especially in the highly vesicular clasts). The evidence for this is as follows. Unit 4 clasts lack large vesicles (or they are present but extremely rare and irregular in distribution, a texture seen in only one clast). The dominant population of moderate-sized vesicles has thick glass walls and yet the shapes of the bubbles are complex and curvilinear. It seems clear that these clasts do not represent a parcel of melt that was quenched by magma:water interaction *prior* to reaching high vesicularity. Rather, the Unit 4 magma seems to have been largely out-gassed at the time of fragmentation, an interpretation supported by existing data on residual contents of magmatic volatiles [*Dunbar et al.*, 1989]. The Phase 4 magma apparently had a longer degassing history than even that of Unit 1, and this history included an extended period where open system degassing occurred to near-surface final pressures and crystallization of microlites had begun. Our preferred option is that the phase 4 melt existed as a shallow plug under the northeastern vent. However, there is a wide range in vesicularity even in single samples, suggesting considerable physical heterogeneity in the magma that was erupted, though, in some cases, recycling of previously erupted moderately-to-highly vesicular pumices cannot be eliminated.

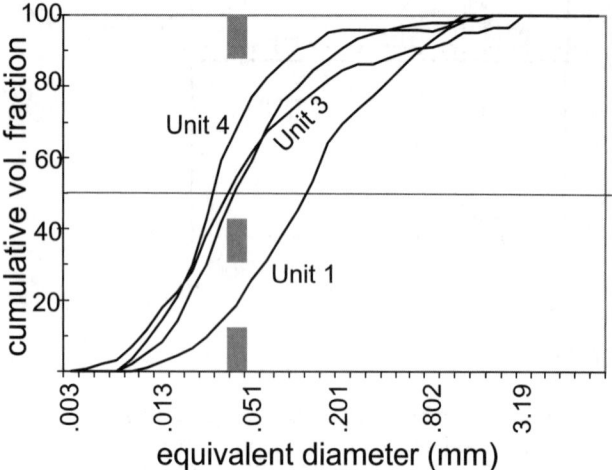

Figure 8. Cumulative volume fractions for the clasts of mean vesicularity from the phreatomagmatic units and subunits at Taupo. Note the displacement of the Unit 1 clast towards higher vesicle diameters and the low abundance of coarse vesicles in the Unit 4 clast. Gray bars correspond to 32–40 μm, the dominant size mode for bubbles in most pumices from the 1.8 ka eruption.

4.2. Model for Fragmentation

Our initial hypothesis was that the availability of external water was the key factor in determining the eruptive style at Taupo, i.e. phreatomagmatism was only possible when lake water could flood into the vent and ceased whenever the vent(s) were above the lake surface (cf. *Walker*, 1981*b*). However, we see *no* textural evidence that water prematurely halted vesiculation during any phase of the 1.8 ka erup-

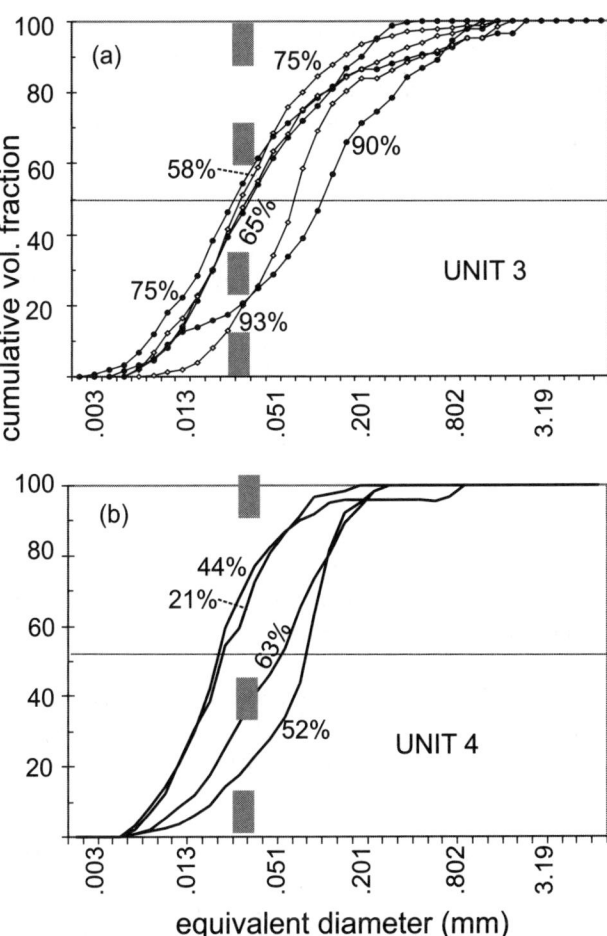

Figure 9. Cumulative volume fractions for all clasts in units 3 and 4. Percentage values are the measured bulk vesicularity of each clast. (a) Unit 3. Clasts from subunit 3C have open diamond symbols and from subunit 3D closed circles. Note the cluster of the four mean and least vesicular clasts around a median diameter of 32-40 μm (gray vertical bars) and the displacement of the two extremely vesicular clasts to higher vesicle diameters. (b) Data from 4 clasts for Unit 4. Note the depletion in large vesicles amongst the Unit 4 clasts.

tion. Instead our textural studies of the juvenile clasts suggest that the phreatomagmatic explosions in both phases 3 and 4 at Taupo probably occurred as a consequence of changing magma rheology in the conduit rather than shifts in the availability of external water.

Phase 1

Magma ascent was relatively slow, but uninterrupted, during conduit opening prior to the weak explosions characterizing phase 1. This led to partial open-system degassing of the magma prior to fragmentation and a mature vesicle population that continued to grow and coalesce after termination of bubble nucleation. When this foam-like magma ascended to shallow levels and began to fragment in came into contact with lake or very shallow groundwater that was ejected along with magmatic volatiles and pumice.

Phase 3

Once conduit conditions were well-established, high rates of magma ascent and discharge characterized plinian phase 2, and probably precluded water from entering the conduit or plume. Phreatomagmatism and the switch to phase 3 were only possible once a shift in vent position occurred and these rates had declined somewhat. During phase 3 the eruption switched between a wetter and less intense mode (represented by subunits 3B and 3D) and a drier, more intense mode (subunit 3C). What caused this shift? A potential answer lies in the tail of denser clasts within the deposits of Unit 3. The vesicle size data imply that these clasts represent a small portion of the melt that was more advanced in its vesiculation and degassing history than that represented by the dominant clast populations. This suggests, in turn, that this portion of the melt had had a slightly longer residence time and/or slower ascent in the conduit. In addition Unit 3 is unusual amongst the fall deposits in containing a significant content of long-tube pumice characterized by sheared and attenuated vesicles. A similar content of long tube pumice is also present in Unit 6, the Taupo ignimbrite, where shearing appears to have accompanied a drastic increase in ascent and eruptions rates. We suggest that this reflects temporary imbalances between the input and output of melt through the shallow conduit and a focusing of the velocity gradient inboard of stagnated material now effectively partially blocking the conduit. Such a partial conduit

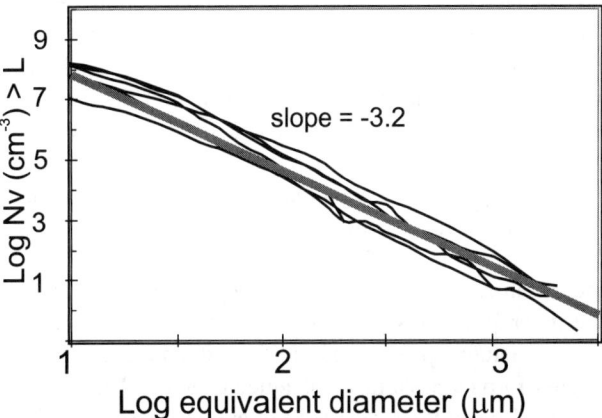

Figure 10. Cumulative number densities for clast of mean vesicularity from each Taupo unit. $N_v (>L)$ is the number of vesicles per cm^3 with diameters greater than L. The gray linear trace has a slope of -3.2.

blockage may have slowed shallow ascent during the eruption of subunits 3B and 3D. This process where supply exceeded discharge possibly was ultimately responsible for the cessation of activity at the southern vent. What was the nature and timing of magma/water interaction during phase 3? It is clear from the preceding discussion that contact of phase 3 magma with external water was a late-stage process- a logical conclusion given the physical difficulty of introducing surface water to any significant depth in the conduit under conditions of high mass-discharge. The VSD information suggests that the majority of the phase 3 melt had vesiculated to a foam-like state, equivalent to or even more advanced than the phase 2 magma, which fragmented due to vesiculation alone. It could be that fragmentation during phase 3 was due to exsolution of magmatic volatiles and external water mixed only *above* the fragmentation surface with a pre-fragmented gas-particle mixture? This raises the intriguing possibility that external water played a far more important role in controlling pyroclast transport and deposition than during fragmentation. The best test of this idea would be in whole deposit grain size data. Unfortunately much of the Taupo 1.8 ka fall units fell offshore and the quality of the available coverage is not adequate to permit comparisons to be made.

Phase 4

A different mechanism is required to explain phreatoplinian phase 4, associated with the opening of a new vent. Unlike phase 3, phase 4 was succeeded (rather than preceded) by a more energetic and dry plinian phase (phase 5). Phases 1 through 3 and phases 5 and 6 involved magma in a foam-like state, which was actively vesiculating and accelerating up the conduit. The phase 4 clasts represent the most degassed melt erupted in the Taupo eruption and also contain a sparse population of microlites, suggesting that the ascent was sufficiently slow to overcome the kinetic inertia that precluded microlite growth during the other pyroclastic phases. Our preferred model is that melt rose and stalled under the northern portion of the vent system relatively early in, or even prior to, the eruption, perhaps accompanying the ascent of magma that eventually vented during phases 1 and 2. Once at shallow level, this melt underwent relatively open-system degassing. We speculate that its eruption was possible only at the close of phase 3, when fractures propagated northward and the phase 4 magma formed a shallow subaqueous dome. This process permitted a series of repeated violent explosions in which small volumes of lake water were brought in contact with the now largely degassed but still hot melt under the northern vent. The nature of the ejecta gives valuable clues to the geometry of this Unit 4 magma body. Unit 4 is almost free of wall rock lithic clasts, except for minor recycled pumice from the earlier pyroclastic phases. This lack of wall rock particles implies that the explosions were centered at a free surface between magma and standing water and not along conduit walls. However, the juvenile clasts have very diverse microtextures suggesting that either the melt was highly heterogeneous over narrow depth intervals or that the fragmentation surface was highly irregular and extended over a significant depth interval.

5. CONCLUSIONS

At Taupo, the vesiculation history of the melt and its ascent history played the dominant roles in switches between "wet" and "dry" or phreatomagmatic and magmatic explosive volcanism. Availability of water was at best a second order determinant. We consider that slowed or staged ascent was essential for large-scale magma/water interaction and that this was favored by repeated shifts in vent position during the 1.8 ka eruption. An issue that remains is: in some phreatoplinian eruptions is the principal role of external water flushing and aggregating fine particles from the plume rather than driving or assisting fragmentation?

Acknowledgments. This research was initially funded by a Marsden Fellowship (96-GNS-ECA-0006) awarded by the Royal Society of New Zealand (BFH, BJH, KC, CJNW) and subsequently by NSF Grant EAR 01-06700 (BFH). We thank Neville Orr for high quality polished thin sections, Alf Harris and Dave Wild for assistance with SEM operation, and Jo Chizmar, Anne Duncan, Rosalind Houghton and Sarah Houghton for clast density measurements and image processing. Jim Gardner and Jocelyn McPhie provided insightful reviews of the manuscript.

REFERENCES

Blower, J. D., Keating, J. P., Mader, H. M. and Phillips, J. C. The evolution of bubble populations in volcanic eruptions. *J. Volcanol. Geotherm. Res., 120,* 1-23, 2002.

Cashman, K. V. and Mangan, M. Physical aspects of magmatic degassing. II. Constraints on vesiculation process from textural studies of eruptive products. *Rev. Mineral., 30,* 447-478, 1994.

Dunbar, N. W., Hervig, R. L. and Kyle, P. R. Determination of pre-eruptive H_2O, F, and Cl contents of silicic magmas using melt inclusions: examples from the Taupo Volcanic Zone, New Zealand. *Bull. Volcanol., 51,* 177-184, 1989.

Fink, J. Structure and emplacement of a rhyolitic obsidian flow: Little Glass Mountain, Medicine Lake highland, northern California. *Geol. Soc. Am. Bull., 94,* 362-380, 1983.

Fink, J.H. and Manley, C. R. Origin of pumiceous and glassy textures in rhyolite flows and domes. *Geol. Soc. Am. Spec. Pap., 212,* 77-88, 1987.

Houghton, B. F. and Wilson C. J. N. A vesicularity index for pyroclastic deposits. *Bull Volcanol., 51,* 451-462, 1989.

Houghton, B. F., Wilson, C. J. N., Smith, R. T. and Gilbert, J. S. Phreatoplinian Eruptions. In H. Sigurdsson, B. F. Houghton, S. McNutt, H. Rymer, and J. Stix (eds), *Encyclopedia of Volcanoes.* Academic Press, San Diego, pp. 513-525, 2000.

Klug, C., Cashman, K. V. and Bacon, C. R. Vesicularity and permeability of pumice from the climactic eruption of Mt Mazama (Crater Lake) Oregon. *Bull. Volcanol., 64,* 486-501. 2002.

Mangan, M. and Sisson, T. Delayed disequilibrium degassing in rhyolitic magma: decompression experiments and implications for explosive volcanism. *Earth Planet. Sci. Lett., 183,* 441-455, 2000.

Pyle, D. M. The thickness, volume and grainsize of tephra fall deposits *Bull. Volcanol., 51:* 1-15, 1989.

Sahagian, D. L. and Proussevitch, A. A. 3D particle size distributions from 2D observations: stereology for natural applications. *J. Volcanol. Geotherm. Res., 84:,* 73-196, 1998.

Self, S. and Sparks, R. S. J. Characteristics of widespread pyroclastic deposits formed by the interaction of silicic magma and water. *Bull. Volcanol. 41,* 196-212, 1978.

Smith, R. T. Models for Units 2 and 3 of the Taupo Eruption. Ph.D. thesis, University of Canterbury, Christchurch, New Zealand, 1998, 1998.

Smith, R. T. and Houghton, B. F. Vent migration and changing eruptive style during the 1800a Taupo eruption: new evidence from the Hatepe and Rotongaio phreatoplinian ashes. *Bull. Volcanol., 57,* 432-439, 1995a.

Smith, R. T. and Houghton, B. F. Delayed deposition of plinian pumice during phreatoplinian volcanism: the 1800a eruption of Taupo, New Zealand. *J. Volcanol. Geotherm. Res., 67,* 221-226, 1995b.

Sparks, R. S. J., Wilson, L. and Sigurdsson, H. The pyroclastic deposits of the 1875 eruption of Askja, Iceland. *Phil. Trans. R. Soc. Lond., A 299,* 241-273, 1981.

Walker, G. P. L. Taupo pumice: product of the most powerful known (ultraplinian) eruption?, *J. Volcanol. Geotherm. Res., 8,* 69-94, 1980.

Walker, G. P. L. The Waimihia and Hatepe plinian deposits from the rhyolitic Taupo volcanic centre. *N. Z. J. Geol. Geophys., 24,* 305-324, 1981a.

Walker, G. P. L. Characteristics of two phreatoplinian ashes and their water-flushed origins. *J. Volcanol. Geotherm. Res., 9,* 395-407, 1981b.

Wilson, C. J. N. The Taupo eruption, New Zealand. II. The Taupo ignimbrite. *Phil. Trans. R. Soc. Lond., A 314,* 229-310, 1985.

Wilson, C. J. N. Stratigraphy, chronology, styles and dynamics of late Quaternary eruptions from Taupo volcano, New Zealand. *Phil. Trans. R. Soc. Lond., A 343,* 205-306, 1993.

Wilson, C. J. N. and Walker, G.P.L. The Taupo eruption, New Zealand. I. General aspects. *Phil. Trans. R. Soc. Lond., A 314,* 199-228, 1985.

K. V. Cashman, Department of Geological Sciences, University of Oregon, Oregon, Eugene, USA

B. J. Hobden and R. T. Smith, Department of Earth Sciences, University of Waikato, Hamilton, New Zealand

B. F. Houghton, Department of Geology and Geophysics, University of Hawaii, Hawaii, USA

C. J. N. Wilson, Institute of Geological & Nuclear Sciences, Wairakei Research Center, Taupo, New Zealand

Submarine Strombolian Eruptions on the Gorda Mid-Ocean Ridge

David A. Clague and Alicé S. Davis

Monterey Bay Aquarium Research Institute, Moss Landing, California

Jacqueline E. Dixon

Rosenstiel School of Marine and Atmospheric Science, University of Miami, Miami, Florida

Compositionally variable limu o Pele occurs in widely distributed sediments collected during ROV *Tiburon* dives along the Gorda Ridge axis. The fragments formed deeper than the critical depth of seawater and are unlikely to be formed by supercritical expansion of seawater upon heating in contact with hot lava. Discharge of CO_2 through erupting lava is the most likely way to make such bubbles at >298 bars pressure. The distribution and composition of limu o Pele fragments indicate that low-energy strombolian activity is a common, although minor, component of eruptions along mid-ocean ridges. Combined dissolved and exsolved volatile contents of N-MORB from the Gorda Ridge with 12.8–15.6% spherical vesicles are about 0.78% CO_2 and 0.18 wt% H_2O and exceed estimates of primary CO_2 of only 0.07 to 0.095 wt% calculated from whole rock Nb concentrations. This discrepancy suggests that the magmas accumulated an exsolved volatile phase prior to eruption. The evidence that a separated volatile phase drives strombolian eruptions on the seafloor also implies that volatile bubbles coalesce during storage or transport to the surface. The combination of large bubbles in otherwise dense magma suggests nearly complete coalescence of small bubbles and is most consistent with accumulation of the exsolved volatile phase, most likely near the tops of crustal magma chambers, prior to upward transport in shallow conduits to the eruptive vents on the seafloor. A portion of this CO_2-rich separated fluid phase is released in brief bursts during eruptions where it becomes part of event plumes.

1. INTRODUCTION

Abundant observations and sampling along mid-ocean ridges [e.g., *Ballard et al.*, 1979; *Perfit and Chadwick*, 1998; *Batiza and White*, 2000] have led to the widely accepted view that submarine basaltic eruptions are effusive. These eruptions produce pillow, lobate, and sheet flows [e.g., *Perfit and Chadwick*, 1998], with the flow type mainly determined by effusion rate [e.g., *Griffiths and Fink*, 1992; *Gregg and Fink*, 1995]. Associated basaltic volcaniclastic deposits are rare, but occur in a variety of tectonic settings and at a range of depths.

Many of these deep-sea volcaniclastic deposits contain common to rare sideromelane bubble-wall fragments in addition to the usually more abundant dense angular sideromelane fragments. Bubble-wall fragments, named limu o Pele or Pele's seaweed, were first described from Kiluaea's ongoing eruption [*Hon et al.*, 1988] where they form as lava entering the ocean produces lava bubbles due to incorporation of seawater into the lava stream and expansion to steam [*Mattox and Mangan*, 1997]. This limu o Pele formed at sealevel in a littoral environment, where the volume change as water expands to steam is large. The expansion of seawater during boiling is greatly diminished with increasing pressure, however, making formation of limu o Pele by this mechanism less likely as eruption depth increases. Calculations of the specific volume of the combined vapor-liquid-solid phases generated during boiling [see Figure 9 in *Clague et al.*, 2000a] indicated that seawater undergoes up to about 20 times volume expansion at roughly 2 km depth, still adequate for limu o Pele formation. However, at supercritical conditions (greater than 407°C and 298 bars or ~2936 m depth [*Bischoff and Rosenbauer*, 1988]), the volume expansion from heating seawater is limited to about a factor of four and is not instantaneous, since no phase change occurs. In order to form limu o Pele, bubbles must grow at rates that exceed the cooling rate of the glass, since the wall must remain plastic to continue to stretch. Expansion of seawater under supercritical conditions is unlikely to be rapid enough to form limu o Pele, as recently proposed for samples from subcritical depths [e.g., *Clague et al.*, 2000; *Maicher and White*, 2001].

Volcaniclastic deposits containing limu o Pele fragments have been documented around Hawaii, on a near ridge seamount off the EPR, on the Azores Platform section of the MAR, and on the Sumisu rift. None have been previously described from a normal mid-ocean ridge. Existing occurrences of limu o Pele are described in more detail below.

Tholeiitic limu o Pele fragments have been recovered from Kiluaea's Puna Ridge in sediment at >5400 m at the base of Puna Ridge, although their dissolved volatile contents indicate they erupted at about 2000 m depth [*Clague et al.*, 1995], and from about 2200 m depth [*Clague et al.*, 2000a]. These are similar to littoral limu o Pele produced from the ongoing eruption of Kilauea, but less vesicular. Limu o Pele was also found in unconsolidated sand deposits on Loihi Seamount between 1150–1950 m depth. The Loihi limu o Pele fragments are mainly dense tholeiitic basalt (81%), although some are transitional basalt (12%) or alkalic basalt (7%) [*Clague et al.*, 2000a]. The curvature of the fragments suggests that the average bubble diameter of tholeiitic limu o Pele from Loihi is 5.9 cm.

Morphologically similar limu o Pele fragments of MORB and hawaiite composition were described from Seamount 6 near the East Pacific Rise from 1600–2000 m depth [*Maicher and White*, 2001]. These fragments formed the basis for a detailed model of limu o Pele formation in the deep sea [*Maicher and White*, 2001] that also relied entirely on the expansion of seawater as it boils [see Figure 9 in *Clague et al.*, 2000a, and Figure 5 in *Maicher and White*, 2001] to produce thin-walled lava bubbles. *Maicher and White* [2001] proposed that seawater is incorporated into thin, fluid, rapidly advancing lava flows that covered water-saturated sediment.

Volcaniclastic deposits of enriched-MORB occur as deep as 1700 m at 37°50' N, 31°32' W on the Mid-Atlantic Ridge [*Fouquet et al.*, 1998; *Eissen et al.*, this volume]. The fragments in these deposits include scoria and pumice interpreted as evidence that exsolution of magmatic volatiles played a large role in magma fragmentation. The increased surface area of the melt in contact with seawater apparently then led to hydrovolcanic explosions and the formation of small plate-like shards and abundant lithic and broken crystal fragments. Based on this evidence, *Fouquet et al.* [1998] argue that the volcaniclastic deposits formed due to submarine explosions involving a combination of magmatic volatile exsolution and bulk/surface steam explosivity, probably accompanied by thermal-contraction fragmentation.

Scoriaceous back-arc basin basaltic breccia occurs at about 1800 m in the Sumisu Basin [*Gill et al.*, 1990]. The glass contains about 1.3% H_2O after vesiculation, which suggests that high initial magmatic H_2O contributed to the explosivity. *Gill et al.* [1990] also argued, based on textural and magnetic data, that the breccia was assembled while hot, despite accumulation in deep water.

Strongly alkalic North Arch lavas, ranging from alkalic basalt to nephelinite, formed volcaniclastic deposits containing vesicular angular and rounded glass fragments, spherical glass droplets [*Clague et al.*, 1990], and thin curved glass fragments interpreted as limu o Pele [*Clague et al.*, 2002] during eruptions at 4160 m, well below the critical depth for seawater. Unfortunately, limu o Pele fragments in these samples were mostly altered to palagonite. These fragment shapes were interpreted to form by fragmentation during eruption, most probably by magmatic volatile exsolution. Several additional samples from about 4320 m collected by the *Shinkai 6500* submersible in August 2002 (dive S704) contain unaltered angular and limu o Pele fragments up to several mm across. The limu o Pele fragments, compared to those from Loihi Seamount [*Clague et al.*, 2000a], include some small vesicles and the fragments tend to be thicker and less regularly curved.

In summary, volcaniclastic deposits from the Mid-Atlantic Ridge [*Fouquet et al.*, 1998; *Eissen et al.*, this volume], the Sumisu Backarc Basin [*Gill et al.*, 1990], and the North Arch volcanic field [*Clague et al.*, 1990, 2002] were interpreted as pyroclastic deposits with fragmentation driven by exsolution of magmatic H_2O in the case of the Sumisu basin and magmatic CO_2 in the other cases. On the other hand, those from Puna Ridge [*Clague et al.*, 2000a], Loihi Seamount [*Clague et al.*, 2000a], and Seamount 6 [*Maicher and White*, 2001] were interpreted as hydrovolcanic deposits with fragmentation driven by steam explosivity.

In this paper, we describe volcaniclastic deposits from a number of locations along the Gorda Mid-Ocean Ridge that contain limu o Pele that are inferred to have formed deeper than the critical depth for seawater. We present evidence that submarine limu o Pele is of pyroclastic, rather than hydrovolcanic, origin. We then argue that submarine limu o Pele and angular sideromelane fragments form during mild strombolian-like eruptions driven by a separated magmatic volatile phase, dominated by CO_2 along mid-ocean ridges. We conclude with an evaluation of volatile budgets for normal MORB and some speculation about a magmatic component to event plumes.

2. VOLCANICLASTIC MORB FROM GORDA RIDGE

2.1. Geologic Setting and Ridge Morphology

The Gorda Ridge is a 300-km-long mid-ocean ridge located 250 to 350 km west of the northern California-Oregon margin (Figure 1). The ridge is similar in structure to the slow-spreading Mid-Atlantic Ridge and is characterized by a deep axial valley [*Clague and Holmes*, 1987] that is divided into 5 offset segments named, from south to north, the Escanaba, Phoenix, Central, Jackson, and Northern segments [*Chadwick et al.*, 1998]. The ridge is bounded on the north and south by the Blanco and Mendocino Fracture Zones, respectively. The neovolcanic zone of the northern four segments consists dominantly of pillow flows with variable, but thin sediment cover [e.g., *Clague and Rona*, 1989]. The sedimentation rate along Gorda Ridge is relatively high, roughly 5–14 cm/ky [*Karlin and Zierenberg*, 1994]. In places, the floor of the neovolcanic zone is broken by faults and fissures, usually oriented parallel to the ridge axis. The neovolcanic zone along Gorda Ridge ranges from 2867 to 3975 m deep, with only 5 volcanic cones in the entire neovolcanic zone shallower than 2936 m. The Escanaba segment is filled with turbidite sediments [e.g., *Zuffa et al.*, 2000] and the neovolcanic zone is characterized by sills intruded into the sediment [*Morton et al.*, 1994, *Normark and Serra*, 2001]. The only lava flows are an extensive one at the NESCA hydrothermal site near 41°N, 127°30'W and a smaller one at the SESCA hydrothermal site at 40° 47'N, 127°31'W [*Ross and Zierenberg*, 1994; *Davis et al.*, 1994]. The northern end of

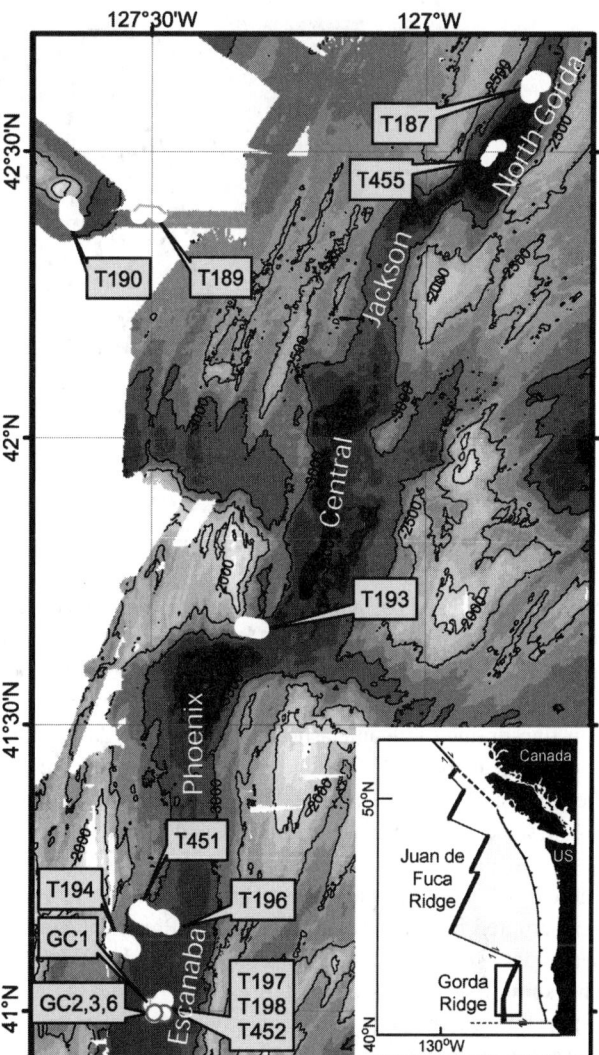

Figure 1. Location map showing *Tiburon* ROV dive locations on Gorda Ridge. Segments of Gorda Ridge axis are labeled. The vesicular MORB samples are from dive T196 and the limu o Pele fragments were recovered in push-cores and sediment scoops during dives T187, T189, T190, T193, T194, T196, T197, T198 in 2000 and during T451, 452, 453, and 455 in 2002 and two gravity cores collected near dives T197 and T198 in 2000.

the Escanaba segment has a group of nearly circular volcanic cones that project above the sediment fill.

2.2. Volcaniclastic Deposits

Volcaniclastic deposits are briefly described from about 3100 m depth at the 1996 eruption site by *Chadwick et al.* [1998]. A bottom photograph (their Figure 3b) shows volcaniclastic black sand on gray talus deposits. They attribute the fine clastic material to fragmentation as pillow lava cas-

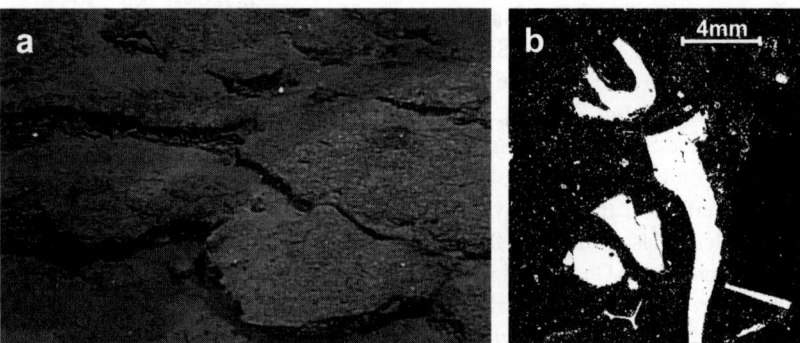

Figure 2. a) Bottom photograph showing fine-grained volcaniclastic deposit at 2883 m depth observed on *Tiburon* dive T190 at top of ridge-parallel fault scarp west of Gorda Ridge axis, field of view is about 1 m wide, b) backscattered electron photomicrograph showing dense fluidal shaped glass fragments in fine-grained matrix consisting of glass fragments to submicron sizes (bright specs).

caded over a fault scarp. The volcaniclastic deposit was not sampled.

We observed and recovered volcaniclastic deposits during two dive programs using MBARI's R/V *Western Flyer* and remotely operated vehicle (ROV) *Tiburon* in August 2000 and July 2002. During dive T190 in 2000, we observed a thin fine-grained volcaniclastic deposit (Figure 2a) at 2883 m depth near the top of a ridge-parallel fault scarp west of the ridge axis. The recovered sample contains dense sand-to-silt sized angular glass fragments and abundant dense, sand-to-silt sized, limu o Pele in a matrix of finely (to submicron sizes) comminuted glass (Figure 2b). The ash may have erupted from the easternmost of the President Jackson Seamounts [*Davis and Clague*, 2000], located to the west. The summit of this seamount presently reaches 1723 m depth, but was probably significantly shallower when the volcano was active and located nearer the spreading center. Due to the uncertainty in the depth of formation of this deposit, we will not discuss it further.

The remaining samples of volcaniclastic materials, and those we will focus on, were collected within the neovolcanic zone along the Gorda Ridge using push-cores and sediment scoops to recover fragments embedded in hemipelagic sediment consisting mainly of mud. At none of the locations were volcaniclastic deposits observed prior to sampling. Sampling sites were selected where sediment cover was thick enough to insert push-cores. Glass fragments were nonetheless recovered in all the samples, except those push-cores collected at the SESCA site in the Escanaba segment. Most of the push-cores yielded only a tiny fraction (a few percent) of glass.

Limu o Pele fragments were recovered in 26 push-cores and sediment scoops during 8 dives using the MBARI ROV *Tiburon* in August 2000, as well as in 2 gravity cores deployed from the R/V *Western Flyer*. Eleven push-cores from 3 additional dives in 2002 also recovered angular sideromelane and limu o Pele fragments. The sites where limu o Pele fragments were recovered are shown in Figure 1 and their distribution as a function of depth is shown in Figure 3. The deepest site is at 3853 m at the southern end of the neovolcanic zone on the Northern segment and the shallowest is off-axis at 2595 m on the west flank of the Escanaba segment.

2.3. Morphology of Fragments

The recovered volcanic fragments are almost entirely unaltered sideromelane glass as dense angular fragments

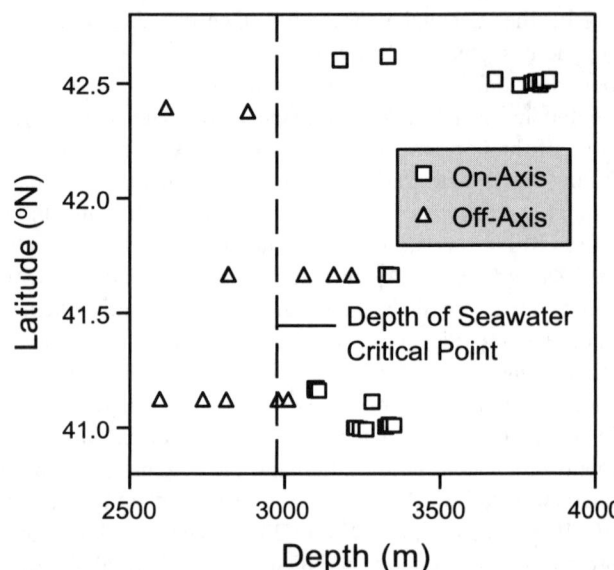

Figure 3. Latitude vs depth of sediment samples from Gorda Ridge that included limu o pele fragments. The samples are divided into axial and off-axis samples.

(Figure 4a) or dense limu o Pele fragments (Figure 4b). Samples from some sites contain plagioclase crystals and glass fragments from other sites have thin white or light tan alteration products or deposits on all sides of the fragments. In most samples, most glass fragments are angular or have concoidal fractures on flat sheets or elongate needles of glass. Most limu o Pele fragments are thin, dense, slightly curved sheets that are smooth on both sides (Figure 4b) or striated on one side (Figure 5b). The thickness is highly variable from nearly colorless transparent sheets a few microns thick to dark brown glass several tens of microns thick (color intensity in Figure 4b is due to variable glass thickness). The striated fragments also have variable thickness, but in elongate patterns. These same fragments com-

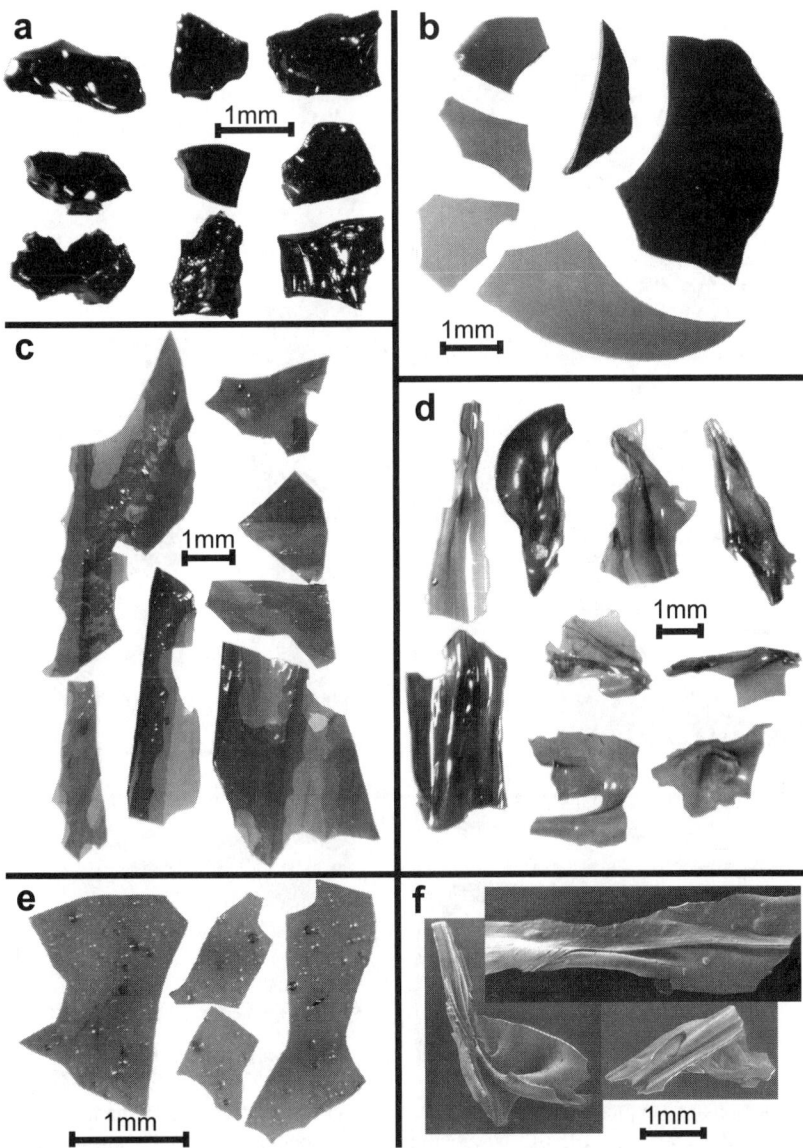

Figure 4. Photomicrographs of samples from *Tiburon* ROV dives on Gorda Ridge in 2000 and 2002. a) dense angular fragments. Pushcore T452-PC67; b) slightly curved limu o Pele sheets that are smooth on both sides. Variable color intensity is due to variable glass thickness. Pushcore T452-PC67; c) limu o Pele fragments that consist of two or even three thin sheets of glass that are tack-welded together. Pushcore T452-PC67; d) limu o Pele fragments that are folded on themselves and tack-welded. Pushcore T196-PC47; e) rare limu o Pele fragments with abundant small vesicles. Pushcore T452-PC67; f) scanning electron microscope images of folded, tack-welded limu o Pele fragments. Left fragment from T193-PC41, top from T196-PC47, and lower right from T193-PC43.

monly contain stretched bubbles with pointed ends (upper right in Figure 5b). Some fragments have a mottled appearance (Figure 5a) due to variable thickness and others contain abundant small vesicles (Figure 4e). The larger of the vesicles in the fragments are broken open, usually on the concave inner wall of the fragment. Other fragments consist of two or even three thin sheets of glass that are tack-welded together (Figure 4c). The layers in these fragments are welded together over most of the surface. Many limu o Pele fragments thicken to one edge (Figure 5d) while others are folded on themselves and tack-welded (Figures 4d and 4f). There are also small fragments of fluidal clasts, which commonly are elongate stretched and twisted glass fragments (Figure 5c). Fluidal clasts range from striated fragments

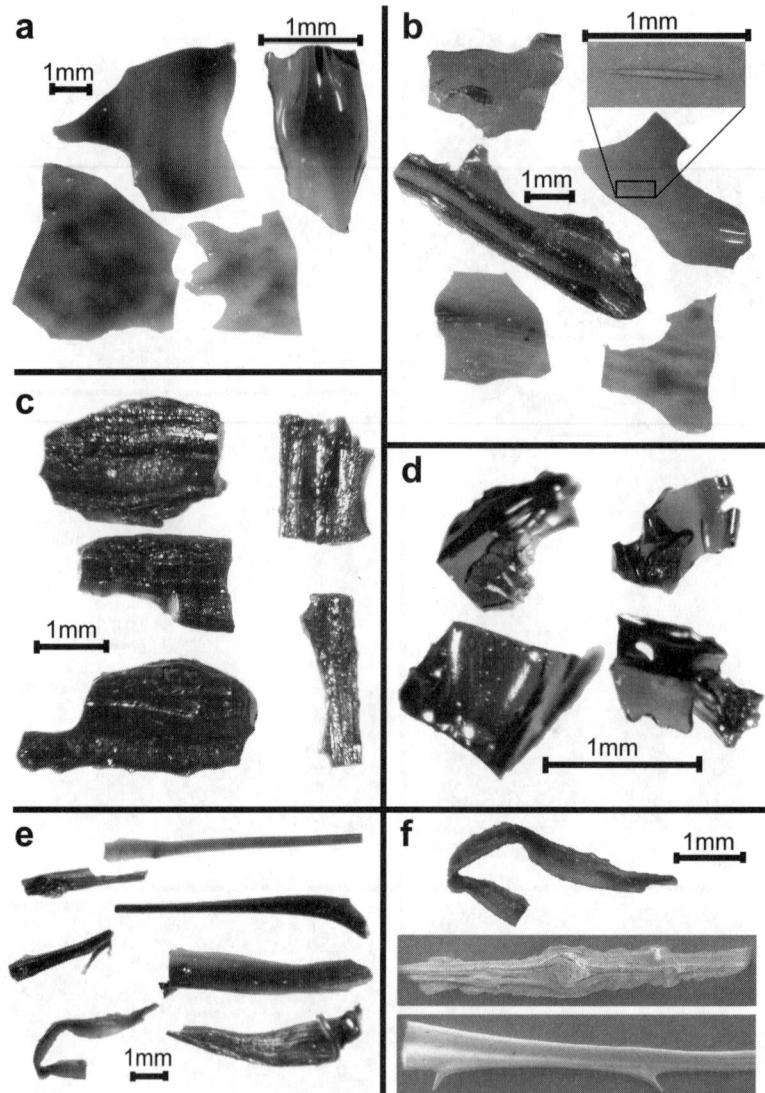

Figure 5. Photomicrographs of samples from *Tiburon* ROV dives on Gorda Ridge in 2000 and 2002. a) three left fragments are mottled limu o Pele, pushcore T452-PC67; upper right fragment has variable thickness, pushcore T452-PC67; b) limu o Pele fragments, pushcore T193-PC47. Upper left fragment has spatter bleb agglutinated to surface, center left is gradational to spatter, other three fragments are striated. Close-up shows that striation in some fragments is due to presence of elongate stretched vesicles; c) small spatter fragments, pushcore T452-PC67, d) limu o Pele that thicken to one edge, pushcore T455-PC57; e) twisted and elongate glass fragments that resemble Pele's hair when extremely elongate, pushcore T194-PC46; f) top image is ribbon spatter enlarged from e); middle is scanning electron microscope image of stretched spatter, T196-PC47; and bottom is scanning electron microscope image of Pele's hair with barbs, T194-PC46.

(left center fragment in Figure 5b) to small stretched rods of glass that resemble Pele's hair when extremely elongate (Figures 5e and 5f). Some of the clasts have a ribbon-like morphology (upper fragment in Figure 5f). There are also rare fragments of limu o Pele with small bits of ribbon-like glass agglutinated to their surfaces (upper left fragment in Figure 5b).

Most limu o Pele fragments are morphologically similar to those from Loihi Seamount described by *Clague et al.* [2000] in having few vesicles or crystals in the glass. One difference is that the curved fragments from Gorda Ridge are rarely regular enough to allow measurements of the initial diameter of the entire lava bubble. The few whose curvature can be determined appear to be from bubbles only a few centimeters in diameter compared to the average 5.9 cm diameter of the measured Loihi bubbles, but similar in size to those reported from Seamount 6 [*Maicher and White*, 2001]. Some rare fragments appear almost flat and represent fragments derived from larger diameter glass bubbles.

2.4. Chemical Compositions of the Volcaniclastic Fragments

We analyzed 241 limu o Pele fragments from the 28 sediment samples recovered in 2000. The fragments range from enriched-MORB to depleted-MORB (Figure 6a). Many have similar compositions to pillow rind glasses from the same dives (Figure 6b), but a number of compositions, including the most enriched-MORB from dive T194, was not sampled as a lava flow. We have also compared the compositional variation of dense angular glass fragments recovered in the same sediment samples and find that there is strong overlap in compositional groups with the limu o Pele fragments (Figure 6c).

Samples from dives T197 and T198 from the NESCA hydrothermal site [*Morton et al.*, 1994], have overlapping glass compositions for pillow rind glasses, dense angular glass fragments and limu o Pele fragments, although the limu o Pele and dense angular fragments extend to somewhat lower MgO than do the sampled pillow rind glasses (6.5% MgO compared to 6.7%). No angular or limu o Pele fragments were recovered at SESCA (dive T199) or an area south of SESCA (dive T200) that are dominated by shallow intrusive sills, rather than surface lava flows. Fragments formed during eruption of the SESCA lava flow may simply

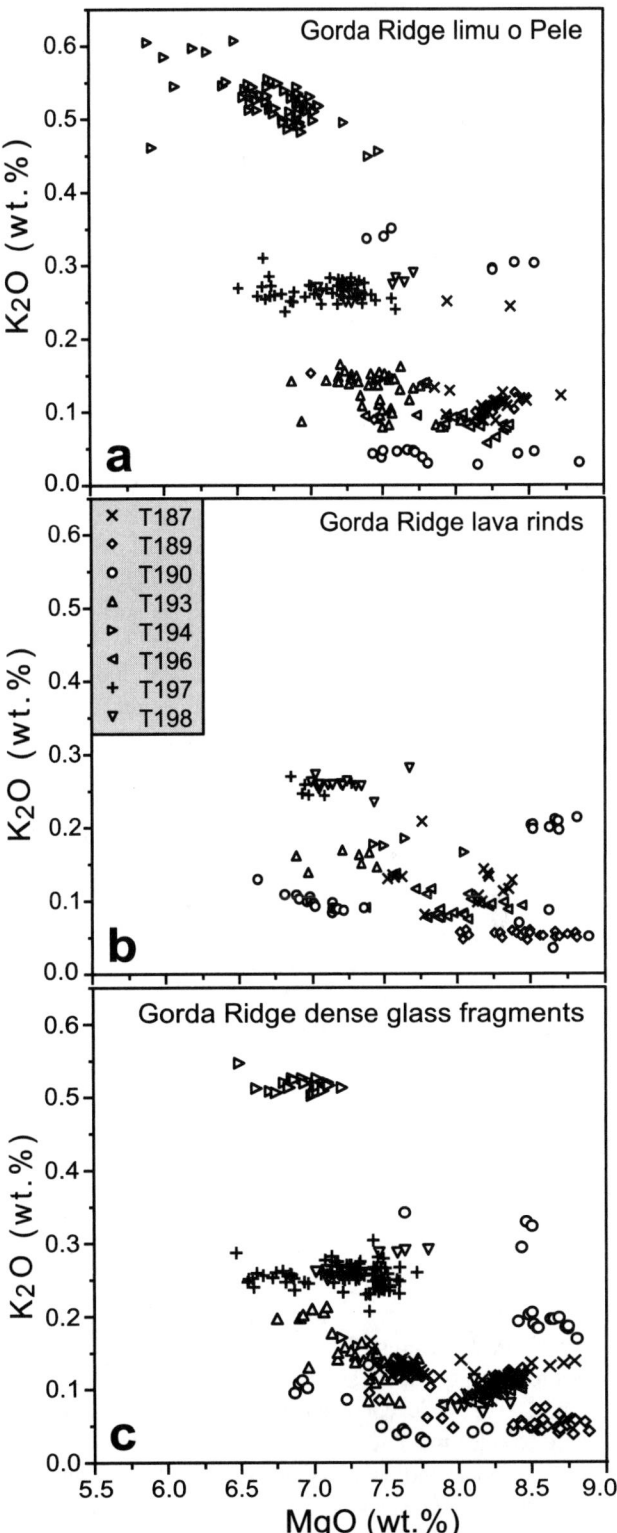

Figure 6. K_2O versus MgO plots for Gorda Ridge glass samples of a) limu o Pele fragments recovered from sediment samples, b) all pillow rinds collected on same dives, and c) angular fragments recovered from same sediment samples as in a).

be too old and covered by sediment too thick to be sampled using 25-cm long push-cores.

2.5. Formation of limu o Pele on the Gorda Ridge

Most of the limu o Pele fragments from the Gorda Ridge was collected at depths greater than the critical point of seawater (Figure 3). If the limu o Pele recovered along the Gorda Ridge from supercritical depths actually formed at those depths, then another mechanism is required to explain formation of limu o Pele. Could it have been produced nearby at depths shallower than the critical point of seawater, transported in the water column, and deposited at the sites where we collected the samples? The short 21-cm cores collected, coupled with regional sedimentation rates of 5-14 cm/ka [*Karlin and Zierenberg*, 1994], suggest that the cores sampled sediment accumulated during the last 2 to 4 k. y. Age estimates based on ^{238}U-^{230}Th disequilibria [*Goldstein et al.*, 1992] indicate that such young eruptions occur within the neovolcanic zone along a narrow band in the center of the axial valley. The entire Gorda Ridge has been mapped by both Seabeam [*Chadwick et al.*, 1998] and high-resolution Simrad EM300 [*MBARI Mapping Team*, 2001] swath systems. These maps show that the only locations along the entire Gorda Ridge where the neovolcanic zone is shallower than 2936 m are the summits of 5 volcanic cones. One is located at the northern end of the Escanaba segment (summit at 2925m), 3 are located near the center of the Phoenix segment (summits at 2867, 2903, and 2930 m), and 1 is located in the center of the Northern segment (summit at 2883 m). Only the summits of these cones rise 11, 69, 33, 6, and 53 m above the 2936 m depth of the critical point of seawater in this area and subcritical conditions for seawater boiling could have occurred only during the final growth of these 5 cones.

If seawater boiling with generation of a vapor phase is responsible for the limu o Pele collected along the Gorda Ridge, then all limu o Pele formed near the summits of these 5 cones and was redistributed along the ridge. The limu o Pele samples recovered at the southern end of the Northern segment are located 33 km from the only shallow cone in the northern half of the Gorda Ridge. The limu o Pele collected at NESCA in the Escanaba segment is located about 23 km from the nearest shallow cone. These are long distances to transport dense angular fragments as large as several mm. The wide compositional range of limu o Pele (Figure 6a) also make it unlikely that all were erupted from these five cones, particularly since our sampling of some of these and other cones along the axis shows that each cone consists of lava with only a small compositional range. Finally, the large lava flow at NESCA is compositionally unique for the entire Gorda Ridge [*Davis and Clague*, 1987, 1990; *Davis et al.*, 1994, 1998; *Rubin et al.*, 1998]. The compositional match of the bubble-wall and angular glass fragments to the flow suggests that the bubble-wall fragments were locally derived. The vent(s) for the NESCA flow are at depths no shallower than 3250 m, well below the critical depth of seawater. These observations demonstrate that most, if not all, limu o Pele from the Gorda Ridge was not only recovered from, but formed at, depths greater than the critical depth of seawater.

If rapid seawater expansion cannot form limu o Pele at supercritical depths, the most likely alternative is that the lava bubbles formed by release of magmatic volatiles exsolved from the magma. Therefore, we propose that limu o Pele at supercritical conditions probably forms during pyroclastic eruptions. The similarity of the fragment morphology and vesicularity of limu o Pele from the Gorda Ridge, where most formed at depths exceeding the critical point of seawater, to that from Loihi Seamount and Puna Ridge [*Clague et al.*, 2000], and Seamount 6 along the East Pacific Rise [*Maicher and White*, 2001], suggests a common mode of formation. Limu o Pele from all these submarine sites differs from the littoral limu o Pele from Kilauea because the fragments are dense rather than highly vesicular glass and the individual fragments tend to be more uniform in thickness. We propose that limu o Pele at all these submarine sites formed during pyroclastic eruptions, regardless of depth, in which exsolved magmatic volatiles disrupted and fragmented magma at vents.

2.6. Eruption Style

What style were these eruptions? Since the eruptions are characterized by release of exsolved magmatic volatiles, they could be either strombolian or hawaiian, or an undescribed style. Strombolian eruptions are characterized by gas and pyroclast discharge in energetic bursts [*Blackburn et al.*, 1976; *Ripepe et al.*, 1993; *Vergniolle and Jaupart*, 1986]. Models of strombolian eruption dynamics fall generally into two classes [*Vergniolle and Mangan*, 2000], those based on collapse of a foam layer in a magma reservoir [*Vergniolle and Jaupart*,, 1990] and bubble coalescence through differential rise speeds of melt and bubbles in the conduit [*Parfitt and Wilson*, 1995]. Both models produce large volatile bubbles in the magma prior to eruption; these bubbles migrate upward in the conduit as a slug flow. Hawaiian fountains, on the other hand, are produced when many small bubbles nucleate and expand in the rising melt, accelerate and disrupt the magma in the conduit, and produce a continuous overshoot of the vent [*Head and Wilson*, 1987, 1989].

Head and Wilson [2003] recently reviewed the theory, landforms, and deposits produced by strombolian and hawaiian eruptions in the submarine environment. *Head and Wilson* [2003] note that deep submarine hawaiian fountains would require such large volatile inventories that basaltic magmas, except for a few highly alkalic and volatile-rich magma types, simply do not contain sufficient volatiles to form such eruptions. The required volatile inventories rise dramatically with increasing pressure, so that a magma at depths as great as 3850 m (depth of the deepest limu o Pele from Gorda Ridge) would have to contain somewhat greater than 13 wt% CO_2 (extrapolated from 3500 m values in their Table 1). Such volatile requirements effectively eliminate the possibility of hawaiian eruptions at these depths. In addition, if the eruptions were hawaiian in character, the angular and the limu o Pele fragments should be highly vesicular and the deposits would most likely be much larger than the observed limu o Pele-bearing deposits. This leaves strombolian as the most likely eruption style to produce limu o Pele in the submarine environment. The eruptions that we propose form limu o Pele are not identical to subaerial strombolian eruptions, but vigorous disruption of the magma surface by slugs of coalesced bubbles are critical in producing pyroclasts in both. We have called these eruptions strombolian, mainly due to this similarity in mechanism. Do the particle types produced tell us anything about how the volatiles accumulated or why subaerial strombolian activity does not produce limu o Pele?

Submarine limu o Pele consists almost entirely of dense glass, as do the associated angular glass fragments. Some exceptions occur, such as the rare fragments described above from Gorda Ridge (Figure 4e), at Loihi Seamount where rare limu o Pele of alkalic basalt composition is moderately vesicular [*Clague et al.*, 2000], and among the strongly alkalic North Arch cones described in the introduction. All the exceptions are only mildly vesicular. The large volatile bubbles that form the submarine sideromelane bubbles (that shatter to form the limu o Pele fragments) therefore discharge through magma that is usually nearly free of small vesicles. As these volatile bubbles rise to the surface of the magma at or near the vent, they create low lava fountains that produce angular glass fragments. In some cases, vigorous bubbling stretches the surface into thin sideromelane bubbles that fragment to become limu o Pele. The absence of small volatile bubbles, but presence of large volatile bubbles, suggests that small bubbles either did not form or have coalesced into large bubbles during magma storage or ascent.

In subaerial strombolian (and hawaiian) eruptions, the small vesicles form during shallow degassing of mainly H_2O and S, with minor contributions from F and Cl. This stage of degassing occurs at very shallow levels (a few hundred m) in the conduit when the magma becomes saturated with these volatiles due to their high solubility in basaltic magma [*Gerlach and Graeber*, 1985]. Carbon dioxide, however, due to its low solubility in basaltic magma [*Dixon et al.*, 1995], exsolves as pressure decreases during rise of magma from the mantle and during storage in sub-axial magma reservoir or lens, where it can accumulate and coalesce into large bubbles. Subaerial strombolian eruptions probably exsolve these volatile components simultaneously, with large coalesced CO_2-rich bubbles triggering eruptive bursts and the shallow-formed H_2O- and S-rich bubbles forming the smaller vesicles in the pyroclasts. In some locations, like Hawaii, where degassing of CO_2 occurs through the summit and degassing of H_2O and S occurs at eruptive vents along the rift zones [*Gerlach and Graeber*, 1985; *Gerlach et al.*, 2002], strombolian activity rarely occurs. Subaerial basaltic eruptions in Hawaii and elsewhere, therefore, almost invariably produce vesicular pyroclasts of spat-

Table 1. Volatile Budgets for E-MORB and N-MORB

Sample	MAR Popping Rocks E-MORB		Gorda Ridge N-MORB T196-R31	
Vesicles (vol%)	17.1		15.6	
Pressure (bars)	383±28		305	
Nb (ppm)	26.2		3.0	
Ce (ppm)	33.5		11.0	
Volatile x	H_2O (wt%)	CO_2 (wt%)	H_2O (wt%)	CO_2 (wt%)
ρ_x at 1000°C	0.0661	0.1453	0.0526	0.1179
$(1-F)M_{x,P}$	0.61	0.60 to 0.82	0.16	0.07 to 0.095
$M_{x,E}$	0.08	−0.46 to −0.22	−0.018	−0.675 to −0.65
$M_{x,D}$	0.51	0.018	0.16	0.015
$M_{x,V}$	0.025	1.04	0.018	0.76

See text for symbols, data sources, and calculation methods. Negative numbers for $M_{x,R}$ indicate that volatile was added to primary melt.

ter and scoria, or reticulite [*Mangan and Vergniolle*, 2000] instead of limu o Pele at primary eruptive vents. These same magmas, after releasing most of their volatiles during transit to the sea, then produce limu o Pele during littoral secondary eruptions.

On the seafloor at depths greater than a few hundred m, only tiny proportions of dissolved H_2O and S exsolve and only a few CO_2-filled bubbles exsolve in the shallow conduit, due to saturation of CO_2 caused by the decrease in pressure as magma migrates upward from the reservoir to the eruptive vent. In mid-ocean ridge basalt, nucleation of bubbles during the final rise to the vent is slow enough that most MORB are supersaturated for their eruption depth [*Dixon et al.*, 1988]. It is because MORB retains its magmatic H_2O and S that it contains so few small vesicles. Discharge of large coalesced CO_2-filled bubbles can stretch the melt and form dense limu o Pele and angular glass fragments. In a later section, we will develop a more complete budget of volatiles for the Gorda Ridge N-MORB.

2.7. Bubble Fragmentation

Fragmentation of the lava bubbles can occur for a variety of reasons. The first is that magmatic gas from the magma continues to inflate the bubble on the lava surface past the point where the thin bubble-wall has become brittle. Some of the limu o Pele fragments that have variable thickness in a mottled pattern probably formed due to over-inflation of the glass bubble. The bubble-wall stretched, became too viscous to stretch evenly, and fragmented. If the bubbles survive intact to this point, they will collapse and fragment as the bubble-wall and the enclosed gas cool in seawater and the pressure inside the bubble decreases below the external pressure. In either case, the initial fragmentation is most likely enhanced by further fragmentation by cooling-contraction granulation of pyroclasts. Some of the limu o Pele fragments that appear to consist of multiple layers of glass (Figure 4c) probably formed during implosion of bubbles, while the glass was still fluid enough to mold against other fragments. A glass bubble could also be blown inside the previous bubble, resulting in double walls.

Implosion of the bubbles during cooling can be quantified if we assume that the volatile phase is entirely CO_2. *Dixon and Stolper* [1995] showed that the exsolved volatile phase in MORB is 95–100% CO_2, so this assumption is a close approximation. In this case, we can calculate the change in density (and volume) as the CO_2 cools inside the bubble using the Peng-Robinson cubic equation of state [*Peng and Robinson*, 1979]. The calculations used a thermodynamic software program, MultiFlash (version 3.0) from Infochem Computer Services Ltd., London. At 250 bars, a magmatic temperature of 1150°C, and a bottom seawater temperature of 2°C, supercritical CO_2 fluid in the bubble with a density of 0.088 gcm^{-3} contracts to liquid CO_2 with a density of 1.031 gcm^{-3} as the glass bubble cools. The 92% volume decrease of CO_2 in the bubble as it cools to ambient temperature leads to collapse and fragmentation of the bubble wall, forming limu o Pele fragments. At 450 bars, magmatic CO_2 with density of 0.152 gcm^{-3} contracts to liquid CO_2 with density of 1.102 gcm^{-3}. At this pressure the volume decrease of the CO_2 in the bubble is still 86%. The increasing density of the magmatic CO_2 as pressure increases indicates that more magmatic CO_2 is required to form the same size glass bubbles at greater depth.

2.8. Transport and Deposition of Fragments

The volcaniclastic deposits along the Gorda Ridge, the summit and upper rift zone on Loihi Seamount [*Clague et al.*, 2000, this issue], the axis of the Puna Ridge, and cones in the North Arch, all contain limu o Pele fragments but are mainly comprised of dense angular glass fragments. These glass fragments have similar compositions to the much less abundant limu o Pele fragments in the same samples. Such angular fragment shapes are commonly attributed to thermal-contraction fragmentation [e.g., *Batiza and White*, 2000]. Many of the fragments have an original glass surface on one side. Particularly along the Gorda Ridge, we have found such dense angular glass fragments, as well as limu o Pele fragments, widely distributed on subdued topography and shallower than nearby lava flows. For example, at NESCA, dense angular glass fragments up to several mm across were recovered in push-cores on a sediment hill 50 m above the surrounding lava flow. These dense fragments had to be ejected from the vent into the water column above the vent at least that high, or ejected to a lower height followed by upward transport in a rising plume of seawater heated at the eruption site, as previously proposed by *Clague et al.* [2000] and *Maicher et al.* [2001]. We propose that these dense fragments also form during mild strombolian-like activity coupled with thermal-contraction fragmentation in much the same way that limu o Pele is formed. The angular fragments may form by discharge of smaller magmatic gas bubbles that do not sufficiently stretch the lava surface to form thin-walled lava bubbles.

Both types of particles are apparently entrained in thermal plumes of hot water rising over the vent and nearby hot flows. *Head and Wilson* [2003] modeled the distribution of pyroclasts in a rising thermal plume. Their calculations were for submarine hawaiian fountains produced by very volatile-rich magma and moderately volatile-rich magma, as well as a case for strombolian eruption from foam accumulated beneath the top of a magma chamber. For the strombolian case, their numerical analysis suggests that the pyroclast/seawater plume should rise only a few hundred m above the erupting vent, so these lower energy eruptions proposed here should form even smaller plumes. Since these eruptions apparently introduce only small amounts of pyroclasts into the water column, the dense particles presumably settle out of the rising plume according to Stoke's law, so we anticipate that particles deposited nearest the source will be more abundant and larger than at greater distances. The height above the bottom, and therefore the distance the particles might be displaced from their source, will vary with the amount of heat input to the thermal plume and the velocity of ocean currents at appropriate depths. Since the initial stage of a thermal or event plume is unlikely to be observed, the size distribution of glass fragments away from a vent might be one way to quantify the time-varying thermal input into the plume.

3. PRIOR EVIDENCE FOR MORB VOLATILE LOSS

Studies over many years of carbon isotope and rare-gas fractionation in MORB [e.g., *Pineau and Javoy*, 1983; *Des Marais and Moore*, 1984; *Mattey et al.*, 1984; *Staudacher et al.*, 1989; *Fisher and Perfit*, 1990; *Javoy and Pineau*, 1991] and CO_2 concentrations in MORB [e.g., *Stolper and Holloway*, 1988; *Gerlach*, 1989; *Dixon et al.*, 1988, 1995; *Dixon and Stolper*, 1995] support the idea that most MORB has degassed a CO_2-rich phase that also acts as a carrier for low solubility rare gases, particularly He. The data in these studies is widely interpreted to indicate that magmatic volatiles dominated by CO_2 are exsolved from MORB magma as it rises from the mantle and the exsolved volatiles may be accumulated and stored as bubbles when the magma is stored, however briefly, in crustal magma chambers [e.g., *Gerlach and Graber*, 1985; *Bottinga and Javoy*, 1989a; *Gerlach*, 1989].

Exsolution of magmatic volatiles and magma buoyancy generated by bubble growth in rising magma is now widely thought to drive magma ascent prior to subaerial eruptions [e.g., *Sparks*, 1978; *Wilson and Head*, 1981] and most probably submarine basaltic eruptions as well [e.g. *Bottinga and Javoy*, 1989a, 1989b, 1990; *Head and Wilson*, 2003]. *Bottinga and Javoy* [1989a] proposed that MORB magmas undergo two stages of bubble nucleation and growth with the first occurring as magma rises through the mantle and the second during eruption. They also proposed that volatiles exsolved during the primary stage of ascent accumulated in magma chambers below the ridge axis and that without the development of this separate volatile rich zone in the magma chamber no eruption occurs. *Sarda and Graham* [1991] discussed the possibility that exsolved volatile accumulation could lead to eruptions where "volatile-charged products may form."

Our analysis shows that N-MORB eruptions produce pyroclasts of limu o Pele during mildly explosive strombolian activity in which large coalesced bubbles of mainly supercritical CO_2 fluid rise through the otherwise dense magma in the vent. Our results are completely in accord with the geochemical studies noted above, but extend those studies by proposing an eruptive mechanism by which the "missing" exsolved volatile phase escapes from MORB magma.

4. MORB VOLATILE EXSOLUTION HISTORY

4.1. Conceptual Framework

Does MORB contain enough magmatic volatiles that can exsolve and accumulate to drive the strombolian-like style of eruptions we have proposed? Our interpretation that MORB limu o Pele from Gorda Ridge forms during vigorous bubbling at the vent and formation of a low pyroclastic eruption column implies that MORB magma contains significantly more gas on eruption than one would infer from the <1% small vesicles contained in typical MORB pillow basalt [e.g., *Moore*, 1979]. We can construct a budget of volatiles by first estimating the Primary volatile contents, $M_{x,P}$ or the mass proportion of volatile x in melt as it separates and rises in the mantle. Since CO_2 exsolves from basaltic magma as it rises through the mantle and while it crystallizes some fraction F in magma chambers, there is a component that we call Exsolved volatiles, or $M_{x,E}$, that may be separated from the melt phase. It is not necessary to independently estimate the percent crystallization F, since the term $(1-F)M_{x,P}$ represents the volatiles in the erupted melts if they had not exsolved any volatiles; these concentrations can be calculated from trace element contents of the erupted magmas, as described below. The volatile component retained in the lava upon eruption is divided into an Vesicle-filling, or $M_{x,V}$ and a Dissolved component, or $M_{x,D}$. A final component consists of volatiles lost during the eruption into the water column; it cannot be determined readily, but may consist mainly of the volatiles separated during rise and storage in a magma reservoir and will be determined as part of the "exsolved, or $M_{x,E}$, component." These volatile components are schematically shown in Figure 7 and are related as follows:

$$(1-F)M_{x,P} = (M_{x,E} + M_{x,V} + M_{x,D}) \quad (1)$$

We use the observation that CO_2 behaves as an incompatible element having a constant ratio to other incompatible

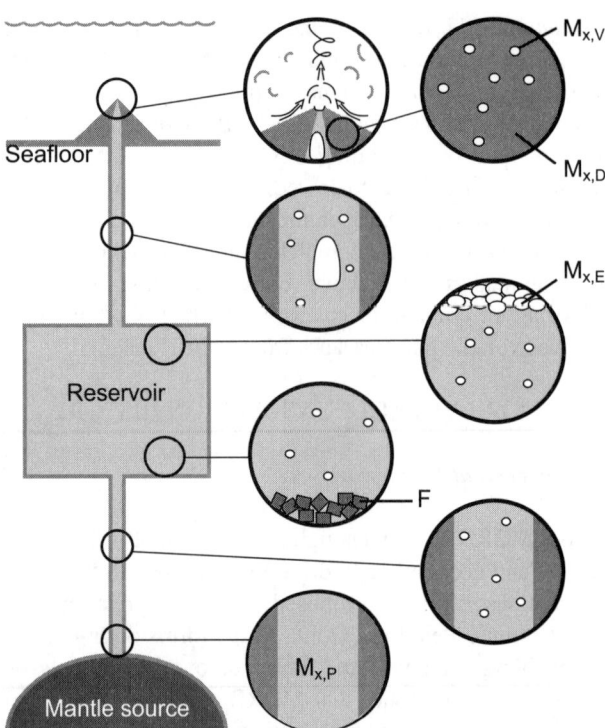

Figure 7. Schematic diagram describing the volatile evolution in MORB magma as it ascends through the upper mantle, pauses in a magma reservoir, ascends to the seafloor, and erupts. White circles and blebs are volatile bubbles, light gray is melt, and dark gray is rock. Upper left exploded view shows formation of limu o Pele during mild strombolian activity with fragments carried upwards in the water column by a rising, heated, turbulent plume of seawater. $M_{x,P}$ is the mass of volatile x in primary magma, $M_{x,E}$ is the mass of volatile x exsolved during transport and storage, $M_{x,D}$ is the mass of volatile x dissolved in the quenched melt, $M_{x,V}$ is mass of volatile x contained in vesicles in the erupted lava, and F is the proportion of fractional crystallization to form the erupted lava from the primary magma. See text for discussion of model and Table 1 for examples of calculated volatile budgets for E-MORB and N-MORB.

elements, such as Nb [*Saal et al.*, 2002; *Hauri et al.*, 2002], and measured Nb concentrations to determine the CO_2 content in the primary magma, corrected for subsequent fractionation, $(1-F)M_{x,E}$ for x=CO_2. The H_2O/Ce=180 for Pacific and south Atlantic MORB [*Michael*, 1995], so we can determine $(1-F) M_{x,E}$ for x=H_2O from the Ce content of the erupted lava. We measure the dissolved $M_{x,D}$ in the glass for both H_2O and CO_2 using FTIR. The vesicle-filling $M_{x,V}$ can be estimated by calculating the volatiles required to form the vesicles retained in the lava sample. In the case of MORB, the mixed volatile phase in the vesicles is 95–100% CO_2 [*Dixon and Stolper*, 1995]; we have used a conservative 95%. The difference between the primary volatile content, corrected for crystallization, and the sum of the dissolved and vesicle-filling volatiles is the volatile content lost during rise from the mantle, storage, and eruption that we have called the exsolved component.

4.2. MAR E-MORB Popping Rocks

One example of this type of reconstruction is based on the so-called "popping rocks" recovered from the Mid-Atlantic Ridge (MAR) [*Hekinian et al.*, 1973; *Pineau et al.*, 1976; *Sarda and Graham*, 1990; *Javoy and Pineau*, 1991; *Gerlach*, 1991; *Graham and Sarda*, 1991]. The popping rocks are enriched- or E-MORB containing 0.63 wt% K_2O and up to 17 volume% vesicles. The samples were recovered from 3770±275 m depth or a pressure of 383±28 bars. Following suggestions from *Gerlach* [1991], *Graham and Sarda* [1991] calculated that the popping rock magma had CO_2 = 0.85 wt% and H_2O =0.59 wt% at 1000°C with a melt density p_m=2.697 gcm^{-3} using non-ideal conditions and assuming that mixing of a binary (H_2O–CO_2) vapor phase is ideal. The 1000°C temperature is preferable to a higher magmatic temperature, since it approximates the temperature when the melt becomes rigid and the vesicles stop contracting as the lava cools from a magmatic temperature of about 1200°C. We have recalculated the content of an ideally mixed supercritical CO_2 and H_2O fluid required to produce the observed volume of vesicles using densities calculated with the Peng-Robinson cubic equation of state [*Peng and Robinson*, 1979], since the ideal gas law is inappropriate for supercritical fluids, such as CO_2 and H_2O at these conditions. The densities of supercritical CO_2 and H_2O at these conditions are listed in Table 1. Simply combining the densities and the volume proportions of the different components allows direct calculation of $M_{x,V}$. The results in Table 1 show that $M_{x,V}$ for x=CO_2 is significantly greater than determined using the ideal gas law.

To calculate the primary volatile contents of the melt $(1-F)M_{x,P}$, we use the Ce content of the popping rocks of 33.5 ppm [*Bougault et al.*, 1988], and H_2O/Ce=183 determined for southern Mid-Atlantic MORB (unfortunately not for the region where the popping rock was recovered) [*Michael*, 1995], and the Nb content of 26.2 ppm [*Bougault et al.*, 1988] and CO_2/Nb covering the range 230±55 [*Saal et al.*, 2002] to 314±125 [*Hauri et al.*, 2002].

The dissolved components, $M_{x,D}$, were measured by FTIR [*J. Dixon*, cited in *Graham and Sarda*, 1991]. Substituting into the equation above, we solve for $M_{x,E}$. The calculations, summarized in Table 1, suggest that the popping rocks have gained some CO_2 during rise and storage.

This increase in CO_2 suggests accumulation of a volatile component consisting mostly of CO_2 supercritical fluid. The imbalance in water probably reflects uncertainty in the H_2O/Ce for this segment of the Mid-Atlantic Ridge.

4.3. Gorda Ridge N-MORB

Gorda Ridge MORB contain between 0.63 and 9.36 ppm Nb and between 5.0 and 16.94 ppm Ce for lavas with glass MgO content ranging from about 7.3 to 9.1 wt% [*Davis and Clague*, 1987, 1990; *Davis et al.*, 1998; *Davis and Clague*, unpublished data). We calculate that $(1-F)M_{x,P}$ for $x=CO_2$ ranges from lows between 0.015 and 0.020 wt% to highs between 0.215 and 0.294 wt% and that $(1-F)M_{x,P}$ for $x=H_2O$ ranges from lows between 0.086 and 0.094 wt% to highs between 0.280 and 0.305 wt%.

Among these many samples, only a few have abundant vesicles that allow reconstruction of $M_{x,V}$. We recovered four vesicular MORB samples from the Gorda Ridge using the MBARI ROV *Tiburon* in August 2000 on dive T296 (Figure 1) [*Clague et al.*, 2001]. The lava fragments were recovered from talus on the inner northern wall of a breached crater in a small volcano located near the northern end of Escanaba Trough at 41.1754° N, 127.5186° W. The most vesicular sample, T196-R31 (Figure 8), contains 15.6 volume% vesicles and was collected at 3006 m depth (305 bars). The top of the cone, the shallowest possible eruption depth, is 2930 m. In contrast to the popping rocks, the Gorda Ridge sample, and several other slightly less vesicular samples from the same location, are depleted N-MORB

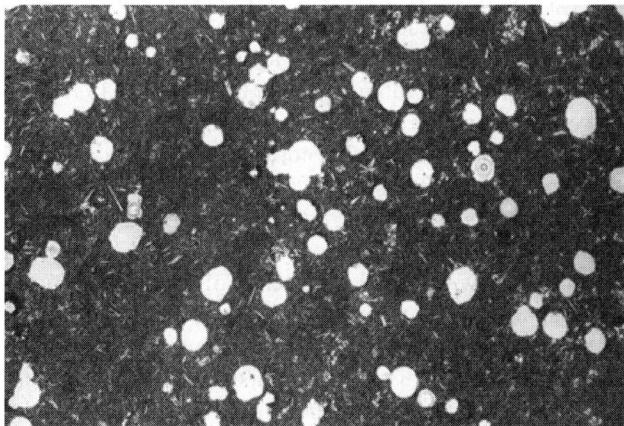

Figure 8. Photograph of thin section (long dimension 2.8 cm) of vesicular N-MORB sample T196-31 collected from 3005 m depth on a small cone at the north end of Escanaba Trough, southern Gorda Ridge. Sample is holocrystalline with no glass rind and has 15.6 volume % vesicles.

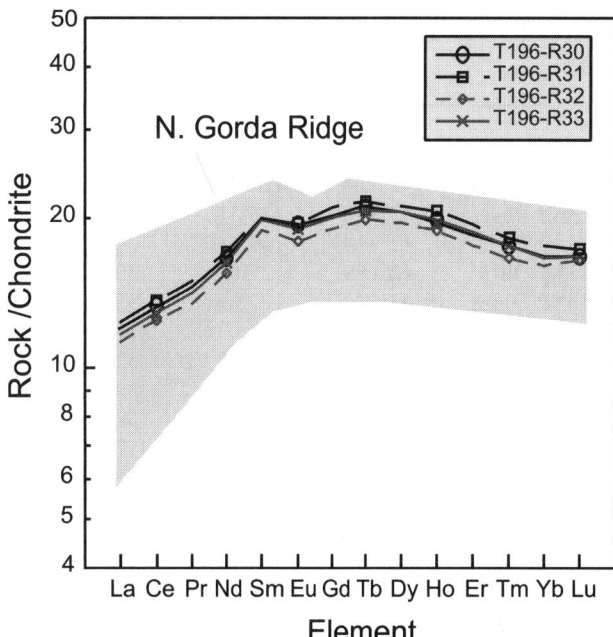

Figure 9. Chondrite-normalized rare earth plot of 4 vesicular basalt samples from Gorda Ridge showing that they are all N-MORB with light-REE-element depleted patterns. Normalizing values from *Boynton* [1984].

as seen by their strongly depleted light-REE patterns (Figure 9).

We repeated the calculations done for the popping rocks for the vesicular N-MORB from Gorda Ridge, except that none of these samples has a glass rind, so $M_{x,D}$ upon eruption and quenching cannot be directly measured using FTIR. However, we can estimate these values using data for 15 other MORB glasses from the axis of Gorda Ridge, determined by FTIR and presented in Table 2. These samples have H_2O (Table 1) that is correlated to Ce [data from *Davis et al.*, 1987] with a $H_2O/Ce=165\pm21$, statistically identical to the ratio of 180 ± 30 proposed by *Michael* [1995] for Pacific MORB. The 15 ridge axis glasses average 147 ppm CO_2, which we have used as a rough estimate of $M_{x,D}$ for $x=CO_2$. The entire volatile budget is summarized in Table 1.

The volatile content of the magma upon eruption is the sum of the dissolved and exsolved components, or 0.18 wt% H_2O and 0.78 wt% CO_2 which contrasts with our calculated $(1-F)M_{x,P}$ for $x=CO_2$ of only 0.07 to 0.095%. The discrepancy indicates that the vesicular samples have accumulated gas exsolved from 8–11 times as much magma. This is direct evidence that $M_{x,P}$, particularly for $x=CO_2$, in mid-ocean ridge magma exsolves during transport and storage, migrates separately from the melt phase, accumulates;

Table 2. Volatile and Trace Element Data for Gorda Ridge MORB Glasses

Sample	Latitude (°N)	Longitude (°W)	Depth (m)	Mg #	Nb (ppm)	Ce (ppm)	H_2O (wt%)	H_2O +/−, N	CO_2 (ppm)	CO_2 +/−, N
KK83NP-D1-8	42°56.6'	126°37.5'	3230	60.4	–	13.8	0.262	0.006, 2	133	11, 2
KK83NP-D3-4	42°54.4'	126°40.3'	3158	62.5	–	12.1	0.234		151	03, 1
KK83NP-D4-15	42°45.4'	126°42.0'	3223	69.4	0.63	5.2	0.088		133	16, 1
KK83NP-D6-12	42°34.3'	126°51.1'	3288	58.8	–	16.5	0.274	0.013, 3	139	09, 3
KK83NP-D7-4	42°28.4'	126°53.3'	3658	62.1	–	15.8	0.280	0.012, 2	156	16, 2
KK83NP-D9-1	42°14.9'	127°04.6'	3048	63.6	–	8.0	0.114	0.000, 2	145	11, 2
KK83NP-D11-8	42°07.2'	127°09.5'	3108	60.9	2.75	12.0	0.215	0.004, 2	164	11, 2
KK83NP-D12-2	42°02.1'	127°09.6'	3425	60.8	–	12.0	0.202		145	14, 1
KK83NP-D13-3	41°48.2'	127°10.1'	3275	64.8	–	8.4	0.126		168	10, 1
KK83NP-D14-8	41°41.7'	127°13.9'	3280	64.1	–	9.9	0.141	0.006, 2	157	15, 2
KK83NP-D16-1	41°31.3'	127°27.4'	3340	56.5	–	16.3	0.278	0.009, 2	134	11, 2
KK83NP-D17-10	41°27.7'	127°13.5'	1703	62.0	–	12.3	0.216		158	12, 1
T196-R31	41°10.5'	127°31.1'	3005	62.8	3.03	11.03	–	–	–	–
L685NC-D36-10	40°59.0'	127°30.2'	3200	60.8	6.8*	16.5*	0.253		122	14, 1
L685NC-D36-11	40°59.0'	127°30.2'	3200	60.8	6.8*	16.5*	0.255	0.017, 2	123	12, 2
L685NC-D31-11	40°46.4'	127°30.8'	3250	65.9	3.9*	8.0*.	0.125		182	15, 1
L585NC-D33-2C	42°45.4'	128°06.1'	~1900	61.2	–	14.0	0.204	0.001, 3	127	07, 3
L585NC-D35-10	42°26.0'	127°40.6'	~1950	69.1	–	6.67	0.119	0.006, 2	161	16, 2

Nb, Ce and Mg# ($100Mg/(Mg+Fe^{2+})$ calculated using $Fe^{2+}=0.9Fe^T$) from *Davis and Clague* [1987], *Davis et al.* [1998], *Davis and Clague* [2000], or *Davis and Clague* [unpublished data]. H_2O and CO_2 by FTIR are the volatile component dissolved in glass. N indicates number of analyses. * indicates trace element data are average for 24 and 6 samples from same sites as L685NC-D36 and –D31, respectively.

it presumably coalesces to form the large bubbles needed to drive mild strombolian activity. This finding is consistent with the widespread view that most MORB has degassed CO_2. In this case, the coalesced exsolved volatile phase may have accumulated in a lava pond in the volcanic cone, since such flat-topped cones are likely to form as overflowing lava ponds [*Clague et al.*, 2000b].

The calculated $(1-F)M_{x,P}$ for $x=CO_2$ of 0.07 to 0.095 wt% minus the $M_{x,D}$ of 0.018 wt% would exsolve to form 1.2–1.7 vol% vesicles, a somewhat greater vesicularity than normally observed in N-MORB from these depths. This discrepancy again supports the idea that supercritical CO_2 liquid exsolves from N-MORB during transport and storage, rather than during shallow devolatization. However, the results for both the MAR popping rocks and the Gorda Ridge N-MORB sample demonstrate that combining vesicle volatile contents with dissolved volatile contents can lead to an overestimation of initial magmatic volatile contents.

5. WATER COLUMN EVENT PLUMES

The exsolved magmatic CO_2 supercritical fluid that forms the limu o Pele bubbles is discharged directly into the seawater above the eruption site. Numerous studies, mainly by a group of researchers at NOAA, have mapped what they call event plumes that accompany seismic swarms along the mid-ocean ridge system [e.g., *Baker et al.*, 1987, 1989, 1998; *Kelley et al.*, 1998; *Lupton et al.*, 1999a, b]. These event plumes are characterized by excess heat and ^3He [*Lupton et al.*, 1999a, 1999b] as well as a wide range of geochemical and particulate markers [e.g., *Baker et al.*, 1998; *Feeley et al.*, 1998; *Kelley et al.*, 1998]. Some tracers in the plumes are conservative (such as ^3He), allowing them to be tracked over long distances [*Lupton*, 1996, 1998].

Interpretation of the origin of these plumes leans heavily toward a massive non-magmatic hydrothermal discharge at the beginning of a submarine eruption [*Baker*, 1998]. In support of a direct eruptive volatile component, as we propose here, *Baker* [e.g., 1995] notes that magmatic volatile input to plumes increases ^3He/heat ratios following eruptions and *Rubin* [1997] suggests that metals and metalloids in plumes may be contributed by magmatic degassing. Based on the widespread occurrence of limu o Pele fragments along at least one mid-ocean ridge, the Gorda Ridge, we propose that one component of event plumes consists of magmatic volatiles discharged directly into the water column during submarine strombolian activity at the erupting vents. The magmatic volatiles would mainly be CO_2, as well as the CH_4, H_2, and ^3He observed in the plumes [*Kelley et al.*, 1998]. The heat in the plume may be mainly derived from conductive cooling of erupted lava, as explored by *Baker* [1998]. The variable ratio of heat/^3He observed in different event plumes [*Lupton et al.*, 1999a, 1999b] may reflect variations in the volume of erupted lava to discharged magmatic volatile phase. In any case, magmatic volatiles are transferred to the oceans by several pathways, one of which is the direct discharge of a separated volatile phase during strombolian activity along mid-ocean ridges.

6. MID-OCEAN RIDGE MAGMA CHAMBERS

There are several implications of the widespread occurrence of submarine strombolian basaltic eruptions since the separated gas phase is first accumulated in the magma chambers and is then discharged into the surrounding seawater during eruptions. It has been widely documented that crustal magma chambers underlying the mid-ocean ridge system produce strong seismic reflections [e.g., *McClain et al.*, 1985; *Detrick et al.*, 1987; *Rohr et al.*, 1988; *Sinton and Detrick*, 1992] that suggest a large velocity contrast between the overlying rock and the underlying magma. If magmatic gas accumulates in magma chambers, it must buoyantly rise and coalesce beneath the rigid lid of the magma chamber. The strong seismic reflections may actually be mapping the distribution of low-density mixtures of magma and separated gas bubbles that grew and coalesced as they rose in the magma. Such a model suggests that the density and therefore the seismic velocity of a thin layer at the top of the magma chamber could be very low. Present seismic models [e.g., *Sinton and Detrick*, 1992] of the size and shape of mid-ocean ridge magma chambers do not consider such a layer.

7. CONCLUDING REMARKS

The results from this study have application to other deep-sea basaltic magmas, including back-arc basin basalts [*Gill et al.*, 1990], ocean island tholeiitic basalt [see *Clague et al.*, this volume], and strongly alkalic basalt [e.g., *Clague et al.*, 1990, 2002]. All of these basalt types contain significantly higher concentrations of primary volatiles than N-MORB which makes it more likely that volatile concentrations required to drive strombolian eruptions [as outlined by *Head and Wilson*, 2003] will be attained by volatile accumulation during transport and storage of these magmas. Pyroclastic basaltic eruptions, similar to subaerial strombolian eruptions because they are driven by release of exsolved and coalesced magmatic volatiles appear to have no depth or compositional limitation in the oceans.

Acknowledgments. The authors thank Captains Mark Vandenberg and Ian Young and the crew of the *Western Flyer*, chief pilots Dale Graves and Buck Reynolds and the entire group of *Tiburon* pilots, and the scientific parties for making the Gorda Ridge cruises so successful. We thank Ed Peltzer for assistance calculating the density of supercritical CO_2 fluids and Jenny Paduan for skillfully preparing many of the figures. The manuscript benefited from reviews by Margaret Mangan, James White and an anonymous reviewer that greatly focused the presentation here. This study was supported by the David and Lucile Packard Foundation through a grant to the Monterey Bay Aquarium Research Institute.

REFERENCES

Baker, E.T., Characteristics of hydrothermal discharge following a magmatic intrusion, *in* Hydrothermal vents and processes, ed. by L.M. Parsons, C.L. Walker, and D.R. Dixon, *Geol. Soc. Spec. Pub. No 87*, 65-76, 1995.

Baker, E.T., G.J. Massoth, and Feeley, R.A., Cataclysmic hydrothermal venting on the Juan de Fuca Ridge, *Nature, 329*, 149-151, 1987.

Baker, E.T., J.W. Lavelle, R.A. Feeley, G.J. Massoth, S.L. Walker, and J.E. Lupton, Episodic venting of hydrothermal fluids from the Juan de Fuca Ridge, *J. Geophys. Res., 94*, 9237-9250, 1989.

Baker, E.T., Patterns of event and chronic hydrothermal venting following a magmatic intrusion: new perspectives from the 1996 Gorda Ridge eruption, *Deep-Sea Res. II, 45*, 2599-2618, 1998.

Ballard, R.D., R.T. Holcomb, and T.H. van Andel, The Galapagos Rift at 86° W. 3. Sheet flows, collapse pits, and lava lakes of the rift valley, *J. Geophys. Res., 84*, 5407-5422, 1979.

Batiza, R., and J.D.L. White, Submarine lavas and hyaloclastite, in *Encyclopedia of Volcanoes*, ed. by H. Sigurdsson, Academic Press, San Diego, 361-381, 2000.

Bischoff, J.L., and R.J. Rosenbauer, An empirical equation of state for hydrothermal seawater, *Am. J. Sci., 285*, 725-763, 1988

Blackburn, E.A., L. Wilson, and R.S.J. Sparks, Mechanisms and dynamics of Strombolian activity, *J. Geol. Soc. London, 132*, 429-440, 1976.

Bottinga, Y., and M. Javoy, MORB degassing: evolution of CO_2, *Earth Planet. Sci. Lett., 95*, 215-225, 1989a.

Bottinga, Y., and M. Javoy, Mid-ocean ridge basalt degassing: bubble nucleation, *J. Geophys. Res., 95*, 5125-5131, 1989b.

Bottinga, Y., and M. Javoy. MORB degassing: bubble growth and ascent, *Chem. Geol., 81*, 255-270, 1990.

Bougault, H., L. Demitriev, J.G. Schilling, A. Sobolev, J.L. Joron, and H.D. Needham, Mantle heterogeneity from trace elements: MAR triple junction near 14°N, *Earth Planet. Sci. Lett., 88*, 27-36, 1988.

Boynton, W.V., Cosmochemistry of the rare earth elements: Meteorite studies, *in Rare earth element Geochemistry*, edited by P. Henderson, Elsevier, N.Y, 63-114, 1984.

Byers, C.D., M.O. Garcia, and D.W. Muenow, Volatiles in basaltic glasses from the East Pacific Rise at 21° N: implications for MORB sources and submarine lava flow morphology, *Earth Planet. Sci. Lett., 79*, 9-20, 1986.

Chadwick, W.W., Jr., R.W. Embley, and T.M. Shank, The 1996 Gorda Ridge eruption: geologic mapping, sidescan sonar, and

SeaBeam comparison results, *Deep-Sea Res. II, 45*, 2547-2569, 1998.

Clague, D.A., A.S. Davis, J.L. Bischoff, J.E. Dixon, and R. Geyer, Lava bubble-wall fragments formed by submarine hydrovolcanic explosions on Loihi Seamount and Kilauea Volcano, *Bull. Volcanol., 61*, 437-449, 2000a.

Clague, D.A., R.T. Holcomb, J.M. Sinton, R.S. Detrick, and M.E. Torresan, Pliocene and Pleistocene alkalic flood basalts on the seafloor north of the Hawaiian Islands, *Earth. Planet. Sci. Lett., 98*, 175-191, 1990.

Clague, D.A., and M.L. Holmes, The geology and mineral potential of the Gorda Ridge, in *Geology and resource potential of the continental margin of western North America and adjacent ocean basins - Beaufort Sea to Baja California*, ed. by Scholl, D.W., Grantz, A. and Vedder, J.G., *Circum-Pacific Council for Energy and Mineral Resources Earth Sci. Series, 6*, 563-580, 1987.

Clague, D.A., J.G. Moore, J.E. Dixon, and W.B. Friesen, Petrology of submarine lavas from Kilauea's Puna Ridge, Hawaii, *J. Petrol., 36*, 299-349, 1995.

Clague, D.A., J.G. Moore, and J.R. Reynolds, Formation of submarine truncated volcanic cones in Hawaii. *Bulletin of Volcanol*ogy , v. 62, p. 214-233, 2000b.

Clague, D.A., and P. Rona, Geology of the GR-14 site on the northern Gorda Ridge, in *Gorda Ridge: A frontier area in the United States Exclusive Economic Zone*, ed. by G.R. McMurray. New York, Springer-Verlag, p. 31-50. 1989.

Clague, D.A., K. Uto, K. Satake, and A.S. Davis, Eruption style and flow emplacement in the submarine North Arch volcanic field, Hawaii, in *Hawaiian Volcanoes: Deep Underwater Perspectives*, ed. E. Takahashi (et al.), Geophysical Monograph 128, Am. Geophys. Union, Washington D.C., 65-84, 2002.

Clague, D.A., R. Zierenberg, A. Davis, S. Goffredi, J. McClain, N. Maher, E. Olsen, V. Orphan, S. Ross, and K. von Damm, MBARI's 2000 expedition to the Gorda Ridge, *RIDGE Events, 11 (3)*, 5-12, 2001.

Davis, A.S., and D.A. Clague, Geochemistry, mineralogy, and petrogenesis of basalt from the Gorda Ridge, *J. Geophys. Res., 92*, 10467-10483, 1987.

Davis, A.S., and D.A. Clague, Gabbroic xenoliths from the northern Gorda Ridge: Implications for magma chamber processes under slow-spreading centers, *J. Geophys. Res., 95*, 10885-10905, 1990.

Davis, A.S., and D.A. Clague, President Jackson Seamounts, northern Gorda Ridge: tectonomagmatic relationship between on- and off-axis volcanism, *J.Geophys. Res., 105*, 27939-27956, 2000.

Davis, A.S., D.A. Clague, and W.F. Friesen, Petrology and mineral chemistry of basalt from Escanaba Trough, southern Gorda Ridge, *U.S. Geol. Surv. Bull., 2022*, 153-170, 1994.

Davis, A.S., D.A. Clague, and W.M. White, Geochemistry of basalt from Escanaba Trough: Evidence for sediment contamination, *J. Petrol., 39*, 841-858, 1998.

Des Marais, D.J., and J.G. Moore, Carbon and its isotopes in mid-oceanic basaltic glasses, *Earth Planet. Sci. Lett., 69*, 43-57, 1984.

Detrick, R.S., P. Buhl, E. Vera, J. Mutter, J. Orcutt, J. Madsen, and T. Brocher, Multichannel seismic imaging of a crustal magma chamber along the East Pacific Rise, *Nature, 326*, 35-41, 1987.

Dixon, J.E., and E.M. Stolper, An experimental study of water and carbon dioxide solubilities in mid-cean ridge basaltic liquids. Part II: Applications to degassing, *J. Petrol. 36*, 1633-1646, 1995.

Dixon, J.E., E.M. Stolper, and J.R. Delaney, Infrared spectroscopic measurements of CO_2 and H_2O in Juan de Fuca Ridge basaltic glasses, *Earth Planet. Sci. Lett., 90*, 87-104, 1988.

Dixon, J.E., E.M. Stolper, and J.R. Holloway, An experimental study of water and carbon dioxide solubilities in mid-cean ridge basaltic liquids. Part I: Calibration and solubility models, *J. Petrol., 36*, 1607-1631, 1995.

Feeley, R.A., E.T. Baker, G.T. Lebon, J.F. Gendron, G.J. Massoth, and C.W. Mordy, Chemical variations of hydrothermal plume particles in the 1996 Gorda Ridge event and chronic plumes, *Deep-Sea Res. II, 45*, 2637-2664, 1998.

Fisher, D.E., and M.R. Perfit, Evidence from rare-gases for magma-chamber degassing of highly evolved mid-ocean-ridge basalt, *Nature, 343*, 450-452, 1990.

Fouquet, Y., J.-P. Eissen, H. Ondreas, F. Barriga, R. Batiza, and L. Danyushevsky, Extensive volcaniclastic deposits at the Mi-Atlantic Ridge axis: results of deep-water basaltic explosive volcanic activity?, *Terra Nova, 10*, 280-286, 1998.

Gerlach, T.M., Degassing of carbon dioxide from basaltic magma at spreading centers: II. Mid-ocean ridge basalts, *J. Volcanol. Geotherm. Res., 39*, 221-232, 1989.

Gerlach, T.M., Comment on "Mid-ocean ridge popping rocks; implications for degassing at ridge crests" by P. Sarda and D. Graham, *Earth. Planet. Sci. Lett., 105*, 566-567, 1991.

Gerlach, T.M., and E.J. Graeber, Volatile budget of Kilauea Volcano, *Nature, 313*, 273-277, 1985.

Gerlach, T.M.., K.A. McGee, T. Elias, A.J. Sutton, and M.P. Doukas, Carbon dioxide emission rate of Kilauea Volcano: implications for primary magma and the summit reservoir, *J. Geophys. Res., 107*, 2189, doi:10.1029/2001JB000407, 2002.

Gill, J., P. Torssander, H. LaPierre, R. Taylor, K. Kaiho, M. Koyama, M. Kusakabe, J. Aitchison, S. Cisowski, K. Dadey, K. Fujioka, A. Klaus, M. Lovell, K. Marsaglia, P. Pezard, B. Taylor, and K. Tazaki, Explosive deep water basalt in the Sumisu backarc rift, *Science, 248*, 1214-1217, 1990.

Goldstein, S.J., M.T. Murrell, D.R. Janecky, J.R. Delaney, and D.A. Clague, Geochronology and petrogenesis of MORB from the Juan de Fuca and Gorda ridges by ^{238}U-^{230}Th disequilibrium, *Earth Planet. Sci. lett., 109*, 255-272, 1992.

Gregg, T.K.P., and J.H. Fink, Quantification of submarine lava flow morphology through analog experiments, *Geology, 23*, 73-76, 1995.

Graham, D., and P. Sarda, Reply to comment by T.M. Gerlach on "Mid-ocean ridge popping rocks: implications for degassing at ridge crests", *Earth Planet. Sci. Lett., 105*, 568-573, 1991.

Griffiths, R.W., and J.H. Fink, Solidification and morphology of submarine lavas: a dependence on extrusion rate, *J. Geophys. Res., 97*, 19729-19737, 1992.

Hauri, E., K. Gronvold, N. Oskarsson, and D. McKenzie, Abundance of carbon in the Icelandic mantle: constraints from melt inclusions, *Trans. Am. Geophys. Union, EOS.*, 2002.

Head, J.W., and L. Wilson, Lava fountain heights at Pu'u O'o,

Kilauea, Hawaii; Indicators of amount and variations of exsolved magmatic volatiles, *J. Geophy. Res., 92*, 13715-13719, 1987.

Head, J.W., and L. Wilson, Basaltic explosive eruptions: Influence of gas release patterns and volume fluxes on near-vent dynamic morphology, and the formation of subsequent pyroclastic deposits (cinder cones, spatter cones, rootless flows, lava ponds, and lava flows), *J. Volcanol. Geotherm. Res., 37*, 261-271, 1989.

Head, J.W., and L. Wilson, Deep submarine pyroclastic eruptions: Theory and predicted landforms and deposits, *J. Volcanol. Geotherm. Res., 121,* 155-193, 2003.

Hekinian, R., M. Chaigneau, and J.-L. Cheminee, Popping rocks and lava tubes from the Mid-Atlantic rift valley at 36° N, *Nature, 245,* 277, 1973.

Hon, K., C.C. Heliker, and J.I. Kjargaard, Limu o Pele: a new kind of hydroclastic tephra from Kilauea Volcano, Hawaii, *Geol. Soc. Am., 20,* A112-A113, 1988.

Javoy, M., and F. Pineau, The volatiles record of a "popping" rock from the Mid-Atlantic Ridge at 14° N: chemical and isotopic compositions of gas trapped in the vesicles, *Earth Planet. Sci. Lett., 107*, 598-611, 1991.

Karlin, R.E., and R.A. Zierenberg, Sedimentation and neotectonism in the SESCA area, Escanaba Trough, *U.S. Geol. Surv. Bull., 2022*, 131-141, 1994.

Kelley, D.S., M.D. Lilley, J.E. Lupton, and E.J. Olson, Enriched H_2, CH_4, and ^3He concentrations in hydrothermal plumes associated with the 1996 Gorda Ridge eruptive event, *Deep-Sea Res. II, 45*, 2665-2682, 1998.

Lupton, J.E., A far-field hydrothermal plume from Loihi Seamount, *Science, 272,* 976-979, 1996.

Lupton, J.E., Hydrothermal helium plumes in the Pacific Ocean, *J. Geophys. Res., 103*, 15853-15868, 1998.

Lupton, J.E., E.T. Baker, R. Embley, R. Greene, and L. Evans, Anomalous helium and heat signatures associated with the 1998 Axial Volcano event, Juan de Fuca Ridge, *Geophys. Res. Lett., 26*, 3449-3452, 1999a.

Lupton, J.E., E.T. Baker, and G.J. Massoth, Helium, heat and the generation of hydrothermal event plumes at mid-ocean ridges, *Earth Planet. Sci. Lett., 171*, 343-350, 1999b.

Maicher, D., and J.D.L. White, The formation of deep-sea limu o Pele, *Bull. Volcanol., 63*, 482-496, 2001.

Mattey, D.P., R.H. Carr, I.P. Wright, and C.T. Pillinger, Carbon isotopes in submarine basalts, *Earth Planet. Sci. Lett., 70*, 196-206, 1984.

Mattox, T.N., and M. T. Mangan, Littoral hydrovolcanic explosions: a case study of lava-seawater interaction at Kilauea Volcano, *J. Volcanol. Geotherm. Res., 75*, 1-17, 1997.

MBARI Mapping Team, MBARI west coast seamounts and ridges multibeam survey, *Monterey Bay Aquarium Research Institute Digital Data Series*, No. 7, 2001.

McClain, J.S., J.A. Orcutt, and M. Burnett, The East Pacific Rise in cross-section: A seismic model, *J. Geophys. Res. 90*, 8627-8640, 1985.

Michael, P., Regionally distinctive sources of depleted MORB: Evidence from trace elements and H_2O, *Earth Planet. Sci. Lett., 131*, 301-320, 1995.

Moore, J.G., Vesicularity and CO_2 in mid-ocean ridge basalt, *Nature, 282*, 250-253, 1979.

Morton, J.L., R.A. Zierenberg, and C.A. Reiss, Geologic, hydrothermal, and biologic studies at Escanaba Trough: an introduction, *U.S. Geol. Surv. Bull., 2022*, 1-18, 1994.

Normark, W.R., and F. Serra, Vertical tectonics in northern Escanaba Trough as recorded by thick late Quaternary turbidites, *J. Geophys. Res., 106*, 13,793-13,802, 2001.

Parfitt, E.A., and L. Wilson, Explosive volcanic eruptions-IX. The transition between Hawaiian-style fountaining and strombolian explosive activity, *Geophys. J. Internat., 121*, 226-232, 1995.

Peng, D.-Y. and D.B. Robinson, The calculation of three-phase solid-liquid-vapor equilibrium using an equation of state, Advances in Chemistry Series 182, American Chemical Society, Washington, D.C. 1979, 185-196, 1979.

Perfit, M.R., and W.W. Chadwick Jr., Magmatism at mid-ocean ridges: Constraints from volcanological and geochemical investigations, in *Faulting and Magmatism at Mid-Ocean Ridges*, Geophyical Monograph 106, Am. Geophys. Union, Washington D.C., 59-115, 1998.

Pineau, F., M. Javoy, and Y. Bottinga, $^{13}C/^{12}C$ ratios of rocks and inclusions in popping rocks of the Mid-Atlantic Ridge: their bearing on the problem of isotopic composition of deep seated carbon, *Earth Planet. Sci. Lett., 29*, 413-421, 1976.

Pineau, F., and M. Javoy, Carbon isotopes and concentrations in mid-ocean ridge basalts, *Earth Planet. Sci. Lett., 62*, 239-257, 1983.

Ripepe, M., M. Rossi, and G. Saccorotti, Image processing of explosive activity at Stromboli, *J. Volcanol. Geotherm. Res., 54*, 335-351, 1993.

Rohr, K., B. Mildreit, and C.J. Yorath, Asymmetric deep crustal structure across the Juan de Fuca Ridge, *Geology, 16*, 533-537, 1988.

Ross, S.L., and R.A. Zierenberg, Volcanic geomorphology of the SESCA and NESCA sites, Escanaba trough, *U.S. Geol. Surv. Bull., 2022*, 143-151, 1994.

Rubin, K., Degassing of metals and metalloids from erupting seamount and mid-ocean ridge volcanoes: Observations and predictions, *Geochim. Cosmochim. Acta, 61*, 3525-3542, 1997.

Rubin, K.H., M.C. Smith, M.R. Perfit, D.M. Christie, and L.F. Sacks, Geochronology and geochemistry of lavas from the 1996 North Gorda Ridge eruption, *Deep-Sea Res. II*, 45, 2571-2597, 1998.

Saal, A.E., E.H. Hauri, C.H. Langmuir, and M.R. Perfit, Vapor undersaturation in primitive mid-ocean ridge basalt and the volatile content of the Earth's upper mantle, *Nature 419*, 451-455, 2002..

Sarda, P., and D. Graham, Mid-ocean ridge popping rocks: implications for degassing at ridge crests, *Earth Planet. Sci. Lett., 97*, 268-289, 1990.

Sinton, J.M., and R.S. Detrick, Mid-ocean ridge magma chambers, *J. Geophys. Res., 97*, 197-216, 1992.

Sparks, R.S.J.,The dynamics of bubble formation and growth in magmas: A review and analysis, *J. Volcanol. Geotherm. Res., 3*, 1-37, 1978.

Staudacher, T., P. Sarda, S.H. Richardson, J.C. Allegre, I. Sagna, and L.V. Dimitriev, Noble gases in basalt glasses from the Mid-

Atlantic Ridge topographic high at 14° N; geodynamic consequences, *Earth Planet. Sci. Lett., 96*, 119-133, 1989.

Stolper, E.M., and J.R. Holloway, Experimental determination of the solubility of carbon dioxide in molten basalt at low pressure, *Earth Planet. Sci. Lett., 87*, 397-408, 1988.

Vergniolle, S., and C. Jaupart, Separated two-phase flow and basaltic eruptions, *J. Geophys. Res., 91*, 12842-12860, 1986.

Vergniolle, S., and C. Jaupart, Dynamics of degassing at Kilauea Volcano, Hawaii, *J. Geophys. Res., 95*, 2793-2809, 1990.

Vergniolle, S. and M. Mangan, Hawaiian and strombolian eruptions, in *Encyclopedia of Volcanoes*, ed. by H. Sigurdsson, Academic Press, San Diego, 447-461, 2000.

Wilson, L., and J.W. Head, Ascent and eruption of basaltic magma on the Earth and Moon, *J. Geophys. Res., 86*, 2971-3001, 1981.

Zuffa, G.G., W.R. Normark, F. Serra, and C.A. Brunner, Turbidite megabeds in an oceanic rift valley recording Jokulhlaups of Late Pleistocene glacial lakes of the Western United States, *J. Geol., 108*, 253-274, 2000.

David A. Clague, Alicé S. Davis, Monterey Bay Aquarium Research Institute, 7700 Sandholdt Road, Moss Landing, CA 95039-9644

Jacqueline E. Dixon, Rosenstiel School of Marine and Atmospheric Science, University of Miami, 4400 Rickenbacker Causeway, Miami, FL 33149

Hyaloclastite From Miocene Seamounts Offshore Central California: Compositions, Eruption Styles, and Depositional Processes

Alicé S. Davis and David A. Clague

Monterey Bay Aquarium Research Institute, California, USA

Hyaloclastite deposits are abundant on mid-Miocene volcanic seamounts offshore central California. The glass compositions are predominantly evolved hawaiite and mugearite, although minor amounts of tholeiitic to alkalic basalt are also present. Textural features give evidence for different eruption styles. For the evolved, alkalic compositions, fragmentation occurred primarily in response to exsolution of magmatic gases as the magma approached and erupted on the seafloor. Textural features of pyroclasts suggest formation of lava fountains of limited size and height, depending on water depth. Monomict, clast-supported hyaloclastite of highly vesicular pyroclasts suggest limited dispersal and deposition near vent sites. Matrix-supported, polymict breccias are reworked and displaced by currents into deeper water. The narrow range of glass compositions extending over multiple layers of volcanic sandstone suggests deposition from a slurry of tephra and water directly related to an eruption. Basaltic hyaloclastite of vesicle-free, angular glass fragments from the deepest site apparently formed from quench granulation. Sulfur contents suggest eruption depths ranging from near sea level to over 2000 m for samples collected from 1300 to over 3400 m depth, implying large amounts of subsidence for these seamounts.

INTRODUCTION

Volcaniclastic deposits consisting largely of basaltic glass fragments have been described from a variety of tectonic settings. Despite a considerable variety of compositions, textural characteristics, and size of particles, all of these rocks may be loosely referred to as hyaloclastite if the main component is glass [*Fisher and Schmincke*, 1984]. Hyaloclastite comprises a large proportion of the material erupted on ocean islands [e.g: *Staudigel and Schmincke*, 1984; *Clague et al.*, 1995; *Moore and Chadwick*, 1995], with the largest amount generated as the volcano grows to near sea level [e.g. *Staudigel and Schmincke*, 1984]. Similar, deep water erupted deposits occur at off-axis seamounts at fast- and slow spreading centers [e.g. *Batiza et al.*, 1984; *Batiza and White*, 2000; *Davis and Clague*, 2000]. They also form a significant component of the sediment record at island arc volcanoes [*Gill et al.*, 1990] and have been reported from nearly flat-lying oceanic flood basalts north of the Hawaiian Islands [*Clague et al.*, 1995, 2002]. No eruption of these deposits in the deep ocean has ever been observed and relatively little is known about them because they are not easily accessible.

This paper describes hyaloclastite samples recovered from Miocene volcanoes located at the continental margin offshore central California. We use the glass compositions and the textural characteristics to interpret the eruptive and depositional processes and compare these hyaloclastites to similar deposits from other locations.

2. GEOLOGIC SETTING

Davidson, Guide, Pioneer, Gumdrop, and Rodriguez seamounts are five volcanic structures located at the continental margin offshore central to southern California (Figure 1), that are unlike intraplate, ocean island volcanoes or near-ridge seamounts [*Davis et al.*, 2002]. Variable in size (~10 to ~50 km in length), these five seamounts are morphologically complex, northeast-southwest trending ridges rather than conical structures with essentially circular bases as is typical for ocean island or near-ridge volcanoes. Unlike most oceanic volcanoes, they lack summit calderas or pit craters. Instead, they consist of a series of cones aligned along parallel ridges that are separated by sediment-filled troughs. This northeast-southwest trending fabric, observed for many seamounts offshore southern California and Baja [e.g. *Lonsdale,* 1991; *Davis et al.*, 1995], reflects the ridge-parallel structure of the underlying ocean crust [*Davis et al.*, 2002]. However, $^{40}Ar/^{39}Ar$ laser fusion ages indicate multiple episodes of volcanism, younger by at least

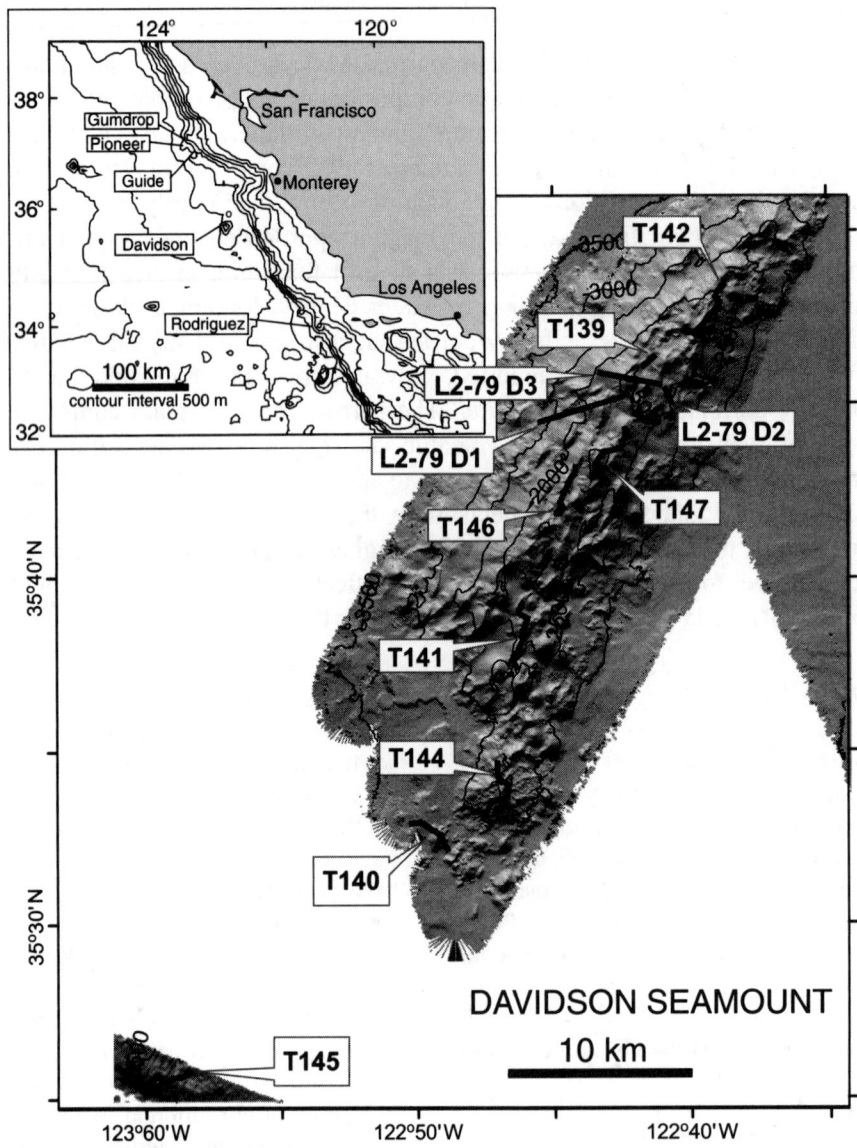

Figure 1A. Maps showing ROV tracks and dredge locations on (A) Davidson Seamount and (B) Guide Seamount. Inset map shows location of volcanic seamounts offshore central California.

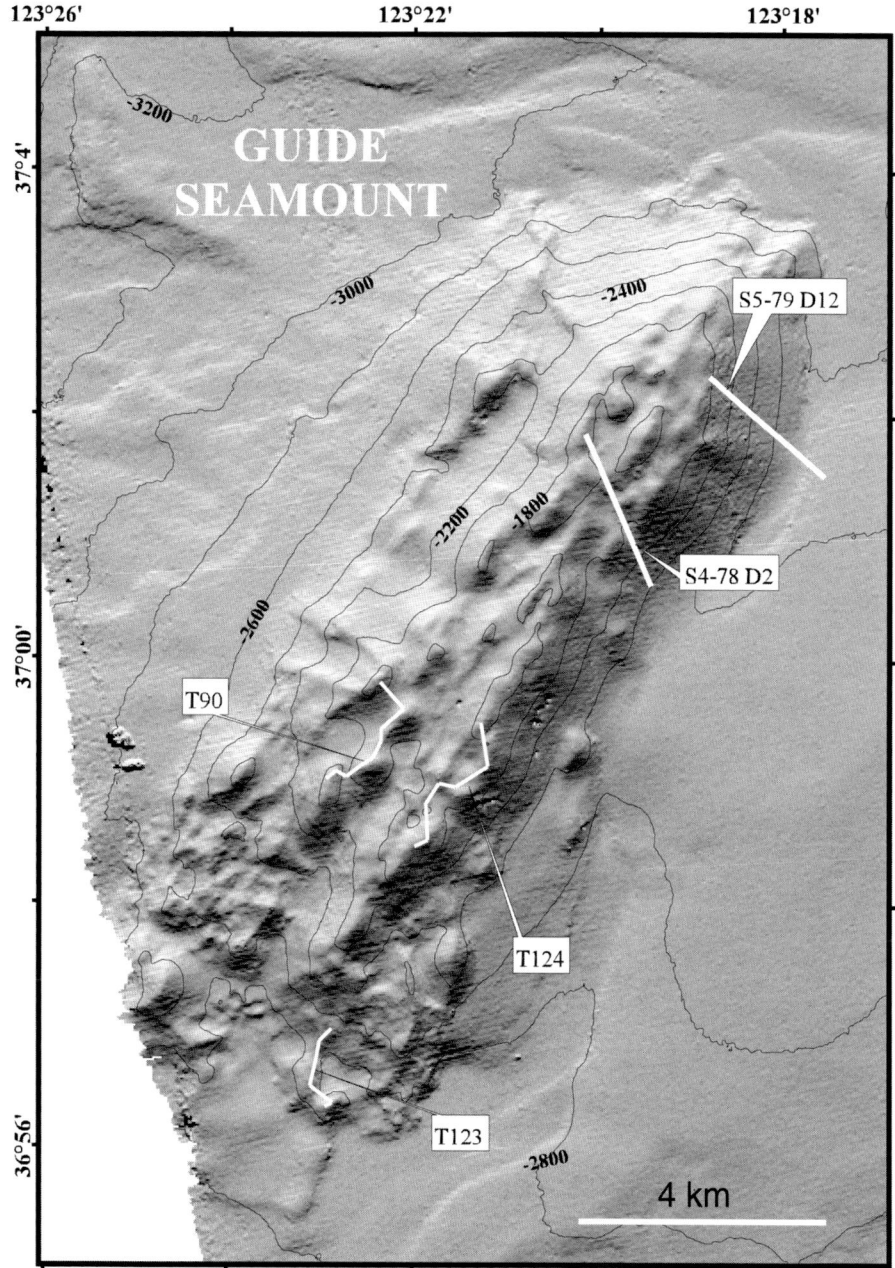

Figure 1B.

7 to 11 million years [*Davis et al.*, 2002] than the underlying ocean crust.

Volcanic rocks recovered from these seamounts are predominantly differentiated alkalic basalt, hawaiite, and mugearite. Isotopic compositions (Sr, Nd, Pb) indicate formation by small percentage of melting of variably enriched MORB-source-type mantle [*Davis et al.*, 2002]. The presence of mantle xenoliths in many samples suggests magma originated in the upper mantle. Xenoliths of alkalic cumulates and megacrysts of amphibole, feldspar, and titanomagnetite suggest extensive crystal fractionation occurred also in the upper mantle. Variable liquid lines of descent suggest small volumes of magma fractionally crystallized and erupted in isolated batches. Episodes of sporadic eruptions (16-10 Ma) presumably resulted from decompression melting of mantle rising along existing zones of weaknesses undergoing extension related to movement along transform-fault systems [*Davis et al.,* 2002].

3. SAMPLING SITES

Hyaloclastite was recovered from each of these seamounts. We will primarily focus on samples from Davidson Seamount (Figure 1a), because they represent the most comprehensive suite, covering the greatest range in collection depths. Several averaged glass analyses for dredged hyaloclastite samples from Davidson and dredge and dive samples from the other seamounts were reported by *Davis et al.* [2002]. We include additional analyses of samples from Guide Seamount (Figure 1b), because the compositionally most diverse polymict hyaloclastite was recovered at that site. Sample locations and water depths are given in Table 1.

Hyaloclastite outcrops, consisting of more or less continuous flat, slabby surfaces (Figure 2a), are found at the tops of each of the seamounts. The non-vesicular hyaloclastite from the deepest dive (~3400 m, T145) occurred as discontinuous, crudely bedded, manganese-covered outcrops (~10 m high), protruding through thick layers of sediment. An outcrop of bedded sandstone, several meters high, was observed on one dive on Davidson Seamount (Figure 2b) but was not sampled. Bedded volcaniclastic rock, consisting of glass sand, was recovered in a dredge that may well be part of this or a similar outcrop.

4. SAMPLING AND ANALYTICAL METHODS

The dive samples were collected using ROV *Tiburon* during two cruises of the Monterey Bay Aquarium Research Institute's RV *Western Flyer* in March and May of 2000. The dredge samples were collected on multiple cruises of the U.S. Geological Survey [for details see *Davis et al.,* 2002].

The hyaloclastite samples were cut for standard and polished thin sections for microscope and electron microprobe analyses, respectively. The glasses were analyzed with a JEOL JX 8900 Superprobe at the U.S. Geological Survey in Menlo Park, using synthetic and natural glass and mineral standards. Details are described by *Davis et al.* [1994]. Volatile analyses were carried out with a Cameca IMS-6F ion probe at Carnegie Institute of Washington, Department of Terrestrial Magnetism, using methods described by *Hauri et al.* [2002].

5. DESCRIPTION OF HYALOCLASTITE

In accord with the broad definition for hyaloclastite [*Fisher and Schmincke,* 1984], all samples are predominantly composed of glass (>70%). The glass may be fresh, or partially to

Table 1. Location and Water Depth for Hyaloclastite Samples

Sample	Latitude (°N)	Longitude (°W)	Depth (m)	Sample type
Davidson Seamount				
T139-R4	35.772	122.691	1611	Monomict breccia
T140-R6	35.547	122.824	3077	Monomict hawaiite sandstone
T141-R8	35.634	122.770	2300	Polymict breccia
T142-R1	35.782	122.659	1871	Monomict breccia
T142-R5	35.785	122.665	1755	Polymict breccia
T142-R8	35.785	122.663	1748	Polymict breccia
T144-R6	35.564	122.786	2624	Monomict breccia
T145-R5	35.427	122.998	3380	Polymict basalt sandstone
T147-R1	35.720	122.724	1337	Polymict breccia
T147-R6	35.723	122.723	1281	Monomict breccia
T147-R8	35.724	122.722	1363	Polymict breccia
T147-R16	35.729	122.713	1580	Polymict volcanic conglomerate
L2-79D1-7	35.749	122.739	2060	Monomict volcanic conglomerate
L2-79D2-H3	35.752	122.680	1910	Bedded sandstone w.lapilli
L2-79D3-1	35.761	122.704	1840	Bedded sandstone
Guide Seamount				
T124-R4	36.974	123.367	1882	Monomict breccia
S4-78D2-15	37.022	123.330	2540	Polymict breccia

Note: Dredge and ROV dive tracks are shown in Figure 1. Locations for dredge samples have been averaged

Figure 2. Photographs from ROV *Tiburon* video showing (A) slabby surface of typical hyaloclastite outcrop and (B) volcaniclastic sandstone outcrop. Field of view for both approximately 5 m.

completely altered to palagonite. We use the textural features in conjunction with the glass compositions to group the samples into six types that reflect their mode of origin. Based on glass chemistry (discussed in the next section), the samples are either mono- or polymict. Based on textural characteristics, we refer to them as breccia, volcanic conglomerate, and bedded or massive volcanic sandstone (Plate 1).

5.1. Monomict Breccia

The most common hyaloclastite samples recovered from each of the seamounts are monomict breccias composed of highly vesicular (to >50%), angular to subrounded glass fragments. The glass fragments (< 1mm to > several cm) are typically crystal rich, with microphenocrysts of euhedral olivine and plagioclase, plus complexly zoned clinopyroxene in some samples. Some of the bigger clasts may show a jigsaw puzzle fit (Plate 1a) indicating in situ brecciation of pillow wedges. In some samples the glass has been completely replaced by palagonite. Others have palagonite rims of variable thickness surrounding fresh glass cores. Virtually all samples contain at least traces of marine sediment and biologic debris (e.g. sponge spicules) in the interstices but some are largely clast-supported whereas others contain a larger proportion of clay matrix (Plate 1b). Minor tachylite is also present and appears to be related to the same flow, based on the phenocryst contents.

5.2 Polymict Breccia

Other, similar looking, hyaloclastite breccias are polymict with more than one glass composition (Table 2). All of these are matrix supported and some contain more diverse and more abundant rock fragments (Plate 1c), including small xenoliths and large xenocrysts of complexly zoned clinopyroxene and plagioclase with resorbed margins, and rare, rounded, and oxidized amphibole megacrysts. The most compositionally diverse glass grains are found in the polymict hyaloclastite (S5-78D2-15) from Guide Seamount (Plate 1c). This sample includes numerous tholeiitic glass spherules (Figure 3a) some of which are totally degassed with respect to sulfur (Table 2).

5.3 Volcanic Conglomerate

Two samples are poorly sorted conglomerates composed of lapilli-sized, sub-rounded to angular clasts that are extremely vesicular (>50%) and nearly crystal free glass (Plate 1d), compared to the breccias above. The glass is virtually unaltered, lacking the palagonite rim observed for many clasts in the breccias. No bedding is observed in these two samples but a third sample consists of a massive layer of lapilli-sized clasts that grades into bedded sand layers (Plate 1e).

5.4 Bedded Volcanic Sandstone

Two dredged samples are bedded volcanic glass sandstone. One has three to five repetitive layers grading from medium to fine sand and silt (Plate 1f). Based on the position of thick Mn crust on the samples, the grading is assumed to be normal. Most of the sand-sized grains are fresh glass but a large proportion of the silt is altered to palagonite and iron oxide. This unit may be from the unsampled, bedded outcrop observed on one of the dives (Figure 2b). The second sample consists of a sand layer

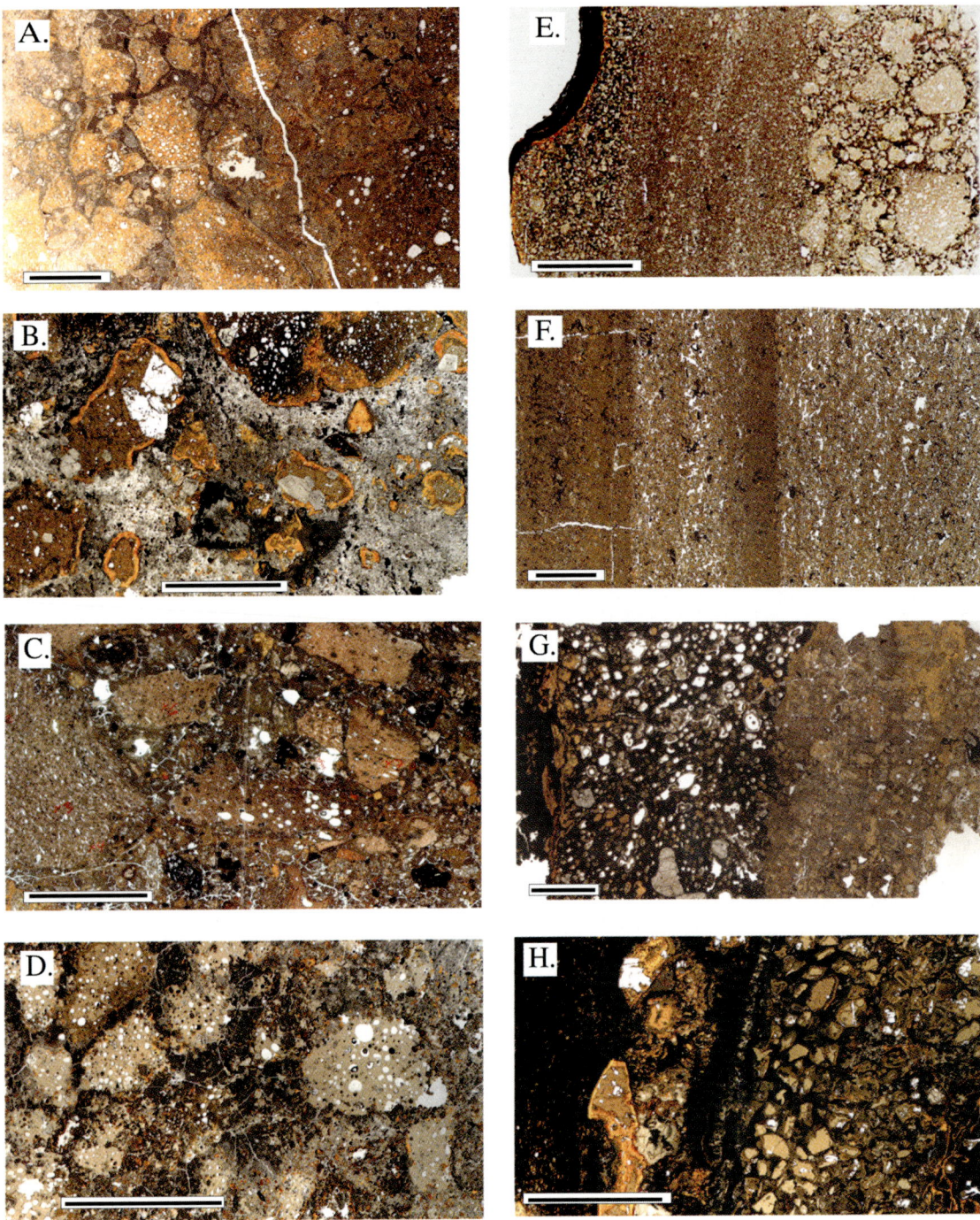

Plate 1. Photomicrographs of hyaloclastite thin sections. (A) monomict, clast-supported breccia (T124-R4), (B) monomict, matrix-supported breccia (T142-R5), (C) polymict breccia with a variety of glassy and lithic fragments (S4-78-D2-15), (D) volcanic conglomerate with highly vesicular, crystal-free pyroclasts (L2-79-D3-1), (E) layer of poorly-sorted lapilli-sized clasts in depositional contact with sand layers (L2-79-D2-H3), (F) bedded sandstone with multiple layers of angular glass shards (L2-79-D3-1), (G) monomict hawaiite sandstone (T140-R6) in contact with highly vesicular lava flow, (H) polymict basalt sandstone (T145-R5), with vesicle-free, angular glass in two layers separated by a band of manganese oxide. Top of section is to the left. Scale bar is 0.5 cm.

Table 2. Representative Glass Compositions

Sample	SiO$_2$	TiO$_2$	Al$_2$O$_3$	FeO	MnO	MgO	CaO	Na$_2$O	K$_2$O	P$_2$O$_5$	SO$_3$	Cl	Total
T139-R4 1	49.46	2.94	18.11	8.47	0.16	4.36	9.01	3.34	2.81	0.87	0.07	0.09	99.65
T139-R4 2	49.53	2.82	18.43	8.40	0.15	4.53	9.16	3.31	2.92	0.88	0.09	0.09	100.29
T140-R6 1	49.16	3.01	18.11	9.40	0.16	4.22	8.79	3.50	2.46	0.84	0.15	0.09	99.87
T140-R6 2	49.45	3.00	18.09	9.55	0.15	4.21	8.73	3.56	2.44	0.80	0.16	0.09	100.20
T140-R6 3	49.36	3.02	18.11	9.49	0.16	4.17	8.81	3.63	2.37	0.76	0.17	0.07	100.13
T141-R8 1	48.78	3.12	19.32	8.87	0.20	3.95	9.13	2.18	2.40	1.15	0.29	0.16	99.52
T141-R8 2	47.76	3.16	19.09	9.05	0.17	3.88	8.95	4.51	3.00	1.02	0.08	0.12	100.75
T141-R8 3	47.54	3.21	19.11	8.85	0.13	4.00	9.09	4.19	2.87	0.94	0.14	0.10	100.16
T141-R8 4	47.90	3.08	19.05	8.83	0.19	3.89	8.93	4.15	2.93	0.99	0.17	0.11	100.18
T142-R5 1	48.65	3.23	18.24	9.61	0.14	3.92	8.91	3.71	2.96	1.03	0.12	0.14	100.62
T142-R5 2	47.74	3.23	18.78	9.29	0.20	3.56	9.27	3.80	2.91	0.90	0.14	0.13	99.91
T142-R5 3	48.01	3.47	18.42	9.53	0.17	3.65	9.13	3.61	3.06	0.95	0.11	0.13	100.22
T144-R6 1	47.64	3.38	17.05	10.02	0.17	4.80	10.08	3.80	2.05	0.74	0.13	0.06	99.91
T144-R6 2	47.57	3.53	16.81	10.03	0.16	4.69	10.05	3.59	1.99	0.72	0.13	0.06	99.32
T144-R6 3	47.51	3.47	16.58	10.31	0.14	4.69	10.04	3.61	2.03	0.75	0.12	0.06	99.30
T144-R6 4	47.28	3.52	16.84	9.97	0.18	4.75	10.02	3.52	2.02	0.77	0.08	0.06	98.98
T145-R5 1	50.05	2.06	16.51	10.24	0.15	6.58	9.61	3.35	0.89	0.35	0.16	0.03	99.97
T145-R5 2	49.60	2.11	16.19	10.64	0.15	6.62	9.56	3.37	0.74	0.28	0.16	0.03	99.44
T145-R5 3	49.84	2.09	16.53	10.25	0.11	6.55	9.67	3.40	0.90	0.33	0.12	0.03	99.81
T145-R5 4	49.24	2.48	16.92	9.56	0.14	5.14	9.74	3.59	1.42	0.44	0.14	0.04	98.85
T145-R5 5	49.87	2.44	17.35	9.56	0.14	5.26	9.98	3.57	1.46	0.48	0.14	0.05	100.27
T147-R1 1	49.36	3.12	18.48	8.62	0.14	4.69	8.97	3.63	2.68	1.00	0.06	0.09	100.82
T147-R1 2	48.77	3.17	18.39	8.61	0.13	4.72	8.51	3.64	2.83	0.97	0.07	0.08	99.85
T147-R1 3	49.11	3.21	18.52	8.54	0.17	4.69	8.85	3.73	2.69	1.03	0.02	0.08	100.61
T147-R6 1	49.26	2.94	18.54	8.25	0.15	4.39	7.88	3.93	3.00	1.20	0.06	0.11	99.68
T147-R6 2	49.25	3.06	18.22	8.19	0.14	4.41	7.98	3.86	2.96	1.22	0.07	0.11	99.43
T147-R6 3	49.42	3.09	18.61	8.09	0.12	4.53	7.99	3.88	2.98	1.30	0.08	0.11	100.16
T147-R6 4	49.41	2.95	18.38	8.30	0.14	4.40	8.00	3.61	2.99	1.24	0.06	0.10	99.55
T147-R8 1	48.77	3.73	17.85	8.77	0.17	3.05	7.62	2.99	3.18	1.57	0.09	0.12	97.86
T147-R8 2	49.60	3.48	17.79	8.42	0.15	3.02	7.30	4.89	3.67	1.28	0.09	0.14	99.78
T147-R8 3	48.94	3.69	17.73	8.60	0.14	3.08	7.38	4.94	3.56	1.40	0.09	0.14	99.65
T147-R16 1	50.53	2.69	19.53	7.70	0.15	4.30	7.43	4.04	2.92	1.03	0.02	0.10	100.40
T147-R16 2	50.44	2.72	19.14	7.59	0.15	4.15	7.40	4.00	2.88	1.02	0.07	0.09	99.62
T147-R16 3	50.24	2.82	19.28	7.86	0.13	4.21	7.36	4.10	2.92	1.05	0.06	0.09	100.09
T147-R16 4	50.37	2.79	19.33	7.89	0.17	4.23	7.35	4.13	2.94	1.04	0.05	0.09	100.36
L2-79D1-7	48.24	3.75	16.83	10.05	0.17	4.70	10.21	2.83	2.20	0.66	0.05	0.06	99.83
L2-79D1-7	49.98	2.49	18.35	7.77	0.13	4.62	8.48	4.12	2.40	0.83	0.03	0.08	99.30
L2-79D1-7	50.39	2.53	18.35	7.70	0.11	4.72	8.58	4.11	2.49	0.86	0.02	0.09	99.98
L2-79D1-7	49.23	2.68	18.28	7.75	0.12	4.66	8.66	4.28	2.68	1.04	0.02	0.09	99.64
L2-79D2-H3	47.44	3.50	16.53	10.34	0.17	4.62	9.84	3.93	2.29	0.70	0.07	0.06	99.48
L2-79D2-H3	48.08	3.39	16.70	10.0	1.14	4.80	10.09	3.95	2.20	0.69	0.08	0.06	100.18
L2-79-D3-1	47.70	3.48	16.82	10.04	0.19	4.66	9.77	3.85	2.14	0.61	0.06	0.10	99.41
L2-79-D3-1	47.53	3.03	16.71	9.74	0.14	4.74	10.24	3.92	2.13	0.59	0.06	0.10	99.22
L2-79-D3-1	47.45	3.59	16.37	9.88	0.15	4.66	10.12	4.23	2.14	0.63	0.06	0.09	99.36
T124-R4 1	48.08	3.21	18.16	8.34	0.17	4.50	8.63	3.77	2.50	1.11	0.07	0.10	98.63
T124-R4 2	48.09	3.23	18.04	8.65	0.15	4.54	8.67	3.97	2.53	1.15	0.09	0.10	99.16
S4D2-15 1	52.09	2.52	16.88	8.82	0.16	4.79	7.92	3.70	1.63	0.57	0.04	0.06	99.16
S4D2-15 2	49.86	2.60	18.04	8.46	0.14	4.61	7.95	4.07	2.11	0.80	0.07	0.07	98.75
S4D2-15 3	51.74	2.81	17.71	8.21	0.19	3.15	5.68	4.38	3.35	0.95	0.02	0.13	98.29
S4D2-15 4	49.86	2.60	18.04	8.46	0.14	4.61	7.95	4.07	2.11	0.80	0.07	0.07	98.75
S4D2-15 5	51.28	2.97	17.02	8.96	0.17	3.73	7.14	4.03	2.80	1.10	0.02	0.10	99.29
S4D2-15 6	50.71	2.34	13.95	12.70	0.23	6.43	9.83	2.59	0.26	0.11	0.01	0.01	99.15
S4D2-15 7	50.60	2.37	13.95	13.27	0.26	6.24	9.94	2.69	0.33	0.22	0.00	0.00	99.87
S4D2-15 8	51.19	2.37	14.12	13.36	0.24	6.27	9.87	2.46	0.45	0.28	0.00	0.01	100.61
S4D2-15 9	50.05	2.44	13.72	13.12	0.23	6.41	10.15	2.61	0.24	0.19	0.00	0.01	99.17

overlying a poorly sorted layer of lapilli-sized clasts (Plate 1e) that was already described. The finer grained layers of this sample contain many small shards that are broken across vesicles, and rare curved fragments that may be bubble-wall fragments (Figure 3b).

5.5 Massive Hawaiite Sandstone

One hyaloclastite sample (T140-R6) is a massive, coarse sandstone of highly angular, vesicular glass shards. The glass is monomict hawaiite, similar in composition to that in the breccias but it is less vesicular and not as crystal rich (Plate 1g). Many glass grains have fluidal shapes, some with parallel, elongated vesicles (Figure 3c). Many of the non-vesicular fragments appear to be broken along vesicle margins (Figure 3c), indicating vesiculation before fragmentation. This fragmental layer is in contact with a highly vesicular, tachylitic lava flow that has the same crystal assemblage as the fragmental material, suggesting it is ejecta from the same eruption.

5.6 Massive Basalt Sandstone

One hyaloclastite sample (T145-R5), collected on the deepest dive (~3400 m), consist mostly of small (mm-size), angular, basalt glass shards. It is vesicle free and almost devoid of crystals (Plate 1h). It has two different glass compositions of transitional and mildly alkalic basalt in two distinct layers that are separated by a mm-thick band of manganese oxide.

6. GLASS COMPOSITIONS

Hawaiite is the most common composition of hyaloclastite glass. Both alkalic basalt and mugearite are present but much less abundant than in the associated lavas [*Davis et al.*, 2002]. Except for two samples, the glass compositions of hyaloclastites from Davidson Seamount (Table 2) closely match those of quenched rinds of lava samples collected nearby. The basalt glass in the sample from the deepest dive (T145-R5) on Davidson Seamount has no comparable flow composition. The most evolved glass composition in the polymict breccia of T147-R8 also has no corresponding lava sample but similar mugearite whole rock compositions were found at other locations on Davidson, as well as on Guide and Rodriguez seamounts [*Davis et al.*, 1995; *Davis et al.*, 2002].

Except for TiO_2, glass compositions of monomict samples are within analytical precision for major elements (Table 2). Variations in TiO_2 reflect different amounts of

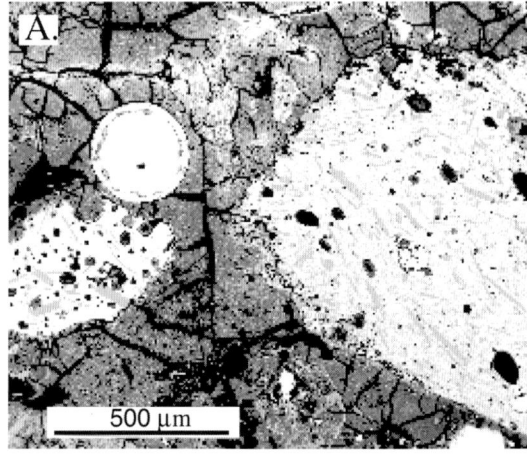

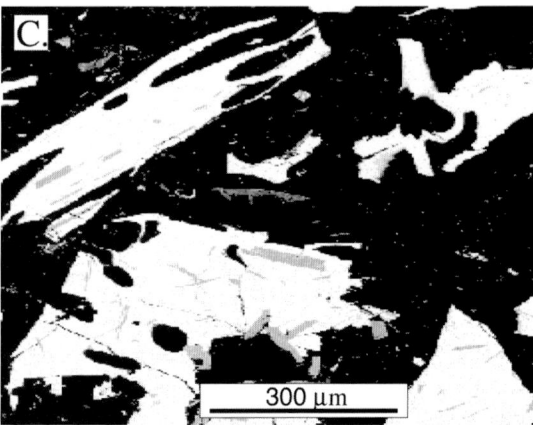

Figure 3. Back scattered electron images of (A) degassed, tholeiitic spherules and crystal-rich pyroclasts in polymict breccia from Guide Seamount, (B) Curved, broken glass shards in fine grained layer of bedded sandstone, (C) fluidal and curved glass fragments in monomict hawaiite sandstone from Davidson Seamount.

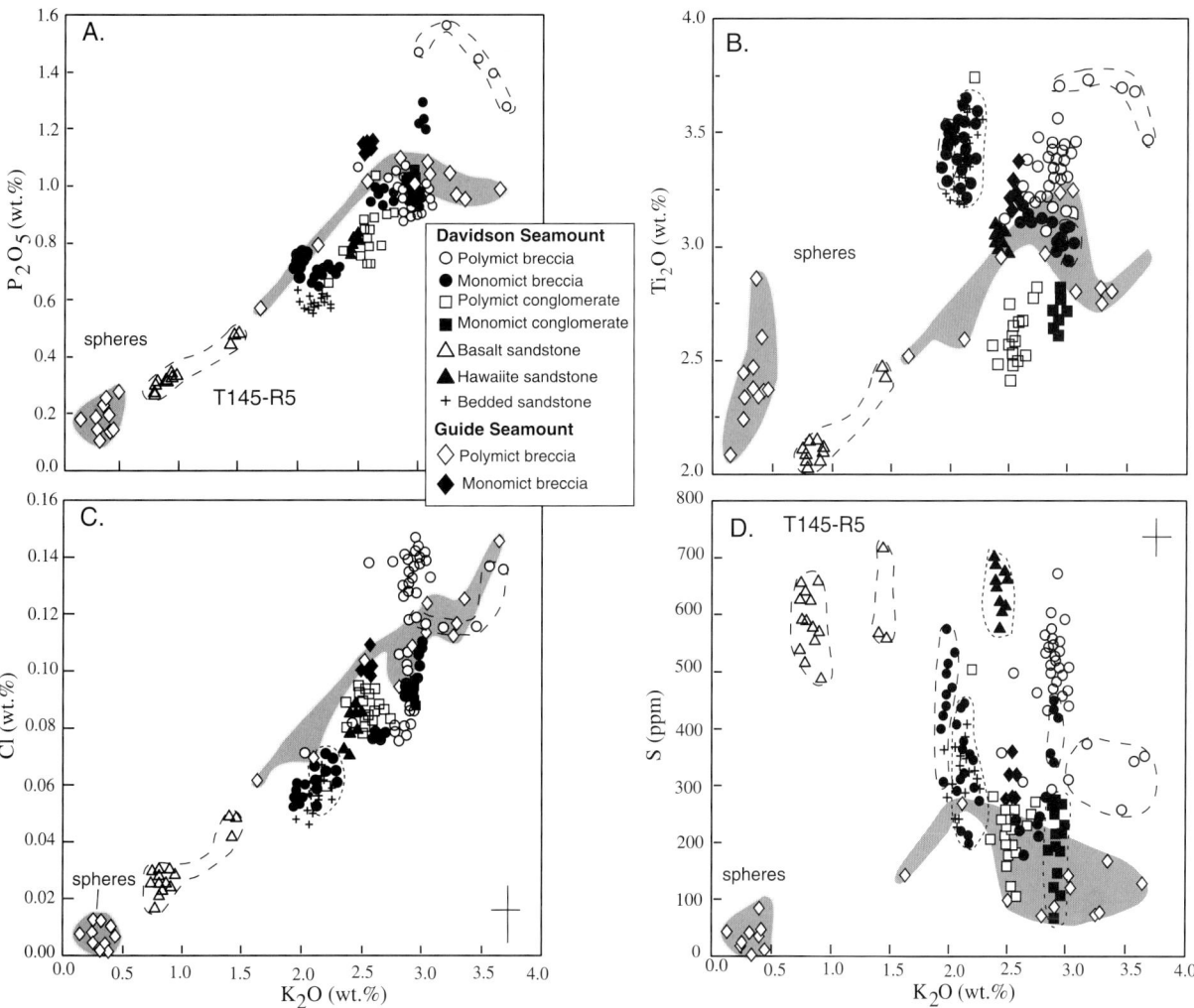

Figure 4. Composition of glass fragments in hyaloclastite samples. (A) K_2O vs. P_2O_5 shows a narrow range for monomict samples. (B) TiO_2 vs. K_2O shows a considerable range of TiO_2 at a given K_2O, reflecting titanomagnetite crystallization. (C) Cl vs. K_2O shows a positive correlation, indicating that Cl behaved as an incompatible element and did not degas significantly. (D) S vs. K_2O shows a large range at a given K_2O, indicating extensive degassing during an eruption. Gray field encloses compositions found in polymict breccia (S5-78D2-15) from Guide Seamount. Dashed lines indicate range of compositions in some other samples.

titanomagnetite fractionation. Because K_2O and P_2O_5 behaved as incompatible elements and are well correlated (Figure 4a), we use these elements as the main indicator to identify mono-or polymict samples. The glasses are generally high in Cl (300 - 1500 ppm). Cl is positively correlated with K_2O (Figure 4c), indicating that it behaved as an incompatible element. S, in contrast, shows a large range in abundance (<100-700 ppm) at a similar K_2O (Figure 4d). S varies by more than 200 ppm (Figures 4d, 5) for grains within single samples that have other elements within analytical precision. The highest values are observed for the non-vesicular basalt glass in the sample collected at ~3400 m water depth. Despite a considerable range in S abundance, there is a positive correlation between the maximum S content of monomict glass samples and collection depth (Figure 5).

The glass from the monomict breccia from Guide Seamount (T124-R4) is compositionally similar to hawaiite from Davidson Seamount. However, the polymict breccia (S5-78D2-15) contains glass fragments of highly diverse compositions, including hawaiite, mugearite, mildly to moderately alkalic basalt, and tholeiitic basalt (Figure 4).

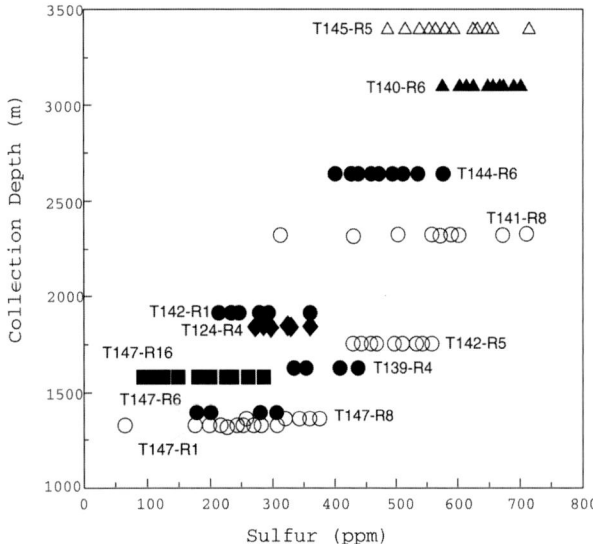

Figure 5. Despite the variation in sulfur abundance within a single sample, maximum sulfur content of monomict samples shows a definite positive correlation with water depth. Symbols as in Figure 4.

Except for the tholeiite, the entire range of compositions observed from Davidson Seamount is represented in this one reworked breccia, indicating a similar diverse lava suite erupted on Guide Seamount as well. The tholeiitic glass spherules are similar to E-MORB and contain <100 ppm S and Cl.

7. ORIGIN OF HYALOCLASTITE

The six types of hyaloclastite identified by a combination of chemistry, grain size, and textural features reflect differences in fragmentation and eruptive style, location relative to the vents, and post-eruptive sedimentary reworking. The types of eruptive products, whether subaerially or subaqueously erupted, are largely controlled by magma composition which determines initial volatile contents, viscosity, and eruptive temperature.

7.1 Fragmentation and Magmatic Degassing

The high vesicle content of the most of the glass indicates extensive exsolution of magmatic volatiles, mainly CO_2, but also H_2O and S. Due to its low solubility, CO_2 is lost from magma residing in shallow chambers and during slow magma ascent to the seafloor. The presence of both mantle xenoliths and alkalic cumulates, that are more evolved than the host lava, suggests that magma originated and crystallized extensively in the upper mantle [Davis et al., 2002]. The absence of calderas and collapse pits on these seamounts suggests lack of magma storage at shallow crustal depth. The build-up of volatile pressure, largely CO_2, may have propelled the hawaiite and mugearite melts, having viscosities at least an order of magnitude greater than basalt [Shaw, 1972], into rapid ascent from deep storage to the seafloor. As the magma neared the seafloor, part of the sulfur and water probably degassed along with the CO_2. Explosive release of volatiles due to decompression and vigorous impact with cold seawater resulted in quenching to glass and extensive fragmentation of vesiculated lava. Fragmentation of magma by volatile exsolution is thought to be restricted under great hydrostatic pressure [e.g. Kokelaar, 1986]. For tholeiitic basalt magma such fragmentation has been proposed to be limited to a few hundred meters [e.g. Moore and Schilling, 1973]. However, at great depths (>3000 m) it appears to be possible for magma more enriched in volatiles, such as those from volcanic arcs [Gill et al., 1991], or strongly alkalic magmas [Clague et al., 2002], including strongly evolved alkalic compositions such as those described here.

The majority of glass compositions from the seamounts had high initial volatile contents. CO_2 is highly incompatible during melting of the mantle and is positively correlated

Table 3. Ion Probe Analyses of Volatiles

Sample	T124-R4	T139-R6	T144-R12	T146-R24
Rock type	Hawaiite	Hawaiite	Alk.basalt	Mugearite
H_2O wt%	0.61	0.69	0.63	0.67
CO_2 ppm	31	51	79	69
F ppm	1791	1139	1179	1582
S ppm	427	446	686	598
Cl ppm	1387	1242	716	1231

Note: only T124-R4 is hyaloclastite, but analysis is not of same grain as for microprobe. Other samples are pillow rims. Analyses are average of 4 to 5 separate analyses.

with Nb [Hauri et al., 2002; Saal et al., 2002]. Based on a constant CO_2/Nb ratio in MORB [~ 273 to 317, Hauri et al., 2002; Saal et al., 2002] these magmas could have had maximum CO_2 contents ranging from 1.7 wt.% for alkalic basalt to >4.2 wt.% for mugearite. The Ce/H_2O ratio is also constant in MORB [174 -223, Hauri et al., 2002] and H_2O contents calculated from Ce concentrations of whole rock samples [Davis et al., 2002] range from 1.1% to >4.2% for alkalic basalt and mugearite, respectively. The estimates for mugearite are minima since Nb and Ce are no longer incompatible elements due to crystallization of Fe-Ti oxides and apatite. Application of these MORB ratios seems reasonable since the isotopic compositions of samples from these seamounts suggest small degrees of partial melting of MORB-like mantle sources [Davis et al., 2002]. Ion probe analyses of several seamount glasses have H_2O ranging from 0.5 to 0.7% and CO_2 <100 ppm (Table 3), clearly indicating extensive exsolution and loss of these volatiles. Dixon [1997] and Dixon et al. [1997], using measured solubilities of CO_2 and H_2O in alkaline melts, have modeled the exsolution of magmatic gases from alkalic magmas and shown that significant amounts of H_2O (and S) are exsolved along with CO_2 as bubbles form. Exsolution of the estimated initial volatiles in these magmas would lead to vesicle contents even higher than those observed, suggesting that some portion of the gas has escaped from the rising magma.

Cl and F are positively correlated with K_2O and apparently did not degas significantly. S, however, undoubtedly degassed to some extent, probably both during formation of vesicles when magma reached shallow depth and during eruption on the seafloor. The decrease in S at a given K_2O (Figure 4d) by about 200 ppm for clasts within nearly every monomict sample, regardless of water depth (Figure 5), reflects at least that much degassing of S occurs during eruption.

7.2. Eruption Style

Comparably gas-rich lava erupted subaerially would have produced significant Hawaiian-style lava fountains [e.g. Head and Wilson, 2002]. The monomict breccias and volcanic conglomerates have highly vesicular (>50%), angular to rounded clasts. They resemble scoria of cinder cones, except they are all glassy indicating quenching in water. The fluidal and curved glass shapes (Figures 3b,c) found in the finer-size fraction of multiple samples appear to be analogous to aerodynamic shapes found in subaerial fire fountains [Vergionelle and Mangan, 2000]. Since no one has observed explosive eruptions in the deep sea, comparison with subaerially erupted ejecta provides the only clues to eruption styles. Although much of the fragmentation may have resulted from shattering of already vesiculated lava, we interpret theses textural features as evidence for Hawaiian-style fountains, probably of low height depending on eruption depth. Pele's hair, characteristic of Hawaiian fountains, was not observed, although its absence may simply indicate that such delicate fragments are unrecognizable after more than 10 m.y. of alteration. Large, irregular, gas cavities in some lava samples indicate that the exsolved gases may have coalesced and escaped in bursts through the vent. In this case, Strombolian activity may have occurred as well. Expansion of steam due to entrapment of water or wet sediment is probably a common occurrence in subaqueous eruptions. Formation of bubble-wall fragments attributed to steam expansion has been described from deep sea eruptions at near-ridge seamounts [e.g. Maicher and White, 2002] and from Loihi seamount [Clague et al., 2000]. The presence of such glass shapes at spreading centers [Clague et al., this volume] and from water depths (>3000 m) that would restrict steam expansion, suggests discharge of a separated gas phase, probably CO_2, may be the driving mechanism. Rare, curved glass fragments, found in some samples (Figure 3b), could be pieces of bubble walls that were broken during transport. However, pyroclast shapes such as spindle bombs or agglutinated spatter that are indicative of Strombolian eruptions were not observed. Rapid quenching in water would probably prevent welding and agglutinating of particles.

Although the mixing of hot rock or magma with water must have led to production of steam, we have found no evidence for large scale phreatic, or phreatomagmatic eruptions on any of the seamounts. Such evidence would consist of lithic fragments of country rock in the hyaloclastite deposits. Extensive phreatic eruptions of basaltic composition commonly accompany caldera or pit crater collapse [e.g., McPhie et al., 1990; Clague et al., this volume], which expose large amounts of hot rock or magma to water. Bathymetric surveys of each of these seamounts show that such collapse structures are notably absent [Davis et al., 2002]. Diverse chemistry of flows and the small volume of individual cones on the seamounts also suggest eruption of small magma volumes thereby providing a limited heat source.

Since episodes of volcanic activity apparently were separated by long time intervals [Davis et al., 2002], the initial eruptions must have had enough gas pressure to open conduits to the surface. Hence the earliest stage of eruption may have started with small-scale, vulcanian-like ejection of country rock clearing the vent. Although some lithic fragments, xenoliths, and broken crystals were observed, especially in the polymict breccias, no unit resembling typical vulcanian deposits was observed. However, the material ejected in an earlier, explosive phase would probably be covered by later lavas and by the fallout from the eruption column.

Despite similar hawaiite composition, the volcanic conglomerates contain generally more vesicular pyroclasts than the breccias but they are virtually free of crystals. Based on the S content, these samples erupted in water depths comparable to or slightly deeper than the crystal-rich breccia samples. Since a high crystal content increases viscosity and inhibits bubble expansion [*Cashman et al.* 2000], these magmas may have had slightly lower viscosity due to more rapid ascent and/or higher effusion rates. Although fragmentation caused by gas exsolution is indicated for the breccias, the jigsaw puzzle fit of some larger clasts suggests that in-situ cooling contraction granulation [*Kokelaar*, 1986] further fragmented some clasts (Plate 1a). The monomict hawaiite sandstone with angular, moderately vesicular clasts erupted in deeper water (3077 m), resulting in lower vesicle and higher S content (Figure 4d). Despite the greater eruption depth, many glass fragments have fluidal shapes and are broken across vesicles, suggesting fragmentation from gas exsolution; they probably formed in Hawaiian-style fountains of limited height. In contrast, the basaltic sandstone from the deepest site (~3400 m) consists of crystal- and vesicle-free, angular glass shards (Plate 1h) that probably resulted mostly from cooling contraction granulation processes. Textures of this sample are similar to hyaloclastite from near-ridge seamounts [e.g. *Smith and Batiza*, 1984; *Davis and Clague*, 2000] that appear to consist mostly of quench granulation fragments and spalled off pillow rinds. Due to a combination of lower initial volatile content of this transitional to mildly alkalic basalt and deeper eruption depth, fragmentation was not explosive. These deposits do not appear to be extensive.

7.3. Eruption Depths and Post-Eruptive Subsidence

S is sensitive to temperature, oxygen and sulfur fugacity, FeO content of the lava, and hydrostatic pressure. The positive correlation of maximum S content with increasing water depths (Figure 5) for comparable hawaiite and mugearite compositions can be attributed to changes in hydrostatic pressure, due to eruption at different water depths. On Davidson Seamount, hyaloclastite samples were collected in water depths ranging from 1300 to 3400 m. Many of these samples apparently erupted in shallow water, as seen by their low dissolved volatile contents and oxidized olivine in some flow samples. Partially rounded pyroclasts with concentric rims of palagonite, including some with reddish brown color, suggest ejection in shallow water; some tephra may have been projected subaerially. Although we are unable with the data at hand to reliably estimate this subsidence, we suspect up to 1000 m of subsidence may have occurred at these seamounts. Evidence for emergence, in the form of wave-cut erosion on Rodriguez Seamount offshore southern California, suggests subsidence of about 700 m since it's formation. Such magnitude of subsidence appears reasonable, based on subsidence documented for sedimentary sequences associated with volcanic rocks from the continental margin [*Wilson et al.*, 2002]. With this interpretation, eruption depths for the hyaloclastite on Davidson Seamount may have ranged from less than a few hundred m to over 2000 m.

The low S contents of tholeiitic glass spherules found in the polymict breccia from Guide Seamount is indicative of subaerial eruption but would imply subsidence of ~1700 m for Guide Seamount. No source vent for these spherules has been identified. The eruption of tholeiitic basalt following the eruption of fractionated alkalic lavas contrasts with the evolution typically observed on ocean islands. It is possible that these spherules were transported here from Miocene, tholeiitic, basaltic volcanoes in the Coast Range of southern California [*Cole and Basu*, 1995].

7.4. Deposition and Post-Eruptive Reworking

The coarseness of the ejecta and the massive, poorly sorted, clast-supported texture of the breccia and conglomerate samples suggest limited dispersal. Most likely these particles are fallout from an eruption column of limited height and were deposited near vent sites, forming the discontinuous slabs observed on the dives. Similar monomict breccia samples that have a larger proportion of finer sediment in the matrix probably were transported somewhat farther from their eruption site.

The polymict, matrix-supported breccias probably are reworked deposits and include a range of glass compositions and a larger variety of lithic clasts as well as abundant biogenic debris, which were entrained when currents displaced them into deeper water. They do not represent single eruptive events. While these polymict hyaloclastites provide less information concerning eruptive processes related to a single event, they are useful in giving an indication of the range of processes and overall compositional range of ejecta present on the seamount. The diverse glass compositions in the polymict breccia (S5-78D2-15) from Guide Seamount clearly demonstrate the great range of compositions erupted, including mildly to moderately alkalic basalt, hawaiite, and mugearite.

The bedded volcanic sandstones have a narrow range of glass compositions spanning multiple layers. The composite sample of lapilli-sized clasts and sand also consists of a common glass composition. They apparently each formed during a single eruptive event. In some respect, they resemble base surge deposits of phreatomagmatic eruptions but apparently were less energetic. No dunes, antidunes, ripples or cross-bedding, common in such deposits [*Fisher and Schmincke*, 1984], were observed. Most likely these samples formed from a slurry of tephra and water, deposited by density currents directly related to volcanic activity [*White*, 2000; *Smellie*, 2001].

8. CONCLUSIONS

Hyaloclastite deposits on mid-Miocene seamounts offshore central California have predominantly evolved alkalic basalt glass compositions. Fragmentation occurred primarily due to exsolution of magmatic gases and interaction of hot magma with cold seawater. Textures suggest formation during mildly explosive eruptions with Hawaiian-style fountains of limited height. Monomict samples with coarse-grained pyroclasts and little or no sediment matrix settled near the vents, whereas polymict, matrix-supported samples were reworked and deposited in deeper water, farther from vent sites. Based on glass composition, the multiple layers of the bedded sandstone formed during a single eruptive event. Less volatile-rich, mildly alkalic basaltic glass in the hyaloclastite from the deepest site apparently formed by quench granulation processes. Eruption depths ranged from near sea level to over 2000 m, suggesting a large amount of subsidence for these seamounts.

Acknowledgments. Support by the David and Lucile Packard Foundation is gratefully acknowledged. J. Paduan helped prepare Figure 1. Helpful reviews by R. Batiza, J. Smellie, and especially D. Maicher improved the manuscript.

REFERENCES

Batiza, R., D.J. Fornari, D.A. Vanko, and P. Lonsdale, Craters, calderas and hyaloclastites on young Pacific seamounts, *J. Geophys. Res., 89*, 8371-8390, 1984.

Batiza, R., and J. D.L. White, Submarine lavas and hyaloclastite, in *Encyclopedia of Volcanoes*, edited by H. Sigurdsson (et al.), Academic Press, San Diego, 361-381, 2000.

Clague, D.A., A.S. Davis, J.L. Bischoff, J.E. Dixon, and R. Geyer, Lava bubble-wall fragments formed by submarine hydrovolcanic explosions on Loihi Seamount and Kilauea Volcano, *Bull. Volcanol., 61*, 437-449, 2000.

Cashman, K.V., B. Sturtevat, P. Papale, and O. Navon, Magmatic fragmentation, in *Encyclopedia of Volcanoes*, edited by H. Sigurdsson, (et al.), Academic Press, San Diego, 421-430, 2000.

Clague, D.A., R.T. Holcomb, J.M. Sinton, R.S. Detrick, and M.E. Torresan, Pliocene and Pleistocene alkalic flood basalts on the seafloor north of the Hawaiian Islands, *Earth. Planet. Sci. Lett., 98*, 299-349, 1995.

Clague, D.A., K. Uto, K. Satake, and A.S. Davis, Eruption style and flow emplacement in the submarine North Arch volcanic field, Hawaii, in *Hawaiian Volcanoes: Deep Underwater Perspectives*, edited by E. Takahashi (et al.), Am. Geophys. Union Monograph 128, Washington D.C., 65-84, 2002.

Cole, R.B., and A.R. Basu, Nd-Sr isotopic geochemistry and tectonics of ridge subduction and middle Cenozoic volcanism in western California, *Geol. Soc. Am. Bull., 107*, 167-179, 1995.

Davis, A.S., and D.A. Clague, President Jackson Seamounts, northern Gorda Ridge: tectonomagmatic relationship between on- and off-axis volcanism, *J.Geophys. Res., 105*, 27939-27956, 2000.

Davis, A.S., D.A. Clague, W.A. Bohrson, G.B. Dalrymple, and H.G. Greene, Seamounts at the continental margin of California: A different kind of oceanic volcanism, *Geol. Soc. Am. Bull., 114*, 316-333, 2002.

Davis, A.S., D.A. Clague, and W.B. Friesen, Petrology and mineral chemistry of basalt from the Escanaba Trough, southern Gorda Ridge, *U.S. Geol Surv. Bull., 2022*, 153-170, 1994.

Davis, A.S., S.H. Gunn, W.A. Bohrson., L.B. Gray, and J.R. Hein, Chemically diverse sporadic volcanism at seamounts offshore southern and Baja California, *Geol. Soc. Am. Bull. 107*, 554-570, 1995.

Dixon, J.E., Degassing of alkalic basalts, *Am. Mineralogist, 82*, 368-378, 1997.

Dixon, J.E., D.A. Clague, P. Wallace, and R. Poreda, Volatiles in alkalic basalts from the North Arch volcanic field, Hawaii: Extensive degassing of deep submarine-erupted alkalic series lavas, *J. Petrol., 38*, 911-939, 1997.

Fisher, R.V., and H.-U. Schmincke, *Pyroclastic rocks*, Springer-Verlag, Berlin, 472 p., 1984.

Gill, J., P. Torssander, H. LaPierre, R. Taylor, K. Kaiho, M. Koyama, M. Kusakabe, J. Aitchison, S. Cisowski, K. Dadey, K. Fujioka, A. Klaus, M. Lovell, K. Marsaglia, P. Pezard, B. Taylor, and K. Tazaki, Explosive deep water basalt in the Sumisu backarc rift, *Science, 248*, 1214-1217, 1990.

Hauri, E., K. Gronvold, N. Oskarsson, and D. McKenzie, Abundance of carbon in the Icelandic mantle: constraints from melt inclusions, *Trans. Am. Geophys. Union, EOS.*, 2002.

Hauri, E.H., J. Wang, J.E. Dixon, P.L. King, C. Mandeville, S. Newman, SIMS analysis of volatiles in silicate glasses, 1: Calibration, matrix effects and comparisons with FTIR, *Chem. Geol.*, in press, 2002.

Head, J.W., III, and L. Wilson, Deep submarine pyroclastic eruptions: Theory and predicted landforms and deposits, *J. Volcano. Geotherm. Res.*, in press, 2002.

Kokelaar, B.P., Magma-water interactions in subaqueous and emergent basaltic volcanism, *Bull. Volcanol., 48*, 275-289, 1986.

Lonsdale, P., Structural patterns of the Pacific floor offshore of peninsular California. *In The gulf and peninsular province of the Californias*, edited by Dauphin (et al.), *Am. Ass. Petrol. Geol. Mem. 47*, 87-125, 1991.

Maicher, D., and J.L. White, The formation of deep-sea Limu o Pele, *Bull. Volcanol., 63*, 482-496, 2001.

McPhie, J., G.P.L. Walker, and R.L. Christiansen, Phreatomagmatic and phreatic fall and surge deposits from explosions at Kilauea Volcano, Hawaii, 1790 A.D.: Keanakakoi Ash Member, *Bull. Volcanol., 52*, 334-354, 1990.

Moore, J.G., and W.W. Chadwick, Jr., Offshore geology of Mauna Loa and adjacent areas, Hawaii, in *Mauna Loa Revealed: Structure, composition, history, and hazards*, edited by J.M. Rhodes and J.P. Lockwood, *Am. Geophys. Union Monograph 92*, 21-44, 1995.

Moore, J.G., and J.G. Schilling, Vesicles, water, and sulphur in Reykjanes ridge basalts, *Contrib. Mineral. Petrol., 41*, 105-118, 1973.

Saal, A.E., E.H. Hauri, C.H. Langmuir, and M.R. Perfit, Vapor undersaturation in primitive mid-ocean ridge basalt and the volatile content of the Earth's upper mantle, *Nature*, in press, 2002.

Shaw, H.G., Viscosities of magmatic silicate liquids: an empirical method of prediction, *Am. J. Sci., 272*, 870-893, 1972.

Smellie, J.L., Lithofacies architecture and construction of volcanoes erupted in englacial lakes: Icefall Nunatak, Mount Murphy, eastern Marie Byrd Land, Antarctica. *Spec. Publs int. Ass. Sediment, 30,* 9-34, 2001.

Smith, T., and R. Batiza, New field and laboratory evidence for the origin of hyaloclastite flows on seamount summits, *Bull. Volcanol., 51*, 96-114, 1989.

Staudigel, H., and H.-U. Schmincke, The Pliocene seamount series of La Palma (Canary Islands), *J. Geophys. Res., 89*, 11,195-11,215, 1984.

Vergniolle, S. and M. Mangan, Hawaiian and Strombolian eruptions in *Encyclopedia of Volcanoes*, edited by H. Sigurdsson (et al.), Academic Press, San Diego, 447-461, 2000.

White, J.D.L., Subaqueous eruption-fed density currents and their deposits, *Precamb. Res., 26,* 87-109, 2000.

Wilson, D. S., P.A. McCrory, and R.G. Stanley, Implications of volcanism in coastal California for the deformation history of western North America, *Tectonics*, in press, 2002.

Alicé S. Davis, David A. Clague, Monterey Bay Aquarium Research Institute, 7700 Sandholdt Road, Moss Landing, CA 95039-9644

Recent MORB Volcaniclastic Explosive Deposits Formed Between 500 and 1750 m.b.s.l. on the Axis of the Mid-Atlantic Ridge, South of the Azores

Jean-Philippe Eissen[1], Yves Fouquet[2], Delphine Hardy[3], and Hélène Ondréas[2]

Extensive scoriaceous volcaniclastic deposits were observed and sampled by submersible and dredging on three segments of the Mid-Atlantic Ridge (MAR) south of the Azores islands between 550 and 1750 mbsl. The deposits are restricted to axial volcanoes at the central part of each of the segments, in close association with zones of major volcanic and hydrothermal activity. The extent and volume of the volcaniclastic deposits decrease significantly with increasing water depth. The deposits generally are composed of well-stratified mm- to cm-scale layers, mainly un-reworked. The layers consist dominantly of juvenile, highly vesicular, scoriaceous, glassy clasts. Accessory lithic clasts are angular non-vesicular and more altered fragments. Thin calcareous sedimentary interlayers are found throughout the deposits, and reach a combined thickness of hundreds of meters. Such a thickness indicates that the deposits formed over long periods of time. The juvenile clasts have E-MORB to alkali basalt geochemistry, and are enriched relative to N-MORBs found elsewhere along the MAR. The primitive volatile content (dominantly CO_2) of this enriched magma remained in solution until magma eruption by dominantly explosive hydrovolcanic activity on the seafloor. The morphology of individual clasts shows that the main mechanism of fragmentation was magmatic (volatile driven) explosivity, followed by quench granulation, bulk magma-water interaction, and accessory surface magma-water interaction. The grain size characteristics of the deposits indicate that the resulting clasts were mainly emplaced by settling of suspended fragments from the water column. Accessory lateral-flowing density currents may have occurred during collapse of convective-settling suspensions. The studied area represents an excellent natural laboratory for additional studies of the fundamental controls of submarine volcaniclastic eruptions.

[1]Institut de Recherche pour le Développement, UR Aléas et Processus Volcaniques, Quito, Ecuador
[2]Institut Français pour l'Exploitation de la Mer, Géologie Marine, Centre de Brest, Plouzané, France
[3]Institut Géologie Albert de Lapparent, Cergy-Pontoise, France and Institut de Recherche pour le Développement, Centre de Bretagne, Plouzané, France

Explosive Subaqueous Volcanism
Geophysical Monograph 140
Copyright 2003 by the American Geophysical Union
10.1029/140GM09

1. INTRODUCTION

Mid-ocean ridge basalts (MORB) recovered along the submarine mid-ocean ridge system are generally emplaced effusively as lava, forming pillowed lava at a low effusion rate, temperature and volume, or sheeted, lobate, or draped flows and lava lakes at progressively higher effusion rate, temperature, or volume [*Ballard and Moore*, 1977; *Francheteau et al.*, 1979; *Bonatti and Harrison*, 1988; *Perfit and Chadwick*, 1998]. The low vesicularity of these lavas (between about 0 and 3 vol.%) [*Moore and Schilling*,

1973; *Moore*, 1979; *Dixon et al.*, 1988] decreases with increasing water depth [*Moore*, 1970].

Unusual, highly vesicular glassy basalts were recovered from the Mid-Atlantic Ridge (MAR) rift valley near 36°N [*Hékinian et al.*, 1973; *Pineau et al.*, 1976] and near 14°N between 3400 and 3620 m depth [*Bougault et al.*, 1988; *Sarda and Graham*, 1990; *Javoy and Pineau*, 1991; 1994]. These rare samples contain uncommonly high vesicle contents (17±1 vol.%), high dissolved H_2O (4160–5300 ppm), and extremely high CO_2 contents (~4000 ppm), along with typical MORB rare gas characteristics [*Pineau and Javoy*, 1994]. The rocks have been called "popping rocks" because of their active "popping" on ship's deck just after their recovery. These exceptional samples represent almost undegassed MORB [*Sarda and Graham*, 1990] and therefore provide insight about parental CO_2 concentrations in MORB's, at least at the level of the magma chamber, which is estimated in this case to be located 2–5 km below the seafloor [*Pineau and Javoy*, 1994]. If retained, such a high volatile content might also have some direct influence on the eruptive style of their host magma, at least at shallow water depth.

However, while basaltic hydroclastic and/or pyroclastic ejecta (hyaloclastites, limu o Pele, …) and spatter deposits are frequently observed on off-axis seamounts of the East Pacific Rise (EPR) [*Batiza et al.*, 1984; *Fornari et al.*, 1984; *Bonatti and Harrison*, 1988; *Smith and Batiza*, 1989; *Batiza and White*, 2000; *Maicher et al.*, 2000; *Maicher and White*, 2001] or on intraplate volcanoes (Loihi [*Clague et al.*, 2000; this volume]; Hawaiian North Arch [*Clague et al.*, 2002]; Macdonald [*Cheminée et al.*, 1991]; Pitcairn [*Binard et al.*, 1992]; Puna Ridge, Kilauea [*Clague et al.*, 2000; *Johnson et al.*, 2002]), such deposits have only rarely been observed along mid-oceanic and back-arc spreading centers. Nevertheless, abundant volcaniclastic deposits have been described from a few limited locations including the Bonin back-arc basin [*Gill et al.*, 1990], the southern Lau basin [*Fouquet et al.*, 1993], the Juan de Fuca Ridge near the Blanco fracture zone [*Batiza*, Pers. Com. 2002], on the Gorda ridge [*Clague et al.*, this volume], and on four segments of the Mid-Atlantic Ridge (MAR) located south of the Azores [*Fouquet et al.*, 1998; *Hékinian et al.*, 2000]. Additional evidence that more abundant volcaniclastic deposits have been produced along the MAR come from the Deep Sea Drilling Project. As an example, a 90 m-thick unit of unconsolidated basaltic volcanoclastics was cored at site 396B (Leg 46) in 13 Ma year old crust [*Dick et al.*, 1978; *Schmincke et al.*, 1978]. Whether or not these volcaniclastic deposits were created by basaltic submarine explosive volcanism, and the role of volatile content in magma fragmentation, are still under debate.

With its three increasingly deeper segments, the area studied hereby is an excellent natural laboratory to evaluate the fundamental controls on the occurrence of submarine volcaniclastic eruptions. Initial work on the area was published by *Fouquet et al.* [1998]. Here we present a more detailed description of different types of volcaniclastic deposits and lava encountered on the three progressively deeper segments of the MAR located just south of the Azores triple junction. Following a brief description of the geological setting of the three deposits, we describe and discuss their extent on the seafloor, their lithology and grain size, as well as the morphology, vesicularity and chemistry of selected clasts. We then discuss the relationship between the chemistry and the role of volatiles in the origin of these volcaniclastic deposits. And finally, we propose an eruptive scenario for their formation and mode of emplacement.

2. GEOLOGICAL SETTING

Around the Azores, the morphology of the MAR and the geochemistry of basalts are strongly influenced by the proximity of the triple junction and the presence of a mantle plume [*Schilling*, 1975; *Bougault and Treuil*, 1980; *Le Douaran and Francheteau*, 1981; *Detrick et al.*, 1995; *Dosso et al.*, 1999]. The spreading axis south of the Azores was mapped during the FARA-SIGMA survey [*Needham et al.*, 1992] using the EM 12 DUAL combined multibeam bathymetrical and seabottom reflectivity tool. The areas located around the axial highs of several segments were studied in detail during the Diva (1994) and Flores (1997) expeditions using the submersible "Nautile", mainly to look for and study hydrothermal activity and deposits [*Fouquet et al.*, 1994; *Ondréas et al.*, 1997; *Fouquet et al.*, submitted.]. As a result, on those cruises the volcaniclastic deposits were secondary targets and their sampling was not optimized.

The three spreading segments studied are located just southwest of the Azores (Figure 1a). The average spreading rate is about 20–25 mm/y [*De Mets et al.*, 1990]. Their development is directly linked to the creation of the Azores platform [*Cannat et al.*, 1999]. The segments have been named, from north to south, respectively: 1) the 38°20′N segment (KP4 segment of *Cannat et al.* [1999]); 2) The Menez Gwen segment; and 3) the Lucky Strike segment; [*Fouquet et al.*, 1994; *Ondréas et al.*, 1997] (Figure 1b). The Azores fracture zone bounds the "38°20′N" segment to the north, the "38°20′N" and Menez-Gwen segments are separated by the Princess Alice fracture zone, and the Menez-Gwen and the Lucky Strike segments are separated by the Pico fracture zone [*Freire Luis et al.*, 1994].

The volcaniclastic deposits are systematically found around the axial topographic high (or "paleo" volcanic highs, see below) of the three ridge segments, where volcanic and hydrothermal activities are also concentrated, probably above a superficial magma chamber. The ends of the segments are marked by increasing water depth and are dominantly affected by tectonic activity [*Ondréas et al.*, 1997; *Fouquet et al.*, 1998].

3. SEGMENT MORPHOLOGY

Following is a brief summary of the segments where the volcaniclastic deposits were found. Detailed descriptions were published elsewhere [*Ondréas et al.*, 1997].

The 38°20′N Segment (called 38N20 hereafter)

This segment is ~45 km long with a minimum depth of ~400 m below sea level (bsl) in its center and a maximum depth of ~2,000 m near its southern limit, the Princess fracture zone. Its center (Figure 2a) lacks the typical deep axial rift valley of the MAR, a morphology that is observed close to the fracture zones. In fact, it is occupied by a circular central volcano ~25 km in diameter and ~1,200 m high relative to the surrounding seafloor. The summit of the volcano is crosscut by an axial graben ~2 km wide, ~800 m long and 500 m deep (between 510 m to 930 m bsl). Using the present spreading rate (20–25 mm/y [*De Mets et al.*, 1990]), the axial graben is estimated to be ~80–100 ka old. Reflectivity images of the seafloor (Figure 2b) and submersible observations show that the inner walls of the graben and the external slopes of the central volcano consist of gently inward-dipping layered volcaniclastic deposits at least 400 m thick. We have identified another area, off axis, ~10 km in diame-

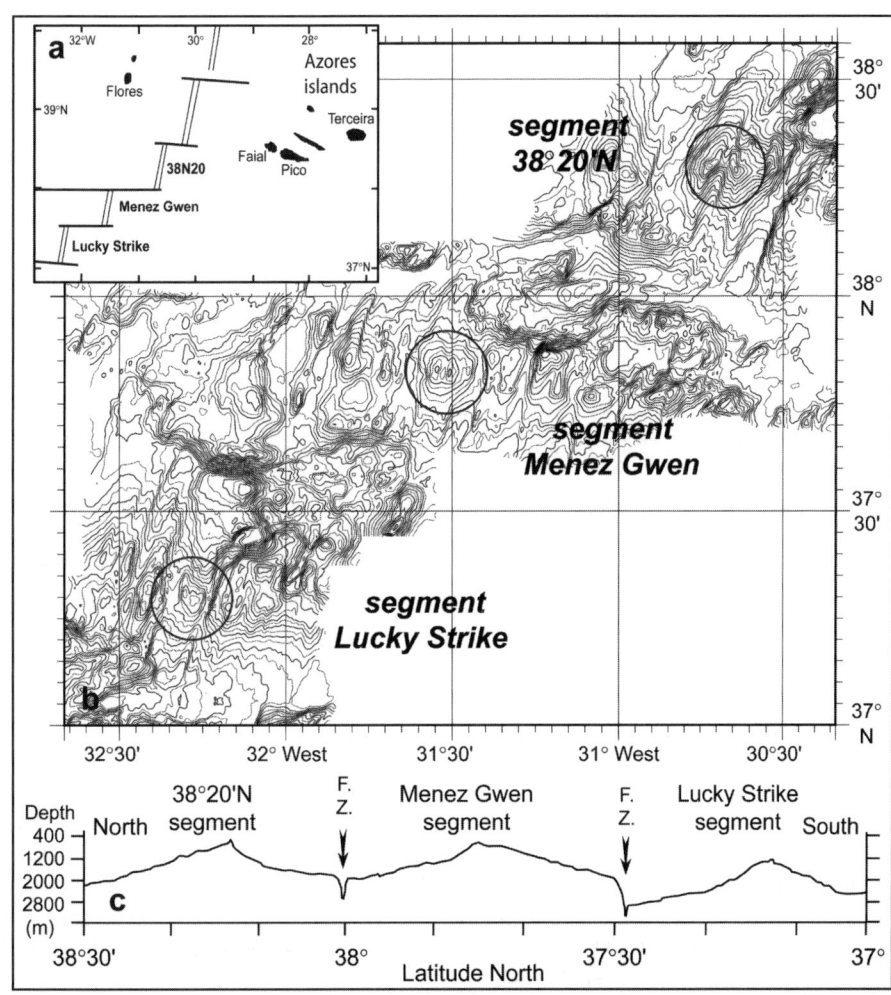

Figure 1. (a) Location of the studied area of the Mid-Atlantic Ridge, just south-southwest of the Azores islands. (b) Bathymetric map of the three successive segments [simplified from *Needham et al.*, 1992]. (c) Along strike, north-south bathymetric section of the three segments along the spreading axis, showing the central high of each segment and the progressive deepening of the segments towards the south (F.Z.: Fracture Zone).

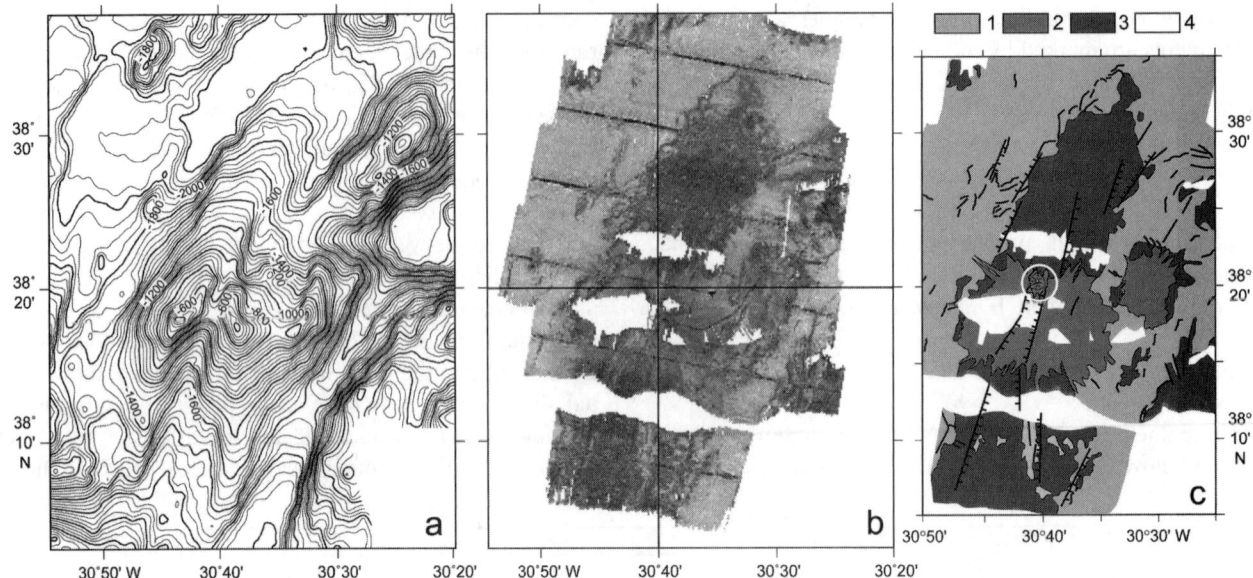

Figure 2. Segment 38N20: (a) Bathymetric map of the central part of the segment; (b) Bottom reflectivity (imagery) of the same area; (c) Geological interpretation compiled from a) and b): 1) pelagic sediments and partially buried pillow lava flows with interconnected sediment pockets (low reflectivity); 2) volcaniclastic and poorly consolidated deposits and breccias, and highly vesicular flows (medium reflectivity); 3) recent volcanics as sheet flows, pillow lavas and lava lakes (high reflectivity); 4) no data; also shown are lineaments, normal faults and one small volcanic cone located within the axial graben. The white circle delineates the diving area. Modified from *Ondréas et al.* [1997].

ter, with a reflectivity similar to that of the central volcano using the bathymetry and the reflectivity image. The area is centered near 38N20 and 30°31′W on a slightly elongated hill, approximately 500 m high (Figure 2c). The structure probably represents axial volcaniclastic deposits emplaced ~400–500 ka ago (a maximum age calculated from the present spreading rate), and rafted off axis by spreading.

The Menez Gwen Segment

This segment is ~60 km long with a minimum depth of ~700 m bsl and a maximum depth of ~2,100 m near its southern limit, the Pico fracture zone, and is morphologically similar to the 38N20 segment. The center of the segment is occupied by a central volcano ~16 km in diameter and ~700 m high. Its summit is cut by an axial graben ~2 km wide, ~6 km long and ~300 m deep (700 m to 1,000 m bsl). The floor of the graben is covered mostly by relatively fresh to very fresh lava flows and pillows, including a ~1,400 m long, 400 m wide and up to 8 m deep lava lake at its deepest point (1,045 m) [*Fouquet et al.*, 1994; 1998]. The age of the axial graben is constrained by the spreading rate to ≤80–100 ka. The volcaniclastic deposits in the Menez Gwen segment are much less extensive than on the 38N20 segment, and based on seafloor reflectivity, are limited to the central part of the segment (Figure 3a). A cross section of the western graben wall, drawn from submersible observations, shows a 240 m-thick volcaniclastic unit overlying a 60 m-thick section of lava flows up to 3 m-thick with columnar jointing and rubbly flow tops (Figure 3c). The contact of the lava flows and the overlying deposits is sharp. A few lava flows are intercalated within the volcaniclastic unit near the top of the eastern wall [*Fouquet et al.*, 1998]. On the reflectivity images, the volcaniclastic deposits are expressed as two fan shaped patches extending east-west over 7 to 8 km, symmetrical about the spreading axis (Figure 3a).

The Lucky Strike Segment

The southernmost and deepest segment is 60 km long with a 15 km wide rift valley (1,570 m in its center to 3,650 m bsl close to its fracture zones). The central part of this axial rift is occupied by a 12 x 8 km central volcano. The summit of the volcano consists of volcanic breccias surrounding a restricted area of layered volcaniclastic deposits and their closely associated highly phyric, highly vesicular basalts which are localized near the center of a 1 km wide and 100 m deep pit crater. The extensive and active Lucky Strike hydrothermal field developed around and south edge of this crater [*Fouquet et al.*, 1994; 1995; submitted.]. The

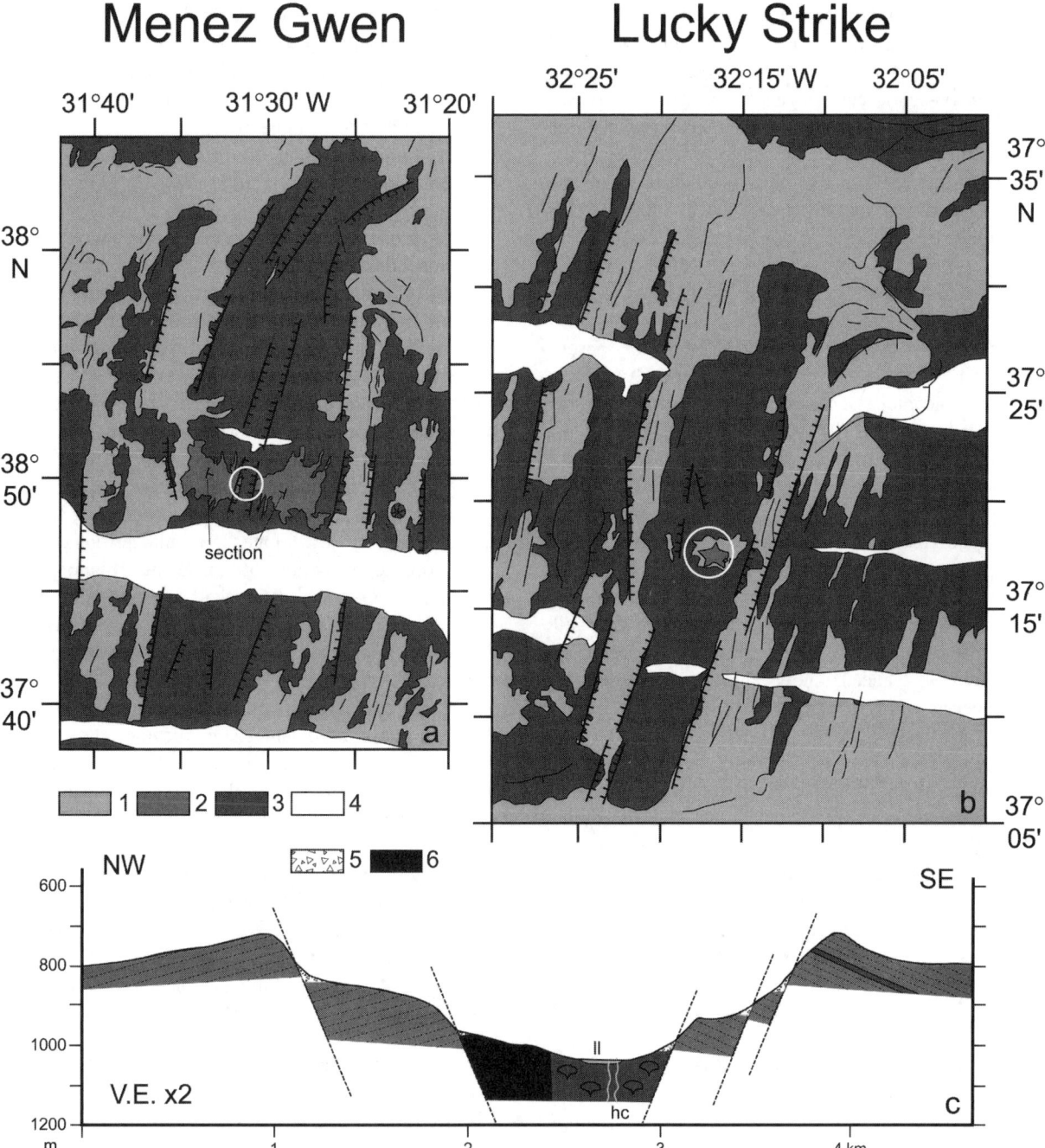

Figure 3. Geological interpretation obtained from the bathymetric and bottom reflectivity maps of the central parts of the Menez Gwen (a) and the Lucky Strike (b) segments. The full bathymetric and imagery data set of these segments are presented by *Ondréas et al.* [1997]. c) WNW – ESE geological cross-section of the central high of the Menez Gwen segment reconstructed from the bathymetry data, the bottom reflectivity data and the submersible observations. Same symbols as Figure 2 plus 5) = talus deposits, 6) massive lava flows, ll = lava lakes; hc = hydrothermal circulations; V.E. = vertical exaggeration.

volcaniclastic deposits are found at around 1700 m bsl. They are massive fragmental units, probably autobrecciated, that grade laterally into in situ brecciated and coherent, highly vesicular -scoriaceous- lava flows and pillow lavas.

Similar scoriaceous volcaniclastic deposits were recently discovered and described from the shallower zones (mainly between 1600 and 1900 m bsl) of another segment of the MAR. The deposits are located at 34°50′N just south of the Oceanographer fracture zone [Hékinian et al., 2000].

4. EXTENT OF THE DEPOSITS

Comparisons of detailed submersible and deep tow observations with EM 12 DUAL seabottom reflectivity images show that the volcaniclastic deposits have a specific acoustic signature [Ondréas et al., 1997]. The volcaniclastics appear gray on the images, intermediate between lava flows (black) and pelagic sediment (white) (Figures 2 and 3). The reflectivity images thus allow an extrapolation of the more spatially limited dive observations. This interpretation shows that the presence of volcaniclastic material is closely restricted to the center of the segments where it forms most of the surface of the volcanic cones, or of the off-axis hill described east of the center of the 38°20N segment. The imagery (Figures 2 and 3) shows that the volcaniclastic deposits cover areas of ~67 km^2, ~15 km^2, and ~2.8 km^2, respectively for the 38N20, Menez Gwen and Lucky Strike segments. Hydrostatic pressure, therefore, has a strong control on the processes of formation. Submersible observations and bathymetrical constraints show that the thickness of the deposits is ≥ 400 m for the 38N20 segment, ≥ 270 m for the Menez Gwen segment, and only a few meters for the Lucky Strike site [Ondréas et al., 1997; Fouquet et al., 1998].

Submersible observations, extrapolated with reflectivity images, reveal the presence of extensive fresh pillow lava, lava flows and, locally lava lakes [Fouquet et al., 1995] around the central high of each segment. The flows formed during a more recent eruptive episode than the volcaniclastic material, and coexist with hydrothermal deposits where there is hydrothermal activity, but with only limited active faulting. In contrast, tectonic activity close to the fracture zones is quite extensive, volcanic flows are older, and hydrothermal and volcaniclastic deposits are absent.

The above observations indicate that, first, there are close geographic relations between extensive volcanic activity, frequent hydrothermal activity and the generation of the volcaniclastic deposits around the central high of each segment. Second, it shows that the formation of the volcaniclastic deposits is linked to the spreading ridge segmentation (i.e. closely related to a morphological high), and mostly independent of the water depth (1750–500 m bsl). Therefore, between the central part and the tips of the ridge segments, there is a strong contrast in volcanic vs. tectonic activity, in effusive vs. explosive volcanic activity, and in presence vs. absence of hydrothermal activity.

5. CONSTRAINTS ON THE DURATION OF ACTIVITY

Abrupt changes in grain size observed between individual thin beds indicates that eruptions were episodic and probably short-lived, perhaps corresponding to isolated explosions. Furthermore, the presence of millimeter-thick, interlayers of fossiliferous calcareous pelagic sediment and calcite-cemented units (Figure 5c and 5d), as well as the thickness of the deposits (see above), indicate that this type of activity probably continued for a long period of time. The sedimentation rate on the Azores platform is well constrained by a study of 50 sediment cores recovered around the ridge axis [Dennielou, 1997; Dennielou et al., 1999]. The sediment is characterized by low biogeneous and terrigenous fluxes with good preservation of carbonates (nanofossils and foraminifers) because of the shallow depth, although some samples have calcite cementation (early diagenesis ?). Near the central high of each segment, the sedimentation rate is very low (~1.6 cm/1000 years). At the 38N20 segment, the very thin sedimentary cover observed allows us to constrain the latest phase of volcaniclastic volcanic activity to be less than 300 years old. However, the total thickness of the deposits demonstrates that this kind of volcanic activity occurred during multiple periods of activity over a minimum of seveval 10^3 to possibly 10^4 years or more, at least for the two northernmost segments. Furthermore, a few cores recovered less than 50 km from the axis contain at least 12 layers of volcaniclastic deposits of the same lithologic characteristics as the ones encountered on the present axis, showing that a similar kind of volcanic activity has been occurring repeatedly since at least 265 ka [Dennielou, 1997].

6. LITHOLOGY OF THE DEPOSITS

Most of the studied samples were recovered by submersible, and therefore are well constrained with respect to environment of deposition. However, some samples recovered by dredges are also included. The origin and nature of the samples are summarized in Table 1.

Volcaniclastic Samples

Some 36 volcaniclastic samples were studied in detail, especially in thin section. The deposits are well-layered

Table 1. Provenance and nature of the studied samples. HPB = highly phyric basalts. DV = DIVA cruise; FL = FLORES cruise; MG = Menez Gwen segment; LS = Lucky Strike segment.

Segment	Number	N Latitude	W Longitude	Depth (m)	Nature of the sample
	Submersible Samples				
38N20	DV-10-1	38°21.036	30°39.824	919	Volcanoclastites
38N20	DV-10-2	38°20.999	30°40.121	849	Volcanoclastites
38N20	DV-10-4	38°20.693	30°40.213	720	Volcanoclastites
38N20	DV-10-5	38°20.396	30°40.223	678	Volcanoclastites
38N20	DV-10-7	38°20.407	30°40.806	543	Volcanoclastites
38N20	DV-10-8	38°19.686	30°40451	843	Volcanoclastites
38N20	DV-10-9	38°19.632	30°40419	845	Volcanoclastites
38N20	DV-11-1	38°17.570	30°40.900	895	Volcanoclastites
38N20	DV-11-3	38°18.390	30°40.080	596	Volcanoclastites altered
38N20	DV-11-5	38°19.040	30°40.130	788	Volcanoclastites
38N20	DV-11-6	38°19.470	30°40.050	822	Volcanoclastites
38N20	DV-11-7	38°19.720	30°40.450	842	Volcanoclastites
38N20	DV-11-8	38°19.720	30°40.450	840	Volcanoclastites altered
	Submersible Samples				
MG	DV-12-3	?	?	?	Volcanoclastites
MG	DV-12-5	37°50.904	31°30.641	1009	Sediment + Volcanoclastites
MG	DV-12-8	37°50.450	31°30.684	883	Volcanoclastites altered
MG	DV-13-4	37°49.880	31°31.892	832	Volcanoclastites
MG	DV-13-5	37°50.900	31°31.640	696	Volcanoclastites
MG	DV-15-3	37°49.800	31°31.080	1003	Volcanoclastites altered
MG	FL-17-03	37°50.445	31°31.312	845	HPB highly vesicular pillow
MG	FL-17-10	37°50.970	31°31.172	955	Altered HPB highly vesicular
MG	FL-17-13	37°51.018	31°31.695	916	HPB vesicular basalt
MG	FL-30-03	37°50.45	31°31.35	843	HPB vesicular basalt
MG	FL-30-04	37°50.45	31°31.35	972	HPB vesicular basalt
MG	FL-30-05	37°49.80	31°30.87	740	Volcanoclastic lapilli
MG	FL-30-06	37°49.80	31°30.76	885	Volcanoclastic lapilli
MG	FL-30-07	37°49.47	31°30.68	820	HPB vesicular basalt
	Dredge Samples				
MG	FL-D9-01	37°49.70	31°30.72	750-900	Volcanoclastic lapilli
MG	FL-D9-02	37°49.70	31°30.72	750-900	Volcanoclastic lapilli
MG	FL-D9-03	37°49.70	31°30.72	750-900	Volcanoclastic lapilli
	Submersible Samples				
LS	DV-4-10	35°17.319	32°16.516	1685	Volcanoclastites
LS	DV-7-4	35°18.561	32°17.994	1650	Volcanoclastites
LS	DV-8-7	35°17.200	32°16.350	1706	Volcanoclastites
LS	DV-8-9	35°17.350	32°16.500	1697	Volcanoclastites encrusted
LS	FL-25-01	37°17.595	32°16.975	1625	Hydrothermalized crust
LS	FL-29-05	?	?	1655	Altered HPB vesicular
LS	FL-29-06	?	?	1660	slab with altered basalt fragments
	Dredge Samples				
LS	FL-D3-05	37°17.70	32°16.90	1600-1730	Slab with large basalt fragments
LS	FL-D4-04	37°17.62	32°16.95	1600-1740	Aphyric pillow
LS	FL-D4-05	37°17.62	32°16.95	1600-1740	Aphyric pillow
LS	FL-D4-06	37°17.62	32°16.95	1600-1740	HPB highly vesicular pillow
LS	FL-D4-07	37°17.62	32°16.95	1600-1740	HPB highly vesicular pillow
LS	FL-D4-08	37°17.62	32°16.95	1600-1740	HPB highly vesicular pillow
LS	FL-D4-09	37°17.62	32°16.95	1600-1740	HPB highly vesicular pillow
LS	FL-D5-01	37°17.53	32°16.55	1620-1720	Volcanoclastites
LS	FL-D5-04	37°17.53	32°16.55	1620-1720	Volcanoclastites
LS	FL-D5-05	37°17.53	32°16.55	1620-1720	Volcanoclastites
LS	FL-D5-06	37°17.53	32°16.55	1620-1720	Volcanoclastites
LS	FL-D5-07	37°17.53	32°16.55	1620-1720	Volcanoclastites
LS	FL-D5-08	37°17.53	32°16.55	1620-1720	Volcanoclastites
LS	FL-D5-09	37°17.53	32°16.55	1620-1720	Volcanoclastic lapilli
LS	FL-D5-10	37°17.53	32°16.55	1620-1720	Volcanoclastites

(Figure 4) over several meters to tens of meters and show small scale variations in grain size (Figure 4 and 5a-5e). Individual layers typically are millimeters to few centimeters thick, consisting of sand- and lapilli-sized clasts. Submersible observations identified very rare meters-thick beds of poorly sorted lapilli or slightly coarser layers, but these have not been sampled. The mechanism of formation of these layers might be different from that of the thinner layers, more related to gravity driven flows.

The deposits are either very poorly consolidated and grain-supported with a large amount of open spaces between the glassy and lithic clasts (Figure 5i, 5j and 5o), or cemented by calcite (Figure 5c and 5d), or lithified by alteration (Figure 5n). The latter frequently results from hydrothermal circulation within the deposits, resulting in strong induration (Figure 5h) or almost total silicification [*Fouquet et al.*, 1994]. But in general, the preservation of granulometrically well-sorted and well-layered ash beds, the very sharp nature of bedding planes between the lapilli and finer ash units and the frequent presence of worm burrows at the tops of some beds (Figure 5a) indicate that most of these deposits have not been reworked. Most of the non-graded layers are restricted to the very center of the segments that could correspond to proximal facies.

Reworked material exists; individual samples show limited signs of gravitational remobilization as small channels (Figure 5a), or restricted intrasedimentary reworked layers near syn-sedimentary slump faults or small tectonic faults, but the extent of these facies is limited. An additional sign of local reworking is provided by a sedimentary sample that

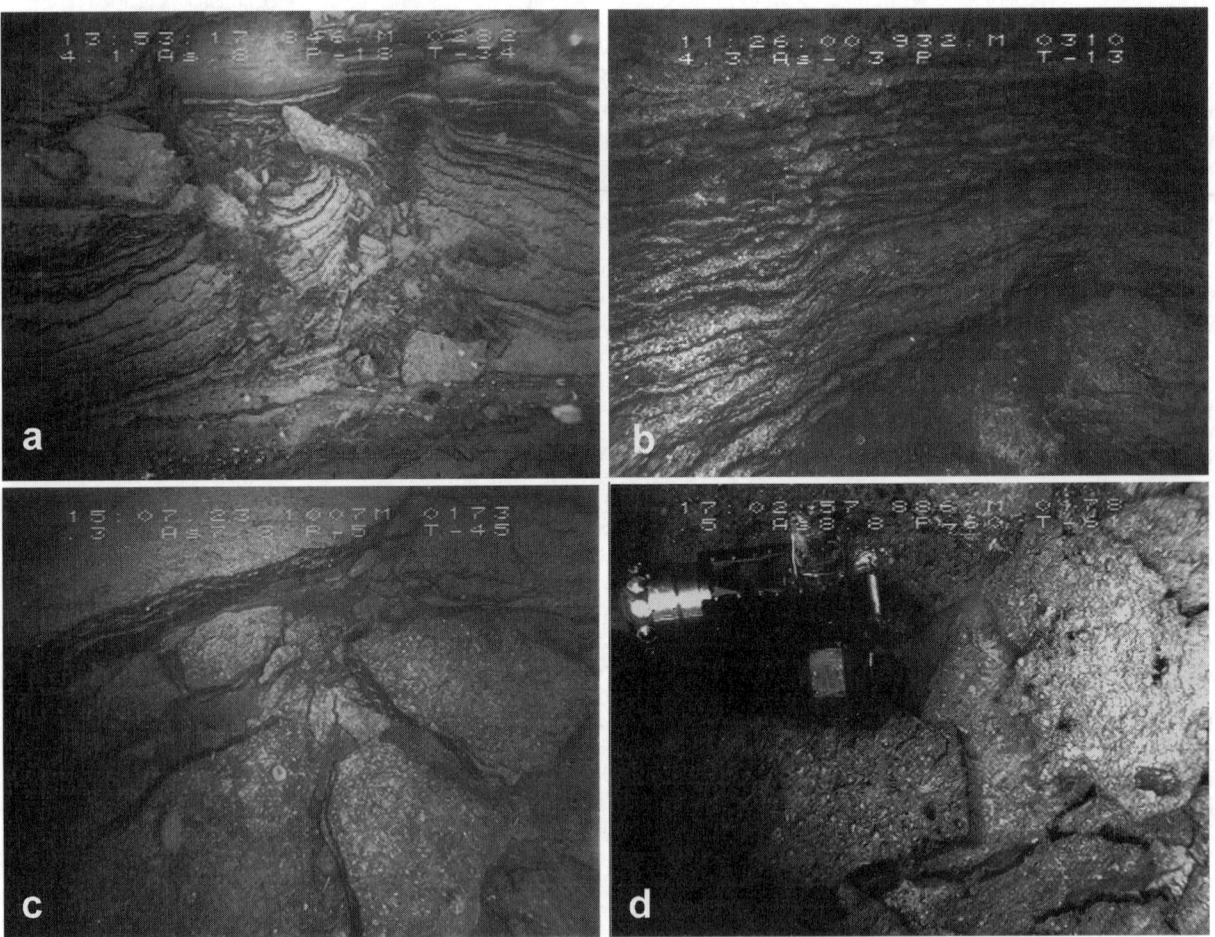

Figure 4. Seafloor photographs of the volcanoclastic deposits taken during the submersible "Nautile" dives during the DIVA cruise. a) layered volcaniclastic deposits of the 38N20 segment, dive 10, depth: 846 m; b) layered volcaniclastic deposits of the Menez Gwen segment, dive 13, depth: 931 m; c) layered volcaniclastic and breccia deposits of the Menez Gwen segment, dive 12, depth: 1007 m; d) recovery of volcaniclastic sample DV-12-8 with the mechanical arm of the submersible "Nautile", Menez Gwen segment, dive 12, depth: 886 m.

contains a 2 cm diameter rounded volcaniclastic "enclave" (Figure 5m). Reworking might be more extensive on the outward diping slopes of the central volcano where distal turbidite flows should exist. Extensive reworking is restricted to very recent deposits formed along the steepest slopes of central neovolcanic cones and inner faulted graben walls where breccias are found. Nevertheless, the volcaniclastic samples represent mostly primary deposits.

The nature of the clasts larger than 0.1 mm found in those deposits are of six different types:

1. Highly vesicular to scoriaceous glassy clasts (Figures 5j and 6), generally closely associated with plagioclase phenocrysts. These correspond to the glassy shard and bubble walls (Figure 5i) found in the common, more fragmented deposits on the 38N20 and Menez Gwen segments. Individually, some of these clasts have shapes that suggest emplacement before quenching in a still fluidal mode, with highly stretched bubbles, sometime with curved bubble wall (Figure 5o), dynamic ejection of stretched clast (Figure 5k) and Pele's hair (Figure 10c). These clasts represent the juvenile clasts of the magma from the eruptions responsible for the generation of these volcaniclastic deposits;

2. Low vesicularity glassy (transparent sideromelane) angular clasts (Figure 6) of sub-aphyric basalt. These clasts are interpreted as fragments of pillow rims of the local basement, incorporated mechanically in the mixture ejected during the eruptions forming the volcaniclastic deposits. They correspond to accessory lithic clasts [*Houghton and Smith*, 1993];

3. Low vesicular devitrified glass to microcrystalline black and angular clasts. These clasts are very similar to the second type, but represent devitrified to microcrystalline pillow interiors. They are also interpreted as accessory lithic clasts extracted from the basement;

4. Highly altered sub-angular to rounded clasts, frequently red to brown in thin sections. These clasts generally have a small diameter relative to the other types of clasts of the same samples, showing that they were more fragile. They are hydrothermally altered rock fragments, whose mineralogy has been completely modified. We interpret them to be basement fragments altered by high temperature fluids;

5. Calcareous sediment is frequently found within the deposits, either as individual unconsolidated or calcitic cemented layers (Figure 5c and 5d), often interspersed into/with the adjacent volcanoclastic-rich layers, or as individual foraminifera dispersed within some samples (Figure 6);

6. About 75% of the samples are free of any fine grained matrix other than sediment (Figure 5i, 5j and 5o). In the remaining samples, the matrix is very fine grained and totally opaque in transmitted light (Figure 6). This opaque matrix is probably made of a mixture of abundant, extremely fine-grained volcanic glassy fragments and calcareous sediment, altered due to its fine grain size.

The relative proportion of the different clast types varies from one sample to another or even from one layer to another within a hand-size sample (Figure 5a-5e), reflecting variations in the nature and intensity of the eruption and the mode of emplacement at the time of formation. Furthermore, these samples are subject to low-temperature weathering (Figure 5n) in seawater and high-temperature hydrothermal alteration (Figure 5h), that transforms the mineralogical and geochemical nature of the clasts. Despite such alteration, their individual morphology is frequently preserved and therefore easy to identify.

Highly Vesicular, Highly Phyric Basalts

Highly vesicular, generally highly plagioclase-phyric basalts (HPB of Table 1; Figure 5f and 5i) are found by submersible or dredging in close association with the volcaniclastic deposits. Identification of these flows from the submersible is difficult, as the basalt frequently forms pillow flows with an external glassy rim, similar in appearance to basalts with low vesicle contents. Only when seen in section, for example, along a fault, in brecciated flow fragments, or within the pit crater at Lucky Strike, is the highly vesicular nature of the interior of these pillows clearly revealed. These HPB (the scoriaceous lava of *Ondréas et al.* [1997]) apparently are absent from the shallower 38N20 segment, are uncommon on the Menez Gwen segment, and are more abundant on the three summit cones at Lucky Strike, around the pit crater which is the focus of the explosive and hydrothermal activities.

Poorly Vesicular Basalts

The most abundant basalt along each of the segments is poorly vesicular basalt, similar to that found along the entire MAR [*e.g. Bougault and Treuil*, 1980]. Locally, generally aphyric (<2% crystals; Figure 5g) to sub-aphyric (2–5% crystals) basalt forms pillow flows, lobate flows, draped flows, sheeted flows, rare massive flows or, locally, volcanic breccias and, in some deep zones, lava lakes [*Ondréas et al.*, 1997].

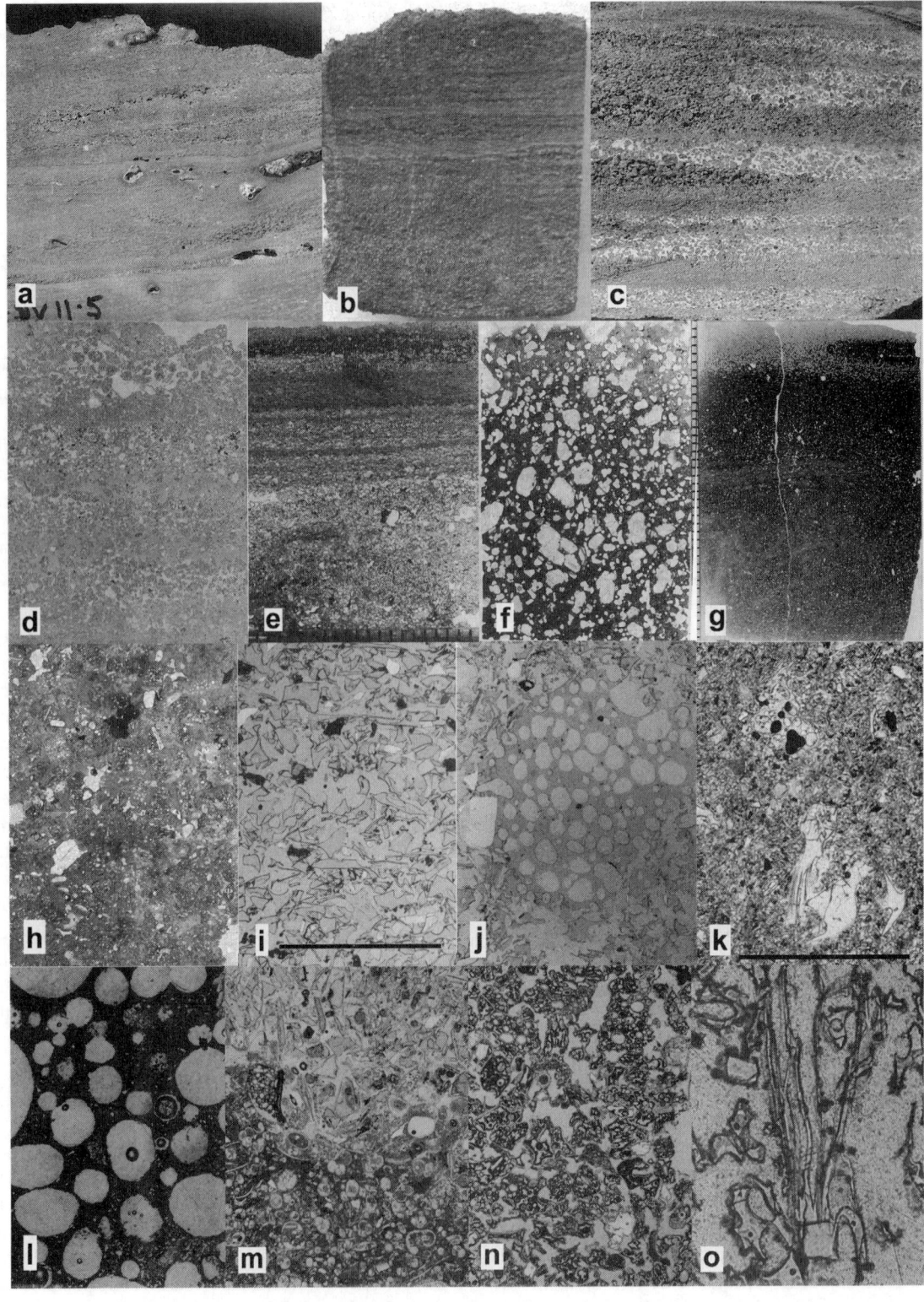

7. GEOCHEMISTRY OF THE DEPOSITS

Geochemical variations are well characterized along the MAR, especially around the Azores platform [*Bougault and Treuil*, 1980; *Dosso et al.*, 1999]. Compositional variations range from depleted N-MORB end-member (the most frequently encountered type) towards an enriched alkali basalt end-member, and include all possible intermediates (essentially E-MORBs). It is even possible to clearly identify, in two independent and extensive data sets [*Langmuir et al.*, 1997; *Dosso et al.*, 1999], depleted and enriched lavas at several localized sites on the MAR, for example, near 32°N, 33°40'N, 34°50'N–35°20'N, and on the Lucky Strike and the Menez Gwen segments.

Glass fragments of about half of our studied samples were analyzed using an electronic microprobe (Table 2). The results are presented using $K_2O\%$ versus Mg# diagrams only (Figure 7), as the purpose of the present paper is not to present an extensive geochemical study. Compositions include a relatively large range in Mg# (74–51) and in $K_2O\%$ from N-MORB (<0.20) to E-MORB (0.20–0.45), and alkali basalt values >0.45. Many of the Lucky Strike glasses are quite primitive, having Mg#>70. However, *all* the highly vesicular clasts from *all* three segments are only of E-MORB or alkali basalt compositions. The latter are more frequent for the Lucky Strike and Menez Gwen vesicular clasts whereas E-MORB compositions are most frequent on 38N20.

The composition of associated massive or dark glass clasts (accessory lithics as described above) is indistinguishable for the 38N20 segment (Figure 7a), much more differentiated (lower Mg#) and less alkalic for the Menez Gwen segment (Figure 7b) and less alkalic for the Lucky Strike segment, where the majority are N-MORB's (Figures 7c and d). These distinctione are also observed at the sample scale, where the various textures and chemistry frequently coexist. The highly vesicular and phyric pillows have compositions similar to the highly vesicular clasts.

Lastly, some glass inclusions analyzed in plagioclase phenocrysts are presented for two segments. For Menez Gwen, the parental magma of the most differentiated E-MORBs (51<Mg#<55) was much more primitive with Mg# up to 73. For Lucky Strike, the analyzed glass inclusions correspond to basalt samples that are mineralogically and chemically unrelated to the scoriaceous clasts, with compositions that plot within the N-MORBs field. These are similar to some dark glass clasts, described as accessory lithic incorporated in the volcanoclastics during the eruption.

Our data set compares closely to two other volcaniclastic deposits, one from the MAR on a segment located near 34°50'N, slightly south of our sector [*Hékinian et al.*, 2000], and one from Seamount Six [*Maicher et al.*, 2000], an off-axis seamount of the East Pacific Rise located near 12°45'N–102°35'W. The extent of the deposits near 34°50'N [*Hékinian et al.*, 2000] is somewhat smaller than those described here, but the lithological and geochemical compositions are more variable (Figure 7c), indicating a wider variety of lava types. Most important, their scoriaceous samples (also co-existing with highly vesicular and highly phyric basalts) are strictly limited to alkalic basalt compositions. This geochemical similarity suggests that the

---

Figure 5. Photographs of the volcaniclastic and associated lava samples.
—Macroscopic photographs: a) sample DV-11-5 showing a change in grain size, a small V-shaped chanel and worm burrows (width = ~6 cm); b) sample FL-D5-4 (width = ~4 cm); c) sample DV-11-6 showing a change in grain size and calcite cemented layers (width = ~6 cm);
—Entire thin section photographs: d) sample DV-11-5 showing a change in grain size and calcite cemented layers (width = 3 cm); e) sample FL-D5-6 showing very fine scale layering and a small change in grain size (width = 2 cm); f) sample FL-30-03, a highly phyric and vesicular basalt (width = 3 cm); g) sample FL-D4-4, an aphyric basalt (width = 3 cm); h) sample DV-11-3, highly altered volcaniclastic samples in which the morphology of individual fragments is well preserved (width = 3 cm);
—Photomicrographs: i) sample DV-10-1, a poorly consolidated volcaniclastic deposit with horizontal layering of most of the elongated clasts and vesicle walls, the presence of black glass (lithic fragment) and plagioclase (white) fragments and the lack of cement (bar = 1 mm); j) sample DV-8-7 showing a large vesicular clast with an attached plagioclase on the left, and both large (primitive) and small (secondary) vesicles (see the text for explanations) (width = ~1 mm); k) sample DV-10-9 showing an elongated comet-like shard with an glomerocryst "head" and stretched vesicles demonstrating that the clast was ejected still in a ductile stage (bar = 1 mm); l) sample FL-D4-5 showing matrix of highly vesicular basalt (width = ~1 mm), and the coexistence of large and small vesicles; m) sample DV-12-5 showing rounded glass-rich volcaniclastic clast (upper part) included within a calcareous sediment (lower part) (width = ~1 mm); n) sample DV-12-8, a completely altered volcaniclastic sample showing that the morphology of the altered glassy clasts is preserved (width = ~1 mm); o) sample DV-4-10, a glassy fragment showing highly streched vesicles with deformed bubble walls (width = ~0.5 mm).

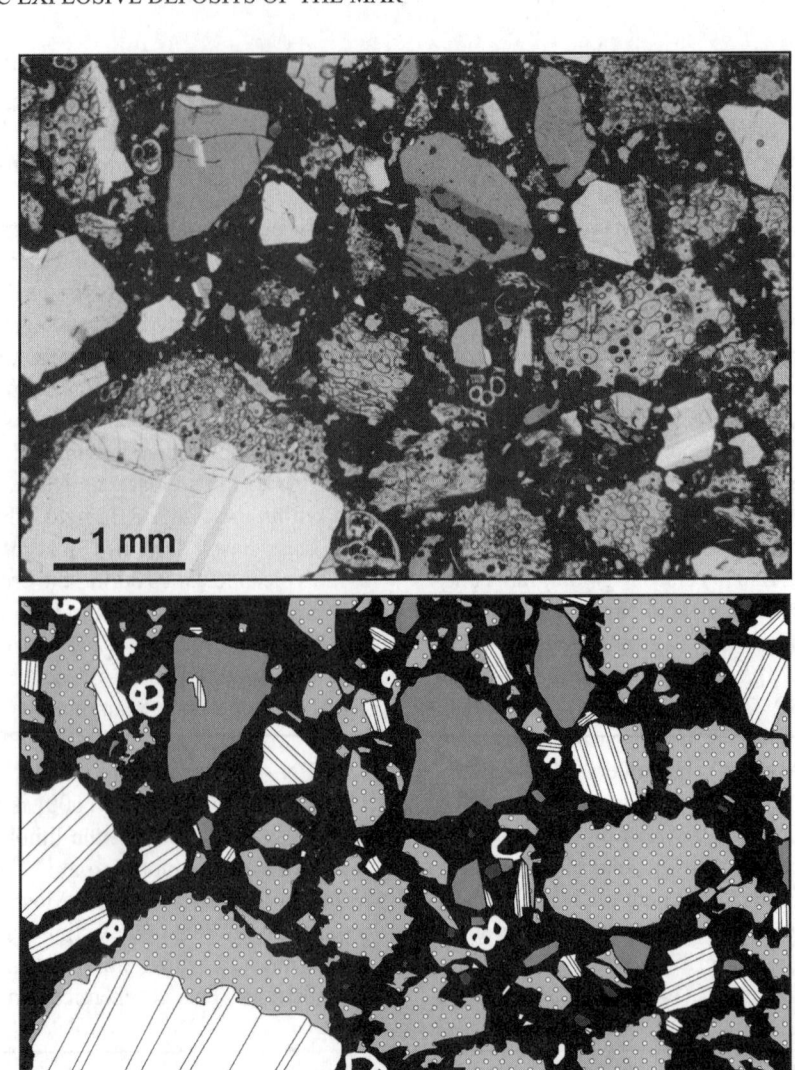

Figure 6. Top: Thin section microphotograph of sample DV-13-5B seen in transmitted light; Bottom: Lithological interpretation of the different clasts observed: 1) scoriaceous basaltic glass; 2) massive basaltic glass; 3) plagioclase crystals, usually broken and attached to vesicular glass; 4) black, undifferentiated, very fine grained mesostasis; 5) hydrothermally(?) altered glassy clast; 6) foraminifera.

mechanisms of formation of these deposits are probably very similar to the one we propose here.

The hyaloclastic deposits have a limited extent on Seamount Six and consist of angular low vesicle content glassy shards and thin, bent plate-like shards called "limu o Pele" [*Maicher et al.*, 2000]. These clasts formed mainly by quench fragmentation and bursting of magma bubbles, formed by vaporization of water entrapped within or below thin lava flows [*Maicher et al.*, 2000]. The extensive range of compositions observed on Seamount Six (Figure 7d) includes a group of N-MORBs and a group of alkalic (their hawaiite group) compositions extending up to 2% K_2O (not all shown on Figure 7). The bubble rich fragments (mainly limu o Pele) have both N-MORB and alkalic compositions. This observation indicates, as proposed by *Maicher et al.* [2000], that the volcanic dynamism responsible for their formation differs from our observations of the MAR volcaniclastic deposits, and is independent of their volatile content. *Clague et al.* [this volume] propose that limu o Pele of N-MORB composition formed deeper than the critical

Table 2. Selected microprobe analyses of basaltic glasses from the MAR volcaniclastic deposits and their associated lavas. Analyses performed using the CAMECA SX 100 "microsonde ouest" of Brest, France using their standard analytical procedure for the main oxides (accelerating voltage = 15 kV; beam current = 20 nA; beam defocused to a spot of ca. 10 μm of diameter) and 60 s of couting for Cl and S. H$_2$O data from *Fouquet et al.* [1998]. Mg# calculated with FeO = 0.9 × FeO^{total}. See the text for explanation on the type of analyzed clasts. The full data set is available upon request from the senior author.

Sple Nber	DV-10-2	DV-10-9	DV-10-9	DV-10-9	DV-10-9	DV-10-9	DV-10-9	DV-12-5	DV-12-5	DV-13-5A	DV-13-5A	DV-13-5B	DV-13-5B	DV-4-10	DV-4-10	DV-7-4	DV-7-4	DV-7-4	DV-7-4	DV-8-7	FL-25-1
segment	38N20	38N20	38N20	38N20	38N20	38N20	38N20	MG	MG	MG	MG	MG	MG	LS	LS	LS	LS	LS	LS	LS	LS
clast average	3 6	1 5	2 5	4 3	11 5	12 5	12 3	7 4	8 4	12 -	13 4	5 4	1 4	5 5	6 6	8 4	3 4	5 3	11 4	3 3	10 5
type	scoria	scoria	massive	scoria	scoria	scoria	massive	scoria	scoria	massive	scoria	scoria	massive	scoria	scoria	scoria	black	scoria	scoria	scoria	scoria
SiO$_2$	50.22	49.67	49.89	49.78	50.39	50.72	51.01	50.98	50.46	51.41	50.01	51.31	48.61	48.42	50.10	50.36	50.32	50.21	50.77
TiO$_2$	1.20	1.66	1.16	1.29	1.20	1.99	1.28	1.09	1.18	1.57	1.26	1.48	1.33	1.29	0.95	0.97	0.95	1.30	1.23
Al$_2$O$_3$	15.64	15.29	15.26	15.07	15.36	14.57	14.30	14.49	15.01	13.99	15.47	14.39	15.13	15.09	14.66	14.50	14.77	15.80	15.13
FeO	8.01	9.06	8.32	8.61	8.80	11.56	9.70	8.86	8.62	10.77	8.29	11.26	7.67	7.67	7.65	7.94	7.80	7.91	7.90
MnO	0.15	0.20	0.12	0.24	0.20	0.21	0.18	0.11	0.12	0.21	0.09	0.16	0.12	0.10	0.18	0.20	0.14	0.16	0.17
MgO	7.95	7.42	8.35	8.47	7.84	5.48	6.73	7.89	7.81	6.38	8.07	6.83	8.36	7.97	8.47	9.28	8.45	8.00	8.02
CaO	13.23	12.18	13.60	13.28	13.57	10.38	11.91	12.91	12.74	10.75	13.33	10.98	13.22	13.24	13.25	13.84	13.16	13.30	13.06
Na$_2$O	2.15	2.53	2.04	2.08	1.96	2.98	2.50	2.36	2.43	2.60	2.42	2.67	2.39	2.37	1.94	1.82	1.94	2.14	2.33
K$_2$O	0.46	0.65	0.39	0.35	0.40	0.70	0.31	0.27	0.50	0.41	0.58	0.36	0.56	0.56	0.31	0.22	0.30	0.45	0.46
P$_2$O$_5$	0.19	0.25	0.18	0.12	0.22	0.35	0.15	0.13	0.21	0.24	0.19	0.18	0.18	0.19	0.15	0.08	0.15	0.21	0.18
Cr$_2$O$_3$	0.06	0.06	0.03	0.05	0.01	0.02	0.03	0.01	0.07	0.04	0.04	0.01	0.06	0.06	0.10	0.13	0.09	0.05	0.02
S (ppm)	594	331	428	580	631	780	970	917	609	1333	376	1280	603	424	553	412	416		535
H$_2$O	0.38				0.39	0.47													
Cl (ppm)	237	142	261	287	292	940	346	102	311	308		311	415	285	134	112	146	287	287
Total	99.05	96.47	99.45	99.48	99.95	98.96	98.08	99.08	99.15	98.38	99.74	99.63	97.63	96.94	97.77	99.33	98.06	99.53	99.28
Mg#	66.3	61.9	66.5	66.1	63.8	48.4	57.9	63.8	64.2	54.0	65.8	54.6	68.4	67.3	68.7	69.8	68.2	66.7	66.8
Cl/K	0.06	0.03	0.05	0.07	0.09	0.16	0.14	0.05	0.08	0.09	0.04	0.10	0.09	0.06	0.05	0.06	0.06	0.08	0.08

Sple Nber	FL-25-1	FL-25-10	FL-25-10	FL-D3-5	FL-D3-5	FL-D5-1	FL-D5-1	FL-D5-4	FL-D5-4	FL-D5-6	FL-D5-6	FL-D5-7	FL-D5-7	FL-D5-8	FL-D5-8	FL-D5-8	FL-D5-10	FL-D5-10
segment																		
clast average	15 4	5 3	6 3	5 5	6 2	5 5	5 5	8 3	9 3	10 6	5 5	6 4	7 4	13 4	14 3	15	12 4	13 5	17 3
type	scoria	scoria	massive	scoria	massive	scoria	scoria	scoria	scoria	scoria	scoria	scoria	scoria	scoria	scoria	black	scoria	scoria	massive
SiO$_2$	50.14	48.85	49.13	50.40	50.21	49.06	49.43	49.27	49.74	49.88	49.71	49.48	49.24	50.33	51.35	49.90	49.24	49.59	49.61
TiO$_2$	1.23	1.17	1.26	1.27	1.25	1.18	1.29	1.28	1.28	1.32	1.23	1.24	1.24	1.23	1.23	1.20	1.22	1.25
Al$_2$O$_3$	14.99	15.29	15.24	15.12	15.25	14.96	14.99	15.41	15.65	15.53	16.00	15.07	15.50	15.59	15.10	15.39	15.09	15.08
FeO	7.99	6.48	6.17	7.90	8.00	6.54	6.59	6.87	6.98	6.88	7.17	6.42	6.97	6.71	6.73	6.00	6.67	7.22	7.24
MnO	0.17	0.11	0.12	0.14	0.21	0.17	0.19	0.08	0.10	0.13	0.15	0.08	0.12	0.15	0.14	0.12	0.18	0.12
MgO	7.85	8.30	8.39	8.24	8.30	8.33	8.41	8.48	8.52	8.70	7.94	8.70	8.33	8.37	8.78	8.40	8.14	8.15
CaO	12.68	13.25	13.34	13.08	13.37	13.62	13.63	13.92	13.98	13.90	13.59	13.74	13.87	13.78	14.84	13.64	12.94	12.95
Na$_2$O	2.26	2.21	2.22	2.32	2.25	2.15	2.25	2.25	2.25	2.20	2.21	2.15	2.24	2.27	2.20	2.23	2.27	2.27
K$_2$O	0.44	0.55	0.58	0.47	0.48	0.56	0.58	0.59	0.56	0.58	0.57	0.56	0.58	0.57	0.18	0.56	0.48	0.49
P$_2$O$_5$	0.14	0.20	0.16	0.17	0.19	0.21	0.16	0.25	0.17	0.15	0.17	0.19	0.23	0.16	0.22	0.23	0.17	0.21
Cr$_2$O$_3$	0.09	0.07	0.06	0.04	0.07	0.06	0.09	0.10	0.05	0.09	0.07	0.04	0.10	0.08	0.01	0.09	0.08	0.12
S (ppm)	585	368	412	558	528	359	279	406	469	458	491	396	495	557	298	323	341	454
H$_2$O																		
Cl (ppm)	128	142	261	287	263	439	258	367	391	447	183	327	311	436	112	285	247	128
Total	97.96	96.68	99.15	99.57	96.83	97.14	98.67	98.63	99.20	99.68	98.06	98.23	98.79	99.26	100.02	97.76	97.37	97.48
Mg#	66.0	72.9	71.7	67.4	67.3	71.6	71.7	70.6	71.0	70.6	71.0	71.2	71.1	71.1	74.4	71.4	69.1	69.0
Cl/K	0.04	0.05	0.07	0.07	0.07	0.10	0.05	0.08	0.08	0.10	0.04	0.07	0.06	0.09		0.06	0.06	0.03

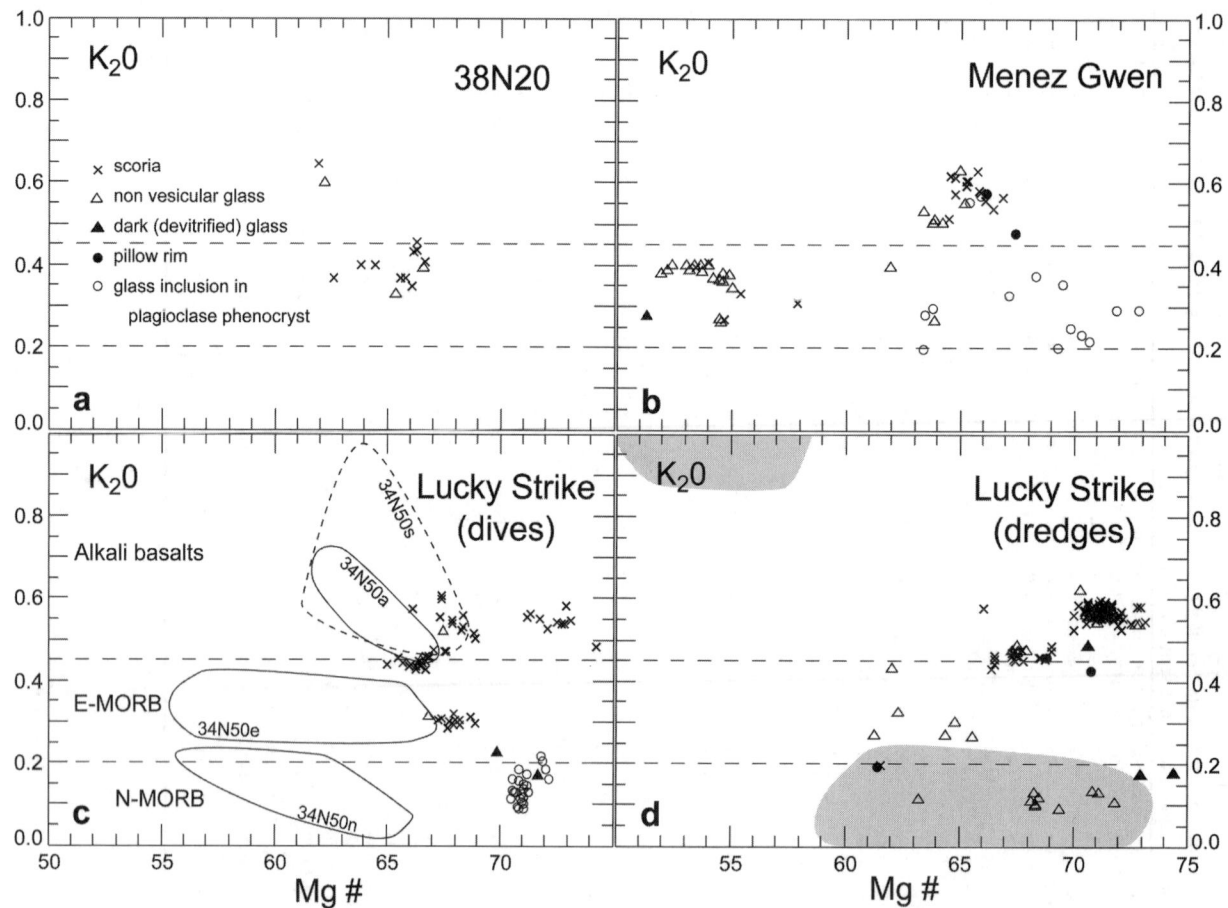

Figure 7. K_2O weight % versus Mg# compositional variations of the basaltic glasses found among the different sample lithologies of the three studied segments. a) 38N20 segment; b) Menez Gwen segment; c) Lucky Strike segment. Samples recovered by submersible. Fields show data of *Hékinian et al.*, [2000]: 34N50n: N-MORB lavas; 34N50e: E-MORB lavas; 34N50a: alkali lavas; 34N50s: scoriaceous samples; d) Lucky Strike segment, samples recovered by dredges. Lightly shaded fields show unpublished data of Doris Maicher from Seamount Six [*Maicher et al.*, 2000]. Horizontal dashed lines mark the geochemical limits between N-MORBs (0–0.2 K_2O %), E-MORBs (0.2–0.45 K_2O %) and alkali basalts (K_2O % > 0.45). *Hékinian et al.* [2000] attributed most of the compositional variation to: 1) variations of partial melting of the more mafic compositions (from 2% from alkali lava; ~5 % from E-MORBs and up to 20 % for N-MORBs), as in the vertical variations observed for the scoria and lava samples of Lucky Strike segment; 2) crystal fractionation leading to horizontal dispersion from high Mg# towards lower Mg# compositions, as in the variations observed between the glass inclusions—more primitive melts (Mg# > 70) and the massive glass of the Menez Gwen segment; and 3) additional variation from some mixing, at the level of the magma chamber, between the different compositions observed.

depth along the Gorda Ridge, have a pyroclastic rather than hydrovolcanic origin, with a similar role for the magmatic volatile phase that we propose here.

8. VESICULARITY

The vesicularity of scoriaceous glassy clasts from 15 samples was determined on thin sections, averaging the vesicularity calculated on several individual clasts (up to 10) of each sample. The vesicularity of the clasts (e.g. Figure 5j) was obtained by drawing them from photomicrographs in order to avoid any misinterpretation, then scanning the drawing. Finally, computer imaging software (NIH Image, National Institute of Health, USA) was used to calculate the vesicularity. The results show that the average vesicularity of the scoriaceous clasts varies from 27 to 65% (Figure 8). No systematic variation with water depth is observed. However, the vesicularity of the clasts from the two shallowest segments (38N20 and Menez Gwen) is strongly *underestimated*. When the vesicularity exceeds 75–80%, the material is totally fragmented, and it is impossible to calculate a lava vesicularity from the clasts (*e.g.* Figure 5i and 5o). Therefore, we propose that, in many cases, the vesicularity of the magma erupted at the shallowest sites was so high (i.e. > 75%) that it was thoroughly fragmented leaving only bubble walls.

9. GRAIN SIZE OF THE DEPOSITS

Grain size data has been obtained from unconsolidated samples and from a few lithified samples whose calcareous cement could be dissolved. We used a standard granulometric column for the fraction over 1 mm and a laser diffraction (Coulter LL130) system for the smaller size fraction. Six representative histograms of full grain size distribution are given as examples (Figure 9). All the clasts are smaller than 8 mm, or even 4 mm, with the highest frequency for a diameter from 0.25 to 0.5 mm. Sample DV-12-03 from the Menez Gwen segment has very fine grain size with a lack of fragments > 0.31 mm. Sample FL-D5-09 shows a bimodal distribution indicating that two individual thin sedimentary horizons were probably sampled, mixed and then measured. The grain size distribution of the volcaniclastic samples (presented in a diagram of *Inman*, 1952), covers a wide range, larger than the fields shown for comparison (Figure 9). They extend from extremely fine grained but very poorly sorted deposits (sample DV-12-03) to more-or-less coarse sand size deposits moderately (FL-D5-08 and DV-8-7) to very well sorted (DV-4-10 and FL-D5-09). The coarser samples, lapilli size on average, are extremely well sorted (FL-30-06), with negative heterogeneity coefficients, a likely result of strong sorting prior to deposition, as might be expected for underwater fall deposits [*Cashman and Fiske*, 1991]. This coarse-rich distribution might also result from syn- or post-depositional elutriation. In summary, the volcaniclastic deposits observed on the studied segments of the MAR are generally well sorted, but this sorting might result from the mode of emplacement, taking into account that elutriation might take place. This point will be discussed later.

10. MORPHOLOGY OF THE CLASTS

The detailed morphology of several hundred individual clasts of unconsolidated volcaniclastic deposits have been

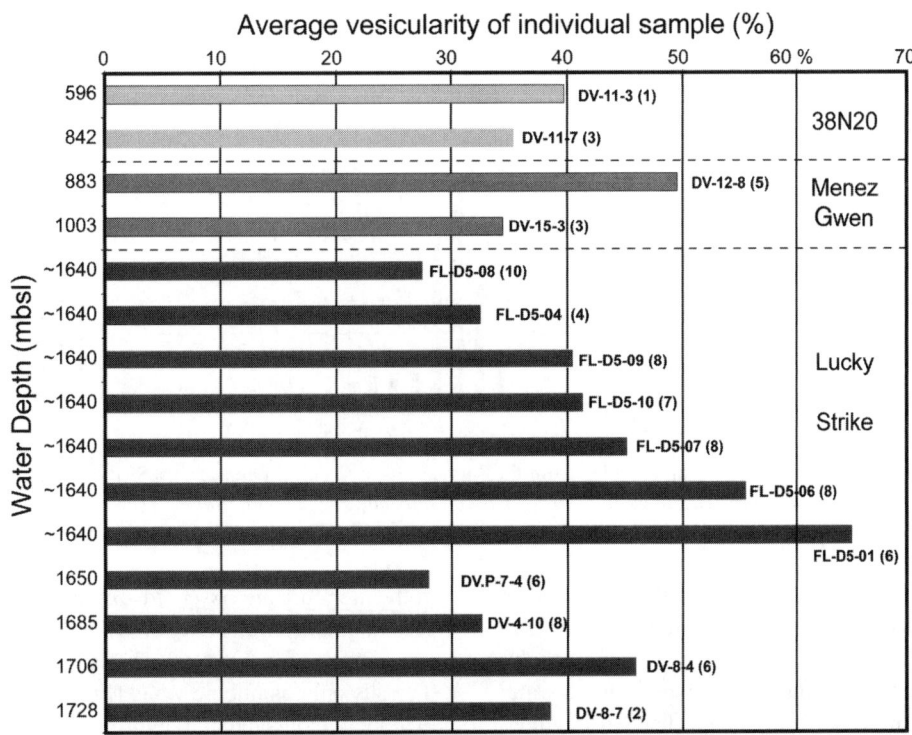

Figure 8. Average vesicularity of individual samples calculated from the vesicularity of several clasts within each sample. Results are ranked by increasing water depth from top to bottom, from 38N20, Menez Gwen and finally Lucky Strike segments. The number given next to the sample number relates to the number of clasts used in the vesicularity calculation. See the text for explanations.

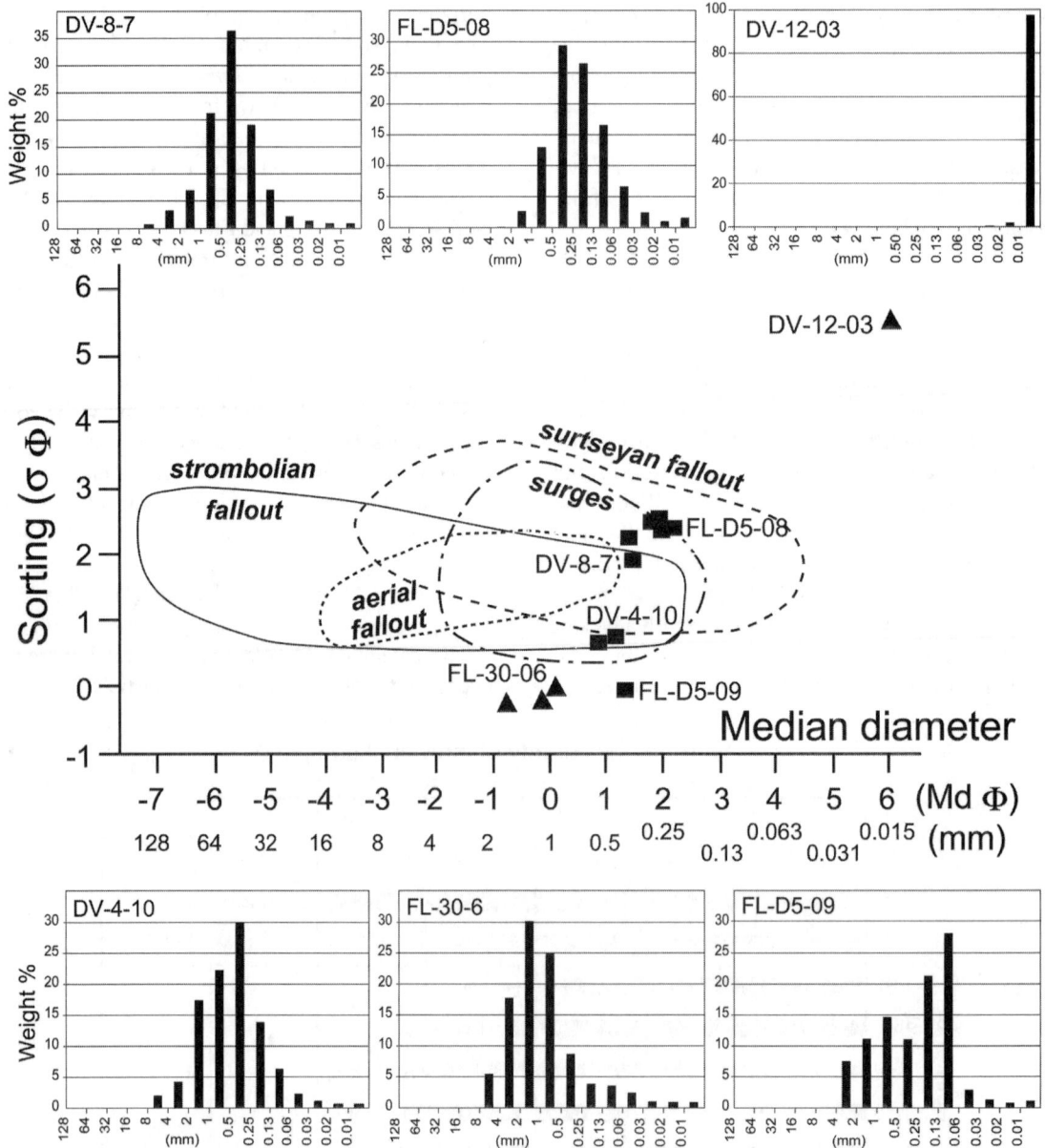

Figure 9. Grain size data obtained on samples from the Menez Gwen (triangle) and Lucky Strike (square) segments, plotted according to *Inman* [1952]. Fields from *Walker and Croasdale* [1972] and *Wohletz* [1983] are given for comparison. Individual results obtained on selected samples are shown in plots surrounding the main figure.

observed using a SEM in order to constrain their mechanism of fragmentation. Four main types of fragmentation have been observed: magmatic explosivity (by volatile exsolution), surface magma-water interaction, bulk magma-water interaction and cooling contraction i.e. mechanical fragmentation during quenching [*Kokelaar*, 1986]. It was shown many years ago that the shape of individual fragmented clasts is closely related to its mode of interaction with water [*Wohletz*, 1983]. Most of the fragmentation processes defined in the literature are observed among the clasts of our sample set. Magmatic explosivity is revealed by the extremely high vesicularity of many clasts that are frequently found in the deposits only as bubble walls (Figure 5i), as clasts showing extremely elongated and deformed vesicles (Figures 5o, 10d and 10h), or as Pele's hair (Figure 10c). More than half the clasts from the volcaniclastic deposits of the MAR result dominantly from this mode of fragmentation, at least for the two shallower segments. Surface magma-water interaction results in the formation of foam-like clasts as illustrated in Figure 10f. This type of clast is rare, and represents no more than 5% of the clasts observed. Massive highly vesicular clasts with sharp edges result generally from bulk magma-water interaction and are frequently found (Figure 10e). Cooling contraction is probably the most common way of fragmentation of a magma during its emplacement in a deep submarine environment as it is frequently observed in the hyaloclastic deposits of submarine seamounts [*Batiza and White*, 2000; *Maicher et al.*, 2000]. Only about 10% of the observed clasts show the typical angular or concave shape created by this mode of fragmentation (Figure 10g).

Recycling of primary magmatic clasts [*Houghton and Smith*, 1993] during explosive eruptions is indicated by the frequent presence of mud coatings on clasts, clearly illustrated in Figures 10a, 10b, 10 h and to a lesser extend 10e. Additional signs of recycling come also from the presence of several rounded clasts, also with mud coatings.

Because of the preservation of an exceptionnaly high volatile content in the E-MORB's and alkali basalts, magmatic fragmentation by exsolution of volatiles is the dominant mechanism of fragmentation at shallow depth followed by bulk magma-water interaction. But the diversity of magma-seawater interactions resulting in a wide range of mechanisms of fragmentation reveals the complexity of the eruptive style involved in the formation of the basaltic volcaniclastic deposits of the MAR.

11. DISCUSSION

Role of the Volatiles

Volatiles are present in MORB's either in the vesicles or as dissolved species. The dominant volatile species are CO_2 and H_2O [*Moore et al.*, 1977; *Pineau and Javoy*, 1983; *Fine and Stolper*, 1986; *Dixon et al.*, 1988; 1995], plus accessory sulfur, chlorine, fluorine [*Moore and Fabbi*, 1971; *Sakai et al.*, 1984; *Michael and Cornell*, 1998] and traces of noble gases and nitrogen [*Craig and Lupton*, 1976; *Marty and Ozima*, 1986; *Marty and Humbert*, 1997]. The average content of dissolved H_2O in MORB glasses varies from 1200 to 3500 ppm and the total (dissolved plus vesicle), CO_2 content is 100–200 ppm [*Pineau and Javoy*, 1983; *Dixon et al.*, 1988; 1991; *Mattey et al.*, 1989]. In fact, due to its low solubility, CO_2 is the first volatile to exsolve during magma ascent in two steps, first at a deep and later at a superficial level, above the magma chamber [*Bottinga and Javoy*, 1989]. Therefore, basaltic magma emplaced onto the ocean floor has lost ~90% of its original carbon content [*Bottinga and Javoy*, 1989; 1990a; 1990b], even if they are CO_2-saturated or even oversaturated when they quench [*Dixon et al.*, 1990].

The presence of extensive scoria-rich volcaniclastic deposits found between 500 and 1750 mbsl (even 1900 mbsl if we take into account the 34°50'N site [*Hékinian et al.*, 2000]) and the presence of "popping rocks" demonstrate that volatile-rich basaltic submarine deposits might not be as exceptional as thought a few years ago: they are present on at least six segments of the MAR, but remarkably absent from the EPR despite extensive exploration. The main characteristic of all the rocks is their high volatile content that is directly reflected in their high to extremely high vesicularity; from ~15% or more for the "popping rocks" to at least 60% or probably more then 80% for some extremely fragmented clasts described here.

Therefore the original volatile content of the magma must have a crucial role in the genesis of these rocks. Another key character of all these rocks is their variable but systematic geochemical enrichment relative to N-MORBs toward E-MORB or even more alkalic compositions. Whatever the reason for this enrichment (i.e. a plume-like deep mantle source or a low partial fusion rate of a shallow mantle source), alkalic magmas are known to have a higher content of volatiles, especially of CO_2. This observation confirms the role of volatiles in the origin of these peculiar rocks.

N-MORB magma generated beneath the spreading ridges is very rich (several thousands of ppm) and saturated in CO_2 at the level of its source [*Bottinga and Javoy*, 1989]. During magma ascent, two generations of bubbles nucleate, a deep one in the mantle and a shallower one above the magma chamber. But during its stay in the magma chamber, magma generally loses most of its first generation CO_2-rich bubbles. Therefore, when it is emplaced on the ocean floor, these magmas, which are usually saturated in CO_2 [*Dixon et al.*, 1990], have in fact lost most of their original carbon content.

Initially, we thought that the general morphology of the ridge axis, with the presence of a central shallower volcano where the scoriaceous volcaniclastic deposits have been found, would play a role in the concentration of the volatiles by simple density segregation toward the shallow part of the segment. But evidence from other studies led us to understand the key role played by CO_2 in the fragmentation of these magmas, without concentration but rather the exceptional preservation of the juvenile volatile content of the magmas.

In fact, the absence of preferential segregation of the large diameter vesicles (first and deep stage of nucleation) from the small diameter vesicles (late and superficial stage of nucleation) [*Sarda and Graham*, 1990] as seen on Figure 10l, means there was 1) no loss of the volatiles contained in the first generation bubbles and 2) no preferential concentration of these bubbles at the chamber's roof. *Sarda and*

Graham (1990) explained this observation by a probable short residence time of the volatile-rich magma within the magma chamber located a few km below the spreading center axis. In fact, this residence time must be short enough to prevent any significant direct volatile losses or any movement of the bubbles either relative to each other as a function of differences of diameters or relative to the magma during their combined ascent.

The presence of a relatively large amount of plagioclase phenocrysts (~5 vol.% with the vesicles; 25–35 vol.% recalculated vesicle free) showing no complex zoning, means, on the other hand, that the magma stayed in the magma reservoir for a sufficient length of time to allow quiet crystalization of phenocrysts to take place. At a high (but observed experimentally [*Clocchiatti and Massare*, 1985]) growth rate of 1 µm/hour, the minimum time needed to grow a 2 mm long plagioclase phenocrysts is ≈ 83 days. Therefore, it seems that residence times on the order of 100 days (to even possibly a small multiple of this) might represent a sufficiently short period of time to prevent any significant bubble segregation as well as any volatiles loss from the magma reservoir. The crystallization of relatively large amounts of plagioclase will also raise the volatile content of their host magma, as volatiles behave as incompatible elements.

Limited volatile data has been gathered on our samples for S, Cl (Table 2) and H_2O [*Fouquet et al.*, 1998]. Sulfur and chlorine contents fall within the range known for submarine basalt with the exception of a few samples showing an enrichment in chlorine, resulting probably from interactions with hydrothermal fluids before eruption, as suggested by their Cl/K ratios higher then 0.8 [*Michael and Cornell*, 1998] (Table 2). Water contents have been determined in different clasts (massive and scoriaceous) of a few samples. Results are quite variable but within the known range of E-MORBs. Therefore, it seems that none of these volatiles play a crucial role in producing the observed vesicularity or the eruptive style which generated these deposits. Unfortunately, no CO_2 content is analysed yet, but by analogy with the scoriaceous deposits at 34°50'N [*Hékinian et al.*, 2000], we assume that the primary carbon contents of the magmas forming the scoriaceous deposits was extremely high, on the order of 20–30 times the content of N-MORBs.

The pressure of the water column normally limits the extent of volatile degassing with depth. Therefore, the *preservation of most of the juvenile volatile content* of some of the enriched magma generated locally on the MAR (along with N-MORBs) is probably the most important factor to explain the generation of deep submarine explosive scoriaceous volcaniclastic basaltic deposits. Additional work is needed to understand why N-MORBs lose most of their original CO_2 content whereas some E-MORBs or alkalic basalts preserve it in some exceptional cases.

Eruptive Style

The information obtained by SEM on the morphology of individual clasts is essential in order to constrain the mechanism of fragmentation involved in the generation of these deposits. Taking into account the previous description and discussion, we propose that the scoriaceous volcaniclastic deposits observed are generated and emplaced by the following mechanisms (Figure 11):

1. In a shallow magma chamber a volume of E-MORBs to alkalic magma with a high content of CO_2-rich bubbles reside for no more than a few months. This magma crystallizes plagioclase and its volatile content increases even more;
2. During the rapid rise of this magma from the magma chamber toward the surface, the first generation of bubbles grows, and the second generation of bubbles nucleates and grows. When water saturation is reached (at a very shallow depth; ~200 m below the sea bottom), some water exsolving from the melt is added to the bubbles, increasing their volume and lowering further the density of the melt, which is rising even faster (Figure 11#1);
3. At shallow depth (<100 m), interaction with hydrothermal circulations might occur and some bulk magma-water interaction might occur locally. The first accidental lithics (hydrothermally altered clasts) are incorporated in the rising flux (Figure 11#2);
4. In the deepest part of the vent, because of the highly fractured volcanic basement rocks (pillow lavas, sheeted flows, ...), more seawater is in contact with the

---

Figure 10. SEM photomicrographs of individual clasts from the volcaniclastic deposits revealing their detailed morphology that can be intrepreted in terms of mechanism of fragmentation [e.g. *Wohletz*, 1983; *Kokelaar*, 1986]. a) sample DV-4-10, mud-coated clast, probably recycled [*Houghton and Smith*, 1993]; b) sample DV-4-10, detail of a mud coating; c) sample DV-12-3, Pele's hair clast; d) sample DV-12-3, clast showing extremely elongated—tubular–deformed vesicles; e) DV-8-7, highly vesicular, equant clast, with a thin mud coating; f) DV-12-3, foam-like clast; g) DV-8-7, arrow = concave clast, result of quench granulation; h) DV-8-7, highly elongated vesicle and mud coating on the clast.

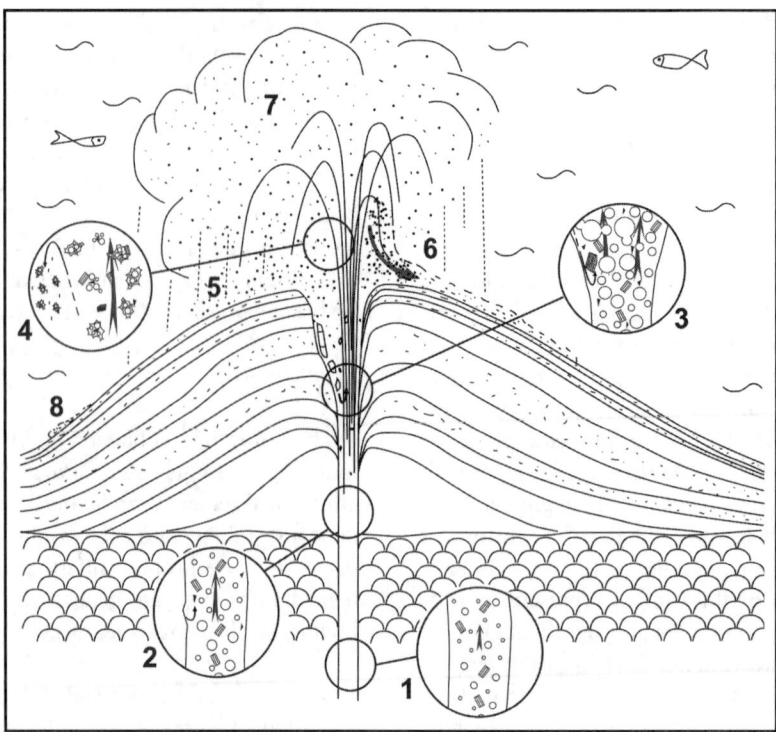

Figure 11. Schematic idealised illustration showing the interpreted processes of eruption, fragmentation and deposition of the scoriaceous deposits formed at the axis of the MAR south of the Azores. 1) Rising magma with plagioclase phenocrysts, and two generations of CO_2 bubbles; 2) Magma rising faster as bubbles grow due to decompression and new bubble nucleation, with plagioclase phenocrysts and accidental lithic clasts from the basement (eventually altered by hydrothermal circulation); 3) Magma rising still faster as bubbles grow and start to coalesce. More accidental lithic clasts from the basement are incorporated and part of the recent volcaniclastic deposits is recycled in the vent. Start of magma fragmentation, mainly by bulk magma-water interaction in a confined environment; 4) Extensive fragmentation of the highly vesiculated magma, injected still hot and ductile in the water column. Fragmentation is dominated by volatile driven magmatic fragmentation by coalescence of bubbles and magmatic volatile expansion. However, surface magma-water interaction and quench granulation strongly enhance fragmentation. Laterally, fallout is the dominant mode of deposition of the fragmented clasts; 5) Extensive and repeated fallout of the clasts takes place all around the vent under the rising column; 6) If the rising mixture of juvenile clasts, crystals, lithics, gas and water become denser than the surrounding sea water, local laterally flowing density currents derived from convective-settling suspensions take place in the proximal areas; 7) Especially in the upper part of the column, elutriation of the clasts is taking place, some of the finer particles being taking away by the sea current; 8) In the distal slopes, lateral gravity driven reworking occurs locally. See the text for explanations.

rising vesiculated magma. Extensive bulk magma-water interaction occurs. Numerous accidental lithics from the basement rocks are incorporated (Figure 11#3);

5. In the upper part of the vent, a larger amount of seawater interacts with the rapidly rising, highly vesicular magma that starts to fragment by magmatic explosivity. Magma-water interaction occurs also with quench granulation, bulk magma-water and surface magma-water interaction. Previously erupted material fall back into the vent and is recycled. Pele's hair and stretched vesicles are also formed by rapid injection of magma in seawater (Figure 11#4);

6. In the plume formed above the vent, quench granulation in direct contact with seawater is the dominant mechanism of fragmentation.

At any stage in the vent, limited surface magma-water interaction occurs.

Emplacement of the Deposits

From the submersible observations, the texture of the samples, their clast morphology, and the grain size data obtained, we can infer that the mechanism of emplacement of these clastic deposits were of four types:

1. Suspended-water fallout deposition dominates (Figure 11#4 and #5). In most outcrops and samples, the absence of sedimentary structures such as bed lenticularity, scouring, clast imbrications and cross stratification argues for a proximal site of emplacement, rather by suspended-fall deposition then by current-emplaced beds [*White*, 2000]. However, normal grading is not always observed as some elutriation of the fines might have occurred (Figure 11#7). Additional arguments for this mode of emplacement come from the lack of grain abrasion (preservation of many clasts with complex and fragile morphology), the frequent subparallel orientation of the elongated fragments (juvenile glassy clasts, accessory lithic clasts, crystals,...; Figures 5, 6 and 10), the lack of mixing of the volcaniclastic mm- to cm-thick layers with mm-thick layers of the pelagic ooze, and the geochemical homogeneity of the juvenile clasts within a single layer. The good grain size sorting of many layers, as it is generally observed in a subaerial environment, is due to a more efficient sorting of clasts within the water column then within the air [*Cashmann and Fiske*, 1991];
2. Accessory laterally flowing density-driven currents derived from the collapse of dense convective-settling suspensions (Figure 11#6) in the very proximal areas around the vents;
3. Relatively limited density-driven reworking emplacement occurs. A few small sedimentary structures, such as erosional channels, have been observed (Figures 5a and 11#8) as well as volcaniclastic-rich clasts within thick pelagic ooze layers at the periphery of the volcaniclastic deposits (Figure 5m). However, their extend is very limited and probably restricted to very local slump-type density flows on the steeper slopes of the hill sides of a growing volcaniclastic pile;
4. Very rarely laterally flowing density current of limited extend extending directly from the vents might have occured in the very proximal areas, resulting in the formation of the thicker and coarser layers observed. But in every case, all the fragments were deposited cold as no welding or ductile deformation was ever observed.

Note however that a more detailed submersible study and sampling devoted specifically to volcanoclastic problems would have brought additional field observations as well as lithological and volcanological informations. The recovery of more distal and proximal facies, that have not been sampled until now, would probably reveal more variations and reworking than described in the presented work.

The highly vesicular and highly phyric pillow lavas are found in close association with the volcaniclastic deposits. They have similar geochemical characteristics and might correspond to the rise of much more limited volumes of magma and at a much lower rising velocity. The emplacement would then occur at a much lower effusion rate as it is the case for low vesicularity pillow lavas.

Similar old deposits have been described in various locations but especially in the Archean volcanic province of Canada [*Mueller et al.*, 1994; *Doucet et al.*, 1994]. The comparison of these old deposits (where often the interior of the vents is exposed) with some of their modern analogs will lead to future refinement of the existing models.

Acknowledgments. This work has been financialy supported by IRD, Ifremer and the MAST3 program of EEC. The DIVA cruise was part of the French-American FARA program. The FLORES cruise was part of the French-Portugese AMORES program. The authors thank the Captains and the crews of the R.V. Nadir and R.V. l'Atalante and the submersible Nautile group that allowed the recovery of the studied samples. We are grateful to Ronan Apprioual (Ifremer), Marcel Bohn (CNRS), Philippe Crassoux (Ifremer) for their help respectively in the confection of the thin sections, in the acquisition of the electronique microprobe analyses and of the MEB images. We are deeply gratefull to Doris Maicher for providing us some unpublished geochemical data of the Seamount Six hyaloclastic deposits and for her comments on a preliminary version of this work. We thank also deeply Kate Bull for her carefull checking of our "frenglish". We acknowledge also Rodey Batiza, and Peter Kokelaar for early discussions and suggestions about this work as well as the editors of the present monograph for organizing a great conference in January 2002 and for their stimulative support. Finally, we acknowledge the constructive reviews and the pertinent comments by Rodey Batiza and Jennifer Reynolds and the fine editorial work of David Clague.

REFERENCES

Ballard, R. D., and J. G. Moore, 1977, *Photographic Atlas of the MAR rift valley*, Springer Verlag, 114 p.

Batiza, R., and White, J. D. L., Submarine lavas and hyaloclastite, in *Encyclopedia of Volvanoes*, ed. by H. Sugurdsson, Academic Press, San Diego, 361-381, 2000.

Batiza, R., Fornari, D. J., Vanko, D. A., and Lonsdale, P., Craters, calderas and hyaloclastites on young pacific seamounts, *J. Geoph. Res., 89*, 8371-8390, 1984.

Binard, N., Hékinian, R., and Stoffers, P., Morphostructural study and type of volcanism of submarine volcanoes over the Pitcairn hot spot in the South Pacific, *Tectonophysics, 206*, 245-264, 1992.

Binard, N., Hékinian, R., Cheminée, J.-L., Searle, R. C., and Stoffers, P., Morphological and structure studies of the Society and Austral hotspot regions in the South Pacific, *Tectonophisics, 186*, 293-312. 1991.

Bonatti, E., and Harrison, C. G. A., Eruption styles of basalt in oceanic spreading ridges and seamounts: Effect of magma temperature and viscosity, *J. Geoph. Res., 93*, 2967-2980, 1988.

Bottinga, Y., and Javoy, M., Degassing: bubble growth and ascent, *Chem. Geol., 81*, 255-270, 1990.

Bottinga, Y., and Javoy, M., Mid-ocean ridge basalt degassing: bubble nucleation. *J. Geoph. Res., 95,* 5125-5131, 1990.

Bottinga, Y., and Javoy, M., MORB degassing: evolution of CO_2, *Earth Planet. Sci. Lett., 93,* 215-225, 1989.

Bougault, H., Dmitriev, L., Schilling, J.-G., Sobolev, A., Joron, J.-L., and Needham, H. D., Mantle heterogeneity from trace elements: MAR triple junction near 14°N, *Earth Planet. Sci. Lett., 88,* 27-36, 1988.

Bougault, H., and Treuil, M., Mid-Atlantic Ridge: zero-age geochemical variations between Azores and 22°N, *Nature, 286,* 209-212, 1980.

Cannat, M., Briais, A., Deplus, C., Escartin, J., Georgen, J., Lin, J., Mercouriev, S., Meyzen, C., Muller, M., Pouliquen, G., Rabain, A., and da Silva, P., Mid-Atlantic Ridge-Azores hotspot interactions; along-axis migration of a hotspot-derived event of enhanced magmatism 10 to 4 Ma ago, *Earth Planet. Sci. Lett., 173,* 257-269, 1999.

Cashman, K. V., and Fiske, R. S., Fallout of pyroclastic debris from submarine eruptions, *Science, 253,* 275-280, 1991.

Clague, D. A., Davis, A. S., Bischoff, J. L., Dixon, J. E., and Geyer, R., Lava bubble-wall fragments formed by submarine hydrovolcanic explosions on Loihi seamount and Kilauea volcano, *Bull. Volcanol., 61,* 437-449, 2000.

Clocchiatti, R. and Massare D., Experimental crystal growth in glass inclusions: the possibilities and limits of the method, *Contrib. Mineral. Petrol., 89,* 193-204, 1985.

Craig, H., and Lupton, J. E., Primordial neon, helium and hydrogen in oceanic basalts, *Earth Planet. Sci. Lett., 31,* 369-385, 1976.

Dennielou, B., Dynamique sédimentaire sur le plateau des Açores pour les derniers 400 ka: distribution, lithologie, flux et processus—implications paléocéanographiques, PhD tesis, Univ. Bretagne Occidentale, Brest, France, pp. 341, 1997.

Dennielou, B., Auffret, G. A., Boelaert, A., Richter, T., Garlan, T., and Kerbrat, R., Control of the Mid-Atlantic ridge and the Gulf Stream on the Quaternary sedimentation on the Azores Plateau, *Comptes Rendus Acad. Sci. Paris, 328,* IIA, 831-837, 1999.

Dick, H. J. B., Honnorez, J., and Kirst, P. W., Origin of the abyssal sand, sandstone and gravel from DSDP hole 396B, Leg 46, Initial Reports of the Deep Sea Drilling Project, ed. by L. Dmitriev, J. Heirtzler, et al., U.S Government Printing Office, Washington D.C., 301-339, 1978.

Dixon, J. E., Clague, D. A., and Stolper, E. M., Degassing history of water, sulfur and carbon in submarine lavas from Kilauea volcano, Hawaii, *J. Geol., 99,* 371-394, 1991.

Dixon, J. E., Stolper, E. M., and , J. R. Delaney, Infrared spectroscopic measurements of CO_2 and H_2O in Juan de Fuca Ridge basaltic glasses, *Earth Planet. Sci. Lett., 90,* 87-104, 1988.

Dixon, J. E., Stolper, E. M., and Holloway, J. R., An experimental study of water and carbon dioxide solubilities in mid-ocean ridge basaltic liquids. Part I: Calibration and solubility models, *J. Petrol., 36,* 1607-1631, 1995.

Dosso, L., Bougault, H., Langmuir, C., Bollinger, C., Bonnier, O., and Etoubleau, J., The age and distribution of mantle heterogeneity along the Mid-Atlantic Ridge (31–41°N), *Earth Planet. Sci. Lett., 170,* 269-286, 1999.

Doucet, P., Mueller, W., and Chartrand, F., Archean, deep-marine, volcanic eruptive products associated with the Coniagas massive sulfide deposits, Quebec, Canada, *Canad. J. Earth Sci., 31,* 1569-1584, 1994.

Fine, G., and Stolper, E. M., Dissolved carbon dioxide in basaltic glasses: concentrations and speciation, *Earth Planet. Sci. Lett., 76,* 263-278, 1986.

Fornari, D. J., Ryan, W. B. F., and Fox, P. J., The evolution of craters and calderas on young seamounts: Insights from Sea Marc 1 and Sea Beam sonar surveys of a small seamount group near the axis of the East Pacific Rise at 10°N, *J. Geoph. Res., 89,* 11069-11083, 1984.

Fouquet, Y., Charlou, J.-L., Costa, I., Donval, J.-P., Radford-Knoery, J., Pellé, H., Ondréas, H., Lourenço, N., Ségonzac, M., and Kingstom Tivey, M., A detailed study of the Lucky Strike hydrothermal site and discovery of a new hydrothermal site: Menez Gwen; Preliminary results of the DIVA 1 cruise, *InterRidge News, 3,* 14-17, 1994.

Fouquet, Y., Ondréas, H., Charlou, J.-L., Donval, J.-P., Radford-Knoery, J., Costa, I., Lourenço, N., and Tivey, M. K., Atlantic lava lakes and hot vents, *Nature, 377,* 201, 1995.

Fouquet, Y., Eissen, J.P., Ondréas, H., Barriga, F., Batiza, R., and Danyushevsky, L., Extensive volcaniclastic deposits at the Mid-Atlantic Ridge axis: results of deep-water basaltic explosive activity *Terra Nova, 10,* 280-286, 1998.

Fouquet, Y., Charlou, J. L., Donval, J. P., Radford-Knoery, J., Ondreas, H., Costa, I., Lourenco, N., Segonzac, M., Tivey, M. K., Barriga, F., Cambon, P., Bougault, H., and Etoubleau, J., Hydrothermal and volcanic processes in shallow hydrothermal fields near the Azores Triple Junction (Lucky Strike and Menez Gwen), *Marine Geol.,* submitted.

Francheteau, J., Juteau T., and Rangin C., Basaltic pillar in collapsed lava-pools on the deep ocean floor, *Nature, 281,* 209-211, 1979.

Gill, J., Torssander, P., Lapierre, H., Taylor, R., Kaiho, K., Koyama, M., Kusakabe, M., Aitchison, J., Cisowski, S., Dadey, K, Fujioka, K., Klaus, A., Lovell, M., Marsaglia, K., Pezard, P., Taylor, B., Tazaki, K., Explosive deep water basalt in the Sumisu backarc rift, *Science, 248,* 1214-1217, 1990.

Hékinian, R., Chaigneau, M., and Cheminée, J.-L., Popping rocks and lava tubes from the Mid-Atlantic rift valley at 36°N, *Nature, 245,* 371-373, 1973.

Hékinian, R., Pineau, F., Shilobreeva, S., Bideau, D., Gracia, E., and Javoy, M., Deep sea explosive activity on the Mid-Atlantic Ridge near 34°50'N: Magma composition, vesicularity and volatile content, *J Volcanol. Geotherm. Res., 98,* 49-77, 2000.

Houghton, B. F., and Wilson, C. J. N., A vesicularity index for pyroclastic deposits, *Bull. Volcanol., 51,* 451-462, 1989.

Houghton, B. F., and Smith, R. T., Recycling of magmatic clasts during explosive eruptions: estimating the true juvenile content of phreatomagmatic volcanic deposits, *Bull. Volcanol., 55,* 414-420, 1993.

Inman, D.L., Measures for describing the size distribution of sediments, *J. Sediment. Petrol., 22,* 125-145, 1952.

Jambon, A., and Zimmermann, J. L., Water in oceanic basalts: evidence for dehydration of recycled crust, *Earth Planet. Sci. Lett., 101,* 323-331, 1990.

Jaupart, C., Physical models of volcanic eruptions, *Chem. Geol., 128,* 217-227, 1996.

Javoy, M., and Pineau, F., The volatiles record of a "popping" rock from the Mid-Atlantic Ridge at 14°N: chemical and isotopic composition of gas trapped in the vesicles, *Earth Planet. Sci. Lett., 107,* 598-611, 1991.

Jendrzejewski, Y., Trull, T. W., Pineau, F., and Javoy, M., Carbon solubility in Mid-Ocean Ridge basaltic melt at low pressures (250-1950 bar), *Chem. Geol., 138,* 81-92, 1997.

Johnson K.T.M., Reynolds, J.R., Vonderhaar, D., Smith, D.K., and Long, L.S., Petrological Systematics of Submarine Basalt Glasses from the Puna Ridge, Hawaii: Implications for Rift Zone Plumbing and Magmatic Processes, in E. Takahashi, P.W. Lipman, M.O. Garcia, J. Naka and S. Aramaki (ed), *Hawaiian Volcanoes: Deep Underwater Perspectives,* AGU Geophys. Monogr. Ser., 128, 143-159, 2002.

Kappel, E.S. and Ryan, W.B.F., Volcanic episodicity and a non steady-state rift valley along the Northeast Pacific Spreading Centers: evidence from Sea-Marc I, *J. Geoph. Res., 91,* 13925-13940, 1986.

Kokelaar, B. P., Magma-water interactions in subaqueous and emergent basaltic volcanism, *Bull. Volcanol., 48,* 275-289, 1986.

Langmuir, C., Humphris, S., Fornari, D., Van, D. C., Von, D. K. L., Tivey, M. K., Colodner, D., Charlou, J. L., Desonie, D., Wilson, C., Fouquet, Y., Klinkhammer, G., and Bougault, H., Hydrothermal vents near a mantle hot spot; the Lucky Strike vent field at 37°N on the Mid-Atlantic Ridge, *Earth Planet. Sci. Lett., 148,* 69-91, 1997.

Maicher D., White J.D.L. and Batiza R., Sheet hyaloclastite: density current deposits of quench and bubble-burst fragments from thin, glassy sheet lava flows, Seamount Six, Eastern Pacific Ocean, *Marine Geol., 171,* 75-93, 2000.

Marty, B., and Humbert, F., Nitrogen and argon isotopes in oceanic basalts, *Earth Planet. Sci. Lett., 152,* 101-112, 1997.

Marty, B., and Ozima, M., Noble gas distribution in oceanic basalt glasses, *Geoch. Cosmochim. Acta, 50,* 1093-1097, 1986.

Mattey, D. P., Exley, R. A., and Philinger, C. T., Isotopic composition of CO_2 and dissolved carbon in glass, *Geoch. Cosmochim. Acta, 53,* 2377-2386, 1989.

Michael, P.J., and Cornell, W.C., Influence of spreading rate and magma supply on crystallization and assimilation beneath mid-ocean ridges: Evidence from chlorine and major element chemistry of mid-ocean ridge basalts, *J. Geophys. Res., 103,* B8, 18325-18356, 1998.

Moore, G., Vennenman, T., and Carmichael, I. S. E., Solubility of water in magma to 2 kbar, *Geology, 23,* 1099-1102, 1995.

Moore, J. G., and Schilling, J. G., Vesicles, water, and sulfur in Reykjanes ridge basalts, *Contr. Mineral. Petrol., 4,* 105-118, 1973.

Moore, J. G., Batchelder, J. N., and Cunningham, C. G., CO_2-filled vesicles in mid-ocean basalt, *J Volcanol. Geotherm. Res., 2,* 309-327, 1977.

Moore, J. G., Vesicularity and CO2 in mid-ocean ridge basalt, *Nature, 282,* 250-253, 1979.

Moore, J. G. and Fabbi, B. P., "An estimate of the juvenile sulfur content of basalts." *Contribution to Mineralogy and Petrology 33:* 118-127, 1971.

Moore, J. G., Water contents of basalts erupted on the ocean floor, *Contr. Mineral. Petrol., 28,* 272-279, 1970.

Mueller, W., Chown, E.H., and Potvin, R., Substorm wave base felsic hydroclastic deposits in the Archean Lac des Vents volcanic complex, Abitibi belt, Canada, *J. Volcanol. Geoth. Res., 60,* 273-300, 1994.

Ondréas, H., Fouquet, Y., Voisset, M., and Radford-Knoery, J., Detailed study of three contiguous segments of the Mid-Atlantic Ridge, South of the Azores (37° N to 38°30' N), using acoustic imaging coupled with submersible observations, *Marine Geophys. Res, 19,* 231-255, 1997.

Peckover, R. S., Buchanan, D. J., and Ashley, D. E., Fuel-coolant interactions in submarine volcanism, *Nature, 245,* 307, 1973.

Perfit, M. R., and Chadwick Jr., W. W., Magmatism at mid-ocean ridges: Constraints from volcanological and geochemical investigations, *Faulting and Magmatism at Mid-Ocean Ridges,* Geophysical Monograph, Washington DC, USA, American Geophysical Union, p. 59-115, 1998.

Pineau, F., and Javoy, M., Carbon isotopes and concentrations in mid-oceanic basalts, *Geochimica Cosmochimica Acta, 46,* 371-379, 1983.

Pineau, F., and Javoy, M., Strong degassing at ridge crests: the behaviour of dissolved carbon and water in basalt glasses at 14°N, Mid-Atlantic Ridge, *Earth Planet. Sci. Lett., 123,* 179-198, 1994.

Pineau, F., Javoy, M., and Bottinga, Y., 13C/12C ratios of rocks and inclusions in popping rocks of the Mid-Atlantic Ridge and their bearing on the problem of isotopic compositions of deep seatted carbon, *Earth Planet. Sci. Lett., 29,* 413-421, 1976.

Pineau, F., Shilobreeva, S., Javoy, M., and Hékinian, R., Deep sea explosive activity on the Mid-Atlantic Ridge near 34°50'N, Part II: evidence of magma origin from stable isotopes (C, O, H), *J Volcanol. Geotherm. Res.,* in prep.

Sarda, P., and Graham, D., Mid-ocean ridge popping rocks: implication for degassing at ridges, *Earth Planet. Sci. Lett., 97,* 268-289, 1990.

Schmincke, H.-U., Robinson, P. T., Ohnmacht, W., and Flower, M. F. J., Basaltic hyaloclastites from hole 396B, DSDP Leg 46, Initial Reports of the Deep Sea Drilling Project, ed. by L. Dmitriev, J. Heirtzler, and et al., U.S Government Printing Office, Washington D.C., 301-339, 1978.

Sheridan, M. F., and Wohletz, K. Hydrovolcanism: basic considerations for review, *J Volcanol. Geotherm. Res., 28,* 257-284, 1983.

Smith, T. L., and Batiza, R., New field and laboratory evidence for the origin of hydroclastite flows on seamount summits, *Bull. Volcanol., 51,* 96-114, 1989.

Walker, G.P.L., and Croasdale R., Characteristics of some basaltic pyroclastics, *Bull. Volcanol., 35,* 303-317, 1972.

Wohletz, K. H., Mechanism of hydrovolcanic pyroclast formation: Grain-size, scanning electron microscopy, and experimental studies, *J Volcanol. Geotherm. Res., 17,* 31-63, 1983.

Zimanowski, B., Froehlich, G., and Lorenz, V., Quantitative experiments on phreatomagmatic eruptions, *J Volcanol. Geotherm. Res., 48,* 341-358, 1991.

Jean-Philippe Eissen, IRD (Institut de Recherche pour le Développement), Whymper 442 y Coruña, A.P. 17-12-857, Quito, Ecuador; now at : IRD (Institut de Recherche pour le Développement), UR "Processus et Aléas Volcaniques", UMR "Magmas et Volcans", 5 rue Kessler, 63038 Clermont-Ferrand, France (Jean.Philippe.Eissen@ird.fr)

Yves Fouquet and Hélène Ondréas, Ifremer (Institut Français pour l'Exploitation de la Mer), Centre de Brest, Géologie Marine, B.P. 70, 29280 Plouzané Cedex, France (Yves.Fouquet@ifremer.fr; Helene.Ondreas@ifremer.fr)

Delphine Hardy, Lafarge Granulats, Compagnie des Sablières de la Seine, Bernières sur Seine, France (delphine.hardy@lafarge.com)

A Cluster Of Surtseyan Volcanoes at Lookout Bluff, North Otago, New Zealand: Aspects of Edifice Spacing and Time

Doris Maicher

University of Otago, New Zealand

Growth of three or more separate small volcanoes, with substantial separation in time, is recorded at Lookout Bluff, which is situated on the east coast of North Otago, South Island, New Zealand. Their deposits now consist of devitrified glassy basaltic volcaniclastic rocks of upper Eocene age, and include well bedded, vesicular-clast tuff breccias, lapillistones and tuffs interbedded with continental shelf siltstones. The deposits formed from subaqueous fall and eruption-fed density current deposits, and each volcano was further shaped by resedimentation events. Glauconitic siltstone deposits separate series of volcaniclastic rocks, indicating that the group of monogenetic cones formed from eruptions separated by hundreds of thousands of years or more. All eruptions are inferred to have taken place on a continental shelf, this part of which remained submerged for most of the early- to mid-Cenozoic. The long gap between eruptions recorded in the modest thickness of siltstone illustrates the need for comprehensive facies analysis within successions including monogenetic volcanic deposits, and the stochastic nature of vent siting within monogenetic volcanic fields. These partly overlapping deposits formed from entirely independent excursions of magma to the surface; though each centre is monogenetic, the group as a whole is not.

INTRODUCTION

Basaltic eruptions initiated in shallow to moderately deep water commonly build cones of clastic material to sea level and above [*Kokelaar*, 1986]. Commonly, this type of monogenetic eruption produces several, closely spaced and sometimes overlapping edifices, erupted from poorly confined and often shifting vent sites, for instance Surtsey and associated satellite vents Surtla, Surtlingur, Jolnir 1963–67 [*Thorarinsson*, 1967; *Thorarinsson et al.*, 1964, *Lorenz*, 1974; *Jakobsson and Moore*, 1982; *Kokelaar and Durant*, 1983], Falcon Island, Tonga [*Hoffmeister et al.*, 1929]; Capelinhos, Azores 1957–58 [*Machado et al.*, 1962; *Camus et al.*, 1981]. In some fields, typically those developed atop larger central volcanoes, multiple monogenetic volcanoes have formed in small areas, and the younger monogenetic volcanoes in part overlap remnants of older ones [e.g. Capelinhos, *Cole et al.*, 1996], resulting in formation of intricate stratigraphic assemblages that consist of the products of multiple monogenetic eruptions, widely spaced in time [*Verwoerd and Chevallier*, 1987].

Good descriptions of spatial relationships among deposits of subaqueous Surtseyan-style volcanoes, however, are rare [e.g. *Sohn*, 1995; *White*, 1996; *Smellie*, 2001], and historical observations offer an incomplete representation of the timescales involved in the development of Surtseyan complexes. Better information is also needed to assess the range of constructional and degradational processes during, and between, eruptions that form the closely spaced edifices of both subaerial [e.g. *Nemeth*, 2001] and subaqueous monogenetic volcanic complexes.

Early Tertiary volcanism in North Otago is recorded by remnants of small basaltic volcanoes in outcrops scattered over an area of ca. 800 km². These comprise locally interbedded pillows and extensive tephra deposits of Surtseyan-style eruptions formed largely or entirely in submarine settings, e.g. at Moeraki Peninsula [*Andrews*, this volume], Bridge Point [*Cas et al.*, 1989], Kakanui, and Cape Wanbrow-Boatmans Harbour [*Mueller et al.*, 2002]. North Otago therefore holds the remnants of an extinct subaqueous volcanic field, in which groups of monogenetic volcanoes were active [*Coombs et al.*, 1986]. At Lookout Bluff, well-exposed remnants of apparently non-emergent Surtseyan volcanoes interfinger with one another and with shelf sediments, thus providing evidence for the nature of both the short-time growth of submerged edifices of a Surtseyan complex, and for development of the complex by eruptions spread over an extended period of time [*Maicher*, 1999].

GEOLOGICAL SETTING

The volcanic deposits are interbedded within a latest Cretaceous to mid Cenozoic transgressive shelf sequence of glauconitic mudstones, locally with thin lenses of sandstone and fossiliferous shallow marine limestones [*Field and Browne*, 1989]. In the late Cenozoic, the coastal region experienced a phase of minor deformation with local folding, related to plate reorganization during oceanic spreading and inception of the current Alpine Fault boundary [*Norris et al.*, 1978]. Today's regional dip is about 10° due east [*Mutch*, 1963].

At Lookout Bluff, bedded tephra forms cliffs up to 40 m high (Fig. 1) The original surface morphology is not preserved, but Lookout Bluff exists because these volcaniclastic rocks are more resistant to erosion than were enclosing marine rocks. Core vent facies are not present in outcrop, and other products of these eruptions extend beyond the mapped area. Numerous clastic dikes are present, and are inferred to have been injected early in the eruptive history. The complex is discordantly capped by effusive lavas formed by late Cenozoic subaerial eruptions, and by loess deposits.

TEPHRA CHARACTERISTICS

Petrographic and depositional features are used to subdivide the sequence into lithofacies units (Table 1, Fig. 1). Sideromelane clasts, of tholeiitic to transitional composition, are variably vesicular and predominantly angular to subrounded, with fracture-bounded surfaces transecting vesicles. Vesicles range in diameter from 10 µm to a few mm, with vesicularities of 20 to 70 vol%. These clast characteristics indicate a coupling of magma fragmentation by volatile exsolution with hydroclastic processes [*Houghton and Wilson*, 1989].

A distinctive clast type at Lookout Bluff comprises pyroclasts with an irregular subequant exterior that have an outer layer of weakly-to-moderately vesicular sideromelane glass that surrounds an older pyroclast core (Fig. 2A). Such cored clasts are interpreted to have formed by recycling of cooled pyroclasts, i.e. by entrainment within coherent melt with subsequent ejection and (re)fragmentation. *Mueller et al.* [2002] described similar features elsewhere in this field, and infered that they may form by slipping and disaggregation of vent–wall material into the active vent during eruption.

Fluidally shaped clasts of lapilli to bomb size (Fig. 2B), similar in appearance to spattery clasts described by *Kokelaar and Durant* [1983], corroborate submarine fountaining [*Mueller and White*, 1992; see also *Cas*; *Kano*; and *Fujibayashi and Sakai*, this volume].

Accidental clasts are dominantly pebble- to cobble-sized, smooth-rimmed clasts of siltstone, schist and basaltic lithics, as well as grains of detrital quartz and glaucony. All of these clasts, derived from the underlying subvolcanic strata, are commonly dispersed within the deposits.

MORPHOLOGICAL RECONSTRUCTION AND EVOLUTION

Shallow marine volcanism at Lookout Bluff produced several small volcaniclastic edifices in close proximity to, and in some cases impinging upon, one another. Individual volcanoes, described first, are recognized by their outward dipping beds and supporting directional indicators in sediment gravity flow deposits. Bedding dips, except within the fault-bounded tuffs Ts, approximate original ones, but with the slight eastward regional tilt superimposed. Stratigraphic relationships and carefully extrapolated bedding dip azimuths, as well as the presence of interbedded siltstones, allow reconstruction of the relative chronology of eruptions. For references to units and lithofacies abbreviations see Table 1.

Northern Area

Tuff breccia centre (TBC). The basal contact with the shelf sediments is not exposed. The lowermost visible part of the edifice consists of a near-vent tuff breccia. At the tip of the promontory, breccia beds lie subhorizontally. Along the flanks of the headland, the beds dip in northwesterly and

Table 1. Lithofacies

	constituents	sedimentary features	interpretation
Tuff breccia (TB)	coarse ash to blocks, mostly blocky; vesicularity highly variable with 2 populations: small, stretched & large oval vesicles; tabular crystalline basaltic blocks with parallel vesicle rows, spatter-like clasts & abundant xenoliths	8–10 m thick coarse beds, often amalgamated units, individual beds massive, coarse-tail graded & clast-to-matrix supported, with trains of large clasts; inter-bedded with 5–10 cm thick well-bedded, planar beds of coarse ash, both poorly sorted	breccia: proximal high-concentration laminar to weakly turbulent, water-supported, cohesionless debris flows; tuffs: suspension fallout
Tuff (Tn)	coarse and fine ash, clasts irregular-shaped, angular; 25–40 % vesicularity, with small, stretched & large, oval vesicles; rarely highly vesicular scoria clasts, isolated cauliflower bombs & angular blocks with parallel vesicle rows; some xenolithic clasts	20–30 cm thick beds, well-bedded & planar to locally scoured bed contacts, base inverse-to-normal graded or massive, normal graded tops, traction structures and down-slope flow indicators, scattered large clasts without bomb sags; coarse beds open framework & poorly sorted, in finer grained beds closely packed grains, low angle cross bedding & density grading	relatively dense gravity flows with +/– deposition from suspension fallout
Tuff (Tm)	medium and fine ash, clasts angular to sub-rounded, closely packed and moderately well sorted; some xenolithic clasts	10–15 cm thick apparently massive beds with density grading of clasts, thinner & finer grained beds with planar laminations, low angle cross stratification, horizons of highly vesicular clasts, isolated ripples & dewatering structures; clastic dikes; top bed bioturbated; inter-bedded non-glauconitic siltstone	dilute density currents combined with suspension fallout deposition; syn- to very early post-volcanic redepostion of volcaniclastic material
Tuff (Ts)	medium to coarse ash, clasts angular, open framework, poorly to moderately well sorted in finer beds; few outsized blocks of structureless siltstones without bomb sags; some xenolithic clasts	well-bedded & massive beds of 15–30 cm thickness, internal low angle cross stratification, scouring parallel dip azimuth & planar stratification, normally and/or density graded; deformed bedding and intense veining	high- to low-concentration density currents with +/– suspension fallout deposition
Lapilli tuff (LT1) (LT2) (LT3)	coarse ash to lapilli sized clasts, angular & mostly blocky; 30–40% vesicularity, oval to slightly stretched vesicles; cauliflower bombs with & without sag structures, spatter-like clasts, rarely xenolithic clasts	alternating 0.5–3 m lapilli-tuff and 10–20 cm coarse tuff, coarse beds open framework & poorly sorted, massive with diffuse bedding and subtle scouring; faint alignment of large clasts, grain size homogeneity of large clasts in individual beds, fine beds moderately well sorted and with down-slope flow indicators	vertical debris falls transformed into lateral high-particle concentration debris flows
Lapilli-stone (LS)	angular to subrounded blocky, equant lapilli & irregular, angular ash; deposits poorly sorted, matrix-supported; large blocks and bombs, composite lapilli, spatter, glauconite grains and few highly vesicular lapilli	planar bedding, massive with very diffuse, gradual bedding boundaries defined by colour changes of fine clasts (greenish to orange brown), m-scale bedding; in finer grained, cm-to-dm thick deposits lensoidal-shaped clusters or trains of large clasts & wedge-shaped beds	voluminous vertical debris falls transformed into lateral high-particle concentration debris flows

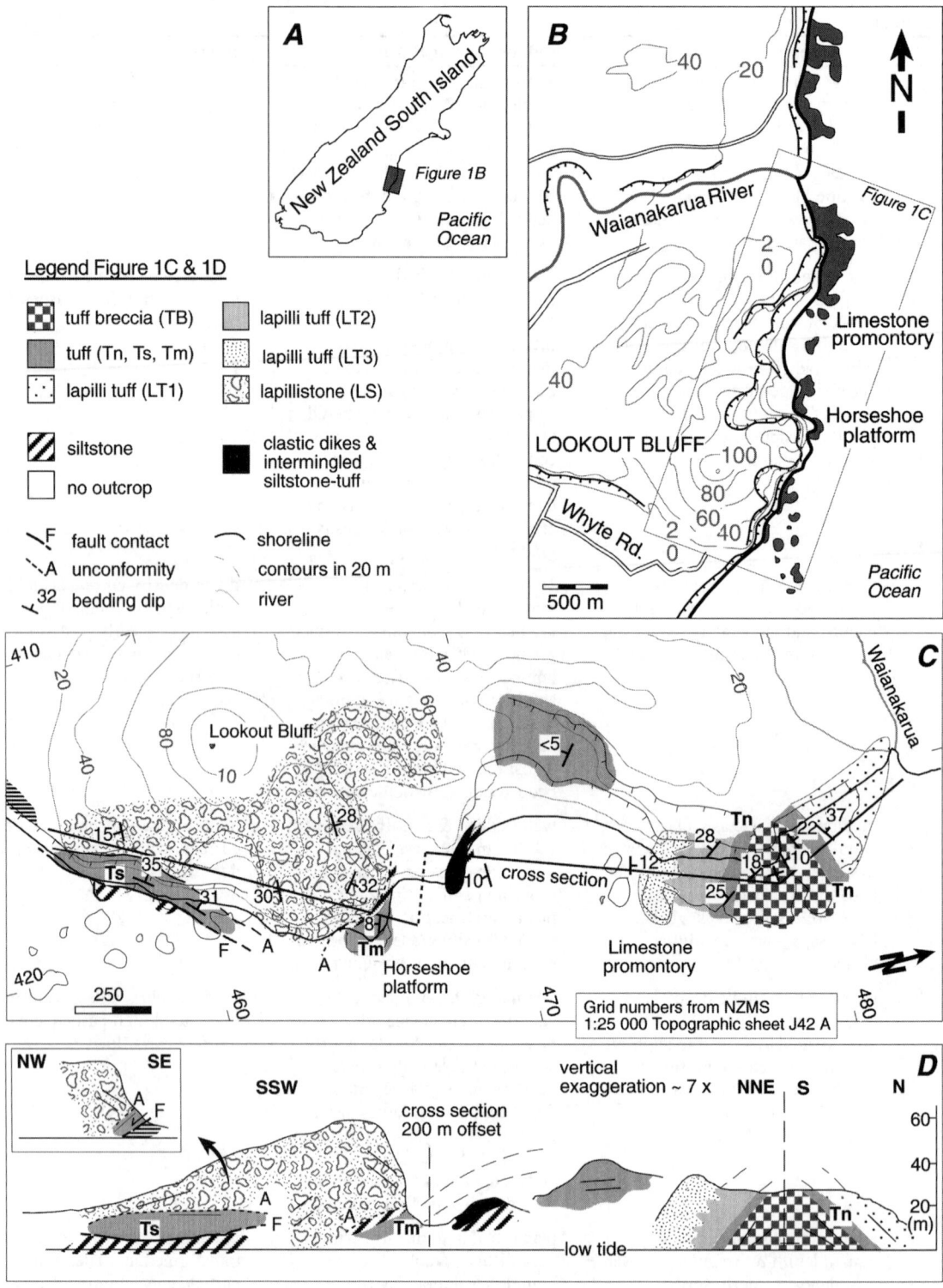

Figure 1. (A) Outline map of South Island of New Zealand with location of Lookout Bluff area; (B) sketch map of the Lookout Bluff area, showing cliffs and wave-cut platforms (shaded), contour lines in metres, definition of informal names; (C) geological map with volcaniclastic lithofacies, siltstone and approximate bedding dips; note location of cross section along the cliff exposure and outcrop gap in the central area; and (D) geological cross sections of the Lookout Bluff area.

southwesterly directions, and at increasingly steeper angles farther from the head of the promontory (Fig. 1C, 1D). The flanks of the edifice are formed by successions of tuffs and lapilli tuffs, mostly conformably overlying the tuff breccia. Their sequences to the N (LT1) and S (LT2) of the promontory can be correlated, although they are slightly differently developed.

The geometry is interpreted as representing outwardly dipping beds of a volcaniclastic edifice. Original bedding dips were in the same sense as present ones, as indicated by down-dip directional flow indicators. Dip magnitudes additionally reflect syn-depositional shear faulting and slumping. Individual beds are subtly lenticular or planar tabular over the scale of available outcrop (meters to tens of meters).

Lapilli tuff units LT1 and LT2 are characterized by decimeter-scale beds with an ash-sized matrix component. Beds show current indicators such as subtle scouring and local clast alignment; amalgamated beds are recognized by clast trains within thicker (up to 3 m) layers. These features are interpreted to result from eruption-fed density currents moving outward from the vent area [*White*, 2000]. Scouring and layer irregularities may also reflect asymmetric spreading of sediment gravity flows induced by directional blasts from the vent, and/or interaction with ocean currents [e.g. *Hoffmeister et al.*, 1929).

Steep-dipping beds are preferential hosts of normal shear faults, and the undulating geometry of the fault planes indicates deformation in the soft state. It is inferred that the edifice flanks were unstable during eruption because of vibration and the rapid input of sediment, causing syn-eruptive slope failure, shearing and faulting. Dips in the stratigraphically higher deposits are steeper than expected for a depositional slope (removing the 10° easterly regional dip yields dips as high as 35° to the west in this segment), particularly one showing planar bedding and depositional characteristics indicative of density current deposition. Tilting of those upper deposits may have been caused by down- and outward failure of the underlying sequence.

Limestone Promontory (LP). Some 600 m south of the tuff breccia center, along continuous outcrop, lies another centre called Limestone promontory (informal name, Fig. 1B). It is composed of a complex sequence of tuffs and lapilli tuffs (LT3, Fig. 3).

Deposits at the base of the promontory consist of a 3.5 m thick sequence of interlayered tuffs and lapilli tuffs with bedding gently dipping N to NE, thus toward TBC. The deposits are strongly sheared, normally faulted, and vary in thickness over 10's of meters from 60 cm to nil. In some, tuff stratification has been destroyed to form homogenized, somewhat, irregular strata. Overlying this sequence is a distinctive lapilli tuff horizon that contains a high concentration of micritic siltstone blocks (unit "S" in Fig. 3), inferred to represent a phase of substantial quarrying into the substrate. This is overlain in turn by a complex sequence of beds indicating transport from the north (LT2, TBC-derived), which alternate with and are deformed by voluminous lapilli tuffs and tuffs derived from the SW (LT3). The most distinctive feature is a "syncline" of well-bedded tuffs, fed from the N. The tips of individual tuff beds are overturned toward the N (Fig. 3, arrows) below the contact with an overlying structureless lapilli tuff beds derived from the SW, which have incorporated a large chunk of the bedded tuff. In the lateral continuation of the tuff, beneath the structureless lapilli tuff unit, the beds are buckled, deformed and squeezed out. Further interlayered deposits from both sources occur higher in the cliff.

This sequence appears to originate from two simultaneously active volcanic centers, a new vent, LP, and the TBC vent which produced the beds underlying this series of intercalated deposits.

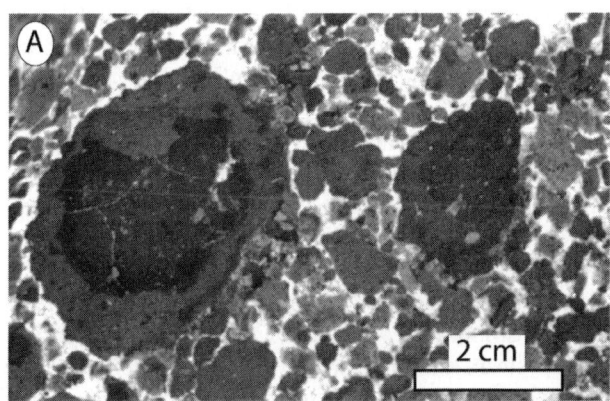
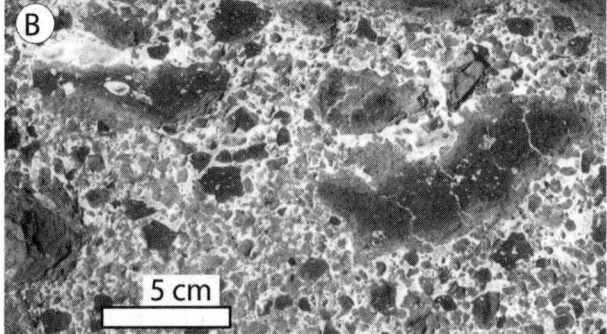

Figure 2. Lapillistone clasts in white calcitic cement; (A) cored clasts with a more dense lapillus as a core and an envelope of moderately vesicular sideromelane; both components have distinctly different vesicle shapes; (B) fluidally shaped clasts with highly altered rims, ubiquitous in massive lapillistone units.

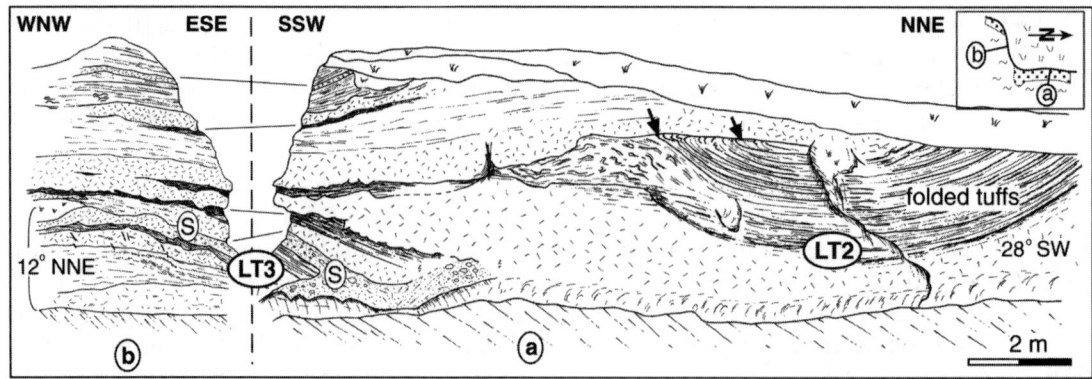

Figure 3. Cliff exposure at Limestone Promontory; sketches at right angles to each other detailing bedding dip changes and bed discontinuity of lapilli tuffs and tuffs of LT3. Folded tuff sequence with tips of beds overturned (arrows) and squeezed out by overriding massive lapilli tuff flow; **S** = layer with occurrence of abundant baked calcareous siltstones.

Central Area

Clastic Dikes and Intermingled Horizons (CD). In the central Lookout Bluff area several volcaniclastic dikes cross-cut tuffs and siltstones and beds of intermingled volcaniclastics and siltstone crop out. The textural characteristics of the volcaniclastic material are those of the tuff lithofacies (Tm; Table 1). Dikes are 2–50 cm wide with sharp, irregular margins and branch into multiple dikelets or end abruptly. Intruded tuff is undisturbed, whereas intruded siltstones appear structureless near the dikes (some 10's of cm). The longest traceable dike emplaced in silt is about 6 m long.

An origin of dike material by phreatomagmatic disruption of magma at some depth within water-saturated siltstones is envisaged. Whether coherent or pre-fragmented volcanic material initially intruded the siltstones cannot be demonstrated directly, e.g. by in situ fragmentation of a progressive sequence from coherent material to volcaniclastics [*Nakamura and Coombs*, 1973; *Hanson and Wilson*, 1993; *Kano et al.*, 1993]. Phreatomagmatic explosions in the subsurface are known to inject volcaniclastic material into country rock as clastic dikes [*Heiken et al.*, 1988]. The poorly consolidated nature of the host sediment is reflected in the ill-defined, branching and rejoining pattern of the dikes, similar to that of clastic dikes previously described from subaqueous settings [*Mayo*, 1976; *Kokelaar*, 1982]. Intrusions were probably related to loading of the substrate, lateral extension and shear zones [*Heiken et al.*, 1988]. Emplacement may have been facilitated by seismic shaking and successive basalt intrusions which liquified or weakened the host (see also *Andrews*, this volume); these processes are also implicated in the edifice instability during growth previously inferred.

Above the dikes, beds of intermingled tephra and siltstone are exposed. The deposits consist of ash to lapilli-sized tephra and totally disrupted silty fragments, which are internally structureless, angular to amoeboid-shaped and vary in size from gravel to microscopic scale. The siltstone clasts exhibit a pervasive colour change from light grey in the centres to greenish darker grey rims. Siltstone clasts are aligned parallel to bedding, and define a coarse-tail grading in the bed. The basal contact of a bed with glauconitic siltstones is sharp and irregular, with rip-up siltstone clasts, flame structures, large load casts and small dikelets protruding down.

The intermingled horizons are thought to have formed by magma intruding into wet, unconsolidated siltstones at shallow levels beneath the seafloor. Formation of volcaniclastic material was followed by intimate mingling and jostling of hot tephra with a granulated-siltstone slurry [cf. *Kokelaar*, 1982; *McClintock and White*, 2002]. Continued pressurization by steam and hot water may have facilitated the advance of the intrusive mixture [*White and Busby-Spera*, 1987], which ultimately breached the sediment-water interface. Subsequent destabilization and slumping of the exposed mixture produced pulses of subaqueous sediment gravity flows.

Southern Area

Horseshoe (Tm). The tuff facies Tm, constituting the Horseshoe platform (informal name, Fig. 1B), is seen as a rather distal deposit relative to the site of clast formation. It consists of a homogeneous sequence of tuff beds interlayered with one non-glauconitic siltstone horizon and one bioturbated tuff horizon (second-to-last of the sequence). The sequence is capped by glauconitic siltstones of at least 50 cm thickness.

Deposition of the tuff is postulated, tentatively, to be from post-eruptive reworking. Although there is no evidence of different generations of alteration in the ash grains and there is no sigificant biogenic input [cf. *Cas et al.*, 1989], the presence of glauconitic siltstone and bioturbated tuff bed suggest passage of substantial amounts of time during deposition of the beds.

Fault-Block (Ts). Tuff subfacies Ts crops out as an isolated block with a faulted contact against the underlying shelf sediments. Near the base, bedding of the succession is subparallel in a wavy pattern with slightly varying dip angles and azimuths. A network of cm-to-dm-sized normal faults dissects the whole sequence and individual beds are pervasively veined by calcite. The underlying glauconitic siltstones are partly deformed and structureless near the contact.

The succession is interpreted as a slumped block, resulting from flank failure of a volcaniclastic edifice located to the E offshore. Downhill displacement involved little rotation of the block, because sedimentary flow indicators indicate downdip flow. Failure must have occurred soon after tuff deposition, when the deposits were still pliable. Further support for slumping is given by the presence of an erosion-resistant sea stack consisting of a m-sized, tightly folded tuff sequence. Evidence for a former cone to the E of the block is given by extensive kelp-covered wave-cut platforms offshore, which are remnants of more erosion-resistant volcanics (similar to the platform S of LP).

Lapillistone Centre (LC). In the southern area, remnants of a large volcaniclastic centre, consisting of lapillistone deposits, is exposed above the deposits of Ts and Tm. Above the Horseshoe platform dip angles of 30–35° are observed, with dip azimuths to the NE and N further inland, and to a southerly direction some 200 m to the S. Assuming a circular edifice, the vent site is proposed to be some 300 m WSW inland from the Horseshoe platform. The central zone between the differently dipping beds appears to be massive and may represent the margin of the collapsed, chaotic vent area of the cone. Toward the S, the angle of dip decreases gradually to about 15° and no significant discordances are recognized in the deposits.

Geometry

The original edifice diameters and heights are inferred geometrically from present-day outcrops (tops not exposed), angle of dip and dip direction to reconstruct the volcanic field (Fig. 4). The northern TBC, with a total

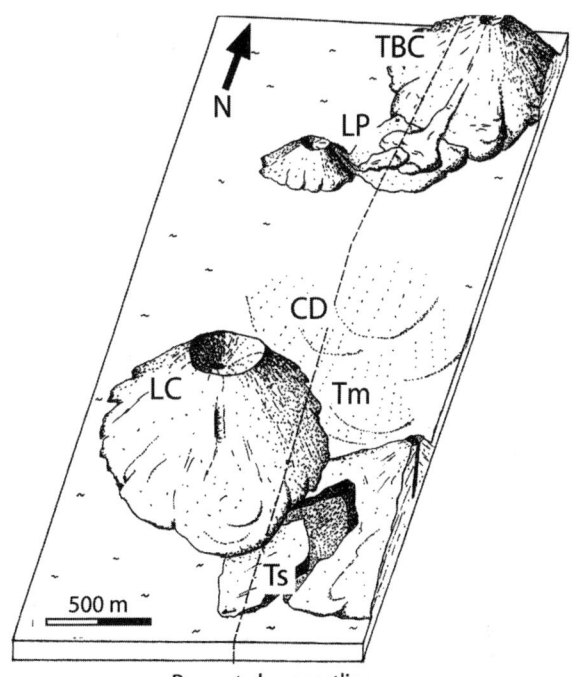

Figure 4. Sketch of Lookout Bluff area in late Eocene times on the submerged continental shelf; stippled lobes represent tuff subfacies Tm; dashed line indicates present-day coastline; site of clastic dikes and intermingled horizons is indicated by lettering (CD); siltstone background sedimentation omitted for clarity.

diameter of 815 m, possibly reached a height of 80 m. The lapillistone beds of the central area, ca. 400 m in diameter, project to a cone about 120 m high. These values neglect post-eruptive removal of material and height reduction by syn-eruptive flank failure and internal shearing and slumping. A cone of this height could have formed entirely subaqueously given the inferred inner to mid-shelf setting.

Reconstructing, tentatively, the eruptive centres of each edifice, the spacing ranges from about 600 m between LP and TBC in the northern sub-complex to 830 m between Ts and LC in the southern sub-complex. The overall distance between the sub-complexes is ca. 2350 m (between Ts and TBC), oriented N-S with only a few 10's of m of lateral offset in an E-W direction.

TIME CONSTRAINTS

Volcanic Periods

Most monogenetic volcanic eruptions are short-lived, with activity commonly lasting from days to a few years. The pre-emergent phase at Surtsey and its satellite vents, as well as at Capelinhos, took several days to a couple of weeks to build the edifice some 120 m from the seafloor up to sea level [*Thorarinsson et al.*, 1964, *Machado et al.*, 1962], with eruptions continuing for a few years in distinctive phases [*Cole et al.*, 1996]. A similar timespan of activity can be envisaged for development of the northern and southern parts at Lookout Bluff. This contrasts strongly with the depositional period required for siltstone, described in the following section.

Non-Volcanic Periods

Siltstone horizons interbedded within the volcaniclastic rocks are used to estimate the duration of volcanic quiescence between eruptive events by using (1) sedimentation rates and (2) growth rates of glauconite.

(1) In a continental margin setting, such as the Cenozoic Canterbury basin, siliciclastic sedimentation rates range from 10–100 m/Ma [*Einsele*, 1992], controlled by basin subsidence and denudation rates, combined with minor biogenic input. For the area considered here, a relatively slow sedimentation rate of 6 m/Ma is calculated from a 60 m thickness of siltstone [Galleon drill site 30 km SE of Lookout Bluff and isopach maps in *Field and Browne*, 1989] deposited over 10 Ma in the middle to upper Eocene. The depositional period includes at least one unconformity, consequently, the actual sedimentation rate during deposition was probably somewhat higher than this.

(2) Glauconitization is the process of impregnating and coating pre-existing grains with potassium-rich smectite, e.g. faecal pellets and rock fragments, and infilling pore spaces in micro-oganism tests. Initial glauconitic smectite, with a growth period of 10^3–10^4 years, matures to glauconite mica in 10^5–10^6 years [*Odin and Matter*, 1981].

Lookout Bluff, situated on a continental shelf during a major transgression at time of volcanic activity, experienced favourable conditions for glauconite formation, as seen by the local background sedimentation. The siltstone capping the Horseshoe tuff sequence (Fig. 1) has, despite the disconformable contact with the overlying lapillistones and recent slope displacement, a minimum thickness of 50 cm, representing an accumulation time of 83 ky (using the above calculated sedimentation rate of 6 m/My, and excluding rapid silt emplacement by slumping). A 15 cm thick horizon of non-glauconitic siltstone within the Horeshoe-sequence however indicates unfavourable conditions for glauconite formation at times, and/or faster sediment accumulation rates. A depositional time span of 25 ky for the deposition of this horizon is inferred. Subsequent fast burial by tuff deposition possibly cut off the exchange between seawater and siltstone, preventing post-tuff-emplacement glauconitization.

The time constraints used above show that fossil ages of the sediment (38-34 Ma) are insufficiently precise to pinpoint the absolute time of volcanic activity. Assuming that the section is complete, i.e. that no silts or tuffs were eroded, the calculations suggests that the entire sequence accumulated in a little over 108 000 y (based on sedimentation rates, in accordance with glauconite formation rates) sometime after 38-34 Ma. The siltstones cover the major length of time, interspersed with very short volcanic episodes.

Sequence of Events

Both of the northern and southern parts of the Lookout Bluff area form well-defined Surtseyan complexes. Correlation of the two is uncertain because of discontinuous outcrops and an absence of marker beds. Extrapolation of bedding dip and directions, however, allows a tentative reconstruction of the sequence of events for the whole area (Fig. 5).

In the northern part of the area, the TBC and LP sections show a clearly defined sequence without breaks in the volcanic activity. After deposition of the lower TBC units, LP as a second eruptive source started feeding volcaniclastic flows, which interfere and alternate with TBC-derived flows (northern monogenetic complex).

The clastic dikes and the sequence of intermingled siltstone-tuff horizons are supposed to have a stratigraphic position below the Tm sequence further south, as indicated by careful extrapolation of bedding orientation of both Tm and intermingled horizons. Consistent with this fact, there are no underlying siltstones exposed at the Horseshoe sequence. Siltstones however, are ubiquitous in the central area and clastic dikes are thought to indicate the earliest stages of volcanic activity. Time-relationship indicators in the northern part of the area are non-specific.

The southern part is composed of 3 major units, the large LC discordantly overlying Tm and Ts. The contact between Tm and lapillistones is separated by glauconitic siltstones of at least 50 cm thickness, which indicates a substantial period of volcanic quiescence. Similar hemipelagic sediments are in contrast absent along the contact between the Ts and lapillistones, thus indicating lapillistone emplacement shortly after flank failure (southern monogenetic complex). Only the relationship between Ts and Tm remains ill-defined.

To unravel the chronological relationship between the southern and northern areas, it is important to determine the source of Tm. Bedding characteristics of the gently SE dipping Tm beds suggest that there was a volcaniclastic source some distance to the N. This *may* have been the same eruptive centres from which LP or TBC were derived. In this case the deposits in the southern part of the area (Ts and LC) would be younger than the northern part of the area.

Alternatively Tm may have been derived from a different source (primary syn-eruptive or redeposited post-eruptive) to the north which is unidentified, in which case the relative age of the northern and southern sub-complexes remain unascertained. Figure 5 shows the preferred sequence of events.

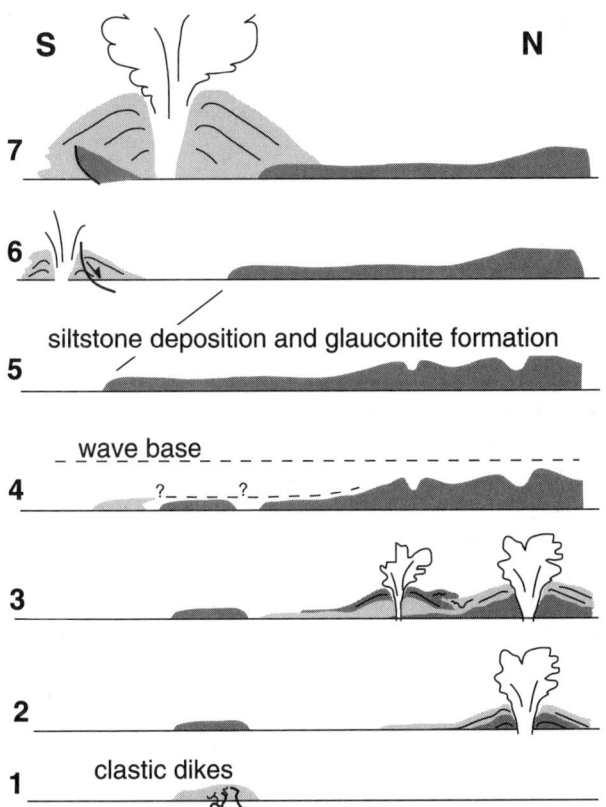

Figure 5. Spatial relationships of eruptive centres and most likely sequence of events in the Lookout Bluff area; grey shaded areas represent all previously deposited material; abbreviations of lithofacies units see Table 1. (1) Emplacement of clastic dikes and intermingled horizons. (2) The northern eruptive centre formed, feeding sediment gravity flows radially outward (tuff breccia, tuffs, LT1 and LT2). (3) Further south at Limestone Promontory, a second, smaller volcaniclastic centre developed contemporaneously, feeding massive lapilli tuffs intermittently from a southwesterly direction. Sediment flow deposits from both vents interfinger (LT3). (4) Deposition of the succession of tuff subfacies Tm from a northerly source, possibly contemporaneous to stage (2) and (3). The interbedded non-glauconitic siltstones represent a time break in the tuff (re-) deposition. (5) Deposition of glauconitic siltstone on top of Tm indicates a long break in the eruptive activity, potentially allowing for substantial erosion of the northern volcaniclastic pile. (6) A tuff cone formed at the southern end of Lookout Bluff. Subsequent flank failure emplaced a large coherent block (Ts) off the eastern flank onto glauconitic siltstones. (7) Submarine eruption of lapillistones (LS) covered discordantly the tuff-slump block (Ts) and the tuff-siltstone sequence (Tm).

DISCUSSION

Subaqueous deposits of Surtseyan eruptions, including the edifices themselves, have attracted much less attention than their subaerial counterparts. This reflects the lack of subaqueous observation of eruptions, and the difficulties of studying still-submerged modern deposits. The subaerial parts of Surtseyan volcanoes, however, have very low preservation potential because loose volcaniclastic material unprotected by later-stage capping lava (e.g. Surtsey) erodes very rapidly in the face of wave attack [*Hoffmeister et al.*, 1929; *Thorarinsson*, 1967]. Often, only the coherent magmatic vent fills are preserved as "sea-stacks" above water.

The Lookout Bluff complex is an exhumed stratigraphic assemblage representing several small monogentic volcanoes upon which a larger one, produced by an entirely separate eruption, was later formed. All eruptions took place in a submarine setting of moderate depth, as indicated by underlying and intercalated fossiliferous, glauconitic shelf siltstones, ubiquitous presence of sideromelane, sedimentary characteristics of subaqueous eruption-fed density flows, water escape structures [*Kano*, 1996] and good density grading of variably vesicular clasts [*Cashman and Fiske*, 1991; *Oehmig and Wallrabe-Adams*, 1993]. A depth somewhat below storm wave base is indicated by the glauconitic mudrock. Moreover, there is no evidence of reworking of the tephra by surface waves, or of wave planation of the older volcanoes [cf. the wave cut platform cut rapidly to 45 m bsl at Surtla, *Kokelaar and Durant*, 1983]. Shelfal depths, rather than deep marine, are indicated again by the glauconitic siltstone, and by moderate but variable vesicularities suggesting phreatomagmatic disruption of a fairly actively vesiculating magma [*Houghton and Wilson*, 1989].

Edifice Instability

Growing piles of water-saturated volcaniclastic rubble are unstable and subject to frequent slumping and internal shearing, [e.g. *Sohn and Chough*, 1992; *Smellie*, 2001]. In addition to syn-eruptive faulting, slumping on the scale of cm to several 10's of metres [*Maicher*, 2000] is an important process to effectively reduce the height of submerged edifices. Mechanisms lowering the stability are loading of the flanks by rapid sediment input and syn-eruptive seismic shaking, both causing liquefaction and ultimate failure of inclined beds. Occurrence of cored clasts, which are discussed to represent recycling of clasts after vent-wall collapse, and volcaniclastic dikes emplaced within zones of weakness further highlight edifice vulnerability.

These mass-wasting processes, expressed by in- and outward collapses during construction, furthermore may alter the original inclination of the cones' flanks, and it is suggested here that height calculations based on edifice remnants should be evaluated carefully.

Spacing and Recurrence

Absolute length and relative duration of syn-eruption periods and periods between eruptions can vary greatly, [e.g. *White*, 1991]. Looking at timing and spacing of the edifices, Lookout Bluff reveals both true monogenetic clusters consisting of multiple edifices in close vicinity formed contemporaneously and, on a slightly larger scale, clusters formed at the same locality but separated by signifiant periods of time.

The small scale spacing of volcano remnants, as described above, is comparable to those of Surtsey and its satellite volcanoes (Jolnir to Surtsey-West 1.7 km, Syrtlingur to Surtsey-West 1.4 km, Surtla to Surtsey-West 2.6 km), which consist of coalescent edifices formed in single, monogenetic eruptive events.

This contrasts strongly with the time elapsed to form the whole of the Lookout Bluff volcanic area, documented by siltstones which separate the volcanic events on a decimetre-scale in space, but by some 100 000 years in time. The recurrence of volcanic activity strongly depends on the geologic and tectonic setting of an area, i.e. mid-ocean ridge-related eruptions (Surtsey, Iceland), plume-related (Azores) and subduction-related (Falcon Island, Tonga arc). Multiple eruptions have occurred at more or less the same spot since historic times, recorded from Falcon Island [*Hoffmeister et al.*, 1929], and since pre-historic times at Capelinhos [*Machado et al.*, 1962]. Longer intervals are recorded from Marion and Prince Edward Islands, which had several volcanically active periods during the Holocene, the Pleistocene and in recent times [*Verwoerd and Chevallier*, 1987], and from other volcanic fields [*White*, 1991].

Similarly, Lookout Bluff's volcanism, probably related to plate reorganization and incipient oceanic spreading [*Norris et al.*, 1978], stretched over about 110 000 y in the late Eocene, including at least one long period of volcanic quiescence. Noticably, any directional progression of time of the volcanoes is lacking, although the centres are distinctly aligned in N-S direction. Siting and timing is comparably scattered, on a regional scale, in North Otago with its many Surtseyan complexes along the coastline, i.e. Bridge Point situated 5 km, Kakanui 8 km and Cape Wanbrow 17 km N of Lookout Bluff; and Moeraki some 9 km to the S.

The term "monogenetic" refers to volcanic edifices formed during single episodes of volcanic activity, without subsequent eruptions [*Sigurdsson*, 2000]. The constellation of several monogenetic clusters, emplaced by effectively unrelated eruptive events and coinciding spatially in very close proximity is clearly different, and previously, no emphasis has been given to the fact that long breaks in eruptive activity can be incorporated within those more complex clusters. A clear distinction to "normal" monogenetic complexes (i.e. *without* a break in time) appears necessary, esp. in view of facies analysis, studies of stratigraphic sequences and volcanic hazard assessment. Whether this distributional pattern is common in Surtseyan complexes elsewhere, most likely depends on the tectonic setting controlling magma supply, which, however, is poorly understood in North Otago.

CONCLUSIONS

At Lookout Bluff, all lithofacies support the interpretation that the depositional setting in which the bulk of the volcanic pile accumulated was subaqueous, and the regional context further contrains the setting to submarine. Architectural features and time components represented in the area are complex and differ widely. The time required to produce a single volcanic edifice, each of which probably accumulated within hours to days, contrasts strongly with the length of volcanically quiet periods (as indicated by interbedded siltstones). Furthermore, the timing of eruptive activity ranges from contemporaneous eruptions from adjacent vents to activity widely spaced in time though coincidentally closely sited, which depicts the sporadic nature of the volcanism. Recurrence of volcanism similar in style and in close vicinity requires steady conditions and a local, long-lived supply of small volumes of magma.

Acknowledgments. This research has been supported by an Otago University doctorate scholarship. Substantial help by JDL White is gratefully acknowledged. Many thanks to C Zink and SR White for assistance in the field. Discussion and constructive comments by C Landis and W Mueller are highly appreciated, and thoughtful revisions by G Heiken, K Kano and J Smellie helped to improve the manuscript considerably.

REFERENCES

Camus, G., P. Boivin, A. de Goer de Herve, A. Gourgaud, G. Kieffer, J. Mergoil, and P.M. Vincent, Le Capelinhos (Faial, Açores) vingt ans après son éruption: le modèle éruptif "surtseyen" et les anneaux de tufs hyaloclastiques, *Bulletin de Volcanologie*, 44, 31-42, 1981.

Cas, R.A.F., C.A. Landis, and R.E. Fordyce, A monogenetic, Surtla-type, Surtseyan volcano from the Eocene-Oligocene Waiareka-Deborah volcanics, New Zealand: a model, *Bulletin of Volcanology*, 51, 281-298, 1989.

Cashman, K.V., and R.S. Fiske, Fallout of pyroclastic debris from submarine volcanic eruptions, *Science*, 253, 275-280, 1991.

Cole, P., A. Duncan, and J. Guest, Capelinhos: The disappearing volcano, *Geology Today*, 68-72, 1996.

Coombs, D.S., R.A.S. Cas, Y. Kawachi, C.A. Landis, W.F. McDonough, and A. Reay, Cenozoic volcanism in North, East, and Central Otago, *Royal Society of New Zealand Bulletin*, *23* (Cenozoic Volcanism in New Zealand), 278-312, 1986.

Einsele, G., *Sedimentary basins: Evolution, facies and sediment budget*, 628 pp., Springer, Berlin, 1992.

Field, B.D., and G.H. Browne, Cretaceous and Cenozoic sedimentary basins and geological evolution of the Canterbury region, South Island, New Zealand, *NZ Geol. Surv. Basin Studies*, *2*, 94, 1989.

Hanson, R.E., and T.J. Wilson, Large-scale rhyolite peperites (Jurassic, southern Chile), *Journal of Volcanology and Geothermal Research*, *54*, 247-264, 1993.

Heiken, G., K. Wohletz, and J. Eichelberger, Fracture fillings and intrusive pyroclasts, Inyo Domes, California, *Journal of Geophysical Research*, *93*, 4335-4350, 1988.

Hoffmeister, J.E., H.S. Ladd, and H.L. Alling, Falcon Island, *American Journal of Science*, *18*, 461-471, 1929.

Houghton, B.F., and C.J.N. Wilson, A vesicularity index for pyroclastic deposits, *Bulletin of Volcanology*, *51*, 451-462, 1989.

Jakobsson, S.P., and J.G. Moore, The Surtsey Research Drilling Project of 1979, *Surtsey Research Progress Report*, *9*, 76-93, 1982.

Kano, K., A Miocene coarse volcaniclastic mass-flow deposit in the Shimane Peninsula, SW Japan: product of a deep submarine eruption?, *Bulletin of Volcanology*, *58*, 131-143, 1996.

Kano, K., T. Yamamoto, and K. Takeuchi, A Miocene island-arc volcanic seamount: the Takashibiyama Formation, Shimane Peninsula, SW Japan, *Journal of Volcanology and Geothermal Research*, *59*, 101-119, 1993.

Kokelaar, B.P., Fluidization of wet sediments during the emplacement and cooling of various igneous bodies, *Journal of the Geological Society of London*, *139*, 21-33, 1982.

Kokelaar, B.P., Magma-water interactions in subaqueous and emergent basaltic volcanism, *Bulletin of Volcanology*, *48*, 275-289, 1986.

Kokelaar, B.P., and G.P. Durant, The submarine eruption and erosion of Surtla (Surtsey), Iceland, *Journal of Volcanology and Geothermal Research*, *19*, 239-246, 1983.

Lorenz, V., Studies of the Surtsey tephra deposits, *Surtsey Research Progress Report*, *7*, 72-79, 1974.

Machado, F., W.H. Parsons, A.F. Richards, and J.W. Mulford, Capelinhos eruption of Fayal volcano, Azores, 1957-1958, *Journal of Geophysical Research*, *67*, 3519-3529, 1962.

Maicher, D., Hyaloclastite beds of shelf and seamount: Roles of exsolution, entrapment and expansion at Lookout Bluff, New Zealand and Seamount Six, Pacific Ocean, PhD thesis, University of Otago, Dunedin, New Zealand, 1999.

Maicher, D., Architecture and development of a shallow marine tuff cone at Lookout Bluff, New Zealand, in *International Maar Conference*, pp. 309-317, Terra Nostra, Daun/Vulkaneifel, Germany, 2000.

Mayo, E.B., Intrusive fragmental rocks directly or indirectly of igneous origin, *Arizona Geological Society Digest*, *10*, 347-430, 1976.

McClintock, M.K., and J.D.L. White, Granulation of weak rock as a precursor to peperite formation: Coal peperite, Coombs Hills, Antarctica. In: Peperites: processes and products of magma-sediment mingling, in *Journal of Volcanology and Geothermal Research*, edited by I. Skilling, J.D.L. White, and J. McPhie, pp. 205-217, 2002.

Mueller, W., and J.D.L. White, Felsic fire-fountaining beneath Archean seas: pyroclastic deposits of the 2730 Ma Hunter Mine Group, Quebec, Canada, *Journal of Volcanology and Geothermal Research*, *54*, 117-134, 1992.

Mueller, W.U., J.D.L. White, and P.L. Corcoran, Fieldtrip guide: Dynamic evolution of the submarine Oamaru volcanic complex: syn-eruptive collapse, unconformities, eruption-fed density currents, and backset strata, in http://www.otago.ac.nz/geology/staff/jdlw/fieldguide_sml.pdf 2002.

Mutch, A.R., Oamaru, *NZ Geol. Surv. Map*, *Sheet 23*, 1:250000, 1963.

Nakamura, Y., and D.S. Coombs, Clinopyroxenes in the Tawhiroko tholeiitic dolerite at Moeraki, North-Eastern Otago, New Zealand, *Contributions Mineralogy and Petrology*, *42*, 213-228, 1973.

Nemeth, K., Phreatomagmatism at the Waipiata Volcanic Field, Otago, New Zealand, University of Otago, Dunedin, 2001.

Norris, R.J., R.M. Carter, and I.M. Turnbull, Cainozoic sedimentation in basins adjacent to a major continental transform boundary in southern New Zealand, *Journal of the Geological Society of London*, *135*, 191-205, 1978.

Odin, G.S., and A. Matter, De glauconiarum origine, *Sedimentology*, *28*, 611-641, 1981.

Oehmig, R., and H.-J. Wallrabe-Adams, Hydrodynamic properties and grain-size characteristics of volcaniclastic deposits on the mid-Atlantic ridge north of Iceland (Kolbeinsey Ridge), *Journal of Sedimentary Petrology*, *63*, 140-151, 1993.

Sigurdsson, H., *Encyclopedia of volcanoes*, 1417 pp., Academic Press, San Diego, 2000.

Smellie, J.L., Lithofacies architecture and construction of volcanoes erupted in englacial lakes: Icefall Nunatak, Mount Murphy, eastern Marie Byrd Land, Antarctica, *Spec. Publ. Int. Asoc. Sed.*, *30*, 9-34, 2001.

Sohn, Y.K., Geology of Tok Island, Korea: eruptive and depositional processes of a shoaling to emergent island volcano, *Bulletin of Volcanology*, *56*, 660-674, 1995.

Sohn, Y.K., and S.K. Chough, The Ilchulbong tuff cone, Cheju Island, South Korea: depositional processes and evolution of an emergent, Surtseyan-type tuff cone, *Sedimentology*, *39*, 523-544, 1992.

Thorarinsson, S., *Surtsey, The new island in the North Atlantic*, 47 pp., The Viking Press, New York, 1967.

Thorarinsson, S., T. Einarsson, G.E. Sigvaldason, and G. Elisson, The submarine eruption off the Westmann Islands 1963-1964, *Bulletin de Volcanologie*, *27*, 1-11, 1964.

Verwoerd, W.J., and L. Chevallier, Contrasting types of Surtseyan tuff cones on Marion and Prince Edward Islands, southwest Indian Ocean, *Bulletin of Volcanology*, *49*, 399-414, 1987.

White, J.D.L., The depositional record of small, monogenetic volcanoes within terestrial basins, in *Sedimentation in Volcanic*

Settings, edited by R.V. Fisher, and Smith, pp. 155-170, SEPM, 1991.

White, J.D.L., Pre-emergent construction of a lacustrine basaltic volcano, Pahvant Butte, Utah (USA), *Bulletin of Volcanology*, *58*, 249-262, 1996.

White, J.D.L., Subaqueous eruption-fed density currents and their deposits, *Precambrian Research*, *101*, 87-109, 2000.

White, J.D.L., and C.J. Busby-Spera, Deep marine arc apron deposits and syndepositional magmatism in the Alisitos Group at Punta Cono, Baja California, Mexico, *Sedimentology*, *34*, 911-928, 1987.

Doris Maicher, Adalbertstr 27, 24106 Kiel, Germany, doris@maicher.net.

Eruptive and Depositional Mechanisms of an Eocene Shallow Submarine Volcano, Moeraki Peninsula, New Zealand

Benjamin Andrews

Department of Geological Sciences, University of Oregon, Eugene, Oregon
Currently at Department of Geology and Geophysics, University of Alaska Fairbanks, Fairbanks, Alaska

Eocene Surtseyan lapilli tuff deposits with maximum stratigraphic thicknesses of at least 175 m record a submarine volcano that underwent two cycles of edifice construction and erosion at Moeraki Peninsula, South Island, New Zealand. Basalt dike fragments, basalt clasts exhibiting fluidal deformation, and angular schist derived xenoliths are present throughout the lapilli tuff units; together with petrographic examination of the lapilli tuff, these rocks indicate that a broad range of eruptive and fragmentary mechanisms took place in the Moeraki volcano. Processes involved probably included phreatomagmatic interaction of magma with pyroclast-mud-seawater slurries in the vent. Data from two measured stratigraphic sections show that the volcano was formed during two distinct phases of eruption, separated by quiescence during which 1.5 m of laminated volcaniclastic sandstones were deposited. Lapilli tuff units below the sandstones are generally massive, while those above are well bedded and often alternate between finer- and coarser-grained lenticular bedsets. The stratigraphy and lithology of rocks at Moeraki indicate that the volcano was constructed through the explosive eruption of lapilli and ash in water no deeper than 400 m. The cone built to a height of at least 100 m above the seafloor by a combination of fall and eruption fed density currents, before eruption ceased and erosion of the volcano occurred. Renewed volcanic activity resulted in a volcano that rose above storm wave base and may even have emerged. As with modern analogues Surtla and Kavachi, erosion of the volcano to storm wave base or below quickly followed cessation of eruption.

INTRODUCTION AND GEOLOGICAL SETTING

Shallow submarine and emergent volcanoes are often short-lived structures quickly planed off to storm wave base unless a carapace of lava is erupted to protect the cone against waves. The erosion of Surtsey's satellite Surtla in 1964 and the many-times-emergent Solomon Islands volcano Kavachi provide examples of how fast erosion can occur [*Thorarinsson*, 1967; *Kokelaar and Durant*, 1983; *Johnson and Tuni*, 1987]. The ephemeral nature of many shallow or emergent cones combined with the difficulty in observing shallow explosive eruptions, has resulted in a relatively small, though rapidly growing body of literature describing this type of activity [e.g. *Maicher*, this volume]. Though numerous papers discuss the eruptive and fragmentary mechanisms of these volcanoes [*Moore*, 1985; *Kokelaar and Moore*, 1987; *White*, 1996], few have tied the eruptive processes to resultant deposits [*White*, 2000; *Smellie*, 2001; *Mueller*, this volume], or discussed the earliest phases of

Explosive Subaqueous Volcanism
Geophysical Monograph 140
Copyright 2003 by the American Geophysical Union
10.1029/140GM11

eruption—i.e. the entirely submarine 1963 eruptions at Surtsey prior to emergence [*Thorarinsson*, 1967].

The northern half of Moeraki Peninsula (Figure 1) is formed by the dissected remnants of an Eocene submarine volcano deposited on marine mudstone of the Kaiatan age Mohiki Formation, one of numerous early Tertiary Surtseyan volcanoes in North Otago [*Coombs et al.*, 1986]. Erosion at Moeraki has occurred such that a cross section of the volcano is exposed along the shore, albeit not through the likely vent. The bulk of the deposits is lapilli tuff with numerous larger clasts and is intruded by basalt and clastic dikes. These exposures allow for detailed study of the stratigraphy at multiple locations, and textural analysis of pyroclasts, which together provide insight into the eruptive and depositional processes of a shallow submarine volcano.

LITHOLOGY

The outcrops at Moeraki can be divided into several distinct lithologies, including basalt lapilli tuff, basalt and clastic dikes, irregular basalt intrusions, and mudstone. Further subdivisions can be made within the lapilli tuff to describe three types of "large" accidental clasts. Petrography and electron microprobe analysis of fresh glass in lapilli and basalt dikes characterizes the basalts as olivine tholeiites.

Basalt Lapilli Tuff

Basalt lapilli tuff is the dominant rock type at Moeraki (Figure 2), and often forms present-day wave-cut benches and bluffs. The units are light tan to gray in color, and particles (typically) range in diameter from <0.1 mm to ~4 cm. Nearly all of the lapilli tuff has been altered to clay, but original textures are preserved. The smallest particles (<0.2 mm) usually occur as low- to nonvesicular shards of sideromelane. The shards have concave morphologies suggestive of shattered vesicle walls. Vesicularities of the larger lapilli (2–40 mm) range from <5 to >40 vol%. These particles also exhibit a wide range in morphology (Figure 3) from blocky and equant to scoriaceous to stretched, and exhibit varying degrees of abrasion. Some lapilli are surrounded partially by a lapillus of an apparently different generation (i.e. one lapillus agglutinated to and partially surrounding another more or less abraded lapillus).

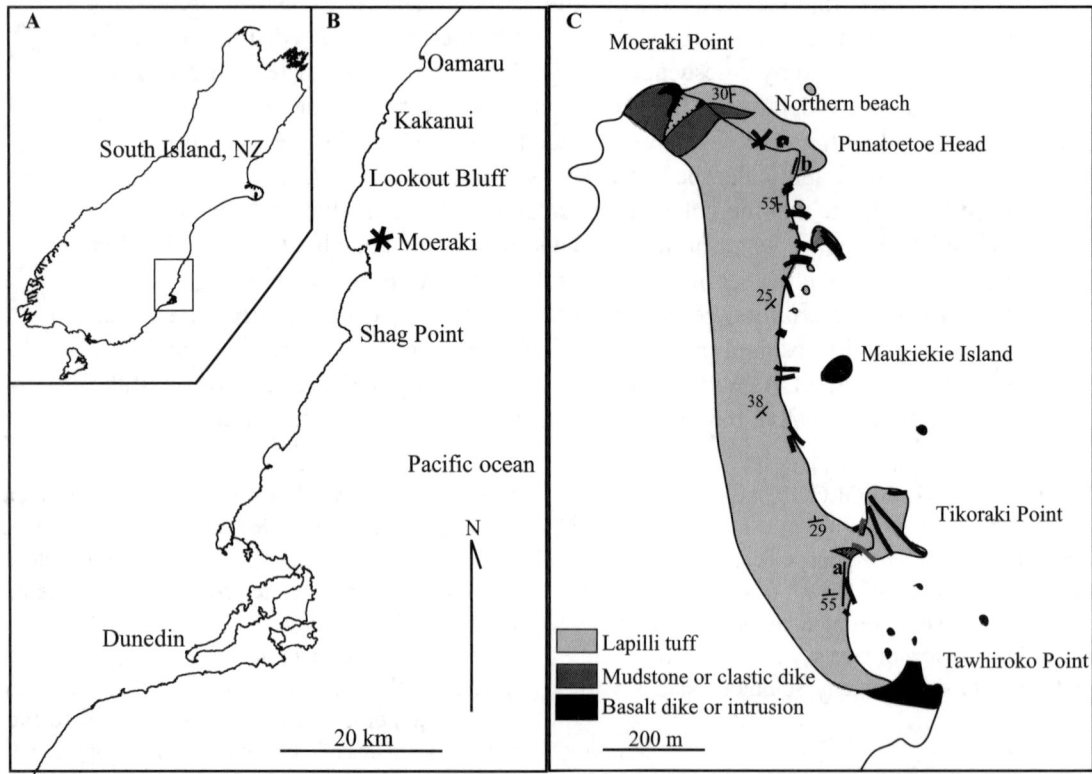

Figure 1. Sketch map of A) South Island, New Zealand showing locations of B) Moeraki Peninsula and other Tertiary volcanic sites in Otago. Map C shows geology of the northern half of Moeraki Peninsula with locations of measured stratigraphic sections "a" at Tawhiroko and "b" at Punatoetoe Head.

The morphologies and vesicularities of the lapilli support an interpretation of their formation under a diverse range of fragmentation processes, consistent with a submarine origin [*Fisher and Schmincke*, 1984; *White*, 1996; *Morrissey et al.*, 2000; *White and Houghton*, 2000]. The abundance of low-vesicularity, blocky lapilli indicates fragmentation dominated by quench processes, whereas scoriaceous lapilli show that exsolution and expansion of magmatic volatiles also played a role in fragmentation [*Honnorez and Kirst*, 1975; *Fisher and Schmincke*, 1984]. Lapilli with moderate vesicularities (20–30%) and stretched morphologies seem to indicate some combination of processes, such as fragmentation by expansion of volatiles terminated by rapid quenching. The different degrees of clast abrasion indicate variable recycling of particles in the vent. This is further supported by the presence of agglutinated lapilli, suggesting reworking of existing pyroclasts, perhaps during vent wall slumping.

Larger Clasts Within Lapilli Tuff

Basalt dike fragments. Angular fragments of basalt dikes occur in nearly all of the lapilli tuff units (Figure 4). These

Figure 2. Photographs of lapilli tuff outcrops, at a) wave cut bench at Tikoraki Point and b) base of bluff on northernmost beach.

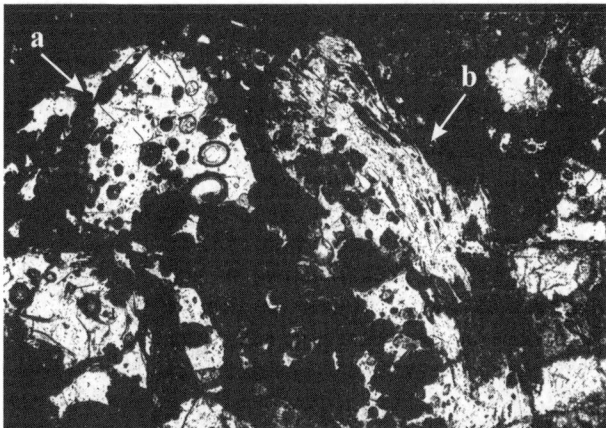

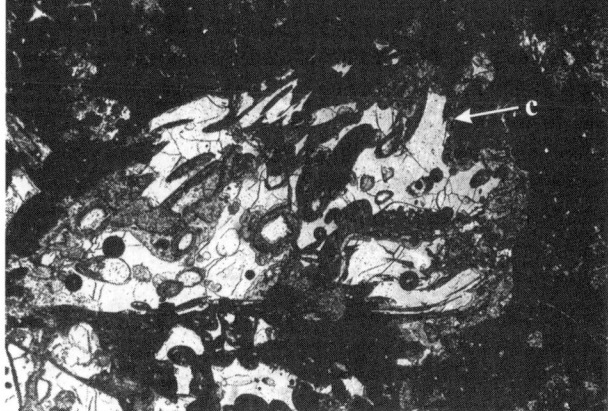

Figure 3. Photomicrographs of fresh sideromelane glass lapilli with a) blocky, b) stretched, and c) angular morphologies. Note also the large range in vesicularity and the extensive alteration to palagonite (dark gray) and clay (black). Vertical axis of pictures is 0.4 mm.

rocks have large diameters (8–30 cm) and blocky morphologies, with alternating bands of high and low vesicularity similar to those observed in numerous basalt dikes intruding the lapilli tuff at Moeraki. Several of these fragments also contain schist xenoliths of approximately the same size and morphology as those observed in basalt intrusions at Moeraki [*Benson*, 1943].

These accidental clasts are interpreted to have formed by the excavation of the vent around existing dikes. The angular morphologies, high crystallinities, and banded textures of the clasts indicate that the dikes were cooled before incorporation by the vent. Excavation occurred as a result of explosive activity removing material from the walls and floor of the vent [McClintock, pers. comm.]. Additionally, the occurrence of the clasts throughout the entire Moeraki outcrop indicates that vent wall excavation occurred in nearly all of the eruptive episodes, and that eruptions were explosive enough to eject the fragments out of the vent.

3:1 to 6:1. Additionally, apparent quenched rims are present on other large clasts, though these rims may be a product of pervasive alteration to clays. Within this group are also possible bombs with less irregular, non-fluidal morphologies. Several of these bombs are bright red suggesting oxidation while still hot.

These clasts were formed by the ejection of molten material from the vent and likely proximal deposition of at least some of the pyroclastic material. Unfortunately, it is difficult to construct a mechanism by which a hot rock in a submarine setting could stay molten long enough to be fluidally deformed. The only plausible processes for their formation are eruption through and transport by a vapor-rich fluid, such as a gas-supported density current [*White*, 2000] or steam cupola [*Kokelaar*, 1986], insulating steam envelope, and/or eruption through the water-air interface, allowing subaerial cooling in a tephra jet [*Thorarinsson*, 1967; and *Kokelaar*, 1986].

Quartz schist xenoliths. Elongate fragments of polycrystalline quartz occur in at least five of the lapilli tuff units (Figure 4). These rocks have aspect ratios of ~4:1 and are usually 2–7 cm long. Occasionally these fragments occur coring larger basalt clasts. The clasts are similar in size and shape to schist xenoliths in basalt dikes [*Benson*, 1943]. In thin section it is apparent that the clasts are nearly monomineralic and are composed of roughly equant quartz grains with an anisotropic fabric oriented parallel to the clast long axes.

The presence of these fragments indicates that the source magma for at least some of the eruptions was the same as that producing the xenolith-bearing dikes and intrusions. Furthermore, the occurrence of schist xenoliths as both isolated fragments and fragments within larger basalt clasts suggest that dikes may have fed eruptions. The variable size and abundance of these clasts, together with data from other clast types, indicate that fragmentation efficiency was variable.

Dikes

Basalt dikes and intrusions. Basalt dikes occur throughout the mapped area and range in size from 10–300 cm in thickness. Although there are several clusters of dikes that appear to have been emplaced radially about loci (south of Punatoetoe head and at Tikoraki Point), there is no clear pattern linking the majority of the dikes at Moeraki. The most notable features of these rocks are the occurrence in some of the larger dikes and intrusions of up to 40% schist xenoliths, and prominent vesicle banding in approximately half of all dikes. This banding is comprised of alternating cm-thick

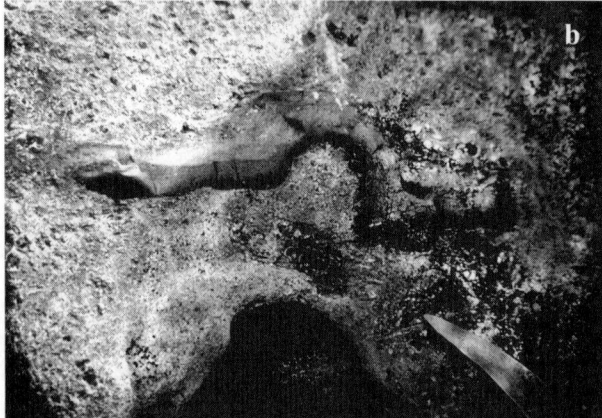

Figure 4. Examples of accidental clast types within lapilli tuff. a) Dike fragment, b) fluidally deformed basalt and c) quartz schist xenolith. Hammer point in b) is 10 cm long.

Fluidally deformed basalt clasts. Basalt clasts with uneven, fluidal morphologies and quenched rims suggestive of fluidal deformation (Figure 4) are present throughout approximately half of all lapilli tuff units at Moeraki. These rocks are commonly 5–30 cm in size with aspect ratios of

bands of high and low vesicularity, symmetrical about the center of each dike.

Several intrusions are also present at Moeraki including an outcrop of minipillows (10 cm in diameter) inshore of Maukiekie Island, and a 3 m amorphous, highly vesicular intrusion south of Punatoetoe Head. Larger irregularly shaped intrusions include a vesicle-banded circular outcrop 5 m in diameter on the northern beach filled with a calcite cemented clast-supported basalt breccia, and Maukiekie Island, formed of columnar basalt.

The presence of schist xenoliths in the dikes, as well as the lapilli tuff, indicates that the magma ascended through the underlying Haast Schist, which it fractured and entrained. This interpretation is consistent with the regional geology and work by *Coombs and Roedder* [1994] on a dolerite intrusion at Tawhiroko establishing a magma chamber depth of 7–14 km. The prominent vesicle banding in the dikes records a history of multiple injections of magma into these dikes. Essentially, as magma was injected, the central regions of the dikes remained warm long enough for more complete vesiculation, subsequent injections occurred through the hotter, more ductile centers of the dikes. The near-spherical shapes of the vesicles indicate that the vesicles experienced little coeval shear. Thus the higher vesicularity bands may have been quite cool (below solidus) and the dikes were mechanically "unzipped" to allow injection of new magma. Presence of other dikes with deformed vesicle bands (i.e. a z-fold with interior bands pinching out in the folds) indicate structural deformation occurred at times. The circular outcrop and breccia on the northern beach is believed to have formed through the partial collapse of a large pillow-like body intruding the lapilli tuff; interaction of the intrusion's hot interior with wet lapilli tuff produced the breccia.

Polyphase clastic dikes. At Tikoraki Point there are several polyphase clastic dikes that are in contact with and appear to stem from basalt dikes (Figure 5). These clastic dikes are composed of fragments of baked cross-laminated mudstones, mudstone breccias, basalt dike, and altered vesicular glass and range in thickness from <5 cm to ~40 cm. Generally, as these dikes are traced away from the basalt dikes they grade into narrow (<5 cm) mudstone dikes.

The polyphase clastic dikes are interpreted to have formed by phreatomagmatic interaction of basalt dikes with the mudstones and sediments immediately underlying the volcanic sequence. Sediments fluidized by the basalt dikes [*Kokelaar*, 1982] were expelled toward the tip of a hydraulically propagating fracture and fragments of lithified mudstone were carried upward into the volcano by these fluids.

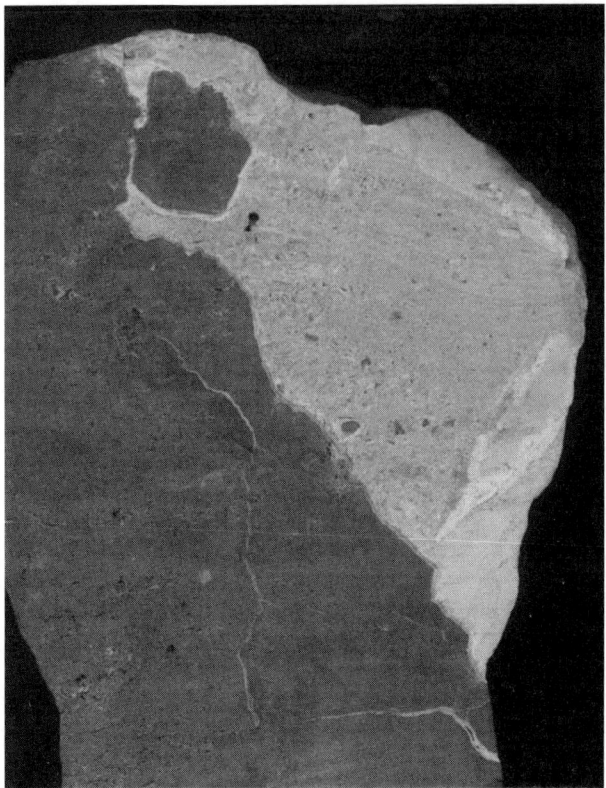

Figure 5. Polyphase clastic dike (light gray) in contact with basalt dike (dark gray) intruded at Tikoraki Point. Note cross-lamination in baked mudstone and fragments spalled from a basalt dike.

Any mudstone remaining in contact with the basalt dikes was baked.

STRATIGRAPHY

Stratigraphic sections have been measured at both Tawhiroko and Punatoetoe Head. Though neither section is bounded above or below by non-volcanic rocks (and thus total thickness is unknown) they record a significant portion of the volcano's history. A submarine origin for the rocks at Moeraki is supported by marine mudstones and sandstones of the Mohiki Formation that immediately underlie the volcanic material. Regional stratigraphy supports a submarine origin for the volcanic rocks at Moeraki: though the marine mudstones and sandstones of the Mohiki Formation are depositionally overlain by volcanic material at Moeraki, elsewhere in Otago, they are overlain by more marine rocks. Furthermore, within the volcanic sequence there is a pillow breccia and intrusive pillows, trough- and planar-cross-laminated sandstones, and sorted and well-graded clasts [*Cashman and Fiske*, 1991].

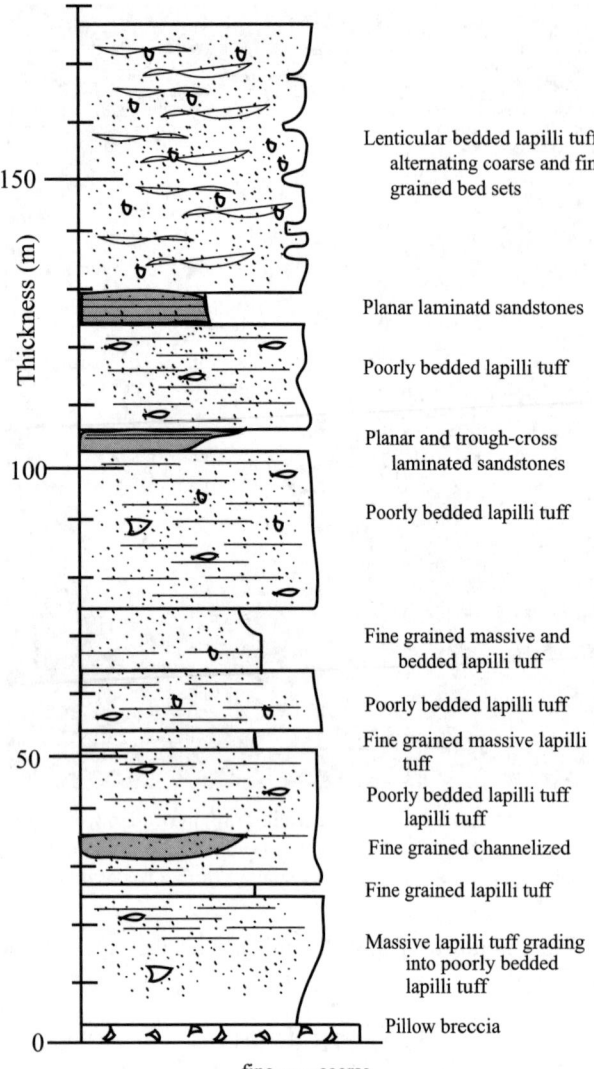

Figure 6. Schematic stratigraphic section measured north of Tawhiroko Point.

Tawhiroko

Beginning with a pillow breccia, and terminating against a present-day raised beach, 175 m of section dipping 60° were measured at Tawhiroko (Figure 6, Table 1). Of particular interest is a ~1.5 m thick volcaniclastic sandstone unit with planar- and cross-laminated beds that occurs between 104th and 106th m of measured section. This sandstone unit separates distinctly different lapilli tuff lithofacies: the lower lapilli tuff units are generally poorly bedded and coarse, whereas the upper units are well-bedded, often occurring as alternating broadly lenticular bedsets of finer- and coarser-grained lapilli tuff. Wavelengths of these lenticular bed forms are on the order of 10 m.

Interpretation of Tawhiroko stratigraphy. The sequence at Tawhiroko records two distinct phases of eruption separated by a quiescent period during which the sandstones were deposited. Differences between the upper and lower sections show that the character of the eruptions and depositional mechanism also changed.

The lowest exposed deposits, a pillow breccia and massive lapilli tuff, indicate that activity recorded in the section was initially effusive and progressed to explosive eruption of large volumes of lapilli and ash. The upward transition into explosive activity may have occurred through construction of the edifice to shallow depths and lower hydrostatic pressures, or through increased effusion rate facilitating fragmentation [*Fisher and Schmincke*, 1984; and *Morrissey et al.*, 2000].

The bulk of the sequence below the sandstones records a period of rapidly repeated eruption and deposition of lapilli and ash. Conceptually, these deposits probably resulted from tephra fall and density currents. The presence of four units of finer grained lapilli tuff within an 85 m sequence of poorly bedded coarse lapilli tuff is interpreted to reflect temporary increases in eruption intensity and fragmentation efficiency. These finer grained deposits are most likely deposits from eruption fed density currents [*White*, 2000].

The laminated sandstones in the middle of the Tawhiroko section were deposited during a period of volcanic quiescence and erosion. These deposits may thus have been derived from wave erosion and fragmentation of lapilli to sand size during planation of the volcano to storm wave base [*Kokelaar and Durant*, 1983]. Transport direction in the sandstones indicates that the summit of the cone, and thus the vent, was located to the northeast of Tawhiroko Point.

The section above the sandstones formed during a phase of renewed volcanic activity. These units are interpreted as reworked deposits of episodic eruptions [*Wright*, 2001]. Its alternating finer- and coarser-grained nature suggests that eruptions were episodic, and punctuated by periods during which erosion of material from farther upslope occurred. The lenticular character of the bedding and erosion of material from upslope is consistent with eruption and deposition in a submarine setting [*Wright*, 2001]. The reworked natures of these beds make determination of the immediately post-eruptive depositional processes difficult. However, based upon the alternating grain sizes of the beds, deposition occurred as both coarse grained fall and density current deposits, as well as finer-grained remobilized deposits winnowed from upslope [*Kokelaar and Durant*, 1983; *Allen and McPhie*, 2000; and *Wright*, 2001]. The units at Tawhiroko were not deposited above storm wave base, but originated from a vent and deposits that were.

Table 1. Description of lithologies comprising the section north of Tawhiroko Point

Level (m)	Lithology	Lapilli size	Bedding	Significant features	Interpretation
127–175	Lapilli tuff	Coarse/fine	Lenticular	Alternating fine and coarse bed sets, fines are ~50% at base, only ~10% at top	Deposits of episodic eruptions and reworked material from above
123–127	Lapilli tuff and sandstone	Fine	Well bedded	Seven reverse-graded lapilli tuff beds interbedded with planar laminated sandstone	storm wave base
112–123	Lapilli tuff	Coarse	Massive to poorly bedded	Bombs and schist fragments (some hosted in basalt clasts)	Eruption and deposition of lapilli and ash Inefficient fragmentation
108–112	Lapilli tuff	Coarse	Poorly bedded	Fining upward, moderate to high vesicularity clasts, schist fragments	Eruption and deposition of lapilli and ash. Efficient fragmentation
105.5–108	Lapilli tuff	Coarse/fine	Poorly bedded	Interbedded coarse and fine beds	Episodic eruption of lapilli and ash
104–105.5	Sandstone	Fine-Medium	Planar and cross-laminated	Volcaniclastic, coarsen up paleoflow from present NE	Erosional deposits from quiescent period
65–75	Lapilli tuff	Medium/fine sand	Poorly bedded	Two beds (65–69 and 69–75), lower with bombs and dike fragments, upper with bombs	Eruption and deposition of lapilli and ash by density currents Efficient fragmentation
53–59	Lapilli tuff	Fine	Massive	Two beds (53–57 and 57–59), lower finer than upper, lower poor in large clasts, upper rich (30% 4–12 cm)	Eruption and deposition of lapilli and ash by density currents
36–38	Lapilli tuff	Fine	Massive	Channelized	Eruption of lapilli and ash
29–31	Lapilli tuff	Fine	Massive	Bombs up to 20 cm long	Eruption of lapilli and ash
19–104	Lapilli tuff	Coarse	Poorly bedded	30% 3–6 cm basalt clasts dike fragments and bombs, four fine grained breaks in unit	High frequency, episodic eruption and deposition of lapilli and ash
1–19	Lapilli tuff	Coarse	Massive	3–30+ cm basalt clasts, isolated schist xenoliths	Eruption and deposition of lapilli and ash
0–1	Pillow breccia	15 cm	Massive	Indeterminate thickness	Effusive eruption

Punatoetoe Head

The basal 20 m of the section at Punatoetoe Head is composed of poorly bedded lapilli tuff largely and covered by modern beach sands. As at Tawhiroko, the sequence is overlain by broadly lenticularly bedded lapilli tuff (Figure 7, Table 2). A green sandstone unit near the base of the section is of particular interest. The unit is 0.5 m thick, massive, composed of fragments of lapilli with a high degree of clay alteration, and contains abundant large bombs up to 30 cm long and burgundy in color. Individual sand grains are highly abraded. Overlying the sandstone is a lapilli tuff unit that also has green alteration to clay. Numerous lapilli tuff units at Punatoetoe Head (meter 7 and between meters 8 and 12) have unimodal and bimodal clast size distribution. Where distribution is bimodal, the large clasts have higher vesicularities than the small clasts.

Interpretation of Punatoetoe Head stratigraphy. The basal unit at Punatoetoe Head represents the upper deposits formed during an early eruptive phase, and is analogous to the units below the central sandstones at Tawhiroko Point. The massive green sandstone near the base of Punatoetoe Head is an erosional unit derived from underlying, upslope strata. The grain size and stratigraphic location of the unit are quite similar to the sandstones at Tawhiroko. Thus the unit might be interpreted similarly and it may represent the same period of reduced volcanic activity or quiescence. However, the lack of bedding, high degree of clay alteration, and presence of bombs indicate a more complex his-

186 ERUPTIVE AND DEPOSITIONAL MECHANISM OF AN EOCENE SHALLOW SUBMARINE VOLCANO

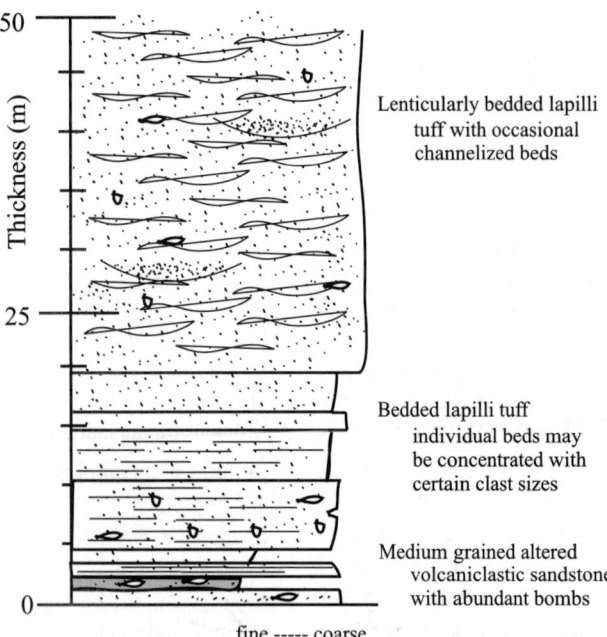

Figure 7. Schematic stratigraphic section measured at Punatoetoe Head.

tory. It is proposed that this unit was initially winnowed from deposits higher on the volcano [*Wright*, 2001], subsequently altered, and then remobilized in a slump or slide. The bombs in the unit may be lag material from previous eruptions exposed at the surface [e.g. *Kokelaar and Durant*, 1983] and reincorporated in the slump or slide.

The overlying bedded lapilli tuff unit with green clay alteration records a period of slower erosion of lapilli tuff from higher volcano flank outcrops. The rest of the sequence is composed of bedded lapilli tuff recording renewed volcanic eruptions. Most of the units are likely the product of eruption-fed density currents [*Morrissey et al.*, 2000; *White*, 2000], indicated by the channelized units above meter 30 in the section. The presence of beds with unimodal and bimodal clast size distributions indicates that deposition occurred as hydraulically sorted fall deposits [*Cashman and Fiske*, 1991; *Oehmig and Wallrabe-Adams*, 1993]. Though reworking by wave and current action cannot be discounted [*Wright*, 2001] the preservation of likely fall deposits indicates that any such reworking has not been pervasive. As with the upper units at Tawhiroko, the lenticular beds at Punatoetoe Head are interpreted as the reworked products of repeated eruptions [*Wright*, 2001]; material from individual eruptive pulses was reworked prior to being covered by the next pulse. The onset of lenticular bedding at Punatoetoe Head indicates growth of the cone into shallower water [*Wright*, 2001] similar to Tawhiroko.

DISCUSSION

Textural and stratigraphic interpretations of rocks at Moeraki suggest the eruptive and depositional history for the Moeraki volcano. These lines of reasoning suggest that Surtla or Kavachi are appropriate modern analogues, where eruptions built a cone above, or to within meters of, sea level before activity ceased and the volcanic edifice was rapidly eroded [*Kokelaar and Durant*, 1983; *Johnson and Tuni*, 1987]. Depending upon the length of the quiescent interval separating the two periods of eruption at Moeraki, and whether or not the volcano can be classified as a single monogenetic cone, a group of monogenetic cones [see *Maicher*, this volume] or a single reactivated cone, then either Surtla (monogenetic) or Kavachi (numerous eruptions this century alone) is the more appropriate analogue. Based upon the sandstones within the lapilli tuff sequences, particularly the altered unit at Punatoetoe Head, it seems likely that a minimum period of several years occurred between the two eruptive phases; such a period is necessary for alteration to occur. Thus Kavachi may provide a better comparison. This contrasts with Bridge and Aorere Points, 30 km north of Moeraki, described by *Cas et al.* [1989] as a "monogenetic Surtla-type volcano."

The rocks at Moeraki also add information on processes occurring in submarine vents [e.g. *Kokelaar*, 1983; *Moore*, 1985; *Kokelaar*, 1986; *Kokelaar and Moore*, 1987; *White*, 1996]. In particular, analysis of the lapilli tuff and larger clasts within it indicates that fragmentation occurred through a combination of expansion of magmatic volatiles and phreatomagmatic processes [*White*, 1996; *Morrissey et al.*, 2000; and *White and Houghton*, 2000]. The presence of bombs and fluidally deformed basalt clasts demonstrates that molten material ejected from the vent did not instantly solidify, but stayed molten long enough to develop stretched and curled morphologies. Furthermore, the occurrence of schist xenoliths in both the lapilli tuff and numerous dikes shows that the eruptions were likely dike-fed, and the presence of dike fragments and recycled lapilli indicates that the vent excavated and incorporated parts of the vent walls. When these interpretations are coupled with the occurrence of polyphase clastic dikes generated through the fluidization and phreatomagmatic interaction of sediments with magma [*Kokelaar*, 1982], it seems likely that a slurry of seawater, mud, and recycled pyroclasts occupied the vent. Magma entering the vent would mingle with the slurry, resulting in phreatomagmatic eruption [*Kokelaar*, 1983; *White*, 1996]. Mingling might occur through Fuel-Coolant-Interaction processes [*White*, 1996]. This mechanism could allow for the larger particles to stay molten long enough to develop fluidal deformation, provided the particles develop an insu-

Table 2. Description of lithologies comprising the section at Punatoetoe Head

Level (m)	Lithology	Lapilli size	Bedding	Significant features	Interpretation
20–50	Lapilli tuff	Coarse	Lenticular	5+ cm clasts present through unit, some 5 cm well-bedded units, some channels	Deposits of episodic eruptions and reworked material from upslope
17–20	Lapilli tuff	Coarse	Poorly bedded		Eruption and deposition of lapilli and ash
16–17	Lapilli tuff	Coarse	Massive	30% of unit is clasts 4+ cm	
12–16	Lapilli tuff	Medium	Poorly bedded	Bi-modal clast size distribution, clasts either <1 or 2–3 cm in diameter	Eruption and deposition of lapilli and ash as fall
7.5–12	Lapilli tuff	Medium	Bedded	Massive beds at base, 5 cm beds at top, some upper beds concentrated with 3 cm clasts	Eruption and deposition of lapilli and ash as density currents (base) and fall (top)
7–7.5	Lapilli tuff	Coarse	Faintly bedded	Concentrated with 2 cm clasts	Deposition as fall
3–7	Lapilli tuff	Medium	Well bedded	10 cm beds, coarsens upward, accidental clast size increases up from 3 to 5 cm	Eruption and deposition of lapilli and ash as density currents
1.5–3	Lapilli tuff	Fine	Bedded	Altered lapilli, green, no large clasts	Derived during quiescent period from upslope
1–1.5	Sandstone	Medium sand	Massive	Composed of lapilli fragments with green clay alteration, abundant bombs, some with burgundy color	Erosional deposit derived altered lapilli higher on volcano during period of volcanic quiescence
0–1	Lapilli tuff	Coarse	Poorly bedded	Abundant 3 cm basalt clasts	Analogous to Tawhiroko lapilli tuff at 103rd m

lating steam envelope. Additionally, it is likely that occasional pyroclasts were ejected above sea level, as evidenced by the red oxidation of bombs at the base of Punatoetoe Head. Oxidation of hot (1000°C) bombs in air may take place in seconds and is therefore feasible for pyroclasts derived from submarine vents that are insulated from the water, ejected into the air, and retain sufficiently high temperatures [personal communication from *J. Gardner, S. Tait, and P. Wallace*].

The stratigraphy at Moeraki records the growth of a volcanic edifice to a height of at least 175 m above the seafloor. Interpretations of the exposed sections suggest that the volcano progressed from effusive eruption of pillows to highly fragmentary explosive activity. These lapilli tuff-producing eruptions occurred as repeated events of varying individual magnitudes, producing deposits ranging from <1 m to 18 m thick for individual beds. The time interval between eruptive pulses is also likely to have varied considerably; the 85 m of poorly bedded lapilli tuff at Tawhiroko was likely formed through high-frequency eruptive pulses, whereas the alternating coarser- and finer-grained deposits comprising the upper section at Tawhiroko Point represent lower frequency explosive ejection of material, allowing for deposition of remobilized material between events [e.g. *Smellie*, 2001]. However, for both sequences it is likely that the entire eruptive, cone-constructing sequence of events lasted on the order of weeks, such as at Surtla (though Surtla grew from the flanks of Surtsey, and not the seafloor) [*Thorarinsson*, 1967].

Estimates of the minimum size of the volcano can be made using a conical volcano reconstruction, an angle of repose for the submarine lapilli tuff of 20°, and the 175 m stratigraphic thickness at Tawhiroko. If the Tawhiroko section is close to the approximate maximum thickness of the volcano, then a cone is implied with a radius of ~480 m and volume of ~0.2 km³. The radius of this cone is far too small to encompass the Moeraki outcrops today, suggesting that several monogenetic cones or a larger single center were present. If, on the other hand, Maukiekie Island is treated as the throat and thus center of a large volcano, the cone may have had a height of ~400 m and volume of ~0.6 km³. This reconstruction is sufficient to cover the entirety of the volcanic deposits at Moeraki. The presence of oxidized pyroclasts suggests that the volcano grew to within meters of the surface where subaerial eruption or flight occurred. As a consequence, the water depth at Moeraki is unlikely to have exceeded 400 m.

CONCLUSIONS

The volcanic deposits comprising the northern half of Moeraki Peninsula are the remnant of a shallow to emergent submarine volcano. The volcano experienced two periods of eruption and edifice construction, separated by a period of degradation that probably lasted a few years, before ultimately being planed off to storm wave base. Fragmentation was accomplished through a combination of processes including exsolution and expansion of magmatic volatiles, quench fracturing, and fuel-coolant interaction. Occasionally, large juvenile particles were ejected while still molten and cooled slowly enough to retain fluidal morphologies; some of these pyroclasts are likely to have been erupted through the water-air interface. Deposition of material occurred as primary fall deposits and eruption fed density currents. Following cessation of the second eruptive period, the volcano was quickly eroded to storm wave base.

Acknowledgments. Financial support and assistance by University of Otago to the author are greatly appreciated. This research would not have occurred without insights and discussions contributed by James White and Murray McClintock, or discussions of oxidation rates with James Gardner, Steve Tait and Paul Wallace. This manuscript was improved significantly by reviews and comments made by Nate Sheldon, Volker Lorenz and John Smellie.

REFERENCES

Allen, S.R., and J. McPhie, Water-settling and resedimentation of submarine rhyolite pumice at Yali, eastern Aegean, Greece, *J. Volcanol. Geotherm. Res., 104*, 237-259, 2000.

Benson, W.N., The basic igneous rocks of eastern Otago and their tectonic environment, *Trans. Roy. Soc. N.Z., 72*, 85-185, 1943.

Cas, R.A.F., C.A. Landis, and R.E. Fordyce, A monogenetic, Surtla-type, Surtseyan volcano from the Eocene-Oligocene Waiareka-Deborah volcanics, New Zealand: a model. *Bull. Volcanol., 51*, 281-298, 1989.

Cashman, K.V., and R.S. Fiske, Fallout of pyroclastic debris from submarine volcanic eruptions, *Science, 253*, 275-280, 1991.

Coombs, D.S., R.A.F. Cas, Y. Kawachi, C.A. Landis, W.F. McDonough, and A. Reay, Cenozoic volcanism in North, East, and Central Otago, in *Cenozoic Volcanism in New Zealand*, edited by I.E.M. Smith, *Roy. Soc. N.Z. Bull. 23*, 278-312, 1986.

Coombs, D.S., and E. Roedder, On the significance of CO_2 Inclusions in plagioclase microphenocrysts in tholeiite from Moeraki, New-Zealand, *Bull. Volcanol. 56*, 23-28, 1994.

Fisher, R.V., and H.-U. Schmincke, *Pyroclastic rocks*, Springer-Verlag Berlin, 472 pp, 1984.

Honnorez, J., and P. Kirst, Submarine basaltic volcanism; morphological parameters for discriminating hyaloclastites from hyalotuffs, *Bull. Volcanol. 39*, 441-465, 1975.

Johnson, R.W., and D. Tuni, Kavachi, an active forearc volcano in the western Solomon Islands; reported eruptions between 1950 and 1982, *Marine geology, geophysics and geochemistry of the Woodlark Basin-Solomon Islands. Circum-Pacific Council for Energy and Mineral Resources, Earth Science Series 7*, 89-112, 1987.

Kokelaar, B.P., Fluidization of wet sediments during the emplacement and cooling of various igneous bodies, *J. Geol. Soc. London, 139*, 21-33, 1982.

Kokelaar B.P., The mechanism of Surtseyan volcanism. *J. Geol. Soc. London, 140*, 939-944, 1983.

Kokelaar, B.P., and G.P. Durant, The submarine eruption and erosion of Surtla (Surtsey), Iceland. *J. Volcanol. Geotherm. Res.*, 19: 239-246, 1983.

Kokelaar, B.P., Magma-water interactions in subaqueous and emergent basaltic volcanism. *Bull. Volcanol. 48*, 275-289, 1986.

Kokelaar, B.P., and J.G. Moore, Structure and eruptive mechanisms at Surtsey Volcano, Iceland; discussion and reply. *Geol. Mag. 124*, 79-86, 1987.

Moore, J.G., Structure and eruptive mechanisms at Surtsey Volcano, Iceland. *Geol. Mag. 122*, 649-661

Morrissey, M.M., B. Zimanowski, K. Wohletz, and R. Buettner, Phreatomagmatic fragmentation, in *Encyclopedia of Volcanoes* edited by H. Sigurdsson, B.F. Houghton, S.R. McNutt, H. Rymer, J. Stix, and D. Ballard, 431-445, 2000.

Oehmig, R., and H.-J. Wallrabe-Adams, Hydrodynamic properties and grain-size characteristics of volcaniclastic deposits on the mid-Atlantic ridge north of Iceland (Kolbeinsey Ridge), *J. Sed. Petrol., 63*, 140-151, 1993

Smellie, J.L., Lithofacies architecture and construction of volcanoes erupted in englacial lakes: Icefall Nunatak, Mount Murphy, eastern Marie Byrd Land, Antarctica, *Spec. Publs. Int. Ass. Sediment., 30*, 9-34, 2001.

Thorarinsson, S., *Surtsey: the new Island in the North Atlantic*, Viking Press, New York, 47 pp., 1967.

White, J.D.L., and B.F. Houghton, Surtseyan and related phreatomagmatic eruptions, in *Encyclopedia of Volcanoes* edited by H. Sigurdsson, B.F. Houghton, S.R. McNutt, H. Rymer, J. Stix, and D. Ballard, 897-912, 2000.

White, J.D.L., Impure coolants and interaction dynamics of phreatomagmatic eruptions. *J. Volcanol. Geotherm. Res,. 74*, 155-170, 1996.

White, J.D.L., Subaqueous eruption-fed density currents and their deposits. *Processes in physical volcanology and volcaniclastic sedimentation; modern and ancient, Precambrian Res., 101*, 87-109, 2000.

Wright, I.C., In situ modification of modern submarine hyaloclastite/pyroclastic deposits by ocean currents; an example from the southern Kermadec Arc (SW Pacific), *Marine Geol, 172*, 287-307, 2001.

B.J. Andrews, Department of Geology and Geophysics, University of Alaska Fairbanks, P.O. Box 755780, Fairbanks, AK, 99775-5780.

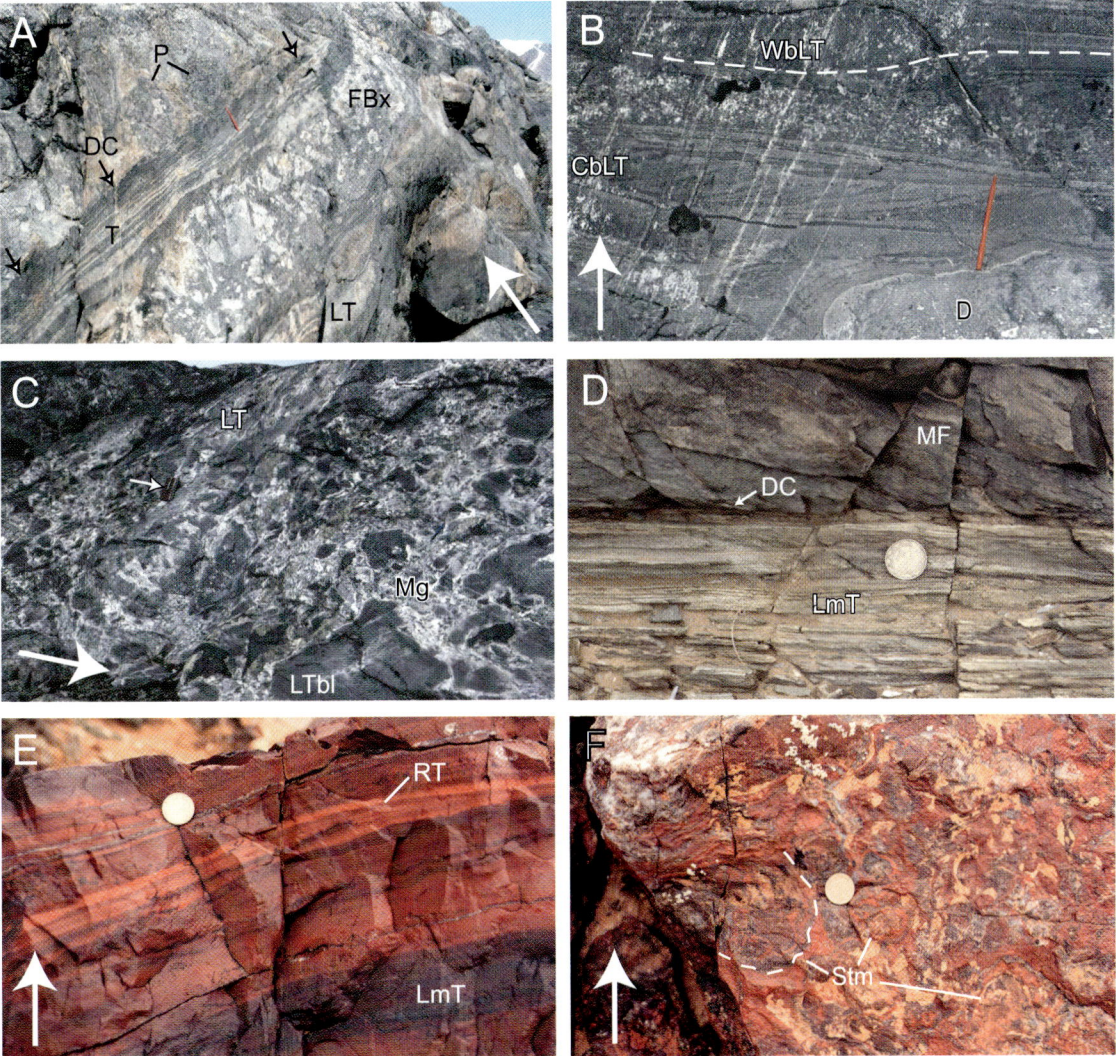

Plate 1. Characteristics of the bounding facies in the Kangerluluk [A-C] and Schakalsberg Mountain areas [D-F]. White arrow indicates top in all photographs. A) Sharp upper depositional contact (DC) between tuff of pyroclastic lithofacies and pillows (P) of volcanic lithofacies. Note inverse graded felsic breccia (FBx) and underlying lapilli tuff (LT). Scale, red pen 16 cm long. B) Synvolcanic dyke (D) intruding crossbedded (CbLT) and wavy bedded lapilli tuff (WbLT). Note chilled dyke margin and cm-scale vesicles filled with calcite-epidote. Scale, pen 16 cm long. C) Peperite composed of lapilli-tuff blocks (LTbl). Magma (Mg) has a sharp contact with lapilli tuff sequence (LT) and LT blocks (LTbl). Scale, field book 23cm long (small arrow). D) Fine-grained laminated tuff (LmT) displays a sharp depositional contact (DC) with a massive basalt flow (MF). Scale, coin 2.5 cm in diameter. E) Laminated (LmT) and rippled tuff (RT) replaced by Fe-rich hydrothermal fluids (jasper). Scale, coin 2.5 cm in diameter. F) Jasper-altered stromatolites (STM). Scale, coin 2.5 cm in diameter.

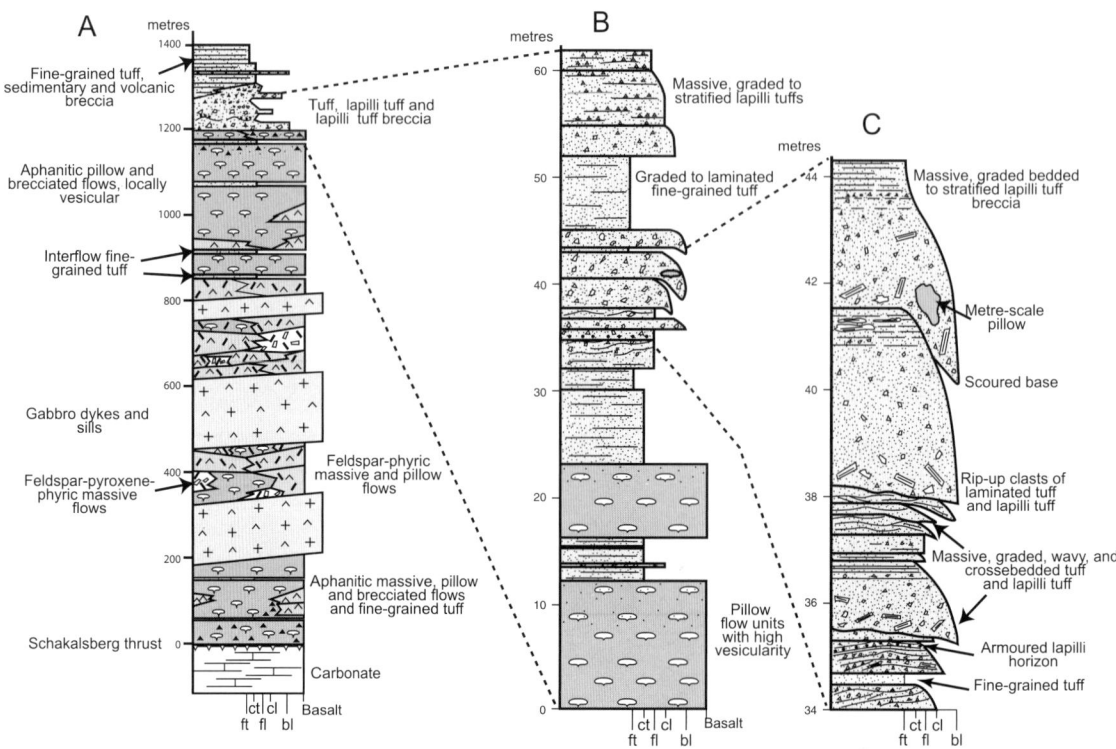

Figure 3. Schakalsberg Mountains stratigraphy and lithological units. A) General stratigraphy with subaqueous basalt flows intruded by low-angle dykes or sills. The top of the sequence features volcaniclastic rocks. B) Volcanic lava flows and pyroclastic deposits of the upper volcaniclastic-dominated segment. C) Detailed section in the volcaniclastic segment.

tuff turbidites capping coarser beds reflect dilute turbidity current and suspension deposits (Bouma T_{bd}, T_d) elutriated from the main body and head during turbulent transport. A time frame of hours is required for the settling of fine- to coarse-grained tuff and may be comparable to slack-water periods in macrotidal zones that generate clay drapes on sandstone foresets to form sigmoidal tidal bundles [Dalrymple, 1992]. Local outsized cognate pyroclasts, have disrupted graded beds by impact and are interpreted as ballistically emplaced bombs.

4.2. Kangerluluk Lapilli Tuff Breccia

A distinct 5–15 m-thick lapilli tuff breccia (Plate 2d) extends ca. 100 metres along strike. The crystal-enriched deposits occur in a series of 10–60 cm-thick massive, graded or laminated beds that are disrupted by subrounded to rounded block-size pyroclasts. The feldspar-phyric pyroclasts with plagioclase phenocrysts up to 2 cm, sag into the fine-grained tuffs similar to sedimentary load casts or impact structures. The volcanic matrix associated with the pyroclasts is rich in broken and euhedral feldspar crystals, minor pyroxene crystals and lithic volcanic fragments. Locally, mm-thick films of fine-grained tuff surround the clasts. The block-size pyroclasts constitute ca. 30% of the deposit.

4.2.1. Interpretation of transport process. The lapilli tuff breccia at Kangerlukuk must reconcile two co-existing transport processes, which are unusual for a subaqueous milieu. The graded bedded tuff-lapilli tuffs resemble deposits of aqueous density currents, but the origin of sag structures associated with large pyroclasts is puzzling. Multiple impact structures (Plate 2d) are observed from small-scale subaerial explosions and have been interpreted as bomb sags [*Fisher and Schmincke,* 1984]. Such structures may also form in water but under dry to moist conditions, as in a subaerial eruption [*White,* 1996]. The observed pyroclasts and their abundance at Kangerluluk are inferred to be ballistically transported to the site of deposition from

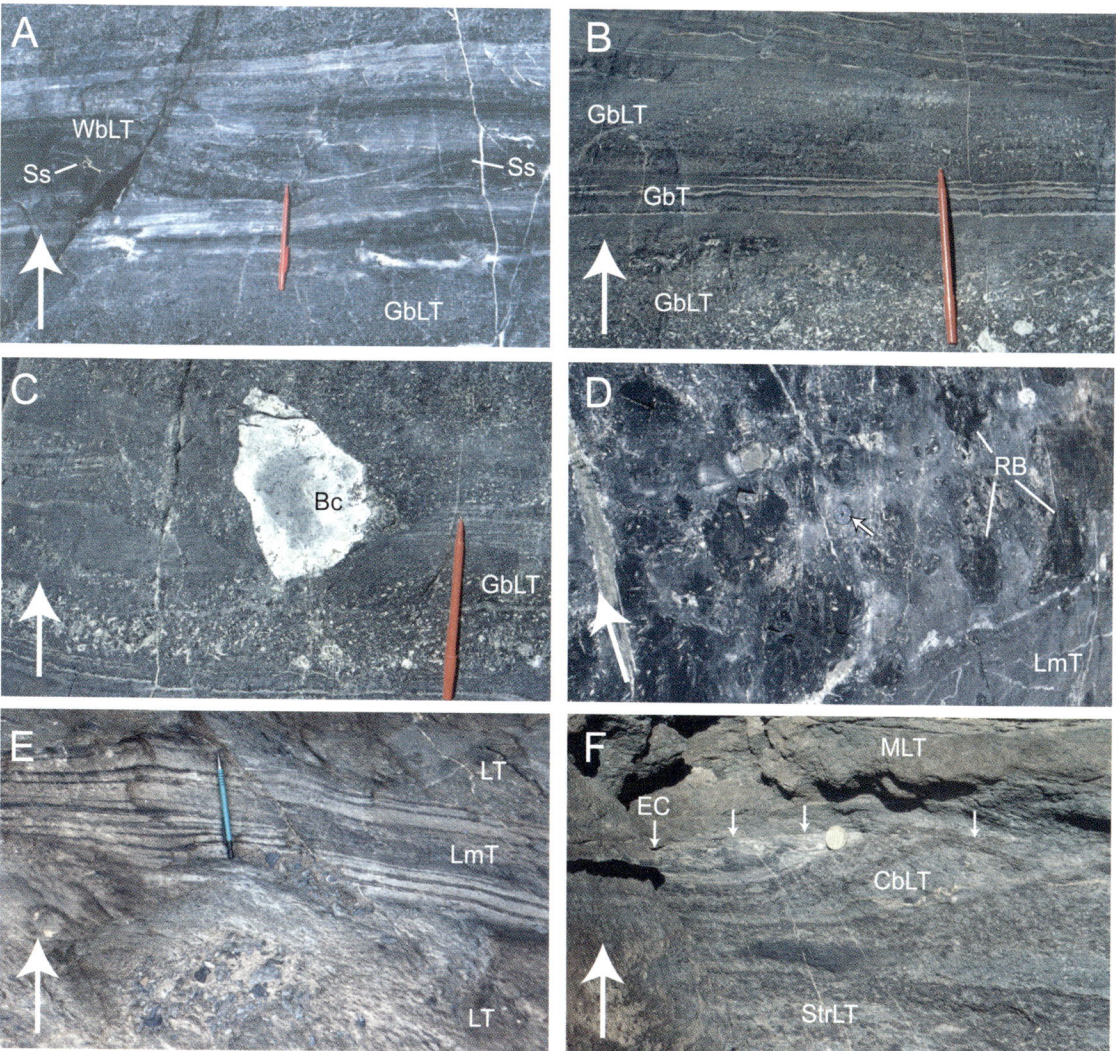

Plate 2. Salient features of the pyroclastic deposits at Kangerluluk [A-D] and Schakalsberg [E-F]. White arrow indicates top in all photographs. A) Graded bedded (GbLT) and wavy bedded tuff and lapilli tuff (WbLT). Low-angle channel-shaped scour (Ss) is filled with wavy-stratified lapilli tuff composed of alternating horizons of fine lapilli and coarse-grained tuff. Scale, pen 16 cm long. B) Two graded bedded lapilli tuffs (GbLT) separated by numerous cm-mm-scale fine-grained graded tuffs (GbT). Two 15–20 cm-thick small-scale fining-upward sequences are present. Scale, pen 15 cm long. C) Block-size clast (Bc) disrupting graded lapilli tuff beds (GbLT). Note abundance of crystals at base of bed. Scale, pen 16 cm long. D) Lapilli tuff breccia unit with rounded bombs (RB) forming sag structures. Scale, coin (arrow), 1.8 cm in diameter. E) Detailed view of fine-grained laminated tuff (LmT) between lapilli tuffs (LT). Scale, pen 13 cm long. F) Crossbedded lapilli tuff (CbLT) composed of fine lapilli and coarse-grained tuff eroded by massive lapilli tuff (MLT). StrLT, stratified lapilli tuff. EC, erosive contact. Scale, coin 2.5 cm in diameter.

a subaqueous eruption. Films of fine-grained tuff around some bombs suggest wet ash adhered to these pyroclasts, allowing classification as armoured lapilli or bombs [*Waters and Fisher,* 1971; *Fisher and Schmincke,* 1984]. The rounded bombs, formed by surface tension of non-consolidated clasts during transport [*Martin and White,* 2001] are considered juvenile magma rather than cognate fragments.

4.3. Schakalsberg Tuff

The 0.10–5 m thick Schakalsberg tuff units (Figure 3c) are interstratified with pillowed and massive basalt flows (Plate 1d) as well as lapilli tuff beds (Plate 2e). A 5m-thick unit represents the stratigraphic top in the study area (Figure 3a). The 0.5–3 cm-thick, fine- to- coarse-grained, mafic tuff beds are composed of cuspate sideromelane, scoria and lithic volcanic particles that display graded bedding and parallel lamination, with minor rippled horizons. Contacts with the overlying lapilli tuff are sharp and locally erosive (Plate 2e), whereas basal contacts are sharp and depositional but non-erosive. Locally, these units have undergone extensive hydrothermal carbonate alteration.

4.3.1. Interpretation of transport process. The Schakalsberg tuffs represent Bouma T_{bd}-T_{cd}, and T_d sequences whereby the former represent dilute turbidity flows and the latter suspension deposits [*Lowe,* 1988; *Mueller et al.,* 1994]. Rapid subaqueous deposition, in comparison with pelagic background sedimentation, is inferred for these syneruptive deposits. Water selectively retained the ash component during subaqueous eruptions causing accumulation and congestion in the water column until turbulence abated [*Lowe,* 1988; *Mueller and White,* 1992]. Their association as caps to lapilli tuff beds in thicker lapilli tuff sequences is consistent with (1) elutriation processes during aqueous transport (surface transformation; Fisher, 1983) and (2) a short residence time in the water column. The tuffs were probably deposited below wave base as suggested by the absence of wave-induced structures and intraformational unconformities (>> 50m; cf. Smellie and Hole, 1997).

4.4. Schakalsberg Lapilli Tuff

Up to 1.50 m of lapilli tuff (Figure 3c) forms a distinct depositional unit composed of planar stratified to crossbedded structures (Plate 2f) and distinct wavy-stratified beds (Plate 3a). The deposit is overlain both conformably by thin-bedded, graded to laminated tuffs and erosively by thick sediment gravity flow deposits (Figure 3c). Individual beds, 10–40 cm-thick, are massive to graded, and stratification is observed in crossbeds and planar beds. Stratification is accentuated by changes in grain-size and hydrothermal alteration. Stratified planar beds may grade laterally into crossbeds, which in turn change back into planar beds. Synsedimentary deformation is local, but scour surfaces are common. The distinguishing feature of these depositional units is the presence of distinct 1–10 mm large spherical to ovoid armoured lapilli (Plate 3b). Lithic volcanic fragments and scoria are abundant, whereas liberated crystals are minor. The armoured lapilli were observed in both the massive to graded as well as wavy beds.

4.4.1. Interpretation of transport process. The lapilli tuffs display similar sedimentary structures to that of the Kangerluluk rocks, and hence have comparable transport processes. Massive to graded beds reflect high concentration density flows with rapid en-masse fallout [*Lowe,* 1982]. Planar stratified to cross-stratified beds with basal scouring are indicative of bedload transport and highly unsteady, turbulent density currents with low to high particle concentrations [*Mueller et al.,* 2000a]. The grain-size changes in graded but stratified beds can be explained by density currents, in which the coarser fragments were deposited from the main body of the flow, and the fine lapilli and coarse-grained tuff represents the tail of the flow [*Martin and White,* 2001]. The stratification within individual beds is consistent with changing internal shear forces operating during flowage [*Yamazaki et al.* 1973; *Doucet et al.,* 1994] and is in part responsible for segregation within the low- to high-concentration, turbulent gravity flows. The armoured lapilli [*Waters and Fisher,* 1971; *Fisher and Schmincke,* 1984; p. 94] or "core-type" accretionary lapilli [*Schumacher and Schmincke,* 1991] necessitate a water-exclusion zone if erupted under water and therefore require special consideration.

5. PETROGRAPHY OF SCHAKALSBERG LAPILLI TUFF

The lapilli tuffs are composed of lapilli-size lithic volcanic fragments, block-size rip-up clasts of tuff and lapilli tuff, eruptive clasts with an amalgamation of shards and scoria, and armoured lapilli (Plate 3b). The matrix of these deposits is a fine-grained shard-rich tuff that locally has been replaced by hydrothermal magnetite. The ubiquitous armoured lapilli are discussed because of implications for eruptive processes. The 1–10 mm armoured lapilli (average size 3–5 mm) may be dispersed or occur in clast-supported pods (thin-section observation). They have thick or thin

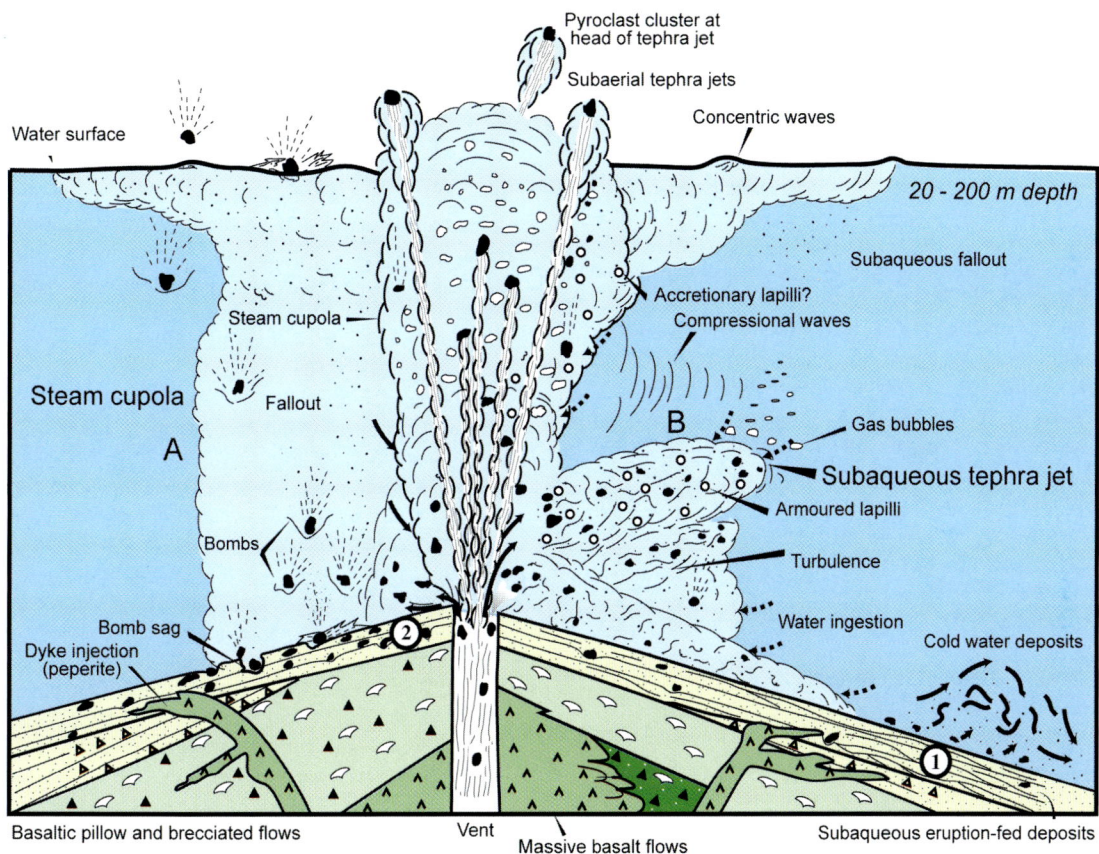

Plate 4. A model for shallow-water, small-volume, Surtseyan-type eruptions with continuous uprush conditions (A) and individual tephra jets (B). (1) lapilli tuff; subaqueous eruption-fed density current deposits. (2) lapilli tuff breccia; ballistically emplaced pyroclasts forming impact structures under a steam cupola. Diagram not to scale.

and is the outgrowth of regional mapping programmes. The logistical support given to the author for these remote areas indicates the high degree of professionalism under which these projects were conducted. GEUS and Kumba Resources are gratefully acknowledged for their permission to publish this information. Constructive reviews of N. Riggs and an anonymous reviewer improved the manuscript considerably. Editor, J. Smellie is thanked for impeccable editing and incisive comments.

REFERENCES

Cashman. K.V., and R.S. Fiske, R.S., Fallout of pyroclastic debris from submarine volcanic eruptions. *Science*, 253, 275-280, 1991.

Chadwick, B., and A.A. Garde, Palaeoproterozoic oblique plate convergence in South Greenland: a re-appraisal of the Ketilidian orogen, in *Precambrian crustal evolution in the North Atlantic region*, edited by T.S, Brewer, *Geol. Soc. London, Spec. Publ.* 112, 179-196, 1996.

Chough, S.K., and Y.K, Sohn, Depositional mechanics and sequences of base surges, Songaksan tuff ring, Cheju island, Korea, *Sedimentology*, 37, 1115-1135, 1990.

Cole, R.B., and R.G. Stanley, 1994, Sedimentology of subaqueous volcaniclastic sediment gravity flows in the Neogen Santa Maria Basin, California. *Sedimentology* 41, 37-54, 1994.

Dalrymple, R.W., Tidal depositional systems, in *Facies Models: response to sea level change*, edited by R.G. Walker and N.P. James, *Geol. Assoc. Can.,* pp. 195-218, 1992.

Doucet, P., W. Mueller, and F. Chartrand, Archean, deep-marine, volcanic eruptive products associated with the Coniagas massive sulfide deposit, Quebec, Canada, *Can. J. Earth Sci.*, 31, 1569-1584, 1994.

Einarsson, T., Der Surtsey Ausbruch: Reykjavik, *Heimskringla,* 23 pp., 1996.

Fisher, R.V., Puu Hou littoral cones Hawaii, *Geol. Rundschau*, 57, 837-864, 1968.

Fisher, R.V., Models for pyroclastic surges and pyroclastic flows, *J. Volcanol. Geotherm. Res.*, 6, 305-318, 1979.

Fisher, R.V., Flow transformations in sedimentary gravity flows. Geology, 11, 273-274, 1983.

Fisher, R.V., and H.U. Schmincke, H., Pyroclastic Rocks, *Springer-Verlag,* New York, 472pp, 1984.

Fiske, R. S., and T. Matsuda, 1964, Submarine equivalents of ash flows in the Tokiwa Formation, Japan, *Am. Jour. Sci.* 262, 76-106, 1964.

Frimmel, H.E., U. Klötzli, and P. Siegfried, New Pb-Pb single zircon age constraints on the timing of Neoproterozoic glaciation and continental break-up in Namibia, *Jour. Geol.*, 104, 459-469, 1996a.

Frimmel, H.E., C.J.H. Hartnady, and F. Koller, Geochemistry and tectonic setting of magamtic units in the Pan-African gariep Belt, Namibia, *Chem. Geol.*, 130, 101-121,1996b.

Frimmel, H.E., R.E. Zartman, and A. Späth, The Richtersveld Igneous Complex, South Africa: U-Pb zircon and geochemical evidence for the beginning of Neoproterozoic continental breakup, *Jour. Geol.*, 109, 493-508, 2001.

Fritz, W. J., and M.F. Howells, A shallow marine volcaniclastic facies model: an example from sedimenatery rocks bounding the subaqueously welded Ordovician Garth Tuff, North Wales, U.K., *Sed. Geol.*, 74, 217-240, 1991.

Gilbert, J.S, and S.J. Lane, The origin of accretionary lapilli, *Bull. Volcanol.,*56, 398-411.

Kokelaar, B.P., The mechanism of Surtseyan volcanism, *Jour. Geol. Soc. London*, 140, 939-944, 1983.

Kokelaar, B.P., Magma-water interactions in subaqueous and emergent basaltic volcanism, *Bull, Volcanol.*, 48, 275-289, 1986.

Kokelaar, B.P., and G.P. Durant, The submarine eruption and erosion of Surtla (Surtsey), Iceland, *Jour. Volcanol. Geotherm. Res.,* 19, 239-246, 1983.

Lowe, D.R., Sediment gravity flows: II. Depositional models with special reference to the deposits of high-density turbidity currents, *Jour. Sed. Petrol.*, 52, 279-297, 1982.

Lowe, D.R., Suspended-load fallout rate as an independent variable in the analysis of current structures. *Sedimentology*, 35, 765-776, 1988.

Martin, U., and J.D.L. White, Depositional and eruptive mechanisms of density current deposits from a submarine vent at the Otago Peninsula, New Zealand, in *Particulate Gravity Currents*, edifted by W.D. Mccaffrey, B.C. Kneller, and J. Peakall, *Spec. Publs. int. Ass. Sediment.*, 31, 245-259, 2001.

Moore, J.G., Structure and eruptive mechanisms at Surtsey volcano, Iceland, *Geol. Mag.*, 122, 649-661, 1985.

Mueller, W., and J.D.L. White, Felsic fire-fountaining beneath Archean seas: pyroclastic deposits of the 2730 Ma Hunter Mine Group, Quebec, Canada, *Jour. Volcanol. Geotherm. Res.*, 54, 117-134, 1992.

Mueller, W., E.H. Chown, and R. Potvin, R, Substorm wave-base felsic hydroclastic deposits in the Archean Abitibi belt, Canada. *Jour. Volcanol. Geotherm. Res.*, 60, 273-300, 1994.

Mueller, W. U., A.A. Garde, and H. Stendal, Shallow-water, eruption-fed, mafic pyroclastic deposits along a Paleoproterozoic coastline; Kangerluluk volcano-sedimentary sequence, Southeast Greenland, *Precamb. Res.*, 101, 163-192, 2000.

Mueller, W.U., J. Dostal, and H. Stendal, Inferred Paleoproterozoic arc rifting along a consuming plate margin: insights from the stratigraphy, volcanology and geochemistry of the Kangerluluk sequence, southeast Greenland, *Intern. Jour. Earth Sci.*, 91, 209-230, 2002a.

Mueller, W. U., J.D.L. White, and P.L. Corcoran, Dynamic evolution of the submarine Oamaru volcanic complex: syn-eruptive collapse, unconformities, eruption-fed density currents, and backset strata, *AGU Chapman Conference on Explosive Subaqueous Volcanism*, Dunedin, New Zealand, Fieldtrip guidebook, pp 1-10, 2002b.

Schumacher, R., and H.-U. Schmincke, Internal structure and occurrence of accretionary lapilli- a case study at Laacher See volcano, *Bull. Volcanol.*, 53, 612-634, 1991.

Schumacher, R., and H.-U. Schmincke, Models for the origin of accretionary lapilli, *Bull. Volcanol.*, 56, 626-639, 1995.

Smellie, J.L., and M.J. Hole, Products and processes in Pliocene-Recent, subvaqueous to emergent volcanism in the Antartic Peninsula: examples of englacial Surtseyan volcano construction, *Bull. Volcanol.*, 58, 628-646, 1997.

Sohn, Y.K., Depositional processes of submarine debris flows in the Miocene fan deltas, Pohang Basin, SE Korea with special reference to flow transformation, *Jour. Sediment. Res.*, 70, 491-503, 2000.

Sohn, Y.K., and S.K. Chough, Depositional processes of the Suwolbong tuff ring, Cheju island (Korea), *Sedimentology*, 36, 837-855, 1989.

Staudigel, H., and H-U. Schmincke, The Pliocene seamount series of La Palma/Canary Islands, *J. Geophys. Res.*, 89, 11195-11215, 1984.

Stendal, H., W. Mueller, N. Birkedal, E.I. Hansen, and C. Østergaard, Mafic igneous rocks and mineralization in the Paleoproterozoic Ketilidian orogen, South-east Greenland: project SUPRASYD 1996, *Geol. Greenland Sur. Bull.* 176, 66-74, 1997.

Stix, J., Flow evolution of experimental gravity currents: implications for pyroclastic flows at volcanoes, *Jour. Geol.*, 109, 381-398, 2001.

Thorarinsson, S., Surtsey. The New Island in the North Atlantic, New York, *The Viking Press*, 47pp, 1967.

Walker, G. P. L., Explosive volcanic eruptions—a new classification scheme, *Geol. Rundschau*, 62, 431-446, 1973.

Waters, A.C., and R.V. Fisher, Base surges and their deposits: Capelinhos and Taal volcanism, *J. Geophys. Res.*, 76, 5596-5614, 1971.

White, J.D.L., Pre-emergent construction of a lacustrine basaltic volcano, Pahvant Butte, Utah (USA), *Bull. Volcanol.*, 58, 249-262, 1996.

White, J.D.L., Subaqueous eruption-fed density currents and their deposits, *Precamb. Res.*, 101, 87-109, 2000.

Wohletz, K.H., Mechanisms of hydrovolcanic pyroclast formation: grain-size, scanning electron microscopy, and experimental studies, *Jour. Volcanol. Geotherm. Res.*, 17, 31-63, 1983.

Yamazaki, T., Kato, I., Muroi, I., and Abe, M., 1973, Textural analysis and flow mechanism of the Donzurubo subaqueous pyroclastic flow deposits, *Bull. Volcanol.*, 37, 231-244, 1973

Wulf Mueller, Sciences de la terre, Université du Québec à Chicoutimi, Chicoutimi, Québec, Canada G7H 2B1, E-mail: wmueller@uqac.uquebec.ca

Basaltic Lava Balloons Produced During the 1998-2001 Serreta Submarine Ridge Eruption (Azores)

João L. Gaspar, Gabriela Queiroz, José M. Pacheco, Teresa Ferreira, Nicolau Wallenstein, Maria H. Almeida, and Rui Coutinho

Centro de Vulcanologia e Avaliação de Riscos Geológicos, Universidade dos Açores, Ponta Delgada, Portugal

In December 1998 a volcanic eruption started about 10 km W of Terceira Island, Azores, on the so-called Terceira Rift. The eruptive vents were located in the Serreta Submarine Ridge, at depths ranging from 300 to 1000 meters. The observed eruption was preceded by a small seismic crisis, and lasted for more than two years. The tectonic setting of the eruption site, the alignment of the eruptive vents and the type of volcanic products point towards a basaltic fissure eruption. The most striking features formed during this eruption were "lava balloons". These hollow structures, spherical to ellipsoidal in shape, are interpreted as the result of puffing up of gas under the plastic surface of lava at vent level. Degassing of a very fluid, gas-rich magma within or beneath submarine lava lakes and/or during lava fountaining episodes is believed to be the process that generates such structures.

INTRODUCTION

Since the settlement of the Azores islands, in the middle of the 15th century, more than twenty volcanic eruptions have been registered in this Atlantic region (*Weston*, 1964; *Van Padang et al.*, 1967) (Fig. 1) along the main WNW-ESE regional tectonic trend (*Machado*, 1959; *Searle*, 1980). The first recorded eruption on land occurred sometime between 1439 and 1443 inside the caldera of Furnas volcano on S. Miguel Island (*Queiroz et al.*, 1995). More recently the 1957–58 Capelinhos eruption, at the northwestern end of Faial Island, became one of the most well-known Azorean eruptions (*Castello Branco*, 1959).

During the past five hundred years several volcanic eruptions have been reported from the Azores Sea. Some of them were described as ephemeral episodes of vapour and gas emissions with minor solid volcanic products, which remained on the sea surface for only short periods of time. This was the case for the 1867 eruption on the Serreta Submarine Ridge (*Zbyszewsky*, 1967). More violent events accompanied growth of temporary islands, as in 1720, on D. João de Castro Bank, between Terceira and S. Miguel islands (*Canto*, 1879; *Chaves*, 1960) and in 1811, during the Sabrina eruption, which occurred about 2 km W of S. Miguel (*Canto*, 1879; *Van Padang et al*, 1967; *Queiroz*, 1997).

Some aspects of the 1998–2001 Serreta Submarine Ridge eruption were unique, and may contribute to the understanding of other reported submarine events that have occurred in historical times. In this paper we discuss the genesis of low-density basaltic lava structures produced during the eruption, which are here referred to as "lava balloons".

DIRECT OBSERVATIONS

Seismic events related to the eruption were first detected on the 23rd of November 1998 by the Azores Seismological

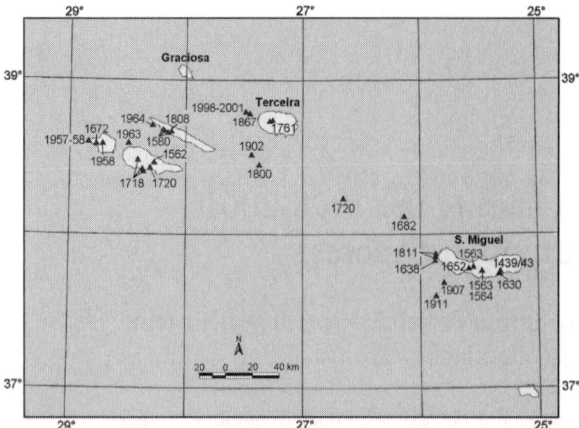

Figure 1. Location of historical eruptions in the Azores region (data from *Weston*, 1964; *Queiroz et al.*, 1995).

Surveillance Network and were located on the Serreta Submarine Ridge, W of Terceira Island (SIVISA, 1998). The activity increased in the following days, reaching a maximum of about four hundred micro-earthquakes on November 29th (Fig. 2). The seismicity then dropped to normal levels, with new peaks of tectonic activity registered only in the middle of December. No seismic events were felt during this precursory crisis.

The first observations of the eruption at the sea surface were made by a group of fishermen on the dawn of December the 18th. The activity was located 10 km W of Terceira Island and was intermittently visible until April 2000. After that date short-lived steam emissions were witnessed in late 2000, and during August 2001.

Early accounts of the volcanic activity referred the presence of floating and steaming dark objects that were first considered to be "whales or tree trunks". Approaching the site, fishermen realised that the observed material consisted of "hot steaming stones whose high temperature caused minor damage to the fishing ropes". In the area a large quantity of dead or injured fish appeared at the sea surface. At night fishermen from another boat, travelling from Graciosa Island to Terceira, reported the presence of "fire coming out from the seawater spreading on the air like sparks of fireworks". Until the end of 1998 only one similar episode was observed from land, on the night of December 23. Scientific missions organised during this period were fruitless due to the discontinuous character of the phenomenon.

The authors took the first images of the event on January 8, 1999, during a helicopter flight. At that time, dozens of lava balloons floating in an area larger than 20.000 m² were releasing thick columns of steam as a result of interaction between seawater and the lava (Fig. 3). This scenario, frequently observed during the following months, was characterised by batches of lava balloons and gas bubbles gently rising towards the surface. As the lava proceeded to cool the steam column started diminishing and sometimes completely disappeared. The lava balloons floated, generally for no more than 15 minutes, before sinking. Sometimes lava balloons exploded, projecting fragments for distances of several tens of metres. Measured temperatures on the interior wall of the lava balloons reached more than 900°C. Sporadically, these fields of lava balloons were pierced by black jets of volcanic ash and water that projected several meters above the sea surface.

The eruption proceeded in periods of different intensity, defined by significant changes in the amount of floating lava balloons and gas bubbles. Clusters of emerging lava balloons and gas bubbles were located using the Global Positioning System and Differential Global Positioning System, and persisted in individual locations for several days, or even weeks, independently of ocean current conditions (Fig. 4). In the early eruptive stages the simultaneous presence of several persistent clusters of floating lava balloons clearly showed an alignment along a NE–SW direction, spanning an area with depths ranging from 300 to more than 1000 metres. After March 2000 some clusters were concentrated on the northern part of the trend, often defining a NW–SE alignment, in an area with an average depth of 400 metres.

SAMPLE CHARACTERISATION

During the development of the eruption hundreds of floating lava balloons were observed at the sea surface. Their longest axes measured from 0.4 up to approximately 3 meters. Regardless of their size, all lava balloons had

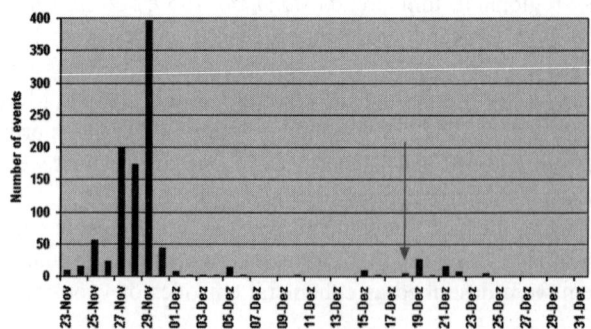

Figure 2. Daily number of micro-earthquakes associated with the Serreta Submarine Ridge volcanic eruption registered by the Azores Seismological Surveillance Network in 1998. The arrow indicates the first observation of the eruption.

spherical to ellipsoidal shapes, and no angular blocks were ever observed.

Several samples were collected, with four lava balloons preserved almost intact (Table 1). The balloon-like structure consists of a thin lava shell surrounding a closed hollow interior (Fig. 5), normally corresponding to a single large vesicle (Fig. 6), or a few large convoluted vesicles (Fig. 7).

Two main layers form the lava balloons' shells: an outer layer (OL) and an inner layer (IL) (Fig. 8). The total thickness of the crust is in general 3 to 8 cm. The OL consists of a thin glassy and very vesicular skin (vesicularity over 65%), brownish to greenish in colour, with well-developed striations in different directions. Because it is very fragile, it is easily detached from the balloons, and was thus absent from many samples. SEM analyses show very fine and delicate fluidal shapes on this outer surface (Fig. 9). Vesicles are elongated and millimetric in size, although, near the interface with the inner layer there are occasionally some larger vesicles that reach several centimetres in diameter. The transition between layers OL and IL is marked, generally, by a thin horizon of white and orange deposits.

Layer IL is less fragile, and is divided into three different sections that differ in their vesicularity. The outermost section (IL1) is usually the thinnest and is very vesicular (vesicularity over 65%). The middle section (IL2) is the least vesicular (vesicularity often less then 40%). The inner section (IL3) is generally the thickest and is also very vesicular (more then 60%). The vesicles of IL3 are the largest of layer IL, and their size often increase towards the interior of the lava balloon.

The petrography is marked by the presence of phenocrysts of olivine (3 mm) and pyroxene (4 mm)—both with opaque inclusions—and plagioclase. Transparent phenocrysts of plagioclase reach dimensions of 10 mm.

Figure 3. Photo of floating lava balloons taken during a helicopter flight on January 8, 1998.

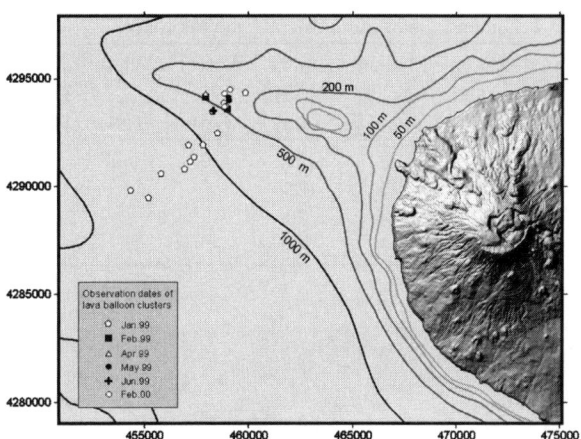

Figure 4. Distribution of lava balloons clusters at sea surface.

Microphenocrysts of olivine, pyroxene, and plagioclase are also present. Within the vitric matrix there are microlites of plagioclase, olivine and pyroxene.

Whole-rock chemical analyses show that the products of this eruption are alkali basalts (Table 2), with a composition similar to that of other lavas erupted along the Terceira Island basaltic chain and during other Azorean historical eruptions (Self, 1974; Self and Gunn, 1976; Flower et al., 1976).

DISCUSSION

The seismic activity registered on late November 1998 is interpreted to have been the result of fracturing due to magma ascent through the submarine volcanic chain that extends towards the W of Terceira Island (Gaspar & Wallenstein, 1999). The low number of seismic events registered during the eruption, as well as their low magnitude, can be explained as the result of the injection of a very fluid magma having become localised along a pre-existing fault system.

The 1998–2001 eruption occurred along the same tectonic lineament as the 1867 submarine volcanic eruption (Zbyszewsky, 1967) but, at that time, a much stronger seismic crisis preceded the event, presumably because more new fracturing was taking place. In addition, fresh basaltic glass was dredged in the same area by an oceanographic vessel a few years ago, and must have been produced during an undetected recent eruption (M. Miranda, Univ. Lisboa, personal communication, 1998). These two facts reinforce the inference that a well-developed volcano-tectonic system already existed in the area at the time of the 1998–2001 eruption.

The clusters of floating lava balloons and gas bubbles represent the sea surface manifestations of actively erupting

Table 1. Dimensions of samples collected unbroken.

Sample	Major Axis (cm)	Medium Axis (cm)	Minor Axis (cm)	Weight (kg)	Volume (m³)	Total Density (g/cm³)
S1	47	32	17	20	0,28	0,7
S2	40	26	25			
S3	43	34	30			
S4	84	38	35			

Figure 5. Sample collected on February 10th, 1999. Note the hollow interior limited by a thin shell that encapsulates the gas.

submarine vents at depth, as deduced from their steady location trough time regardless of the oceanic currents' directions. Thus, the well-defined NE–SW and NW–SE trends traced by the distribution of such clusters are believed to reflect the main tectonic directions. The existence of simultaneously active vents along such volcano-tectonic alignments, and the magma composition, are consistent with a basaltic fissure eruption.

The most noteworthy features observed during the Serreta Submarine Ridge eruption were the floating basaltic lava balloons. The occurrence of floating basaltic blocks has been reported associated with lava flows entering the sea or, less commonly, related with submarine eruptions (Siebe *et al.*, 1995; Ken Hon, Univ. Hawaii, personal communication, 2002). The formation of floating lava blocks, where lava enters the sea, results from the incorporation of seawater into the lava, either (1) by the entrapment of small amounts of water as the lava pours onto a beach with waves, enhancing the development of vapour bubbles in the lava blocks, or (2) by the entrapment of water in a lava tube as waves break over the subaerial portion of the tube. In this case, given the right conditions, the vaporised water can be transported in the submarine portion of the tube, down to the lava front, where pillows are being formed, with the result that

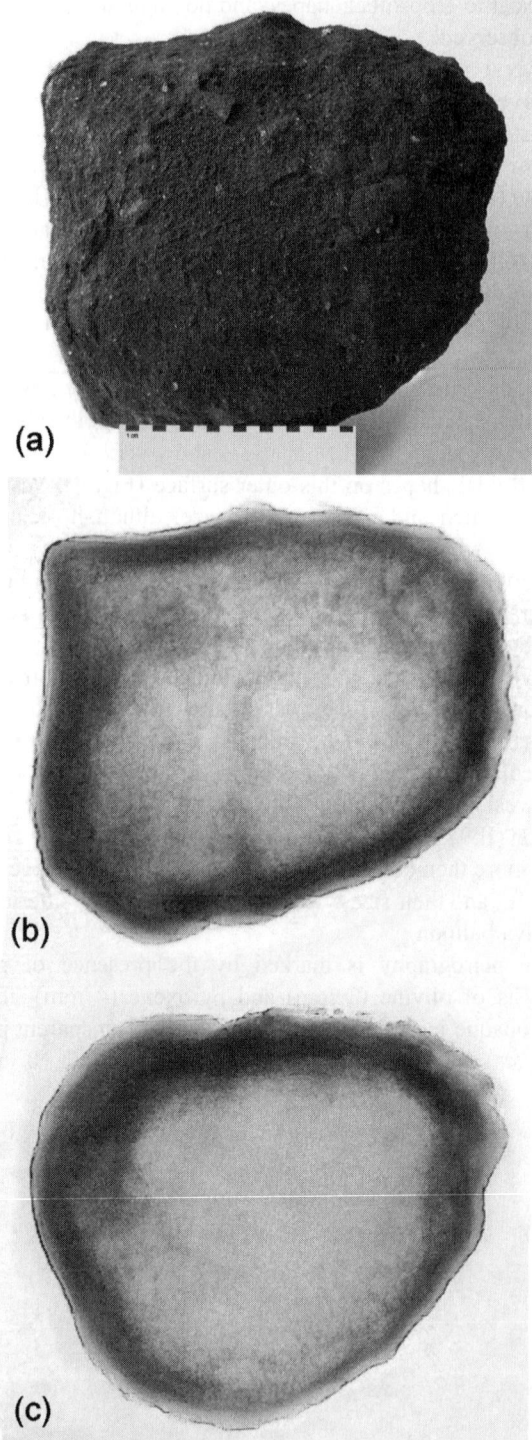

Figure 6. Photograph (a) and XR images (b, c) of sample S3. XR images are perpendicular sections and darker areas are the most opaque to XR. The interior of the sample is made of one single large hollow. (Scale is in cm).

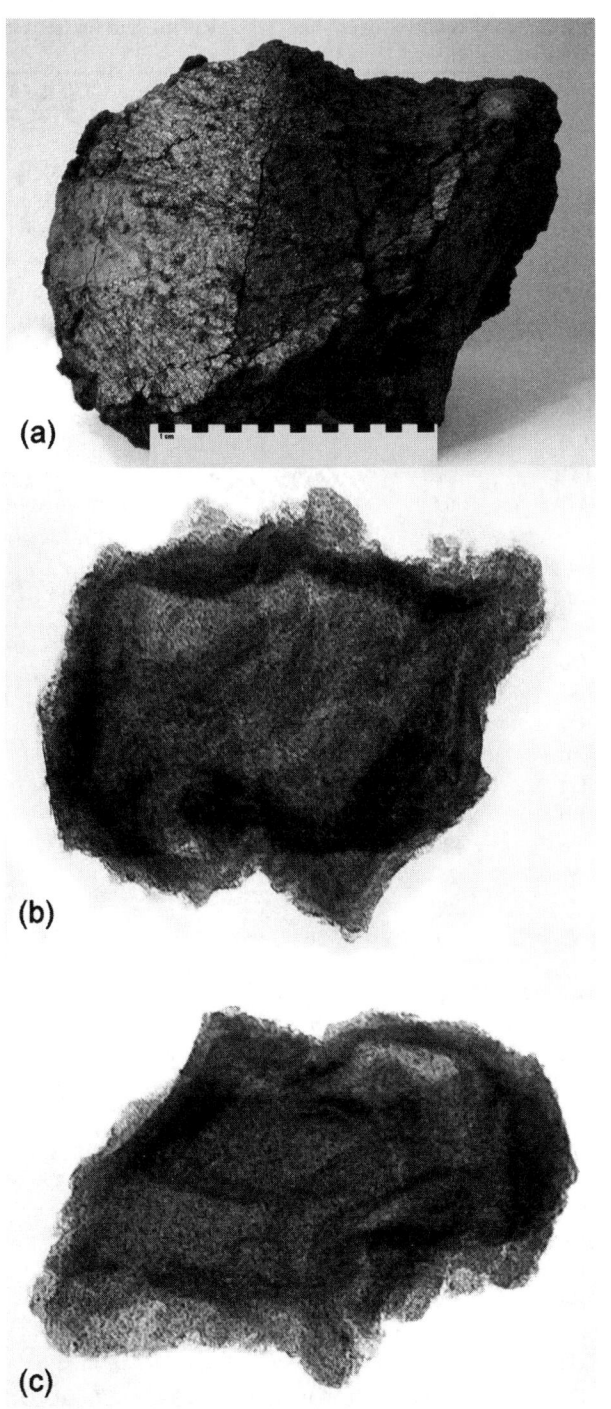

Figure 7. Photograph (a) and XR images (b, c) of sample S2. XR images are perpendicular sections and darker areas are the most opaque to XR. The interior of the sample presents several septa interpreted as vesicle walls. (Scale is in cm).

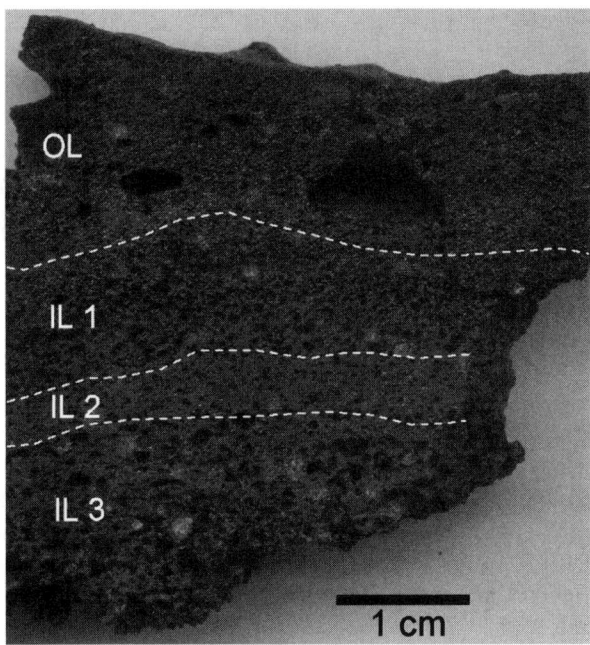

Figure 8. Balloon lava shell section with outer layer OL and inner layers IL1, IL2 and IL3

the vapour bubbles are included in the pillows, which may became buoyant (Ken Hon, Univ. Hawaii, personal communication, 2002).

The production of floating basaltic blocks in a submarine environment was reported during the 1993–94 eruption, near Socorro Island, Mexico (Siebe et al., 1995). Floating scoria and reticulite blocks formed during this event were interpreted as being related to lava fountaining episodes from vents located at 30 and 210 m depth (Siebe et al., 1995). These scoria blocks had an external ropy texture, were highly vesicular, with vesicles up to 10 cm, and commonly had reticulite masses in their cores.

During the Serreta Submarine Ridge eruption, lava balloons were produced away from the coastal land/sea interface, in a submarine setting at depths exceeding 300 meters. The steady location of the rising gas bubbles and the basaltic floating lava balloons throughout most of the eruption, suggests that they were originated at stable vents and not at the fronts of extensive lava flows. The hydrostatic pressures at the depths of the Serreta Submarine Ridge eruption range from 3 to about 10 MPa. At these pressures vesiculation of magma is readily achieved (Wilson et al., 1980; Sparks, 1978).

Considering that the magma involved in the Serreta Submarine Ridge eruption was very fluid and gas-rich, as inferred from the surface texture and gas content of the lava

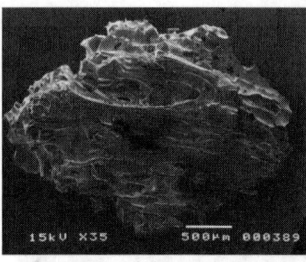
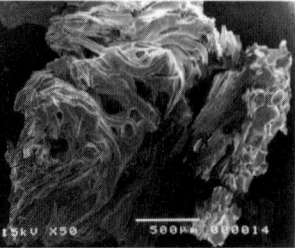

Figure 9. SEM photos of fragments from lava balloon's thin glassy crust. Note the deformation and fine scale fluidal shapes of the magma.

balloons, the segregation and accumulation of gas under a cooler lava crust at vent level would lead to the development of large gas bubbles, as is frequently observed at subaerial lava lakes. At a critical point of accumulation, these large gas bubbles can form blisters that, in a subaqueous setting, detach from the vent as swollen lava balloons that rise by flotation (buoyancy). Submarine lava fountaining can also occur at these depths (Batiza et al., 1984; Smith & Batiza, 1989; Clague et al., 1990 in Clague et al., 2000; Siebe et al., 1995; Clague et al., 2000; Simpson & McPhie, 2001) generating gas-rich lava fragments that can rise by flotation as they degas progressively inwards from their outer surfaces.

Under both conditions, the quenching of the lava at the interface with the seawater results in a thin, initially pliant crust, which acts as an insulating skin allowing high internal temperatures to be maintained and preventing the magmatic gases in the interior from escaping. The magma forming the lava balloons can continue to degas as the balloons rise; the exsolving volatiles further inflate the balloons, which are expanding primarily as a consequence of decompression. This expansion leads to enlargment of the balloons' surface area, and hence the formation of new skin; the well-developed striations observed on the crust of the fragments are considered to have formed as a result of this process. It is the presence of still-molten lava inside the lava balloon that prevents complete rupture of the outer solid crust and allows the formation of the new skin.

When the lava balloons reached the sea surface they floated for several minutes, releasing steam as a result of interaction between seawater and the hot lava crust (Fig.10). Some lava balloons continued to float after the external surface had cooled down, when no more vapour was produced (see Fig.3). Lava blocks sank only when fissures in their crusts allowed the trapped magmatic gas to escape and the seawater to penetrate and fill the balloon cores. The entrance of seawater into the hot interiors of the lava balloons was sometimes an explosive process that projected fragments for distances of several meters.

Table 2. Representative major- and trace-element data for Serreta whole-rock samples.

	TSER-01	TSER-02	TSER-03	TSER-04
Major elements (wt%)				
SiO_2	48.74	47.99	47.46	47.94
TiO_2	2.80	2.70	2.44	2.44
Al_2O_3	14.91	14.94	14.36	14.29
Fe_2O_3	11.66	11.16	10.87	10.66
MnO	0.16	0.16	0.16	0.16
MgO	8.94	8.80	10.52	10.19
CaO	10.39	10.37	10.97	10.92
Na_2O	2.81	2.94	2.62	2.73
K_2O	0.84	0.72	0.67	0.69
P_2O_5	0.41	0.36	0.32	0.30
LOI	-0.76	-0.76	-0.64	-0.81
TOTAL	100.89	99.38	99.74	99.52
Trace elements (ppm)				
V	218	224	207	235
Cr	277	256	383	401
Co	30	41	36	46
Ni	103	89	91	136
Cu	23	33	29	38
Zn	38	86	66	32
Ga	15	18	16	18
Rb	14	15	13	15
Sr	480	447	417	437
Y	25	26	23	25
Zr	200	160	141	119
Nb	26	26	23	25
Ba	193	176	176	178
Hf	3.6	4.1	3.7	3.3
Ta	1.7	2	1.8	1.9
Th	1.9	2.4	2.2	2.1
U	0.8	0.8	0.7	0.8
La	21.0	22.0	20.1	20.7
Ce	45.0	46.3	43.0	43.8
Pr	5.6	5.8	5.3	5.5
Nd	24.0	25.4	23.4	23.8
Sm	5.6	5.9	5.4	5.6
Eu	2.2	2.2	2.0	2.1
Gd	4.8	5.9	5.4	5.7
Tb	0.8	1.0	0.9	0.9
Dy	4.7	5.2	4.7	4.9
Ho	1.0	0.9	0.9	0.9
Er	2.6	2.5	2.4	2.5
Tm	0.3	0.3	0.3	0.3
Yb	1.8	1.9	1.8	1.8
Lu	0.2	0.3	0.3	0.3

Figure 10. Steaming lava balloon just after it reached the sea surface.

CONCLUSIONS

The 1998–2001 Serreta Submarine Ridge eruption took place in a well-developed volcano-tectonic system where other historical events have occurred. The lava-balloon phenomenon presented characteristics of a basaltic fissure eruption, which developed along NE–SW and NW–SE faults with vents located at depths ranging from 300 to more than 1000 meters.

During the eruption a singular type of volcanic product was formed, here referred as "lava balloons". These are closed, hollow, structures, spherical to ellipsoidal in shape, with a thin vesicular crust. It is proposed that the lava balloons originate at the erupting vents, and are related to lava lake crustal dynamics and/or lava fountaining activity.

Despite elevated confining pressures at depth in submarine environments, the abundance of volatiles (juvenile gases and seawater) and low viscosity of alkaline magmas were sufficient to allow the formation of these buoyant lava balloons. While floating lava blocks are commonly associated with subaerial and shallow submarine eruptions, the lava balloons of the 1998–2001 Serreta eruption provide clear evidence that they can be produced by deeper submarine eruptions.

Acknowledgments. This work was supported by the Azores Regional Service of Civil Protection trough Project SIMOVA2. We thank T. Mattox, D. Maicher, J. White, A. Duncan, C. Siebe and S. Self for reviewing the manuscript and R. Batiza and K. Hon for their additional comments and contributions. N. Oskarson performed the ICP chemical analyses at the NVI. SEM analyses were performed at CIRN. The geological surveys were possible thanks to the Portuguese Air Force, the Portuguese Navy and the crew of Maria Medina fishing-boat.

REFERENCES

Batiza, R., Fornari, T.J., Vanko, D.A. and Lonsdale, P., Craters, calderas and hyaloclastites on young Pacific seamounts, *J. Geophys. Res.,* 89, 8371-8390, 1984.

Blackburn, E.A., Wilson, L. and Sparks, R.S.J., Mechanisms and dynamics of strombolian activity, *J. Geol. Soc. Lond.,* 132, 429-440, 1976.

Canto E., Vulcanismo nos Açores desde a época da descoberta até ao presente, In: *Archivo dos Açores,* vol.I, III, IV, V, IX e XI, Typ. do Archivo dos Açores, Ponta Delgada, 1879.

Castello Branco, A., *Le Volcanisme de l'Île de Faial et l'Éruption du Volcan de Capelinhos*, Serviços Geológicos de Portugal, memória nº4, 99p, 1959.

Chaves, F.A., Erupções submarinas nos Açores. Informações que os navegantes podem prestar sobre tal assunto, *Açoreana*, v.5, nº5, 50p., 1960.

Clague, D.A., Davis, A.S. and Bischoff, J.L., Lava bubble-wall fragments formed by submarine hydrovolcanic explosions on Lo'ihi seamount and Kilauea volcano, *Bull. Volcanol.,* 61, 437-449, 2000.

Flower, M.F.J., Schmincke, H.-U. and Bowman, H. Rare earth and other trace elements in historic azorean lavas, *J. Volcanol. Geoth. Res.,* 1, 127-147, 1976.

Gaspar, J.L. and Wallenstein, N., Submarine eruption west of Terceira Island (Azores), Preliminary report, Departamento de Geociências da Universidade dos Açores, in: *Bull. Global Volcanism Network*, vol.24, nº1, 1999.

Machado, F., Submarine pits of the Azores plateau, *Bull. Volcanol.,* serie II, tome XXI, 109-116, 1959.

Queiroz, G., *Vulcão das Sete Cidades (S. Miguel, Açores): História Eruptiva e Avaliação do Hazard*, Tese de doutoramento no ramo de Geologia, especialidade de Vulcanologia, Universidade dos Açores, Departamento de Geociências, 1997.

Queiroz, G., Gaspar, J.L., Cole, P., Guest, J., Wallenstein, N., Duncan, A. and Pacheco, J., Erupções vulcânicas no Vale das Furnas (ilha de S. Miguel, Açores) na primeira metade do século XV, *Açoreana*, VIII(1), 159-168, 1995.

Searle, R, Tectonic pattern of the Azores spreading centre and triple junction, *Earth Planet. Sci. Lett.,* 51, 415-434, 1980.

Self, S., *Recent volcanism on Terceira, Azores*, PhD thesis, London Univ., Imperial College, 1974.

Self, S. and Gun, B.M., Petrology, volume and age relation in alkaline and peralkaline volcanics from Terceira, Azores, *Contrib. Mineral. Petrol.,* 54, 293-313, 1976.

Siebe, C., Komorowski, J.-C., Navarro, C., McHone, J., Delgado, H. and Cortés, A., Submarine eruption near Socorro Island, Mexico: Geochemistry and scanning electron microscopy studies of floating scoria and reticulite, *J. Volcanol. Geoth. Res.,* 68 (4), 239-271, 1995.

Simpson, K. and McPhie, J., Fluidal-clast breccia generated by submarine fire fountaining, Trooper Creek Formation, Queensland, Australia, *J. Volcanol. Geoth. Res.,* 109, 339-355, 2001.

SIVISA, *Boletim Sísmico Preliminar. Meses de Novembro e Dezembro*. Ed. Instituto de Meteorologia e Centro de Vulcanologia da Universidade dos Açores, 1998

Smith, T.L. and Batiza, R., New field and laboratory evidence for the origin of hyaloclastite flows on seamount summits, *Bull. Volcanol.*, 51, 96-114, 1989.

Sparks, R.S.J., The dynamics of bubble formation and growth in magmas: a review and analysis, *J. Volcanol. Geoth. Res.*, 3, 1-37, 1978

Van Padang, M.N., Richards, A.F., Machado, F., Bravo, T., Baker, P.E. and Le Maitre, R.W., *Atlantic Ocean, Catalog of active volcanoes of the world*, IAVCEI, Rome, 128p., 1967.

Weston, F.S., List of recorded volcanic eruptions in the Azores with brief reports. *Bol. Mus. Lab. Min. Geol. FCUL*, 10(1), 3-18, 1964.

Wilson, L., Sparks, R.S.J. and Walker, G.P.L., Explosive volcanic eruptions—IV. The control of magma properties and conduit geometry on eruption column behaviour, *Geophys. J. R. Astr. Soc.*, 63, 117-148, 1980.

Zbyszewski, G., As observações de F. Fouqué sobre o vulcanismo dos Açores, *Bol. Núcleo Cultural da Horta*, Sep. V.4, n°2-3, 17-95, 1967

Gabriela Queiroz, João L. Gaspar, José M. Pacheco, Maria H. Almeida, Nicolau Wallenstein, Rui Coutinho, and Teresa Ferreira, Centro de Vulcanologia e Avaliação de Riscos Geológicos, Universidade dos Açores, Ponta Delgada, Portugal.

Subaqueous Pumice Eruptions and Their Products: A Review

K. Kano

Institute of Geoscience, Geological Survey of Japan, AIST, Tsukuba, Japan

Subaqueous pumice eruptions include mainly 1) subaqueous plinian-type eruptions, 2) subaqueous flow-generating eruptions, 3) explosive bulk interactions of vesicular magma and water, and 4) non-explosive release of pumice clasts from the thermally and mechanically brecciated surface of growing vesicular domes. Subaqueous plinian-type eruptions form normally graded deposits as the umbrella region is gravitationally inverted to a density current. The column of subaqueous flow-generating eruptions may expand by water-ingestion and heating the water to steam, but eventually collapses with increasing water-ingestion to produce poorly sorted, inversely-to-normally graded or massive, and partly stratified density-current deposits. Pumice eruptions driven by bulk interaction of a well-vesicular magma with the ambient water recurs as long as the magma is supplied to the watery setting. Resulting dilute currents form thinly-bedded sequences of inversely to normally graded, stratified, and well-sorted deposits. Pumice blocks and lapilli can be carried to the surface by buoyancy during non-explosive thermal and mechanical brecciation of pumiceous lava. Fallout of these pumices produces fairly well sorted, inversely graded or massive units. Magmatic pumice eruptions are plausible down to a water depth of 500–1000 m. Phreatomagmatic pumice eruptions occur mainly at domes growing in water shallower than 200-300 m, or slightly emergent. Growth of a lava dome to shallow water lessens the confining pressure and allows explosive magma-water interaction or magma vesiculation to form a pyroclastic cone on the dome.

1. INTRODUCTION

Pumice is a highly vesicular material in terms of vesicularity index [*Houghton and Wilson,* 1989], and is commonly produced by catastrophic fragmentation of vesicular magma or lava mainly of silicic composition. Pumice eruptions, which eject large volumes of pumice, can occur subaqueously. There is also a large volume and variety of subaqueous pumice deposits formed from subaerial eruptions that discharged their products directly into water. Subaqeous eruptions of pumice, however, are not well understood. This is because 1) observations are limited mainly to the emergent or shallow water eruptions, 2) an expanding, turbulent eruption mass prevents close observation, and 3) subsequent recovery of the products has not allowed for complete reconstruction of eruptive processes. Models of subaqueous pumice eruptions have been thus proposed mainly through close examination of the ancient deposits now exposed on land. Documented cases of subaqueous pumice eruptions are relatively rare and incomplete, thus theoretical and experimental approaches are required to fill the gap of our knowledge between the observations of eruptions and the products. In this context, this paper reviews the direct observation of subaqueous pumice

Explosive Subaqueous Volcanism
Geophysical Monograph 140
Copyright 2003 by the American Geophysical Union
10.1029/140GM14

eruptions and the occurrence of pumice deposits emplaced in water, and discusses theoretically plausible eruptive mechanisms and the factors that control the behavior of such eruptions.

2. OBSERVATION OF ERUPTIONS

Observations of subaqueous pumice eruptions are limited to two cases. The first case is phreatomagmatic eruption at shallow depths, and the second case is silent upwelling of pumice blocks and lapilli, discolored hot water, and bubbles from deep water (Table 1). Brief descriptions of typical cases are given below.

2.1. 1952-1953 Eruption of Myojinsho

Myojinsho is a submarine volcano located on the northeast rim of a submerged, 8-km x 10-km caldera, which is 420 km south of Tokyo. The present shallowest point of the summit is 43 m below sea level.

The 1952–1953 eruption of Myojinsho involved dome growth, submarine explosions, and emergent black tephra jets and ash plumes [*Niino and Kumakori,* 1953; *Morimoto and Ossaka,* 1955; *Fiske et al.,* 1998]. Dacitic domes 300–500 m across and 200 m high emerged above sea level three times and collapsed by subsequent phreatomagmatic explosions. Submarine explosions were recognized with upwelling of discolored water, bubbles, and sparse tephra. Tephra jets intermittently projected into the air and collapsed to form surges and ash plumes. With intermittent black tephra jets (Figure 1), pumice lapilli and ash and minor pumice blocks were directly projected into the sea and flowed down the submarine flank of Myojinsho, or floated on the sea and drifted away with ocean currents [*Fiske et al.,* 1998]. Surges traveling over the sea and ash plumes contributed additional fall of pumice lapilli and ash into the sea.

2.2. 1985-1986 Eruption of Fukutoku-Oka-no-Ba

Fukutoku-Oka-no-Ba is a submarine volcano located 1,300 km south of Tokyo. The basal diameter is 5–6 km and the elevation from the basement is approximately 400 m. The present vent is located 14 m below sea level on a dome 250 m above the basement. The summit has emerged above sea level three times, in 1904, 1914 and 1985–1986 with repeated submarine explosive eruptions.

According to the joint party of the *Hydrographic Department, Maritime Safety Agency, Japan, Faculty of Engineering, Tokyo Institute of Technology, and Faculty of Science, Okayama University* [1986], the 1985–1986 eruption started with a submarine ash plume, followed by emergent black tephra jets. Dacitic pumice lapilli and ash were spouted from the submarine vent and accumulated around the vent, or remained afloat on the sea and drifted away with the ocean currents [*Kato,* 1988]. At the most active stage of the eruption, repeated phreatomagmatic explosions built a tuff ring partly emergent with a submerged crater 300 m across (Figure 2). This tuff ring was eroded by wave action as the activity waned.

The 1904 and 1914 eruptions were similar in eruption mode, but were larger in scale than the 1985–1986 eruption [*Ossaka,* 1991]. The 1904 eruption produced a partly submerged, asymmetrical tuff ring above sea level and 4.5 km across. The 1914 eruption also produced a tuff ring 130 m above sea level and 900 m by 1300 m in diameter. These tuff rings were subsequently eroded below sea level by wave action.

2.3. 1984 Eruption of Kaitoku Seamount

Kaitoku Seamount is a submarine volcano located 1,000 km south of Tokyo. The basal diameter is 40 km and the elevation from the basement is approximately 2000 m. This seamount comprises two major edifices, Higashi-Kaitoku-Ba and Nishi-Kaitoku-Ba. The summit areas of the two edifices are 200 to 400 m below sea level with the shallowest points at ca. 100 m.

The 1984 eruption occurred at Higashi-Kaitoku-Ba from March to July. During the eruption, discolored warm water intermittently spouted, occasionally with pumice clasts [*Hydrographic Department, Maritime Safety Agency, Japan and Faculty of Engineering, Tokyo Institute of Technology,* 1984; *Tsuchide et al.,* 1985]. Sparse black pumice clasts up to 1–3 m in length were observed when discolored water spouted every 5 to 10 minutes. They were afloat with steaming for 5 minutes or less and sunk into the sea as steaming ceased. A subsequent bathymetric survey by *Hydrographic Department, Maritime Safety Agency, Japan* [1994] disclosed a cone 0.7 km by 1.1 km in diameter and 150 m high with a summit 95 m below sea level and a slope angle of 20–25°. This cone is located just beneath the center of spouting water, and is likely to comprise the ejecta from the 1984 eruption. The pumice clasts that reached to the sea surface appear quite small in volume and many of the vesicles are isolated. These observations, and that eruptions recurred at short time intervals [*Tsuchide et al.,* 1985], collectively suggest a phreatomagmatic eruption.

2.4. 1934-1935 Eruption of Shin-Iwojima, Kikai Caldera

Kikai submarine caldera is located 100 km south of Kagoshima, Kyushu, southwest Japan. Satsuma-Iwojima,

Table 1. Subaqueous pumice eruptions and their deposits

No	Eruption name or site, or deposit	Water depth (m)	Rock composition	Clast size	Clast morphology	Deposit	Eruption mode	References
1	1986 Fukutoku-Oka-no-Ba eruption off Minami-Iwojima, Izu-Ogasawara Arc	around sea level	Dacite	Mainly ash to pumice lapilli	Not described	Pumice rafts	Tephra jets mainly breached the sea surface from the dome top	Hydrographic Department, Maritime Safety Agency, Japan et al., 1986
2	A seamount in the area of the South Sandwich Islands	27 m	Rhyolite	Mainly pumice lapilli	Not described	Pumice rafts	Upwelling of pumice	Gass et al., 1963
3	1952–1953 Myojinsho eruption, Izu-Ogasawara Arc	Shallower than 100–200 m	Rhyolite	Ash to lapilli, with minor blocks	Not described	Not recovered	Tephra jets mainly breached the sea surface from the dome top	Niino and Kumakori, 1953; Morimoto and Ossaka, 1955; Fiske et al., 1998
4	Devonian Bunga Beds, southern Australia	100–200 m	Rhyolite	Ash to lapilli, with minor blocks	Not described	Stratified pumice breccias and crystal-rich tuff	Phreatomagmatic eruption from the dome top	Cas et al., 1990
5	Wakamiko caldera, Kagoshhna Bay, SW Japan	160–200 m	Rhyolite	Ash to lapilli, with minor blocks	Platy, bubble wall shards, and blocky pumice	Stratified to inversely graded deposits	Subaqueous large-volume current-forming eruption	Kano et al., 1996
6	Upper Miocene-Pliocene Shirahama Group, Japan	Above the wave base	Silicic, moderately vesicular	Ash to small lapilli	Platy, bubble wall shards, and blocky pumice	Massive or graded to stratified deposits with well sorted tephra	Subaqueous plinian eruption	Cashman and Fiske, 1991; Tamura et al., 1991
7	Yali, eastern Agean, Greece	15–200 m	Rhyolite and minor dacite	Ash to block	Platy, bubble wall shards, and blocky pumice	Fines-poor pumice breccia, and graded stratified lapilli tuff to tuff	Subaqueous phreatomagmatic eruption and non-explosive syn-eruptive dome collapse	Allen and McPhie, 2000
8	1984 Kaitoku Seamount eruption, Izu-Ogasawara Arc	95–240 m	Dacite	Black pumice up to 1–3 m in length	Not described	Not recovered	Upwelling of discolored water occasionally with pumice, and growth of a cone 0.7 km by 1.1 km in diameter and 150 m high	Hydrographic Department, Maritime Safety Agency, Japan and Faculty of Engineering, Tokyo Institute of Technology, 1984; Tsuchide et al., 1985; Hydrographic Department, Maritime Safety Agency, Japan, 1994
9	2002 eruption west of Vava'u Islands, Tonga	Shallower than 200–300 m	Not described	Pumice and ash	Not described	Not recovered	Subaqueous phreatomagmatic eruption	Taylor, 2002
10	1934–1935 Shin-Iwojima eruption in the Kikai caldera, SW Japan	Shallower than 300 m	Rhyolite	Pumice up to 10 m in length	Not described	Pumice rafts and fallout	Upwelling of pumice, followed by phretomagmatic eruption and lava effusion	Tanakadate, 1935
11	1953–1957 Tuluman eruption in the northern Bismarck Sea		Rhyolite	Ash to block	Not described	Pumice rafts and fallout	Upwelling of pumice, followed by phretomagmatic eruption and lava effusion	Reynolds et al., 1980
12	Torishima caldera or Minami-Sumisu caldera, Izu-Ogasawara Arc	300–500 m or shallower	Rhyolite	Ash to lapilli, with minor blocks	Platy, bubble wall shards, and blocky pumice	Fines-depleted poorly sorted massive to inversely graded pumice lapillistone to lapilli tuff	Subaqueous large-volume current-forming eruption	Nishimura et al., 1991
13	1924 submarine eruption off the Iriomote Island Rhyukyu Arc	400–500 m	Rhyolite	Pumice up to 2 or 3 m	Not described	Pumice rafts and perhaps fallout	Upwelling of pumice	Kato, 1991
14	Myojin Knoll caldera. Izu-Ogasawara Arc	500–900 m	Rhyolite	Lapilli to blocks (mainly 4 mm to 40 cm)	Angular to subangular pumice	Fines-depleted, massive to stratified pumice lapillistone and pumice breccias	Subaqueous large-volume current-forming eruption	Yuasa, 1995; Fiske et al., 2001

Table 1. (continued)

No	Eruption name or site, or deposit	Water depth (m)	Rock composition	Clast size	Clast morphology	Deposit	Eruption mode	References
15	Archean Hunter Mine Group Quebec, Canada	Sub-wave base, deeper than 200 m	Felsic, moderately vesicular	Ash to block	Fluidal to blocky moderately vesicular	Massive breccias, diffusely stratified matrix-rich and -poor lapilli tuffs, and intebedded tuff and lapilli tuff	Subaqueous lava fountain	*Mueller and White*, 1992
16	Late Triassic to Eraly Jurassic Vandever Mountain Tuff, Mineral King, California	Deeper than 150 m	Rhyolite	Ash to small lapilli	Mainly bubble wall shards, and vesicular lapilli	Massive welded pumice-lapilli tuff with many mega- and meso-blocks, bedded tuff, lapilli tuff and tuff breccia, and tuff	Subaqueous large-volume current-forming eruption	*Kokelaar and Busby*, 1992
17	Tokiwa Formation, Japan	Shallower than 150–500 m	Dacite and rhyolite	Ash to lapilli	Platy, bubble wall shards, and blocky pumice	Massive or graded, stratified deposits	Subaqueous plinian eruption or large-volume current-forming eruption	*Fiske and Matsuda*, 1964
18	Healy caldera, southern Kermadec Arc	550–1000 m	Rhyolite	Ash to lapilli	Not described	Not described	Subaqueous large-volume current-forming eruption	*Wright et al.*, 2003
19	Okinawa Trough 100 km west of Okinawa Island, SW Japan	1500–1600 m	Silicic dacite	Pumice blocks 1-3 m long and 1 m or less thick	Not described	Pumice blocks occurs sparsely or in swarm on the upper slopes of presumed lava domes	Subaqueous vulcanian or non-explosive dome collapse	*Kato*, 1987
20	Early Miocene Tayu Tuff, Shimane Peninsula, SW Japan	200–1000 m or shallower	Rhyolite and minor andesite	Ash to block with pumice up to 2 or 3 m	Not described	Inversely to normally graded, stratified with basal lithic breccia	Subaqueous vulcanian?	*Kano*, 1996

Takeshima, and other post-caldera volcanoes lie along the periphery of the submerged caldera floor [*Matsumoto*, 1943].

According to a detailed description by *Tanakadate* [1935a, b], the 1934–1935 eruption started at the northern periphery of the caldera floor at a depth of 300 m, adjacent to the emergent, active rhyolitic domes of Satsuma-Iwojima and Takeshima Islands. At the beginning of the submarine eruption, rhyolitic pumice blocks of meter size repeatedly rose in swarms to the sea surface with doming of water to a height of 1–2 m above sea level. The floating pumice blocks heated the ambient seawater and white steam rose in plumes up to a height of 800-1000 m from the pumice swarm (Figure 3). Some floating pumice blocks up to 30 m^3 in volume sunk abruptly upon invasion of water into the hot interiors through cracks opened by water-cooling and internal gas-expansion. However, many pumice blocks and lapilli remained afloat on the sea and drifted away from the source with the ocean currents (Figure 4). The eruption was gentle and calm with little sound. Ships were able to advance through the near-vent area that was covered with pumice rafts. This mode of eruption subsequently changed to phreatomagmatic eruptions with growth of an emergent pyroclastic cone, and further changed to lava effusion over the emerged edifice of Shin-Iwojima Island.

2.5. 1953-1957 Eruption of Tuluman Volcano

Tuluman volcano is located on the southern extension of Lou Island in the northern Bismarck Sea. This volcano consists of 7 cones built by the 1953–1957 eruption of an alkali-

Figure 1. Tephra jets rising to a height of 410 m from Myojinsho at 13:12 JST, September 23, 1952. Soon after this photo was taken by J. Ossaka, the jets collapsed to form a surge running over the sea and a plume, from which ash fell.

Figure 2. Tephra jets from the tuff ring of Fukutoku-Oka-no-Ba. This photo was taken at 9 o'clock, January 21, 1986 by J. Ossaka from a plane of the Maritime Safety Agency, Japan.

rich rhyolitic magma. The eruption started under the sea with upwelling of vesicular lava or pumice clasts of meter size, followed by phreatomagmatic explosions and lava effusion with emergence of the volcanic centers and degassing. These eruption modes described by *Reynolds et al.* [1980] are similar to the 1934–1935 eruption of Shin-Iwojima, Kikai caldera. The water depths at which the eruptions of 7 cones started are not described in the literature.

2.6. 1924 Eruption off the Iriomote Island, SW Japan

The gentle but continuous rise of rhyolitic pumice clasts up to 2 or 3 m across was also observed during the 1924 eruption off Iriomote Island in the Ryukyu Islands of Japan [*Kato*, 1991]. The source vent has not yet been identified, but is most likely a mound 400–500 m below sea level in the reported area of pumice emergence [*Kato*, 1991]. Many pumices remained afloat on the sea and drifted to the north along the coast of Japan for over 1 year.

2.7. 2001 Eruptions West of the Vava'u Islands

According to a summary by *Taylor* [2002], an unnamed submarine volcano west of the Vava'u Islands, Tonga was active from September to November of 2001, being accompanied with upwelling of pumice, discolored water and bubbles. Emergence of an island 3 km long with an ash plume was reported. Subsequently, aircraft observations showed a new submarine bank 2.4 km across at 18.358°S and 174.346°W, a location different from the reported island, and pumice clasts 1-20 cm across stranded along the coast of Kadav and on the south coast of Viti Levu, Fiji.

Figure 3. Pumice swarm rising from a submarine eruption center close to Satsuma-Iwojima Island, with a plume of steam heated from the seawater in direct contact with pumice blocks [*Tanakadate*, 1935a]. The diameter of the swarm is not described, but is likely less than 1 km. This photo was taken in the middle of September, 1934 by the Osaka Mainichi Shinbun.

Some pumice rafts were reported to be more than 100 m in diameter. The reported island was, therefore likely to be a pumice raft. The water depth of the volcano summit prior to this eruption was 200–300 m. Post-eruptive depths have not yet been determined.

3. ERUPTION PRODUCTS

Subaqueous pumice deposits produced by subaqueous eruptions are generally depleted in fines relative to their subaerial counterparts, commonly contain platy or blocky non- to well-vesicular glass shards, and show upward coarsening of pumice. These deposit features can be related mainly to the different settling velocities of particles in air and water, behavior of hot pumice in water, and interaction between hot magma and the ambient water. A wide spectrum of subaqueous pumice deposits has been recognized so far, but they can be described by three end-members; types 1, 2 and 3 classified according to the relative differences in unit bed thickness, grain-size population, internal structures, and constituents.

3.1. Type 1 Deposits

Type 1 deposits are thickly bedded, poorly sorted, inversely-to-normally graded or massive, and internally, partly stratified. They are sometimes accompanied by a lithic breccia or lithic-rich bed at the base and a relatively well sorted, stratified, and finer-grained bed at the top. Primary volcaniclastic deposits of this type are thought to be subaqueous counterparts of pyroclastic flows, as represented by the Shinjima Pumice exposed on Shinjima Island, Kyushu, southwest Japan [*Kano et al.*, 1996].

Shinjima Island, located immediately outside of the submerged Wakamiko caldera at a water depth of 200 m, is known to have emerged in 1780 from the 100–200 m deep seafloor, perhaps by intrusion of the An'ei lava effused from Sakurajima volcano. The Shinjima Pumice, rhyolitic in composition, was produced by the eruption that formed Wakamiko caldera and emplaced in water. Absence of similar deposits on adjacent land suggests that the eruption occurred at a water depth of 100–200 m and never breached the sea surface.

Multiple type 1 beds constitute the sequence of Shinjima Pumice with a total thickness over 40 m. Individual beds are

Figure 4. A pumice block on the beach of Satsuma-Iwojima Island [*Tanakadate*, 1935a]. Note prismatic open cracks on the surface. The size of this block was not described, but the pumice blocks landing on the shore of the Satsuma-Iwojima Island were reported to reach a maximum of 10 m in length.

1–10 m thick, and are inversely graded and diffusely stratified with upward coarsening of pumice (Figure 5). They are similar in internal structures and in grain-size distribution to high-density turbidites [*Lowe*, 1982], and are better sorted than the majority of subaerial pyroclastic flow deposits (Figure 6). Constituent pumice blocks, however, have polyhedral or prismatic joints of thermal contraction origin (Figure 7). Pumice shards are commonly blocky and many glass shards are platy (Figure 8); these morphological features are diagnostic of explosive magma-water interaction [*Heiken and Wohletz*, 1985].

Type 1 deposits of rhyolitic composition have also been found at ODP site 788 on a 1000 m-deep part of the Shichito-Iwojima Ridge, Izu-Ogasawara (Bonin) Arc [*Nishimura et al.*, 1991, 1992]. The Pliocene-to-Pleistocene pumice deposits at this site constitute a bedded sequence 250 m thick. Individual beds are 20–50 meters or less in thickness, and commonly show inverse grading of coarse pumice clasts (Figure 9). They are poor in fine ash but commonly contain blocky, polyhedral ash-sized pumice and variably vesicular clasts [*Nishimura et al.*, 1991] indicative of an interaction between magma and water. The source of the pumice deposits is unknown but could be one of the nearby submarine calderas, Torishima caldera 40 km to the south [*Nishimura et al.*, 1991] or Minami Sumisu caldera 30 km to the north.

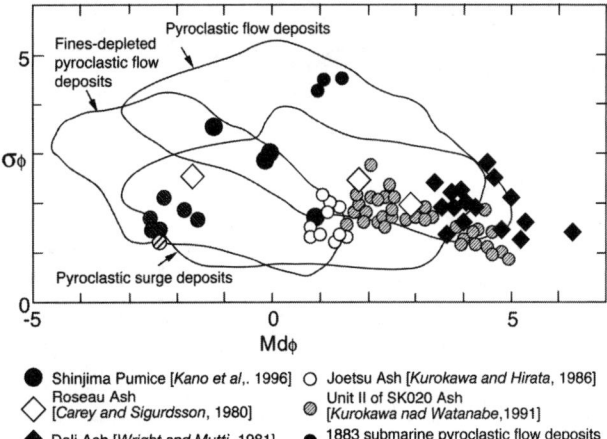

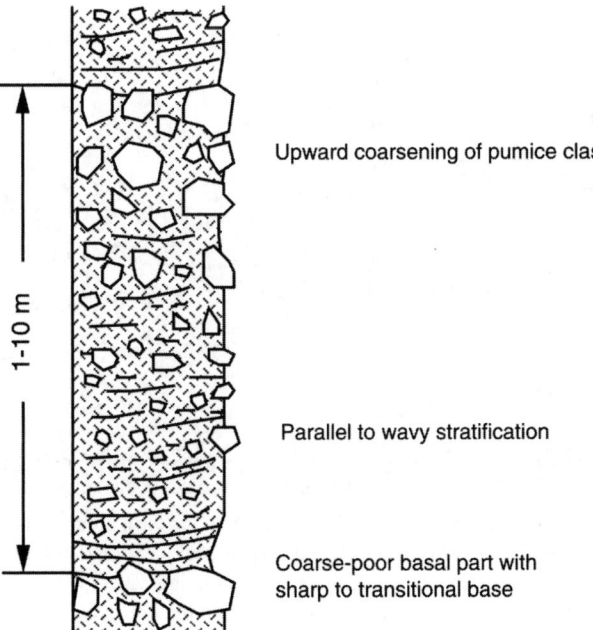

Figure 5. Idealized single flow unit of the Shinjima Pumice. Reprinted from *Kano et al.* [1996], with permission from Elsevier Science.

Figure 6. Plot of median size (Md_ϕ) vs. graphic standard deviation (σ_ϕ) in phi scale for the Shinjima Pumice, the Yali Pumice Breccia, and other volcaniclastic deposits. Boundaries of pyroclastic flow, fines-depleted pyroclastic flow and pyroclastic surge are adopted after Walker (1983). Reproduced from *Kano et al.* [1996] with permission from Elsevier Science. Data from *Allen and McPhie* [2000] are added to the original.

Submersible observations by *Yuasa* [1995] and *Fiske et al.* [2001] show that type 1 rhyolitic pumice deposits are exposed on the wall of the Myojin Knoll caldera located 100 km north of the Minami Sumisu caldera on the same Shichito-Iwojima Ridge of the Izu-Ogasawara Arc. They are 200–300 m thick with a crude upward coarsening of pumice and depletion in fines. Fines-depleted, massive pumice lapillistone is exposed on the lower wall, and stratified pumice lapillistone is exposed on the upper wall with a topmost deposit of large pumice blocks 1–2 m in diameter. The overall architecture of these deposits is similar to the Shinjima Pumice.

A Miocene or Pliocene pumice deposit at Dogashima, Izu Peninsula, Japan, which is likely to have been directly fed by an eruption [*Tamura et al.*, 1991], consists of a lower, inversely to normally graded lithic-rich tuff breccia, a middle, diffusely stratified lithic-pumice lapilli tuff, and an upper normally graded pumice lapilli tuff [*Cashman and Fiske*, 1991]. This deposit possesses characteristics of classical subaqueous pyroclastic flow deposits [*Fiske and Matsuda*, 1964] but appears somewhat different from the Shinjima Pumice. The difference is, however, attributed to the density contrast and grain-size population of the constituents. The Dogashima deposit is interpreted to have settled from a turbulent suspension of a mixture of pumices, lithics and other clasts. Lithics are, therefore, concentrated in the lower part whereas pumice is concentrated in the

Figure 7. Pumice block in the Shinjima Pumice. Note the polyhedral contraction cracks. Lens cap is 6 cm across. Reprinted from *Kano et al.* [1996], with permission from Elsevier Science.

upper through the middle parts. Absence of upward coarsening of pumice implies that pumice settled after being fully saturated with water in the suspension mass, suggesting a fallout origin of the normally graded main part of the Dogashima deposits [*Cashman and Fiske,* 1991].

3.2. Type 2 Deposit

Type 2 deposits are thinly bedded, internally inversely to normally graded and stratified, and better sorted than type 1 deposits. They are similar in depositional features to pyroclastic surges, as represented by the Yali Pumice Breccia exposed on the Yali Island in the eastern Agean Sea [*Allen and McPhie,* 2000].

The Yali Pumice Breccia is over 150 m thick and comprises mainly multiple type 2 beds. The pumice clasts settled to the seafloor, presumably 150–200 m deep. Juvenile materials are dominated by pumice; non-vesicular materials are a minor component. Polyhedral or prismatic joints are well developed in pumice blocks and coarse lapilli.

Individual beds of type 2 deposits are 0.1 to 3 m thick at most. They are internally parallel to wavy stratified, and are inversely to normally graded with relatively continuous, parallel to wedge-shaped bedforms (Figure 10). These thin beds range in grain size mainly from lapilli to coarse ash and are intercalated with well-sorted, massive, planar beds of pumice blocks and coarse lapilli. They are similar in sedimentary structures and constituents to the type 1 deposits of the Shinjima Pumice, but are dominated by traction structures. In addition, they are more depleted in fines than the type 2 Shinjima Pumice (Figure 6). The coarse deposits intercalated with these type 2 deposits are more depleted in fines and are interpreted to be fallout deposits [*Allen and McPhie,* 2000], namely of type 3.

3.3 Type 3 Deposits

Type 3 deposits are fairly well sorted, inversely graded or massive pumice breccias or pumice blocks scattered in lacustrine or marine sediments. A representative occurrence is the isolated or partly stacking pumice blocks enclosed in a laminated ash bed of Sierra La Primavella caldera, Mexico [*Mahood,* 1980; *Clough et al.,* 1981]. The pumice blocks have radial joints and are rounded to some extent, perhaps representing thermal contraction of the surface and spalling off the angular corners by intense thermal contraction and cracking. Individual pumice blocks are interpreted to have settled in sequence through the caldera lake water. Laminae of the host ash bed are warped downward immediately below the pumice blocks and discordantly abut on them (Figure 11). This plastic deformation of the laminae might have resulted from the load of pumice and the rotation of pumice with subsequent upward doming of the caldera floor [*Clough et al.,* 1981].

Another type 3 deposit is found in the Okinawa Trough on the backarc side of Ryukyu Arc, SW Japan, where dacitic pumice blocks 1–3 m long and 1 m or less thick occur locally stacking on the flank of a knoll 1500–1600 m below sea level, together with blocks of cognate dacitic, poorly vesicular lava [*Kato,* 1987]. They have a coarse fibrous texture with extremely elongate vesicles, giving an appearance of wood, and are therefore called woody pumice by *Kato*

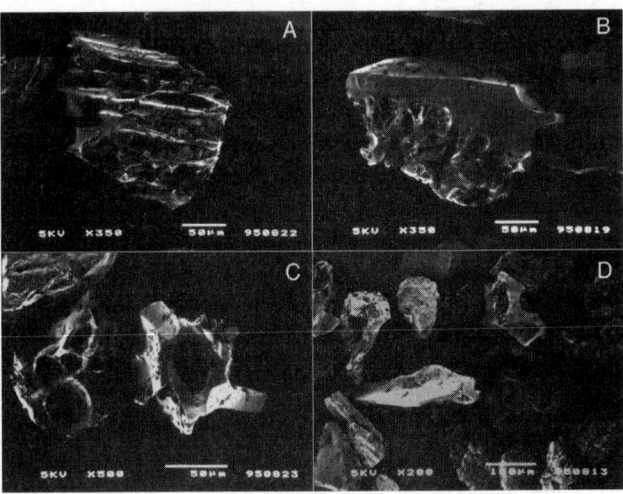

Figure 8. Scanning electron photomicrographs of ash components in the Shinjima Pumice. (A) Blocky equant pumice with long-tube vesicles and curviplanar fracture surfaces cutting across the vesicles. (B) Blocky, vesicular clasts with walls thicker than 10 μm. (C) Pumice clasts with large vesicles. Thick walls are fractured with curviplanar surfaces. (D) Platy or curved angular glass shards. Reproduced from *Kano et al.* [1996], with permission from Elsevier Science. Arrangement of photos and inset letters is slightly modified.

[1987]. The fibrous texture indicates stretching of the vesicular viscous lava during its effusion or flowage.

4. MODES OF ERUPTION

Observations of subaqueous pumice eruptions and even recovery of the products of observed eruptions are rare. Consequently, attempts have been made to theoretically link observed eruptions to deposit types, mainly based on the sedimentary features of deposits and the morphological fea-

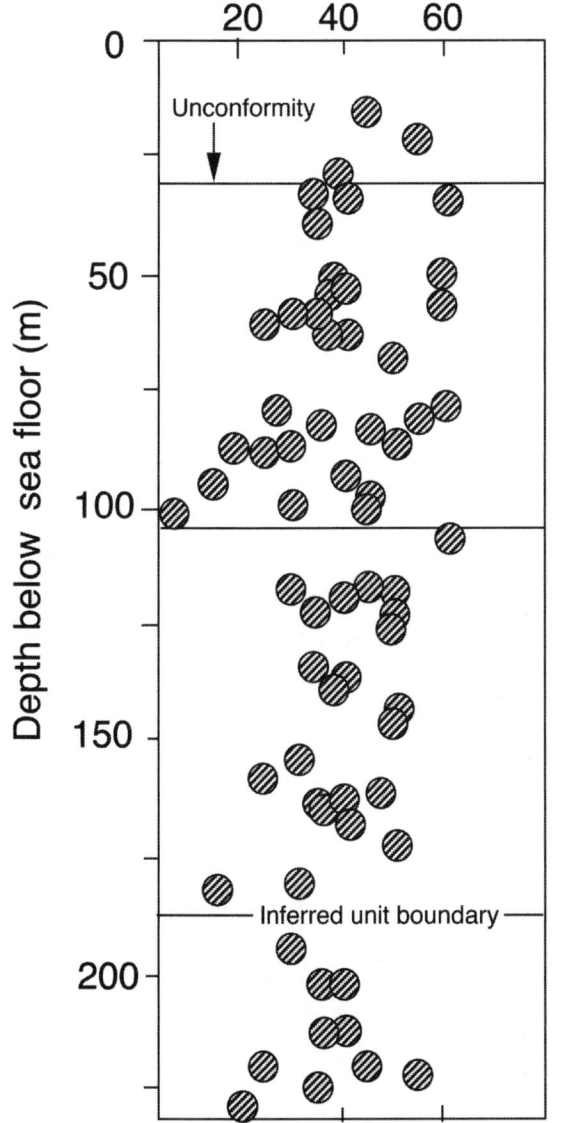

Figure 9. Variations in maximum grain size of pumice lapilli of the pumice deposits at ODP site 788 [modified from *Nishimura et al.,* 1991].

Figure 10. Bedded sequence of the Yali Pumice Breccia, covering a height of 18 m. Coarser pumices are contained in the thicker beds. Photo courtesy of S. R. Allen.

tures and compositions of ejected materials. Three modes of subaqueous pumice eruptions have been proposed so far: 1) magmatic eruption, 2) phreatomagmatic eruption, and 3) syn-eruptive, non-explosive dome collapse.

4.1. Magmatic Eruption

Magmatic eruption, which is driven mainly by magma vesiculation, ejects fragmented magma and volcanic gas. We have poor knowledge of what types of magmatic eruptions occur in water, but they could be similar to subaerial counterparts with similar eruptive processes.

When it thrusts into water, a gas-supported flux of pumice and comminuted clasts likely forms an eruption column [*Cashman and Fiske,* 1991; *Kokelaar and Busby,* 1992; *Kano et al.,* 1996]. The subaqueous eruption column, if large in scale, may contain a basal gas-thrust region, a tur-

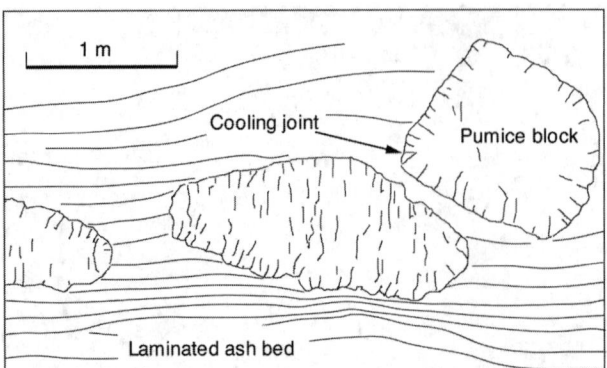

Figure 11. Pumice blocks in the laminated ash deposits of Sierra La Primavella caldera, Mexico. Sketched from a photo (Figure 13.46a) of *Cas and Wright* [1987].

bulent region where the flux of gas-particle mixture ingests ambient water, a buoyancy-driven convection region, and an umbrella region at the water surface, similar to subaerial plinian eruption [*Cashman and Fiske, 1991*]. A large volume of particle suspension in the umbrella region likely transforms into a turbulent density current by density inversion [*Soh et al., 1989; Carey, 1997; Fiske et al., 1998*], resulting in settling, mainly forming a normally graded deposit. *Cashman and Fiske* [1991] applied this model to account for the thick, normally graded pumiceous part and overlying stratified part of the Dogashima deposit. This type of eruption is similar to plinian eruptions in terms of the dynamic behavior of the eruption column and may be termed subaqueous plinian eruption.

A large-volume eruption column in water is expected to ingest the ambient water by its turbulent, rapid flow regime [*Kano et al., 1996*]. The water incorporated in the eruption column plays an important role in determining the subsequent fate of the eruption column. If the volume of incorporated water is enough to totally condense the contained gas (mainly steam), the eruption column collapses immediately. On the other hand, if the volume of incorporated water is small enough to be vaporized by the internal heat, the eruption column expands with further quench-fragmentation of juvenile materials, and continues to rise. The former case exclusively occurs when the heat source is limited to the internal steam (Figure 12A). The latter case occurs when the heat source is derived from both steam and pyroclasts (Figure 12B). The expanding column, however, further ingests ambient water and finally collapses (Figure 12B) with the production of a density current and an ascending flow of buoyant clasts (Figure 13). This column collapse may be accompanied by a gas-supported hot pyroclastic flow [*Kokelaar and Busby, 1992*]. The flow is driven by the internal pressure and gravity potential of the gas thrust

region, but will become water-logged in a short distance by ingestion of ambient water (Figure 13) if the flow volume is small. This type of eruption is similar to ignimbrite-forming eruptions and is here called a subaqueous flow-generating eruption. Subaqueous column collapse is, however, accompanied by a well-developed buoyancy-driven convective flow of vesicular clasts (Figure 13), different from the subaerial counterparts.

A subaqueous flow-generating eruption is inferred for the Shinjima Pumice [*Kano et al., 1996*] because the eruption column is thought to have collapsed in water. The deposits show features of high-density turbidites. They are thick, poorly sorted, inversely-to-normally graded, and partly stratified, but monomictic, similar to pyroclastic flow deposits. Polyhedral or prismatic joints of the constituent pumice blocks indicate that the blocks were hot and slowly cooled down relative to the associated finer clasts, as observed in the 1934–1935 eruption of Shin-Iwojima, Kikai

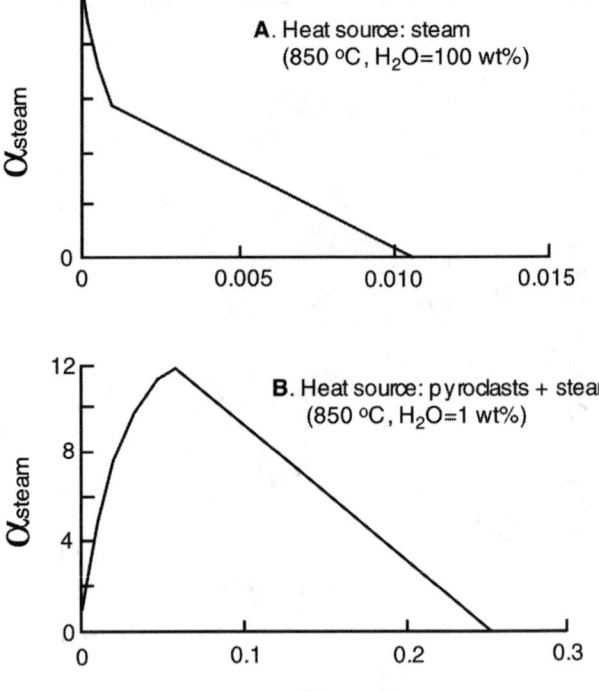

Figure 12. Variations of the volume ratio, α_{steam}, of the total steam after the heat exchange to the original steam in the plume with the mixing volume ratio, V_{water}/V_{plume}, of the entrained water to the plume at 1 MPa for a rhyolitic magma containing an exsolved water (steam) of 1 wt.% [*Kano et al., 1996*]. A: the case where heat source is limited to the internal steam. B: the case where heat is derived from both the internal steam and pyroclasts. Reprinted from *Kano et al.* [1996], with permission from Elsevier Science.

caldera [*Tanakadate*, 1935a,b]. Upward coarsening of pumice clasts implies that they remained buoyant while internal heat maintained a sufficient volume of steam in the interiors. At temperatures much higher than the boiling point of the ambient water, steam films are developed on the surfaces of hot pumices to prevent invasion of water into the interiors [*Whitham and Sparks*, 1986; *Kano et al.*, 1996]. The steam films collapse with decreasing temperatures, but steam remains in the interiors until the pumice clasts are cooled down to the boiling points. Thus the larger the pumice clasts are, the later they settle. Blocky pumice shards and platy glass shards likely indicate quench fragmentation of hot particles in direct contact with ingested water in the eruption column. The Shinjima Pumice is, thus, interpreted to have been erupted and emplaced in water. Subaqueous flow-generating eruptions involve all the processes inferred above (Figure 13). Inversely graded multiple units of the Shinjima Pumice suggest repeated deposition from an unsteady large plume or multiple plumes repeatedly built by a sequence of eruptions. The submarine eruption of the Shinjima Pumice produced the Wakamiko caldera, 6.5 km by 5.5 km in diameter, with an estimated eruption volume of 12 km^3 [*Kano et al.*, 1996].

The source of the pumice deposits at ODP site 788 is also inferred to be a submerged caldera, and similar deposits are exposed on the wall of the Myojin Knoll caldera, 6 km by 7 km in diameter [*Yuasa et al.*, 1991; *Fiske et al.*, 2001]. These observations also collectively imply large-scale eruptions for type 1 deposits.

Hawaiian and strombolian eruptions of silicic magma are not common because high viscosity prevents rapid movement and coalescence of bubbles. In deep water, vesiculation is, however, suppressed to keep magma fluidal and this effect may enable hawaiian and strombolian eruptions of silicic magma. Fire fountains of silicic magma envisaged for the deposits of slightly to moderately vesicular, partly plastically deformed glassy volcaniclasts [*Müeller and White*, 1992] likely represent a hawaiian type eruption in water. Quench fragmentation dominated, perhaps due to incorporation of water into the eruption column along the turbulent surface [*Müeller and White*, 1992].

Strombolian eruption of silicic magma in water has not yet been described, though it is plausible if the magma is sufficiently fluidal to allow coalescence of bubbles beneath the thin, fragile quench crust developed over the molten lava. This type of eruption may account for the 1984 eruption of Kaitoku Seamount [*Hydrographic Department, Maritime Safety Agency, Japan and Faculty of Engineering, Tokyo Institute of Technology*, 1984; *Tsuchide et al.*, 1985; *Hydrographic Department, Maritime Safety Agency, Japan*, 1994].

Vulcanian silicic eruptions are thought to be common in water. When silicic magma slowly effuses in water, a large lava dome may grow over the vent, or the conduit may be plugged with the solidified lava in direct contact with water. In either case, a gas pocket will be produced beneath the lava crust or plug by accumulation of the gas exsolved from the magma and, if highly pressurized, can explode with ejection of blocks and finer clasts of the lava.

Vulcanian eruption from the conduit may recur as long as the magma head is kept in direct contact with water. If the lava crust at the magma head is thin or fragile, the eruption will also eject a certain volume of molten vesicular lava clasts. In this way, strombolian and vulcanian eruptions may have common sources.

Vulcanian eruptions of lava domes originate mainly from the pumiceous zone developed beneath the lava crust [*Fink and Manley*, 1989], and therefore the deposits commonly contain pumice clasts, including large pumice blocks and significant amounts of non- to poorly-vesicular lava clasts

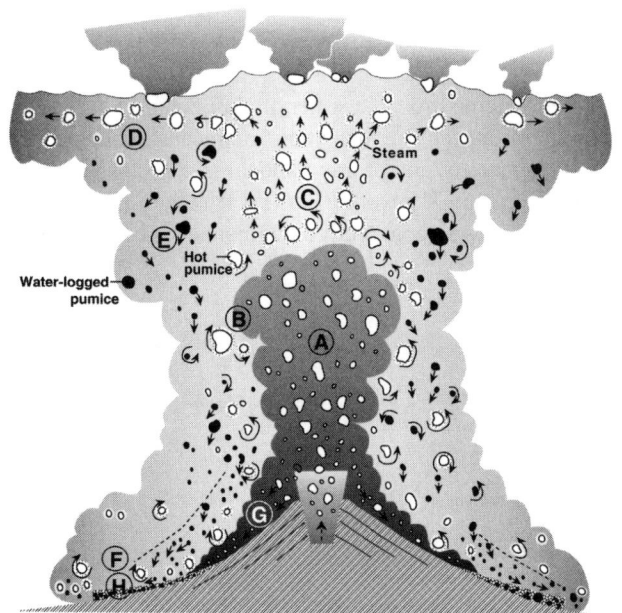

Figure 13. A model of subaqueous flow-generating eruption. A=gas-supported eruption plume. B=mixing zone between eruption plume and water. C= buoyancy-driven convective plume carrying pumice clasts and ash. D=suspension cloud of pumice clasts and ash. E=fallout of water-logged pumice clasts and other dense materials. F=gas- and water-logged density current. G=gas-supported hot pyroclastic flow. H=pumice deposit. Hot pumice clasts are buoyant as the vesicles are filled with steam. At high temperatures, steam films form on the surfaces and prevent the pumices from rapid cooling and water-invasion. Reproduced from *Kano et al.* [1996], with permission from Elsevier Science. Half-tone gray color and diagonal pattern are added to the original.

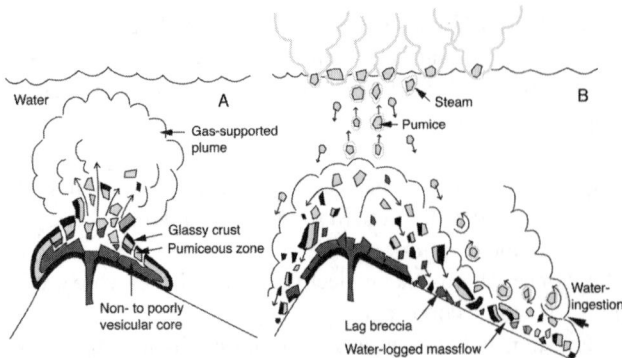

Figure 14. Vulcanian eruption of a pumiceous dome (A) and generation of a mass flow of coarse pumice (B). Reprinted from *Kano* [1996], with permissions from Springer-Verlag.

at the base. *Kano* [1996] envisaged a submarine vulcanian eruption of a large rhyolitic lava dome to explain a thick bed mainly composed of coarse pumice with basal lithic breccia of lava crusts and dominated by traction structures (Figure 14).

4.2. Phreatomagmatic Eruption

Allen and McPhie [2000] proposed this mode of eruption for the type 2 Yali Pumice Breccia. The Yali deposits are thinly bedded and internally stratified with sedimentary features characteristic of low-density turbidites or grain-flow deposits [*Middleton and Hampton*, 1973; *Lowe*, 1982], similar to pyroclastic surge deposits. Pumice blocks have polyhedral or prismatic joints produced by thermal contraction. Polyhedral ash-sized clasts indicate explosive water-magma interaction, and multiple thin beds represent repetition of small-scale eruptions with occasional production of dilute density currents.

Explosive magma-water interaction is limited in scale by the area of the interface (the vent aperture or the area of exposed molten surface), and is likely to occur consecutively as long as the newly exposed surface of the magma remains hot and remains in direct contact with water. Resulting explosions are small in scale relative to subaqueous plinian eruptions and subaqueous flow-generating eruptions (Figure 15), and if produced, the flows must be dilute, similar to pyroclastic surges. Intermittent cessation of explosions leads to cooling and solidification of the magma, and the solidified surface may be broken by a combination of quenching and successive ascent of magma, or by accumulation of gas beneath the lava surface (vulcanian eruption). This process of fragmentation of pumiceous lava could produce pumice blocks, and some isolated pumice blocks still remaining hot would ascend through the water, drift on the surface, saturate with water, and then settle to the bottom to produce type 3 deposits (Figure 16). Repetition of these processes requires intermittent extrusion of lava to form a dome and eventual explosive magma-water interactions in the conduit or lava extrusion (Figure 15), as envisaged by *Allen and McPhie* [2000]. Pumice fallout likely represents temporal interruption of these eruption activities.

Continuous growth of a lava dome to shallow water may allow explosive magma-water interaction or magma vesiculation to form a pyroclastic cone on the dome top [*Cas et al.*, 1990]. This is the case observed at the emergent phase of the 1934-1935 eruption of Shin-Iwojima, Kikai caldera [*Tanakadate*, 1935a,b].

Dome-top explosions by magma-water interaction are perhaps common, including the 1952–1953 eruption of Myojinsho and the 1985–1986 eruption of Fukutoku-Oka-no-Ba. Direct deposition by tephra jets onto the slopes of the domes and density inversion of the suspension fed by continuous fallout generate relatively dilute density currents [*Fiske et al.*, 1998], perhaps resulting in type 2 deposits.

4.3. Syn-Eruptive, Non-Explosive Dome Collapse

When a dome grows in water, it may collapse by a combination of quench fragmentation and mechanical fragmentation (Figure 16) and/or gravitational instability.

As inferred by *Kano et al.* [1991], growth of a lava dome in water results in consecutive fragmentation by rapid ther-

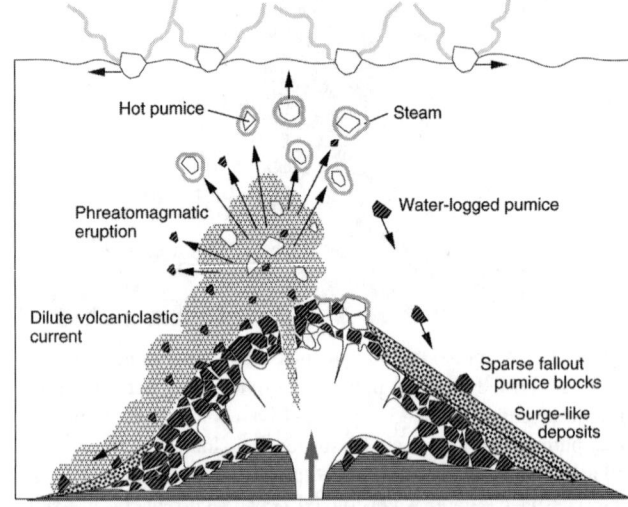

Figure 15. Phreatomagmatic eruption from the interior of a pumiceous dome and resulting dilute density flow. Hot pumice blocks may ascend to the water surface, drift for a while, and settle separately from other finer clasts when saturated with water. Some ascending pumice blocks will be disintegrated by thermal contraction and internal gas expansion.

mal contraction and successive flow of the internal molten part, accumulation of the syn-eruptive breccias around the growing front, and further ascent of the lava through the breccias. When the lava is pumiceous and produced pumice clasts are sufficiently large to retain heat until they reach the water surface, a plume of pumice clasts could be observed as observed during the initial phase of the 1934–1935 eruption of Shin-Iwojima, Kikai caldera [*Tanakadate*, 1935a,b]. This eruption ejected large pumice blocks of meter size. Some pumice blocks were observed to sink around the source while others floated away. Subsequent submersible surveys revealed isolated emplacement of large pumice blocks [*Nakamura et al.*, 1986]. This mechanism may account for an origin of type 3 deposits.

5. FACTORS CONTROLLING SUBAQUEOUS PUMICE ERUPTIONS

Explosive subaqueous eruption of silicic magma occurs by magma vesiculation, magma-water interaction, or a combination of both. The mode and scale of the eruption depends on the extent of magma vesiculation [*Houghton and Wilson*, 1989] and water-ingestion by magma [*Wohletz and McQueen*, 1984; *Wohletz*, 1986; *Kano et al.*, 1996], which in turn relate mainly to the confining pressure (water depth), volatile contents in magma, and magma properties such as viscosity and temperature [*McBirney*, 1963; *Fisher*, 1984; *Kokelaar*, 1986]. In addition, there may be transitions

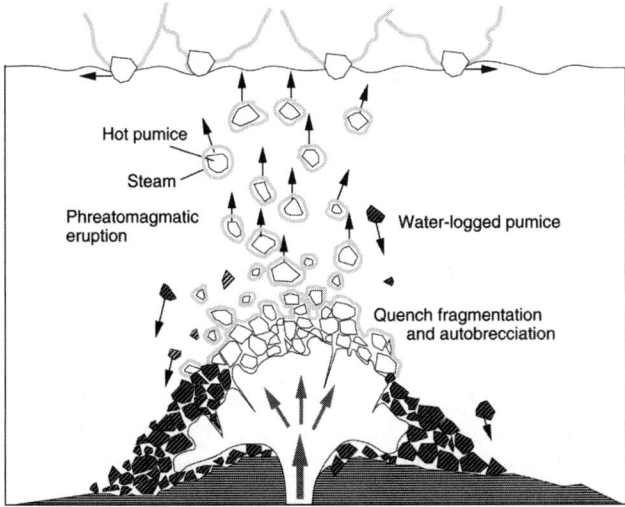

Figure 16. Quench fragmentation and mechanical fragmentation with outward extrusion of the interior of a subaqueous pumiceous dome. Isolated hot pumice blocks ascend to the water surface, drift for a while, and settle to the water bottom. Ascending pumice blocks may be disintegrated by thermal contraction and internal gas expansion.

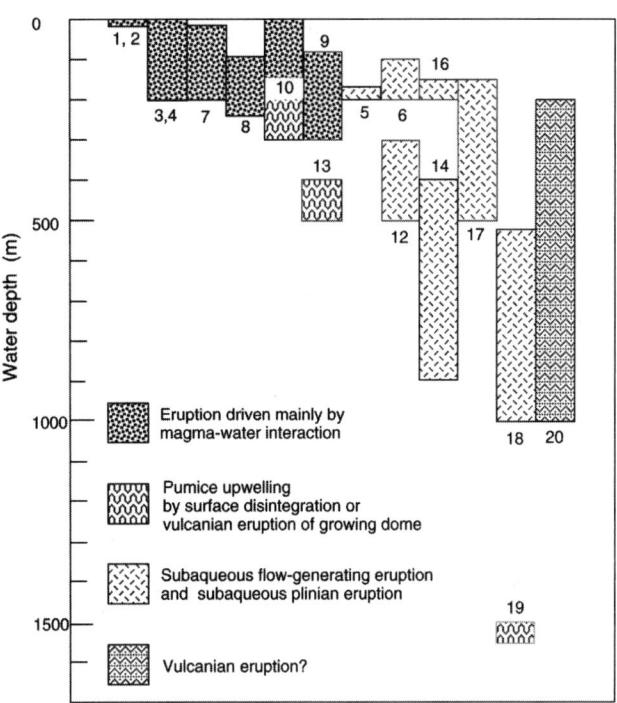

Figure 17. Relationship between the water depth and mode of pumice eruption. Numerals identify deposits listed in Table 1.

from explosive to effusive activity, or the reverse, during an eruption episode. Such transitions between explosive and non-explosive eruption depend on the magma flow rate in the permeable conduit or the extent of internal pressure building in the magma chamber [*Jaupart and Allégre*, 1991; *Woods and Koyaguchi*, 1994]. A silicic lava dome may, therefore, form by effusion of the magma that degassing during or prior to its ascent. The lava may foam during ascent and could be emplaced as a dense unit following collapse of the foam [*Eichelberger et al.*, 1986]. In many cases, a significant amount of gas, however, has escaped prior to its effusion. Subsequent exsolution of volatile in the interior of the effused lava forms an inner pumiceous zone [*Fink et al.*, 1992], and may cause secondary explosions from the highly pressurized pumiceous zone [*Fink and Manley*, 1989].

Figure 17 shows the relationship between water depth and eruption mode for silicic magma. Individual eruption modes are inferred from direct observations and/or deposit features, and water depths are taken as reported or inferred from the literatures (Table 1). A subaqueous flow-generating eruption occurred at a water depth of 100–200 m in submarine Wakamiko caldera [*Kano et al.*, 1996]. Myojin Knoll caldera was produced by a catastrophic rhyolitic eruption at a water depth of 400–500 m, presumably after an explosive pumice eruption at a water depth of 500–900 m

[*Fiske et al.*, 2001]. Similar submarine rhyolitic calderas lie on the same Shichito-Iwojima Ridge, Izu-Ogasawara Arc have wide floors 2–10 km across and rims down to 700 m below sea level [*Yuasa et al.*, 1991]. According to *Wright et al.* [2003], Healy caldera, southern Kermadec Arc was produced by a large-volume explosive eruption of silicic magma even at an inferred water depth of 550–1000 m.

Theoretically, explosive vesiculation of rhyolitic magma is plausible even at a water depth of 1000–1300 m, with a water content of 6 wt.% and a temperature of 850°C (Figure 11 in *Wright et al.* [2003]). On the other hand, a rhyolitic magma at 850°C with 1 wt.% H_2O cannot explode by magma vesiculation even at a water depth of 200 m [*Kano et al.*, 1996]. *Burnham* [1979] argued that volatiles concentrate in the residual melt during crystallization of anhydrous minerals. This process, called retrograde or second boiling, is accompanied by a drop in temperature that contributes to magma explosion in deep water. Most of the ejecta are, however, poor in phenocrysts, suggesting that explosive eruptions normally occur at an incipient stage of crystallization.

In the 1934–1935 eruption of Shin-Iwojima in Kikai caldera, large rhyolitic pumice clasts rose gently to the sea surface from a water depth of 200-300 m. This eruption is minor, and is likely attributed to quench fragmentation and mechanical fragmentation of a growing pumiceous dome, a mechanism discussed earlier. The 1934 pumice of Shin-Iwojima contains only 1 wt.% H_2O in the melt, which is significantly lower than the H_2O content of 3 wt.% in the melt of the pumice produced in the caldera-forming eruption of Kikai caldera [*Saito et al.*, 2001].

The 1924 eruption off Iriomote Island erupted large pumice clasts to the sea surface, similar to the 1934–1935 eruption of Shin-Iwojima. The eruption mechanism remains uncertain. Notable earthquakes were, however, few during the upwelling of pumice and no caldera features have been found in the inferred source area [*Kato*, 1991]. The eruption, therefore, might have occurred in a manner similar to the initial phase of the 1934–1935 eruption of Shin-Iwojima.

The origin of the woody pumice blocks from water depths of 1500–1700 m in the Okinawa Trough is also ambiguous. Either vulcanian or gravitational collapse of a growing dome is a plausible explanation for the local occurrence of pumice blocks with cognate poorly vesicular blocks on the flank of a submarine knoll.

Explosive magma-water interaction includes two mechanisms, contact surface interaction (fuel-coolant interaction) and bulk interaction [*Wohletz*, 1986; *Kokelaar*, 1986]. Gas-release and resulting vesicles increase magma viscosity and likely prevent fuel-coolant interaction [*Zimanowski et al.*, 1991]. Bulk interaction of enclosed water with magma, however, may allow explosions. The process includes engulfing wet sediments by magma intrusion [*Hunns and McPhie*, 1999], surface perturbation of magma by drain back [*Yamamoto et al.*, 1991], and sudden intrusion of water into the interior of magma through contraction cracks [*Allen and McPhie*, 2000]. Vesicular, silicic magma is normally too highly viscous to effectively intermingle with the ambient water or watery sediments. A primary mechanism for vesicular, silicic magma is bulk interaction of water convoluted in the interior of the magma, a process envisaged for the production of pumiceous peperite [*Hunns and McPhie*, 1999].

The maximum water depth at which explosive bulk interaction takes place is uncertain, but could never exceed 2200 m or 3100 m, the critical point of water or seawater [*Kokelaar*, 1986]. The volume ratio of steam to liquid water increases exponentially as pressure decreases [*McBirney*, 1963], and therefore, explosive magma-water interaction is more likely to occur at shallower water depths.

The pumice eruptions driven mainly by magma-water interaction observed so far occurred mainly on the tops of submerged to slightly emerged domes in water shallower than 200 m (Table 1), where water expands by a factor of a hundred to more than a thousand in being transformed to steam. Growth of the volatile-poor Shin-Iwojima dome to shallow water depths facilitated explosive magma-water interaction with decreasing confining pressure, as it did during the 1952–1953 eruption of Myojinsho and the 1986 eruption of Fukutoku-Oka-no-Ba. Low volatile contents of these three domes perhaps suppressed explosive magma vesiculation of the domes even at shallow depths, whereas brecciated, ragged surfaces of growing domes allowed access of ambient water into the interiors, leading to explosive bulk interaction between the internal viscous lava and the external water.

The 1984 eruption of Kaitoku Seamount is thought to have produced a cone by repeated explosions at water depths between 240 m and 95 m. One of the pumice clasts contains 62 wt.% SiO_2 [*Hydrographic Department, Maritime Safety Agency and Faculty of Engineering, Tokyo Institute of Technology*, 1984] and is therefore close to andesite in composition and perhaps sufficiently fluid to mix vigorously with ambient water without forming a dome. To test this working model for the 1984 eruption, direct observation of the cone and recovery of the constituents are necessary.

The 2001 eruption of an unnamed submarine volcano west of Vava'u produced a huge submarine bank by ejection of pumice. The bank-like landform is 2.4 km across and likely to be less than 200–300 m high from the summit of the submarine volcano, and is, therefore, morphologically similar to a tuff ring. The relatively small size of pumice suggests the eruption was explosive, and the morphology of the newly produced 'bank' suggests a phreatomagmatic eruption.

Observed cases of subaqueous pumice eruptions are fairly rare and our search is perhaps inadequate to cover all modes of explosive eruptions over a wide range of water depth and magma composition.

6. SUMMARY

Observations of subaqueous pumice eruptions and even recovery of the products of observed eruptions are rather limited. The modes of eruption are commonly inferred, mainly based on the sedimentary features of the deposits, and morphological features and compositions of ejected materials. Subaqueous pumice eruptions observed or inferred from deposits include plinian-type eruptions, flow-generating eruptions, lava explosions by delayed vesiculation (vulcanian eruptions), lava fountaining, explosive bulk interaction of vesicular magma and water, and non-explosive release of pumice clasts isolated from the thermally and mechanically brecciated surface of a growing vesicular dome.

In the case of magmatic eruptions, a subaqueous column ingests ambient water, grows large with heating of the ingested water to steam, collapses with increasing ingestion of water, and thus produces a density current. The products are poorly sorted, inversely-to-normally graded or massive, and partly stratified deposits. They are sometimes accompanied by a lithic breccia or lithic-rich bed at the base and a relatively well sorted, stratified, and finer-grained bed at the top with upward coarsening of pumice. Plinian-type eruptions in water likely form an umbrella region sustaining particles on the top of the eruption column with subaqueous density inversion to form a density current. Settling of this material may form a normally graded. subaqueous fallout deposit.

Pumice eruptions driven by bulk interaction of a well vesicular magma with water recurs as long as magma is supplied to the watery setting. Resulting dilute currents form a thinly bedded sequence of internally inversely to normally graded and stratified, and well sorted deposits. These explosive pumice eruptions are accompanied by upwelling and subsequent fallout of pumice blocks, as observed in explosive or non-explosive collapse of subaqueous pumiceous lava. Resulting fallout pumice deposits are fairly well sorted, inversely graded or massive. Upwelling of pumice from deep water is a sign of eruption, but the mode of eruption cannot be specified based only on this phenomenon.

Subaqueous flow-generating pumice eruptions can be sufficiently large to form calderas. Caldera formation is more likely with increasing volatile content and magma temperature and decreasing water depth. The water depth at which such catastrophic eruptions occur is likely to range to a depth of 500–1000 m.

Subaqueous pumice eruptions driven mainly by magma-water interactions occurred mainly at growing domes in water shallower than 200–300 m or slightly emergent, even though explosive bulk interaction between magma and water is plausible down to a depth of 2200 m in water and 3100 m in seawater. Continuous growth of a lava dome to shallow water lessens the confining pressure and may allow explosive magma-water interaction or magma vesiculation to form a pyroclastic cone on the dome.

Acknowledgments. R. Fiske, S. Carey, and D. Clague critically reviewed this paper. Discussion with T. Yamamoto, I. Miyagi, and M. Yuasa were helpful to finish this work. J. Ossaka and S.R. Allen kindly permitted to use their photos. M. Pino'on, a librarian of the Geological Survey of Papua New Guinea and P. W. Taylor kindly sent literatures I requested. J. White and B. Houghton encouraged writing this paper on the occasion of the Chapman Conference on Explosive Subaqueous Eruption. I thank all these persons.

REFERENCES

Allen, S.R., and J. McPhie, Water settling and sedimentation of submarine rhyolitic pumice at Yali, eastern Aegean, Greece, *J. Volcanol. Geotherm. Res.,* 95, 285-307, 2000.

Burnham, C.W., The importance of volatile constituents, in *The Evlution of Igneous Rocks,* edited by H.S. Yorder Jr., pp. 39-482, Princeton Univ. Press, Princeton, 1979.

Carey, S.N., Influence of convective sedimentation on the formation of widespread tephra fall layers in the deep sea, *Geology,* 25, 839-842, 1997.

Carey, S.N., and H. Sigurdsson, The Roseau Ash: deep-sea tephra deposits from a major eruption on Dominica, Lesser Antilles Arc, *J. Volcanol. Geotherm. Res.,* 7, 67-86, 1980.

Cas, R.A.F., and R.L. Allen, *Volcanic Successions*, 528pp., Allen & Unwin, London, 1987.

Cas, R.A.F., R.L. Allen, S.W. Bull, B.A. Clifford, and J.V. Wright, Subaqueous, rhyolitic dome-top tuff cones: a model based on the Devonian Bunga Beds, southern Australia and a modern analogue, *Bull. Volcanol.*, 52, 159-174, 1990.

Cashman, K.V., and R.S. Fiske, Fallout of pyroclastic debris from submarine volcanic eruptions, *Science,* 253, 275-280, 1991.

Clough, B.J., J.V. Wright, and G.P.L. Walker, An unusual bed of giant pumice in Mexico, *Nature*, 289, 49-50, 1981.

Eichelberger, J.C., C.R. Carrigan, H.R. Westrich, and R.H. Price, Non-explosive silicic volcanism, *Nature*, 323, 598-602, 1986.

Fink, J.H., and C.R. Manley, Explosive volcanic activity generated from within advancing silicic lava flows, in *Volcanic Hazar,*

IAVCEI Proceeding in Volcanology 1, edited by Latter, J.H., pp.169-179, Springer-Verlag, Berlin Heidelberg, 1989.

Fink, J.H., S.W. Anderson, and C.R. Manley, Textural constraints on effusive silicic volcanism: beyond the permeable foam model, *J. Geophys. Res.*, 97, 9073-9083, 1992.

Fiske, R.S., K.V. Cashman, A. Shibata, and K. Watanabe, Tephra dispersal from Myojinsho, Japan, during its shallow submarine eruption of 1952–1953, *Bull. Volcanol.*, 59, 262-275, 1998.

Fiske, R.S., J. Naka, K. Iizasa, M. Yuasa, and A. Klaus, Submarine silicic caldera at the front of the Izu-Bonin arc, Japan: Voluminous seafloor eruptions of rhyolite pumice, *GSA Bull.*, 113, 813-824, 2001.

Fiske, R.S., and T. Matsuda, Submarine equivalents of ash flows in the Tokiwa Formation, Japan, *Am. J. Sci.*, 262, 76-106, 1964.

Fisher, R.V., Submarine volcaniclastic rocks, in *Marginal Basins Geology; Volcanic and Associated Sedimentary Processes in Modern and Ancient Marginal Basins, Geol. Soc. London, Spec. Publ.*, 16, edited by B.P. Kokelaar and M.F. Howells, 5-27, 1984.

Gass, I.G., P.G. Harris, and M.W. Holdgate, Pumice eruption in the area of the South Sandwich Islands, *Geol. Mag.*, 100, 321-330, 1963.

Heiken, G., and K.H. Wohletz, *Volcanic Ash*, 246pp., Univ. California Press, Berkley, 1985.

Houghton, B.F., and C.J.N. Wilson, A vesicularity index for pyroclastic deposits, *Bull. Volcanol.*, 51, 451-462, 1989.

Hunns, S.R. and J. McPhie, Pumiceous peperite in a submarine volcanic succession at Mount Chalmers, Queensland, Australia, *J. Volcanol. Geotherm. Res.*, 88, 239-254, 1999.

Hydrographic Department, Maritime Safety Agency, Japan, Bathymetric survey at Kaitoku Sea Mount, *Report of Coordinating Committee for Prediction of Volcanic Eruptions*, 58, 76-77 (in Japanese), 1994.

Hydrographic Department, Maritime Safety Agency, Japan, and Faculty of Engineering, Tokyo Institute of Technology, Submarine volcanic activities of the Kaitoku Seamount, *Report of Coordinating Committee for Prediction of Volcanic Eruptions*, 37, 5-69 (in Japanese), 1984.

Hydrographic Department, Maritime Safety Agency, Japan, Faculty of Engineering, Tokyo Institute of Technology, and Faculty of Science, Okayama University, Volcanic activities of the Hukutoku-Oka-no-Ba, *Report of Coordinating Committee for Prediction of Volcanic Eruptions*, 31,66-73 (in Japanese), 1986.

Jaupart, C., and C. Allégre, Gas content, eruption rate and instabilities of eruption regime in silicic volcanoes, *Earth Planet. Sci. Lett.*, 102, 413-429, 1991.

Kano, K., A Miocene coarse volcaniclastic mass-flow deposit in the Shimane Peninsula, SW Japan: product of a deep submarine eruption?, *Bull. Volcanol.*, 58, 131-143, 1996.

Kano K., K. Takeuchi, T. Yamamoto, and H. Hoshizumi, Subaqueous rhyolite block lavas in the Miocene Ushikiri Formation, Shimane Peninsula, SW Japan, *J. Volcanol. Geotherm. Res.*, 46, 241-253, 1991.

Kano, K., T. Yamamoto, K. Ono, Subaqueous eruption and emplacement of the Shinjima Pumice, Shinjima (Moeshima) Island, Kagoshima Bay, SW Japan, *J. Volcanol. Geotherm. Res.*, 71, 187-206, 1996.

Kato, Y., Woody pumice generated with submarine, *J. Geol. Soc. Japan*, 77, 193-206, 1987.

Kato, Y., Gray Pumices Drifted from Fukutoku-oka-no-ba to the Ryukyu Islands, Bull. Volcanol. Soc. Japan, 2nd Ser., 33, 21-30 (in Japanese with English abstract), 1988.

Kato, Y., The 1924 submarine eruption off Iriomotejima, *Gekkan Chikyu (Monthly Earth Science)*, 13, 644-649 (in Japanese), 1991.

Kokelaar, B.P., Magma-water interactions in subaqueous and emergent basaltic volcanism, *Bull. Volcanol.*, 48, 275-289, 1986.

Kokelaar, B.P., and C. Busby, Subaqueous explosive eruption and welding of pyroclastic deposits, *Science*, 257, 196-201.

Kurokawa, K., and I. Hirata, Grain-size characteristics of the Joetsu Ash (Unit 1), a subaqueous ash flow turbidite of early Pleistocene, central Japan., *Mem. Fac. Educ. Niigata Univ.*, 28, 15-24, 1986.

Kurokawa, K., and T. Watanabe, The SK020 (Uonuma Pink) Ash and its grain-size characteristics: the early Pleistocene subaqueous ash containing accretionary lapilli in the Niigata region, central Japan, *Mem. Fac. Educ. Niigata Univ.*, 32, 75-121, 1991.

Lowe, D.R., Sediment gravity flows: II. Depositional models with special reference to the deposits of high-density turbidity currents. *J. Sediment. Petrol.*, 52, 279-297, 1982.

Mahood, G.A., Geological evolution of a Pleistocene rhyolitic center—Sierra La Primavera, Jalisco, Mexico, *J. Volcanol. Geotherm. Res.*, 8, 199-230, 1980.

Mandeville, C.W., S. Carey, H. Sigurdsson, and J. King, Paleomagnetic evidence for high-temperature emplacement of the 1883 subaqueous pyroclastic flows from Krakatau Volcano, Indonesia. *J. Geophys. Res.*, 99, 9487-9504, 1994.

Matsumoto, T., The four gigantic caldera volcanoes of Kyushu. *Japan. J. Geol. Geograph.*, 19, 1-57, 1943.

McBirney, A.R., Factors governing the nature of submarine volcanism, *Bull. Volcano.*, 26, 455-469, 1963.

Middleton, G.V., and M.A. Hampton, Sediment gravity flows: mechanics of flow and deposition, in *Turbidites and Deep Water Sedimentation*, edited by G.V. Middleton, and A.H. Bouma, pp.1-38, SEPM Pacific Sect., 1973.

Morimoto, R., and J. Ossaka, The 1952–1953 submarine eruption of the Myojin reef near the Beyonnaise rocks, Japan (I), *Bull. Earthq. Res. Inst., Tokyo Univ.*, 33, 221-250, 1955.

Müeller, W.M., and J.D.L. White, Felsic fire-fountaining beneath Archean sea: pyroclastic deposits of the 2730 Ma Hunter Mine Group, Quebec, Canada, *J. Volcanol. Geotherm. Res.*, 54, 117-134, 1992.

Nakamura, K-I, K. Sakaguchi, T. Nagai, Marine geological survey in the Kikai caldera with special references to the occurrence of giant pumices, the detailed survey on topography and the measurement of seawater temperature, *Tech. Rep. Japan Marine Sci. and Technol. Center, Spec. Issue of the 2nd Symposium on Deep Sea Research Using the Submersible "Shinkai 2000" System*, pp.137-155 (in Japanese with English abstract), 1986.

Niino, H., and T. Kumakori, Report on the submarine eruption of Myojin-sho, *J. Tokyo. Univ. Fisheries,* 40, 1-32, 1953.

Nishimura, A., K.M. Masaglia, K.S. Rodolfo, A. Colella, R.N. Hiscott, K. Tazaki, J.B. Gill, T. Janecek, J. Firth, M. Isiminger-Kelso, Y. Herman, R.N. Taylor, B. Taylor, K. Fujioka, and Leg 126 Scientific Party, Pliocene-Quaternary submarine pumice deposits in the Sumisu Rift area, Izu-Bonin Arc, in *Sedimentation in Volcanic Settings, SEPM Spec. Publ.,* 45, edited by R.V. Fisher, and G.A. Smith, pp.201-208, 1991.

Nishimura, A., K.S. Rodolffo, A. Koizumi, A. Gill, and K. Fujioka, Episodic deposition of Pliocene-Pleistocene pumice from the Izu-Bonin Arc, in *Proc. ODP, Scientific Results*, 126, B. Taylor, K. Fujioka, et al., pp.3-21, Ocean Drilling Program, College Station, Texas, 1992.

Ossaka, J., Eruptions of submarine volcanoes in the proximity of Japan, 300pp. (in Japanese), Tokai University Press, 1991.

Reynolds, M.A., J.G. Best, and R.W. Johnson, 1953-57 Eruption of Tuluman volcano: Rhyolitic volcanic activity in the northern Bismarck Sea, Geol. Surv. Papua New Guinea Memoir 7, 1980.

Saito, G., Kazahaya, K., Shinohara, H., Stimac, and Y. Kawanabe, Variation of volatile concentration in a magma system of Satsuma-Iwojima volcano deduced from melt inclusion analyse, *J. Volcanol. Geotherm. Res.,* 108, 11-31, 2001.

Soh, W., Taira, A., Ogawa, Y., Taniguchi, H., Pickering, K.T., and Stow, D.A., Submarine depositional processes for volcaniclastic sediments in the Mio-Pliocene Misaki Formation, Miura Group, central Japan, in *Sedimentary Facies in the Active Plate Margin*, edited by A. Taira and F. Masuda, Terra Scientific Publ, Tokyo, 619-630, 1989.

Tamura, Y., M. Koyama, and R.S. Fiske, Paleomagnetic evidence for hot pyroclastic debris flow in the shallow submarine Shirahama Group (Upper Miocene-Pliocene), Japan, *J. Geophys. Res.,* 96, 21,779-21,787, 1991.

Tanakadate, H., Preliminary report of the eruption of Iwojima, Kagoshima Prefecture, *Bull. Volcanol. Soc. Japan, Ser. 1,* 2, 188-209 (in Japanese), 1935a.

Tanakadate, H., Evolution of a new volcanic islet near Io-zima (Satuma Prov.). *Proc. Imperial Academy,* 11, 152-154, 1935b.

Taylor, P. W., Volcanic hazards assessment following the September-October 2001 eruption of a previously unrecognized submarine volcano W of Vava'U, kingdome of Tonga, *Australian Volcanological Investigations Occasional Rept.,* 02/01, pp. 1-7, 2002.

Tsuchide, M., S. Kato, A. Uchida, H. Sato, N. Konishi, J. Ossaka, and J. Hirabayashi, Submarine volcanic activity at the Kaitoku Seamount in 1984, *Rep. Hydrographic Res,.* 20, 47-82 (in Japanese), 1985.

Walker, G.P.L., Ignimbrite types and ignimbrite problems, *J. Volcanol. Geotherm. Res.,* 17, 65-88, 1983.

Whitham, A.G., and R.S.J. Sparks, Pumice, *Bull Volcanol.,* 48, 209-223, 1986.

Wohletz, K.H., Explosive magma-water interactions: thermodynamics, explosion mechanisms, and field studies, *Bull. Volcanol.,* 48, 245-264, 1986.

Wohletz, K.H., and R.G. McQueen, Experimental studies of hydromagmaic volcanism: Inception, evolution, and hazards. *Studies in Geophysics*, pp.158-169, National Academy Press, Washington D.C., 1984.

Woods, A.W., and T. Koyaguchi, Transitions between explosive and effusive eruptions of silicic magmas, *Nature*, 370, 641-644, 1994.

Wright, J.V., and E. Mutti, The Dali Ash, Islands of Rhodes, Greece: A problem in interpreting submarine volcanic sediments, *Bull. Volcanol.,* 44, 153-167, 1981.

Wright, I.C., J.A. Gamble, and P.A.R. Shane, Submarine silicic volcanism of the Healy Caldera, southern Kermadec Arc (SW Pacific): I—volcanology and eruption mechanisms, *Bull. Volcanol.,* 65, 15-29, 2003.

Yamamoto, T., T. Soya, S. Suto, K. Uto, A. Takada, K. Sakaguchi, and K. Ono, The 1989 submarine eruption off eastern Izu Peninsula, Japan: ejecta and eruption mechanisms, *Bull. Volcanol.,* 53, 301-308, 1991.

Yuasa, M., Myojin Knoll, Izu-Ogasawara arc: Submersible study of submarine pumice volcano, *Bull. Volcanol. Soc. Japan*, 40, 277-284 (in Japanese with English abstract), 1995.

Yuasa, M., F. Murakami, E. Saito, and K. Watanabe, Submarine topography of seamounts on the volcanic front of the Izu-Ogasawara Bonin) arc, *Bull. Geoll. Surv. Japan*, 42, 703-743, 1991.

Zimanowski, B., G. Frohlich, and V. Lorenz, Quantitative experiments on phreatomagmatic explosions, *J. Volcanol. Geotherm. Res.*, 48, 341-358, 1991.

K. Kano, Institute of Geoscience, Geological Survey of Japan, National Institute of Advanced Industrial Science and Technology, Tsukuba Central 7, 1-1 Higashi 1-chome, Tsukuba, Ibaraki 305-8567, Japan.

Submarine Silicic Calderas on the Northern Shichito-Iwojima Ridge, Izu-Ogasawara (Bonin) Arc, Western Pacific

Makoto Yuasa

Geoinformation Division, Geological Survey of Japan, AIST, Tsukuba, Ibaraki Japan

Kazuhiko Kano

Institute of Geoscience, Geological Survey of Japan, AIST, Tsukuba, Ibaraki Japan

Eight submarine silicic calderas have been discovered on the northern Shichito-Iwojima Ridge, Izu-Ogasawara Arc. They have typical caldera structures with a circular to elliptical rim 2–10 km across with a flat or slightly concave floor 1–7 km across and a wall rising 0.1–1.1 km above the floor and gently inclining, mainly at an angle of 14–30°. They are exclusively developed on submarine stratovolcanoes with aspect ratios 1:20 and comparable in size to either Nigorikawa or Crater-Lake calderas. Gravity anomalies of the submarine calderas are not low, due to central lava domes or major mass of the stratovolcanoes underlying the calderas. Magnetic anomalies are low with narrow wavelength, presumably reflecting complex growth of stratovolcanoes. Acoustic reflection profiles show stratified layers on the flanks and within these calderas, and recent submersible surveys have disclosed a thick pumiceous sequence on the wall of Myojin Knoll caldera. Pumice deposits recovered from ODP sites 788–791 close to Minami-Sumisu and Sumisu calderas commonly contain not only pumice lapilli and blocks but also blocky or platy, variably vesicular glass shards, suggesting a submarine eruption by magma vesiculation and water-ingestion into the eruption plume. Catastrophic submarine silicic eruption and massive collapse of the eruption column by ingestion of the ambient water are likely to have resulted in thick accumulation of pumice deposits. Stratification of the deposits might represent recurrent eruption or recurrent emplacement from a suspension mass unsteadily loaded by a sustained eruption.

1. INTRODUCTION

Izu-Ogasawara (or Izu-Bonin) Arc is situated along the eastern edge of the Philippine Sea Plate in the western Pacific Ocean. The size of the whole arc including the submarine parts is about 1200 km long from Oshima to Minami-Iwojima islands and 400 km wide. From west to

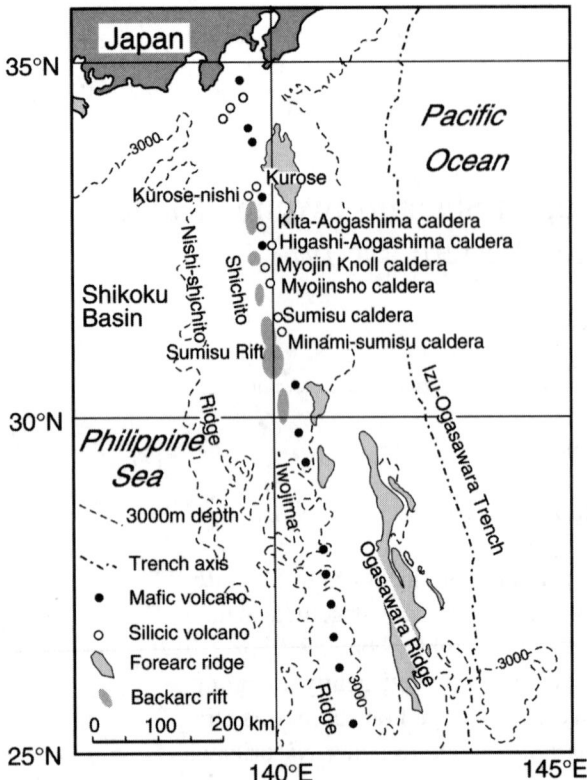

Figure 1. Topographic and geologic features of the Izu-Ogasawara Arc.

east, the arc consists of the Nishi-shichito, Shichito-Iwojima and Ogasawara ridges (Figure 1). The Shichito-Iwojima Ridge is the volcanic front of the Izu-Ogasawara Arc, comprising a chain of 13 volcanic islands from Oshima to Minami-Iwojima islands and 15 major volcanic seamounts. Most of these are composite volcanoes composed mainly of tholeiitic mafic rocks with subordinate amounts of silicic rocks, as shown by the surface geology [*Ono et al.*, 1981] and dredge samples [*Yuasa and Nohara*, 1992]. Seismic profiles and submarine topographic data have disclosed 10 submarine calderas on the northern Shichito-Iwojima Ridge [*Murakami and Ishihara*, 1985; *Yuasa et al.*, 1991; *Klaus et al.*, 1992; *Research group of the area between the Hachijoshima Is. and Aogashima Is.*, 1995]. *Yuasa et al.* [1991] have described morphological and geophysical features of nine of these submarine calderas and interpreted that seven were produced by silicic pyroclastic eruption with thick accumulation of pumice, and that the remaining two were produced by either effusion or pyroclastic eruption of mafic magma, similar to Oshima volcano on the same Shichito-Iwojima Ridge. Subsequent surveys mainly by the Shinkai 2000 submersible of the Japan Marine Science and Technology Center, however, discovered one additional submarine caldera [*Klaus et al.*, 1992; *Research group of the area between the Hachijoshima Is. and Aogashima Is.*, 1995] and allowed more detailed observations, which led to several controversial views on the origin of the submarine calderas [*Yuasa*, 1995; *Murakami*, 1997; *Sakamoto et al.*, 2000; *Fiske et al.*, 2001; *Ueda et al.*, 2001]. This paper reviews some details of the morphological and other features of the submarine calderas, and discusses the caldera features in relation to the mode of eruption in water.

Submarine silicic calderas have also been discovered on the Kermadec Ridge [*Wright and Gamble*, 1999] and its northern extension [*Worthington et al.*, 1999]. Many submarine silicic calderas plausibly remain undiscovered in submerged volcanic arcs, and we may know only a small number of the calderas. Description of the morphology and geology is the first step to understanding their origin. There is economic as well as volcanological interest in the submarine silicic calderas. Myojin Koll caldera and other submarine calderas accommodate Kuroko-type massive sulfide deposits [*Iizasa et al.*, 1999]

Table 1. Morphlogical and geophysical features of the caldera volcanoes on the northern Shichito-Iwojima Ridge, Izu-Ogasawara Arc.

Caldera name	Caldera rim			Caldera floor		Slope of wall (degree)	Central cone	Gravity anomaly	Magnetic anomaly
	Minimum depth (m)	Diameter (km)	Relative relief (m)	Maximum depth (m)	Diameter (km)				
Kurose Hole	107	5 x 7	500	760	2	14–20	absent	high	weak
Kurose-Nishi Hole	400	5 x 6	1100	1500	2	20	absent	–	–
Kita-Aogashima	280	5 x 5	160	530	2 x 3	6–22	present	–	weak
Higashi-Aogashima	180	5 x 10	100–200	805	4 x 7	14–20	absent	high	weak
Myojin Knoll	366	5 x 7	700–900	1300	5 x 6	20–30	present	high	weak
Myojinsho	10 a.s.l	7 x 9	650–900	1120	5 x 6	30–90	present	high	weak
Sumisu	136 a.s.l	6 x 9	600–700	969	5 x 6	32	present	high	weak
Minami-Sumisu	269	2 x 4	300–500	842	1	20–40	absent	–	–

2. DESCRIPTION OF SUBMARINE CALDERAS

The Shichito-Iwojima Ridge contains at least ten submarine calderas; nine calderas in the northern part and one caldera in the southern part. Eight of the nine calderas in the northern part are commonly accompanied by silicic volcaniclasts, which are low alkali, low-K tholeiitic dacite to rhyolite, as defined by *Gill* [1981] on the alkaili-silica discrimination diagram [*Yuasa and Nohara*, 1992; *Sakamoto and Tanimoto*, 1996]. The remaining two appear associated with mafic lava and/or volcaniclasts, and will not be described further in this paper. The morphological and geophysical features of the silicic calderas on the northern Shichito-Iwojima Ridge are summarized in Table 1.

2.1. Kurose Hole Caldera

Kurose Hole is a nearly circular caldera 5–7 km in diameter (Figures 2 and 3) located at 33°24′N and 139°41′E on Kurose Bank, 30 km north of Hachijojima Island. Kurose Bank is 107 m below sea level and 28 x 32 km in diameter. The caldera floor is 2 km across and 600–700 m below sea level. The floor inclines to the north and reaches a maximum water depth of about 760 m. The relief of the caldera wall from the caldera floor is about 500 m and the slope of the wall is 14–20°. The caldera rim is flat-topped, shallower than 200 m depth, and increases in depth to about 300 m toward the northern part where a canyon-like feature is developed. The canyon runs across the upper wall of the caldera rim and has steep-sided walls, perhaps

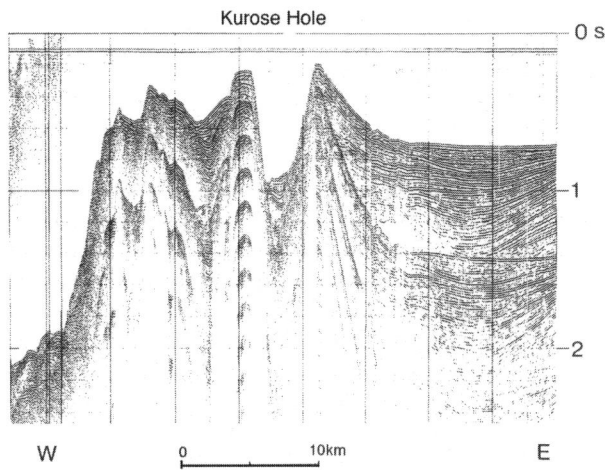

Figure 3. Acoustic profile across Kurose Hole caldera west to east [*Yuasa et al.*, 1991].

representing syn- or post-eruptive erosion by volcaniclastic mass flows.

An acoustic reflection profile in the east to west direction (Figure 3) shows 1) the caldera fill appears layered with slump folds and 2) the surface of the caldera rim comprises multiple layers dipping outward. Pyroxene-hornblende dacite pumice was recovered from the caldera wall and outer slope [*Yuasa and Nohara*, 1992]. A submersible survey found many fragments of moderately vesicular silicic pumice and limestone scattered on the floor and wall and also found limestone exposed on the lower slope of the wall [*Iwabuchi et al.*, 1989]. The limestone is poorly consolidated with a ragged, brownish surface and contains a mud components of 5–10 wt% [*Iwabuchi et al.*, 1989], suggesting a bioclastic origin.

The free-air gravity anomaly is high on Kurose caldera, but the central part is lower than the surroundings [*Ishihara and Yamazaki*, 1991]. No magnetic anomaly is recognized in this caldera [*Yamazaki et al.*, 1991], indicating that silicic rocks and less magnetic other rocks constitutes the caldera rims.

2.2. Kurose-Nishi Hole Caldera

Kurose-Nishi Hole is located at 33°18′N and 139°33′E, 20 km southwest of Kurose caldera (Figure 2). The caldera rim is about 5–6 km in diameter with the shallowest part 400 m below sea level. The floor is as deep as 1500 m, and the slope of the caldera wall is mainly about 20°. An acoustic reflection profile in the south to north direction shows 1) the caldera fill is layered with slight down-warping and 2) the surface of the caldera rim comprises layers dipping outward (Figure 4),

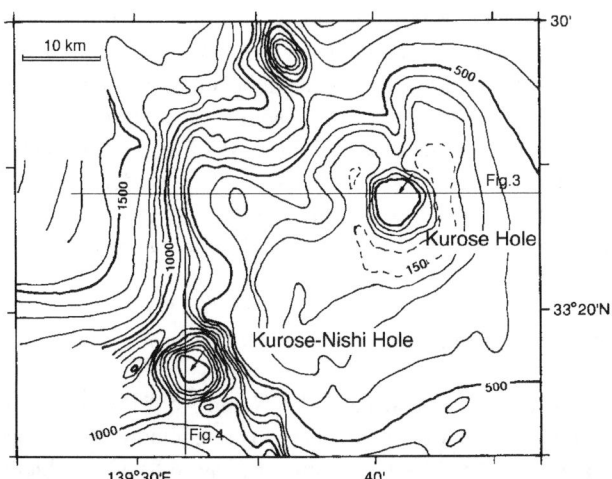

Figure 2. Submarine topography of Kurose Hole and Kurose-Nishi Hole calderas [*Saito et al.*, 1988]. Solid lines indicate locations of acoustic profile shown in Figures 3 and 4. Arrows show topographic depression.

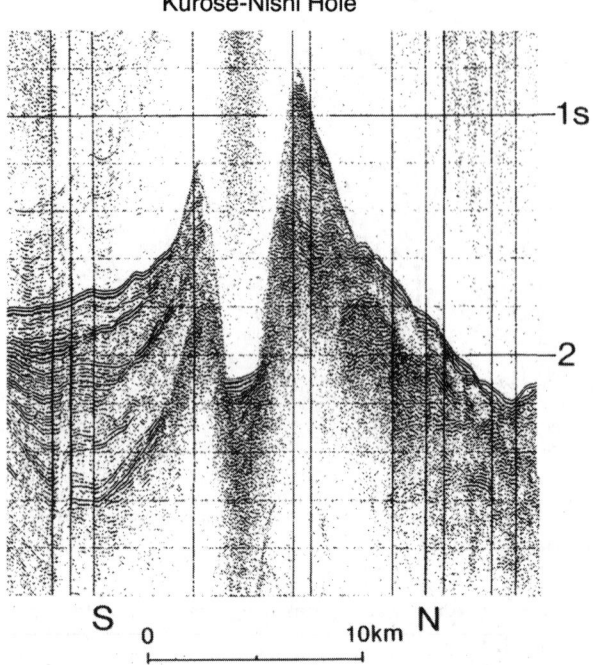

Figure 4. Acoustic profile across Kurose-Nishi Hole caldera west to east [*Yuasa et al.*, 1991].

similar to adjacent Kurose caldera. A submersible survey showed that the wall comprises interbedded volcaniclastic deposits mainly of rhyolite with collapses by slope failures or faulting in places [*Fujioka and Saito*, 1992].

2.3. Kita-Aogashima Caldera

Kita-Aogashima caldera [*Klaus et al.*, 1992] is located at 32°42'N and 139°45'E, on the broad bank south of Hachijojima Island [*Research group of the area between the Hachijoshima Is. and Aogashima Is.*, 1995], which has a poorly defined base 13 x 15 km in diameter (Figure 5). This caldera has an asymmetrical rim 5 x 5.5 km in diameter with a relief of 100 m from the caldera floor, which is 1.5 x 2 km in diameter and 500–540 m below sea level. A cone 0.9–1.4 km across and 100 m high occurs in the central southern part. A submersible survey showed 1) breccia of rhyolite lava and large blocks of rhyolite pumice cover central cone, 2) pumice and sand-to-silt-sized volcaniclasts mantle the caldera floor, and 3) a rhyolite lava dome occurs on the eastern rim [*Sakamoto and Tanimoto*, 1996; *Sakamoto et al.*, 2000]. Many pumice fragments were recovered from the caldera rim. They are variably vesicular with contraction cracks. Pumice blocks on the slope of the central cone are giant with a length up to 3–5 m and appear woody with a fibrous texture [*Sakamoto et al.*, 2000]. Such large blocks of pumice could be produced by quenching and mechanical fragmentation of a subaqueous growing dome [*Kano*, this volume].

2.4. Higashi-Aogashima Caldera

This caldera is located at 32°27'N and 139°54'E, 12 km east of the Aogashima Island. It rests on a NNE-SSW trending seamount 1723 m in height and 33 x 35 km in diameter with two peaks, Daini-Higashi-Aogashima Knoll to the north and Daisan-Higashi-Aogashima Knoll to the south (Figures 6 and 7). The shape of the caldera is elliptical with an E-W trending short diameter of 5.4 and N-S trending long diameter of 9.9 km. The caldera floor is 4–7 km across and 600–800 m below sea level. The maximum water depth of the floor is 805 m at the southern part. The caldera rim includes the two knolls at 180 m depth on the north and 275 m depth on the south. The western ridge of the caldera rim deepens to 600–700 m, at which the relative height from the caldera rim to the floor becomes only 100–200 m. The slope of the caldera wall is 14–20°, characteristically small. The caldera fill is layered and locally folded or gently inclined toward the caldera wall, and the surface of the deposits and the caldera rim appears mantled with well-stratified deposits (Figure 7). Samples recovered from the caldera wall and outer slope (southern and northern flanks of Daini-Higashi-Aogashima Knoll) include fragments of plagioclase-porphyritic orthopyroxene, clinopyroxene andesite lava, orthopyroxene dacite pumice, and altered volcanic breccia [*Yuasa and Nohara*, 1992].

The Bouguer gravity anomaly is highest at the central southern part of the caldera [*Ishihara and Yamazaki*, 1991]. A remarkable dipole magnetic anomaly, negative to the north and positive to the south, is centered on the southern part of the caldera [*Yamazaki et al.*, 1991]. This, together with the rocks recovered from the caldera rim and the asymmetrical topographic features, suggests that the Higashi-Aogashima caldera formed on a mafic composite volcano.

2.5. Myojin Knoll (Kita-Bayonnaise) Caldera

This caldera is located at 32°06'N and 139°51'E, NNW of the Bayonnaise Rocks and lies on a seamount 950 m in height and 19 x 22 km in diameter (Figure 8). The caldera rim is nearly circular 6 x 7 km in diameter (Figure 9). The caldera floor is about 1400 m below sea level and 5–6 km across. The floor is separated into the northwestern and southeastern parts by a central cone 1.8–2.2 km across and

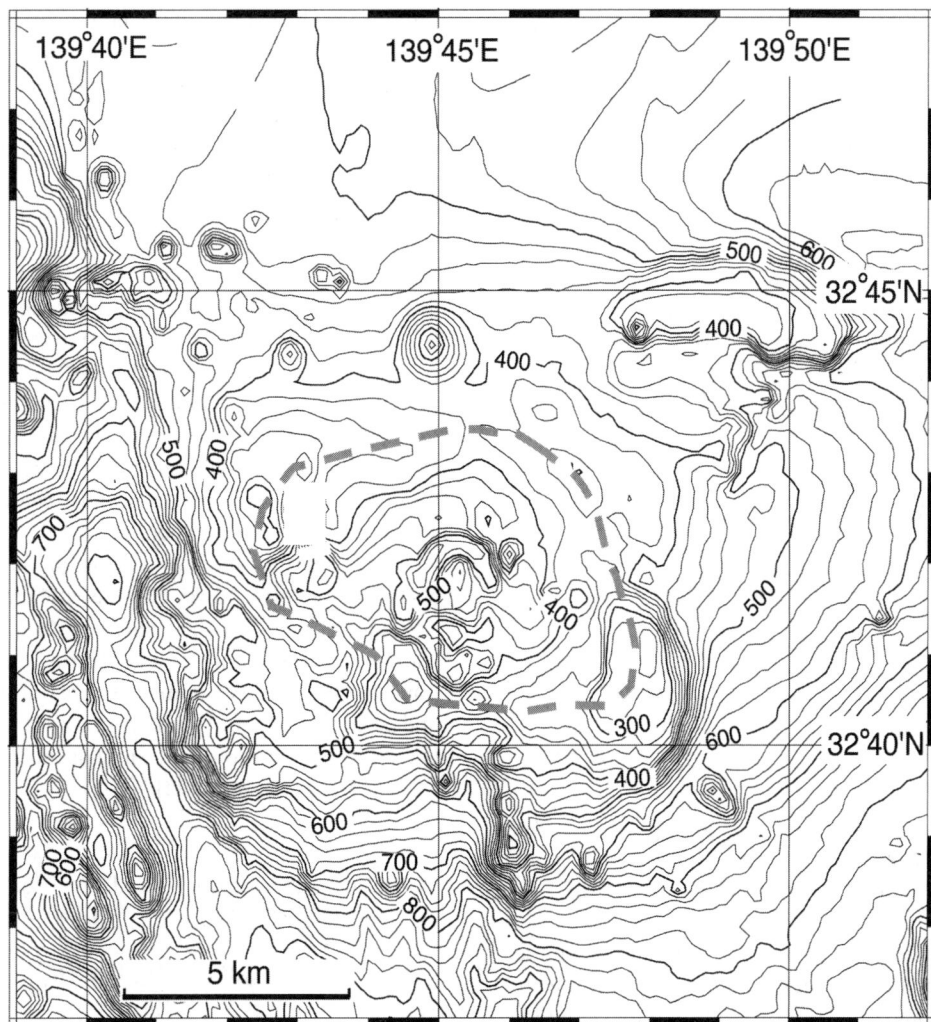

Figure 5. Submarine topography of Kita-Aogashima caldera [*Sakamoto et al.,* 2000].

250 m above the caldera floor. The caldera rim is flat-topped and generally 600–700 m deep, except a shallow point named Myojin Knoll that is 366 m deep. Relief of the caldera wall is about 500–900 m and the inner slope is as steep as 20–30°.

An acoustic profile (Figure 10) shows that this caldera volcano is composed of upper multi-layered deposits and an underlying poorly layered, vaguely resolved mass. The upper unit mantles the caldera rim and thins out at distal areas of the outer slope. This unit appears to drape even the upper slope of the caldera wall and is thus thickest on the crest of the caldera rim.

Submersible surveys [*Yuasa,* 1995; *Naka et al., 1995*; *Fiske et al.,* 2001] show a sequence of rocks exposed on the caldera wall; rhyolite lava and indurated volcaniclastic rocks in the lower part, slightly-indurated, fines-poor massive pumice breccias interbedded with thinner fine pumice breccia and lapilli tuff in the middle part, and loosely-packed, fines-poor pumice breccia to lapilli tuff in the upper part. The middle part is 200–300 m thick and the upper unit is 150–200 m thick. The upper unit appears massive, but has local stratifications represented by different grainsize populations and contains 5–15 vol.% of rhyolititic lithic fragments in its upper half. Constituent pumices in the middle and upper units are rhyolites, chemically indistinguishable from one another [*Yuasa and Nohara,* 1992; *Fiske et al.,* 2001]. Sparse rhyolite lithics in the upper unit are petrographically similar to the rhyolite lava of the lower unit and are likely to have been derived from it with widening of the vent [*Fiske et al.,* 2001]. The upper multi-layered acoustic

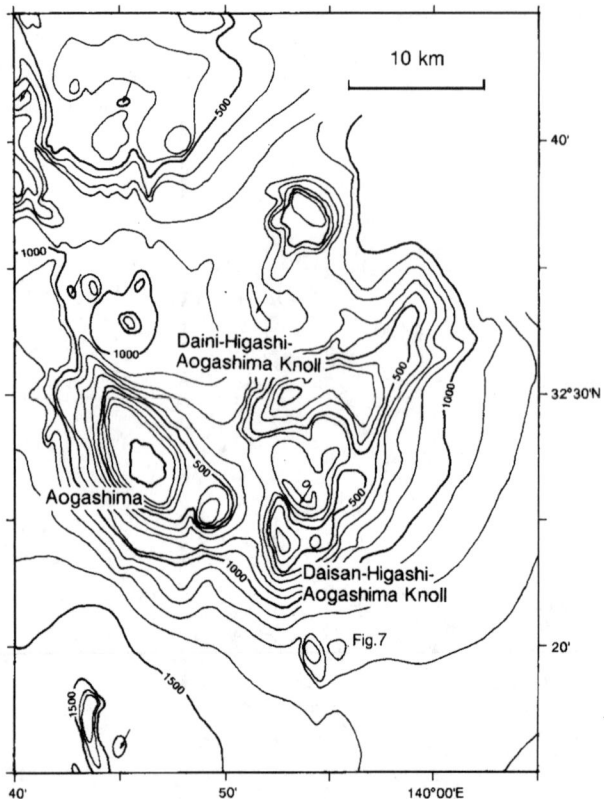

Figure 6. Submarine topography of Higashi-Aogashima caldera [*Saito et al.*, 1988]. Solid line indicates location of acoustic profile shown in Figure 7. Arrows show topographic depression.

unit likely corresponds to the upper and middle pumice units, and the lower acoustic unit perhaps represents the pre-caldera rhyolite complex. The relatively indurated middle pumice unit is, however, interpreted by *Fiske et al.* [2001] to have been produced prior to the caldera formation. The original edifice covered with the middle pumice unit was likely at a water depth of 500–900 m as extrapolated from the upper depositional surface [*Fiske et al.*, 2001].

The central cone is a rhyolite dome protruded through the debris covering the caldera floor. A low gravity anomaly is observed on the eastern outer slope, whereas the central part of the caldera is relatively high. No remarkable magnetic anomaly is recognized in this caldera. The high gravity anomaly is thus attributed to the central dome of low magnetic intensity [*Murakami*, 1997].

2.6. Myojinsho Caldera

This caldera is located at 31°53′N and 139°59′E, and lies on a seamount 1410 m in height and 28 x 32 km in diameter. The Bayonnaise Rocks are a peak on its western caldera rim (Figure 8). The caldera rim is horseshoe-shaped, open to the north, and 7–9 km across (Figure 11). Myojinsho, after which this caldera is named, is an active dacitic volcano on the northeastern caldera rim. The northern part of the caldera rim deepens to 700 m depth. The diameter of the caldera floor is 6.5 km in the east to west direction and 5 km in the north to south direction. The deepest point of the floor is 1120 m in the western part, whereas the northeast part shoals to 600 m depth because of the growth of Myojinsho volcano. The relief of the caldera is 650 to 900 m. The slope of the caldera wall is steep and inclines 30°. A large cone with a top 330 m deep occupies the central part of the caldera floor. Acoustic profiles (Figure 12) show multiple layers that mantle the caldera rim and fill the caldera. The caldera floor deepens away from Myojinsho volcano, demonstrating some contribution of the post-caldera volcanic products to the caldera floor. No samples of caldera fill have been recovered.

Free-air gravity anomaly and magnetic anomaly are high at the Bayonnaise Rocks and the volcanoes on the caldera rim, and are low in the caldera with a local subtle high anomaly corresponding to the central cone. Model calculations for the gravity anomaly suggest a dense silicic mass underlies the caldera [*Ueda et al.*, 2001]. The central cone is a dacite dome protruded through the debris covering the caldera floor [*Iizasa*, 1992].

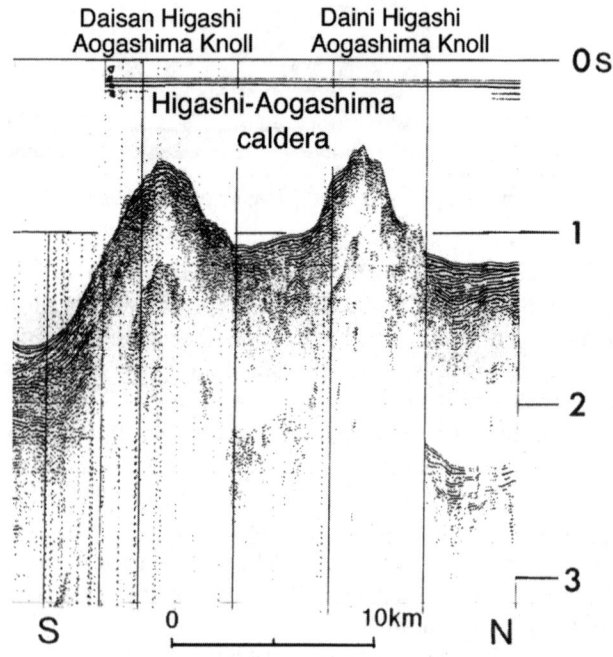

Figure 7. Acoustic profiles across Higashi-Aogashima caldera south to north [*Yuasa et al.*, 1991].

2.7. Sumisu Caldera

Sumisu caldera is located at 31°19'N and 140°03'E, 6 km north of the Sumisujima Island (136 m above sea level), which is a peak on its caldera rim. This caldera lies on a large seamount with a basal diameter of about 20 km (Figure 13) and has a nearly circular rim about 6–9 km across (Figure 14). The caldera floor is 5–6 km across and 800–969 m. Relief of the caldera is 600 to 700 m. Small cones 1–2 km across and 100–200 m high lie along the rim and center of the caldera floor, and a large cone, Shirane, with a relief of 700 m and a diameter of 2–3 km occupies the eastern part of the floor. The caldera wall is steep with an average angle of 32° (Figure 14). Acoustic profiles show layers dipping outward on the outer slope, whereas acoustic layers are warped upward in the central and marginal parts of the caldera floor, perhaps due to intrusion of lava domes into the caldera fill (Figure 15). Dredge samples from the caldera wall, the upper outer slope, and the eastern large cone are andesitic and rhyolitic lithics, dacite pumice, and poorly vesicular dacite blocks [*Yuasa and Nohara*, 1992]. These collectively suggest that the acoustic layers covering the caldera rim comprise mainly pumice, and the eastern cone is a dacite lava dome. A mudstone recently recovered from the caldera wall by a submersible survey contains nannofossils, which suggest that the caldera is younger than 0.16 Ma (Iwabuchi, 1999).

Sumisu caldera has a gravity anomaly about 25 mGal (Bouguer) higher than the surrounding area [*Ishihara and Yamazaki*, 1991] and a strong dipole magnetic anomaly, negative to the north and positive to the south [*Yamazaki et al.*,

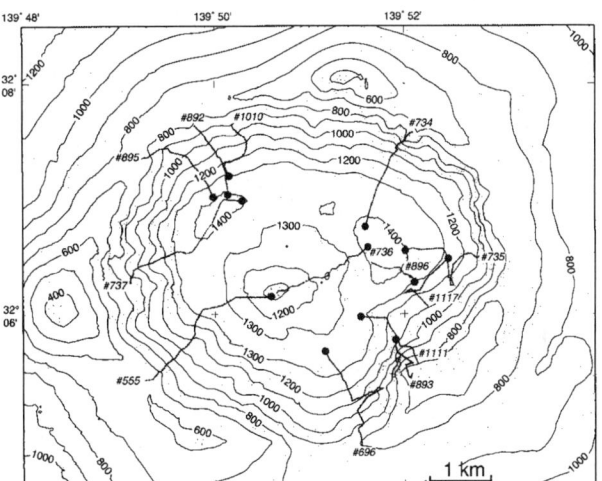

Figure 9. Submarine topography of Myojin Knoll caldera [*Fiske et al.*, 2001]. Contour interval is 100 m. Diving routes and numbers are shown.

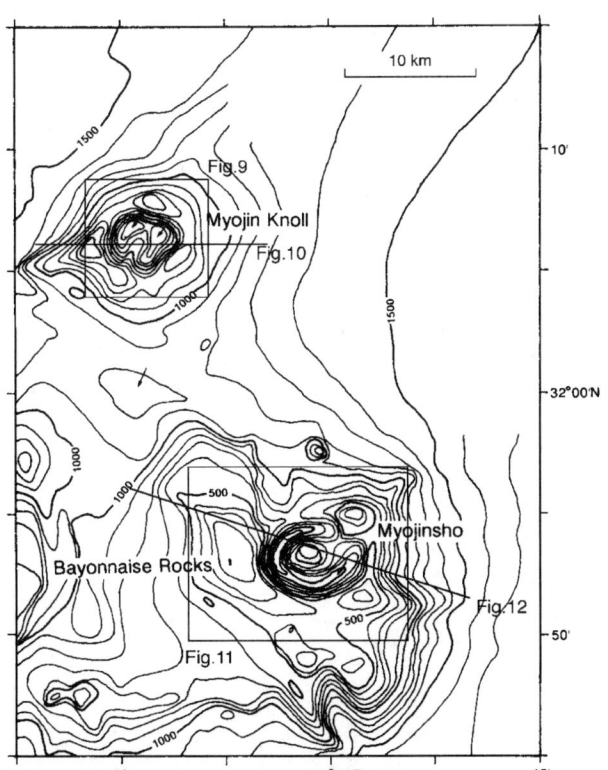

Figure 8. Submarine topography of Myojin Knoll and Myojinsho calderas, and their environs [*Saito et al.*, 1988]. Solid lines indicate locations of acoustic profile shown in Figs. 10 and 12. Boxes show locations of Figures 9 and 11. Arrows show topographic depression.

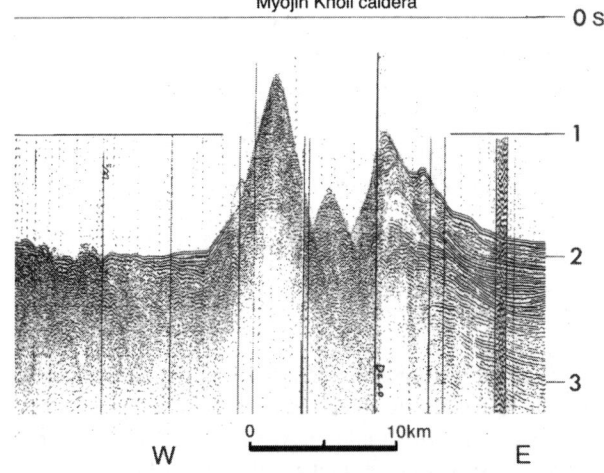

Figure 10. Acoustic profile across Myojin Knoll caldera west to east [*Yuasa et al.*, 1991].

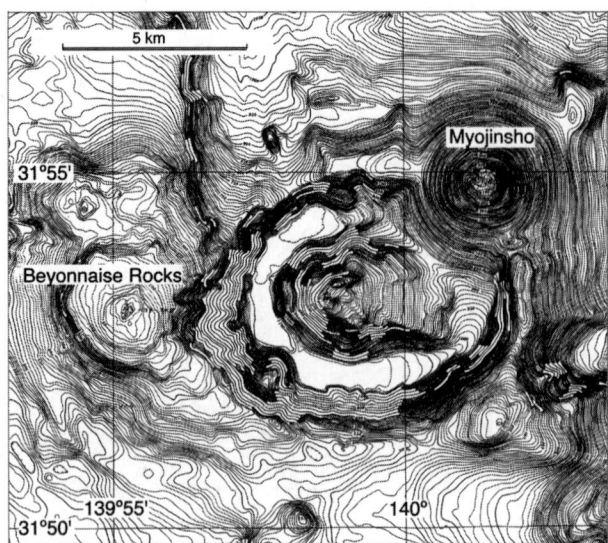

Figure 11. Submarine topography of Myojinsho caldera [*Japan Coast Guard*, 2000]. Contour interval is 10 m.

1991]. These geophysical features suggest that the seamount is composed mainly of mafic rocks and the mass deficiency attributed to caldera formation is small relative to the total mass of the seamount.

2.8. Minami-Sumisu Caldera

This small caldera 2–4 km in diameter is located at 31°16'N and 140°04'E on Daisan-Sumisu Knoll, which is 10 km south of the Sumisujima Island and 850 m in height and 17 x 19 km in diameter (Figure 13). The caldera floor is 1 km across with a maximum depth of 842 m. The summit of the caldera rim is 300–500 m below sea level, although a peak rises to 269 m in the northwestern part. The slope of the caldera wall is 20–40°. Acoustic profiles show poorly resolved layers dipping parallel to the outer slope and infilling the caldera (Figure 16). Dredge samples from the caldera wall include rhyolite volcanic breccia and dacite pumice [*Yuasa and Nohara*, 1992].

3. DISCUSSION

3.1. Caldera Structures

The submarine silicic calderas on the northern Shichito-Iwojima Ridge are smaller than or comparable to Crater Lake caldera, in size (Table 1). Acoustic profiles show stratified deposits mantle the flanks and partly fill the insides of the calderas. Locally recovered samples suggest the deposits are mainly silicic pumice. Some calderas have active volcanoes and similar landforms on the rims and lava domes centered on the floors. These features of the northern Shichito-Iwojima submarine calderas are common to collapse calderas [*Smith*, 1960; *Smith and Bailey*, 1968; *McBirney*, 1990; *Lipman*, 1997].

Silicic calderas are filled mainly with loosely packed vesicular fragments that produce low gravity anomalies, and mafic calderas are filled mainly with dense lava that produces high gravity anomalies [*Yokoyama*, 1974, 1983; *Yokoyama and Ohkawa*, 1986]. Magnetic anomalies are, therefore, low in silicic calderas and high in mafic calderas. The submarine calderas on the northern Shichito-Iwojima Ridge are associated with high gravity anomalies and in many cases, have a low magnetic intensity. Close examination at Myojin Knoll caldera [*Murakami*, 1997] and Myojinsho caldera [*Ueda et al.*, 2001], however, suggest that the high gravity anomalies arise from the central lava dome or an underlying large mass of mafic or dense lava. Higashi-Aogashima and Sumisu calderas have a density of 2.4 g/cm^3 whereas adjacent non-caldera volcanoes have a density of 2.6-3.0 g/cm^3 [*Ishihara*, 1987]. This implies that the caldera fills comprise a significant amount of loosely packed fragments or vesicular materials.

Caldera structures could be produced by 1) massive collapse of seamounts with eruption of large amounts of magma, 2) explosive erosion and retrogressive collapse of the conduits, or combination of these. The former process has been widely accepted for large calderas of Crater Lake type [*Lipman*, 1997], piecemeal calderas [*Branney and Kokelaar*, 1994], or downsag calderas [*Branney*, 1995]. The latter process has been proposed by *Yokoyama* [1974, 1983], *Yokoyama and Ohkawa* [1986], and *Aramaki* [1983, 1984], and might be applied to Nigorikawa caldera [*Ando*, 1981,

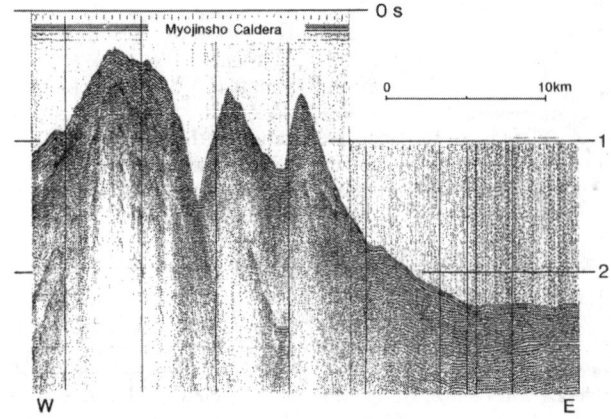

Figure 12. Acoustic profile across Myojinsho caldera west to east [*Yuasa et al.*, 1991].

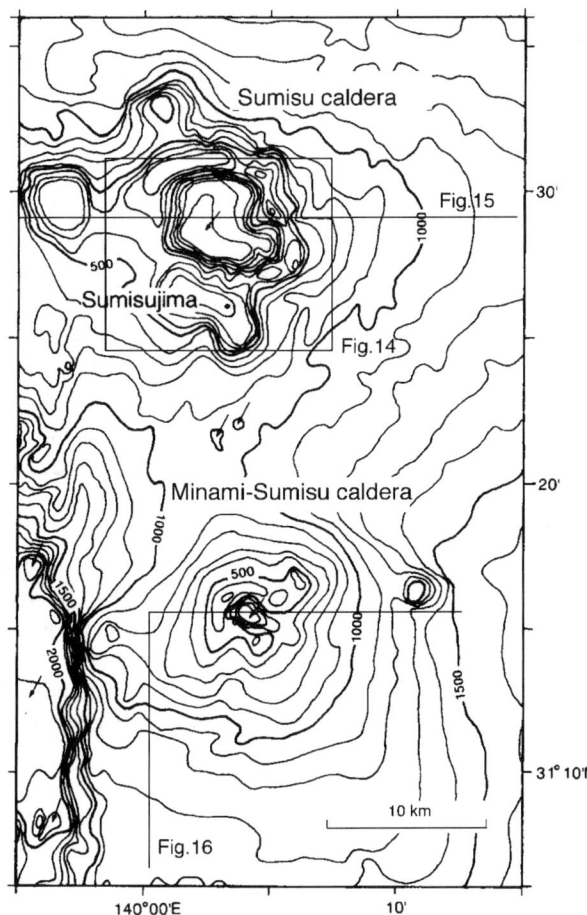

Figure 13. Submarine topography of Sumisu and Minami-Sumisu calderas, and their environs [*Saito et al.*, 1988]. Solid lines indicate locations of acoustic profile shown in Figures 15 and 16. Box shows location of Figure 14. Arrows show topographic depression.

1983], Sunakohara caldera [*Yamamoto*, 1992], Novarupta [*Eichelberger and Hildreth*, 1986], Shiotani conduit [*Kano et al.*, 1997], and other relatively small calderas or large funnel-shaped eruption conduits. *Aramaki* [1983, 1984] envisaged the internal structure of small calderas to be funnel-shaped like an explosion crater, mainly based on the drilling results in Nigorikawa caldera [*Ando*, 1981, 1983] and named this type of small caldera as Haruna type after Haruna volcano, central Japan.

Murakami [1997] suggested that either process is plausible for Myojin Knoll caldera because of mass deficiency attributable to the caldera formation is small. *Ueda et al.* [2001] envisaged the former process for Myojinsho caldera because the observed high gravity anomaly over the Myojinsho caldera is attributable to the central lava dome.

The submarine calderas on the Shichito-Iwojima Ridge have diameters smaller than or comparable to Crater-Lake-type calderas, and have smaller inner slopes of 10–40° than those expected for the ring faults of Crater-Lake-type calderas. The slopes are also smaller than the angle of repose of gravitationally collapsed hard rocks, and are not simply attributed to retrogressive slumping of caldera rims. The caldera wall of Myojinsho caldera has an upper gentle slope and a lower steep slope. The slope changes from 30° to 90° toward the caldera floor [*Ueda et al.*, 2001], similar to a funnel. However, the topographical and structural criteria for distinction between the Crater Lake type and the Nigorikawa type are also ambiguous; even in the Nigorikawa type calderas, ring faults may be present at depths of the funnel-shaped conduit under the cover of volcaniclastic materials [*Lipman*, 1997].

Sakamoto et al. [2000] argued that Kita-Aogashima caldera represents a down sag structure by effusion of a large volume of rhyolite lava. This is because the caldera is quite shallow and rhyolite lava occupies the major area of the caldera. Pumices and volcaniclastic sands, however, spread over the caldera floor. No acoustic profile to examine the internal structure.

3.2. Caldera-Forming Submarine Pumice Eruptions

A large-volume submarine pumice eruptions are envisaged for the Shichito-Iwojima submarine silicic calderas. Silicic

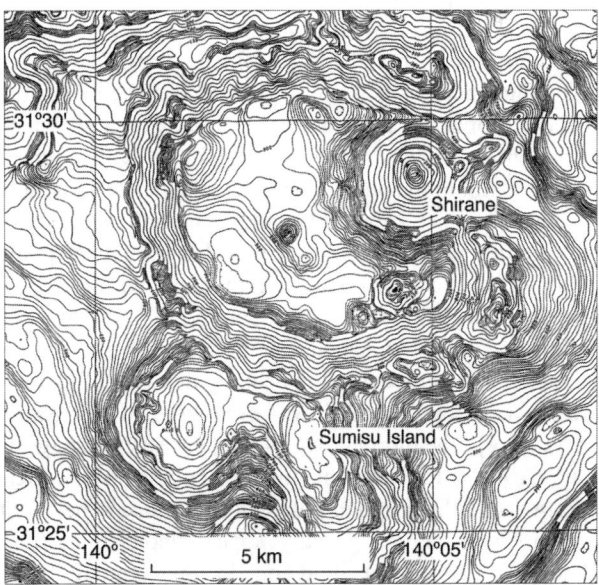

Figure 14. Submarine topography of Sumisu caldera [*Maritime Safety Agency, Japan.*, 1997]. Contour interval is 10 m.

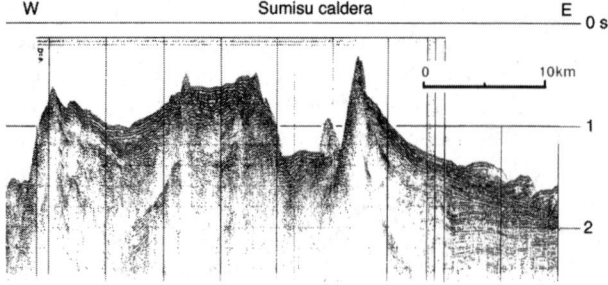

Figure 15. Acoustic profile across Sumisu caldera west to east [*Yuasa et al., 1991*].

pumices have been dredged from seven submarine calderas and/or directly observed by submersible surveys. Acoustic layers mantle the outer slopes. These observations collectively demonstrate thick accumulation of pumice in the calderas and even distal flanks. *Fiske et al.* [2001] estimated an eruption volume exceeding 40 km³ from Myojin Knoll caldera, only 6–7 km across. Other submarine calderas of similar size are also likely to have individually ejected similar volumes for a total volume in excess of 10^2 km³ in total. This large volume of ejecta appears to be dispersed over the Shichito-Iwojima Ridge and in adjacent Sumisu rift and other backarc rift basins [*Nishimura et al., 1991, 1992*].

The detailed features of the deposits from the Shichito-Iwojima calderas have not been examined, but the Plio-Pleistocene rhyolitic pumice deposits at ODP site 788 on a 1000 m-deep part of the Shichito-Iwojima Ridge, Izu-Ogasawara (Bonin) Arc [*Nishimura et al., 1991, 1992*] may be representative of some of the features. The source of the pumice deposits is unknown but could be one of the adjacent submarine calderas, Torishima caldera 40 km to the south [*Nishimura et al., 1991*] or Minami-Sumisu caldera 30 km to the north. The pumice deposits constitute a bedded sequence 250 m thick and commonly contain not only pumice lapilli and blocks but also blocky or platy, variably vesicular glass shards. Individual beds are 20–50 meters or less in thickness, and commonly show inverse grading of coarse pumice clasts, one of the characteristics proposed to be inherent to the primary deposits of submarine pumice eruption-fed flows [*Kano et al., 1996; Kano,* this volume]. The morphology of the constituents suggests a submarine eruption by magma vesiculation and water-ingestion into the eruption plume. The pumice deposits of Myojin Knoll caldera visually appear depleted in fines [*Fiske et al., 2001*], but could contain some amount of ash components, similar to other subaqueous deposits emplaced directly from subaqueous eruption-fed flows [*Kano,* this volume].

Submarine pumice eruptions of this kind could arise mainly from catastrophic magma vesiculation. The depths of the eruptions on the Shichito-Iwojima Ridge remain open to question, but could be above the caldera floors 600–1300 m deep and even above the minimum water depths of the caldera rims, which are between 100 and 400 m except where post-caldera volcanoes occur on the rims (Table 1). At the Myojin Knoll caldera, the pre- and syn-caldera pumice eruptions are inferred to have occurred at water depths of 500–900 m and 300 m by extrapolating the upper slopes of the deposit surfaces toward the center [*Fiske et al., 2001*]. *McBirney* [1963] proposed that explosive vesiculation of rhyolite magma with a low volatile content occurs mainly at water depths shallower than 500 m. At higher volatile contents, explosive magma vesiculation is plausible even in deeper water. On the Kermadec Ridge north of Taupo volcanic zone, New Zealand, the submarine silicic caldera Healy is 2–2.5 km across with a caldera floor 1660–1690 m deep [*Wright and Gamble, 1999*]. This caldera was produced by a large volume explosive eruption of silicic magma even at an inferred water depth of 550–1000 m. The resulting rhyolitic pumice contains 6 wt.% H_2O, which allows development of an eruption column with pyroclasts having a vesicularity of 80–85 vol.%, at the magma temperature of 850°C estimated by pyroxene thermometry [*Wright et al., 2002*].

In subaqueous settings, water pressure suppresses explosive magma vesiculation but magma may interact with water-saturated sediments or water at shallow sub-bottom levels and widen the vent with less thermal energy than in eruptions by magma vesiculation. A correlation of crater or caldera diameter to eruption volume (or eruption energy) shows a one or two-order-larger energy is required to produce the same crater or caldera diameter in magmatic eruption than in phreatomagmatic eruption [*Sato and Taniguchi, 1997*], perhaps because magmatic eruptions occur at a level deeper than where phreatomagmatic eruptions occur. In addition, phreatomagmatic eruptions recur so long as fresh magma is

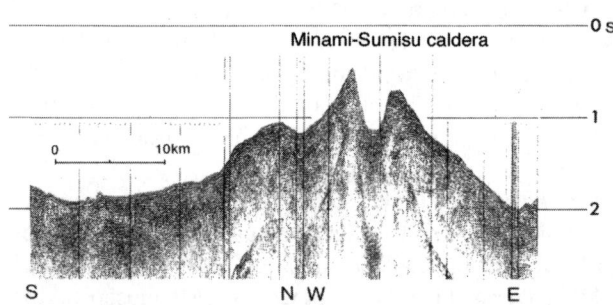

Figure 16. Acoustic profile across Minami-Sumisu caldera west to east [*Yuasa et al., 1991*].

exposed directly to water. The resulting explosion crater may thus widen to a size of small crater. Immediate collapse of the eruption plume by water-ingestion results in thick massive accumulation of pumice in and around the eruption center [*Kano, this volume*], which produces a landform like a shallow-concave caldera or tuffring, or tuffcone [*Yuasa et al., 1991*]. Pumice is emplaced with elutriation of fines from the suspension of the collapsed eruption plume, and the deposits are relatively depleted in fines [*Fiske et al., 2001*], whereas the deposits from eruption-fed flows are less sorted and commonly contain ash components [*Nishimura et al., 1992*].

The kinetic energy to destroy wall rocks may not be so large to produce explosion craters or calderas with diameters of several kilometers [*Scandone, 1990*]. Nevertheless, there is a tuffring or shallow concave caldera structure 2 km across and 600 m deep produced with repeated phreatomagmatic silicic eruptions in the Oga Peninsula, northeast Japan [*Kano et al., 2002*]. This occurrence suggests that recurrent pumice eruptions plausibly enlarge the caldera structure. Locally stratified pumice deposits in the Myojin Knoll caldera acoustic layers perhaps represent multiple submarine pumice eruptions or repetitive emplacements from a sustained eruption. Four units of submarine pumice flow at ODP site 788 span a time interval of two or three million years from Late Pliocene to Pleistocene [*Nishimura et al., 1992*] and suggest that eruptions recur on average every 10^5 years, provided the four units had a single source.

4. CONCLUSIONS

There are at least eight submarine silicic calderas on the Shichito-Iwojima Ridge, Izu-Ogasawara (Bonin) Arc. They are likely to have been produced by submarine explosive pumice eruptions. The morphological, geological and geophysical features are comparable to small calderas presumed to have either funnel-shaped vent-conduit systems or ring fault systems, though we have no conclusive evidence at present for their internal structures. Water depths where caldera-forming pumice eruptions occurred may be equal to or shallower than 500–1000 m. Catastrophic vesiculation of silicic magma is plausible perhaps even below this water depth with higher volatile contents, with fragmentation enhanced by magma-water interaction. Collapse of the eruption plume occurs by ingestion of ambient water leading to massive emplacement of pumice in and around the submarine calderas.

Acknowledgments. Discussion with R.S. Fiske was stimulating. Thoughtful reviews by J. Gill and C. Busby and editorial advice by D. Clague were helpful to improve the manuscript. The Geological Survey of Japan financially supported this research. Japan Marine Science and Technology Center kindly approved submersible observations by M. Yuasa.

REFERENCES

Ando, S., An example of the structure of Crater Lake-type caldera, Nigorikawa caldera, southwest Hokkaido, Japan, Abstracts, 1981 IAVCEI Symposium—Arc Volcanism, pp. 9-10, 1981.

Ando, S., Structure of the Nigorikawa caldera interpreted from borehole data, *Gekkan Chikyu (Earth Monthly)*, 5, 116-121 (in Japanese), 1983.

Aramaki, S., Some problems on the genesis of Japanese calderas, *Mining Geol. Special Issue*, 11, 139-154 (in Japanese with English abstract), 1983.

Aramaki, S., Formation of Aira caldera, southern Kyushu, 22,000 years ago, *J. Geophys, Res.*, 89, 8485-8501, 1984.

Branney, M.J., Downsag and extension at calderas. New perspective on collapse geometries from ice-melt, mining, and volcanic subsidence, *Bull, Volcanol.*, 57, 303-318, 1995.

Branney, M.J. and P. Kokelaar, Volcanic faulting, soft state deformation, and rheomorphism of tuffs during development of piecemeal caldera, English Lake District, *Geol. Soc. Am. Bull.*, 106, 507-530, 1994.

Eichelberger, J.C. and W. Hildreth, Research drilling at Katmai, Alaska, *EOS, Trans. Am. Geophys, Union*, 67, 778-780, 1986.

Fiske, R.S., J. Naka, K. Iizasa, M. Yuasa, and A. Klaus, Submarine silicic caldera at the front of the Izu-Bonin arc, Japan: Voluminous seafloor eruptions of rhyolite pumice, *Geol. Soc. Am. Bull.*, 113, 813-824, 2001.

Fujioka, K. and S. Saito, Is Kurose-nishi Hole cauldron ?, *Japan Mar. Sci. Technol. Center Tech. Rept. Deepsea Res.*, 8, 221-228 (in Japanese with English abstract), 1992.

Gill, J., *Orogenic Andesites and Plate Tectonics*, Springer-Verlag, Berlin-Hiderberg-New York, 390p., 1981.

Iizasa, K., Hydrothermal activities in the floor of submarine Myojinsho caldera, Izu-Ogasawara Arc, *Gekkan Chikyu (Earth Monthly)*, 158, 499-505 (in Japanese), 1992.

Iizasa, K., R.S. Fiske, O. Ishizuka, O., M. Yuasa, J. Hashimoto, J. Ishibashi, J. Naka, Y. Horii, Y. Fujiwara, A. Imai, A., and S. Koyama, A Kuroko-type polymetallic sulfide deposit in a submarine silicic caldera, *Science*, 283, 975-977, 1999.

Ishihara, T., Gravimetric determination of densities of seamounts along the Bonin Arc, in *Seamounts, islands, and atolls*, edited by B. H. Keating *et al.*, *Geophys. Monogr. Series*, 43, 97-113, 1987.

Ishihara, T. and T. Yamazaki, Gravity anomalies over the Izu-Ogasawara (Bonin) and northern Mariana Arcs, *Bull. Geoll. Surv. Japan*, 42, 687-702, 1991.

Iwabuchi, Y., Sumisu Caldera, *Japan Mar. Sci. Technol. Center J. Deep Sea Res.*, 15, 83-94 (in Japanese with English abstract), 1999.

Iwabuchi, Y., J. Ashi, and K. Fujioka, Geological and morphological survey of the Kurose Hole, north of the Hachijo Island, *Japan Mar. Sci. Technol. Center Tech. Rept. Deepsea Res.*, 5, 37-45 (in Japanese with English abstract), 1989.

Japan Coast Guard, The Basic Map of the Sea in Coastal Waters (1: 50,000), Beyonesu Retugan: Submarine structural chart and report of survey, pp. 1-32 (in Japanese), Japan Coast Guard, Tokyo, 2000.

Kano, K., Subaqueous pumice eruptions and their deposits, in *Subaqueous explosive volcanism*, edited by J. White, D. Clague, and Smellie, *AGU Monograph*, this volume.

Kano, K., H. Matsuura, and S. Yamauchi, Miocene rhyolitic welded tuff infilling a funnel shaped eruption conduit, Shiotani, southeast of Matsue, SW Japan, *Bull. Volcanol.*, 59, 125-135, 1997.

Kano, K., T. Ohguchi, S. Hayashi, K. Uto, and T. Danhara, Toga volcano: An alkali-rhyolite tuff-ring in the western end of Oga Peninsula, NE Japan, *Bull. Volcanol. Soc. Japan*, 47, xxx-xxx (in Japanese with English abstract), 2002.

Kano, K., T. Yamamoto, and K. Ono, Subaqueous eruption and emplacement of the Shinjima Pumice, Shinjima (Moeshima) Island, Kagoshima Bay, SW Japan, *J. Volcanol. Geotherm. Res.*, 71, 187-206, 1996.

Klaus, A., B. Taylor, G.F. Moore, F. Murakami, and Y. Okamura, Back-arc rifting in the Izu-Bonin Arc: Structural evolution of Hachijyo and Aogashima Rifts, *The Island Arc*, 1, 16-31, 1992.

Lipman, P. W., Subsidence of Ash-flow calderas: relation to caldera size and magma-chamber geometry, *Bull. Volcanol.*, 59, 198-218, 1997.

Maritime Safety Agency, Japan, The Basic Map of the Sea in Coastal Waters (1: 50,000), Sumisu Sima: Bathymetric chart, submarine structural chart and report of survey, pp. 1-34 (in Japanese), Maritime Safety Agency, Tokyo, 1997.

McBirney, A.R., Factors governing the nature of submarine volcanism, *Bull. Volcano.*, 26, 455-469, 1963.

McBirney, A.R., An historical note on the origin of calderas, *J. Volcanol. Geotherm. Res.*, 42, 303-306, 1990.

Murakami, F., The formation mechanism of the submarine caldera on Myojin Knoll in the northern part of the Izu-Ogasawara (Bonin) Arc, *J. Geograph.*, 106 (in Japanese with English abstract), 70-86, 1997.

Murakami, F. and T. Ishihara, Submarine calderas discovered in the northern part of Izu-Ogasawara Arc, *Gekkan Chikyu (Earth Monthly)*, 7, 638-646 (in Japanese), 1985.

Naka, J., R.S. Fiske, A. Taira, F. Yamamoto, K. Iizasa, K., and M. Yuasa, Submarine geology of Myojin Knoll, northwest of Myojinsho, Izu Islands, *Japan Marine Sci. and Technol. Center Symp. on Deep Sea Res. Rept.*, 11, 323-331 (in Japanese with English abstract), 1995.

Nishimura, A., K.M. Masaglia, K.S. Rodolfo, A. Colella, R.N. Hiscott, K. Tazaki, J.B. Gill, T. Janecek, J. Firth, M. Isiminger-Kelso, Y. Herman, R.N. Taylor, B. Taylor, K. Fujioka, and Leg 126 Scientific Party, Pliocene-Quaternary submarine pumice deposits in the Sumisu Rift area, Izu-Bonin Arc, in *Sedimentation in Volcanic Settings, SEPM Spec. Publ.*, 45, edited by R.V. Fisher, and G.A. Smith, pp.201-208, 1991.

Nishimura, A., K.S. Rodolffo, A. Koizumi, A. Gill, and K. Fujioka, Episodic deposition of Pliocene-Pleistocene pumice from the Izu-Bonin Arc, in *Proc. ODP, Scientific Results*, 126, B. Taylor, K. Fujioka, et al., pp.3-21, Ocean Drilling Program, College Station, Texas, 1992.

Ono, K., T. Soya, and K. Mimura, eds., *Volcanoes of Japan, Second Edition*, Geological Survey of Japan, 1 sheet map, 1981.

Research group of the area between the Hachijoshima Is. and Aogashima Is., Geological and geophysical study of the area between the Hachijoshima Is. and Aogashima Is. (Report of 1990 to 1992 expeditional cruises by Tokai Daigakumaru II), *J. School Mar. Sci Technol,, Tokai Univ.*, 40, 41-62 (in Japanese with English abstract), 1995.

Saito, E., K. Watanabe, J. Miyazaki, J., and F. Murakami, Bathymetry of submarine calderas in the Izu-Ogasawara Arc, in *1987 F.Y. report for Investigation of heavy metal resources accompanied with submarine hydrothermal activity*, Geol. Surv. Japan, pp. 10-15, 1988.

Sakamoto, I. and H. Tanimoto, Rhyolitic pumice dredged from the bank located between Hachijyoshima Is. and Aogashima Is., *J. School Mar. Sci Technol,, Tokai Univ.*, 41, 139-156 (in Japanese with English abstract), 1996.

Sakamoto, I., S. Wu, T. Sato, and M. Ishida, Topographical and geological characteristics of 'South-Hachijyo Bank', located between Hachijyojima Island and Aogashima Island, *Japan Mar. Sci. Technol Center J. Deep Sea Res.*,16, 69-85 (in Japanese with English abstract), 2000.

Sato, H. and H. Taniguchi, Relationship between crater size and ejecta volume of recent magmatic and phreatomagmatic eruptions; Implications for energy partitioning, *Geophys. Res. Lett.*, 24, 205-208, 1997.

Scandone, R., Chaotic collapse of calderas, *J. Volcano. Geotherm. Res.*, 42, 285-302, 1990.

Smith, R.L., Ash flows, *Geol. Soc. Am. Bull.*, 71, 795-842, 1960.

Smith, R.L. and R.A. Bailey, Resurgent cauldrons, *Geolo. Soc. Am. Mem..* 116, 83-104, 1968.

Ueda, Y., K. Onodera, Y. Ootani, and A. Suzuki, Geophysical structure of the Myojin-sho caldera and its volcanological interpretation, *Bull. Geol. Soc. Japan*, 46, 17-185 (in Japanese with English abstract), 2001.

Wortington, T.J., M.R. Gregory, and V. Bondarenko, The Denham Caldera on Raoul Volcano: dacitic volcanism in the Tongna-Kermadec arc, *J. Volcanol. Geotherm. Res.* 90, 29-48, 1999.

Wright, I.C. and J.A. Gamble, Southern Kermadec submarine caldera arc volcanoes (SW Pacific): caldera formation by effusive and pyroclastic eruption, *Mar. Geol.*, 161, 207-227, 1999.

Wright, I.C., J.A. Gamble, and P.A. Shane, "Deep-water" submarine pyroclastic volcanology of the Healy Caldera, southern Kermadec Arc (SW Pacific), *AGU Chapman Conference on Explosive Subaqueous Volcanism, Dunedin, New Zealand, Proceedings*, pp.33-34, AGU, Washington D.C., 2002.

Yamamoto, T., The middle Pleistocene explosive volcanism in Sunagohara cakdera volcano, Aizu, Japan: evidence from non-marine volcaniclastic facies of the Todera formation, *J. Geolo. Soc. Japan,* 99, 721-737 (in Japanese with English abstract), 1992.

Yamazaki, T., T. Ishihara, and F. Murakami, Magnetic anomalies over the Izu-Ogasawara (Bonin) Arc, Mariana Arc and Mariana Trough, *Bull. Geoll. Surv. Japan,* 42, 655-686, 1991.

Yokoyama, I., Calderas and their formation, *Assoc. Geol. Collab. Japan, Monogr.,* 18, 41-53 (in Japanese), 1974.

Yokoyama, I., Gravimetric studies and drilling results at the four calderas for study of their formation, in *Arc Volcanism: Physics and Tectonics,* pp. 29-41, Terra Sci. Publ. Co., Tokyo, 1983.

Yokoyama, I. and S. Ohkawa, Subsurface structure of Aira caldera and its vicinity in southern Kyushu, Japan, *J. Volcanol. Geotherm. Res.,* 30, 253-282, 1986.

Yuasa, M., Myojin Knoll, Izu-Ogasawara arc: Submersible study of submarine pumice volcano, *Bull. Volcanol. Soc. Japan,* 40, 277-284 (in Japanese with English abstract), 1995.

Yuasa, M., F. Murakami, E. Saito, and K. Watanabe, Submarine topography of seamounts on the volcanic front of the Izu-Ogasawara Bonin) arc, *Bull. Geoll. Surv. Japan,* 42, 703-743, 1991.

Yuasa, M., and M. Nohara, Petrographic and geochemical along-arc variations of volcanic rocks on the volcanic front of the Izu-Ogasawara (Bonin) Arc, *Bull. Geol. Surv. Japan,* 43, 421-456, 1992

M. Yuasa, Geoinformation Division, Geological Survey of Japan, AIST, Tsukuba Central 7, 1-1 Higashi 1-chome, Tsukuba, Ibaraki 305-8567, Japan.

K. Kano, Institute of Geoscience, Geological Survey of Japan, AIST, Tsukuba Central 7, 1-1 Higashi 1-chome, Tsukuba, Ibaraki 305-8567, Japan.

Submarine, Silicic, Syn-Eruptive Pyroclastic Units in the Mount Read Volcanics, Western Tasmania: Influence of Vent Setting and Proximity on Lithofacies Characteristics

Jocelyn McPhie

Centre for Ore Deposit Research and School of Earth Sciences, University of Tasmania, Hobart, Australia

Rodney L. Allen

Centre for Ore Deposit Research; University of Tasmania, Hobart, Australia; and Volcanic Resources Limited, Boledin, Sweden

Lithofacies characteristics of submarine, silicic, syn-eruptive pyroclastic units in the Cambrian Mount Read Volcanics, western Tasmania, have been used to infer the source vent setting and proximity. The submarine, syn-eruptive pyroclastic units typically consist of one or more, massive to graded, very thick (a few metres to >100 m) beds and are laterally extensive (>10 km along strike). They are composed of variable proportions of rhyolitic or dacitic pumice, crystals (mainly quartz and feldspar), shards, volcanic lithic clasts and non-volcanic sedimentary clasts. Although the dominant components are juvenile pyroclasts and the units are very thick, there is no textural evidence for hot emplacement. In some cases, pumice clasts define a bedding-parallel foliation formed during diagenetic compaction. Three types of units can be defined on the basis of lithofacies characteristics: (1) graded pumice-lithic breccia–shard-rich sandstone, probably generated by submarine explosive eruptions and deposited in relatively deep water at medial to distal sites; (2) very thick, graded to massive pumice breccia–shard-rich sandstone, considered to be the proximal, mainly below-wave-base submarine record of a large-magnitude (>10 cubic km), explosive eruption from an intrabasinal submarine vent; (3) very thick, crystal-rich volcanic sandstone; the high crystal abundance in this facies is attributed to interaction between subaerial pyroclastic flows and seawater, and formation of crystal-enriched, submarine, water-supported, gravity currents; in this case, it is most likely that the source vent was subaerial, and that deposition took place at proximal to medial locations in relatively shallow water (probably <200 m).

INTRODUCTION

Submarine silicic pyroclastic units are common in ancient volcanic successions worldwide [e.g., *Fiske*, 1963; *Fiske and Matsuda*, 1964; *Niem*, 1977; *Wright and Mutti*, 1981; *Cas*, 1983; *Yamada*, 1984; *Cashman and Fiske*, 1991;

McPhie and Allen, 1992; and *Allen et al.*, 1997]. Submarine silicic pyroclastic deposits have also been recognised in drill core recovered from modern oceans [e.g., *Nishimura et al.*, 1991] and on the modern ocean floor [e.g., *Mandeville et al.*, 1996; and *Fiske et al.*, 2001] at a wide range of water depths. Broadly defined, submarine pyroclastic facies can include: (1) submarine, hot-gas-supported, pyroclastic flow deposits that are primary and essentially identical in origin and characteristics to subaerial pyroclastic flow deposits; (2) syn-eruptive, water-supported gravity current and water-settled deposits composed of freshly erupted juvenile pyroclasts; and (3) pyroclast-rich deposits generated by post-eruptive reworking and/or resedimentation of non-welded pyroclastic deposits.

In many submarine volcanic successions, the syn-eruptive category dominates but units of this type are by far the most difficult to interpret in terms of transport and depositional processes and source setting. The transport and depositional processes are likely to be varied. In relatively simple situations, submarine gravity currents and suspensions may be fed directly from a single, contemporaneous, short-lived (hours to days?) eruption. The records of more prolonged eruptions (days to months?) are likely to be more complicated in including water-settled fall and gravity-current deposits from repeated syn-eruptive resedimentation events. It is also apparent that buoyant pyroclasts, especially ash, may be temporarily stored in particle-rich layers in the water column before delivery to the sea floor via vertical gravity currents [e.g., *Carey*, 1997]. Field criteria that allow recognition of different genetic varieties of syn-eruptive pyroclastic facies have yet to be established. Nevertheless, the broad category of syn-eruptive pyroclastic facies is distinctive and characterised by the high abundance of unmodified juvenile pyroclasts such as pumice, crystals and shards that have a uniform composition or narrow range in composition.

Moderate- to large-magnitude explosive eruptions are capable of dispersing pyroclasts over hundreds to thousands of square kilometres [e.g., *Wilson et al.*, 1995], so depositional sites may be far removed from source vents. Because submarine environments are depocentres, they are repositories for silicic syn-eruptive pyroclastic deposits originating not only from submarine intrabasinal vents, but also from subaerial basin-margin or island vents. One of the major challenges confronting facies analysis of submarine pyroclastic successions is distinguishing between syn-eruptive facies from subaerial versus submarine sources. In both cases, the interstitial fluid involved in pyroclast transport undergoes transitions between hot magmatic gas, steam and liquid water. However, in detail, the rate and complexity of the transitions can vary widely, resulting in units with a very broad range in lithofacies characteristics. The submarine, silicic, syn-eruptive pyroclastic units in the Mount Read Volcanics, western Tasmania, display just such textural diversity. Within that diversity, gravity current deposits are by far the dominant kind of syn-eruptive pyroclastic facies. Three main types of syn-eruptive, pyroclastic gravity current units can be recognised, primarily on the basis of lithofacies characteristics. These types are interpreted to reflect contrasting source vent settings, identified as relatively deep submarine, or subaerial to shallow submarine. Another important influence on the lithofacies characteristics of the three types may have been proximity to the source vent. In this paper, examples of the three types are described; they provide a framework for distinguishing source setting and proximity using deposit characteristics.

Naming the lithologies in submarine syn-eruptive pyroclastic units can be difficult because they originate through a blend of volcanic and non-volcanic surface processes. The volcanic processes are principally responsible for the production of particles, in this case, explosive fragmentation. It is therefore appropriate to identify the particles as pyroclasts [*Fisher*, 1966]. The non-volcanic surface processes are important in final transport and deposition. Hence, terms for primary pyroclastic aggregates (ignimbrite, tuff, lapilli-tuff, tuff-breccia etc) are unsuitable. In the absence of better terms, we use "breccia" (grain size >2 mm, angular clasts), "sandstone" (average grain size 0.0625-2 mm) and "mudstone" (grain size < 0.0625 mm) for the pyroclastic aggregates, qualified by terms indicating the principal components (e.g., crystals, shards, pumice). Most of the submarine syn-eruptive pyroclastic units in the Mount Read Volcanics consist of pumice breccia, pumice-lithic breccia, crystal-rich volcanic sandstone and shard-rich mudstone.

MOUNT READ VOLCANICS

The Mount Read Volcanics consist of compositionally and texturally diverse, mainly Middle Cambrian, submarine lavas, intrusions and volcaniclastic facies, interbedded with subordinate non-volcanic or mixed-provenance sedimentary facies [*Corbett*, 1992; and *McPhie and Allen*, 1992]. They presently occupy an area some 200 km long (north-south) and up to 20 km wide (Figure 1). The original thickness ranged from <1 km to more than 3 km. The succession was affected by Cambrian deformation and hydrothermal alteration, followed by Devonian regional deformation and metamorphism that produced greenschist facies metamorphic assemblages, major faults, folds and variably developed cleavage. Although any original volcanic glass has devitrified, primary textures in both coherent and clastic volcanic facies are generally well preserved. Exceptions occur in strongly cleaved domains and where hydrothermal alteration is intense, close to massive sulfide ore deposits and prospects that occur throughout the belt. The Mount Read Volcanics are well exposed along major rivers and roads, and around dams and mines. Elsewhere, outcrop is poor. There are, however, numerous diamond-drill-core sections through the succession, especially close to the principal ore deposits. The combination of outcrop and drill-core data has provided good control on the vertical and lateral facies variations within the succession.

Four lithostratigraphic units have been recognised [*Corbett*, 1992]: the Central Volcanic Complex, the Eastern quartz-phyric sequence, the Western volcano-sedimentary sequences and the Tyndall Group (Figure 1). Of these, only the Tyndall Group has been formally defined [*Corbett et al.*, 1974; and *White and McPhie*, 1996]. There are also several, informally named successions dominated by andesite and basalt [e.g., Anthony Road Andesite; *Corbett*, 1992; Figure 1]. Stratigraphic relationships among the Central Volcanic Complex, Eastern quartz-phyric sequence and Western volcano-sedimentary sequences are complicated but the three units are considered to be more or less contemporaneous [*Corbett*, 1992; Figure 2]. Four isotopic dates (U-Pb in zircon and ^{40}Ar/^{39}Ar in hornblende) indicate emplacement of this part of the succession at 502.6 ± 3.5 Ma [*Perkins and Walshe*, 1993]. The Tyndall Group is probably younger than the other lithostratigraphic units [*White and McPhie*, 1997] because it: (1) unconformably overlies Cambrian granite that intrudes the Central Volcanic Complex and Eastern quartz-phyric sequence near Mount Darwin; (2) contains a slightly younger fossil assemblage than the other lithostratigraphic units [*Corbett*, 1992]; (3) has an isotopic age of 494.4 ± 3.8 Ma [U-Pb in zircon; *Perkins and Walshe*, 1993]; and (4) locally has a conformable, gradational contact with the overlying, Late Cambrian-Early Ordovician, siliciclastic Owen Conglomerate.

Each of the lithostratigraphic units consists of variable proportions of volcanic and sedimentary facies. The principal volcanic facies are lavas, domes, syn-volcanic intrusions and diverse volcaniclastic facies, including thick and extensive syn-eruptive pyroclastic units and in situ and resedimented hyaloclastite [*McPhie and Allen*, 1992; and *Gifkins and Allen*, 2001]. The lavas, syn-volcanic intrusions and syn-eruptive volcaniclastic facies are predominantly rhyolitic and dacitic. Sections up to several hundred metres thick composed of andesitic and basaltic facies are present throughout the succession [e.g., Sterling Valley Volcanics, Lynch Creek Basalts, Que-Hellyer Volcanics; *Corbett*, 1992] but are not laterally extensive (<10 km). Magma compositions were mainly calc-alkaline; some of the oldest basalts and andesites are tholeiitic and some of the youngest are shoshonitic [*Crawford et al.*, 1992]. The volcanic facies are interbedded with a sedimentary facies association that comprises black mudstone, micaceous mudstone and quartz-lithic sandstone of non-volcanic or mixed provenance. The non-volcanic components are predominantly metamorphic and vein-quartz fragments derived from adjacent Precambrian basement terranes. Mudstone units are massive or delicately parallel laminated. Sandstone beds have sharp bases and are graded, showing complete or partial Bouma sequences. Middle Cambrian trilobites and other marine fossils are sparsely distributed in the sedimentary facies [*Corbett*, 1992].

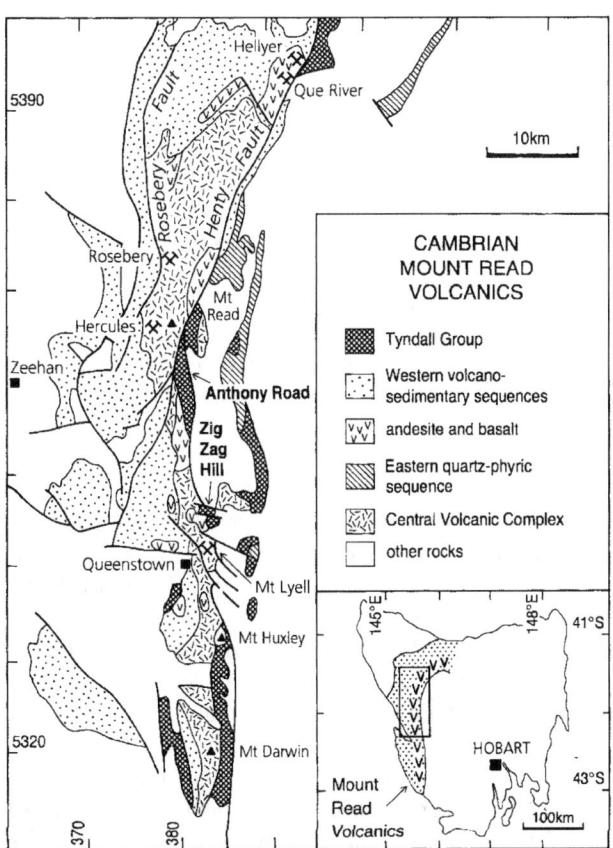

Figure 1. Geological setting and principal lithostratigraphic units of the Cambrian Mount Read Volcanics in western Tasmania, modified from Corbett [1992]. Submarine syn-eruptive pyroclastic units occur in the Western volcano-sedimentary sequences, the Central Volcanic Complex and the Tyndall Group. The inset map shows the full extent of the Mount Read Volcanics in western Tasmania.

DEPOSITIONAL SETTING OF THE MOUNT READ VOLCANICS

The depositional setting of the Mount Read Volcanics is best constrained by the character of the non-volcanic or

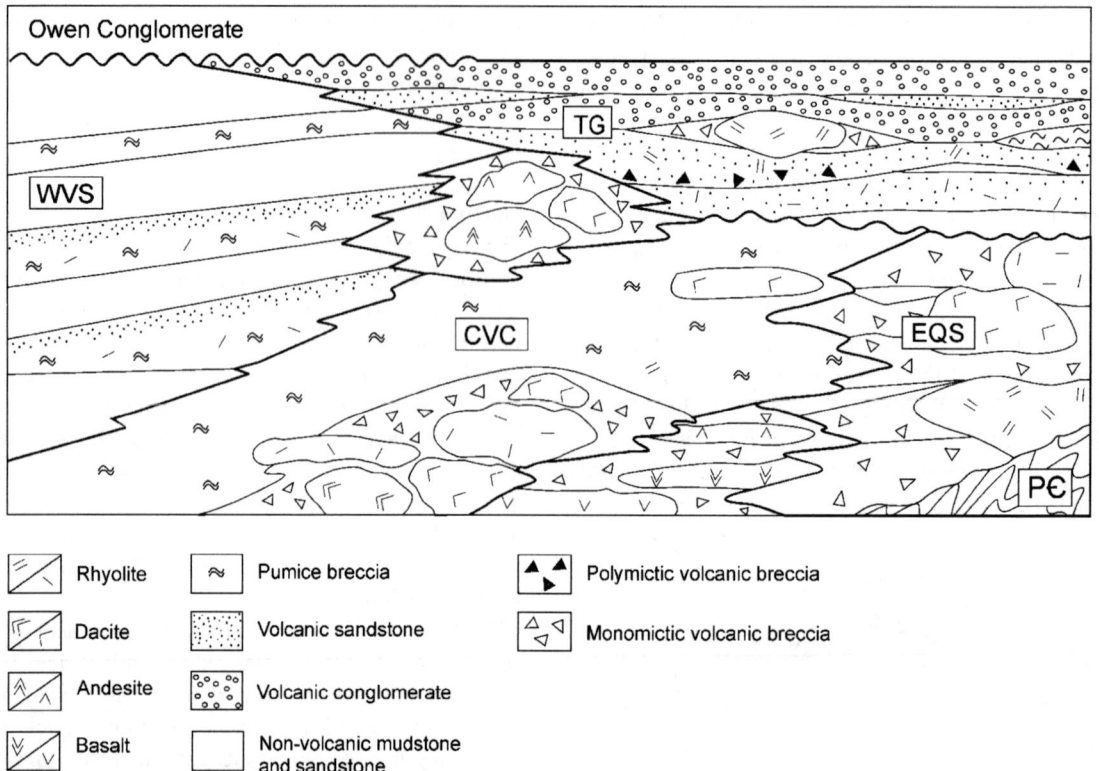

Figure 2. Simplified stratigraphic relationships among the principal lithostratigraphic units in the Mount Read Volcanics. The Central Volcanic Complex (CVC) is composed of rhyolitic and dacitic lavas and syn-volcanic intrusions, and thick, rhyolitic pumice breccia. The Eastern quartz-phyric sequence (EQS) is dominated by thick, massive, rhyolites and dacites. The Western Volcano-sedimentary sequences (WVS) consist of rhyolitic, pumice- and crystal-rich volcaniclastic units interbedded with black mudstone and non-volcanic turbidites. The Tyndall Group (TG) comprises a lower part mainly composed of crystal-rich sandstone and an upper part composed of volcanic conglomerate. The silicic units are locally interleaved with andesite- and/or basalt-dominated intervals. The volcanic succession overlies Pre-Cambrian metamorphic rocks and is overlain by the siliciclastic Owen Conglomerate.

mixed-provenance sedimentary facies: the presence of tabular sandstone beds with well-developed Bouma sequences, very thick intervals of laminated black mudstone and a variety of marine fossils imply submarine deposition below wave base, probably in relatively deep water. This setting is well established for the Western volcano-sedimentary sequences that include very thick, continuous sections of the sedimentary facies association. In the Central Volcanic Complex and the Eastern quartz-phyric sequence, the sedimentary facies association is far subordinate but nevertheless reliable in indicating a below-wave-base depositional setting. This setting is also consistent with the presence of very thick, graded to massive, volcaniclastic units in the Central Volcanic Complex. Both the Western volcano-sedimentary sequence and the Central Volcanic Complex are host to polymetallic massive sulfide deposits which typically form in submarine, relatively deep water settings [e.g., *Lydon*, 1988].

The depositional setting of the Tyndall Group was submarine, as indicated by in situ fossiliferous limestone near the base [*Jago et al.*, 1972; and *Halley and Roberts*, 1997] and marine fossils in correlative volcaniclastic units [*Corbett*, 1992]. In addition, much of the Tyndall Group consists of very thick, massive to graded, volcaniclastic units interbedded with minor thin intervals of sandstone turbidites and hemipelagic mudstone, suggesting a below-wave-base setting. However, the in situ limestone at the base includes a shallow-marine fossil assemblage [*Jago et al.*, 1972]. Also, well-rounded pebbles and cobbles are common in the volcaniclastic units [*White and McPhie*, 1996]. We infer that deposition occurred in proximity to very shallow or subaerial environments, and probably in shallower water than the rest of the Mount Read Volcanics. A relatively shallow depositional environment is also consistent with the presence of welded ignimbrite in the lower part of the group [*White and McPhie*, 1997], and the inferred setting of the syn-genetic Au-rich Henty ore deposit that occurs in the lower Tyndall Group at Henty [*Large et al.*, 2001].

CHARACTERISTICS OF SUBMARINE, SILICIC, SYN-ERUPTIVE PYROCLASTIC UNITS

Submarine, silicic, syn-eruptive pyroclastic units inherit their major textural and lithofacies characteristics from the source explosive eruptions. Such explosive eruptions can generate very large volumes (>10 km^3) of relatively fine particles (mm-cm) in just a few hours to days [e.g., *Self et al.*, 1984; and *Ninkovich et al.*, 1978]. The particles (pyroclasts) are characteristically angular or ragged in shape, and dominated by juvenile pumice lapilli and crystal- or shard-rich ash. In most cases, non-juvenile pyroclasts are also present but not abundant. Hence, submarine, silicic, syn-eruptive pyroclastic units can have the same range of components and dimensions as their subaerial equivalents.

Transport and depositional processes are also comparable in both settings, involving combinations of gravity currents and suspension fallout, although there is a significant contrast in the nature of the dominant interstitial fluid, being hot gas (subaerial) versus water (submarine). Beds deposited from submarine, water-supported, pyroclastic gravity currents are typically very thick, massive to graded, and widespread [e.g., *Fiske*, 1963; and *Fiske and Matsuda*, 1964], reflecting the large volumes and high rates of supply. In many examples, there is a lower, massive or weakly graded, thicker and coarser division and an upper, diffusely thinly bedded, thinner and finer division. The differences may relate to an initial high-particle-concentration, waxing phase of the current (lower division) followed by a more dilute, possibly turbulent, waning phase [e.g., *Fiske and Matsuda*, 1964]. Grading and relatively good sorting [e.g., *Kano et al.*, 1996; and *Wright and Mutti*, 1981] appear to be features that help distinguish submarine, water-supported, pyroclastic gravity current deposits from subaerial non-welded pyroclastic flow deposits. In many cases, the submarine gravity currents pass across wet-mud or wet-sand substrates and incorporate an admixture of non-juvenile, intrabasinal sedimentary clasts.

Submarine, syn-eruptive gravity current deposits dominated by silicic pumice lapilli are relatively common. Because most dry silicic pumice clasts are less dense than water, their incorporation in water-supported gravity currents implies that they were already waterlogged. The behaviour of pumice clasts in water depends on the size and temperature of the clasts, the interconnectedness of vesicles in the pumice, and the environment in which the clasts cool (whether air or water) [*Whitham and Sparks*, 1986; *Kano et al.*, 1996; and *Fiske et al.*, 2001]. Hot juvenile pumice lapilli (less than ~16 mm) immersed in water readily become water logged and sink rapidly [*Whitham and Sparks*, 1986; and *Fiske et al.*, 2001], making them available to syn-eruptive gravity currents. Hence, we infer that a high abundance of juvenile pumice lapilli in gravity current deposits reflects direct delivery of hot pyroclasts into the submarine environment from submarine vents, rather than from subaerial vents.

Coarse hot pumice clasts (larger than ~16 mm) can be temporarily buoyant [*Kano et al.*, 1996], and cold pumice clasts cooled in air can remain buoyant for very long periods of time [e.g., *Simkin and Fiske*, 1983]. Buoyant pumice and fine ash suspended in the water column will eventually settle to form suspension fallout deposits. There are few descriptions of such deposits in ancient successions [e.g., *Niem*, 1977; *Cashman and Fiske*, 1991; *White et al.*, 2001; *Doyle and McPhie*, 2001; and *Stewart and McPhie*, in press]. Submarine fallout deposits may show strong bimodality, expressed either by combinations of coarse pumice with much finer lithic clasts [*Cashman and Fiske*, 1991], or coarse pumice clasts and fine ash [*Stewart and McPhie*, in press].

SUBMARINE, SILICIC, SYN-ERUPTIVE PYROCLASTIC UNITS IN THE MOUNT READ VOLCANICS

The vast majority of syn-eruptive pyroclastic units in the Mount Read Volcanics were derived from rhyolitic or dacitic source magmas, as indicated by the crystal assemblages dominated by quartz and feldspar, the presence of quartz- and feldspar-phyric pumice and dense juvenile fragments, and the whole-rock geochemical characteristics of the juvenile pumice-rich units. Most units are considered to be deposits from submarine water-supported gravity currents: they have sharp bases and mainly consist of a coarser, graded to massive lower portion overlain by a finer, thinner stratified upper portion. The upper portion probably includes deposits from suspension fallout as well as from dilute gravity currents. Components that were originally glassy (pumice and shards) have devitrified and/or been altered to fine-grained quartz-feldspar-phyllosilicate-carbonate assemblages, but clast shapes are still well preserved and can be distinguished easily in most cases.

The submarine, silicic, syn-eruptive pyroclastic units in the Mount Read Volcanics do not preserve textural evidence for hot emplacement: most pumice clasts and shards are uncompacted, reflecting the early infilling of vesicles and replacement of the glass by secondary feldspar (or a precursor phase such as zeolite) [*Allen and Cas*, 1990; and *Gifkins and Allen*, 2001]. Other indicators of high-temperature emplacement, such as high-temperature crystallisation textures (spherulites, micropoikilitic texture), columnar joints, and features indicating the presence of hot gas (gas-escape structures), are absent. Many pumice breccia units composed of well-preserved uncompacted pumice clasts also contain bedding-parallel phyllosilicate-rich lenses. The lenses were probably also pumice but compacted as a result of early alteration of the originally glassy and porous pumice to a mechanically "weak" mineralogy (e.g., clay) [*Allen and Cas*, 1990; and *Gifkins and Allen*, 2001].

The submarine, silicic, syn-eruptive pyroclastic units in the Mount Read Volcanics vary widely in thickness, lateral extent, internal textural variations and components. Within that broad spectrum, three distinctive types can be identified: (1) graded pumice-lithic breccia—shard-rich sandstone—; (2) very thick, graded to massive pumice breccia—shard-rich sandstone, and (3) very thick, crystal-rich volcanic sandstone.

Graded Pumice-Lithic Breccia—Shard-Rich Sandstone

These units are common throughout the Western volcano-sedimentary sequences [e.g., Yolande River Sequence, Southwell Subgroup, White Spur Formation; *Corbett*, 1992] and are interbedded with substantial thicknesses of black mudstone and sandstone turbidites. They are typically a few metres to a few tens of metres thick with gradational upper contacts (Figures 3 and 4). Some examples in the Southwell Subgroup and the White Spur Formation are laterally extensive and have been traced for more than 13 km along strike [*McPhie and Allen*, 1992].

The juvenile pyroclastic components are pumice clasts (most <2 cm, up to ~10 cm), crystals (1-2 mm) and shards (<1 mm) (Figures 4 and 5). The principal non-juvenile components are non-volcanic mudstone or sandstone clasts and variable amounts and types of volcanic lithic clasts. Although poorly sorted overall, the sedimentary clasts and large dense lithic clasts are typically concentrated near bases of beds, pumice lapilli dominate in the middle portions, and fine-grained components (especially shards and fine pumice clasts) are concentrated in the upper stratified intervals (Figures 3, 4 and 5). In some cases, the matrix is a heterogeneous mixture of fine juvenile (shards and crystals) and non-juvenile components, the latter being derived from disintegration of sedimentary clasts. The larger pumice clasts commonly show bedding-parallel alignment due to early compaction that accompanied diagenesis.

The relatively deep submarine depositional setting of these units is well constrained by the thick interbedded intervals of black mudstone and sandstone turbidites. The principal components are pyroclasts and clasts derived from intrabasinal sediment, suggesting that the outflow paths of the gravity currents were exclusively submarine. The incorporation of abundant silicic pumice clasts in the water-supported gravity currents requires that the pumice clasts were already water-logged. This condition can be achieved most efficiently if the hot pumice is erupted directly into the submarine environment, suggesting that source vents were submarine, rather than subaerial.

These units are generally better sorted than most subaerial pyroclastic flow deposits and lack abundant fine matrix in the lower lithic-rich and middle pumice-rich parts. The sorting characteristics could reflect the ease with which fine components were separated from coarse components in water-supported gravity currents, and the selection of pumice lapilli below a threshold size by the water-logging process, as described above. However, the sorting characteristics are complex because most units of this type contain an admixture of fine sediment (mud and sand) derived from the outflow path or from disintegration of sedimentary clasts.

The units are relatively thin but laterally extensive, and show only very slight lateral variations in grain size or texture. They are typically interbedded with non-volcanic sedimentary facies (mudstone and sandstone) and are not known to be associated with proximal volcanic facies such as lavas, domes, cryptodomes, or coarse, thick volcanic breccias. Hence, they are considered to represent medial to distal settings with respect to submarine source vents. Each unit may have been generated by a single explosive eruption involving substantial magma volumes, probably in the order of several to 10 km^3. This submarine facies is analogous to subaerial medial to distal outflow ignimbrite facies [e.g., *Hildreth*, 1983], in being the remote product of a substantial explosive eruption.

Very Thick, Graded to Massive Pumice Breccia— Shard-Rich Sandstone

Parts of the Central Volcanic Complex are dominated by this facies, the essential characteristics of which are extreme thickness and almost exclusively juvenile composition (that is, poor in lithic clasts and intrabasinal sedimentary clasts). Allen [1994a] described a pumice breccia succession in the Rosebery-Hercules area that is more than 800 m thick, and extends more than 14 km along strike. This succession comprises a small number of amalgamated depositional units, each of which is tens to >100 m in thickness (Figure 6). The lower part of each unit consists of a very thick, massive interval of moderately sorted pumice breccia. The lower part grades up into a thinner upper part composed of

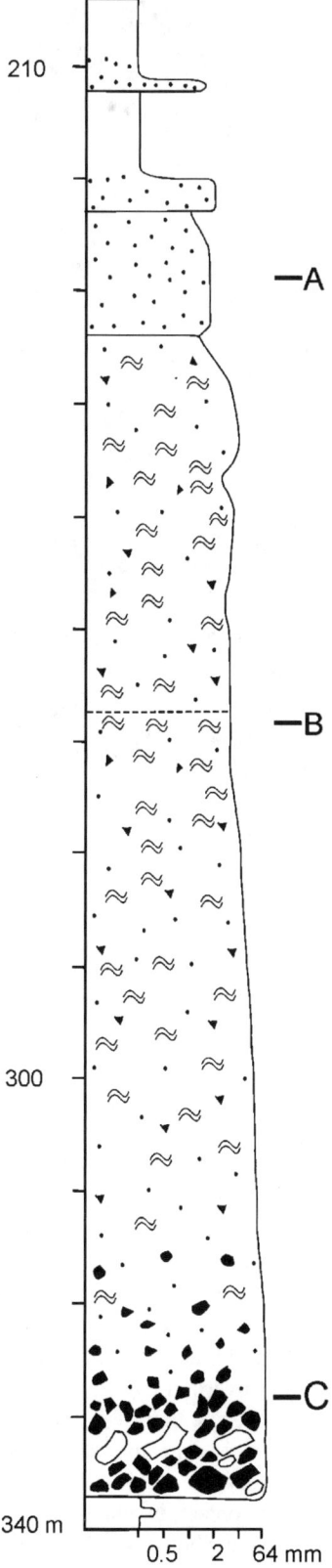

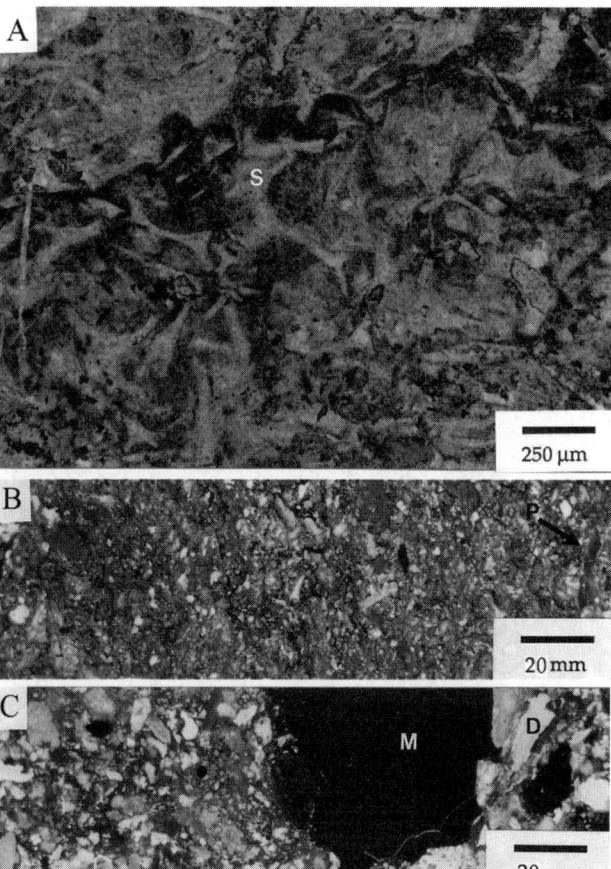

Figure 4. Three samples from the very thick, graded to pumice lithic breccia–shard-rich sandstone unit shown in Figure 3. Polymictic volcanic breccia at the base (sample C, 331.1 m) consists of angular and ragged dacitic (D) and other volcanic clasts, together with black mudstone clasts (M). The interior of the unit is dominated by massive volcanic sandstone–granule breccia (sample B, 291.6 m) composed of feldspar crystal fragments, dacitic clasts, ragged pumice clasts (P) and fine shard-rich matrix. Shard-rich sandstone (sample A, 228.6 m; photomicrograph, plane polarised light) occurs at the top of the unit. Relic bubble-wall shards (S) are well preserved.

Figure 3. Graphic log of an example of graded to pumice-lithic breccia–shard-rich sandstone in the Western volcano-sedimentary sequences. This unit occurs in diamond drill core HP 2 located at High Point, ~5 km southwest of Hellyer (Figure 1). The graded to volcaniclastic unit is ~120 m thick and interbedded with black mudstone. Samples from three depths (A, B, C) are shown in Figure 4.

Figure 5. Textures in samples of a graded to pumice-lithic breccia–shard-rich sandstone unit in the Western volcano-sedimentary sequences on the Anthony Road (Figure 1). This unit is ~50 m thick and grades upward from crystal-lithic granule breccia (C) to diffusely stratified, shard-rich, fine sandstone (B) and shard-rich mudstone (A). **A.** The shard-rich mudstone at the top (~40 m above the base) is dominated by poorly preserved relic shards with scattered crystal fragments (c) (photomicrograph, plane polarised light). **B.** The middle part of the unit (~20 m above the base) is composed of undeformed platy and cuspate relic shards (s) and angular crystal fragments (c) (photomicrograph, plane polarised light). **C.** The lowermost part of the unit (~2 m above the base) consists mainly of crystal (Fl) and volcanic lithic (L) fragments together with minor relic shards (s) (photomicrograph, plane polarised light).

thin, graded pumice breccia beds, and/or diffusely thinly stratified fine pumice breccia, shard-rich sandstone and shard-rich mudstone. The pumice breccia intervals commonly show a bedding-parallel foliation defined by diagenetically compacted pumice clasts [*Allen and Cas*, 1990; Figure 7]. The Rosebery and Hercules massive sulfide ore bodies occur in the topmost part of the pumice breccia succession [*Allen*, 1994b; and *Large et al.*, 2001].

The entire >800-m-thick succession is dominated by texturally and mineralogically uniform rhyolitic, feldspar-phyric pumice clasts. Subordinate components include feldspar crystals, shards and minor dense volcanic lithic clasts. Although composed almost exclusively of juvenile pyroclasts, there is no textural evidence for hot emplacement such as welding compaction, columnar jointing or gas-escape structures [*Allen and Cas*, 1990]. The base of the succession is not exposed. Relationships at the top of the succession vary from place to place. There are two main types of relationships [*Allen*, unpublished data]. In one, the pumice breccia has a normally graded top and is overlain either by black mudstone and thinly bedded sandstone turbidites (Figure 6), or by up to 30 m of stratified, moderately to well sorted, lithic-crystal-rich volcanic sandstone and pebbly sandstone and then black mudstone. In the other type of contact relationship, the normally graded top is missing and pumice breccia is overlain by either a thick, normally graded pumice-lithic breccia unit of different composition, or the lithic-crystal-rich volcanic sandstone described above, or black mudstone.

The first type of top contact, in which the pumice breccia has a graded top, is attributed to deposition in paleo-topographic low areas where the entire pumice breccia succession accumulated in deep water (below wave base). The second type of top contact is an erosion surface that developed in at least two different situations. In the cases where a graded pumice-lithic breccia unit overlies the pumice breccia, currents have scoured the upper tens of metres of the pumice breccia succession, removing the normally graded top of the succession. The lithofacies characteristics of the overlying, normally graded pumice-lithic breccia unit indicate that its deposition, and probably also erosion of the underlying pumice breccia, involved below-wave-base gravity currents. The other cases where the pumice breccia succession is truncated and overlain sharply by stratified, lithic-crystal-rich sandstone or black mudstone are more difficult to interpret. These contacts indicate that at least several tens of metres of the pumice breccia succession have been eroded, but there is no directly overlying gravity-current deposit to help constrain the setting or agent of erosion. It is possible that such contacts reflect paleo-topographic high areas where the pumice succession filled the basin to near sea level. Wave erosion could then have rapidly stripped the unconsolidated pumice breccia down to wave base before further subsidence returned the eroded contact to deeper-water conditions.

The very substantial preserved thickness of the pumice breccia strongly suggests deposition close to the source vent(s) and a large-magnitude (>10 km^3), explosive, silicic eruption. Consequently, the depositional basin has been interpreted to be a caldera [*Green et al.*, 1981; and *Allen and Cas*, 1990]. The setting of the eruptive vent(s) is not well constrained but, given the proximal character of the succession, the source was probably within the depositional basin and either submarine or near sea level. In addition, the high abundance of pumice lapilli suggests that hot pumice clasts were rapidly waterlogged. Rapid water logging is favoured by direct delivery of hot pumice clasts to the submarine setting [e.g., *Whitham and Sparks*, 1986; and *Fiske et al.*, 2001]. In addition, the pumice breccia units lack components that record outflow through subaerial or shallow-marine environments. This facies can thus be considered as the proximal submarine record of a synchronous, large, explosive eruption. The very thick, graded to massive pumice breccia—shard-rich sandstone facies in the Mount Read Volcanics is broadly comparable in composition, componentry and dimensions to intracaldera facies in other ancient submarine volcanic successions [e.g., *Busby-Spera*, 1986; and *Morton et al.*, 1991]. However, submarine intracaldera facies elsewhere may include genuine pyroclastic flow deposits that show evidence for hot deposition [e.g., welding in the Vandever Mountain tuff; *Kokelaar and Busby*, 1992], and some examples are well bedded, each bed showing systematic variations in composition and texture [e.g., Mattabi succession in the Sturgeon Lake caldera, *Morton et al.*, 1991].

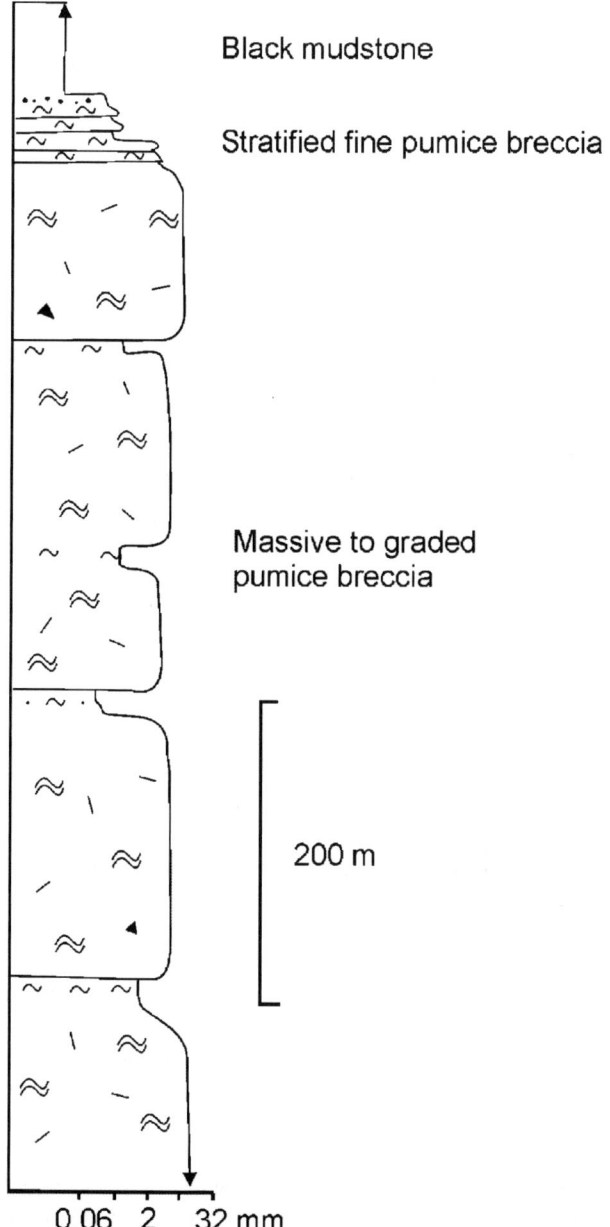

Figure 6. Graphic log of very thick, graded to massive pumice breccia—shard-rich sandstone units that form a >800-m-thick interval in the footwall to the Rosebery and Hercules massive sulfide ore bodies (Figure 1). The pumice breccia beds are very thick and composed of feldspar-phyric tube pumice clasts, feldspar crystals and relic shards, together with very sparse volcanic lithic fragments.

Very Thick, Crystal-Rich Volcanic Sandstone

Crystal-rich volcanic sandstone is an important component of the Tyndall Group, the youngest lithostratigraphic

Figure 7. Outcrops (upper) of the very thick, graded to massive pumice breccia–shard-rich sandstone commonly show a bedding-parallel foliation defined by phyllosilicate-rich lenses formed during diagenetic compaction of pumice clasts. The paler domains between the dark lenses consist of uncompacted, feldspar-altered pumice clasts. The hand sample (lower) shows the early compaction foliation and the later, steeply dipping regional tectonic foliation. The early diagenetic foliation is crenulated by, and transposed into, the regional tectonic foliation. Samples from the Hercules Mine Haulage Road at Hercules (Figure 1).

unit in the Mount Read Volcanics. The Tyndall Group comprises a diverse collection of submarine felsic volcanic and sedimentary facies [*White and McPhie*, 1996]. The crystal-rich volcanic sandstone dominates sections in the lower part of the Tyndall Group (Comstock Formation) throughout its extent, and is most commonly associated with thin intervals of laminated mudstone and turbidite sandstone, quartz-feldspar-phyric rhyolitic lavas and intrusions, and welded rhyolitic ignimbrite. The upper part of the Tyndall Group (Zig Zag Hill Formation) consists mainly of very thickly bedded, polymictic volcanic conglomerate and pebbly volcanic sandstone.

The crystal-rich volcanic sandstone occurs in very thick (1 m to tens of metres), normally graded beds, and very thick (tens of metres to >100 m), apparently massive units [*White and McPhie*, 1996; Figure 8]. The graded beds have coarse, volcanic lithic clast-rich bases (Figure 9), thick intervals of massive crystal-rich sandstone in the middle, and thin intervals of diffusely stratified, crystal-rich or shard-rich mudstone at the top. Grading is defined by a gradual upward decrease in size and abundance of the volcanic lithic clasts. The massive units of crystal-rich volcanic sandstone are remarkably uniform in texture and composition, varying only slightly and locally in crystal and lithic clast abundance. The crystal-rich volcanic sandstone has been affected by a distinctive style of alteration involving pink albite and dark green chlorite. These alteration minerals occur in patchy albite- or chlorite-rich domains or in bands that are commonly, but not everywhere, parallel to bedding (Figure 10). The albite and chlorite are mainly confined to the matrix in a framework composed of crystals.

The principal components in the crystal-rich volcanic sandstone are crystals, primarily angular fragments and euhedral crystals of quartz and feldspar (mainly plagioclase) (Figure 10), together with subordinate titanomagnetite, clinopyroxene and hornblende. Crystals account for 35-70 modal percent of the rocks. The crystals are typically 1-2 mm and form a grain-supported framework (Figure 10). The matrix now comprises a fine-grained quartz-albite-chlorite assemblage; however, in some samples, relic bubble-wall shards are present, suggesting that originally, much of the matrix was composed of glass shards. Chlorite-altered, wispy pumice clasts and angular quartz-feldspar-phyric juvenile clasts are very minor. The lithic clast population is dominated by purple and red, quartz-feldspar-phyric felsic lava clasts, but purple and red welded ignimbrite clasts, trilobite-bearing limestone clasts and non-volcanic basement-derived quartzite clasts are present in some sections. The ignimbrite clasts are typically angular whereas the other lithic clast types vary from angular to rounded.

The presence of in situ, marine fossil-bearing limestone below the crystal-rich sandstone indicates that the depositional setting for the crystal-rich sandstone was submarine. The massive to graded, very thick, tabular beds suggest that deposition occurred below wave base. The widespread distribution and apparent continuity of this distinctive facies are also consistent with deposition below wave base. However, the in situ limestone contains a relatively shallow

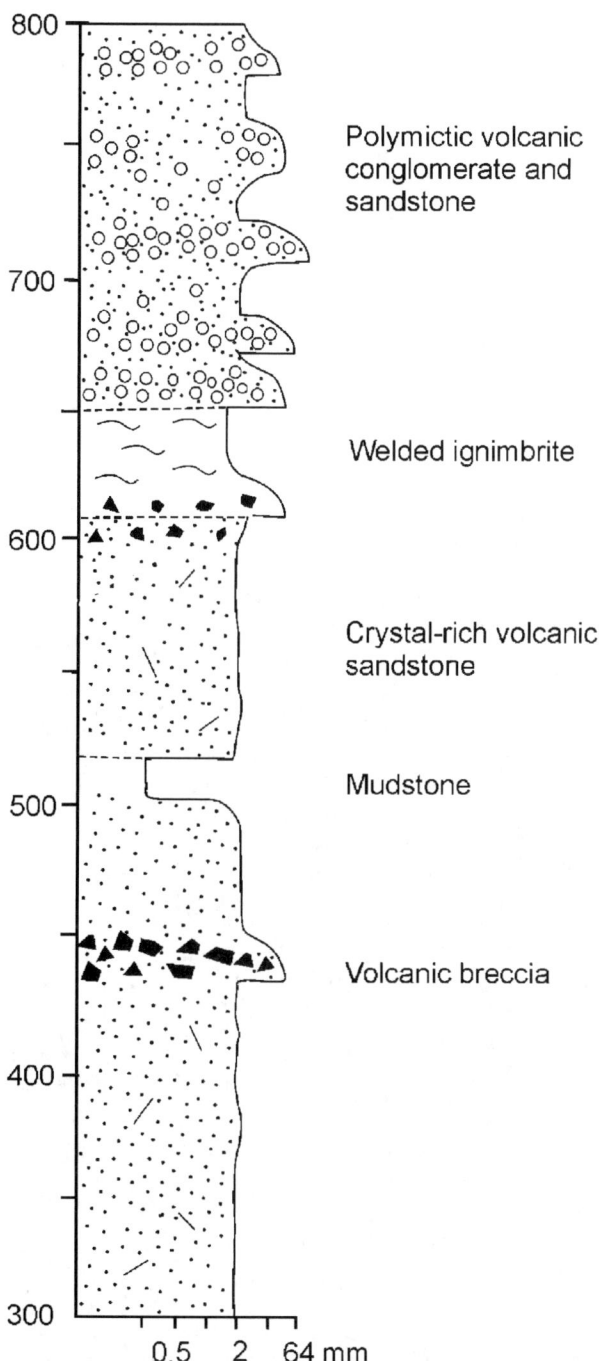

Figure 8. Graphic log of the Tyndall Group at Zig Zag Hill (Figure 1). The lower two-thirds of the section (Comstock Formation) consists mainly of crystal-rich volcanic sandstone interbedded with subordinate volcanic lithic breccia and laminated mudstone or sandstone. Welded ignimbrite occurs at the top of the crystal-rich volcanic sandstone. The top part of the section consists of polymictic volcanic conglomerate and sandstone (Zig Zag Hill Formation).

water fossil assemblage [*Jago et al.*, 1972], and there is a paucity of the "deep-water" sedimentary facies, such as laminated black mudstone and sandstone turbidites, that are common elsewhere in the Mount Read Volcanics. In addition, there is a close association between the crystal-rich volcanic sandstone, and in situ, purple and red, welded ignimbrite and coarse breccia composed of similar welded ignimbrite clasts [*White and McPhie*, 1997]. Although welded ignimbrite can occur in deep water, this facies is far more common in subaerial and very shallow-water settings [*Cas and Wright*, 1991]. Hence, the water depth during deposition of the crystal-rich sandstone facies may have been shallower than for the other syn-eruptive pyroclastic facies described above, and possibly no more than about 200 m [*White and McPhie*, 1997]. It is plausible that, in some sections, deposition of the crystal-rich sandstone facies resulted in appreciable shoaling because intervals of this facies are texturally uniform for at least tens of metres and interbeds of other facies are uncommon and thin, implying continuous, probably rapid aggradation.

Given the relatively shallow depositional setting for the crystal-rich volcanic sandstone, the source vent(s) can reasonably be constrained as subaerial or very shallow marine. The crystal-laden currents contained appreciable amounts of well-rounded lithic clasts and angular welded ignimbrite clasts, both of which were collected en route. These clasts were most likely formed in a subaerial or above-wave-base marine setting, implying that the source of the currents was also subaerial or, if submarine, then in even shallower water than that at the site of deposition.

The distinctive feature of the crystal-rich volcanic sandstone facies is the dominance of crystals through continuous sections more than a hundred metres in thickness. The crystal abundance in this facies exceeds that typical of most rhyolitic ignimbrites and lavas. This large volume of euhedral crystals and angular crystal fragments of uniform composition was probably supplied by a large-magnitude explosive eruption involving a porphyritic magma [*White and McPhie*, 1997], and, given the constraints on the setting outlined above, the vent is inferred to have been subaerial. Crystals may have been concentrated to extreme proportions by the combined effects of: (1) subaerial transport as pyroclastic flows involving preferential elutriation of glassy ash [e.g., *Walker*, 1972]; (2) interaction of gas-supported pyroclastic flows with seawater at and beyond the coast, leading to secondary explosions and further loss of glassy ash [e.g., *Freundt*, 2003]; (3) separation of pumice lapilli, either in pumice rafts formed by cold pumice and/or by rapid water-logging and deposition of hot pumice at the shoreline, and (4) sorting during submarine transport in water-supported, probably turbulent currents, resulting in

Figure 9. Coarse lithic breccia at the base of very thick (~8 m), graded crystal-rich volcanic sandstone beds on the Anthony Road (Figure 1). The breccia consists of poorly sorted, dark grey, angular, felsic volcanic clasts in a paler, crystal-rich sandstone matrix.

separation of light components (mainly glassy ash and fine pumice clasts) into trailing suspensions, rapid near-source deposition of the densest components (mainly coarse dense lithic clasts), and formation of crystal-concentrated, more-or-less steady, gravity currents.

The crystal-rich volcanic sandstone facies is widespread, occurring throughout the extent of the Tyndall Group and correlative units, a strike extent of some 200 km. The substantial thickness and local presence of abundant, relatively coarse (up to 1 m), dense lithic clasts suggest the preserved sections are proximal to medial with respect to source vents. This facies is thus considered typical of the proximal to medial, submarine record of a subaerial, or very shallow submarine, large-magnitude explosive eruption.

CONCLUSIONS

Volcaniclastic units composed of pumice, crystals, and shards are common in the late Middle Cambrian Mount Read Volcanics in western Tasmania. These units are broadly similar in composition (rhyolitic or dacitic), and further characterised by their high abundance of juvenile pyroclasts, graded to massive, very thick, tabular beds, and lack of evidence for hot emplacement. They are direct submarine records of explosive volcanism that differ from their subaerial counterparts because final transport and deposition involved water-supported (rather than hot-gas supported) gravity currents. The distinction between hot gas- versus water-supported transport modes is significant, even though both may be carrying similar populations of juvenile, texturally unmodified pyroclasts. The transport and depositional modes strongly affect the deposit bed forms, internal bed organisation and sorting. As a result, the processes accompanying eruption, transport and deposition of pyroclasts in submarine settings are complicated by the presence of water.

The lithofacies characteristics of the syn-eruptive pyroclastic units in the Mount Read Volcanics vary enormously. We conclude that much of the textural variation can probably be related to the proximity of the source vent, its setting and the depositional environment. Three types have been identified: (1) Graded to pumice-lithic breccia–shard-rich sandstone; these volcaniclastic units are typically interbedded with thick intervals of black mudstone and turbidites. They were deposited from water-supported pyroclast-rich gravity currents generated by submarine explosive eruptions and represent medial to distal, relatively deep-water

Figure 10. Polished slab (upper) and photomicrograph (lower) of the crystal-rich volcanic sandstone on the Anthony Road (Figure 1). The crystal-rich volcanic sandstone shows diffuse, alternating paler albite-rich versus darker chlorite-rich bands. The bands are sub-parallel to bedding. The albite and chlorite have mainly replaced former matrix between the quartz and feldspar crystals. Angular crystals and crystal fragments form a grain-supported framework that accounts for about 60 modal%.

depositional settings. (2) Very thick, graded to massive, pumice breccia—shard-rich sandstone; this facies dominates a single proximal section more than 800 m thick and was generated by a large-magnitude, submarine explosive eruption [*Allen*, 1994a]. Deposition occurred mainly in relatively deep water. (3) Very thick, crystal-rich volcanic sandstone—shard-rich mudstone; the extreme concentration of crystals and relatively good sorting in this facies reflect a complex transport and depositional history, involving subaerial (or very shallow marine) explosive eruptions, interaction between pyroclastic flows and seawater, and final submarine deposition from water-supported crystal-rich gravity currents.

Acknowledgments. This research was funded by the Australian Research Council's Special Research Centre program. We are grateful for discussions with Keith Corbett, Cathryn Gifkins and Greg Ebsworth. Critical reviews by James White, John Smellie and two anonymous referees led to substantial improvement of earlier versions.

REFERENCES

Allen, R.L., Volcanic facies analysis indicates large pyroclastic eruptions, sill complexes, synvolcanic grabens, and subtle thrusts in the Cambrian "Central Volcanic Complex" volcanic centre, western Tasmania, in *Contentious Issues in Tasmanian Geology: A Symposium,* edited by D.R. Cooke and P.A. Kitto. Geological Society of Australia, Abstracts 39: 41-43, 1994a.

Allen, R.L., Synvolcanic subseafloor replacement model for Rosebery and other massive sulfide ores, in *Contentious Issues in Tasmanian Geology: A Symposium,* edited by D.R. Cooke and P.A. Kitto. Geological Society of Australia, Abstracts 39: 107-108, 1994b.

Allen R.L. and Cas R.A.F., The Rosebery controversy: distinguishing prospective submarine ignimbrite-like units from true subaerial ignimbrites in the Rosebery—Hercules ZnCuPb massive sulphide district, Tasmania. *Geological Society of Australia Abstracts* 25: 31-32, 1990.

Allen R.L., Weihed P. and Svenson S.Å., Setting of Zn-Cu-Au-Ag massive sulfide deposits in the evolution and facies architecture of a 1.9 Ga marine volcanic arc, Skellefte district, Sweden. *Economic Geology* 91: 1022-1053, 1997.

Busby-Spera C.J., Large-volume rhyolite ash flow eruptions and submarine caldera collapse in the lower Mesozoic Sierra Nevada, California, *Journal of Geophysical Research* 89, B10: 8417-8427, 1986.

Carey S., Influence of convective sedimentation on the formation of widespread tephra fall layers in the deep sea, *Geology* 25: 839-842, 1997.

Cas R.A.F., 1983 Submarine 'crystal tuffs': their origin using a Lower Devonian example from southeastern Australia. *Geological Magazine* 120: 471-486, 1983.

Cas R.A.F. and Wright J.V., Subaqueous pyroclastic flows and ignimbrites: An assessment, *Bulletin of Volcanology* 53: 357-380, 1991.

Cashman K.V. and Fiske R.S., Fallout of pyroclastic debris from submarine volcanic eruptions, *Science* 253: 275-279, 1991.

Corbett K.D., Stratigraphic-volcanic setting of massive sulfide deposits in the Cambrian Mount Read Volcanics, Tasmania. *Economic Geology* 87: 564-586, 1992.

Corbett K.D., Reid K.O., Corbett E.B., Green G.R., Wells K. and Sheppard N.W., The Mount Read Volcanics and Cambro-Ordovician relationships at Queenstown, Tasmania, *Journal of the Geological Society of Australia* 21: 173-186, 1974.

Crawford A.J., Corbett K.D. and Everard J.L., Geochemistry of the Cambrian volcanic-hosted massive sulfide-rich Mount Read Volcanics, Tasmania, and some tectonic implications. *Economic Geology* 87: 597-619, 1992.

Doyle M.D. and McPhie J., Shallow-water microbialite-volcaniclastic facies association in the Cambro-Ordovician Mt Windsor Subprovince, Australia, *Australian Journal of Earth Sciences* 48: 815-831, 2001.

Fisher R.V., Rocks composed of volcanic fragments and their classification, *Earth Science Reviews* 1: 287-298, 1966.

Fiske R.S., Subaqueous pyroclastic flows in the Ohanapecosh Formation, Washington, *Geological Society of America Bulletin* 74: 391-406, 1963.

Fiske R.S. and Matsuda T., Submarine equivalents of ash flows in the Tokiwa Formation, Japan. *American Journal of Science* 262: 76-106, 1964.

Fiske R.S., Naka J., Iizasa K., Yuasa M., Klaus A., Submarine silicic caldera at the front of the Izu-Bonin arc, Japan: Voluminous seafloor eruptions of rhyolite pumice. *Geological Society of America Bulletin* 113: 813-824, 2001.

Freundt A., Entrance of hot pyroclastic flows into the sea: experimental observations. *Bulletin of Volcanology* 65: 144-164, 2003.

Gifkins C.C. and Allen R.L., Textural and chemical characteristics of diagenetic and hydrothermal alteration in glassy volcanic rocks: Examples from the Mount Read Volcanics, Tasmania, *Economic Geology* 96: 973-1002, 2001.

Green G.R., Solomon M. and Walshe J.L., The formation of the volcanic-hosted massive sulfide ore deposit at Rosebery, Tasmania, *Economic Geology* 76: 304-338, 1981.

Halley S.W. and Roberts R.H., Henty: A shallow-water gold-rich volcanogenic massive sulfide deposit in western Tasmania, *Economic Geology* 92: 438-447, 1997.

Hildreth W., The compositionally zoned eruption of 1912 in the Valley of Ten Thousand Smokes, Katmai National Park, Alaska, *Journal of Volcanology and Geothermal Research* 18: 1-56, 1983.

Jago J.B., Reid K.O., Quilty P.G., Green G.R. and Daily B., Fossiliferous Cambrian limestone from within the Mt Read Volcanics, Mount Lyell mine area, Tasmania, *Journal of the Geological Society of Australia* 19: 379-382, 1972.

Kano K., Yamamoto T. and Ono K., Subaqueous eruption and emplacement of the Shinjima Pumice, Shinjima (Moeshima) Island, Kagoshima Bay, SW Japan, *Journal of Volcanology and Geothermal Research* 71: 187-206, 1996.

Kokelaar P. and Busby C. Subaqueous explosive eruption and welding of pyroclastic deposits, *Science* 257: 196-201, 1992

Large R.L., McPhie J., Gemmell J.B., Herrmann W. and Davidson G.J., The spectrum of ore deposit types, volcanic environments, alteration haloes, and related exploration vectors in submarine volcanic successions: Some examples from Australia, *Economic Geology* 96: 913-938, 2001.

Lydon J.W., Volcanogenic massive sulphide deposits, Part 2: Genetic Models. *Geoscience Canada Reprint Series* 3: 155-182, 1988.

Mandeville C.W., Carey S. and Sigurdsson H., Sedimentology of the Krakatau 1883 submarine pyroclastic deposits, *Bulletin of Volcanology* 57: 512-529, 1996.

McPhie J. and Allen R.L., Facies architecture of mineralised submarine volcanic sequences: Cambrian Mount Read Volcanics, western Tasmania, *Economic Geology* 87: 587-596, 1992.

McPhie J., Doyle M.J. and Allen R.L., Volcanic Textures: A guide to the interpretation of textures in volcanic rocks. CODES, University of Tasmania, Hobart, 196 pp, 1993.

Morton R.L., Walker J.S., Hudak G.J. and Franklin J.M., The early development of an Archean submarine caldera complex with emphasis on the Mattabi Ash-flow Tuff and its relationship to the Mattabi massive sulfide deposit, *Economic Geology* 86: 1002-1011, 1991.

Niem A.R., Mississippian pyroclastic flow and ash-fall deposits in the deep-marine Ouachita flysch basin, Oklahoma and Arkansas, *Geological Society of America Bulletin* 88: 49-61, 1977.

Ninkovich D., Sparks R.S.J. and Ledbetter M.T., The exceptional magnitude and intensity of the Toba eruption, Sumatra: An example of the use of deep-sea tephra layers as a geological tool, Bulletin Volcanologique 41: 286-298, 1978.

Nishimura A., Marsaglia K.M., Rodolfo K.S., Colella C., Hiscott R.N., Tazaki K., Gill J.B., Janacek T., Firth J., Isiminger-Kelso M., Herman Y., Taylor R.N., Taylor B., Fujioka K. and Leg 126 Scientific Party, Pliocene-Quaternary submarine pumice deposits in the Sumisu Rift area, Izu-Bonin Arc, in *Sedimentation in Volcanic Settings* edited by R.V. Fisher and G.A. Smith. SEPM (Society for Sedimentary Geology) Special Publication No. 45: 201-208, 1991.

Perkins C. and Walshe J.L., Geochronology of the Mount Read Volcanics, Tasmania, Australia, *Economic Geology* 88: 1176-1197, 1993.

Self S., Rampino M.R., Newton M.S. and Wolff J.A., Volcanological study of the great Tambora eruption of 1815, *Geology* 12: 659-663, 1984.

Simkin T. and Fiske R.S., *Krakatau 1883: The volcanic eruption and its effects*, Smithsonian Institution Press, Washington, D.C., 464 p, 1983.

Stewart A.L. and McPhie J. An Upper Pliocene coarse pumice breccia generated by a shallow submarine explosive eruption at Milos, Greece, *Bulletin of Volcanology*, in press, 2003.

Walker G.P.L., Crystal concentration in ignimbrites, *Contributions to Mineralogy and Petrology* 36: 135-146, 1972.

White J.D.L., Manville V., Wilson C.J.N., Houghton B.F., Riggs N.R. and Ort M., Settling and deposition of AD 181 Taupo pumice in lacustrine and associated environments. *Special Publications International Association of Sedimentologists* 30: 141-150 (2001).

White M.J. and McPhie J., Stratigraphy and palaeovolcanology of the Cambrian Tyndall Group, Mt Read Volcanics, western Tasmania. *Australian Journal of Earth Sciences* 43:147-159, 1996.

White M.J. and McPhie J., A submarine welded ignimbrite-crystal-rich sandstone facies association in the Cambrian Tyndall Group, western Tasmania, Australia. *Journal of Volcanology and Geothermal Research* 76: 277-295, 1997.

Whitham A.G. and Sparks R.S.J., Pumice. *Bulletin of Volcanology* 48: 209-224, 1986.

Wilson C.J.N., Houghton B.F., Kamp P.J.J. and McWilliams M.O., An exceptionally widespread ignimbrite with implications for pyroclastic flow emplacement, *Nature* 378: 605-607, 1995.

Wright J.V. and Mutti E., The Dali Ash, Island of Rhodes, Greece: a problem in interpreting submarine volcanogenic sediments. *Bulletin of Volcanology* 44: 153-168, 1981.

Yamada E., Subaqueous pyroclastic flows: their development and their deposits, in *Marginal Basin Geology*, edited by B.P. Kokelaar and M.F. Howells. Geological Society Special Publication 16: 29-36, 1984.

Rodney L. Allen, Volcanic Resources Limited, Guldgaten 11, S-93632 Boledin, Sweden.

Jocelyn McPhie, Centre for Ore Deposit Research and School of Earth Sciences, University of Tasmania, Private Bag 79, Hobart, Tasmania. 7001, Australia.

Vesiculation and Eruption Processes of Submarine Effusive and Explosive Rocks From the Middle Miocene Ogi Basalt, Sado Island, Japan

Norie Fujibayashi

Department of Geology, Faculty of Education and Human Sciences, Niigata University, Niigata, Japan

Umio Sakai

Hokuriku Branch, Nittoc Construction Co. Ltd., Toyama, Japan

The Middle Miocene Ogi Basalt, formed in a back-arc basin, contains products of a variety of eruption processes, from extrusion of pillow lava (Sawasaki pillow lava) associated with formation of hyaloclastite to explosive formation of fluidal spatter and scoriaceous pyroclasts (Sawasaki pyroclastic rocks). The vesiculation textures of juvenile samples were examined to determine eruption style. The Sawasaki pillow lava, hyaloclastite, and pyroclastic rocks are characterized by their high vesicularities: 22–44% for the pillow lava and hyaloclastite, and 19–64% for the pyroclastic rocks. The Sawasaki pyroclastic rocks are characterized by their high contents of small vesicles (<0.2mm across) that show an abrupt increase in vesicle size-frequency patterns. Large vesicles, up to 1 cm in diameter, are minor in the spatter clasts. Population densities calculated for the small vesicles are similar to those of Kilauea lava fountain products. Vesiculated magma that rose slowly erupted effusively to make pillow lava and then hyaloclastite. Subsequently, accelerated magma, rich in small bubbles, must have erupted in lava fountain style through the water column and produced spatter clasts. Those were hydroclastically disintegrated into scoria lapillis, and formed thick scoriaceous pyroclastic rocks, instead of a formation of clastogenic lava on land.

INTRODUCTION

Explosive basaltic products from the deep sea environments were recently reported from the Sumisu back-arc rift [*Gill et al*, 1990], Mid-Ocean Ridges [*Siebe et al.*, 1995; *Eissen, et al.*, this volume; *Clague et al.*, this volume], and New Ireland basin [*Herzig, et al.*, 1998], indicating that basaltic explosive eruptions can take place even in a deep sea environment. Though high volatile content of magmas might be a reason for explosive eruptions [eg. *Gill et al.*, 1990], fragmentation processes have not been evaluated for any of these samples.

Vesicularities reported for subaqueous explosive basaltic pyroclasts are usually lower than 75%: 30–40% for pyroclasts in the Sumisu Back-arc Rift [*Gill et al.*, 1990] and 50–80% for those from La Palma in the Canary Islands [*Staudige and Schmincke*, 1984]. This suggests that their fragmentation process is unlikely to be the "maximum packing of bubbles" [e.g., *Sparks*, 1978; *Wilson and Head*, 1981; *Gerlach*, 1986; *Greenland et al.*, 1988; *Sparks et al.*, 1994; *Mader et al.*, 1994; 1996]. Recently, two alternative processes have been proposed for basaltic explosive eruptions. One is the "annular or slug flow model connected with a physical collapse of the foam layer at the top of a conduit" [*Vergniolle and Jaupart*, 1986; 1990; *Jaupart and Vergniolle*, 1988; 1989; *Proussevich et al.*, 1993]. The other proposal is the "vesiculation burst at high degree of volatile supersaturation" [*Cashman and Mangan*, 1994; *Mangan and Cashman*, 1996; *Cashman et al.*, 2002]. Based on detailed study of vesiculation textures of juvenile clasts, *Cashman and Mangan* [1994] and *Mangan and Cashman* [1996] suggested that fragmentation of basaltic magma is caused by acceleration of magma upwelling due to this vesiculation burst even with vesicularity less than 75%, which is considered to be the lower vesicularity limit for fragmentation of magma by internal overpressuring of bubbles.

Middle Miocene Ogi Basalt includes pyroclastic rocks that contain fluidal spatter clasts. This paper describes vesiculation textures, population densities and vesicularity of juvenile clasts in pyroclastic rocks and compares them with those of associated pillow lavas and hyaloclastite breccia from the same horizon. Based on these data, we propose an explosive eruption process of basaltic magma at intermediate- to deep-sea environment.

THE OGI BASALT

The Ogi Basalt [*Yamakawa and Chihara*, 1968] is an assemblage of basaltic to andesitic pillow lavas and volcaniclastic rocks (Figure 1). It formed a seamount in a marginal basin, the Japan Sea (Figure 1A), in the Middle Miocene (15.5 to 10.7 Ma [*Tsunakawa, et al.*, 1983; *Shinmura, et al.*, 1995]). Basaltic to andesitic pillow lavas and volcaniclastic rocks are intercalated with acidic tuff, tuffaceous mudstone and/or mud- to siltstone of the Tsurushi Formation, which was deposited in an intermediate- to deep-sea environment [*Matsunaga*, 1963]. *Research Group of Ogi Basalt* [1977] and *Komatsu and Takegawa* [1985] reported subaqueous bombs of explosive origin from the volcaniclastic rocks.

The Ogi Basalt is subdivided into four eruptive and two intrusive units (Figures 1B and 2). The lowermost Itsubo andesitic basalt consists of close-packed pillow lava [*Yamagishi*, 1994] grading upward to pillow fragment-bearing hyaloclastite breccia [*McPhie et al.*, 1993]. The rocks have olivine phenocrysts with spinel inclusions, often accompanied by microphenocrysts of olivine, plagioclase, and clinopyroxene, and xenocrysts of plagioclase and clinopyroxene (Table 1).

The lower-middle Sawasaki basalt is composed of basal pillow lava and directly overlying pyroclastic rocks. The pillow lava covers the Itsubo pillow lava and hyaloclastite breccia discordantly. It is close-packed pillow lava that grades horizontally into massive lava (Figure 2), and sometimes contains thin fine hyaloclastite beds < 1m thick. The pyroclastic rocks (Sawasaki pyroclastic rocks) are scoriaceous and associated with fluidal spatter clasts and are described in the next section. Both pillow lava and pyroclastic rocks are olivine basalt that contains olivine phenocrysts with spinel inclusions, and microphenocrysts of olivine and plagioclase (Tables 1 and 2).

The upper-middle Etsumi basalt comprises basal angular fragment–dominant hyaloclastite breccia, middle close-packed pillow lava and upper epiclastic hyaloclastite breccia (Figure 2). The rocks are characterized by large phenocrysts of olivine, plagioclase and clinopyroxene, and abundant xenocrysts of olivine, plagioclase, clinopyroxene and orthopyroxene (Table 1).

The uppermost Konagase andesite is composed mainly of hyaloclastite breccia associated with minor close-packed pillow lava (Figure 2). The hyaloclastite breccia contains abundant isolated pillow lobes and angular pillow fragments. The rocks are almost aphyric and often contain microphenocrysts of plagioclase, clinopyroxene and orthopyroxene (Table 1).

The Shiraki picritic dolerite sheet intrudes into the Sawasaki pillow lava at the boundary with underlying Itsubo pillow lava. The Itsubo-Shiroyama basaltic andesite sill intrudes into tuffaceous mudstone of the Tsurushi Formation and hyaloclastite breccia of the Konagase andesite. In addition, there are numerous contemporaneous basaltic dikes [< 2 m in width] intruding the Ogi Basalt, not shown in Figure 1.

THE SAWASAKI PYROCLASTIC ROCKS

The Sawasaki pyroclastic rocks compose the top of the Sawasaki basalt unit covering the Sawasaki pillow lava (Figure 3A). The rocks are scoria agglomerate, scoria lapilli tuff and scoria tuff (Figure 3B). Roughly estimated, total thickness is about 100 m. Both scoria agglomerate and lapilli tuff contain water-chilled spatter clasts with fluidal shape (Figures 3 C, D and E).

The scoria agglomerate is thickly stratified with varying amounts of spatter (Figure 3B). The largest spatter observed is around 30 cm in diameter. Each bed is poorly sorted (Figure 3C). The matrix is composed of scoria lapilli with a wide variation in size. Accumulated spatter clasts occur along the coastline between Sawasaki and Fukaura (Figures 3D and E). The facies is depleted in fine matrix. Spatter

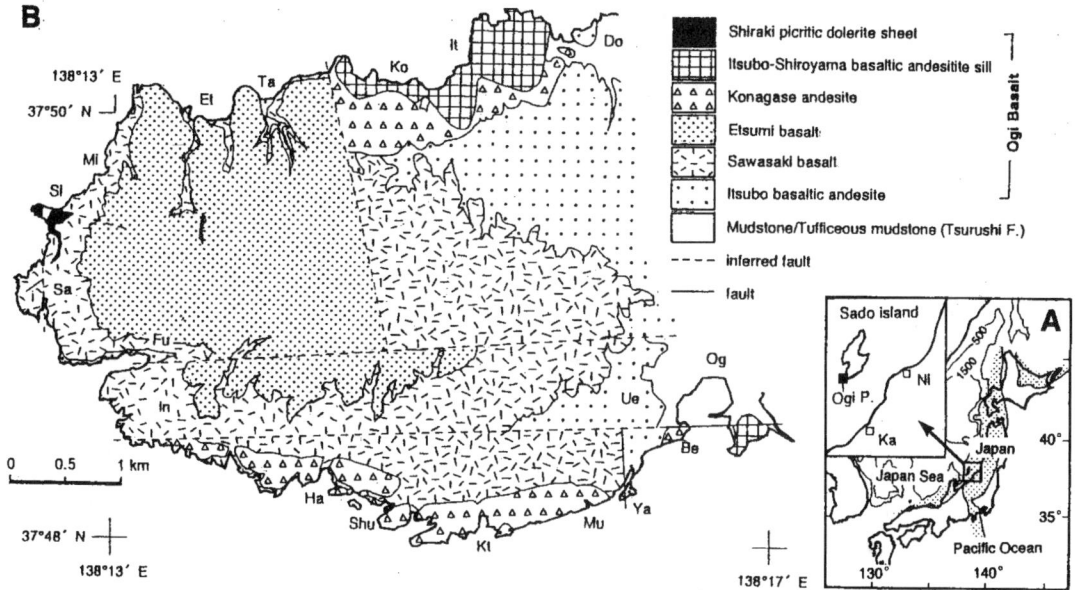

Figure 1. Location (A) and geological map (B) of the Ogi Peninsula. Ni: Niigata, Ka: Kashiwazaki, Ogi P.: Ogi Peninsula, Do: Dougama, It: Itsubo, Ko: Konagase, Ta:Tanoura, Et: Etsumi, Mi:Mitsuya, Si: Shiraki, Sa: Sawasaki, Fu: Fukaura, In: Inugamidaira, Ha: Haginoura, Shu: Shukunegi, Kt: Kotoura, Mu:Mushiya, Ya: Yajima, Ue: Ueno, Be: Bentenzaki, and Og: Ogimachi.

clasts are not flattened and not welded to each other (Figure 3E). Scoria is unoxidized.

The scoria lapilli tuff is characterized by thinly bedded (cm to ten cm scale) stratification (Figure 3F). Blocks of broken spatter are supported by scoria lapilli breccia and ash. Some lithic fragments derived from the Itsubo basaltic andesite are included.

The scoria tuff overlies the scoria agglomerate and scoria lapilli tuff. The thickness varies from around 30cm at Mitsuya (Figure 3B) to 10 cm at Kowashimizu (Figure 3G). It is composed of silt-size basaltic glass shards and fine-grained scoria clasts with laminae structure. The shards and scoria clasts have angular shapes.

VESICLES

Vesicles observed microscopically in thin sections have a wide variation both in size and shape. The pillow lavas and hyaloclastites usually contain spherical to ellipsoidal vesicles in the glassy crust of a pillow lobe (Figure 4A). Amoeboidal vesicles are common in the inside of pillow lobes in which intersertal texture is formed [Figure 4B]. Inside of a lobe, even early-formed large vesicles show irregular forms having small apophyses. Vesicles become larger, up to 1 cm in diameter, toward the center of a pillow lobe.

Spherical to ellipsoidal vesicles developed in the pyroclastic rocks, even in the center of spatter clast, mostly range from 0.5 to 5 mm in diameter (Figure 4C). Based on macro-scale observation, only a few vesicles up to 1 cm in diameter are sparsely distributed in spatter clasts. These large vesicles are spherical to ellipsoidal in shape even with a matrix of intersertal to hyalo-ophitic texture (Figure 4C), in contrast to the vesicles of irregular form inside the pillow

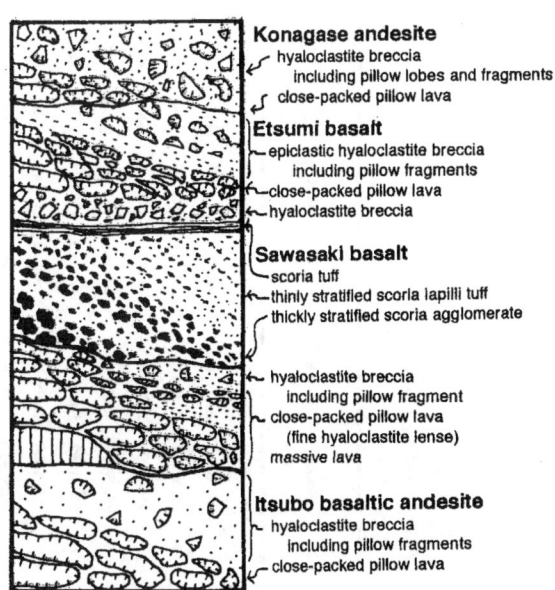

Figure 2. Schematic stratigraphic column of the Ogi Basalt.

Table 1. Modal compositions and vesicularities of the pillow lavas and hyaloclastites from the Ogi basalts. Modal compositions are recalculated into vesicle-free basis.

No.	Sample	Geological Unit	Locality	lithology	Occurrence	Analyzed part	Texture	Phenocrysts OL	SP	PL	Cpx	Micro-phenicrysts OL	PL	CPX	OPX	Xenocrysts OL	PL	CPX	OPX	Qz	GM	Vsph (%) >0.2 mm	<0.2 mm	Virr (%) >0.2 mm	<0.2 mm	Vesicularity (%)
1	96103002	Itsubo ba.	Itsubo	pillow l.	lobe	rind	V	1.1	-	-	-	6.7	-	-	-	-	-	-	-	-	92.1	7.9	-	-	-	7.9
2	96103105	Itsubo ba.	Itsubo	pillow l.	lobe	rind	V	2.1	-	-	-	5.7	tr.	tr.	-	-	-	-	-	-	92.2	8.9	-	-	-	8.9
3	96051902	Itsubo ba.	Itsubo	pillow l.	lobe	inside	I	tr.	-	-	-	-	tr.	-	-	-	-	-	-	-	100.0	-	-	19.2	19.4	38.5
4	95080901	Itsubo ba.	Kotoura	pillow l.	lobe	inside	I	-	-	-	-	-	1.4	tr.	-	-	2.9	-	-	-	95.7	-	-	27.1	12.5	39.6
5	95073105b	Itsubo ba.	Fukaura	pillow l.	lobe	inside	I	-	-	-	-	0.4	1.3	tr.	-	-	tr.	0.9	-	-	96.9	-	-	2.0	8.8	10.8
6	95081702	Itsubo ba.	Sawasaki	pillow l.	lobe	rind	V	6.6	tr.	-	-	2.1	9.4	tr.	-	-	-	-	-	-	81.8	4.8	3.8	-	-	8.6
7	96041406	Itsubo ba.	Sawasaki	pillow l.	lobe	rind	V	6.5	tr.	tr.	-	1.0	6.7	0.2	-	-	1.4	-	-	-	84.0	5.3	0.9	-	-	6.2
8	96052802a	Itsubo ba.	Ueno-kita	pillow l.	lobe	rind	V	0.5	-	-	-	1.0	1.0	1.0	-	-	tr.	tr.	-	-	96.3	8.7	1.6	-	-	10.4
9	96052802b	Itsubo ba.	Ueno-kita	pillow l.	lobe	inside	I	0.8	-	-	-	2.0	4.0	1.6	-	-	-	-	-	-	90.8	-	-	15.3	12.4	27.7
10	96052803	Itsubo ba.	Ueno-kita	hyalo.	matrix	clast	V	0.7	-	-	-	1.0	1.1	1.0	-	-	tr.	-	tr.	-	96.0	8.5	1.8	-	-	10.4
11	95080703a	Itsubo ba.	Itsubo	hyalo.	lobe	rind	V	1.7	tr.	-	-	1.0	-	-	-	-	6.4	-	tr.	-	90.9	12.1	4.0	-	-	16.1
12	95080703b	Itsubo ba.	Itsubo	hyalo.	lobe	rind	V	2.0	tr.	-	-	tr.	-	-	-	-	tr.	-	-	-	97.2	13.7	4.5	-	-	18.2
13	96051707	Itsubo ba.	Tanoura	hyalo.	lobe	rind	V	0.6	-	-	-	1.7	0.0	1.4	-	-	1.2	-	-	-	95.1	3.5	1.2	-	-	4.6
14	96052703	Itsubo ba.	Ogimachi	hyalo.	matrix	clast	V	-	-	-	-	1.7	4.3	3.4	-	-	-	-	-	-	90.5	-	0.9	-	-	0.9
15						clast	V	-	-	-	-	3.5	0.9	1.8	-	-	-	-	-	-	93.8	-	-	-	-	0.0
16	95073005 β	Sawasaki bs.	Sawasaki	pillow l.	lobe	inside	I	21.1	tr.	-	-	1.8	0.8	-	-	-	-	-	-	-	76.4	-	-	10.6	17.3	27.9
17	95082003	Sawasaki bs.	Sawasaki	pillow l.	lobe	inside	I	5.8	tr.	-	-	3.0	0.3	-	-	-	-	-	-	-	91.0	-	-	21.0	14.6	35.5
18	95082001	Sawasaki bs.	Sawasaki	sheeted l.	lava	inside	H	6.0	tr.	-	-	0.7	3.5	-	-	-	-	-	-	-	89.8	-	-	17.8	19.2	37.0
19	96041407	Sawasaki bs.	Yajima	pillow l.	lobe	rind	V	tr.	tr.	-	-	tr.	tr.	-	-	-	-	-	-	-	100.0	41.7	0.7	-	-	42.4
20						inside	feather	1.5	tr.	-	-	tr.	tr.	-	-	-	-	-	-	-	97.6	-	-	24.5	12.9	37.4
21	96041403	Sawasaki bs.	Ueno-kita	pillow l.	lobe	rind	V	4.2	tr.	-	-	3.1	1.2	-	-	-	-	-	-	-	91.5	42.6	1.4	-	-	44.0
22						inside	H	2.7	tr.	-	-	0.4	0.0	-	-	-	-	tr.	-	-	96.9	-	-	38.7	10.7	49.4

V: vitrophyric, I: intersertal, H: hyaloophitic OL: olivine, Sp: spinel, PL: plagioclase, CPX: clinopyroxene, OPX: orthopyroxene, QZ: quartz, GM: groundmass, tr.: trace, -: none, Vsph: vesicularity for spherulitic vesicles, and Virr: vesicularity for irregular vesicles, ba.: basalt, bs.: basaltic andesite, bs.: basalt, and and.: andesite.

Table 1. (continued)

No.	Sample	Geological Unit	Locality	lithology	Occurrence	Analyzed part	Texture	Phenocrysts OL	SP	PL	Cpx	Micro-phenicrysts OL	PL	CPX	OPX	Xenocrysts OL	PL	CPX	OPX	Qz	GM	Vsph (%) >0.2 mm	<0.2 mm	Virr (%) >0.2 mm	<0.2 mm	Vesicularity (%)
23	96052805b	Sawasaki bs.	Ueno-kita	hyalo.	matrix	clast	V	0.8	tr.	-	-	1.6	0.8	-	-	-	-	-	-	-	96.8	18.6	4.5	-	-	23.1
24	96052805a	Sawasaki bs.	Ueno-kita	hyalo.	matrix	clast	V	6.0	tr.	-	-	1.9	-	-	-	-	-	-	-	-	92.1	20.1	2.0	-	-	22.0
25	96052806	Sawasaki bs.	Ueno-kita	hyalo.	lobe	rind	V	0.9	tr.	-	-	2.8	-	-	-	-	-	-	-	-	96.3	31.1	2.5	-	-	33.7
26	96052806	Sawasaki bs.	Ueno-kita	hyalo.	matrix	clast	V	3.6	tr.	-	-	2.5	1.0	-	-	-	-	-	-	-	92.9	41.5	0.5	-	-	42.0
27						clast	V	3.0	tr.	-	-	3.1	1.2	-	-	-	-	-	-	-	92.7	19.1	3.4	-	-	22.5
28	96052807	Sawasaki bs.	Ueno-kita	hyalo.	matrix	clast	V	3.2	tr.	-	-	5.8	3.2	-	-	-	-	-	-	-	87.8	29.7	1.1	-	-	30.8
29						clast	V	0.6	tr.	-	-	3.4	tr.	-	-	-	-	-	-	-	94.9	40.8	0.0	-	-	40.8
30	96053001b	Sawasaki bs.	Haginoura	hyalo.	matrix	clast	V	6.0	tr.	-	-	1.2	8.4	-	-	-	-	-	-	-	84.3	27.6	1.0	-	-	28.6
31						clast	V	9.2	tr.	-	-	4.2	tr.	-	-	-	-	-	-	-	79.8	40.5	1.2	-	-	41.7
32	95081504	Sawasaki bs.	Sawasaki	hyalo.	matrix	rind	V	4.4	tr.	-	-	tr.	tr.	-	-	-	-	-	-	tr.	95.6	36.9	0.5	-	-	37.5
33	96051711	Esumi bs.	Esumi	pillow l.	lobe	inside	I	2.0	tr.	1.4	0.5	1.0	2.4	0.7	-	4.0	12.2	4.6	1.3	-	74.1	-	-	17.3	11.4	28.8
34	96051803	Konagase and.	Konagase	pillow l.	lobe	rind	V	0.2	tr.	-	-	tr.	0.4	tr.	tr.	-	3.1	0.2	0.4	-	95.6	0.9	0.2	-	-	1.1
35	96051805	Konagase and.	Konagase	pillow l.	lobe	inside	I	tr.	-	-	-	-	0.3	tr.	0.3	-	3.6	0.6	0.3	-	94.9	-	-	6.3	13.8	20.1
36	94050801a	Konagase and.	Konagase	pillow l.	lobe	inside	I	0.4	-	-	-	-	tr.	tr.	tr.	-	2.2	0.4	tr.	-	97.1	-	-	7.0	0.4	7.4
37	96041504	Konagase and.	Dougama	pillow l.	lobe	rind	H	0.7	-	-	-	1.1	1.1	-	-	-	1.1	0.4	-	-	95.5	4.1	0.4	-	-	4.5
38	00102001	Konagase and.	Mushiya	pillow l.	lobe	inside	I	tr.	-	-	-	1.0	0.5	1.0	tr.	-	2.4	1.5	tr.	-	93.7	-	-	12.4	5.2	17.6
39	95080511	Konagase and.	Shukunegi	hyalo.	matrix	clast	V	1.9	tr.	-	-	1.1	tr.	tr.	tr.	-	tr.	-	-	-	98.0	10.0	8.0	-	-	18.0
40	96051707	Konagase and.	Tanoura	hyalo.	matrix	clast	V	1.0	tr.	-	-	0.9	0.4	tr.	tr.	-	tr.	-	-	-	97.7	16.0	5.5	-	-	21.5
41	95080801a	Konagase and.	Bentenzaki	hyalo.	lobe	inside	I	1.2	tr.	-	-	1.6	tr.	tr.	tr.	-	2.2	-	tr.	-	95.0	-	-	2.9	15.3	18.2
42	95080801b	Konagase and.	Bentenzaki	hyalo.	lobe	rind	V	2.2	tr.	-	-	1.1	tr.	tr.	tr.	-	tr.	-	-	-	96.7	12.9	10.6	-	-	23.5
43	95080802	Konagase and.	Bentenzaki	hyalo.	lobe	rind	V	1.0	tr.	-	-	1.0	0.3	tr.	tr.	-	1.6	-	-	-	96.1	15.5	5.3	-	-	20.8
44						inside	H	1.8	tr.	-	-	0.6	0.6	0.6	tr.	-	1.8	-	-	-	95.2	-	-	19.3	4.6	23.9
45	00102113	Konagase and.	Konagase	hyalo.	lobe	rind	H	0.6	tr.	tr.	-	0.3	0.9	1.2	-	-	0.9	-	tr.	-	96.1	-	-	8.5	9.0	17.6

Table 2. Modal compositions and vesicularities of the Sawasaki pyroclastic rocks.

No.	Sample	lithology	Occurrence	Analyzed part	Texture	Phenocrysts				Micro-phenicrysts				Xenocrysts					Modal Compositions (%) GM	Vsph (%) >0.2 mm	Vsph (%) <0.2 mm	Virr (%) >0.2 mm	Virr (%) <0.2 mm	Vesicularity (%)
						OL	SP	PL	Cpx	OL	PL	CPX	OPX	OL	PL	CPX	OPX	Qz						
1	95073006b	Scoria agg.	lapilli	rind	V	0.9	tr.	tr.	-	1.9	2.3	-	-	-	-	-	-	-	94.9	25.4	10.3	-	-	35.7
2	95080502	Scoria lpt.	matrix	clast	V	2.8	tr.	-	-	tr.	tr.	-	-	-	tr.	tr.	tr.	-	97.2	20.3	25.0	-	-	45.3
3				clast	V	0.9	tr.	-	-	1.3	0.4	3.1	-	-	-	-	-	-	94.2	11.5	30.1	-	-	41.6
4	96053006b	Scoria lpt.	matrix	clast	V	2.9	tr.	-	-	tr.	tr.	tr.	-	-	-	-	-	-	97.1	12.3	6.9	-	-	19.1
5				clast	V	7.0	tr.	-	-	1.2	tr.	-	-	-	-	-	-	-	91.9	32.7	17.6	-	-	50.3
6	96053101b	Scoria lpt.	matrix	clast	V	10.9	tr.	-	-	tr.	-	-	-	-	-	-	-	-	89.1	31.7	-	-	-	31.7
7				clast	V	3.0	tr.	-	-	0.6	-	-	-	-	-	-	-	-	96.3	14.2	17.7	-	-	31.9
8	95080503	Scoria lpt.	lapilli	inner	I	-	-	-	-	0.6	tr.	-	-	-	-	-	-	-	98.2	-	-	46.2	3.5	49.7
9	96052904	Scoria lpt.	spatter	rind	V	3.7	tr.	-	-	0.0	0.7	-	-	-	-	-	-	-	95.6	28.1	0.3	-	-	28.4
10	96053105	Scoria agg.	matrix	clast	V	9.5	tr.	-	-	1.7	tr.	-	-	-	-	-	-	-	88.8	9.1	30.7	-	-	39.8
11				clast	V	4.8	tr.	-	-	1.9	tr.	-	-	-	-	-	-	-	93.3	17.9	17.9	-	-	35.8
12	96053005	Scoria agg.	matrix	clast	V	6.7	tr.	-	-	3.3	tr.	-	-	-	-	-	-	-	90.0	38.9	11.1	-	-	50.0
13				clast	V	2.3	tr.	-	-	1.1	tr.	-	-	-	-	-	-	-	96.6	25.6	6.4	-	-	32.0
14				clast	V	2.1	tr.	-	-	tr.	tr.	-	tr.	-	-	-	-	-	89.4	35.2	18.7	-	-	53.8
15	96053003b	Scoria agg.	matrix	clast	V	3.0	tr.	-	-	2.6	0.8	-	-	-	-	-	-	-	93.6	44.1	10.4	-	-	54.6
16				clast	V	7.9	tr.	-	-	1.8	1.2	-	-	-	-	-	-	-	89.0	40.7	15.6	-	-	56.3
17	96053002	Scoria agg.	matrix	clast	V	-	-	-	-	0.9	0.6	tr.	-	-	0.9	0.3	-	-	97.3	14.1	26.3	-	-	40.4
18				clast	V	2.8	tr.	-	-	tr.	-	-	-	-	-	-	-	-	97.2	30.5	26.8	-	-	57.3
19	96053103b	Scoria agg.	matrix	clast	V	7.9	tr.	-	-	0.6	-	-	-	-	-	-	-	-	91.6	47.2	16.8	-	-	64.0
20				clast	V	-	-	-	-	tr.	-	-	-	-	-	-	-	-	100.0	37.9	23.3	-	-	61.2
21	96052914	Scoria agg.	matrix	clast	V	1.5	tr.	-	-	tr.	-	-	-	-	-	-	-	-	98.5	9.8	38.2	-	-	48.0
22				clast	V	1.7	tr.	-	-	1.7	-	-	-	-	-	-	-	-	96.6	25.9	35.4	-	-	61.2

agg.: agglomerate, and lpt: lapilli tuff.

Table 2. (continued)

No.	Sample	lithology	Occurrence	Analyzed part	Texture	Phenocrysts				Micro-phenicrysts				Xenocrysts					Modal Compositions (%)	Vsph (%)		Virr (%)		Vesicularity (%)
						OL	SP	PL	Cpx	OL	PL	CPX	OPX	OL	PL	CPX	OPX	Qz	GM	>0.2 mm	<0.2 mm	>0.2 mm	<0.2 mm	
23	96050406	Scoria lpt.	matrix	clast	V	7.8	tr.	tr.	-	7.8	6.0	3.0	-	-	-	-	-	-	83.1	21.6	12.0	-	-	33.7
24				clast	V	2.2	tr.	tr.	-	2.2	1.8	2.7	-	-	-	-	-	-	93.3	16.6	7.6	-	-	24.2
25	96052901	Scoria lpt.	matrix	clast	V	1.4	tr.	-	-	2.9	-	-	-	-	-	-	-	-	95.7	33.6	9.4	-	-	43.0
26				clast	V	2.9	tr.	-	-	1.0	-	-	-	-	-	-	-	-	96.1	21.2	25.0	-	-	46.2
27	96052903	Scoria lpt.	matrix	clast	V	tr.	tr.	-	-	tr.	-	-	-	-	-	-	-	-	100.0	10.8	9.5	-	-	20.3
28				clast	V	9.7	tr.	-	-	4.8	-	-	-	-	-	-	-	-	85.5	30.1	18.4	-	-	48.5
29				clast	V	5.8	tr.	-	-	tr.	-	-	-	-	-	-	-	-	94.2	18.3	27.8	-	-	46.1
30	96052910	Scoria lpt.	matrix	clast	V	5.1	tr.	-	-	tr.	-	-	-	-	-	-	-	-	94.9	27.4	12.9	-	-	40.3
31				clast	V	2.8	tr.	-	-	tr.	-	-	-	-	-	-	-	-	97.2	12.2	40.5	-	-	52.7
32	96052911	Scoria tuff	matrix	clast	V	-	-	-	-	-	-	-	-	-	-	-	-	-	100.0	29.5	26.2	-	-	55.7
33				clast	V	-	-	-	-	-	-	-	-	-	-	-	-	-	100.0	24.5	11.3	-	-	35.8
34	96053104	Scoria lpt.	matrix	clast	V	1.4	tr.	-	-	1.4	-	-	-	-	-	-	-	-	97.3	50.0	5.6	-	-	55.6
35				clast	V	18.3	tr.	-	-	2.8	-	-	-	-	1.4	-	-	-	77.5	27.9	29.5	-	-	57.4
36	96052909	Scoria lpt.	matrix	clast	V	3.2	tr.	-	-	1.1	tr.	-	-	-	-	-	-	-	95.7	10.8	29.1	-	-	39.9
37				clast	V	0.8	tr.	-	-	tr.	0.8	-	-	-	-	-	-	-	98.3	12.0	48.7	-	-	60.7
38				clast	V	0.7	tr.	-	-	0.7	0.7	-	-	-	tr.	-	-	-	97.8	19.9	40.7	-	-	60.5
39	96053007	Scoria lpt.	matrix	clast	V	9.3	tr.	-	-	tr.	-	-	-	-	-	-	-	-	90.7	47.0	12.0	-	-	59.0
40				clast	V	1.6	tr.	-	-	0.5	-	-	-	-	-	-	-	-	97.9	47.7	9.1	-	-	56.8
41	96053102	Scoria lpt.	matrix	clast	V	-	-	-	-	-	tr.	tr.	-	-	-	-	-	-	100.0	18.0	19.5	-	-	37.6
42				clast	V	5.6	tr.	-	-	3.7	tr.	tr.	-	-	-	-	-	-	90.7	29.9	6.5	-	-	36.4

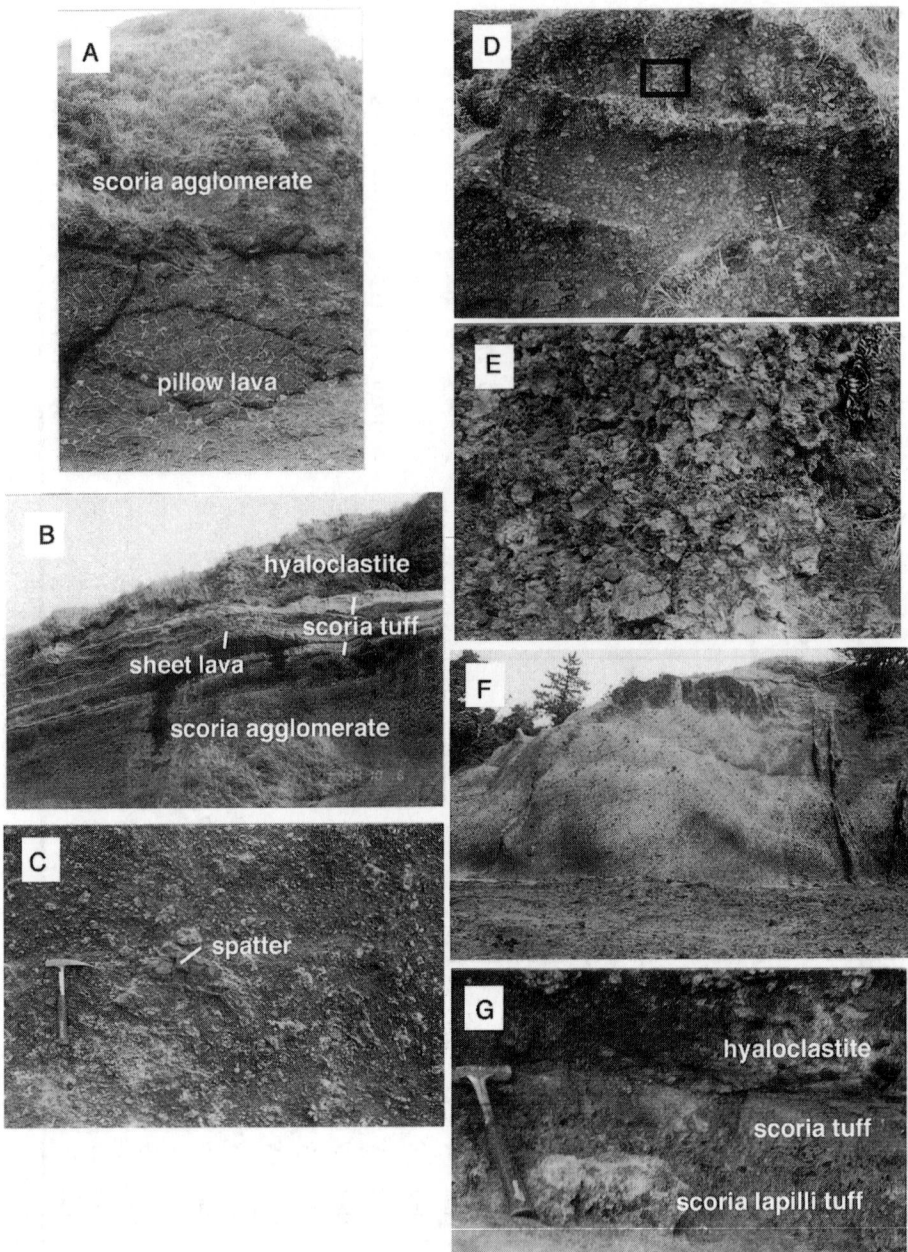

Figure 3. Photographs of the Sawasaki pyroclastic rocks.
A: Sawasaki pillow lava and overlying Sawasaki pyroclastic rocks from the cliff beneath the Sawasaki Lighthouse.
B: Scoriaceous agglomerate (more than 50 m thick) and scoriaceous tuff (around 30 cm thick) from the Sawasaki pyroclastic rocks (Sawasaki basalt unit) from the cliff in Mitsuya. The scoriaceous tuff is intercalated with thin sheet lava and overlain by the hyaloclastite breccia of Etsumi basalt unit.
C: Scoriaceous agglomerate from Inugamidaira containing spatter with irregular shape in a scoriaceous matrix.
D: Spatter-concentrated agglomerate from the Sawasaki Fishery Harbor. The area surrounded by the black border is shown in F. The largest spatter clast in the left-bottom is around 30cm in diameter.
E: Accumulated water-chilled spatter clats with irregular shapes. The spatter clasts are not flattened or welded.
F: Thinly bedded scoria lapilli tuff from Shukunegi Coast.
G: Scoria tuff from Kowashimizu-zawa (stream), overlying scoria lapilli tuff. The hyaloclastite breccia is of Etsumi basalt unit.

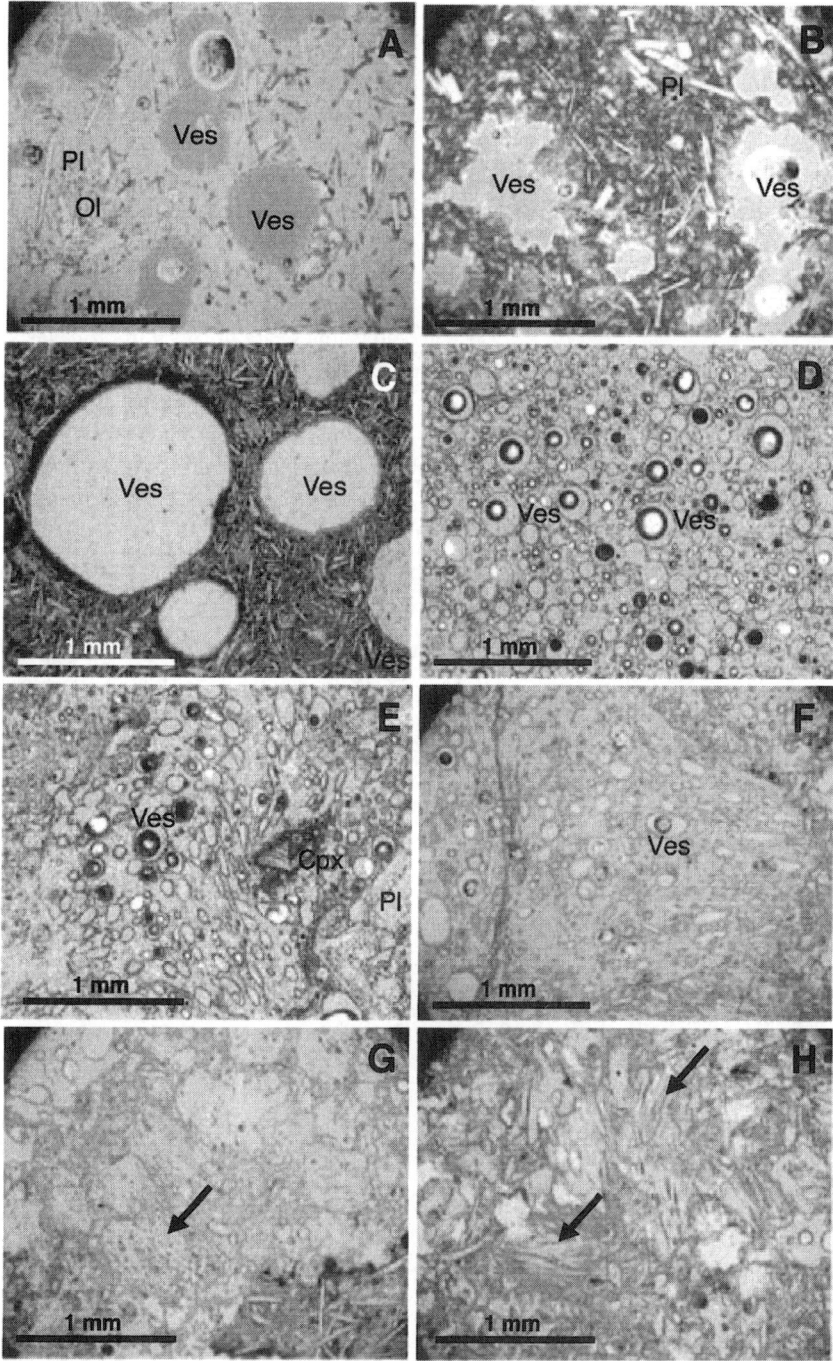

Figure 4. Vesicle shapes in pillow lava and pyroclasts. A: spherical vesicles in the glassy crust of a pillow lobe (96041403), B: amoeboidal vesicles in the inside of a pillow lobe (96041403), C: largely expanded vesicles in the center of a water-chilled spatter clast (95080404), D: a typical glassy clast in scoria lapilli tuff, rich in small spherical vesicles (95080502), E: vesicles spherical to ellipsoidal on the left-hand side of the clast are elongated into tube-like shapes on the right-hand side showing a variation in elongation direction (95080502)_F: vesicles elongated into tube-like shapes on the right-hand side of the central clast (96053002), G: fragmented fine pyroclasts in a matrix of lapilli tuff (96092910), and H: glass shards with flattened vesicles in scoria tuff (96092911).

lobes (Figure 4B). Chilled margins of spatter clasts and glassy shards, both with vitrophiric texture, usually have small vesicles of spherical to ellipsoidal shape (Figure 4D).

Some ash size fragments have tube-like vesicles (Figures 4E, F, and G). Gradation from ellipsoidal to tube-like shape is observed in those clasts, often coupled with a change in elongation direction (Figures 4E and F). Flattened vesicles also exist in the finer fragments less than 0.5 mm in length in scoria lapilli tuff (Figure 4G), and are typical in scoria tuff (Figure 4H).

VESICLE SIZE DISTRIBUTION AND POPULATION DENSITY IN GLASSY SAMPLES

Vesicle size varies from the outer rim toward the center of a spatter clast and a pillow lobe, as mentioned previously.

Quenched glassy samples with vitrophyric texture record vesiculation textures at the time of eruption. Therefore, vesicle size and number of vesicles per unit area were measured for representative glassy samples with vitrophyric texture. Measurements were carried out manually on binary images of thin sections. Because the quenched rims of pillow lobes and spatter clasts are thin [less than 5 mm thick], and the clasts with vitrophyric texture in the matrix of pyroclastic rocks and hyaloclastites are smaller than 5 mm in length, analyzed areas are not large (6.1 to 25.6 mm^2 for pillow lobes, about 5.2 mm^2 for clasts in hyaloclastites, and 1.4 to 2.7 mm^2 for clasts in pyroclastic rocks).

Vesicle size frequencies are shown in Figure 5 as plots of vesicle number per unit area versus diameter length. Glassy clasts in pyroclastic rocks are enriched in small vesicles in the range less than 0.2 mm diameter with strong peaks at

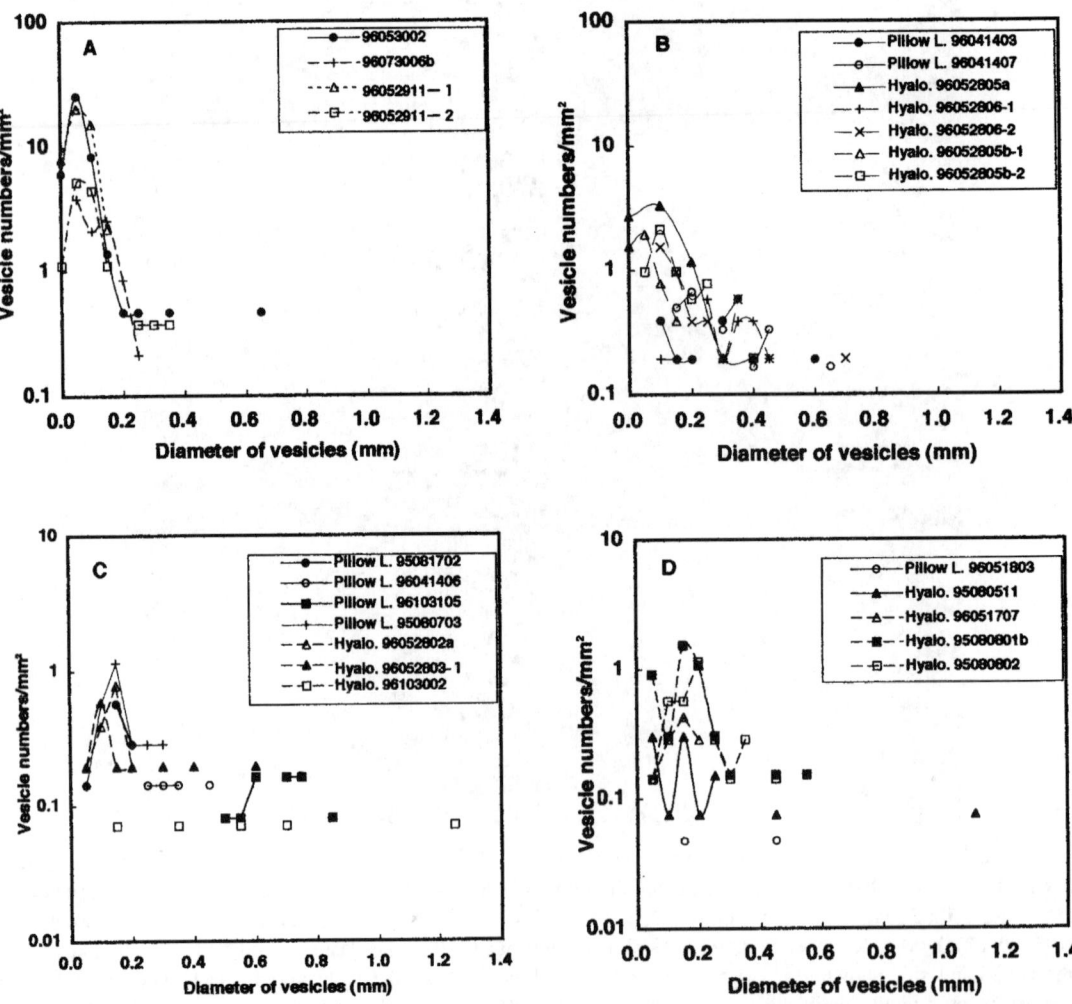

Figure 5. Vesicle numbers per unit area (mm^{-2}) as a function of diameter. A: Sawasaki pyroclastic rocks, B: Sawasaki pillow lava and hyaloclastite breccia, C: Itsubo pillow lava and hyaloclastite breccia, and D: Konagase pillow lava and hyaloclastite. All the analyzed samples are glassy.

0.05 mm (Figure 5A). On the other hand, pillow lavas and hyaloclastites of effusive origin show only weak peaks in that range (Figures 5C, D, and E). In a range larger than 0.2 mm diameter, both effusive samples and pyroclastic rocks show similar flat patterns.

Population density, n, was calculated by the method previously applied for vesiculation studies [*Sarda and Graham*, 1990; *Mangan et al.*, 1993]. The two-dimensional distribution of vesicles in thin section was converted to the three-dimensional distribution, using the method of *Saltikov* [1967] that was introduced for the analysis of "popping rock" by *Sarda and Graham* [1990]. The results are shown in Figure 6 as a plot of *ln* n (population density) versus vesicle diameter (D). Although, analyzed areas are small, some glassy clast (96053002 and 96092911-1) show almost linear vesicle size distributions (VSDs) in a range smaller than 0.2 mm diameter (Figure 6A). Larger vesicles plot above those straight lines. Pillow lava and hyaloclastite do not have linear correlations (Figure 6B).

When steady state bubble nucleation and growth is postulated, the relationship between *ln* n and D is represented by the following equation (1).

$$ln\ n = ln\ n_0 - (1/Gt)D \quad (1)$$

n_0 is the nucleation density, n is the population density by vesicle diameter, G is the growth rate assumed to be independent of size, t is the residence time of magm, and D is the vesicle diameter [e.g. *Sarda and Graham*, 1990]. The total population densities for all vesicle size classes less than 0.2 mm diameter are estimated to be 5.5×10^5 cm^{-3} for 96053002 and 5.7×10^5 cm^{-3} for 96092911-1. These values are similar to those for vesicles smaller than 0.2 cm in diameter in Kilauea lava fountains [1.9×10^4 to 1.8×10^5 cm^{-3} after *Mangan and Cashman*, 1996].

From the straight line drawn for the sample 96053002, the slope [1/Gt] calculated (36.1 mm^{-1}) is larger than for the Kilauea lava fountain products [50.0 to 104.9 cm^{-1}, after *Mangan and Cashman*, 1996], reflecting smaller vesicle size in the Sawasaki pyroclastic rocks.

VESICULARITY

The vesicularity was measured in two-dimensions using standard point counting techniques on thin sections. Spherical to ellipsoidal vesicles (Vsph) were distinguished from irregular vesicles (Virr) observed in the inside of the pillow lobes. Virr tends to be higher than Vsph for the rocks from the same unit, except for the Sawasaki basalt unit in which hollows are observed in some pillow lobes.

As shown in Tables 1 and 2, both the Sawasaki pyroclastic rocks and pillow lava are characterized by high Vsphs with wide variation ranged from 19 to 64% and from 22 to 44%, respectively. The other units have Vsph less than 21%. When vesicles were subdivided into those smaller than 0.2 mm and those larger than 0.2 mm in size, the Sawasaki pyroclastic rocks are distinctive for large amounts of the smaller vesicles (Figure 7). The amounts of large vesicles are similar between the Sawasaki pyroclastic rocks and effusive rocks (maximum 55% and 45%, respectively).

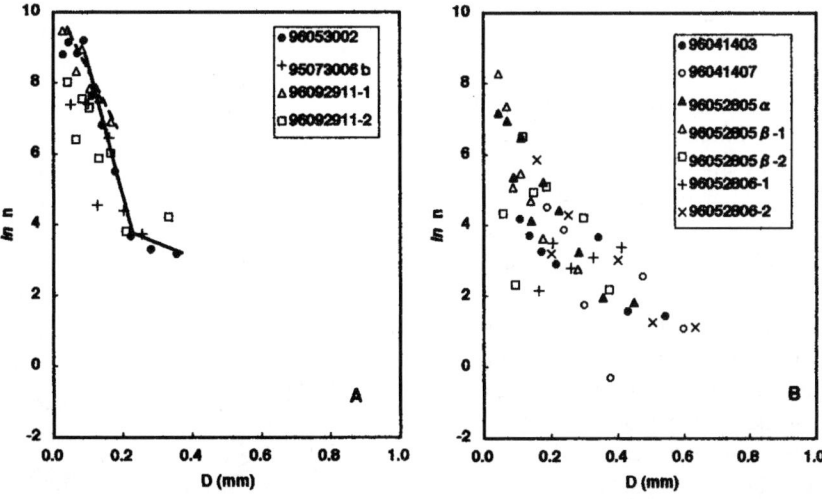

Figure 6. Vesicle size distributions as plots of ln n vs. Diameter (mm), where n is the population density calculated by the method of Sarda and Graham [1990]. A: Sawasaki pyroclastic rocks and B: Sawasaki pillow lava and hyaloclastite breccia.

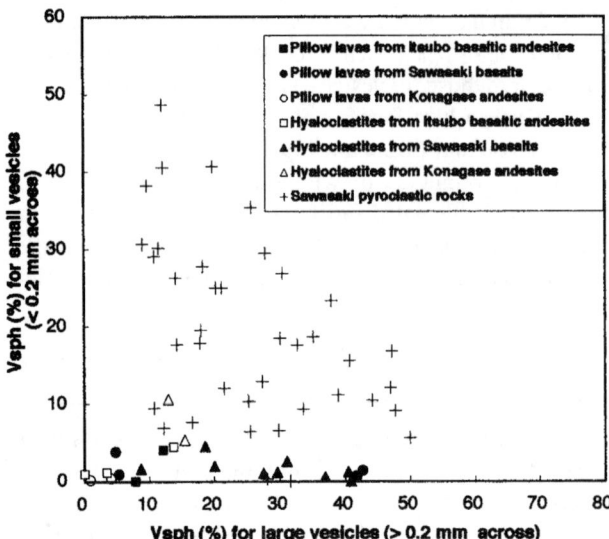

Figure 7. Spherical vesicularities (Vsphs) for large vesicles (>0.2 mm in diameter) vs. those for small ones in each samples from the Ogi basalts. All the samples shown have spherical vesicles in glassy groundmass.

ASCENT OF MAGMA FOR SAWASAKI PILLOW LAVA AND PYROCLASTIC ROCKS

The Sawasaki pillow lava and the Sawasaki pyroclastic rocks have common petrological features, and the latter directly overlies the former. The Sawasaki pillow lava is an effusive precursor of explosive eruption of the Sawasaki pyroclastic rocks. Basaltic eruptions tend to begin with a quiet rise of lava and develops into spectacular fountaining, as the vent becomes a widening fissure [e.g. *Bardintzeff and McBirney*, 2000]. The Sawasaki pillow lava and pyroclastic rocks are consistent with such an eruptive sequence.

As mentioned previously, the vesiculation texture characteristic of the Sawasaki pyroclastic rocks is the abrupt increase in small vesicles (<0.2 mm in diameter) (Figure 5). This increase implies that formation of small bubbles is directly correlated with the evolution of eruption style from effusive to explosive. The abrupt increase in these small vesicles suggests two vesiculation stages during the magma upwelling.

Based on numerical study, it was suggested that bubbles tend to nucleate within a narrow range of depth, and thereafter grow diffusively with decompressional expansion in a magma upwelling with a steady ascending rate [*Toramaru*, 1989]. Nucleation of small bubbles should be caused by some change in conditions of ascending magma, such as an acceleration of magma upwelling caused by injection of new magma into the chamber. Even a small amount of magma, injected into a magma chamber, may lead to high supply rate of magma to a narrow vent and may accelerate magma upwelling. Contemporaneous expansion of vent fissure could decompress magma and also cause new bubble nucleation. It is also possible that the small vesicles form as H_2O becomes saturated at shallow depth after the vesiculation of CO_2 at greater depth. However, this is not likely to be the case, because active formation of small vesicles is not observed in the effusive precursor. Another alternative is a secondary boiling by crystallization. This, however, is not supported because the crystallinity is similar for the rim of Sawasaki pillow lava and the glassy clasts of Sawasaki pyroclastic rocks (Tables 1 and 2). Therefore, we conclude that injection of a new magma into the chamber is the most likely cause of a second vesiculation of magma for the Sawasaki pyroclastic rocks (Figure 8).

The buoyancy generated is estimated to be about 0.23 per unit volume, if the difference in vesicularity is estimated at 9% (by the subtraction of average vesicularity of effusive rocks (34%) from that of pyroclastic rocks (45%)) and the density of basaltic magma is 2.6 [kg/m^3]. The magma will be forced to ascend faster with acceleration of about 0.09 (m/s^2). This value is not low in comparison with the ascending rate, 0.6–0.7 m/s, estimated for explosive magma of Mt. St. Helens based on a seismic movement [e.g. *Rutherford and Gardner*, 2000].

Accordingly, it is considered that the magma that formed Sawasaki pyroclastic rocks was accelerated in the conduit by the volumetric difference between the wide chamber and the thin conduit, decompressional widening of the vent, and vesiculation of magma when new magma injected into the chamber (Figure 8).

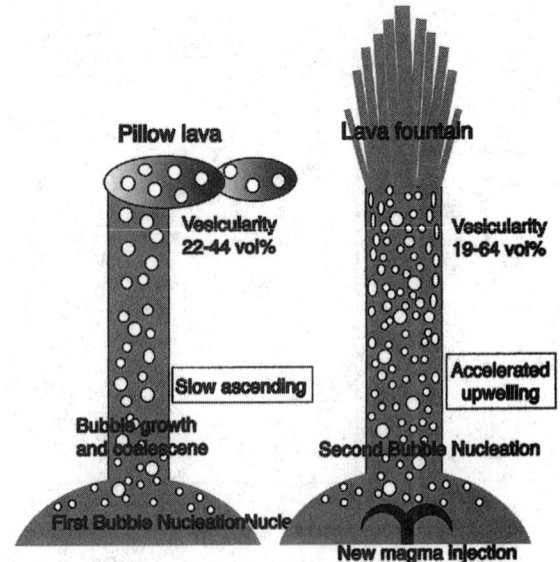

Figure 8. Diagram illustrating the vesiculation processes for the Sawasaki pillow lava and the Sawasaki pyroclastic rocks.

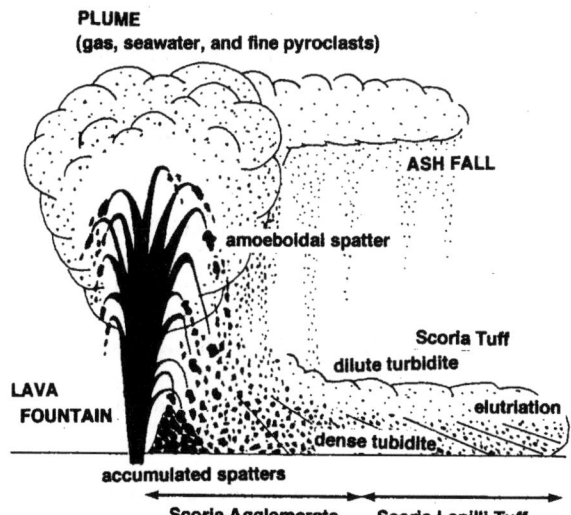

Figure 9. Schematic model of a subaqueous lava fountain eruption of the Sawasaki pyroclastic rocks [modified from Mueller and White_1992].

FRAGMENTATION OF THE SAWASAKI PYROCLASTIC ROCKS

At Kilauea, *Cashman and Mangan* [1994] and *Mangan and Cashman* [1996] suggested that a vesiculation burst might contribute to the fragmentation process based on vesiculation textures of juvenile clasts from the lava fountains. They proposed that fragmentation occurs by ductile deformation of magma caused by the acceleration of magma due to the vesiculation burst and breakup in the liquid jets of accelerated magma. In the present study, the population density of bubbles for the Sawasaki pyroclastic rocks was shown to be similar to the Kilauean clasts described above. We showed that even large vesicles dispersed in the inner part of spatter clasts are less than 1 cm and that the amount of these bubbles is quite low. These observations suggest that there was not significant bubble growth and coalescence to produce large gas pockets or a gas column in the magma [e.g. *Vergniolle and Jaupart*, 1986; 1990; *Proussevich et al.*, 1993]. Accordingly, a vesiculation burst apparently played an important role in fragmentation of magma.

In a submarine environment, fluidal spatter clasts were fragmentized into scoria breccia through hydroclastic fragmentation (Figure 9). Larger amounts of scoria lapilli should have been formed hydroclastically than in a subaerial environment.

The scoria clasts were transported as high-concentration sediment gravity flows and formed stratified scoria agglomerate and scoria lapilli tuff (Figure 9). Thinly stratified scoria lapilli tuff must be a distal facies. Accumulated spatter clasts may represent a near-vent facies, probably the equivalent of a spatter rampart on land. The spatter clasts, however, are unwelded. They have amoeboidal shape and are not flattened (Figure 3E). Due to rapid cooling in subaqueous environment, neither welding of spatter clasts nor formation of clastogenic lava flows took place.

Fine glass shards with flattened vesicles might have originated at the vent margin through shearing between accelerated magma and vent wall (Figure 8). The shards were suspended in a warm water plume over lava fountains and deposited, covering scoria agglomerate and lapilli tuff as scoria tuff (Figure 9). The thin laminae structure of scoria tuff suggests that fine glass shards formed dilute turbidity currents after they settled. The shards may have partly mixed with elutriated fine particles from the underlying scoria agglomerate and lapilli tuff, because the tuff contains some tiny scoria clasts with spherical vesicles. The whole eruption column remained below sea level, because no oxidized pyroclasts have been observed.

Acknowledgments. We are greatly indebted to D. A. Clague, K. V. Cashman, and P. Wallace for their reviews and valuable comments on the manuscript. K. Kano, K. Kurokawa and H. Yamagishi are also gratefully acknowledged for their discussions and encouragement.

REFERENCES

Bardintzeff, J-M., and A. R. McBirney, *Volcanology*, Jones and Bartlett Pub. Inc., French, 268p, 2000.

Cashman K., and M. T. Mangan, Physical aspects of magmatic degassing II. Constraints on vesiculation processes from textural studies of eruptive products, *Reviews in Mineralogy*, 30, 447-478, 1994.

Cashman, K., B. Sturtevant, P. Papale, and O. Navon, Magmatic fragmentation, in *Encyclopedia of Volcanoes*, edited by H. Sigurdsson, et al., pp. 421-430, Academic Press, New York, 2000.

Clague, D. A., A. S. Davis, and J.E. Dixon, Submarine strombolian eruptions on the Gorda Mid-ocean Ridge, this volume, 2003.

Eissen, J-P., Y. Yves Fouquet, D. Hardy, and H. Ondrlas, Volcaniclastic Basaltic deposits between 400 and 2000 m.b.s.l. at the axis of the Mid-Atlantic Ridge south of the Azores, this volume, 2003.

Gerlach, T. M., Exsolution of H_2O, CO_2, and S during eruption episodes at Kilauea volcano, Hawaii. *Jour. Geophys. Res.*, 91, 12177-12185, 1986.

Gill, J., P. Torssanander, H. Lapierre, R. Taylor, K. Kaiho, M. Koyama, M. Kusakabe, J. Aitchison, S. Cisowski, K. Dadey, K. Fujioka, A. Klaus, M. Lovell, K. Marsaglia, P. Pezard, B. Taylor, and K. Tazaki, Explosive deep water basalt in the Sumisu backarc rift. *Science*, 248, 1214-1217, 1990.

Greenland L. P., A. T. Okamura, and J. B. Stokes, Constraints on the mechanics of the eruption. *US Geol. Surv. Prof. Paper, 1463*, 155-164, 1988.

Herzig, P. M. and shipboard scientific party, Volcanism, Hydrothermal Processes and Biological Communities at Shallow Submarine Volcanoes of the New Ireland forearc (PNG), *Technical Cruise Report of R/V Sonne Cruise SO-133*, Freiberg University, 1998.

Jaupart, C., and S. Vrgniolle, Laboratory models of Hawaiian and Strombolian eruptions. *Nature, 331*, 58-60, 1988.

Jaupart, C., and S. Vrgniolle, The generation and collapse of a foam layer at the roof of a basaltic magma chamber, *Jour. Fluid Mech., 203*, 347-380, 1989.

Komatsu, M., and Y. Takegawa, Petrologocal study of submarine basaltic rocks (Miocene age) in the Ogi Peninsula, Sado Island, Central Japan. 1. Morphology of subaqueous volcanic bomb, *Sci. Rep. Niigata Univ., Ser. E, 6*, 17-35, 1985.

Mader, H.M., Y. Zhang, J.C. Phillips, R. S. J. Sparks, B. Sturtvant, and E. Stolper, Experimental simulations of explosive degassing of magma, *Nature, 372*, 85-88, 1994.

Mader, H.M., J.C. Phillips, R. S. J. Sparks, and B. Sturtvant, Dynamics of explosive degassing of magma: Observations of fragmenting two-phase flows, *Jour. Geophys. Res. 101*, 5547-5560, 1996.

Mangan, M.T., and K.V. Cashman, The structure of basaltic scoria and reticulite and inferences of vesiculation, foam formation, and fragmentation in lava fountains, *Jour. Volcanol. Geotherm. Res., 73*, 1-18, 1996.

Mangan, M. T., K. V. Cashman, and S. Newman, Vesiculation of basaltic magma during eruption, *Geology, 21*, 157-160, 1993.

Matsunaga, T., Benthonic smaller foraminifera from oil fields of northern Japan, *Sci. Rep. Tohoku Univ., 2nd ser., 35*, 67-122, 1963.

Mcphie, J., M. Doyle, and R. Allen, *Volcanic textures, a guide to the interpretation of textures in volcanic rocks*, CODES Key Center, Univ. Tasmania, Ttasmania, 198p, 1993,.

Mueller, W., and J. D. L. White, Felsic fire-fountaining beneath Archean seas: pyroclastic deposits of the 2730 Ma Hunter Mine Group, Quevec, Canada, *Jour. Volcanol. Geotherm. Res., 60*, 273-300, 1992.

Proussevitch, A. A., D. L. Sahagian and V. A. Kutolin, Stability of foams in silicate melts, *Jour. Geptherm. Res., 59*, 161-178, 1993.

Research Group of Ogi Basalt, Occurrence of basaltic rocks of the Ogi peninsula, Sado Island, Niigata prefecture, central Japan, *Pub. Sado Museum, 7*, 3-19, 1977. (in Japanese)

Rutherford M. J., and J.E. Gardner, Rates of magma ascent, in *Encyclopedia of volcanoes*, edited by H. Sigurdsson, et al., Academic Press, New York, 207-217, 2000.

Saltykov, S. A., The determination of the size distribution of particles in an opaque material from measurement of the size distribution in their section, in *Stereology, Proc. Second Int. Congr. Stereology*, Edited by H. Helias, Springer, New York, p163, 1967.

Sarda, P., and D. Graham, Mid-oceanic ridge popping rocks: implications for degassing at ridge crests. *Earth Planet. Sci. Lett., 97*, 258-289, 1990.

Shinmura, T., Y. Kobayashi, Y. Arakawa, and T. Itaya, K-Ar ages of Neogene basaltic rocks in the Ogi Peninsula, Sado Island, *Jour Mineral. Petrol. Econ. Geol. Japan, 90*?403-409, 1995. (in Japanese with English abstract)

Siebe, C., J.-C. Komorowski, C. Navarro, J. McHone, H. Delgado, and A. Cortes, Submarine eruption near Socorro Island, Mexico: Geochemistry and scanning electron microscopy studies of floating scoria and reticulite, *Jour. Volcanol. Geotherm. Res., 68*, 239-271, 1995.

Sparks, R. S. J., The dynamics of bubble formation and growth in magmas. *Jour. Volcanol. Geotherm. Res., 3*, 1-37, 1978.

Sparks, R. S. J., J. Barclay, C. Jaupart, H. M. Mader, and J. C. Phillips, Physical aspects of magma degassing I. Experimental and theoretical constraints on vesiculation. *Reviews in mineralogy, 30*, 413-405, 1994.

Staudigel H., and H.-U. Schmincke, The Pliocene seamount series of La Palma/Canary Islands, *Jour. Geophys. Res., 89 (B13)*, 11195-11215, 1984.

Toramaru, A., Vesiculation process and bubble size distributions in ascending magmas with constant velocities, *Jour. Geophys. Res., 94(B12)*, 17523-17542. 1989.

Tsunakawa, H., A. Takeuchi, and K. Amano, K-Ar ages of dikes in Northeast Japan, *Geochem. Jour., 17*, 269-273, 1983.

Vergniolle, S., and C. Jaupart, Separated two-phase flow and basaltic eruptions, *Jour. Geophys. Res., 91*, 12842-12860, 1986.

Vergniolle, S., and C. Jaupart, The dynamics of degassing at Kilauea volcano, Hawaii, *Jour. Geophys. Res., 95*, 2793-2809, 1990.

Wilson, L., and J. W. Head, Ascent and eruption of basaltic magma on the earth and moon, *Jour. Geophys. Res., 86*, 2971-3001, 1981.

Yamagishi, H., *Subaueous volcanic rocks*, Hokkaido Univ. Press, Sapporo, 195p, 1994.

Yamakawa, M., and K. Chihara, Petrological study of the Ogi basalts, Sado Island, *Contr. From Dept. Geol. Miner., Niigata Univ., 2*, 41-80, 1968. (in Japanese with English abstract).

N. Fujibayashi, Department of Geology, Faculty of Education and Human Sciences, Niigata University, Ikarashi ni-no-cho 8050, Niigata, 950-2181, Japan

U. Sakai, Nittoc Construction Co. Ltd., Hokuriku Branch, Honchou, 6-20, Toyama, 930-0916, Japan

The Submarine Record of a Large-Scale Explosive Eruption in the Vanuatu Arc: ~1 Ma Efaté Pumice Formation

Alison M. Raos

Centre for Ore Deposit Research and the School of Earth Sciences, University of Tasmania, Hobart, Australia
Present address: Esso Australia Pty Ltd, Melbourne, Australia

Jocelyn McPhie

Centre for Ore Deposit Research and the School of Earth Sciences, University of Tasmania, Hobart, Australia

The Efaté Pumice Formation (EPF) is the record of a major explosive eruption that occurred in the Vanuatu arc, southwestern Pacific, at about 1 Ma. The EPF is the oldest stratigraphic unit of the Efaté Island Group and consists of a succession of non-welded, trachydacitic pumice breccia and shard-rich sand and silt beds with a minimum thickness of ~500 m and a minimum bulk volume of approximately 85 km^3. The lower part (Efaté Pumice Breccias) of the EPF comprises very thick beds composed almost exclusively of glassy, trachydacitic, pumice fragments with ragged terminations. In contrast, the upper part (Rentabau Tuffs) consists of up to 70 m of well-bedded and well-sorted shard-rich sand and silt. The clast population of this upper part comprises >95 % glassy or formerly glassy shards, but fossil foraminifera are a ubiquitous and important non-volcanic component. Some glass shards have blocky, equant shapes and arcuate fracture surfaces, features typically associated with the influence of external water during fragmentation, but most are cuspate and platy bubble-wall shards. Pyroclast morphologies indicate that the Efaté Pumice Breccias were largely generated by magmatic-volatile-driven ("dry"), explosive fragmentation processes, and lithofacies characteristics indicate deposition in below-storm-wave-base environments, from eruption-sourced, water-supported density currents of waterlogged pumice. The Rentabau Tuffs are interpreted to represent a change to hydromagmatic activity in response to waning discharge that allowed ingress of water (presumably seawater) to the vent(s).

INTRODUCTION

High-silica magmas in island arcs may produce highly explosive eruptions in both subaerial and submarine environments. In island arcs in particular, the best-preserved record of such events is likely to be submarine. Submarine deposits from explosive eruptions have been recognised in both modern settings and throughout the rock record [*e.g.*

274 LARGE-SCALE EXPLOSIVE ERUPTION IN THE VANUATU ARC

Fiske, 1969; *Sparks et al.*, 1980; *Busby-Spera*, 1986; *Cashman and Fiske*, 1990]. Because modern examples are difficult to study, understanding of facies and transport and depositional mechanisms depends largely on ancient examples now well exposed on land [*e.g. Soh et al.*, 1989; *Mángano and Buatois*, 1997; *Stow et al.*, 1998; *Allen and McPhie*, 2000].

The non-welded, pumice and shard-rich deposits of the Efaté Pumice Formation in the immature Vanuatu island arc are >500 m-thick and currently cover 160 km². These deposits were generated by a large-scale eruption of evolved trachydacite magma and rapidly emplaced in a submarine environment. Textures and depositional structures in the Efaté Pumice Formation are very well preserved in scattered cliff outcrops on Efaté and nearby islands. In this paper we describe the Efaté Pumice Formation and use the lithofacies and textural characteristics to constrain the eruption and depositional processes and setting.

GEOLOGICAL SETTING

The Vanuatu arc forms part of a chain of Tertiary to Recent island arcs that mark the boundary between the Pacific plate and the Indo-Australian plate in the southwestern Pacific (inset, Figure 1). Subaerial pyroclastic deposits associated with large-scale, explosive eruptions occur at several caldera centers in the Vanuatu arc [*e.g. Robin et al.*, 1993; *Robin et al.*, 1994a; *Robin et al.*, 1994b; *Robin et al.*, 1995], and other possible calderas probably occur offshore, for example, in the region immediately north of Efaté [*Crawford et al.*, 1988].

The thick pyroclastic sequence on Efaté is distinctive in two key ways: (1) the Efaté deposits are exclusively trachydacitic; basaltic and basaltic andesite products dominate at all other centers, and (2) the Efaté pyroclastic sequences were deposited exclusively in submarine environments whereas other exposed modern and recent Vanuatu arc pyro-

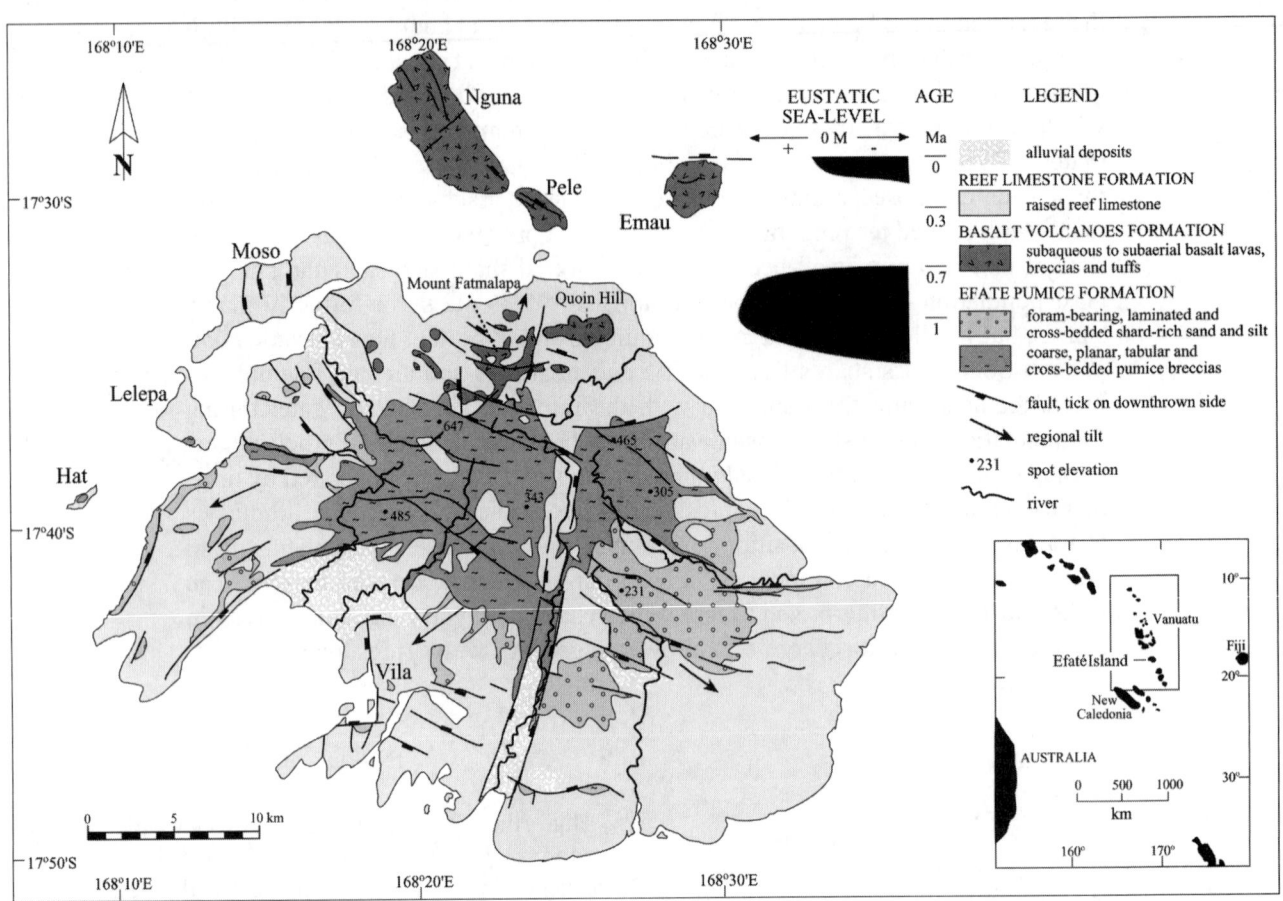

Figure 1. Geology and structure of the Efaté Island Group, Vanuatu arc, southwestern Pacific. Modified from *Ash et al.* [1978], and incorporating eustatic sea-level curves from *Haq et al.* [1988].

clastic sequences that resulted from large-scale eruptions, were all emplaced subaerially. Interestingly, episodes of hydromagmatic activity have been common in many subaerial explosive eruptions in the active Vanuatu arc [e.g. *Robin et al.*, 1993; *Robin et al.*, 1994b].

THE GEOLOGY OF EFATE

Efaté consists of two major volcanic formations overlain by reef-forming limestone [Figure 1, *Mawson*, 1905; *Obellianne*, 1958; *Ash et al.*, 1978]. The oldest formation on Efaté is the ~1 Ma Efaté Pumice Formation (EPF) exclusively comprising trachydacitic pumice breccia and shard-rich sand and silt facies. The Pleistocene to Recent (~0.7-0 Ma) Basalt Volcanoes Formation (BVF) unconformably overlies the EPF and is restricted to the north of Efaté, and islands offshore (Figure 2). Lithofacies include submarine to subaerial basaltic lavas, and associated fine- to coarse-grained volcaniclastic facies. Deposition of in situ biogenic and detrital facies of the Reef Limestone Formation (RLF) began in the late Pleistocene (~0.3 Ma) and this unit unconformably onlaps both the EPF and the BVF. Reef growth is recorded in a series of terraces that developed during broadly domal uplift of Efaté, Hat, Lelepa and Moso islands (Figure 2).

EFATE PUMICE FORMATION

The EPF is exposed in central Efaté, and on the nearby islands of Lelepa and Hat (Figure 1). Beds are essentially flat-lying and the deposits are only poorly consolidated. The base of the formation is not currently exposed and the upper contact with onlapping RLF is erosional everywhere. The EPF is informally divided into lower (Efaté Pumice Breccias) and upper (Rentabau Tuffs) members that differ in their principal facies characteristics (Table 1). The contact between the upper and lower units is apparently conformable, non-erosive and knife-sharp (Figure 2).

Efaté Pumice Breccias

The Efaté Pumice Breccias have a minimum stratigraphic thickness of ~350 m and a minimum bulk volume of 65–80 km³. They are primarily composed of angular to ragged-ended, elongate, tube-pumice clasts (60–80%) and less abundant round-vesicle pumice clasts (5–20%), with subordinate (5–10%) perlitic obsidian and porphyritic lava clasts in a minor matrix (5–15%) of non-abraded, Y-shaped, cuspate and platy bubble-wall shards, crystals and crystal fragments (Figure 3a). Pumice clasts have vesicularities,

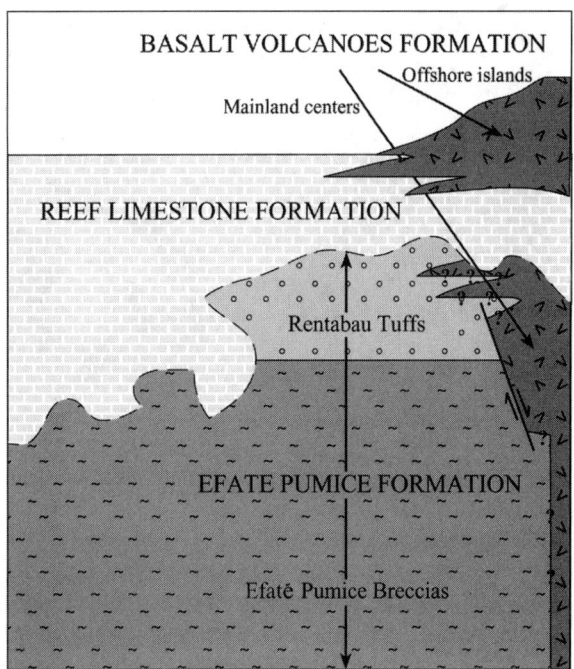

Figure 2. Schematic diagram showing the relationships between the three formations on Efaté Island. Dashed lines represent an unconformable contact, thicker lines are faulted contacts, and ? indicate that these contact relationships are obscured or not exposed.

determined by analysis of oriented thin sections, in the range 65–85%. The glassy components are uniformly trachydacitic and the crystal population comprises plagioclase (andesine), clino- and orthopyroxene, Ti-magnetite and accessory apatite. Rare clasts occurring in lithic-rich intervals include abraded limestone fragments, well-rounded scoria clasts and elongated intraclasts of shard-rich sand and silt.

The Efaté Pumice Breccias comprise a coarse and fine facies association (Table 1). The coarse facies are dominated by planar-tabular, structureless to normally graded and internally stratified, thick to very thick beds (up to 5 m) of clast-supported, moderately sorted, 1–6 cm pumice fragments (MPB, SPB, XPB; Table 1). Pumice breccia beds have sharp bases and gradational to erosional upper contacts (Figure 3b). These beds commonly have an inversely graded crystal-rich basal layer up to 5 cm thick, and are locally interbedded with well-sorted, shard-rich facies. The fine facies include thin to very thin, tabular, shard-rich sand and silt interbeds with textures varying from structureless to normally graded, to planar- and locally cross-stratified (MAS, SAS, XAS, Table 1, Figure 3c). Shard-rich sand and silt beds have sharp bases and locally display flame struc-

Figure 3. Components and facies of the Efaté Pumice Formation (EPF). (a)-(c) Efaté Pumice Breccias: (a) photomicrograph of clast-supported fabric in pumiceous sand; note the abundance of elongate tube pumice clasts, sample AR064, Lelepa Island. (b) Very thick beds of clast-supported massive pumice breccia with interbeds of shard-rich sand and silt, person for scale, Lelepa Island. (c) Fine-grained interval consisting of fine laminated sand, interbedded with angle-of-repose cross-bedding and massive shard-rich silt lenses and drapes (pencil is 14.5 cm long, 8 mm wide), Lelepa Island. (d)-(f) Rentabau Tuffs: (d) scanning electron microscope image of platy shards from a very fine sand, sample AR016, south-central Efaté. (e) Massive (below pencil) and laminated (above pencil) shard-rich sand beds (pencil is 14.5 cm long, 8 mm wide), Forari, SE Efaté. (f) Plastically-deformed (convolute) bedding in laminated shard-rich sand, note thin (3 cm) indurated bed of massive very fine grained shard-rich sand, central Efaté.

Table 1. Summarised facies descriptions for the principal facies of the Efaté Pumice Formation.

LITHOFACIES	BED THICKNESS	GENERAL DESCRIPTION	DEPOSITIONAL PROCESSES
Efaté Pumice Breccias			
massive pumice breccia (MPB)	0.5–5.0 m (thick to very thick)	clast-supported, moderately sorted, fines-poor, pumice breccia; structureless to normally or inverse graded close to bed contacts; planar, tabular, laterally continuous beds with sharp and erosional, to gradational bases; may have lithic-rich basal layer locally	pumice-rich, water-supported, density currents
stratified pumice breccia (SPB)	0.2–5.0 m (medium to very thick)	clast-supported, moderately to well sorted, fines-poor pumice breccia; planar stratified and locally normally and inversely graded, planar, tabular, laterally continuous beds with sharp bases	traction sedimentation from pumice-rich, water-supported density currents
cross-stratified pumice breccia (XPB)	0.5–5.0 m (thick to very thick)	clast-supported, moderately sorted, fines-poor, pumice breccia; internally stratified and cross-stratified discontinuous beds with sharp generally erosive bases; oversteepened foreset structures and isolated scour and fill structures occur locally	traction sedimentation from pumice-rich, water-supported density currents
massive shard-rich sand and silt (MAS)	thin to very thin	shard-rich, well sorted, structureless silt and fine sand; some planar, tabular beds with sharp bases, commonly forms draping beds and may show injection structures and soft-sediment deformation features at upper contacts; indurated and locally carbonate veined	suspension fallout deposition including fallout of the finest particles from shard-rich, water-supported density currents
stratified shard-rich sand and gravelly sand (SAS)	thin to very thin	shard-rich, well sorted gravelly sand to silt; planar, tabular beds with sharp bases; planar stratified and laminated; commonly contain single grain thickness layers; crystal-rich layers common	traction plus suspension sedimentation in shard-rich, water-supported density currents
cross-stratified shard-rich sand and gravelly sand (XAS)	thin to very thin	shard-rich, well sorted gravelly sand to silt; planar, tabular beds with sharp bases; isolated or locally grouped asymmetric ripple bedforms; commonly contain granular pumice stringers or single-grain thickness layers; crystal rich layers common	traction sedimentation in shard-rich, water-supported volcaniclastic density currents
Rentabau Tuffs			
massive shard-rich sand and silt (MAS)	very thin to thick	shard-rich, well sorted sand to silt; planar, tabular, laterally continuous, structureless beds with sharp, conformable contacts; sole marks and load casts occur locally; may contain fossil foraminifera and rare outsize pumice clasts	shard-rich, water-supported density currents
laminated shard-rich sand and silt (SAS)	0.1–1.2 m (medium to very thick)	shard-rich, well sorted sand to silt; planar, tabular internally thickly laminated beds with sharp conformable contacts; may contain rare low-angle truncation surfaces; may contain fossil foraminifera and rare outsize pumice clasts	traction plus suspension sedimentation in shard-rich, water-supported density currents and bottom currents
cross-stratified shard-rich sand and silt (XAS)	0.1–0.5 m (medium to thick)	shard-rich, well sorted sand to silt; planar, tabular laterally continuous or rarely lenticular beds with sharp conformable contacts; bed bases may be undulose; ripple bedforms are typically asymmetric and isolated, but climbing ripples and trough cross-laminae occur locally; may contain fossil foraminifera and rare outsize pumice clasts	traction sedimentation in shard-rich, water-supported density currents and bottom currents
convolute bedded shard-rich sand and silt (CAS)	0.3–1 m (medium to thick intervals)	shard-rich, well sorted sand to silt; contorted bedding with irregular folds; intervals have discordant sharp bases and tops; may contain fossil foraminifera and rare outsize pumice clasts	soft-sediment slumping

tures and load casts, and upper contacts are gradational or erosional. Tractional sedimentary structures, including planar and cross-stratification, internal scouring and low-angle bedding truncations occur in both the fine and the coarse facies (Table 1, Figure 4a).

Rentabau Tuffs

The Rentabau Tuffs occur in sections up to 70 m thick, and have a minimum bulk volume of 3–5 km³. The Rentabau Tuffs are compositionally very similar to the Efaté Pumice Breccias, comprising >90% tube pumice, bubble-wall, cuspate and blocky, glassy (or formerly glassy) shards. Glass shards (where preserved) are chemically identical to glassy components in the underlying Efaté Pumice Breccias and the majority are unabraded. Pumice shards have both ragged and fracture-bounded terminations and blocky shards have curviplanar margins (Figure 3d). Euhedral crystals and angular crystal fragments of plagioclase (andesine), clino- and orthopyroxene, and Ti-magnetite account for up to 10%

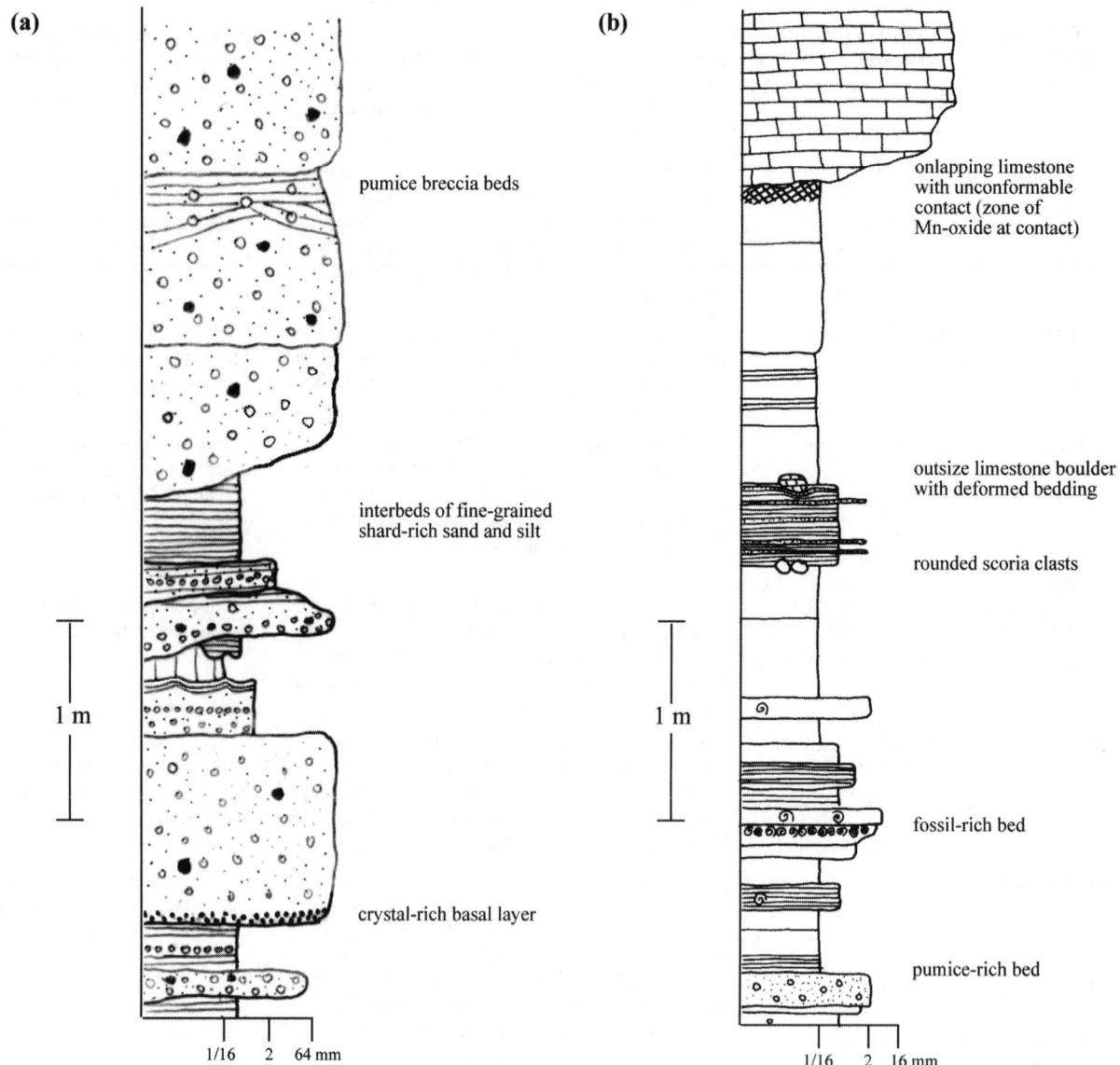

Figure 4. Representative graphic logs from the EPF. (a) Thick to very thick breccia beds and interbedded shard-rich sand and silt from the Efaté Pumice Breccias, Lelepa Island. (b) Shard-rich, fossil bearing sand and silt beds from the Rentabau Tuffs, at Forari, southeastern Efaté.

of components and are locally concentrated within crystal-rich laminae. Non-volcanic clasts include a ubiquitous but minor (3–5 vol.%) population of marine fossil foraminifera.

Bedding in the Rentabau Tuffs is remarkably uniform (Figure 4b). Thin to medium beds with planar, tabular geometries and sharp, conformable contacts are typical (Figure 3e). Beds are dominantly massive or internally laminated and locally ripple-laminated, and well-sorted (MAS, SAS, XAS, Table 1), although rare, randomly scattered, outsize pumice clasts (up to 3 cm) may be present. Intervals of convolute bedding are locally important (CAS, Table 1, Figure 3f). Many sections of the Rentabau Tuffs are substantially altered to halloysite clays.

DISCUSSION

Fragmentation Processes and Source of Pyroclasts

The EPF consists almost exclusively of highly vesicular pumice and cuspate, blocky and bubble-wall glass shards and crystals. Such clasts are typically produced by major explosive eruptions of vesiculating magma. The majority of glassy pumice clasts in the Efaté Pumice Breccias have ruptured bubble walls, and are relatively coarse (1–6 cm), indicating that fragmentation by "dry" magmatic-volatile- driven explosions dominated [*Sparks*, 1978; *Heiken and Wohletz*, 1991; *Cashman and Mangan*, 1994]. Curviplanar surfaces on perlitic obsidian clasts probably represent original macroperlite fracture traces.

In contrast, many of the pumiceous shards in the Rentabau Tuffs have subplanar fracture-bounded surfaces, and some shards have equant, blocky shapes with arcuate clast margins. These features are commonly generated by interaction of magma with external water [*Sheridan and Wohletz*, 1981; *Wohletz*, 1983]. In addition, the consistently fine grain size of the Rentabau Tuffs is consistent with hydromagmatic fragmentation mechanisms being important in the generation of these pyroclasts [*Self and Sparks*, 1978; *Wohletz et al.*, 1989].

Transport and Depositional Mechanisms

Deposition of the Efaté Pumice Breccias at ~1 Ma coincides with a period of high eustatic sea level and a period of rapid volcaniclastic and hemipelagic sedimentation in the central Vanuatu arc [Figure 1, *Haq et al.*, 1998; *Goud Collins*, 1994]. A marine setting for deposition of the Efaté Pumice Breccias is consistent with these conditions and is supported in part by the presence of marine fossils in the conformably overlying Rentabau Tuffs. In addition, reef limestone unconformably overlies both the Efaté Pumice Breccias and the Rentabau Tuffs, and there is no evidence for hot emplacement or gas-supported transport in any facies of the Efaté Pumice Breccias. Instead, depositional structures, including very thick structureless and stratified, planar, tabular pumice beds, suggest the dominance of water-supported volcaniclastic density currents during transport and deposition. The facies characteristics, together with the absence of wave-generated bedforms are consistent with a below-wave-base environment [*c.f. Einsele*, 1991]. Stratification, cross-stratification and grading developed as particle-rich currents lost capacity, allowing traction sedimentation to dominate over suspension fallout sedimentation [*c.f. Hiscott*, 1994]. The generally fines-poor nature of the pumice breccias probably reflects efficient elutriation of fine pyroclasts during transport [*c.f. Cousineau*, 1994; *Druitt*, 1995]. Elutriated shards and crystals would have created suspensions that eventually generated fine-grained density currents from which the shard-rich interbeds were deposited.

Foraminifera tests are widely distributed but sparse (<2 vol.%) throughout the Rentabau Tuffs and include planktic species typical of deep sea or open water settings [*Ash et al.*, 1978; *Raos*, 2001]. Hence foraminifera in the Rentabau Tuffs probably settled from suspension, together with shards and crystals. The presence of structureless and laminated beds, and ripple-laminated intervals implies that deposition was dominated by density current and suspension fallout processes [*c.f. Lowe*, 1982]. The monotonous and extremely regular bedding of the Rentabau Tuffs, and the dominance of fine grain sizes, plus the lack of erosional contacts with the underlying Efaté Pumice Breccias, is consistent with a relatively deep, non-channelised setting. The convolute bedded intervals probably resulted from slumping of unconsolidated sediment on the submarine slope. Earthquakes accompanying active volcanism, and high sedimentation rates were probably important mechanisms for triggering sediment slumping in the Rentabau Tuffs [*c.f. Niem*, 1977; *Stow*, 1994].

Vent Setting

The Efaté Pumice Formation was clearly deposited in a submarine environment but the vent or vents for the eruption are not preserved or not exposed and their precise location remains unknown. However, sparse palaeocurrent indicators and systematic variations in grain-size point to a source offshore to the north of Efaté [*Raos*, 2001]. The vent setting is also difficult to constrain. Although clast morphologies in the Efaté Pumice Breccias suggest dominantly "dry" explosive fragmentation, this does not preclude eruption from submarine vents: energetic discharge may prevent water gaining access to vents, and further, submarine eruption columns may remain protected from interaction with seawater by a steam carapace [*e.g. Kano et al.*, 1996]. For the Rentabau Tuffs, the overwhelming dominance of glass shards <2 mm, and the presence of shards bounded by arcuate fracture surfaces, are typical of pyroclasts resulting from hydromagmatic activity. At Efaté the external water may have been the sea, implying that the vent was submerged. In addition, no regional ash layer correlated with the Efaté Pumice Formation has been recognised within the Vanuatu arc. A subaerial eruption of this style and scale would be expected to produce a widespread ash bed [*e.g. Schmincke and van den Bogaard*, 1991], implying that in this case, vents were probably submarine; although the eruption column may have breached the sea surface without forming a high plume (Figure 5a).

Charred plant remains, terrestrial fossils or accretionary lapilli, which would positively indicate that the eruption occurred from a subaerial vent, are notably absent in beds of the EPF. Such particles are commonly preserved in other submarine successions from subaerial eruptions, even up to 250 km from source [*e.g. Carey and Sigurdsson*, 1980; *Soh et al.*, 1989; *Stow et al.*, 1998]. Hence the available evidence favours a submarine eruption for the EPF.

MODEL FOR ERUPTION AND EMPLACEMENT OF THE EFATE PUMICE FORMATION

Stage I – Efaté Pumice Breccias

Vesiculating trachydacite magma was explosively ejected from shallow submarine vents located to the north of Efaté (Figure 5a). Turbulent mixing with seawater and overload-

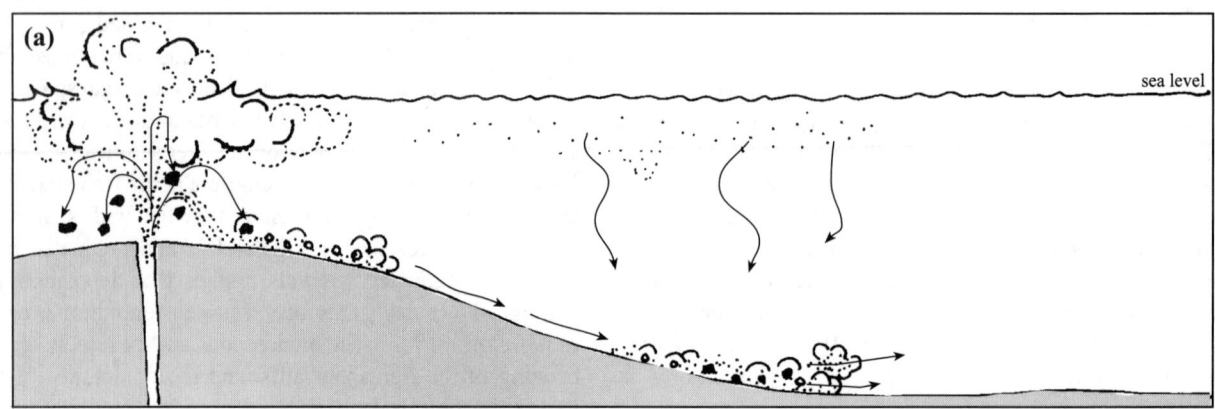

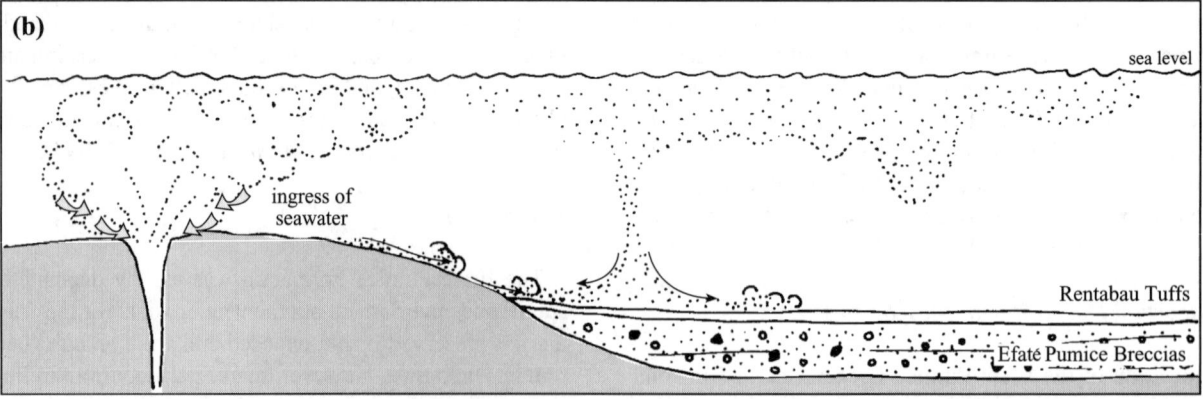

Figure 5. Model for eruption and emplacement of the EPF. (a) Explosive eruption of vesiculating magma from a submerged vent(s) generated abundant highly vesicular pumice clasts and finer glass shards. Turbulent mixing of pyroclasts with seawater generated pumice-rich density currents that deposited thick pumice breccia beds in nearby depocentres. The finest particles were deposited from separate fine-grained, currents produced as by-products of the coarse pumiceous currents, and from suspension settling. As the eruption intensity waned, (b) seawater gained access to the vesiculating magma resulting in hydromagmatic explosions. Abundant fine pyroclasts were incorporated in eruption-sourced shard-rich density currents and descending vertical plumes that rapidly deposited remarkably uniform shard-rich sand and silt beds. See text for details.

ing by pyroclasts led to continuous collapse of the submarine eruption column. This collapsing column and continued explosions initiated a succession of volcaniclastic density currents of waterlogged pumiceous debris (Figure 5a, *Fiske and Matsuda*, 1964; *Cashman and Fiske*, 1991]. Hot pumice clasts, in particular the highly permeable tube pumice, rapidly ingested seawater and lost their initial buoyancy to become incorporated in these eruption-sourced, cold (or cool), water-supported, volcaniclastic density currents [*e.g. Fiske*, 1969; *Whitham and Sparks*, 1986]. Older lavas and domes were shattered by the explosions generating perlitic obsidian and dense porphyritic lithic blocks that became incorporated into the density currents. The finest particles were segregated from the pumice-rich density currents during current generation, due to their lower densities compared with the waterlogged pumice clasts, and during transport by elutriation [*e.g. Cousineau*, 1994; *Fisher*, 1983]. Separate shard-rich currents, generated as by-products of the coarser density currents and from suspension in the water column, locally deposited interbeds of laminated, cross-laminated, and structureless, shard-rich sand and silt. The dominance of mechanically unmodified pyroclasts in the deposits indicates that abrasion following fragmentation was minimal. The Efaté Pumice Breccias may have rapidly accumulated in local depocentres, leading to a very thick and regionally confined succession [*e.g. Cousineau*, 1994].

Stage II – Rentabau Tuffs

As the eruption progressed and magma discharge rates were reduced, interaction of the vesiculating magma with seawater led to a change in eruption dynamics. Hydromagmatic explosions occurred (Figure 5b), and generated mainly fine pyroclasts while the magma was still near the point of peak vesication [*Houghton and Wilson*, 1989; *Houghton et al.*, this volume]. Deposition of the pyroclastic debris occurred mainly from cold (or cool) water-supported, density currents generated directly at submarine vents after turbulent mixing of gas-pyroclast dispersions with seawater. Density currents also originated from descending vertical plumes developing in the upper parts of the particle-laden water column [*e.g. Carey*, 1997; *Fiske et al.*, 1998; *Stow et al.*, 1998]. Probably, seismic activity related to the volcanic eruption, caused local slumping and the development of convolute-bedded intervals within the Rentabau Tuffs. Such seismic events may have triggered further volcaniclastic density currents by remobilisation of this unconsolidated sediment [*Niem*, 1977; *Stow*, 1994]. This unit comprises ~98 % pyroclasts and lacks any significant interbeds of hemipelagic sediment, indicating that deposition occurred contemporaneously with explosions, or rapidly following the eruption.

CONCLUSIONS

The pumice and shard-rich deposits of the EPF were generated by cold (or cool), water-supported, volcaniclastic density currents that were *directly* fed from contemporaneous, (shallow) submarine, explosive eruption(s). Magmatic-volatile-driven explosive fragmentation processes and energetic discharge dominated in the early phases of the eruption, producing the Efaté Pumice Breccias. As the eruption waned, magma interaction with seawater caused a change to hydromagmatic fragmentation mechanisms. The fine-grained pyroclastic debris generated in these explosions was also deposited principally by eruption-sourced, water-supported, volcaniclastic density currents, forming the Rentabau Tuffs. Rapid emplacement and volcanic seismicity following deposition initiated slumping in the poorly consolidated pumiceous sediments.

Acknowledgments. This research was funded in part by an Australian Research Council Large Grant awarded to Professor Tony Crawford for studies in the Vanuatu arc and by the Australian Research Council Special Research Centres program. AMR acknowledges a School of Science and Technology scholarship from the University of Tasmania that enabled her PhD research on Efaté Island.

REFERENCES

Allen, S.R., and J. McPhie, Water-settling and resedimentation of submarine rhyolitic pumice at Yali, eastern Aegean, Greece, *Journal of Volcanology and Geothermal Research*, 95, 285-307, 2000.

Ash, R.P., J.N. Carney, and A. Macfarlane, Geology of Efaté and Offshore Islands, pp. 49, *New Hebrides Condominium Geological Survey*, 1978.

Busby-Spera, C.J., Depositional features of rhyolitic and andesitic volcaniclastic rocks of the Mineral King submarine caldera complex, Sierra Nevada, California, *Journal of Volcanology and Geothermal Research*, 27, 43-76, 1986.

Carey, S., Influence of convective sedimentation on the formation of widespread tephra fall layers in the deep sea, *Geology*, 25 (9), 839-842, 1997.

Carey, S.N., and H. Sigurdsson, The Roseau Ash: deep-sea tephra deposits from a major eruption on Domenica, Lesser Antilles Arc, *Journal of Volcanology and Geothermal Research*, 7, 67-86, 1980.

Cashman, K.V., and R.S. Fiske, Fallout of pyroclastic debris from submarine volcanic eruptions, *Science*, 253, 275-280, 1991.

Cashman, K.V., M.T. Mangan, and S. Newman, Surface degassing and modifications to vesicle size distributions in active basalt flows, *Journal of Volcanology and Geothermal Research, 61,* 45-68, 1994.

Cousineau, P.A., Subaqueous pyroclastic deposits in an Ordovician fore-arc basin: an example from the Saint-Victor Formation, Quebec Appalachians, Canada, *Journal of Sedimentary Research, A64,* 867-880, 1994.

Crawford, A.J., H.G. Greene, and N.F. Exon, Geology, petrology and geochemistry of submarine volcanoes around Epi Island, New Hebrides Island Arc, in *Geology and Offshore Resources of Pacific Island Arcs - Vanuatu Region,* edited by H.G. Greene, and F.L. Wong, pp. 301-327, Circum-Pacific Council for Energy and Mineral Resources, Huston, Texas, 1988.

Einsele, G., Submarine mass flow deposits and turbidites, in *Cycles and Events in Stratigraphy,* edited by G. Einsele, W. Ricken, and A. Seilacher, pp. 313-339, Springer-Verlag, Berlin, Heidelberg, 1991.

Fisher, R.V., Flow transformations in sediment gravity flows, *Geology, 11,* 273-274, 1983.

Fiske, R.S., Recognition of pumice in marine pyroclastic rocks, *Geological Society of America Bulletin, 80,* 1-8, 1969.

Fiske, R.S., K.V. Cashman, A. Shibata, and K. Watanabe, Tephra dispersal from Myojinsho, Japan, during its shallow submarine eruption of 1952-53, *Bulletin of Volcanology, 59,* 263-275, 1998.

Fiske, R.S., and T. Matsuda, Submarine equivalents of ash flows in the Tokiwa Formation, Japan, *American Journal of Science, 262,* 76-106, 1964.

Goud Collins, M.R., Volcaniclastic sediments of the North Aoba Basin: depositional processes and geologic history, in *Proceedings of the Ocean Drilling Program, Scientific Results, Leg 134,* edited by H.G. Green, J.-Y. Collot, L.B. Stokking, et al. pp. 97-107, ODP, College Station, TX, 1994.

Haq, B.U., J. Hardenbol and P.R. Vail, Mesozoic and Cenozoic chronostratigraphy and eustatic cycles, in *Sea-level Changes: an integrated approach,* edited by C.K. Wilgus, B.S. Hastings, H. Postmentier, et al, pp. 71-108, SEPM Special Publication No. 42, 1988.

Heiken, G., and K. Wohletz, Fragmentation processes in explosive volcanic eruptions, in *Sedimentation in Volcanic Settings,* edited by R.V. Fisher, and G.A. Smith, pp. 19-26, SEPM Special Publication No. 45, 1991.

Hiscott, R.N., Loss of capacity, not competence, as the fundamental process governing deposition from turbidity currents, *Journal of Sedimentary Research, A64* (2), 209-214, 1994.

Houghton, B.F., and C.J.N. Wilson, A vesicularity index for pyroclastic deposits, *Bulletin of Volcanology, 51,* 451-462, 1989.

Kano, K., T. Yamamoto, and K. Ono, Subaqueous eruption and emplacement of the Shinjima Pumice, Shinjima (Moeshima) Island, Kagoshima Bay, SW Japan, *Journal of Volcanology and Geothermal Research, 71,* 187-206, 1996.

Lowe, D.R., Sediment gravity flows: II. Depositional models with special reference to the deposits of high-density turbidity currents, *Journal of Sedimentary Petrology, 52* (1), 279-297, 1982.

Mángano, M.G., and L.A. Buatois, Slope-apron deposition in an Ordovician arc-related setting: the Vuelta de Las Tolas Member (Suri Formation), Famatina Basin, northwest Argentina, *Sedimentary Geology, 109,* 155-180, 1997.

Mawson, D., The geology of the New Hebrides, *Proceedings of the Linnean Society of NSW, 30,* 400-485, 1905.

Niem, A.R., Mississippian pyroclastic flow and ash-fall deposits in the deep-marine Ouachita flysch basin, Oklahoma and Arkansas, *Geological Society of America Bulletin, 88,* 49-61, 1977.

Obellianne, J.-M., *Contribution à la connaissance geologique de l'archipel des Nouvelles-Hébrides (îles Vaté, Pentecote, Maewo, Santo).,* 76 pp., Foundation Scientifique de la Geologie et de ses Applications, Nancy, 1958.

Raos, A.M., The volcanic and geochemical evolution of a trachydacite-dominated island arc centre: Efaté Island Group, Vanuatu arc, SW Pacific, Unpublished thesis, University of Tasmania, Hobart, 2001.

Robin, C., J.-P. Eissen, and M. Monzier, Ignimbrites of basaltic andesite and andesite compositions from Tanna, New Hebrides Arc, *Bulletin of Volcanology, 56,* 10-22, 1994a.

Robin, C., J.-P. Eissen, and M. Monzier, Mafic pyroclastic flows at Santa Maria (Gaua) Volcano, Vanuatu: the caldera formation problem in mainly mafic island arc volcanoes, *Terra Nova, 7,* 436-443, 1995.

Robin, C., M. Monzier, A.J. Crawford, and S.M. Eggins, *The geology, volcanology, petrology-geochemistry and tectonic evolution of the New Hebrides Island Arc, Vanuatu,* 86 pp., IAVCEI Pre-conference Excursion Guide A5, 1993.

Robin, C., M. Monzier, and J.-P. Eissen, Formation of the mid-fifteenth century Kuwae caldera (Vanuatu) by an initial hydroclastic and subsequent ignimbritic eruption, *Bulletin of Volcanology, 56,* 170-183, 1994b.

Schmincke, H.-U., and P. van den Bogaard, Tephra layers and tephra events, in *Cycles and Events in Stratigraphy,* edited by G. Einsele, W. Ricken, and A. Seilacher, pp. 393-429, Springer-Verlag, Berlin, 1991.

Self, S., and R.S.J. Sparks, Characteristics of widespread pyroclastic deposits formed by the interaction of silicic magma and water, *Bulletin of Volcanology, 41* (3), 196-212, 1978.

Sheridan, M.F., and K.H. Wohletz, Hydrovolcanic explosions: the systematics of water-pyroclast equilibration, *Science, 212,* 1387-1389, 1981.

Soh, W., A. Taira, Y. Ogawa, H. Taniguchi, K.T. Pickering, and D.A.V. Stow, Submarine depositional processes for volcaniclastic sediments in the Mio-Pliocene Misaki Formation, Miura Group, central Japan, in *Sedimentary Facies in the Active Plate Margin,* edited by A. Taira, and F. Masuda, pp. 619-630, Terra Scientific Publishing Company (TERRAPUB), Tokyo, 1989.

Sparks, R.S.J., The dynamics of bubble formation and growth in magmas: a review and analysis, *Journal of Volcanology and Geothermal Research, 3,* 1-37, 1978.

Sparks, R.S.J., H. Sigurdsson, and S.N. Carey, The entrance of pyroclastic flows into the sea, II. Theoretical considerations on subaqueous emplacemant and welding, *Journal of Volcanology and Geothermal Research, 7,* 97-105, 1980.

Stow, D.A.V., Deep sea processes of sediment transport and deposition, in *Sediment transport and depositional processes*, edited by K. Pye, pp. 257-291, Blackwell Scientific Publications, Oxford, 1994.

Whitham, A., and R.S.J. Sparks, Pumice, *Bulletin of Volcanology*, *48*, 209-223, 1986.

Wohletz, K.H., Mechanisms of hydrovolcanic pyroclast formation: grain-size, scanning electron microscopy, and experimental studies, *Journal of Volcanology and Geothermal Research*, *17*, 31-63, 1983.

Wohletz, K.H., M.F. Sheridan, and W.K. Brown, Particle size distributions and the sequential fragmentation/transport theory applied to volcanic ash, *Journal of Geophysical Research*, *94* (B11), 15,703-15,721, 1989.

J. McPhie. Centre for Ore Deposit Research and the School of Earth Sciences, University of Tasmania, Private Bag 79, Hobart TAS Australia 7001.

A.M. Raos. Esso Australia Pty Ltd, 12 Riverside Quay, Southbank VIC Australia 3006.

Products of Explosive Subaqueous Felsic Eruptions Based on Examples From the Hellenic Island Arc, Greece

S.R. Allen and A.L. Stewart

Centre for Ore Deposit Research, University of Tasmania, Hobart, Australia

We review the outflow products of three explosive felsic eruptions derived from submarine vents in the Hellenic Island Arc, Greece. The facies characteristics vary due to the differences in eruption volume (>0.3–150 km^3), water depth (tens to hundreds of metres), depositional setting (subaerial or submarine), plume dynamics and inferred eruption intensity. Phreatoplinian eruptions are only known from vents that have been sufficiently shallow (a few tens of metres) to optimise magma/water interaction and produce subaerial plumes. Even then, phreatoplinian activity is suppressed in favour of exclusively "dry" activity at sufficiently high eruption rates. In contrast, explosive felsic eruptions sourced from deeper vents (>~200 m) produce large quantities of pumice (reaching metres across) and ash, with subordinate lithic clasts in plumes that are at least dominantly submarine. Transport in water preserves the quenched margins on the pumice clasts. The eruption rate and incorporation of dense clasts influences the plume dynamics. Eruptions with significant vent erosion expel dense lithic clasts that generate submarine gravity currents. Pumice clasts that are sufficiently large to remain hot and buoyant rise through the water column until they become waterlogged and settle depositing a reversely size and density graded bed. Lower intensity eruptions with very little vent erosion, are dominated by coarse proximal pumiceous water-settled fallout that is resedimented down slope of the growing pumice cone.

INTRODUCTION

Modern island arcs, such as the Hellenic Island Arc in Greece, are characterised by numerous volcanic centres, many of which are located in shallow to moderate water depths or become inundated during eruptions. Many of these volcanoes produce explosive eruptions that are commonly of felsic composition. Explosive, submarine felsic eruptions have occurred in the seas around Japan [e.g., 1952–1953 Myojinsho dome eruption, *Fiske et al.*, 1998; and 1989 Teishi Knoll eruption, *Yamamoto et al.*, 1991]. In addition, diverse products of such eruptions have been discovered by recent sea floor exploration [e.g., Myojin Knoll caldera, *Yuasa*, 1995; *Fiske et al.*, 2001; Southern Kermadec arc, *Wright and Gamble*, 1999; and *Wright et al.*, in press]. Furthermore, deposits from subaqueous, water-supported, pumiceous gravity currents [e.g., *Fiske*, 1963; *Fiske and Matsuda*, 1964; *Kano*, 1996; and *Kano et al.*, 1996] and water-settled fall [e.g., *Cas et al.*, 1990; and *Cashman and Fiske*, 1991] have been linked to explosive felsic eruptions from submarine vents. However, the deposits of explosive rhyolitic eruptions from submarine vents can be difficult to interpret due to poor preservation of uplifted successions. In addition, although the depositional environment of the deposit can be determined, there are severe limitations in constraining the vent setting, which can only be assumed to be shallower than the deposit. Furthermore, very little is known about the behaviour of subaqueous eruption plumes derived from pyroclastic eruptions due to the significant lack of numerical and theoretical modelling of submarine eruptions.

Volcanological studies on the deposits of subaqueous eruptions suggest that very shallow subaqueous plumes (less than a few tens of metres), can have sufficient momentum for the gas thrust region to breach the water surface and become subaerial. Such eruptions have been proposed for the deposition of fallout and pyroclastic flows during the Minoan eruption, Santorini [*Bond and Sparks*, 1976; and *Heiken and McCoy*, 1984], the 1800 YR-BP Taupo eruption, New Zealand [*Wilson*, 1985], and the Oruanui eruption, New Zealand [*Wilson*, 2001]. These very shallow eruptions are also accompanied by magma/water interaction involving "fuel-coolant" explosions. In comparison, in deeper water (>~200 m), the eruption plume is more likely to be wholly or at least predominantly subaqueous. Such eruptions are proposed to generate water-supported, pumiceous gravity currents and water-settled fall similar to those that deposited the Wadaira Tuff, Japan [*Fiske and Matsuda*, 1964], Shirihama Group, Japan [*Cashman and Fiske*, 1991], and the Shinjima Pumice, Japan [*Kano et al.*, 1996]. Hence, it appears that submarine pyroclastic facies will reflect a combination of eruption style and water depth of the vent.

The Hellenic Island Arc has a variety of products derived from submarine explosive eruptions, including Pliocene to Quaternary submarine deposits that have been uplifted due to volcano-tectonics. The deposits from three explosive eruptions that were sourced from submarine vents have been well studied: the Kos Plateau Tuff [*Allen and Cas*, 1998; *Allen et al.*, 1999; and *Allen*, 2001], the Yali Pumice Breccia [*Allen and McPhie*, 2000], and the Filakopi Pumice Breccia [*Stewart and McPhie*, in press].

The Kos Plateau Tuff is an example of a moderate to large volume (>150 km^3), very shallow marine, caldera-forming eruption that deposited phreatoplinian fallout and non-welded ignimbrites on land. The Filakopi Pumice Breccia was deposited from a small to moderate volume (>1.8 km^3 bulk volume), explosive eruption from a shallow marine vent (~200 m). The products include water-supported gravity current deposits and water-settled fallout. The Yali Pumice Breccia was generated by a small-volume (>0.3 km^3), explosive submarine eruption at moderate water depths (200–400 m) located on the floor of a submarine caldera. It comprises syn-eruptive, downslope resedimented water-settled pumice fall deposits.

These three units vary in bulk volume by three orders of magnitude, and show differences in eruption style, depositional setting, and transport mechanisms, allowing comparisons to be made between the deposit type, eruption style and source water depth. We use these three well-exposed deposits, along with several other well-documented examples, to create working facies models for the outflow products of felsic pyroclastic eruptions generated from subaqueous, shallow to moderate water (<400 m) vents.

Geological Setting

The Hellenic Island Arc formed on the Aegean microplate in response to the collision between the African and Eurasian plates [*McKenzie*, 1972; and *Le Pichon and Angelier*, 1979]. The modern Hellenic Island Arc (Figure 1A) extends from the Greek mainland (west) across the Aegean Sea to the Kos-Nisyros complex (east) [*Keller*, 1982]. The arcuate line defined by these volcanoes lies above a 130–150 km deep, broadly north-dipping Benioff Zone about 200 km from the Hellenic Trench [*Dewey and Sengör*, 1979, and *Keller*, 1982].

The lithospheric stress field for the subduction zone differs from west (2 cm/yr) to east (4–5 cm/yr) due to variations in the linear rate of underthrusting at the trench [*LePichon and Angelier*, 1979]. Fytikas et al. [1984] recognised that in the west, convergence is directed orthogonal to the plate boundary and volcanism is dominated by the extrusion of domes and lavas. Towards the eastern sector (with sub-parallel convergence), explosive activity and calderas are dominant. Milos lies near the centre of the arc and mostly consists of small to moderate volume, rhyolitic and dacitic pyroclastic deposits, domes and lavas. The Kos-Nisyros complex (east) includes a 20 km-wide caldera, a composite cone with summit caldera (Nisyros), and the intracaldera lava and pumice cone of Yali.

MODERATE TO LARGE VOLUME, VERY SHALLOW MARINE EXPLOSIVE ERUPTION

The >150 km^3 Kos Plateau Tuff is the largest known Quaternary eruption from the Hellenic Island Arc. It is dominated by thick (5 to >15 m) ignimbrites (units D and E) that are extensively preserved on the Greek Dodecanese islands and Turkish peninsulas within 40 km of the source (Figure 2A). Correlation amongst the various exposures has been constrained by textural and lithofacies characteristics, and geochemistry [*Allen et al.*, 1999]. The entire deposit is non-welded. There are no significant submarine deposits of the Kos Plateau Tuff exposed. Subaerial deposits comprise

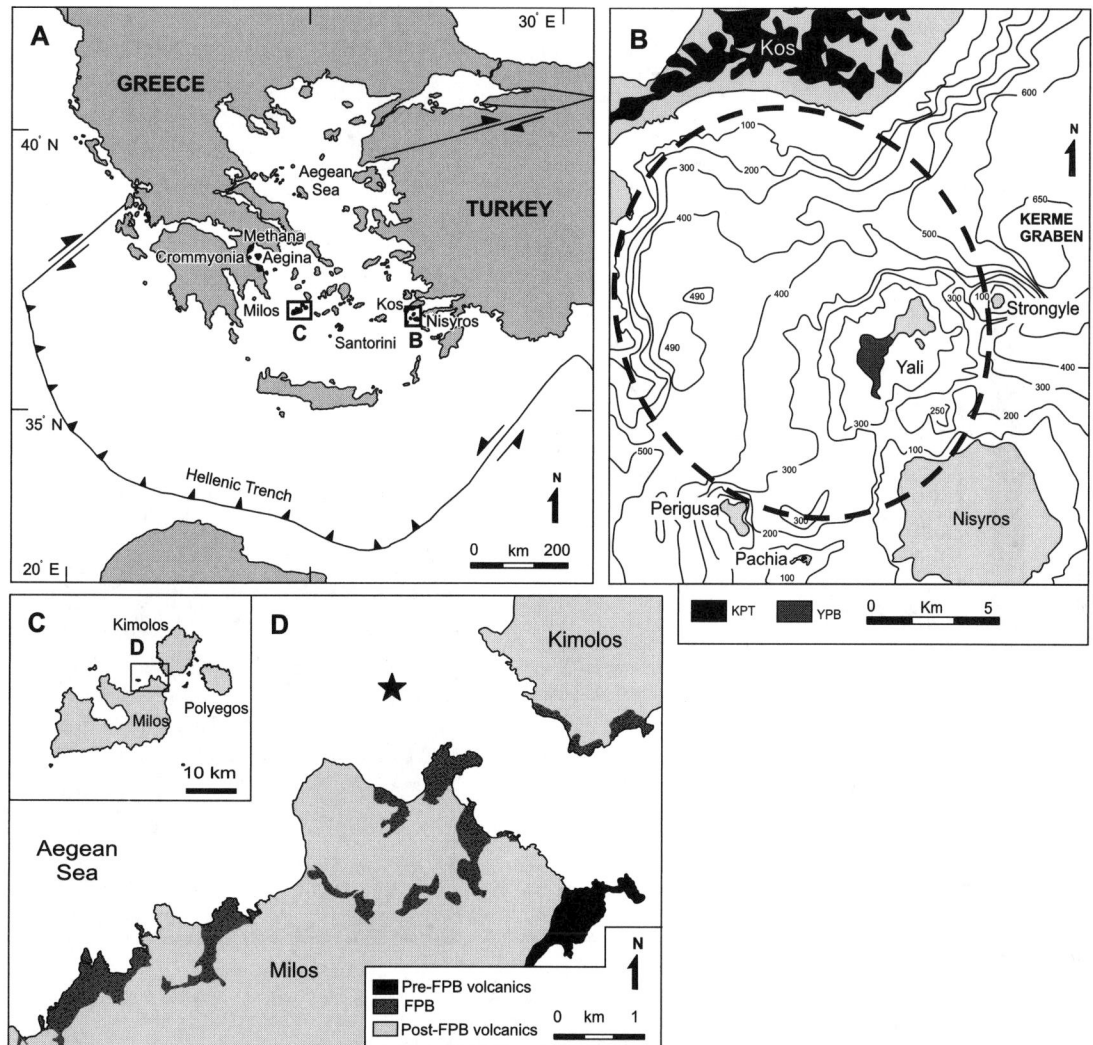

Figure 1. A. Map showing the location of the Kos-Yali-Nisyros complex, Milos and other volcanoes (Crommyonia, Methana, Aegina, Santorini along the Hellenic Island Arc, Greece (solid black fills). **B.** Present-day bathymetry of the area south of central Kos to Nisyros showing the inferred margin of the Kos-Nisyros caldera [from *Allen*, 2001]. Yali is located within the submerged caldera. The distribution of Kos Plateau Tuff on Kos (KPT, black), and the Yali Pumice Breccia (YPB, dark grey) are shown. **C.** Location of Milos, Kimolos and Polyegos. **D.** Distribution of the Filakopi Pumice Breccia (FPB) along the northeastern coast of Milos and southwestern Kimolos. The black star to the north of Milos is the inferred source location [from *Stewart and McPhie*, in press].

lower phreatoplinian units (A and B), overlying "dry" magmatic ignimbrites (units D and E) and a final phreatomagmatic subunit, Fs (Figures 2, 2B).

The eruption occurred 161 000 years ago [Ar-Ar dating, *Smith et al.*, 1996] from a source between the islands of Kos and Nisyros. A ~20 km-wide, 400–600 m-deep submarine caldera was produced and has been modified by the younger volcanoes of Nisyros, Yali and Strongyle (Figure 1B). Palaeo-shoreline reconstructions by Allen and Cas [2001] based on sea level changes, bathymetry and uplift patterns, suggested that at the time of the eruption, sea level was 60-80 m lower than present. Hence, an extensive landmass that included Kos, Kalymnos, Pserimos and Bodrum is inferred for the northern outflow environment (Figure 2A). In comparison, to the southeast and east (Datça Peninsula and Tilos), there were wide expanses of sea. Variations in the textural features of ignimbrites to the north compared with those in southeast-east are discussed in Allen and Cas [2001].

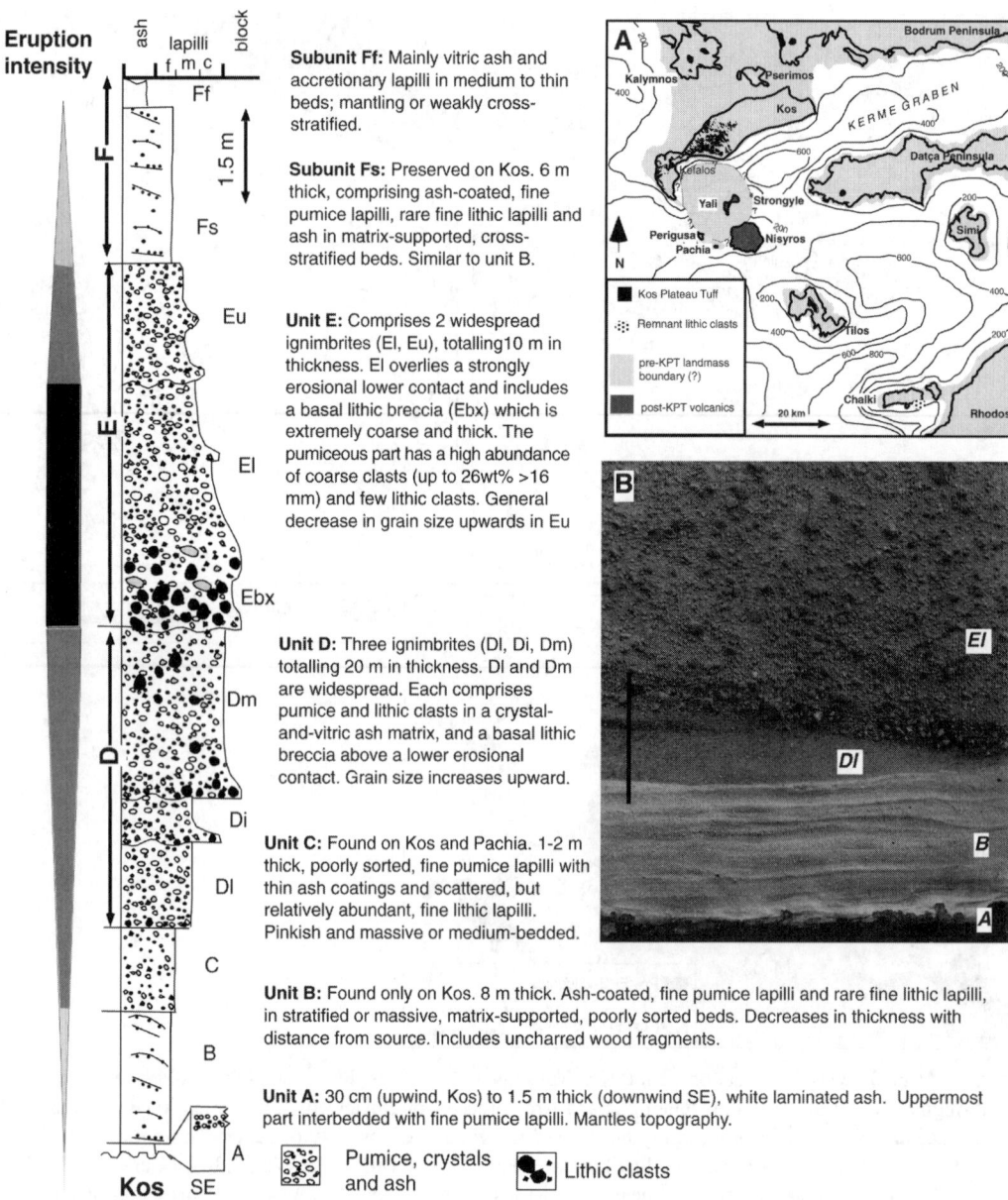

Figure 2. Graphic stratigraphic log of the Kos Plateau Tuff on Kos [modified from *Allen and Cas*, 2001; and *Allen*, 2001] showing the six units and information of their textural characteristics. Eruption intensity (left) was highest during eruption of unit E. The lower intensity phases were phreatomagmatic. **A.** Present-day distribution of the Kos Plateau Tuff and estimated shoreline position during the eruption [from *Allen and Cas*, 2001]. **B.** Phreatomagmatic units (A and B) overlain by thin ignimbrite Dl. Ignimbrite El (top) is rich in coarse pumice lapilli and has an erosional lower contact. Scale bar, 1 m.

Vent Setting

The Kos Plateau Tuff includes a large volume (~3 km³) of vent and conduit-derived andesitic lithic clasts that are texturally and petrologically similar to andesite forming the small islands of Pachia and Perigussa on the southwestern rim of the Kos-Nisyros caldera. The large volume and petrological similarities of the andesite led Allen [2001] to suggest that at the time of the eruption, the area now occupied by at least the southeastern part of the caldera included an andesitic volcano. Hence, even though the source was located within the western submarine extension of the

Neogene—Quaternary, E-W trending Kerme graben, the presence of the andesitic volcano suggests that the vent was much shallower. The volume of the initial phreatoplinian units (A and B) amounts to some 22.5 km³. At the estimated eruption rate of 10^5 m³/s, Allen and Cas [1998] calculated that a large volume of water was required to sustain the phreatoplinian activity. Therefore, it is most likely that seawater had direct, unimpeded access to the vent during this phase and hence, the vent was very shallowly submerged, or at least partly so.

Transport Mechanisms

Transport mechanisms for the Kos Plateau Tuff ranged from subaerial fallout at the beginning (unit A) and end of the eruption (subunit Fs) to pyroclastic density currents that varied in particle concentration. The transporting system that deposited units B, C and subunit Fs were interpreted by Allen and Cas [1998] to have been transitional in particle concentration between a dilute pyroclastic surge and a high-particle-concentration pyroclastic flow. The unidirectional bedforms suggest turbulent depositional boundary conditions whereas thickness (up to 8 m) and distribution (>19 km from source), imply a relatively high-particle-concentration transport system. In comparison, the transporting system that deposited units D and E were more typical pyroclastic flows and formed massive, pumice-rich ignimbrites with basal lithic breccias above erosional lower contacts. However, the ability of these pyroclastic flows to traverse tens of kilometers of water led Allen and Cas [2001] to suggest that they were relatively dilute and stratified, with lower, denser, lithic-rich zones and upper, expanded, pumice-rich zones.

Implications for Eruption Style

A fine grain size, combined with the widespread nature and mantling form of unit A led Allen and Cas [1998] to infer deposition by fallout from a 20–30 km-high phreatoplinian eruption plume. Unit B also has a phreatomagmatic character with thickly ash-coated pyroclasts and low-temperature emplacement indicated by preservation of uncharred wood. It was probably generated from partial? collapse of the phreatoplinian column. Unit C is transitional, with only thin fine-ash coatings on pyroclasts, indicating that moisture was present within the current, whereas the pinkish colour suggests thermal oxidation and a relatively hot temperature of emplacement.

The ignimbrites (units D and E) have matrices dominated by vitric shards and crystals that have no adhering fine ash, indicating dry conditions during eruption and outflow. The variations in grain size, volume and extent suggest that the D ignimbrites reflect waxing eruption intensity that culminated during eruption of ignimbrite El when caldera collapse was initiated [Figure 2; *Allen*, 2001]. Ignimbrite Eu then records the waning phase. The predominance of "dry" explosive conditions during the ignimbrite-forming phases appears to correlate strongly with eruption intensity.

Presence of cohesive ash within subunit Fs indicates the presence of water during renewed phreatomagmatic activity as the eruption continued to wane. The lack of heavy components such as crystals and lithic clasts in this unit suggests that it is co-ignimbrite ash that settled out at the end of the eruption.

Facies Model

The Kos Plateau Tuff provides an example of the outflow deposits of a moderate to large volume, felsic, caldera-forming eruption for which the inferred source environment was shallowly (less than tens of metres) submerged, and the eruption plume and depositional setting was predominantly subaerial. A characteristic feature of this style and setting of eruption is the phreatomagmatic opening phase which produced a widespread ash-rich deposit. Similar facies also occur within the Oruanui Formation [*Self and Sparks*, 1978; *Self*, 1983; and *Wilson*, 2001], 1800 YR-BP Taupo deposits [*Wilson* 1985; and *Smith and Houghton*, 1995], and the 1875 Askja deposits [*Self and Sparks*, 1978], all sourced within lakes. Facies characteristics of the Kos Plateau Tuff are most similar to the 1800 YR-BP Taupo deposits in which initial phreatomagmatic units are followed by non-welded, relatively thin (less than tens of metres thick), but widespread ignimbrite(s) from "dry" eruption and transport conditions. In addition, ignimbrites in both the Kos Plateau Tuff and Taupo deposits are completely non-welded, and include very coarse pumice clasts (more than 30 cm in diameter) within the ignimbrite associated with the eruption climax.

SMALL TO MODERATE VOLUME, SHALLOW MARINE EXPLOSIVE ERUPTION

The 45 m-thick Filakopi Pumice Breccia comprises a coarse lithic-rich basal breccia (unit A) and a stratified pumiceous middle part (unit B) that grades into an upper coarse pumice breccia and shard-rich mud (unit C) (Figure 3). Pumice clasts are moderately to highly vesicular (30–85 vol. %). Large pumice clasts have are prismatic with finely vesicular margins and normal joints, and clast interiors exhibit polyhedral joints. Small pumice clasts are polyhedral, with smooth, planar to curviplanar surfaces that cross-cut vesicle boundaries.

Unit C: upper coarse pumice breccia and shard-rich mud. 6.5–20 m-thick, tabular bed of reversely graded, well sorted grain supported pumice in a finer grained matrix. Dominated by clast-supported, large (64 mm-6.5 m) prismatic pumice clasts. Lacks lithic clasts. Matrix grades internally upward from fine (1-2 cm), well-sorted, diffusely stratified pumice clasts grading from unit B, to shard-rich, laminated mud. Laminae are deformed and contorted around large pumice clasts. Mud matrix comprises glass shards (<1 mm) with lesser amounts of crystal (<1 mm) and fine pumice (<2 mm) fragments. <1 m thick interval of shard-rich mud at the top. Planar and sharp upper contact.

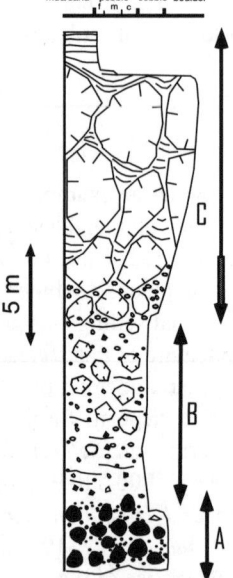

Unit B: Fines-poor, stratified, pumice breccia. 16–17 m thick; internally massive or diffusely stratified, very thick (0.9-4 m), wedge-shaped, poorly defined beds. Predominantly comprises 2 mm- to 1.5 m, angular to subangular pumice clasts (50-98 vol. %) in grain support. Upward decrease in size and abundance of scattered obsidian and lithic clasts. Parallel to wavy diffuse stratification is commonly defined by alignment of coarse pumice clasts, especially in the lower parts of beds. Single beds display normal grading of dense lithic clasts, and reverse grading of coarse pumice clasts.

Unit A: fines-poor, poorly sorted, grain-supported basal lithic breccia. 4-8 m thick, normally graded with a reversely graded base. Tabular bed dominated by dense dacitic/andesitic lithic clasts and sedimentary intraclasts up to several m across. Rare basement clasts. Subrounded coarse lithic clasts (>20 cm across), angular smaller lithic clasts. Sparse, typically large, poorly to moderately vesicular pumice clasts (2-5 vol.% ~40 cm across). Sharp, often highly erosional lower contact. Upper contact is gradational and diffusely stratified.

Figure 3. Graphic stratigraphic log of the Filakopi Pumice Breccia [from *Stewart and McPhie*, in press] and descriptions of the three units. **A.** Prismatic coarse pumice clasts within unit C, have normal joints along the margins and internal polyhedral joints. **B.** Wedge-shaped, medium beds of unit B at the base are overlain by the tabular, upward coarsening unit C that dips gently to the south. A lenticular, 1–2 m-thick, diffusely stratified bed lacking coarse pumice clasts occurs locally within unit C. Scale bar, 5 m. **C.** Reversely to normally graded coarse lithic breccia (unit A). The lithic breccia has a sharp, planar lower contact. The base comprises a fine grained, pumiceous matrix. An upper, 5 m-thick, coarser grained, lithic clast-rich part includes andesitic and dacitic lithic clasts up to 2.5 m in diameter.

An area of at least 18 km² is covered by the pumice breccia including northeastern Milos and southwestern Kimolos (Figures 1C, 1D). The minimum bulk volume is approximately 1.8 km³. Bioturbated and fossiliferous marine siltstone enclose the pumice breccia. In situ and intact bivalve shells and burrows in the underlying siltstone suggests a moderate to shallow marine environment. Overlying sedimentary and volcanic formations record a transition from this relatively shallow but dominantly below wave base setting to a subaerial environment, recording subsequent deposition and uplift. Hence, the pumice breccia was deposited in shallow to moderate water depths (~150 to 400 m). Dating of overlying facies by Fytikas et al. [1986] implies that the pumice breccia is Upper Pliocene in age.

Vent Setting

Grain-size and thickness variations within the pumice breccia measured by Stewart and McPhie [in press], indicate that the source was located offshore to the north of the present northeastern coast of Milos (Figure 1D). The lithic clasts are significantly coarser close to source whereas the overall thickness and pumice clast size both increase to a maximum away from the inferred source and then decrease. Textural and facies characteristics suggest the vent was submerged but shallower than the depositional setting, and was presumably shallow marine (~200 m water depth).

Transport Mechanisms

The poor sorting, concentration of lithic clasts, reverse to normal grading, tabular geometry and presence of very large intraclasts in unit A suggest deposition from a high-particle-concentration, lithic-rich gravity current [*Stewart and McPhie*, in press]. The presence of pumice clasts together with dense lithic clasts suggests that these pumice clasts must have been waterlogged, a condition achieved by hot pumice [e.g., *Whitham and Sparks*, 1986]. Poor sorting in unit B and division into multiple, very thick, pumice-rich, wedge-shaped beds with basal concentrations of dense clasts, are consistent with deposition from a series of density stratified, pumiceous gravity currents. In order for the pumice clasts to be transported by such currents, they must have been denser than water and hence, already waterlogged. However, the coarser pumice clasts in unit B have intact quenched margins formed when they were initially suspended in the water column. These pumice clasts probably settled from suspension up-current and were collected by the gravity currents, and/or settled directly into the active currents. Their concentration in the upper parts of beds is attributed to buoyant lift forces that acted preferentially on coarse clasts [*Stewart and McPhie*, in press].

The three distinct grain-size classes in unit C reflect different transport and depositional processes [*Stewart and McPhie*, in press]. The reversely graded, well-sorted, grain-supported framework of coarse pumice clasts with intact quenched margins was probably generated by passive settling through the water column. Being hot and large they would have been at least temporarily buoyant. The progressive upward increase in their size suggests that the rate of waterlogging was strongly size dependant, similar to those on the Shinjima Pumice [*Kano et al.*, 1996]. An intermediate (1–2 cm) pumice component forms the matrix within the basal part of unit C. It is poorly sorted and diffusely stratified, reflecting deposition from relatively dilute, weak traction currents presumably derived from waning currents that formed unit B. The shard-rich mud matrix at the top presumably settled last as it infills the framework in the upper part of unit C. Shards have slow settling velocities [e.g., *Cashman and Fiske*, 1991], but their concentration in the pumice breccia indicates that deposition was relatively rapid, such as could be redistributed by ocean currents. Settling of the finest pyroclasts through the water column may have been accelerated by convective instabilities in one or more high-particle concentration layers in the water column [cf. *Carey*, 1997].

Eruption Style

The abundance of coarse, highly vesicular pumice and vent- and conduit-derived lithic clasts provide strong evidence for an explosive, open-vent style eruption. The gradational contacts between the three units indicate that a single eruptive event was involved. The normal margin joints and internal polyhedral joints of the large pumice clasts are typical of hot juvenile clasts that have been quenched, consistent with a submarine eruption [e.g., *Kano*, 1996; and *Kano et al.*, 1996]. Given that hot pumice is more rapidly waterlogged than cold pumice [*Whitham and Sparks*, 1986], and the strong evidence for continuous aggradation, Stewart and McPhie [in press] concluded that the Filakopi Pumice Breccia was syn-eruptive.

The high percentage of dense volcanic lithic clasts in unit A reflects the initial vent clearing and conduit erosion. In terms of geometry, stratigraphic position, grain size, componentry and thickness, unit A closely resembles coarse lithic breccias present at the base of some other submarine volcaniclastic deposits believed to have been generated directly from explosive eruptions [e.g., Shirahama Group, *Cashman and Fiske*, 1991; and Tayu Volcaniclastic Bed E, *Kano*, 1996). In subaqueous eruption plumes, the basal gas-thrust region may be suppressed by a combination of mixing with seawater and confining hydrostatic pressure [*Cashman and Fiske*, 1991; and *Kano*, 1996). Dense lithic clasts and smaller pumice clasts that were quickly waterlogged fed density-stratified gravity currents derived more or less directly from the basal part of the plume. The largest hot pumice clasts and fine ash remained buoyant and presumably contributed to a buoyant plume [c.f., *Fiske et al.*, 2001]. Both eventually settled to the seafloor. Given the shallow water-depth inferred for the Filakopi vent (~200 m), it is possible that the plume breached the seawater interface.

The deposition of the shard-rich mud led Stewart and McPhie [in press] to suggest that the pumice breccia was erupted from an area enclosed by islands, possibly a partly submerged caldera, that was sheltered from both wind and ocean currents.

Facies Model

The Filakopi Pumice Breccia provides an example of a small to moderate volume, pyroclastic eruption sourced in shallow water. The pumice breccia is strongly density sorted, with a dense lithic-rich base overlain by pumice and ash dominated middle and upper parts. The lower units were deposited from gravity currents generated directly from the eruption plume whereas the upper part is the product of water settling of buoyant, suspended pyroclasts.

The Filakopi Pumice Breccia shows similar textural and lithofacies characteristics to several other formations in the submarine succession on Milos and to the products of other small to moderate volume, submarine explosive eruptions. The diffuse bedding and poor sorting of unit B are similar to the Shinjima Pumice of Kano et al., [1996], and the coarse lithic breccia at the base resembles similar facies in the Tayu Volcaniclastic Bed E of Kano [1996] or the Shirahama Group of Cashman and Fiske [1991]. In addition, the upper shard-rich mud also shows similarities to the upper-most layers of the Tayu Volcaniclastic Bed E and Shirahama Group, but neither of the units include very large pumice clasts.

SMALL VOLUME, MODERATE DEPTH, MARINE EXPLOSIVE ERUPTION

The small volume (>0.3 km³) Yali Pumice Breccia is exposed on the western part of Yali, a small island within the Kos-Nisyros caldera (Figure 1B). Volcano-tectonic uplift has exposed the pumice breccia at elevations up to 150 masl. The pumice breccia remains undated, but is older than a rhyolite lava to the northeast [4 ka, fission track; *Wagner et al.*, 1976] and younger than the Kos Plateau Tuff (161 ka). Units overlying the pumice breccia show an upward progression (fossiliferous limestone, bioclastic sandstone, brown sandy silt, subaerial tephra fall) that record a progressive transition from shallow marine to subaerial environments [*Allen and McPhie*, 2000].

This submarine pumice breccia comprises highly vesicular, pebble to boulder-sized pumice clasts with very little matrix or lithic clasts (Figure 4). The base is not exposed and there are no internal reworked horizons. The most characteristic features are good size and density sorting that resembles pyroclastic fall deposits (Figures 4A, 4C), the quenched margins on the larger clasts (Figure 4B), and tabular beds of large pumice clasts within diffusely stratified intervals of smaller pumice clasts (Figure 4D).

Vent Setting

The lack of tephras related to the Yali Pumice Breccia on any of the surrounding islands suggests that the eruption plume did not significantly breach the water surface. The pumice breccia was erupted from a vent within the central part of the Kos Plateau Tuff caldera and was presumably deposited on the floor of the caldera in approximately 400–600 m of water. The great thickness of the pumice breccia (>150 m) suggests that the vent may have been somewhat shallower at moderate water depths of ~200–400 m.

Transport Mechanisms

The lack of fine matrix and presence of good density sorting in the pumice breccia are consistent with deposition by means of water-settling from suspension [e.g., *Cashman and Fiske*, 1991]. The tabular, massive beds of large pumice clasts are consistent with this interpretation. However, the bedforms, edge modification and presence of dense lithic clasts within the accumulations of small pumice clasts also suggest lateral transport. The wedge-shaped beds are consistent with deposition from subaqueous gravity flows (modified? grain flows) of waterlogged, cohesionless pumice. Mixed facies of small and large pumice clasts probably resulted from downslope resedimentation of unstable, near-source accumulations of pumice and from water-settling of coarse pumice on to, and incorporated within active grain flows. The grain size of the mixed facies reflects the clast population prior to remobilisation, and/or the dominance of either resedimentation or suspension settling during deposition. The textural and lithofacies characteristics of the pumice breccia are consistent with features of proximal submarine fall deposits where the growing pumice cone is subjected to downslope resedimentation [e.g., *Cas et al.*, 1990].

Implications for Eruption Style

Allen and McPhie [2000] discussed the possibility that the large pumice clasts were the product of submarine, open-vent style, explosive eruptions. Large pumice clasts had previously been documented only for subaqueous dome eruptions [e.g,. *Mahood*, 1980; *Reynolds et al.*, 1980; *Wilson and Walker*, 1985; *Kano*, 1996; and *Fiske et al.*, 1998]. However, recent submersible surveys of calderas within the Izu-Ogasawara Arc, offshore from Japan [e.g., *Yuasa*, 1995], and the southern Kermadec Arc [*Wright et al.*, in press] have also found thick sequences of large pumice clasts within caldera margins, presumably from explosive eruptions. No caldera related to the eruption has been identified [*Papanikolaou et al.*, 1998].

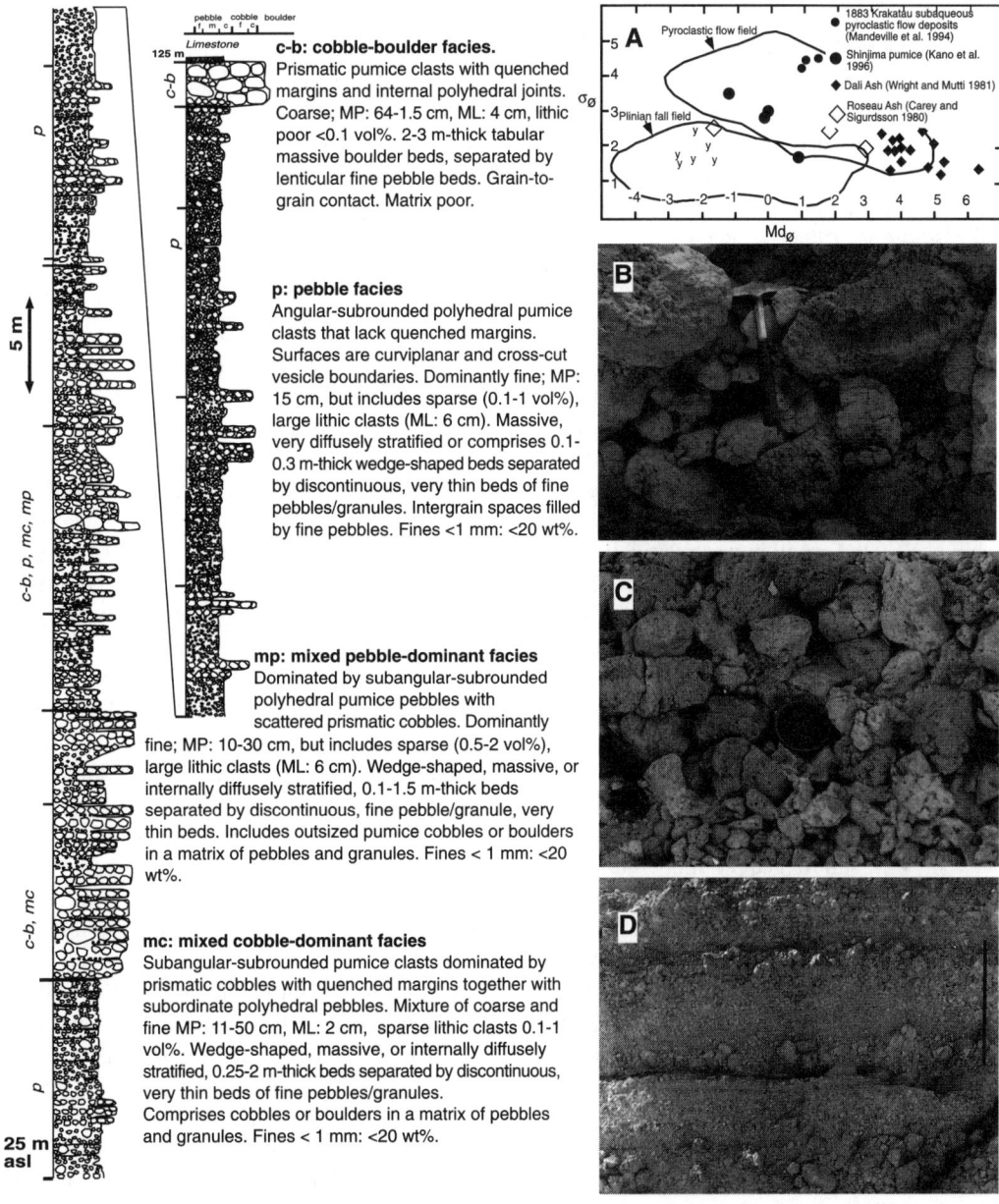

Figure 4. Graphic stratigraphic log of the Yali Pumice Breccia [from *Allen and McPhie*, 2000] and descriptions of the four facies. Five units up to 50 m thick consist of one or more facies. All facies are dominated by highly vesicular pumice clasts. **A.** Plot of median grain size versus sorting showing pyroclastic flow and fall fields of Walker (1983). Pebble facies grain size analyses (Y) are plotted along with the deposits of subaqueous turbulent mass flows [Krakatau subaqueous pyroclastic flow deposits, Shinjima Pumice, Dali Ash, and Roseau Ash; from *Kano et al*. 1996]. Photos of cobble-boulder facies (B), pebble facies (C), and interbedded pebble and mixed cobble dominant facies (D). Scale bar, 2 m.

The lack of reworked horizons suggests that the eruption was relatively continuous. Although proximal, all existing exposures of Yali pumice breccia were deposited at some distance from the source. The vent stratigraphy may be different, including higher concentrations of dense clasts that occur scattered within the resedimented facies.

The pumice breccia lacks products related to collapse of an eruption plume [cf., *Fiske,* 1963; *Fiske and Matsuda,* 1964, and *Kano et al.,* 1996]. However, given that the base of the pumice breccia is not exposed, it is possible that the start of the eruption produced plume-fed gravity currents. For subaqueous eruption plumes, the buoyant rise of pumice above the gas-thrust region is dependent on those clasts remaining gas-charged and less dense than water. The larger the pumice size, the slower the rate of water-logging and hence, the longer time period to remain buoyant. Fine pumice lapilli and ash would also have been buoyant, and incorporated into upwelling plumes of heated seawater and carried away by near-surface currents [e.g., *Cashman and Fiske,* 1991; and *Fiske et al.,* 1998].

We interpret thick, coarse, reversely graded, dominantly pumiceous water-settled fall deposits to be the products of buoyant subaqueous plumes. However, at Yali, coarse pumice clasts are not just at the top of the unit but occur throughout and much of the unit comprises resedimented fallout pumice. Very little is known about the mechanisms required for creating an exclusively or predominantly buoyant plume in the subaqueous environment. A low density plume is thought to be important [*Cashman and Fiske,* 1991]. Presumably at Yali, the eruption was relatively weak with no major vent-widening episodes as it was cone-building and lithic clasts are few and scattered. This suggests that the discharge rate (eruption intensity) was less than that required for plume collapse.

Facies Model

The Yali Pumice Breccia provides an example of a small volume, open-vent, relatively low intensity explosive eruption where the vent was submerged at moderate water depths (~200–400 m) and the eruption plume was at least dominantly submarine. The product of this eruption was a very thick succession of stratified, well-sorted, markedly fines-poor, pumice deposited by water-settled fall and modified grain flows. The coarse nature and wedge-shaped bedforms reflect the proximal nature of the pumice breccia and resedimentation of the water-logged pumice clasts downslope. Proximal water-settled fall deposits generated from submarine explosive eruptions also occur within two submarine rhyolitic tuff cone sequences in the Bunga Beds, Australia [*Cas et al.,* 1990].

DISCUSSION

Very little is known about the eruption style and dynamics of explosive subaqueous eruption plumes. There are no rigorous methods proven to calculate plume height or eruption intensity (discharge rate) and links between deposit characteristics and eruption style have not been established. However, by analogy with the equivalent eruption style in subaerial settings, the greatest velocity increase probably occurs at the level of fragmentation. Magma with water contents of 4–6 wt.% will fragment when the vesicularity reaches a value of approximately 75 vol% which is predicted to occur at pressures of 10–15 MPa [*Shaw,* 1974; *Sparks,* 1978; and *Klug and Cashman,* 1994]. Such pressures equate to crustal depths of ~470–750 m or water depths of 1–1.5 km. Subaerial felsic explosive eruptions have peak eruption intensities that range from 10^6–10^9 kg/s, with most around 10^7–10^8 kg/s [*Carey and Sigurdsson,* 1989]. In "dry" subaerial eruption plumes, height of the gas-thrust region varies between 0.5–4 km but is dependent on temperature, gas content, velocity and vent radius [*Woods,* 1998]. Gas-thrust regions in subaqueous plumes are predicted to decelerate much faster than in subaerial plumes due to the confining pressure, density of the overlying water column and rapid cooling by water [e.g., *Cashman and Fiske,* 1991; and *Koyaguchi and Woods,* 1996]. Hence subaqueous gas-thrust regions will reach lower heights than in the subaerial environment. The Yali Pumice Breccia, Filakopi Pumice Breccia Shinjima Pumice [*Kano et al.,* 1996] and Wadaira Tuff [*Fiske and Matsuda,* 1964] are all interpreted to be the products of submarine eruption plumes suggesting that the thrust heights related to these eruptions were not much greater than the water depth (i.e., <~200–400 m).

Comparisons of the various products derived from subaqueous felsic explosive eruptions at similar water depth estimates to those presented here and others in the literature suggest that magnitude (eruption volume) and facies type vary depending on eruption intensity.

Explosive magma-water interactions are strongly controlled by optimal magma/water ratios, and hence depend on the availability of water, and the eruption intensity [e.g., *Wohletz,* 1986]. In the subaqueous environment, optimal conditions for phreatomagmatic activity appear to be strongly controlled by water depth. For example, there are no documented phreatoplinian eruptions thought to have involved wholly subaqueous eruption plumes or vents at moderate water depths. On the contrary, it appears that phreatoplinian eruptions require suitably (very) shallow water depths for optimum fragmentation conditions and so that the gas-thrust region can penetrate above the water surface (Figure 5). Koyaguchi and Woods [1996] predicted that at the eruption rates required for plinian eruptions, the eruption column height will progressively decrease with increasing incorporation of surface water. They also found that "wet" eruption columns remain buoyant with addition of a maximum of 15–20 wt% mass fraction of water.

The second factor that influences optimal phreatomagmatic activity in the subaqueous environment, is eruption intensity (volume of magma ejected). For example, only the opening and/or lower intensity phases of the Taupo and Kos Plateau Tuff eruptions were phreatomagmatic. Once the eruption rates reached 10^8–10^9 kg/s [*Walker et al.,* 1980;

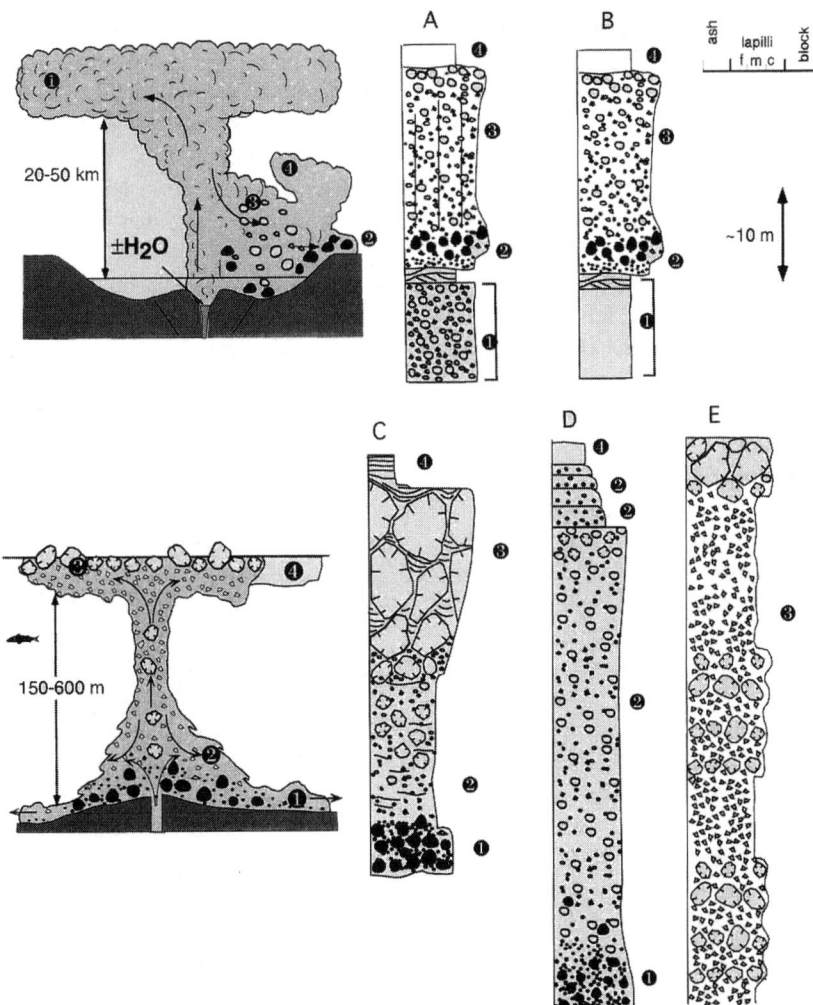

Figure 5. Eruption styles and facies models for the outflow products of open-vent explosive felsic eruptions sourced from submarine vents. Top left hand side; schematic cartoon of a plinian/phreatoplinian eruption plume. 1. Fallout, 2. Basal lithic breccia, 3. Pumiceous ignimbrite, 4. Co-ignimbrite ash. Bottom left hand side; schematic cartoon of a subaqueous pyroclastic eruption plume. Density segregated gravity current with a lithic-rich lower part (1) and pumiceous upper part (2), large buoyant pumice (3), and shards (4). **A.** Facies produced by a subaerial plinian eruption for comparison [Sparks et al., 1973]. **B.** High intensity, moderate-large volume, phreatoplinian/plinian, caldera-forming eruption sourced in very shallow water (les than a few tens of metres) [modified from *Walker*, 1980; and *Allen*, 2001]. **C.** Shallow water (~200 m), small to moderate-volume, moderate intensity eruption [modified from *Cashman and Fiske*, 1991; *Kano et al.*, 1996; and *Stewart and McPhie*, in press]. **D.** Shallow water (~200 m), moderate-large volume, moderate intensity eruption [modified from *Fiske and Matsuda*, 1964]. **E.** Moderate water depth (~200-400 m), small volume, low intensity eruption [modifed from *Allen and McPhie*, 2000].

and *Allen and Cas,* 1998], "dry" explosive conditions prevailed, even though the source was still presumably submerged. The ignimbrites produced cannot be readily distinguished from ignimbrites generated from subaerial vents. Futhermore, changes from "wet" to "dry" eruption styles for the Minoan eruption, Santorini [*Heiken and McCoy,* 1984] and 22 000 YR eruption from Aira caldera, Japan [*Aramaki,*

1984] are also interpreted to be the result of increasing eruption intensity and/or lower rates of available water. In other cases, such as the ~1170 km^3 (bulk volume) Oruanui eruption, magma-water interaction continued throughout the eruption and most deposits are fine-grained and rich in accretionary lapilli [Wilson, 2001]. Hence, the facies model for the products of explosive felsic eruptions from very shallow (less than a few tens of metres) subaqueous vents is fine-grained fallout and flow deposits generated during phreatomagmatic activity that forms a relatively low buoyant plume. However, moderate to large volume explosive felsic eruptions from very shallow vents that increase in eruption intensity or exclude water from the vent area, have facies models broadly similar to that for subaerial caldera-forming eruptions (Figure 5A). In this case, the facies model is an initial fine-grained phreatomagmatic deposit overlain by coarse, non-welded ignimbrites (Figure 5B)

In comparison, deposits from submarine, felsic, explosive eruptions sourced in somewhat deeper water (i.e., ~200 m), have facies characteristics very different from plinian eruptions in subaerial settings. Single-event, small to moderate volume eruptions [e.g., Shirahama Group *Cashman and Fiske,* 1991; Shinjima Pumice, *Kano et al.* 1996; and Filakopi Pumice Breccia] commonly produce basal, poorly sorted, lithic-rich gravity current deposits overlain by pumiceous gravity current deposits, both of which are fines-poor. Density-sorted pumiceous and vitric ash deposited by water-settled fall are also preserved in some examples (Figure 5C). Larger volume eruptions [e.g., Wadaira Tuff, *Fiske and Matsuda, 1964*] produce thick, massive units comprising normally graded dense lithic clasts and reversely graded pumice clasts overlain by successively finer and thinner, graded beds (Figure 5D) and are exclusively derived from syn-eruptive gravity currents. Small volume, lower intensity eruptions at moderate (~200–400 m) water depths (e.g., Yali Pumice Breccia) produce thick, fines-poor laterally confined facies dominated by water-settled coarse pumiceous fallout and their resedimented equivalents (Figure 5E).

Coarse Pumice Clasts

The pumice clasts within the Filakopi and Yali Pumice Breccias are very coarse (up to 6.5 m across) compared with pumice clasts in subaerial plinian deposits. The vesicularities (abundance and morphology) are similar to pumice clasts produced by magmatic-volatile-driven explosive eruptions [*Houghton and Wilson,* 1989]. These coarse pumices survived intact due to their eruption and emplacement in water. Most of the large pumice clasts have well-preserved quenched margins reflecting cooling from the surface inwards. Transport in water greatly reduced the effects of clast collision and abrasion. Large pumice clasts were also created during the Kos Plateau Tuff and Taupo eruptions, but were significantly reduced in size and abraded during transport within pyroclastic flows.

The coarsest pumice clasts occur at the top of both the Yali and Filakopi Pumice Breccias. The progressive upward increase in size results from delayed settling of the largest pumice clasts due to slower waterlogging. Hence, for submarine deposits from explosive eruptions, the stratigraphy is influenced by the ambient fluid being water rather than air. The stratigraphic position of the pyroclasts does not necessarily correlate with the order of eruption. In this way, coarse pumice clasts that were generated early in the eruption may be deposited last. The range of clast types present in submarine pyroclastic deposits does not necessarily reflect the range of clasts erupted, because transport involves very efficient hydraulic sorting.

CONCLUSIONS

Submarine explosive eruptions of felsic magmas appear to be a relatively common phenomenon in several island arc settings. The eruption dynamics, transport, and depositional processes of submarine explosive eruptions are strongly influenced by the presence of water. Distinctive facies are generated depending on eruption magnitude, water depth, and inferred eruption intensity. At very shallow water depths (less than a few tens of metres), the eruption plumes breach the sea-water interface, generating subaerial pyroclastic flows, and phreatomagmatic activity is typical, particularly for the initial or lower-intensity eruptive phases of moderate to large caldera-forming eruptions. At water depths sufficiently deep to suppress the gas-thrust region of the plume (>~200 m of water), wholly or predominantly submarine eruption plumes are generated.

Small to moderate volume, moderate-intensity eruptions produce water-supported, lithic-rich and pumiceous, gravity-current deposits derived from the collapsing plume and overlain by water-settled pumice and ash that were initially buoyant. The products comprise a basal lithic breccia and/or stratified pumice breccia overlain by an interval of reversely graded pumice and/or shard-rich mud. Small volume, low intensity eruptions dominantly produce cones of water-settled pumice that are subjected to downslope resedimentation. The products are very thick, ash and lithic poor, rich in coarse pumice clasts and stratified with a relatively small geographic distribution.

Eruption plume dynamics, transport and depositional processes are clearly different in submarine settings compared to subaerial environments. These facies models provide a starting point for interpretation of the style of submarine explosive eruptions of varying magnitude and water depths.

Acknowledgments. This manuscript greatly benefited from the helpful reviews of Cathy Busby, Ian Wright and Jocelyn McPhie.

The research was funded by the Australian Research Council's Special Research Centres Program. SA also acknowledges funding from an Australian Research Council Post-Doctoral Research Fellowship, and AS, an Australian Postgraduate Research Award.

REFERENCES

Allen, S. R., Reconstruction of a major caldera-forming eruption from pyroclastic deposit characteristics: Kos Plateau Tuff, eastern Aegean Sea. *J. Volcanol. Geotherm. Res. 105,* 141-162, 2001.

Allen, S. R., and R. A. F. Cas, Rhyolitic fallout and pyroclastic density current deposits from a phreatomagmatic eruption in the eastern Aegean Sea, Greece. *J. Volcanol. Geotherm. Res. 86,* 219-251, 1998.

Allen, S. R., and R. A. F. Cas, Transport of pyroclastic flows across the sea during the explosive, rhyolitic eruption of the Kos Plateau Tuff, Greece. *Bull. Volcanol., 62,* 441-456, 2001.

Allen, S. R., and J. McPhie, Water-settling and resedimentation of submarine rhyolite pumice at Yali, eastern Aegean, Greece. *J. Volcanol. Geotherm. Res., 95,* 285-307, 2000.

Aramaki, S., Formation of the Aira caldera, southern Kyushu, ~22 000 years ago. *J. Geophys. Res., 89,* 8485-8501, 1984.

Bond, A., and R. S. J. Sparks, The Minoan eruption of Santorini, Greece. *J. Geol. Soc. London, 132,* 1-16, 1976.

Carey, S., Influence of convective sedimentation on the formation of widespread tephra fall layers in the deep sea. *Geology, 25(9),* 389-842, 1997.

Carey, S., and H. Sigurdsson, The Roseau Ash: deep-sea tephra deposits from a major eruption on Dominica, Lesser Antilles arc. *J. Volcanol. Geotherm. Res., 7,* 67-86, 1980.

Carey, S., and H. Sigurdsson, The intensity of plinian eruptions. *Bull. Volcanol. 51,* 28-40, 1989.

Cas, R. A. F., R. L. Allen, S. W. Bull, B. A. Clifford, and J. V. Wright, Subaqueous, rhyolitic dome-top tuff cones: a model based on the Devonian Bunga Beds, southern Australia and a modern analogue. *Bull. Volcanol., 52,* 159-174, 1990.

Cashman, K. V., and R. S. Fiske, Fallout of pyroclastic debris from submarine volcanic eruptions. *Science 253,* 275-280, 1991.

Dewey, J. F., and A. M. C. Sengör, Aegean and surrounding regions: complex multiplate and continuum tectonics in a convergent zone. *Geol. Soc. Am. Bull., 90,* 84-92, 1979.

Fiske, R. S., Subaqueous pyroclastic flows in the Ohanapecosh Formation, Washington. *Geol. Soc. Am. Bull., 74,* 391-406, 1963.

Fiske, R. S., and T. Matsuda, Submarine equivalents of ash flows in the Tokiwa Formation, Japan. *Am. J. Sci., 262,* 76-161, 1964.

Fiske, R. S., K. V. Cashman, A. Shibata, K. Watanabe, Tephra dispersal from Myojinsho, Japan, during its shallow submarine eruption of 1952-1953. *Bull. Volcanol., 59,* 262-275, 1998.

Fiske, R. S., J. Naka, K. Iizasa, M. Yuasa, and A. Klaus Submarine silicic caldera at the front of the Izu-Bonin arc, Japan: Voluminous seafloor eruptions of rhyolite pumice. *Geol. Soc. Am. Bull., 113,* 813-824, 2001.

Fytikas, M., P. Innocenti, R. Manetti, R. Mazzuoli, A. Peccerillo, and L. Villari, Tertiary to Quaternary volcanism in the Aegean region, in *The geological evolution of the Eastern Mediterranean,* edited by J. E. Dixon and A. H. F. Robertson, *Geol. Soc. London Spec. Publ., 17,* 687-699, 1984.

Fytikas, M., F. Innocenti, N. Kolios, P. Manetti, R. Mazzuoli, G. Poli, F. Rita, and L. Villari, Volcanology and Petrology of volcanic products from the island of Milos and neighboring islets. *J. Volcanol. Geotherm. Res., 28,* 297-317, 1986.

Heiken, G., and F. McCoy, Caldera development during the Minoan eruption. Thira. Cyclades, Greece. *J. Geophys. Res. 89,* 8441-8462, 1984.

Houghton, B. F., and C. J. N. Wilson, A vesicularity index for pyroclastic deposits. *Bull. Volcanol., 51,* 451-462, 1989.

Kano, K., A Miocene coarse volcaniclastic mass-flow deposit in the Shimane Peninsula, SW Japan: product of a deep submarine eruption? *Bull. Volcanol., 58,* 131-143, 1996.

Kano, K., T. Yamanamoto, and K. Ono, Subaqueous eruption and emplacement of the Shinjima Pumice, Shinjima (Moeshima) Island, Kagoshima Bay, SW Japan. *J. Volcanol. Geotherm. Res., 71,* 187-206, 1996.

Keller, J., Mediterranean island arcs, in *Andesites,* edited by R. S. Thorpe, pp. 307-325, J. Wiley & Sons, 1982.

Klug, C., and K. V. Cashman, Permeability development in vesiculating magmas: implications for fragmentation. *Bull. Volcanol., 58,* 87-100, 1996.

Koyaguchi, T., and A. W. Woods, On the formation of eurption columns following explosive mixing of magma and surface-water. *J. Geophys. Res., 101,* 5561-5574, 1996.

LePichon, X., and J. Angelier, The Hellenic arc and trench system: a key to the neotectonic evolution of the eastern Mediterranean area. *Tectonophysics, 60,* 1-42, 1979.

Mahood, G., Geological evolution of a Pleistocene rhyolitic centre–Sierra La Primavera, Jalisco, Mexico. *J. Volcanol. Geotherm. Res., 8,* 199-230, 1980.

McKenzie, D. P., Active tectonics of the Mediterranean region. *Geophys J. R. Astron. Soc. 30(2),* 109-185, 1972.

Mandeville, C.W., S. Carey, H. Sigurdsson, and J. King, Paleomagnetic evidence for high-temperature emplacement of the 1883 subaqueous pyroclastic flows from Krakatau Volcano, Indonesia. *J. Geophys. Res., 99,* 9487-9504, 1994.

Papanikolaou, D.J., P. Nomokou, and V. Lykousis, 6. Submarine reconnaissance, in *Newsletter of the European Centre on Prevention and Forecasting of Earthquakes,* pp. 23-26, ECPFE and EPPO, Athens, 2, 1998.

Reynolds, M. A., J. G. Best, and R. W. Johnson, 1953-57 eruption of Tuluman volcano: rhyolitic volcanic activity in the northern Bismarck Sea, pp. 1-44, Geol. Surv. Papua New Guinea, Mem. 7, 1980.

Self, S., Large-scale phreatomagmatic silicic volcanism: a case study from New Zealand. *J. Volcanol. Geotherm. Res., 17,* 433-469, 1983.

Self, S., and R. S. J. Sparks, Characteristics of widespread pyroclastic deposits formed by the interaction of silicic magma and water. *Bull. Volcanol., 41,* 196-212, 1978.

Shaw, H. R., Diffusion of H2O in granitic liquids, 1. Experimental data, in *Geochemical transport kinetics,* edited by A. W. Hoffman, B. J. Giletti, H. S. Yoder, and R. S. Yund, Carnegie Inst. Washington Publ., 634, 139-170, 1974.

Smith, R.T., and B. F. Houghton, Vent migration and changing eruptive style during the 1800a Taupo eruption: new evidence from the Hatepo and Rotongaio phreatoplinian ashes. *Bull. Volcanol., 57,* 432-439, 1995.

Smith, P.E., D. York, Y. Chen, and N. M. Evensen, Single crystal 40Ar-39Ar dating of a Late Quaternary paroxysm on Kos, Greece: Concordance of terrestrial and marine ages. *Geophys. Res. Lett. 23,* 3047-3050, 1996.

Sparks, R. S. J. The dynaimcs of bubble formation and growth in magmas: a review and analysis. *J. Volcanol. Geotherm. Res., 3,* 1-37, 1978.

Sparks, R. S. J., S. Self, and G. P. L. Walker Products of ignimbrite eruptions. *Geology, 1,* 115-118, 1973.

Stewart, A. L., and J. McPhie, An Upper Pliocene coarse pumice breccia generated by a shallow submarine explosive eruption at Milos, Greece. *Bull. Volcanol.,* in press 2003.

Wagner, G. J., D. Storzer, and J. Keller Spaltspurendatierungen quartärer Gesteinsgläser aus dem Mittelmeerraum. *N. Jb. Miner. Mh.,* 84-94, 1976.

Walker, G.P.L., The Taupo pumice: product of the most powerful known (ultaplinian) eruption. *J. Volcanol. Geotherm. Res., 8,* 69-94, 1980.

Whitham, A., and R. S. J. Sparks, Pumice. *Bull. Volcanol., 48,* 209-223, 1986.

Wilson, C. J. N., The Taupo eruption, New Zealand II. The Taupo Ignimbrite. *Philos. Trans. Royal Soc. London, Ser. A, 314,* 229-310, 1985.

Wilson, C. J. N., The 26.5 ka Oruanui eruption, New Zealand: and introduction and overview. *J. Volcanol. Geotherm. Res., 112,* 133-174, 2001.

Wilson, C. J. N., and G. P. L. Walker, The Taupo eruption, New Zealand. I: General aspects. *Philos. Trans. R. Soc. London, Ser. A, 314,* 199-228, 1985.

Woods, A. W., The fluid dynamics and thermodynamics of eruption columns. *Bull. Volcanol. 50,* 169-193, 1998.

Wohletz, K. H., Explosive magma-water interactions: thermodynaimcs, explosion mechanisms, and field studies. *Bull. Volcanol., 48,* 245-264, 1986.

Wright, I. C., and J. A. Gamble, Southern Kermadec submarine caldera arc volcanoes (SW Pacific): Caldera formation by effusion and pyroclastic eruption. *Marine Geology, 161,* 207-227, 1999.

Wright, I. C, J. A. Gamble, and P. A. Shane, Submarine silicic volcanism of the Healy caldera, southern Kermadec arc (SW Pacific): 1-volcanology and eruption mechanisms. *Bull. Volcanol.,* in press.

Wright, J.V., and E. Mutti, The Dali Ash, Island of Rhodes, Greece: a problem in interpreting submarine volcanigenic sediments. *Bull. Volcanol., 44,* 153-167, 1981.

Yamamoto, T., S. Tatsunori, S. Shigeru, U. Kozo, T. Akira, S. Keiichi, and O. Koji, The 1989 submarine eruption of eastern Izu Peninsula, Japan: ejecta and eruption mechanisms. *Bull. Volcanol., 53,* 301-308, 1991.

Yuasa, M., Myojin Knoll, Izu-Ogasawara Arc: submersible study of submarine pumice volcano. *Bull. Volcanol. Soc. Japan. 40,* 277-284 (in Japanese with English abstract), 1995.

S. R. Allen and A. L Stewart, Centre for Ore Deposit Research, University of Tasmania, GPO Box 252-79, Hobart, Tasmania 7001, Australia

Miocene Submarine Fire Fountain Deposits, Ryugazaki Headland, Oshoro Peninsula, Hokkaido, Japan: Implications for Submarine Fountain Dynamics and Fragmentation Processes

R. A. F. Cas[1], H. Yamagishi[2], L. Moore[1], and C. Scutter[1]

Three in situ Miocene submarine basaltic andesite fire fountain deposits are preserved at Ryugazaki Headland, Oshoro Peninsula, Hokkaido, Japan. Emplacement occurred below storm wave base based on the absence of shallow water tractional reworking structures. The basal facies is a mass flow resedimented pillow clast and spatter clast breccia, indicating early effusive pillow lava and fire fountain eruptive activity upslope from the depositional site. The overlying fire fountain deposits consist of a porphyritic spatter clast breccia facies [oldest], an aphyric spatter micro-breccia facies, and a spheroidal clast breccia facies [youngest]. Only the youngest has a feeder dike exposed. The aphyric spatter micro-breccia facies indicates a high submarine fountain, fed by a high magma discharge, low viscosity magma source. Abundant blocky matrix clasts, indicate pervasive quench fragmentation in the high fire fountain. The fountains for the porphyritic spatter breccia and spheroidal clast breccia were lower, less water was able to enter the fountain, producing less quench fragmentation, and coarser deposits. In submarine fire fountains, the primary fragmentation process is shearing and extensional stretching of the magma fountain as it flares above the vent and decelerates. Quench fragmentation is an important second-order fragmentation process, especially at the margins of the fountain. The low vesicularity of Ryugazaki spatter [generally < 30%] indicates that explosive magmatic vesiculation was not an important process, and there is no direct evidence for phreatomagmatic influences. Spatter debris was deposited by fallout, with only minor evidence for re-sedimentation.

1. INTRODUCTION

Fire fountaining is a low explosive intensity, low dispersal explosive eruption style typically associated with relatively high magma discharge eruptions of low to moderate volatile bearing, low viscosity magma [basalt, phonolite], such as seen on Hawaii [*Macdonald*, 1972; *Walker*, 1973]. Its products are marked by fluidal spatter clasts or bombs, and often agglutinates. Although fire fountaining is usually associated with subaerial settings, reports of deep-water submarine fire fountain deposits or deposits containing fluidal clasts from both modern seafloor settings [e.g. *Kokelaar and Durant*, 1983; *Smith and Batiza*, 1989; *Wright et al.*, 1996; *Clague et al.*, 2000] and the Phanerozoic to Archean geological record [e.g. *Carlisle*, 1963; *Mueller and White*, 1992; *Simpson and McPhie*, 2001; *Fujibayashi and Sakai*, 2002, and this volume] indicate that fire fountaining is also a dynamic subaqueous

[1]School of Geosciences, Monash University, Australia
[2]Department of Environmental Science, Niigata University, Japan.

Explosive Subaqueous Volcanism
Geophysical Monograph 140
Copyright 2003 by the American Geophysical Union
10.1029/140GM20

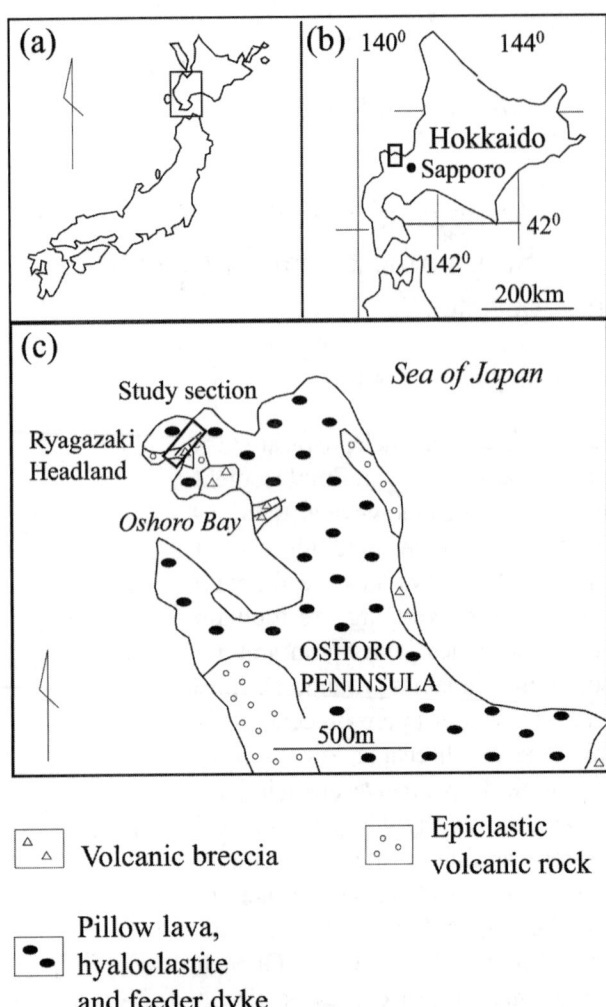

Figure 1. Setting of the Miocene fire fountain deposits of the Oshoro Peninsula [a] in Japan, and [b] in southwest Hokkaido. [c] The simplified geology of Oshoro Peninsula [after *Yamagishi*, 1982].

process. In this study we report on a succession of Miocene fire fountain deposits from the Ryugazaki headland, Oshoro Peninsula, southwest Hokkaido, Japan [Figure 1]. We consider the characteristics of the deposits, the implications for fountain dynamics, and the fragmentation processes associated with submarine fire fountaining.

Ryugazaki headland lies at the northern end of Oshoro Peninsula, which is located 40 km west of Sapporo in southwest Hokkaido, Japan (Figure 1). It is also called Kabuto Cape, but to avoid confusion with another well - known volcanic locality that is also called Kabuto Cape, the alternative local name of Ryugazaki headland is used here. The Ryugazaki succession is part of an extensive belt of Miocene submarine volcanic and sedimentary rocks that extends along the western margin of the Japanese islands,

known as the Green Tuff Belt. This belt is a relic of an extensional basin associated with the separation of Japan from mainland Asia and the opening of the Sea of Japan. The Green Tuff Belt is the principal host for Kuroko - type volcanic hosted massive sulphide (VHMS) deposits in Japan [e.g. *Ohmoto and Skinner*, 1983; *Urabe*, 1987]. An understanding of the eruption processes and facies characteristics of the submarine volcanics may therefore improve exploration concepts for other VHMS deposits [e.g. *Cas*, 1992].

Aspects of the geology of the Oshoro Peninsula region have been documented previously by Yamagishi [1982, 1985, 1987, 1991]. The principal lithofacies include basalt and basaltic andesite pillow and sheet lavas and associated hyaloclastite, spatter deposits, feeder dikes, and re-sedimented volcaniclastic breccias (Figure 2). The succession considered here has been examined only briefly in previous studies by Yamagishi [1987] who recognised the presence of water-chilled spatter bombs, redeposited pillow fragment breccias and considered that the eruption style was analogous to submarine Strombolian activity. The presence of pillow lavas, re-deposited pillow breccias and hyaloclastite indicates a subaqueous setting, which is consistent with the marine nature of the Green Tuff Belt regionally [*Ohmoto and Skinner*, 1983; *Urabe*, 1987].

2. STRATIGRAPHY AND FACIES

The Ryugazaki headland coastal cliffs (Figures 2 and 3) preserves approximately 35 m of variably weathered, bedded, volcaniclastic rocks and a cross-cutting, dike. Dips are up to 20° north with an east-west strike. The geochemistry indicates a reasonably uniform basaltic andesite composition (Table 1) with extreme enrichment of FeO with respect to MgO, indicating tholeiitic affinities. The stratigraphic succession has been subdivided into six principal facies, including the cross-cutting dike (Figure 2):

6) mixed spheroidal clast and spatter clast breccia facies [youngest]
5) basaltic andesite dike [contemporaneous with 4)
4) spheroidal clast breccia facies
3) aphyric spatter clast micro - breccia facies
2) porphyritic spatter clast breccia facies
1) pillow and spatter clast breccia facies (oldest)

The size-texture terms used in the facies names and facies descriptions are descriptive, adopting aggregate terms regularly used by earth scientists and civil engineers to avoid any initial genetic connotations [see *Cas and Wright*, 1987; *McPhie et al*, 1993]. "Breccia" is used for any clastic aggregate in which fragments are mostly greater than 2 to

3 cm in diameter and are angular to irregular in shape. "Micro-breccia" is used for breccias in which most clasts are 2 to 3 cms or less in diameter. Genetic terms such as "lapilli", "tuff", and "agglomerate" are used only after a primary pyroclastic fragmentation *and* depositional origin has been established for a facies [*Fisher*, 1966; *Cas and Wright*, 1987; *McPhie et al.*, 1993].

2.1. Pillow and Spatter Clast Breccia Facies

2.1.1. Description. The basal facies at Ryugazaki is a 12 m succession of crudely bedded pillow and spatter clast breccias, divided into two sub-facies. Sub-facies A is a pillow fragment dominant breccia, whereas sub-facies B is a mixed spatter and angular block breccia. These sub-facies represent a spectrum and so the description of features will be combined. There are approximately 15 identifiable beds which are from 0.5 to 2 m thick, and generally tabular, but variations in clast content and size along some horizons over distances of 5 to 15 m or greater, also define a lensing

Figure 3. Photo of Ryugazaki succession, highlighting the diffusely stratified, tabular to slightly wedging geometry of the lower pillow and spatter clast breccia facies, overlain by the younger stratigraphic units at the top of the cliff. Note L. Moore for scale in upper centre of photo.

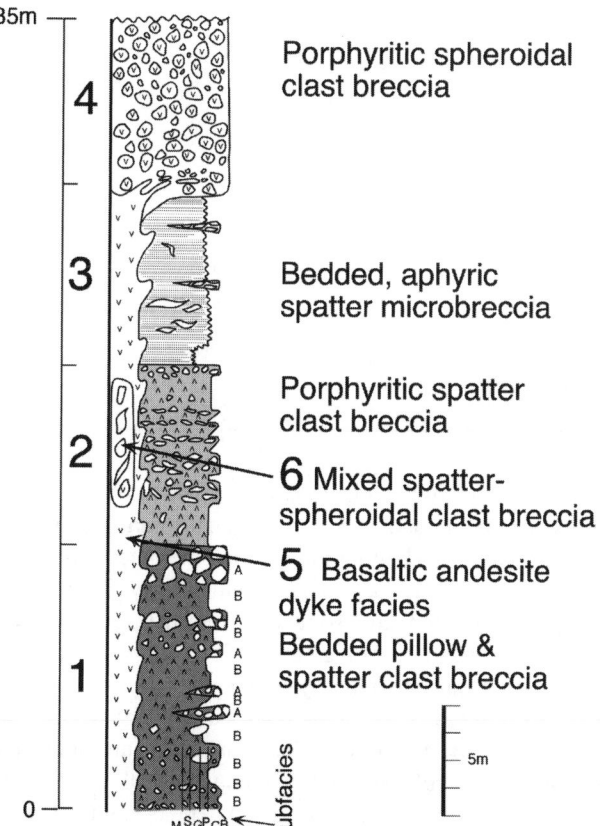

Figure 2. Schematic representation of the stratigraphy and facies architecture of the rock succession exposed at Ryugazaki headland, Oshoro Peninsula.

or wedge geometry (Figure 3). Bedding contacts are diffuse and are defined by changes in the size of the largest clasts, large clast abundance variations and relative proportions of clast types (Figure 3). Beds are massive and non-graded to weakly reverse graded, and sorting is extremely poor. No cross-bedding has been found. Beds are conformable to each other and are overlain conformably by the porphyritic spatter breccia facies. The base is not exposed. All beds are matrix-supported, with a large clast content varying from 20% to 60%. At the top of the reverse graded layers and within the pillow clast dominant breccia horizons, clasts are locally in contact. The matrix is an aggregate of angular pebble sized blocky clasts of the same litho-types as the framework.

Table 1. Geochemical analyses of juvenile clasts and feeder dike in different facies at Ryugazaki, Oshoro Peninsula, Japan. RYU-T[A], [B], [C] and RYU-V are weathered to varying degrees. RYU-U appeared to be freshest in the field.

Sample No: Facies:	RYU-T[A] Matrix of spatter micro-breccia facies	RYU-T[B] Broken bomb, spatter micro-breccia facies	RYU-T[C] Bomb, spatter micro-breccia facies	RYU-U Clast, spheroidal clast breccia	RYU-V Feeder dike
SiO_2	53.10	52.00	50.07	54.93	50.43
TiO_2	0.87	0.88	0.93	0.70	0.77
Al_2O_3	15.46	15.69	17.14	18.64	20.20
Fe_2O_3	12.41	12.37	11.91	8.80	9.58
MnO	0.18	0.16	0.14	0.15	0.15
MgO	3.73	4.04	3.81	3.18	3.36
CaO	6.61	7.09	8.87	9.41	10.79
Na_2O	2.55	2.48	2.32	2.40	2.24
K_2O	0.84	0.72	0.44	0.97	0.45
P_2O_5	0.12	0.10	0.09	0.19	0.18
SO_3	405	428	2091	128	238
Cl	3039	2101	254	581	763
Cr	14	19	20	18	24
Ba	179	241	132	250	148
Sc	35	33	44	26	33
Ce	35	24	36	23	24
Nd	20	14	18	17	14
V	278	268	300	218	271
Co	46	53	50	44	42
Cu	70	61	56	43	47
Zn	96	95	112	81	80
Ni	11	14	12	11	9
Ga	16	16	20	15	13
Zr	53	51	61	69	50
Y	19	15	26	26	26
Sr	219	233	294	284	315
Rb	11	10	9	19	5
Pb	7	8	12	12	3
As	30	17	0	0	10
Loss	3.10	2.69	1.97	0.90	1.29
TOTAL	99.44	98.58	98.04	100.44	99.64

Clast types include whole pillow clasts or pillow fragments up to a metre in maximum dimension, concentric shell jointed spheroidal to elongate clasts, irregular spatter clasts, broken fragments of all of these, and some hydrothermally altered clasts of basaltic andesite (Figure 4). The proportions of clast types vary from bed to bed. Pillow fragments include the round ends of pillow lobes truncated by brittle fracture surfaces, and variably shaped parts thereof, including pie-shaped polygonal clasts. Pillow fragment sizes vary from 15 cm to 1 m. Pillow fragments contain up to 15% plagioclase crystals up to 3 mm long, and up to 5% pyroxene up to 2 mm long. Vesicularity of pillow clasts varies from 5% to 30%. Vesicles include spherical, radial pipe vesicles, and elongate, flow-stretched vesicles. Some preserved surfaces have tortoise shell cooling cracks, ropey wrinkles and chilled margins [cf. *Yamagishi*, 1985, 1994]. Pillow clasts commonly show radial polygonal jointing and some have a concentric shell jointing around the outer margins.

Spheroidal clasts have shapes which vary from nearly spheroidal to elongate, sometimes consisting of several linked spheroids. They range from 25 to 40 cm in maximum dimension. Concentric sheet jointing generally occurs in the outer 50% of the clasts, but pervades almost the whole clast when the clasts are <15 cm. Shells are 2–10 mm thick, and the outermost shells are often incomplete, with pieces having spalled off. The phenocryst content is dominated by 15–20% plagioclase up to 3 mm long, with minor pyroxene crystals. Vesicularity is generally less than 20%.

Irregular spatter clasts or fragments range from 15 to 25 cm long, if whole, down to 1 cm for broken fragments.

Irregular spatter clasts occur in all horizons in varying proportions, from less than 5% in horizons lower in the stratigraphy, to 20% high in this facies interval. There are two types of spatter clasts, one with a relatively high phenocryst content [cf. porphyritic spatter facies described below] and one with low phenocryst content (cf. spatter clast microbreccia facies, below). The crystal poor spatter clasts have <3% phenocrysts of plagioclase up to 2 mm maximum dimension and sparse pyroxene minerals, < 10% vesicles, and they are often oxidised internally to a red-brown colour with yellowish palagonitised or black chilled glassy margins. The crystal rich spatter clasts have up to 20% plagioclase and minor pyroxene phenocrysts, and < 20% vesicles up to 2–3 mm in diameter.

Figure 4. Variations in clast types in the basal pillow and spatter clast breccia. Rounded and blocky fragments are pillow fragments [Pb]; irregular clasts are spatter clasts [Sp]; spheroidal clasts have an outer concentric shell jointing [Sph].

The matrix constitutes 20–80% of this facies. It is composed of blocky clasts ranging from granule to 2–3 cm diameter, but mostly about 1 cm in size. Clast shapes vary from blocky to triangular and curved blade shapes (? spalled concentric shells). Some clasts in the matrix are identifiably concentric shell joint fragments and some blocky spatter clast fragments.

2.1.2. Interpretation. The massive, matrix-supported nature of the beds is consistent with deposition by grain rich, cohesionless debris flows, as is the weak development of reverse grading in some beds [*Lowe*, 1979, 1982]. The diffuse bed boundaries and the relative uniformity of the clast populations are consistent with a reasonably rapid sequence of depositional events. Pillow clasts were formed by either quench fracturing, or gravitational collapse-induced disaggregation. Quench fragmentation of hot erupting pillow lava would have produced a range of grainsizes from boulder to sand size, with clast morphologies varying from curvi-planar spalled fragments to blocky clasts with curvi-planar surfaces [*Cas and Wright* 1987; *Cas*, 1992]. Gravitational collapse would have caused extensional fracturing during initial collapse from the source lava, as well as perhaps abrasion impact fracturing during down slope movement. These processes are also likely to have formed much of the matrix debris that is petrographically similar to the pillow clasts.

Spheroidal, concentric, shell-jointed clasts also appear to have contributed to the matrix by spalling of concentric shells leading to the addition of blade-shaped and fine blocky debris to the matrix, during re-sedimentation. The spatter clasts are comparable to the spatter fragments in the porphyritic spatter clast breccia and the bedded spatter clast micro-breccia facies. Spatter clasts are characteristically formed by fire fountaining (see below for discussion). It thus appears that the pillow and spatter clast breccia facies consists of a mixture of debris from four different deposits, sourced from a succession upslope, mixed and then redeposited downslope by debris flows. Given that fragile spatter clasts are preserved it is likely that the transport distance was very short. Furthermore, since the clasts were not rounded by tractional reworking, derivation from the source[s] and mixing appears to have been caused by gravitational collapse of a wholly submarine succession of multiple facies [pillow lavas and breccias, different spatter deposits], with further mixing during mass flow.

Sub-facies A (pillow fragment dominant breccia) occurs in the upper half of the facies interval and interdigitates with the more polymictic Sub-facies B (spatter clast dominant) at its base, indicating a shift to a more pillow dominant source toward the top of the bedded succession. It is likely that re-

Figure 5. Porphyritic spatter facies, represented by multiple reverse graded depositional layers, each representing fallout from increasing magma discharge, fire fountain, eruption pulses.

near the tops of some beds, but represent only a few percent near the finer bottoms of beds. Spatter clasts vary in shape from oval and "cow-pat " (Figure 6) to irregularly elongate and twisted. Most have a black glassy chilled margin several mm thick, and grey interiors; some have broken margins, with truncated chilled rims. Phenocrysts represent 15–20% and are overwhelmingly 1–3mm long plagioclase phenocrysts; pyroxene crystals represent 5–10%. Glomerocrysts of plagioclase are common. The glassy groundmass has numerous small plagioclase laths. The vesicularity of the spatter clasts varies from 15–25%, based on thin section estimates. The matrix consists of granule to pebble sized, blocky debris, texturally similar to the spatter clasts.

2.2.2. Interpretation. The predominance of fragile fluidal spatter clasts suggests a syn-eruptive derivation for this facies involving fire fountaining of a porphyritic basaltic andesite magma. The reverse graded beds have two possible origins: vent controlled processes, such as increasing magma discharge and fountain height, or mass flow re-sedimentation involving the development of reverse grading through a grain flow mechanism. The absence of internal layering or low angle scours as well as the preservation of delicate bombs, suggests that abrasive grain flow was not responsible for the reverse grading. A fire fountain style of eruption and subaqueous, near-vent fallout, is considered to be the cause of this facies [*Cas et al.*, 1996; cf. *Mueller and White*, 1992; *Simpson and McPhie*, 2001]. The eruption was marked by multiple, short-lived pulses of increasing magma discharge and fountain height, producing increasing dispersal of spatter clasts and reverse size grading. This facies is therefore a primary in situ pyroclastic deposit (cf. pillow and spatter clast breccia facies) and can be given the genetic name, porphyritic spatter bomb lapilli tuff.

sedimentation occurred penecontemporaneously with eruptive activity because there are no signs of lithification of the source materials (e.g. blocks of agglutinated or cemented spatter or breccia), and apart from some apparent alteration in some blocky clasts, all clasts appear to have been first cycle volcanic debris.

2.2. Porphyritic Spatter Clast Breccia Facies

2.2.1. Description. This facies occurs as an 8 m thick interval of 60–80 cm thick beds, which are tabular and coarse-tail reverse graded (Figure 5). The facies is monomictic, consisting of porphyritic basaltic andesite spatter clasts in a matrix of similar blocky granule to pebble size debris. Whole to almost whole spatter clasts constitute up to 50% of the facies

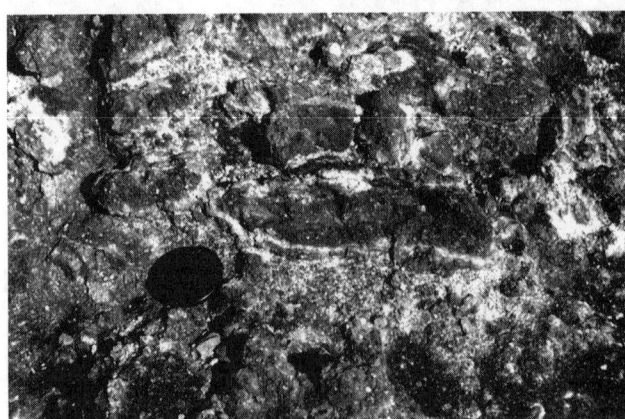

Figure 6. Irregular "cow-pat " form of spatter clasts in the porphyritic spatter clast breccia facies.

2.3. Aphyric Spatter Clast Micro-Breccia Facies

2.3.1. Description. The aphyric spatter clast micro—breccia facies is pervasively thinly planar bedded to laminated (Figure 7). It is approximately 8 m thick, and moderately well sorted except for isolated spatter clasts or trains of spatter clasts within bedding. Strata are generally 3–5 cm thick but vary from millimetre scale laminae to beds 15–20 cm thick. Internally layers are reverse to normally graded. Strata boundaries are diffuse and are marked by gradational grain size change. Occasionally there are low angle truncations in the succession mantled by wedge shaped to planar layers. The package is conformable with the succession above and below. Rare cross-stratification is present, and this is generally low angle (<20°).

Spatter clasts are matrix-supported and constitute only about 15% of the facies, the rest consisting of the finer granule to fine pebble size matrix, hence the textural term used here, micro-breccia. No agglutinated aggregates or deposits are present. Spatter clasts are generally elongate and 10–15 cm long, but range up to 60 cm. They have irregular, fluidal, delicate shapes and wrinkled margins, and are often zoned with black chilled rims and brown-red interiors [Figure 8; cf. *Yamagishi*, 1987]. Clast long axes are usually aligned either parallel to or sub-parallel to bedding, but there are exceptions, including clasts with long axes perpendicular to bedding. Many spatter clasts have internal polygonal to blocky jointing with a spacing of about 1 cm (cf. blocky clasts in matrix). Some spatter clasts are broken and the brittle fractures truncate clasts and their chilled margins (Figure 9). They have generally less than 2–3% phenocrysts, represented only by plagioclase. Vesicularity is about 10% in the chilled margins of spatter clasts but this increases inward, reaching 30–40% generally, but up to 60% in some cases.

Figure 7. Overview of the aphyric, spatter clast micro-breccia facies. Note the generally planar stratification. Notebook for scale.

Figure 8. Irregular, fluidally shaped spatter bomb with black chilled rim and more highly vesiculated core in a matrix of granule to lapilli sized blocky shaped clasts, some of which preserve segments of black chilled rims.

Large multilobate vesicles occur in the centre of many spatter clasts, and many large vesicles are elongate, parallel to the long axis of the spatter clasts. Microscopic vesicles are generally spherical. Some palagonitisation of clast margins is noted.

Matrix grains vary from 1 cm to coarse sand size, and occur in moderately well sorted finer and coarser layers. Fine silt- and clay-sized grains are notably absent. Fragments within the bedded matrix are angular, blocky shaped, and poorly vesiculated. Many of the fragments have chilled margins truncated by brittle fractures. The blocky morphology of matrix grains is similar to the fracture bounded components of the fractured spatter clasts. There is also a similarity of the glass groundmass and the vesicle sizes, morphology and abundance in matrix grains as in the spatter clasts. This facies therefore consists entirely of spatter clasts and fragments thereof.

2.3.2. Interpretation. The abundance of spatter and spatter debris indicate that subaqueous fire fountaining of low viscosity fluidal magma was the dominant eruption and primary fragmentation process for this facies [*Cas et al.*, 1996; cf. *Mueller and White*, 1992; *Simpson and McPhie*, 2001]. The low crystal content indicates eruption at near liquidus temperature, consistent with a low magma viscosity. The spatter bombs formed as very fluidal, irregular magma clots with wrinkled surfaces, which were fragmented through plastic extension and shear in a rapidly discharging fire fountain. Chilled rims demonstrate rapid chilling of spatter in the fountain. Spatter clasts are often concentrated in horizons, suggesting a pulsating eruption, and are usually aligned within bedding. Some however, are sub-vertical,

Figure 9. Broken spatter clast with internal polygonal fractures in a matrix consisting of identical polygonal blocky fragments of spatter debris.

cross cutting several layers consistent with vertical fallout as bombs. The large multilobate vesicles in the centre of many spatter clasts indicate that vesicle coalescence occurred and that vesicle expansion continued in the interiors after chilled margins formed. The elongate vesicles suggest that vesiculation occurred early, prior to clast formation and elongation. The microscopic spherical vesicles are consistent with late vesiculation inside the spatter clasts.

Broken spatter clasts indicate a secondary brittle fracturing process, either in the fountain or upon impact with the depositional surface. The matrix is composed of broken blocky fragments of spatter, which also indicate a secondary brittle fracturing process. Quench fragmentation, or cooling contraction granulation [*Kokelaar*, 1986] of spatter by seawater or steam at the margins of the fountain are considered to be the most likely origin of the brittle fracturing. The pervasive stratification in this facies is consistent with water settling of a nearly continuous, but pulsating, rain of fine, blocky, quenched spatter debris onto the depositional surface by near vent fallout. Planar diffuse layering, transitional grainsize changes from reverse to normal and back again, and good sorting are consistent with progressive aggradation throughfallout. This may also be the explanation for the fines depleted nature of the matrix. That is, whatever fine ash size debris was formed in the fountain or subaqueous column may have been separated from the coarser debris in the fallout column, and transported by turbulence into the upper column and then dispersed laterally by oceanic currents. Discrete beds with reverse grading, low angle truncations and wedging, and rare cross-bedding indicate some down-slope transportation or re-sedimentation, perhaps even syn-fallout generated turbidity current forming as dense packages of fallout debris rain onto

the seafloor and continue to move down-slope under the influence of the momentum acquired during fallout [cf. *Carey*, 1997]. However, since the bulk of the facies is a primary fire fountain fallout deposit, an appropriate genetic name is aphyric spatter bomb lapilli tuff, with the qualifying "re-sedimented" applied to beds that have been re-sedimented.

2.4. Spheroidal Clast Breccia Facies

2.4.1. Description. The basaltic andesite spheroidal clast breccia is monomictic, poorly sorted, massive, and at least 8 m thick. It is connected to a subjacent dike [feeder] and conformably overlies the aphyric spatter clast micro - breccia facies. The facies is matrix-supported and consists of 50-60% spheroidal, irregular and angular blocky clasts, ranging from pebble to boulder size, with 50–40% matrix consisting of angular pebble to granule size debris (Figure 10). Silt and clay size matrix is absent. Spheroidal clasts commonly have both marginal concentric shell jointing and internal polygonal jointing. Many spheroidal clasts are broken, truncated by brittle fracture surfaces. Angular clasts are bounded by planar to curviplanar surfaces and many appear to be derived from disintegration of polygonally jointed larger clasts. Some clasts have triangular to curved blade-like shapes. Clasts contain up to 30% 2–3 mm long plagioclase, plus 10% pyroxene. Glomerocrysts of plagioclase are common. Clast vesicularity ranges from 0–10% and vesicles are spherical to stretched in shape. Pipe vesicles are absent. The groundmass is fresh and glassy.

The facies grades down into a dike of basaltic andesite at one locality. Coherent basaltic andesite in the dike grades up into in situ, jigsaw-fit textured, basaltic andesite breccia to

Figure 10. Spheroidal bombs of the spheroidal clast breccia. Note the concentric shell jointing and blocky nature of the matrix, which consists of fragments spalled off spheroidal clasts.

lobes or apophyses that upwards intrude or cross cut the spheroidal clast breccia, to supra-dike breccia consisting of necked-off lobes and spheroidal clasts of basaltic andesite. The dike is described further below.

2.4.2. Interpretation. The eruption mode for this facies appears to have varied from limited fire fountaining to dike fed effusion, and perhaps even explosive bursts. Some of the spheroidal clasts are interpreted as bombs comparable to subaerial spindle/fusiform bombs. There are also spheroidal clasts and a number of lobes/pillows, which are necked off by quenching from the subjacent dike. Both bombs and lobes experienced quench fracturing. Direct evidence of this process is observed where lobes branch off the top of the dike, extend into the body of the clastic deposit and are necked off and have in situ jigsaw fracture textures. Quenching produced concentric shell jointing (thermal shock exfoliation), polygonal fracture systems (cooling contraction fractures) and production of angular pebble sized blocky matrix clasts.

Emplacement modes include an initial low fire fountain with near vent fallout producing an agglomerate of spheroidal clasts, with high depositional settling rates suppressing sorting leading to massive deposits. Quench fracturing of bombs, while still hot, both during their trajectory through the water and after deposition, led to concentric shell jointing, breaking, and spalling of debris from the bombs, producing matrix. Late in the eruption intrusion from the feeder dike into the agglomerate produced lobes and necked off quench-fragmented debris above the feeder dike.

2.5. Basaltic Andesite Dike

2.5.1. Description. A major dike cross cuts all of the stratigraphy up to the top unit and merges into this unit, the mafic spheroidal clast breccia (Figure 11). The dike is vertical, trends north- east to south-west, and is between 1 to 2 metres wide. It is preserved over a distance of approximately 100 metres, with exposure of both ends terminating at coastal cliffs and rock platforms. The north-east to south-west trend is parallel to other nearby dikes on Oshoro Peninsula suggesting a tectonic control on dike orientation.

Although the dike is a vertical tabular body, its margins are highly irregular in detail and at one locality its trace has a dextral jog of 0.5 metres. Both margins vary from planar to undulating to lobate. Lobes or apophyses protrude up to 1 metre into the enclosing strata. Lobes are up to 1 metre long, but most are about 0.5 metres long. In addition, small faults displace the lobate margins by up to 20 cms. The top of the dike is gradational into the overlying spheroidal clast

Figure 11. Feeder dike to the spheroidal clast breccia, cross-cutting all other facies. The margins of the dike are irregular, marked by intrusive lobes and apophyses, indicating that the whole succession was unconsolidated at the time of intrusion. Note: H. Yamagishi for scale.

breccia (Facies 4) and irregular (see above). The margins of the dike everywhere have outward facing chilled margins up to 10 cms wide, which are recognised by a sharp outer glassy tachylite layer, grading inward into the grey basaltic andesite of the dike interior. The dike rock is massive to flow banded, with flow banding parallel to the dike margins. In addition, *within* some parts of the dike, there are two outward facing chilled margins preserved, each grading from an outer glassy margin to a more crystallised interior. Black glassy clasts of the chilled margin are mixed into the adjacent host volcaniclastic at numerous places along the margin.

The internal architecture of the dike is complex. It consists of two sub-facies:

a) coherent basaltic andesite, and
b) jigsaw fit, monomictic basaltic andesite breccia

In addition, an enclave of mixed spheroidal clast and spatter clast breccia occurs within the dike (see below).

The coherent basaltic andesite contains up to 30% plagioclase and up to 10% pyroxene phenocrysts, with phenocryst size up to 4 mm. Glomerocrystic plagioclase aggregates are common. Vesicularity is variable reaching up to 30%. The jigsaw-fit, monomictic, basaltic andesite breccia sub-facies is randomly distributed throughout the length of the dike. Domains of this facies vary in size from local patches as little as 20 cm in length to zones metres in extent. This breccia grades into coherent dike rock. Local clast rotation of smaller clasts occurs. This breccia is noted for the development of palagonite particularly in the smaller grain sizes. Petrologically, clasts are identical to the coherent dike rock. Clast size is mostly 5–15 cm, but finer grain sizes also occur.

2.5.2. Interpretation. The planar nature of the dike may be controlled by basement tectonic fractures along which the magma injected. However, the irregular lobate margins indicate that the host stratified volcaniclastic facies succession already described had limited strength and cohesion, was not fully lithified, and that dike intrusion was contemporaneous with volcanism and deposition. But the enclosing strata had some strength, because of the generally linear trace of the dike and its generally uniform thickness along its length.

At least two generations or phases of magma intrusion are evidenced by two, outward facing chilled margins in some places. Glassy black clasts along the dike margin have spalled from the dike margin indicating some quench fragmentation and the presence of water in the unconsolidated enclosing sediment.

The composition and texture of the dike rock is similar to clasts in basaltic andesite spheroidal clast breccia at the top of the succession and the gradational contact between the two indicate the dike was a feeder dike and vent to the fire fountain produced spheroidal clast breccia.

The jigsaw fit nature of the monomictic basaltic andesite breccia sub-facies in the dike indicates in situ quench fragmentation [*Cas and Wright*, 1987]. The generally blocky to granular nature of fragments is consistent with quench fragmentation. Irregular distribution of domains of this breccia throughout the dike indicates access of external water into dispersed parts of the core of the dike, presumably resulting from localised quench fragmentation along the margins causing cracks to form that allowed water to penetrate from the sides and/or from above. The mixing of dike margin chilled glassy fragments in the enclosing host stratified facies, also indicates the host was unconsolidated at the time of dike intrusion. Fluidisation mixing of dike clasts laterally into the host volcaniclastic succession occurred through heating and convective mixing of pore water in the volcaniclastics when the magma intruded through the volcaniclastics [cf. *Kokelaar*, 1982]. This is thus a good example of a peperite with a pene-contemporaneous juvenile clast matrix.

2.6. Mixed Spheroidal Clast and Spatter Clast Breccia Facies [Youngest]

2.6.1. Description. The mixed spheroidal clast and spatter clast breccia occurs in one locality as an internal domain, within the dike. This domain has a window-like shape (Figure 12) on the 30° sloping rock outcrop. The domain is about 4 m long and 1.5 m wide and is sharply bounded by coherent basaltic andesite dike all around (Figure 12). The margin of the domain is an *inward* facing chilled margin of the dike. At this locality, the outer margins of the dike also have normal outward facing chilled margins. In three dimensions therefore the domain represents an oblique section through a sub-vertical chimney-like structure filled with the mixed clast breccia facies. It occurs at the stratigraphic level of the top of the pillow and spatter clast breccia and the bottom of the porphyritic spatter clast breccia.

The facies consists of a mixture of two clast populations. The first consists of porphyritic basaltic andesite spheroidal clasts and angular blocky fragments of the same lithology. This lithology is identical to the dike rock and the overlying spheroidal clast breccia to which the dike is a feeder. The second clast population consists of irregular aphyric spatter clasts with chilled rims, and fragments of these, and is identical to the spatter debris of the aphyric spatter clast microbreccia facies. These two clast populations occur in approximately equal proportions but distribution patterns are heterogeneous. The spheroidal clasts are generally larger, up to 25 cm long, whereas the largest spatter fragments are 15 cm long. Large clasts of these lithologies are supported in a matrix of finer, angular debris of the two lithologies. Sorting in this breccia is poor and there is no indication of tractional abrasion of clasts.

2.6.2. Interpretation. The continuous nature of the inward facing chilled margin that bounds the domain containing the breccia deposit indicates that the breccia in-filled a cavity within the dike. Alternatively the domain represents a part of volcanic stratigraphy described here that was intruded by and engulfed by the dike, isolating it from its parent stratigraphic horizon. However, there are two arguments against the latter interpretation. First there are no clasts [inclusions/xenoliths] of the breccia in the surrounding dike. Secondly, the clast population is not similar to the immediately adjacent stratified succession outside the dike. The two principal clast types were derived from the two uppermost stratigraphic units, the aphyric basaltic andesite micro-breccia facies, and the spheroidal clast breccia, which were mixed and transported downwards. The envisaged, if speculative, scenario for the origin of this isolated breccia domain in the dike therefore is:

(a) first, intrusion of the dike and continuous fire fountaining along the dike produced the basaltic andesite spheroidal clast breccia.

(b) secondly, local waning of fountaining and draw down of magma within the dike produced a vertical, chimney shaped cavity inside the dike and open to the seafloor at the top. Chilling of magma lining the cavity produced inward facing glassy chilled margins.

(c) finally, gravitational collapse by rockfall from the upper walls of the dike conduit led to infilling of the cavity by clasts from the aphyric basaltic andesite micro-breccia facies and the basaltic andesite spheroidal clast breccia facies higher up in the walls of the dike (Figure 13).

3. DISCUSSION

3.1. Depositional and Palaeogeographic Setting

There is an absence of intercalated ambient sedimentary facies that could indicate depositional setting in all Oshoro Peninsula successions However, all the known facies are submarine volcanics marked by abundant pillow lavas and hyaloclastites [e.g. *Yamagishi*, 1987]. The presence of pillow lava clasts in the volcaniclastics interpreted as mass flow deposits in the lower part of the Ryugazaki succession indicates derivation from a submarine volcanic source or succession further upslope. Preservation of delicate surface features on pillow-clasts (glassy chilled margins, ropey wrinkles, tortoise shell jointing) as well as irregular spatter indicate no tractional abrasion occurred prior to, during or after deposition of any of the facies. The absence of abundant tractional depositional structures in the Ryugazaki volcanic succession indicate that deposition occurred below storm wave base in relatively deep water, although the exact depth is unknown.

The presence of spatter and spatter deposits indicates fire fountaining and a relatively close proximity to source vents based on (a) preservation of delicate shapes and surface features and (b) preservation of large spatter bombs up to 25 cm maximum dimension.

Although fire fountaining is generally associated with sub-aerial volcanism (e.g. Hawaii), fire fountaining deposits have been recorded from the modern deep seafloor [e.g. *Smith and Batiza*, 1989; *Wright*, 1996] and the geological record [e.g. *Mueller and White*, 1992; *Simpson and McPhie*, 2001]. In sub-aerial settings fire fountain deposits are dominated by large spatter and irregular, ragged scoriaceous debris. The finer character of the Ryugazaki spatter succession and the blocky and fine nature of matrix debris is attributed to quench fragmentation of spatter clasts in the subaqueous eruption fountain through interaction with seawater [cf. *Mueller and White*, 1992]. The absence of agglutinate is also notable, indicating spatter and matrix debris were too

Figure 12. Polymictic breccia in a pocket within the dike. This pocket is bound by an inward facing chilled margin, whereas the margins of the dike are outward facing.

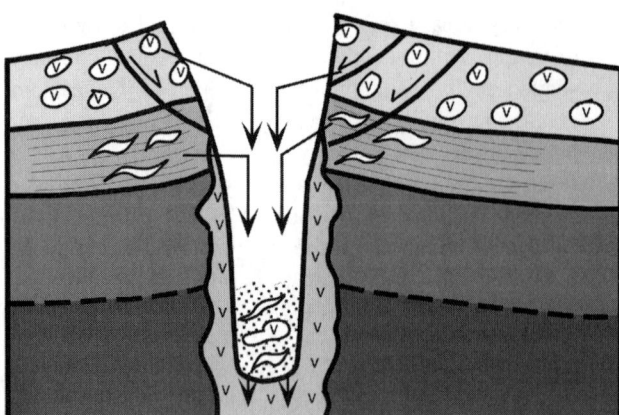

Figure 13. Schematic scenario for the formation of the isolated domain of mixed spheroidal clast and aphyric clast breccia facies preserved in a chimney-like cavity in the basaltic andesite dyke. During final stages of fire fountaining that produced the spheroidal clast breccia deposit [uppermost unit], magma draw down in one localised part of the dyke produced a vertical chimney-like cavity open to the seafloor. Collapse from the upper walls of the cavity incorporated debris from the aphyric spatter breccia [second top unit] and spheroidal porphyritic clast breccia horizons, which fell down into, and filled, the cavity.

cold upon deposition to agglutinate (cf. subaerial spatter deposits).

The absence of interbedded basinal sediments and intraclasts of these in the Ryugazaki succession implies that the eruptions occurred from vents elevated above the normal seafloor, implying the existence of an elevated volcanic edifice. Deposition of the preserved succession must also have occurred on the flanks of this edifice. Given the widespread occurrence of mafic pillow and sheet lavas and feeder dikes, this edifice was probably a submarine lava shield seamount. Eruptions that produced the Ryugazaki succession probably occurred from fissure vents on the flanks of the edifice.

3.2. Eruption Styles: Fire Fountaining or a More Explosive Eruption Style?

The lowest facies of the Ryugazaki succession, the pillow and spatter clast breccia, preserves debris mixed from two different eruption styles, pillow lava forming and spatter forming fire fountaining. The fragmented nature of both components and the massive to diffusely stratified nature of the succession indicate mass flow re-sedimentation from sources and a vent system upslope.

By contrast, the three succeeding facies (porphyritic spatter clast breccia, aphyric spatter clast micro-breccia facies, spheroidal clast breccia , basaltic andesite dike) are all primary, in situ, fire fountain deposits. Two magma types were involved, one producing an almost aphyric population of spatter, and a second producing two deposits of coarsely porphyritic debris. In the absence of any mixing of clast types in these three discrete deposits it is logical to consider the two magma types having been erupted from spatially separated fissure vent systems. Facies characteristics are consistent with fallout deposition and only a limited degree of downslope re-sedimentation, indicating that the vents were nearby. The vent for the uppermost facies, the spheroidal clast breccia is visible in the preserved succession as the dike.

Subaerial Hawaiian fire fountaining is usually classified as the least explosive eruption style [e.g. *Walker*, 1973]. The products range from variably vesiculated, very fluidal, irregular clots of spatter or fragments of spatter to fluidally shaped glassy achnelith fragments, and variably vesiculated scoria.

The principal mechanism producing fragmentation appears to result from the high velocity jetting of the fluidal, highly pressurised basaltic magma out of a narrow conduit or vent. As the magma column leaves the high confining pressure domain of the lithosphere into the low pressure domain of the atmosphere it expands laterally or flares. As a consequence magma extends and stretches and begins to disintegrate, through plastic stretching, into irregular magma clots or spatter fragments. Differential velocity in the fountain from the conduit margins to the centre also causes shearing of the magma. At the upper margins of the fountain, finer debris is air quenched and may be subject to brittle fracturing as a result of quenching and impact upon landing. The high degrees of vesiculation of the interiors of spatter fragments relative to poorly vesicular to chilled margins indicate that much if not most of the vesiculation and/or vesicle growth in fire fountain spatter occurs late, after fragmentation. Gas exsolution and explosive bursting of gas bubbles appears to play a significantly lesser role in the actual fragmentation of fire fountain deposits than in other more energetic explosive eruption styles.

High velocity fountaining of rising basalt magma results from its fluidal nature and high internal fluid pressure. The high fluid pressure can be attributed in the first instance to a combination of high lithostatic and magma-static pressures imposed upon the magma in the source area in the mantle. Secondly, vesiculation, even as little as 10–20% microvesiculation, will cause a volume expansion that triggers immediate acceleration of the magma upwards within the conduit toward the vent [e.g. *Sparks et al.*, 1994].

In subaqueous environments the hydrostatic pressure (10^5 Pa/10 m) will act to retard physical growth of vesicles with increasing water depth. At increasing water depths the potential for explosive expansion of vesicles should be significantly less than from a subaerial vent. In the Ryugazaki case minor vesiculation, generally up to 30%, but locally up to 60%, for the aphyric spatter clast micro-breccia, indicates that gas exsolution and expansion played some role in the formation of a high velocity magma fountain from a seafloor vent. In addition to fragmentation resulting from expansion of the magma column and shearing and stretching of the magma to form fluidal clots, the entry of the fountain into a cold water mass should also have produced significant quench fragmentation of the magma and the spatter fragments in the fountain.

In the aphyric spatter micro-breccia facies dominant fine blocky fragments that were derived by break-up of internally fractured spatter clasts indicate that the spatter clasts must have quickly formed a chilled margin after leaving the vent and then must have been subject to quench cooling and fracturing in the fountain leading to a rain of blocky debris around the vent. The abundance of blocky matrix debris indicates that external water had continuous access to most parts of the fountain.

If fragments had been formed by explosive vesicle growth, significant irregularly shaped scoria should occur, which it doesn't . If phreatomagmatic explosive fragmentation was a major influence in fragmentation, then blocky shaped clasts may have formed [e.g. *Heiken and Wohletz*, 1985], but this would not explain the common occurrence of chilled margins on one side of matrix clasts. Phreatomagmatic explosive interaction usually produces abundant very fine ash debris, of which we find no evidence. In addition, some accretionary lapilli might be expected from such activity, but none were found. Finally, phreatomagmatic explosions commonly produce ballistic clasts derived from disruption of the vent walls, but these are not apparent. Nonetheless some degree of phreatomagmatic explosiveness cannot be completely ruled out.

The spatter clasts occur either in distinct concentrations along particular bedding horizons or as isolated outsized clasts with a matrix of bedded blocky fragment lapillistone. The concentration of spatter clasts within specific horizons reflects increased magma effusion rates and rapid fallout from the lower part of the fountain and rapid burial by spatter debris so limiting the quench fragmentation effects of the ambient water mass. The isolated clasts probably also represent fallout from the lower part of the fountain and rapid burial by finely quenched debris from the upper part of the fountain. A fallout origin for the large spatter clasts is also supported by the vertical long axis orientation of many of these spatter fragments. The majority have long axis orientation parallel to bedding indicating that upon impact they toppled over.

Kokelaar and Durant [1983] accounted for the spatter clasts in the deposits of the wholly submarine cone of Surtla, a satellite of Surtsey, as being a product of shallow submarine explosive activity which generated a steam cupola above the vent into which magma spatter clots were ejected, cooled and deposited. The Surtsey eruption was witnessed and was clearly explosive. *Smith and Batiza* [1989] proposed fire fountaining at depths of about 2,000 m at Seamount 6 near the East Pacific Rise to account for fluidal spatter clasts, but did not comment on the role of volatile induced vesiculation in the fragmentation process. *Gill et al.* [1990] proposed that explosive deep-water fragmentation produced vesiculated basaltic debris and spatter fragments at depths of about 3,000 m in the Sumisu Basin south of Japan. *Clague et al.* [2000] similarly attributed vesiculated basaltic deposits with associated fluidal spatter clasts at the deep-water Lo'ihi centre [> 1,000 m] to explosive fragmentation. *Mueller and White* [1992] advocated the *Kokelaar and Durant* [1983] steam cupola model to account for the formation of rhyolitic spatter clasts and associated debris in Archean deposits at the Hunter Mine in the Abitibi Greenstone belt of Canada. They attributed associated finer blocky debris to hydrovolcanic fragmentation of magma spatter at the margins of the submarine fire fountain and steam cupola. It is unclear if they meant explosive, or quench fragmentation, or both. *Cas et al.* [1996] suggested that explosive fragmentation was not necessary to account for the Ryugazaki Headland succession, suggesting that fire fountaining and quench fragmentation at the margins of the fountain, as proposed in this paper, was responsible for generation of the spatter and associated debris. Although some component of submarine phreatomagmatic explosive activity may have contributed to the formation of the blocky matrix debris at Ryugazaki, the evidence is not compelling, as discussed above. *Simpson and McPhie* [2000] also attributed the generation of blocky debris associated with basaltic-andesite fire fountain deposits of Cambro-Ordovician age in the Trooper Creek Formation in Queensland, Australia, to quench fragmentation at the margins of the submarine fire fountain.

Although some element of a steam cupola may have formed at the margins of the fire fountain, as proposed for the Hunter Mine succession by *Mueller and White* [1992], perhaps this is not necessary to produce spatter subaqueously. As a fountain flares above a vent and the magma column begins to expand sideways, stretch and disintegrate into spatter, in the first instance it is likely that spatter clots will be separated by domains of rarefied gas exsolving from the magma. Only at the margins of the fountain are spatter clasts likely to interface with water or steam.

3.3. A Fire Fountain Model for the Ryugazaki Spatter Succession

The Ryugazaki Headland spatter deposit succession preserves three in situ fire fountain deposits, but for only the youngest, the spheroidal clast breccia, is the vent preserved in the form of the basaltic andesite dike. This dike is laterally continuous and therefore appears to have fed a fissure eruption producing a laterally continuous fountain. There is no evidence that this dike was the vent for the lower two spatter deposits, the porphyritic and aphyric spatter deposits. However, as *Kokelaar and Durant* [1983] have pointed out, fire fountain deposits are likely to be deposited only within close proximity (tens of metres) from vent, so the vents for the other two spatter deposits are likely to have been close to the presently preserved outcrops. This is supported by the generally coarse nature of some of the spatter, the thick accumulations preserved, and the primary fallout characteristics of the deposits. Therefore two older, closely spaced vents were active nearby, and we infer that they were probably sub-parallel fissure vent systems. All vent systems were located on the flanks of a large submarine volcanic edifice such as a lava shield.

We adopt a generally similar fountain eruption model to that proposed by *Mueller and White* [1992]. However, this can be refined for the Ryugazaki case. The differences in the different spatter deposits suggest that the fire fountains producing them varied in their dynamics and characteristics (Figure 14).

The model we envisage for the aphyric spatter micro-breccia facies (Figure 14a) involves the high velocity jetting fountain of low viscosity, aphyric, high (liquidus) temperature, fluidal basaltic andesite magma into the ambient water mass. The highly fluidal nature of the low viscosity magma induced first order primary extensional stretch and shear induced fragmentation of the fountaining magma as it flared out of the vent to produce relatively small spatter clasts, usually no more than 20 cm in maximum dimension (lower part of fountain in Figure 14a). As the magma fountain disintegrated immediately above the vent, spatter clasts in the centre of the fountain were enclosed in discontinuous domains of gas exsolved from the magma. At the margins of the fountain there was probably a gradation from gas in and at the immediate magma fountain margins to a mixture of gas and supercritical steam to a zone of supercritical steam to heated water to the colder water of the oceanic water mass. If the hydrostatic pressure in the water mass, controlled by water depth, was greater than the pressure regime in the physically heterogenous fountain, with its dispersed gas pockets, water would ingress into the fountain in irregular ways (Figure 14a) and be variably converted to steam. Alternatively, and/or in addition, turbulence at the margins of the fountain could cause mixing in of water. The fountain would have accelerated up the conduit above the level of volatile exsolution resulting from volume expansion up the conduit, and increased buoyancy and fluid overpressure resulting from vesiculation (see schematic graphical representation of density contrast, Dr, acceleration, dv/dt, and fluid overpressure, DP, Figure 14a). Upon flaring above the vent, deceleration and rapidly declining overpressure in the fountain led to fallout of debris around the vent, leading to accumulation of a near-vent fire fountain fallout deposit (Figure 14a).

Spatter entering the cooler marginal regimes of steam and water would be subject to chilling and quenching. The relatively high fountain would have had a large surface area so enhancing the capacity of the external water mass to interact with it and cause quite pervasive and efficient second order quench fragmentation. Abundant blocky matrix debris and the occurrence of internal brittle fractures in the whole, intact spatter fragments indicate that spatter fragments were rapidly chilled by contact with cold water. Disaggregation must have occurred in the fountain and/or in the water mass leading to fallout deposition producing diffusely stratified deposits. The diffuse stratification could be due to a pulsing fire fountain, and/or pulsing or cyclical ingress of water into the fountain producing pulsing quenching fronts at the margins of the fountain. Some grainflow re-sedimentation may also have occurred. Some fragmentation of spatter may also have occurred on impact as a third order fragmentation process. However, this appears to account for less debris. Subsequent syn-eruptive grain flow may also have disrupted such clasts.

A significant inferred difference exists in the nature of the fire fountain that produced the underlying package of reverse graded beds of porphyritic spatter. The porphyritic spatter fragments are in general larger and more abundant than the aphyric spatter fragments, and some porphyritic bombs are spheroidal. We attribute this difference to variation in magma properties and fountain height. The porphyritic debris (up to 30% phenocrysts) indicate a sub-liquidus eruption temperature whereas the aphyric spatter would have been erupted at near-liquidus temperature. Sub-liquidus temperature would have caused the magma viscosity to be relatively high. The higher viscosity of the magma is likely to have reduced the magma discharge rate, leading to a lower fountain. Secondly, the higher viscosity is likely to have reduced the ease with which fragmentation due to plastic stretching and shearing could have occurred. It is likely that larger primary clasts are formed as a result of

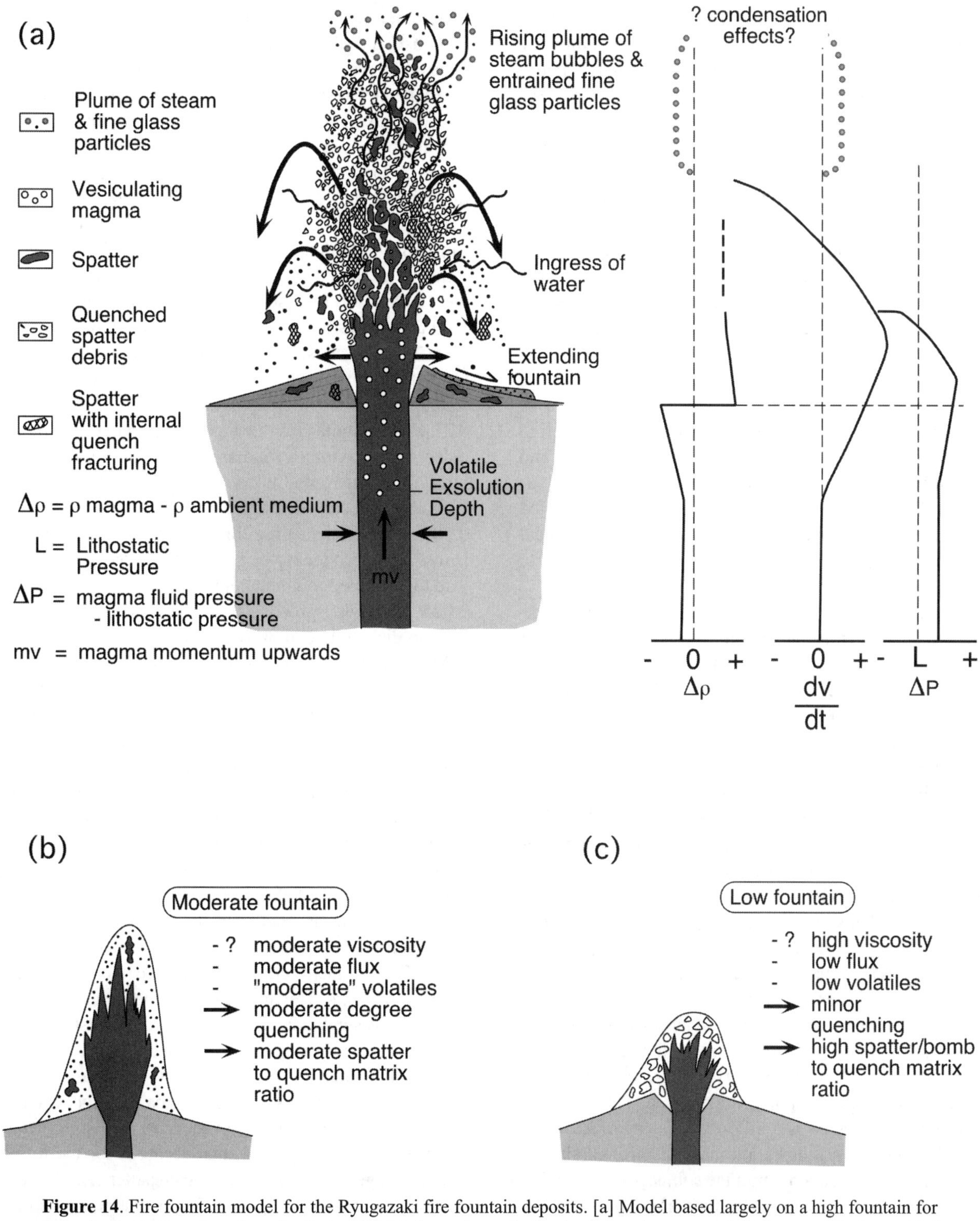

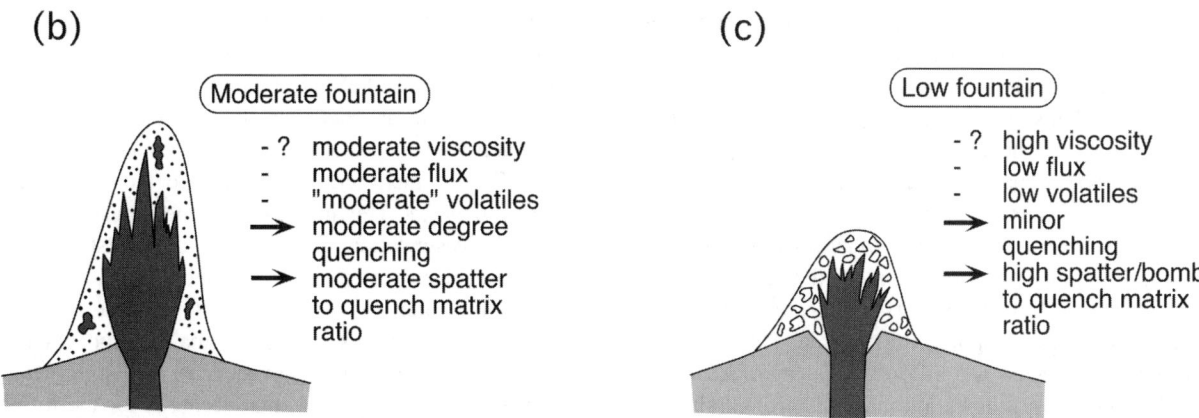

Figure 14. Fire fountain model for the Ryugazaki fire fountain deposits. [a] Model based largely on a high fountain for the aphyric spatter micro-breccia facies. [b] Moderate fountain for the more viscous porphyritic spatter facies and [c] Very low fountain for the spheroidal clast breccia.

this. The overall coarser nature of the porphyritic spatter deposits indicates less effective quench fragmentation by the enclosing water mass compared with the aphyric spatter micro-breccia facies. We therefore consider that the porphyritic spatter facies resulted from a lower fountain eruption (Figure 14b), whereby the time available and the efficiency of quench fragmentation were limited. Rapid fallout from a low column around vent would lead to preservation of large spatter bombs.

The reverse grading in some intervals of the porphyritic spatter facies can be attributed to changing magma effusion rates and efficiency of quench fragmentation. As for the aphyric spatter micro-breccia facies the finer grained base of these reverse graded beds is interpreted to result from relatively high fountains and efficient quenching of the debris in these fountains. Conversely the coarser grain tops of these reverse graded beds is interpreted to reflect a lowering of the fountain showing waning fountain eruption and reduced effectiveness of quench fragmentation. The repetition of several such reverse graded beds indicates repeated resurgence or pulsing of fire fountaining. This spheroidal concentric shell jointed bombs appear to be analogous to subaerial fusiform or spindle bombs.

The fountain for the spheroidal clast breccia is interpreted to have been even lower than for the porphyritic spatter breccia fountain (Figure 14c). The overall grainsize is coarser, and the proportion of matrix is less. In part this is also due to the proximity of the eruption point. However, the gradational contact between the feeder dike and this facies suggests that magma discharge rate was not great, especially at the end of the eruption. Even the larger bombs, however, show evidence of quenching as reflected by widespread preservation of a concentric shell jointing, interpreted to be an exfoliation, cooling jointing. Some shell sheets at the margins of the clasts are truncated by the brittle fractures indicating that some debris has spalled off the margins during eruption and deposition. In addition, adjacent to some clasts, obvious spalled debris is preserved. This facies is least like a fire fountain deposit of all the facies described because clasts are predominantly spheroidal to elongate with multiple small spheroidal connected nodes, and the vesicularity is low (<20%). It superficially resembles pillow lava and pillow clast hyaloclastite breccia. However, no clasts are interconnected as a pillow tube system. All clasts are smaller than most pillows and have spheroidal shapes, and the deposit is gradational into the subjacent feeder dike. It is therefore interpreted as a low fountain deposit, resulting from the minimum magma discharge required to produce a fountain.

Although *Head and Wilson* [1987] considered that subaerial basaltic fire fountain heights were predom-nantly determined by volatile content, *Parfitt et al.* [1995] have demonstrated that magma volume flux and amount of degassed lava re-entrainment (e.g. due to drain back of fused, degassed spatter magma back into the vent after eruption) are also significant factors in determining lava fountain height, based on an analysis of the 1983–1986 fire fountain eruption interval on Pu'u 'O'o, in Hawaii. Furthermore they found that contrary to the model of *Vergniolle and Jaupart* [1990], carbon dioxide concentration was not the force driving fountains from the conduit, since measured ratios of exsolved gases from erupting Hawaiian magmas were 85% H_2O, 12% SO_2 and 3% CO_2. Maximum exsolved H_2O content varied between 0.07 wt% to 0.37 wt%. *Parfitt et al.* [1995] consider that fragmentation of the magma fountain occurred when vesicle content reached a volume fraction of 75%. Given an average exsolved gas content of 0.32 wt%, magma rise rates of greater than 0.087 m/s coincided with maximum fountain heights because under these high flow rate conditions gas bubble coalescence was unlikely to occur. At magma rise rates between 0.024 and 0.087 m/s sufficient time for some gas bubble coalescence is available, producing large decoupled gas pockets, which lead to a reduced fountain height. At magma rise rates below 0.024 m/s widespread gas coalescence should lead to Strombolian style explosive activity.

More recently *Head and Wilson* [2003] have analysed and modelled the factors determining the dynamics of deep water explosive eruptions, and in particular deep water fire fountain eruptions. They recognise that in deep water settings, with high ambient hydrostatic pressures, the role of carbon dioxide as a driving explosive volatile fraction may be dominant over magmatic water because it exsolves before water does, at higher ambient pressures. They consider that fragmentation of the erupting magma in submarine fountains will occur at vesicularity levels of about 75%, at which stage bubble coalescence will occur. However, the vesicularity of both subaerial and subaqueous fire fountain pyroclasts is often well below 75%, which certainly applies to the Ryugazaki fire fountain deposits. This indicates that mechanisms other than explosive vesiculation to levels of 75% must occur. Two processes are suggested here. First, in both subaerial and subaqueous fire fountains of low viscosity magma, the development of shear stresses in the magma jet, and extensional stresses as it flares above the vent must be a significant first order fragmentation process that accounts for the formation of many large magma fragments and spatter clasts. Secondly, in subaqueous settings, quench fragmentation of magma and magma fragments at the margins of the fountain at least, must contribute significantly to second order fragmentation. For the aphyric spatter facies, it is thought that this secondary process accounts for the large volume of relatively fine blocky vitric debris. The low vesicularity of this debris raises doubts as to whether or not explosive vesiculation played any major role in the actual fragmentation of the magma fountain. Vesiculation however clearly would have caused the rising magma to accelerate up the conduit to generate the fountain.

At Ryugazaki Headland, we do not know the original nor exsolved water and carbon dioxide content. However, hydrostatic pressure would have reduced the degree of gas exsolution relative to a subaerial counterpart. Vesiculation levels are well below the 75% cited by *Parfitt et al.* [1995] for fountain fragmentation. Fountain height was probably controlled by both magma discharge rate and exsolved volatile content. The aphyric spatter micro-breccia facies contains the most highly vesiculated spatter [25-30%], whereas the spheroidal clast breccia facies contains clasts with very low vesicularities [<10%], accounting for our interpretation of differing fountain heights for the three spatter forming eruptions (Figure 14).

4. CONCLUSIONS

(1) The Miocene submarine volcanic succession preserved at Ryugazaki Headland, Oshoro Peninsula, Hokkaido, Japan is dominated by the products of basaltic submarine fire fountaining.

(2) Differences in clast compositions and textures indicate that two magma sources and three spatially separated vent systems, probably all fissures, were active sequentially and in close proximity.

(3) Inferred variations in magma effusion rates during fire fountaining have been identified from detailed facies analysis of the fire fountain deposits. Low magma discharge rates and low fountain heights have been inferred from coarse clast dominant, porphyritic spatter and spheroidal clast deposits. Relatively high magma discharge rates and fountain heights have been inferred from bedded, finely fragmented lapilli deposits of aphyric spatter. The finer grainsize of the aphyric spatter deposit is attributed to more efficient quench fragmentation of spatter in a high fountain.

(4) Variations in magma properties such as viscosity and vesicularity influence fire fountain dynamics and pyroclast properties, especially magma discharge rate, column height and clast morphology.

(5) Comparison between the characteristics of subaerial fire fountaining and submarine fire fountaining deposits suggest that a significant difference could be higher degrees of fragmentation and finer grained deposits near vent in subaqueous settings due to the quenching effects of the ambient water mass. Although quench fragmentation is usually associated with subaqueous effusive lava forming events and emplacement of syn-sedimentary sills, quenching has been an important fragmentation influence during more dynamic subaqueous fire fountaining at Ryugazaki.

(6) An important primary fragmentation mechanism associated with fire fountaining, perhaps in both subaerial and subaqueous, is shearing and extensional stretching of the fountain as it flares form the vent. Although violent, explosive gas vesicle bursting may be an important fragmentation mechanism in many fire fountains, evidence for it is lacking in the Ryugazaki succession. In subaqueous settings, quench fragmentation may be an important secondary fragmentation mechanism of the fountain and its hot debris.

(7) The one feeder dike propagated through the bedded sequence when that sequence was still at least semi-unconsolidated. The occurrence of an internal chimney-like cavity within the dike, filled with fall-in debris derived from above, indicates that pulsing and periodic draw down marked the magma discharge history during the final preserved fire fountain event.

Acknowledgments. RAFC acknowledges the Australian Research Council for award of a Large Grant to support this research. Draga Gelt and Steve Morton are thanked for assistance with figure preparation. Kirsty Simpson and Wulf Mueller are thanked for their thorough and helpful reviews, and David Clague for his patience as an editor.

REFERENCES

Carey, S., Influence of convective sedimentation on the formation of widespread tephra fall layers in the deep sea. *Geology*, 25, 839-842, 1997.

Carlisle, D., Pillow breccias and their aquagene tuffs, Quadra Island, British Columbia. *J. Geol.*, 70, 48-71, 1963.

Cas, R.A.F., Submarine volcanism: eruption styles, products, and relevance to understanding the host-rock successions to volcanic-hosted massive sulphide deposits. *Economic Geology*, 87, 511-541, 1992.

Cas, R.A.F., Moore, C.L., Scutter, C. and Yamagishi, H., Fragmentation processes and water depth significance of Miocene submarine fire fountain deposits, Ryugazaki, Hokkaido, Japan. Abstract, *AGU Western Pacific Geophysics Meeting*, Brisbane, p. W 126, 1996.

Cas, R.A.F. and Wright, J.V., *Volcanic successions, modern and ancient.* Allen and Unwin, London. 528pp, 1987.

Clague, D.A., Davis, A.S., Bischoff, J.L., Dixon, J.E. and Geyer, R., Lava bubble wall fragments formed by submarine hydrovolcanic explosions on Lo'ihi seamount and Kilauea volcano. *Bull. Volcanol.* 61, 437-449, 2000.

Fisher, R.V., Rocks composed of volcanic fragments and their classification. *Earth Sci. Rev.*, 1, 287-298, 1966.

Fujibayashi, N. and Sakai, U., Subaqueous fire fountains of basalt. Abstract *Amer. Geophys. Union Chapman Conference Explosive Subaqueous Volcanism*, Dunedin, New Zealand, p. 19, 2002.

Gill, J., Torssander, P., Lapierre, H., Taylor, R., Kaiho, K., Koyama, M., Kusakabe, M., Aitchison, J., Cisowski, S., Dadey, K., Fujioka, K., Klaus, A., Lovell, M., Marsaglia, K., Pezard, P., Taylor, B., and Tazaki, K., Explosive deep—water basalt in the Sumisu backarc rift. *Science, 248*, 1214-1217, 1990.

Head, J. and Wilson, L., Lava fountain heights at Pu'u' O'o, Kilauea, Hawai'i : indicators of amount and variations of exsolved magma volatiles. *J. Volc. Geotherm. Res.*, 92, 13715-13719, 1987.

Head, J.W. III and Wilson, L., Deep submarine pyroclastic eruptions: theory and predicted landforms and deposits. *J. Volc. Geotherm. Res.*, 121, 155-193, 2003.

Kokelaar, B.P. and Durant, G.P., The submarine eruption and erosion of Surtla [Surtsey], Iceland. *J. Volcanol. Geotherm. Res.*, 19, 239-246, 1983.

Kokelaar, B.P., Magma—water interactions in subaqueous and emergent volcanism. *Bull. Volcanol.*, 48, 275-290, 1986.

Lowe, D.R., Sediment gravity flows: their classification and some problems of application to natural flows and deposits. In Doyle, L.J and Pilkey, O.H. [eds] *Geology of continental slopes.* Soc. Econ. Pal. Mineral. Spec. Publ., 27, 75-82, 1979.

Lowe, D.R., Sediment gravity flows: II, depositional models with special referebce to the deposits of high densith turbidity currents, *J. Sed. Petrol.*, 52, 279-297, 1982.

Macdonald, G.A., *Volcanoes.* Prentice-Hall, New Jersey, 510 pp, 1972.

McPhie, J., Allen, R. and Doyle, M., *Volcanic Textures.* CODES, University of Tasmania. 196 pp, 1993.

Mueller, W. and White, J.D.L., Felsic fire fountaiing beneath Archean seas. *J. Volcanol. Geotherm. Res.*, 54, 117-134.

Ohmoto, H. and Skinner, B.J. [eds] 1983. The Kuroko and related volcanogenic massive sulfide deposits. *Economic Geology Monograph 5*, 604 pp, 1992.

Parfitt, E., Wilson, L. and Neal, C.A., Factors influencing the height of Hawaiian lava fountains: implications for the use of fountain height as an indicator of magma gas content. *Bull. Volcanol.*, 57, 440-450, 1995.

Simpson, K. and McPhie, J., Fluidal-clast breccia generated by submarine fire fountaining, Trooper Creek Formation, Queensland, Australia. *J. Volc. Geotherm. Res.*, 109, 339-355, 2001.

Smith, T.L. and Batiza, R., New field and laboratory evidence for the origin of hyaloclastite flows on seamount summits. *Bull. Volcanol.*, 51, 96-114, 1989.

Sparks R.S.J., Barclay, J., Jaupart, C., Mader, H.M. and Phillips, J.C., Physical aspects of magma degassing 1, Experimental and theoretical constraints on vesiculation, *in* Carroll, M.R. and Jolloway J.R. [eds] Volatiles in Magmas, *Reviews in Mineralogy*, 30, 413-445, 1994.

Urabe, T., Koroko deposit modelling based on magmatic hydrothermal theory. *Mining Geology*, 37, 159-176, 1987.

Vergniolle, S. and Jaupart, C., Dynamics of degassing at Kilauea volcano, Hawaii. *J. Geophys. Res.*, 95, 2793-2809, 1990.

Walker, G.P.L., Explosive volcanic eruptions—a new classification scheme. *Geol. Rundsch.*, 62, 431-446, 1973.

Wright, I., Volcaniclastic processes on modern submarine arc stratovolcanoes: sidescan and photographic evidence from the Rumble IV and V volcanoes, southern Kermadec Arc [SW Pacific]. *Marine Geology*, 136, 21-39, 1996.

Yamagishi, H., Miocene subaqueous volcaniclastic rocks of the Oshoro Peninsula, Southwest Hokkaido, Japan. *Jour. Geol. Soc. Japan*, 88, 19-29, 1982.

Yamagishi, H., Growth of pillow lobes—evidence from pillow lavas of Hokkaido, Japan, and North Island, New Zealand. *Geology, 13,* 499-502, 1985.

Yamagishi, H., Studies on the Neogene subaqueous lavas and hyaloclastites in Southwest Hokkaido. *Rep. Geol. Surv. Hokkaido*, 59, 55-117, 1987.

Yamagishi, H., Morphological and sedimentological characteristics of the Neogene submarine coherent lavas and hyaloclastites in southwest Hokkaido, Japan. *Sediment. Geol.*, 74, 5-23, 1991.

Yamagishi, H., *Subaqueous volcanic rocks.* Hokkaido University Press, Sapporo, Japan, 1994.

R. A. F. Cas, L. Moore and C. Scutter, School of Geosciences, Monash University, P.O. Box 28E, Victoria, Australia, 3800.

H. Yamagishi, Department of Environmental Science, Faculty of Science, Niigata University, Igarashi 2-no-chou 8050950-2181, Niigata, Japan.

An Archean Submarine Pyroclastic Flow Due to Submarine Dome Collapse: The Hurd Deposit, Harker Township, Ontario, Canada

C. R. Scott

Sciences de la Terre, Université du Québec à Chicoutimi, Chicoutimi, Québec, Canada

D. Richard and A. D. Fowler

Department of Earth Sciences, University of Ottawa, Ottawa, Ontario, Canada

The Hurd Volcaniclastic Deposit lies within the Kinojevis Assemblage, a group of submarine volcanic rocks dominated by Mg-and Fe-rich metamorphosed pillow-basalt. The volcaniclastic deposit can be subdivided into two facies, a lower unorganized massive facies and an upper imbricated facies. The lower facies is composed of cm-to-m-scale flow-banded dacite blocks interspersed with angular to curved-to-blocky and equant mm-to-cm-scale fragments of altered glass; it has a sharp but undulatory basal contact with underlying massive to lobate dacitic lavas. Some blocks are partly composed of in-situ breccia, and agglomerations or curvilinear arrays of spherulites. This lithofacies contains well-developed columnar joints that terminate at the sharp contact with the ~2m thick upper imbricated facies. With the exception of numerous large imbricated tabular clasts, the imbricated facies is dominated by blocky to angular mm-to-cm-scale fragments. The presence of pillows, chert, jig-saw fit breccias, and microscopic quench textures are evidence of a subaqueous environment of deposition. Similarly, the presence of columnar joints, plastically deformed clasts, quench breccias, and possible gas escape structures indicate hot emplacement, perhaps by a gas-supported pyroclastic flow. Together, the massive and imbricated facies represent one flow unit that was quickly emplaced. The lower facies was deposited by a high-density flow isolated from the aqueous environment by the development of a steam carapace, whereas the overlying facies was emplaced by a turbulent, aqueous flow. The onset of magmatic fragmentation may have been the result of sudden decompression possibly triggered by the gravitational collapse of a submarine dacite dome.

1. INTRODUCTION

The nature of transport within, and deposition from, subaqueous pyroclastic flows, either those of subaerial origin [*Whitham*, 1989; *Mandeville et al.*, 1996; *Kokelaar and Königer*, 2000] or originating from completely submarine eruptions [*Kokelaar*, 1992; *Fritz & Stillman*, 1996], remains a contentious issue [*Cas and Wright*, 1991]. The continued debate stems from uncertainty related to the extrinsic conditions and mechanics of subaqueous eruptions, and is fueled by the paucity of irrefutably primary pyroclastic flow deposits emplaced in a submarine environment. Direct observations of entirely subaqueous eruptions are absent and only recently have pyroclastic flows entering the sea have been well documented, from Soufriere Hills, Montserrat [*Young et al.,* 1997], though some information is also available from historical events such as the eruption of Mt. Pelée, Martinique [*Lacroix,* 1904]. Subaerially produced pyroclastic flows that reach the coast may only have limited interaction with water. For instance, they may glide over the water surface as observed from video of pyroclastic flows flowing out of Tar River Valley in Montserrat [e.g. *Allen,* 2001], or segregate into an overriding turbulent cloud with an advancing underwater subaqueous density current [*Carey & Sigurdsson,* 1980; *Mandeville et al.,* 1996].

Typically a submarine pyroclastic flow will transform from an initial gas-supported, high-density, dominantly laminar current to a water-supported, high-density turbulent aqueous current due to water ingestion [*Fisher,* 1983; *White,* 2000]. Resulting deposits tend to be better sorted due to the viscosity contrast between gas and water [e.g. *Stix,* 1991] forming stratified beds that can be doubly graded [*Fiske & Matsuda,* 1964]. Critical to the unambiguous recognition of a subaqueous pyroclastic flow deposit is the confirmation of a hot-state of deposition from a gas-supported flow and documentation of a submarine setting, based on bounding lithofacies [*White and McPhie* 1997]. Possible indicators of high temperature emplacement for ancient deposits include welding, columnar jointing, and gas segregation pipes [e.g. *Cas and Wright,* 1991], as well as supportive microscopic evidence such as perlitic fractures and spherulitic crystallization [*McArthur et al.,* 1998].

The few reported examples of subaqueous primary pyroclastic flow deposits are typically associated with large volume, caldera-forming eruptions, either subaqueous [*Busby-Spera,* 1984; *Kokelaar & Busby,* 1992] or initially subaerial. Products of these eruptions flowed into shallow, intertidal settings [*Howells et al.,* 1985; *Kokelaar & Königer,* 2000], smaller subaqueous eruptions produced submarine lava fountaining [*Mueller & White,* 1992], and some subaerial eruptions have emplaced relatively thin welded ignimbrites in subaqueous environments [*White & McPhie,* 1997]. All these deposits lack one or more of the diagnostic features given above [*Cas and Wright,* 1991] typically used to support subaqueous deposition by a primary pyroclastic flow, rather than deposition of localized hot fragments. The one Archean example [*Mueller and White,* 1992] includes agglomerated welded clasts, but lacks columnar joints.

In this paper, we examine a thin Archean subaqueous volcaniclastic sequence (known locally as the Hurd property after the late Mr. D. Hurd who exposed the outcrops for prospecting) composed of a massive basal facies and overlying stratified facies. The evidence for both hot and subaqueous emplacement conditions leads us to the interpretation that the volcaniclastic sequence was deposited by a submarine pyroclastic flow.

1.1 Regional Geology

Rocks in the vicinity of the volcaniclastic sequence belong to the Kinojevis Assemblage (Figure 1) deposited about 2701 Ma [*Ayer,* 2002]. These rocks are associated with metakomatiite and are dominated by thick successions of tholeiitic metabasalt, mostly pillow basalt and related facies, with very subordinate amounts of interflow metasediment. The Kinojevis rocks are interpreted to be analogous to those found in modern mid-oceanic ridge environments [*Jackson et al.,* 1994]. The degree of metamorphism experienced by rocks within the area was mild, ranging from prehenite-pumpellyite to lower greenschist facies. Indeed, it is not too unusual to find fresh forsterite preserved in the ultramafic successions. The rocks are weakly deformed except in proximity to the major faults or intrusions of the region. Delicate volcanic textures, for instance spherulites, spinifex texture, shards and perlitic fractures have experienced little to no deformation as they apparently preserve their original geometry. Although the rocks have all been metamorphosed, for the sake of brevity, from here onward we drop the prefix "meta."

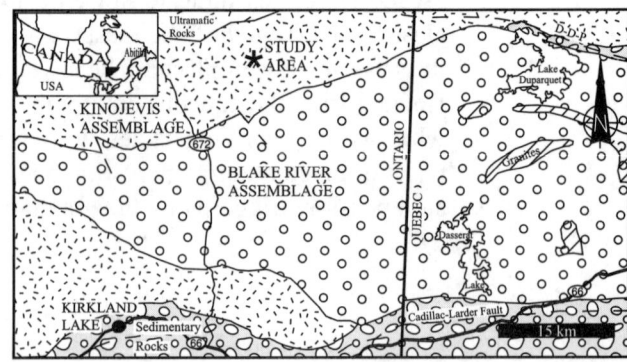

Figure 1. Simplified regional geology map depicting the general location of the study area. D-D-P is the Duparquet-Destor-Porcupine Fault. Modified from mapping by the Ontario Geological Survey.

1.2 Terminology

Subaqueous pyroclastic flows are defined as hot, high-particle concentration density currents that are gas-supported, explosively produced, and deposited in a subaqueous setting [e.g. *White*, 2000]. They are at least in part effectively isolated from water, and thereby retain their high temperatures and gas-supported nature. Increasing hydrostatic pressure with water depth limits volatile expansion and may restrict large volume eruptions to shallow depths of <1 km. Usage of the term volcaniclastic follows the non-genetic definition of *Fisher* [1961] that is, a rock composed of a mixture of volcanic and non-volcanic fragments regardless of specific mode of volcanic fragmentation and transportation. Volcaniclastic deposits are described using the granulometric scheme of *Fisher* [1966], whereby a tuff is lithified ash fragments <2 mm, lapilli fragments are 2–64 mm, and breccia contains fragments mostly >64 mm.

2. STRATIGRAPHY OF THE VOLCANO-SEDIMENTARY SEQUENCE

Locally, the section is composed of a southward younging volcano-sedimentary sequence comprising massive to pillowed basalt, massive and lobate dacite, and a dacite volcaniclastic lapilli tuff to tuff breccia (Figure 2).

Compositions were determined using REE and other trace elements as well as the rather immobile oxides TiO_2, FeO and Al_2O_3 [*Jones*, 1992]. The pillowed basaltic lithofacies facilitates identification of the depositional environment, but emphasis is here placed on the dacitic volcaniclastic lithofacies well exposed in a 75 m x 45 m stripped outcrop (Figures 2 and 3) and exposed in trenches over a total distance of >500m.

2.1 Mafic Lavas and Cherty Horizon

These lithofacies are located lower in the stratigraphy than the volcaniclastic deposit and immediately underlie the massive and lobate dacite lava flows. A cherty horizon, composed of very fine-grained silica, was observed over 500 m as a discrete band or as discontinuous pods. This chert horizon is overlain by a sequence of massive to pillowed mafic lavas (Figure 2).

2.2 Massive Dacitic Lavas

A sequence of dacitic lavas and subordinate interstratified fragmental lithofacies form the base of the outcrop mapped in detail (Figures 3 and 4A). These lavas can be subdivided into three probable flow units based on contact relationships and textural features.

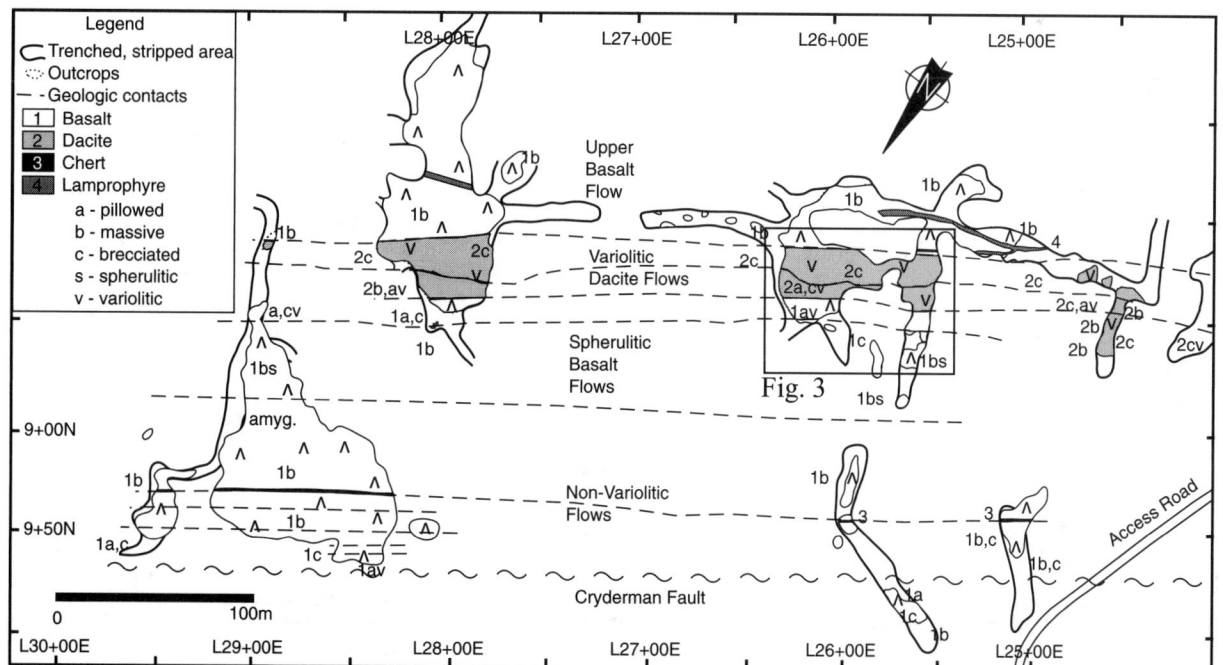

Figure 2. Local geology of study area with pillowed flows and interstratified chert indicating a deep marine environment. Area of interest highlighted. Modified from *Jones* [1992].

The lower flow unit is aphyric and almost non-vesicular (~2%, visual estimate) with possible lobate structures towards its upper contact. Overlying this flow unit is a thin brecciated lithofacies composed of lapilli- to tuff-sized monolithic fragments that appear to be compositionally similar to the underlying lava. Fragments commonly contain individual mm-sized spherulites and perlitic fractures. This brecciated facies typically grades into a closely packed, jigsaw pattern fragmental unit of angular to subangular fragments with quenched margins.

The subsequent lava flow is relatively thin and homogeneous. Its upper contact is defined by a linear band of mm-sized spherulites at the base of the next lava flow.

The overlying massive flow unit is also relatively homogeneous, but with several flow bands outlined by mm-sized spherulites. Large irregular carbonate pockets near the upper contact fill possible lithophysae cavities. Vesicularity decreases up section, from approximately 5%, 2mm diameter vesicles at the base, to 2%, 1mm diameter ones at the top. Its upper contact is irregular, but sharp, and varies from being in contact with a volcaniclastic lithofacies to the east versus with a lobate dacite lithofacies to the west (Figures 3 and 4A).

This sequence of lavas and intercalated brecciated facies records several individual flows, with the interstratified lapilli tuffs interpreted as flow top breccias produced by quench fragmentation during interaction with external water [e.g. *Batiza and White,* 2000]. The in-situ and jigsaw character of lapilli tuffs demonstrate that little or no reworking occurred following their formation.

2.3 Lobate Dacite Facies

The lobate facies is restricted to the western side of the main outcrop and measures up to 5 m in thickness (Figure 4A). Individual lobes vary from 0.5 to 2 m in width and form flattened to more classic convex-upward shapes, which probably attests to variations in viscous deformation. Nevertheless, all lobes close towards the contact with the overlying volcaniclastic facies, in a generally easterly direction. We observed products of in-situ brecciation inside lobes, tortuous flow banding and numerous spherulites

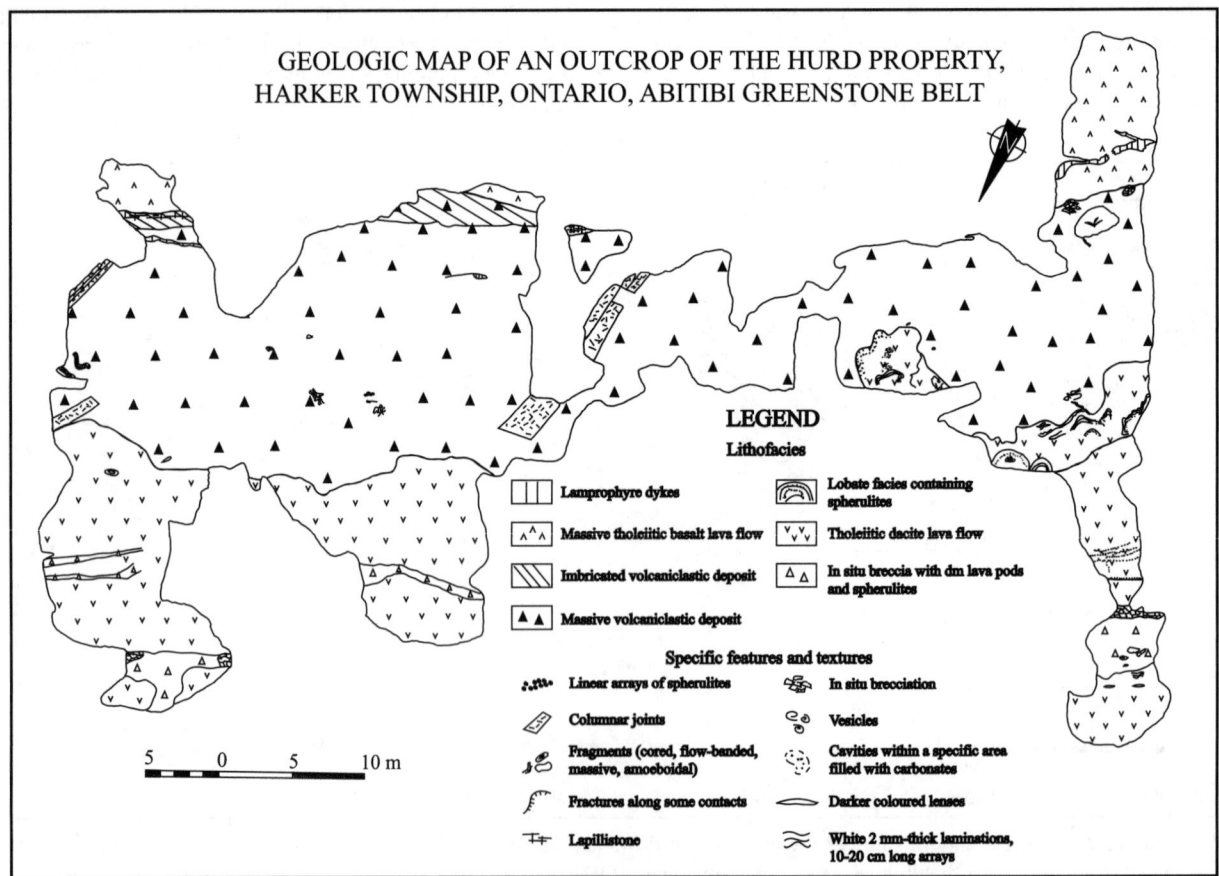

Figure 3. Detailed geology map of the main area of the Hurd volcaniclastic deposit.

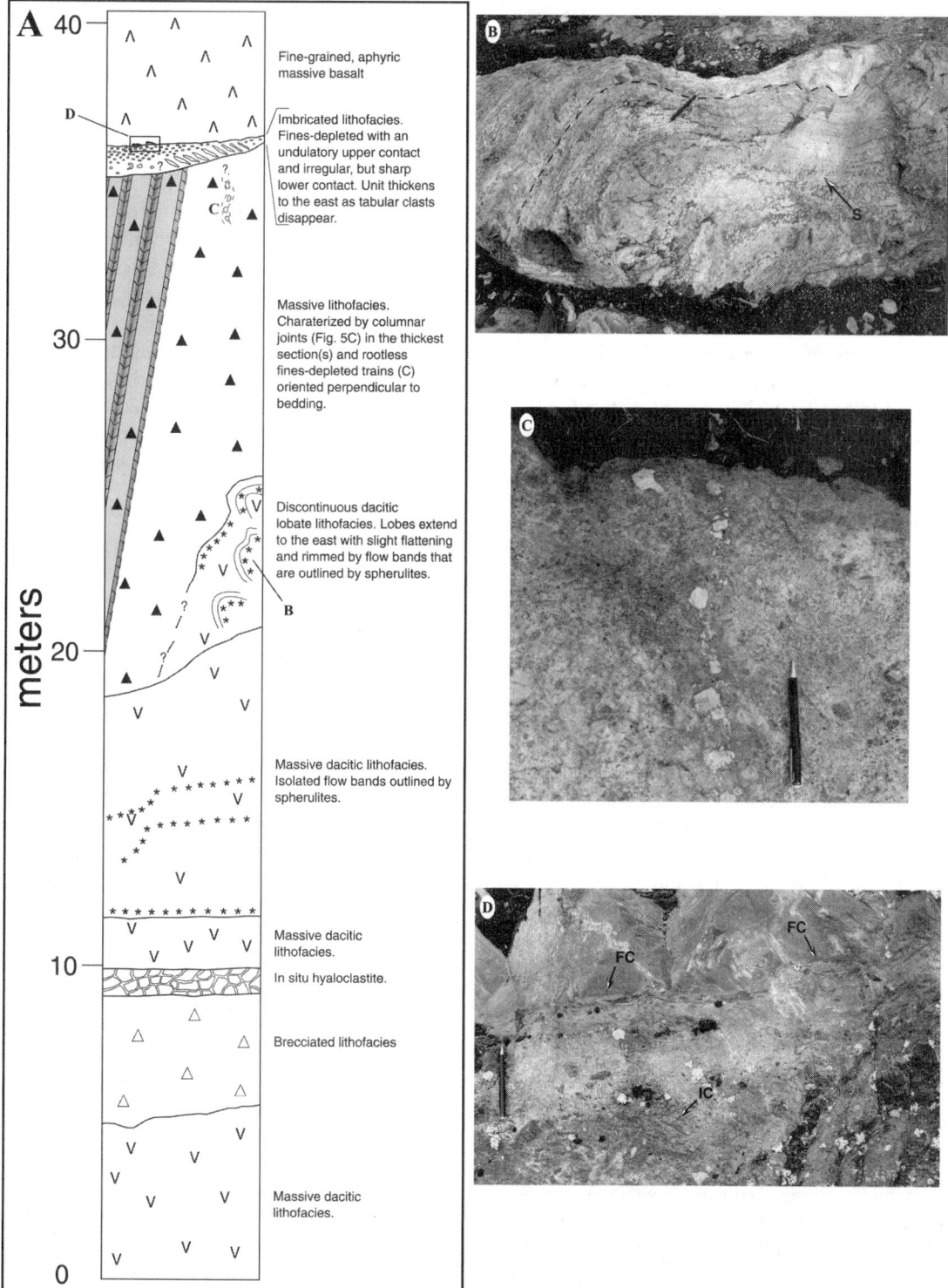

Figure 4. A) Stratigraphic column of mapped area illustrating contact relationships and local features. B) Individual lobe that is slightly flattened. Flow bands are defined by spherulites (*) and outlined by dashed line. Pencil is 12 cm long and points in the younging direction. C) Train of altered clasts oriented perpendicular to bedding. Pencil is 12 cm long and pointing in younging direction. D) Contact relationship between imbricated lithofacies and overlying massive basalt. Contact is characterized by fluidal textures at the base of the basalt (FC). Several imbricated clasts are also visible (IC). Pencil is 12 cm long and pointing in younging direction.

either in agglomerations or in linear arrays (Figure 4B). The lower contact of the lobate facies with the more massive dacitic lava flow is sharp and relatively straight (Figures 3 and 4A) and defined by the appearance of lobate structures. The upper contact with the overlying volcaniclastic deposit is sharp, but irregular and commonly disturbed.

2.4 Volcaniclastic Lithofacies

The volcaniclastic lithofacies forms a continuous stratigraphic horizon > 500 m in an east-west direction with a reasonably consistent thickness of approximately 15 m (Figure 2). This fragmental unit is composed exclusively of volcanic clasts. Stratigraphic and physical volcanologic relationships permit the subdivision of this lithofacies into a massive lower part and an imbricated upper division.

2.4.1 Massive lower lapilli tuff breccia. The roughly 10-14 m-thick lower unit forms a massive, non-graded, poorly sorted lapilli tuff breccia composed of volcanic debris of identical composition to the overlying imbricated division. The variable thickness is a function of the uneven lower contact with the dacite lobate and/or massive flows (Figure 3) and development of the overlying imbricated division. Contact with the lobate facies is highly contorted and disturbed, to the point that it appears to 'intrude' into the volcaniclastic facies (Figures 3 and 4A). However, no peperitic textures are observed. Contact with the overlying imbricated division is sharp and commonly associated with a mm-thick, dark green alteration band. Furthermore, the massive part is altered a dark gray-green as opposed to the lighter gray alteration of the upper division.

Volcanic fragments include former vitrophyric shards, gray flow banded clasts, and subordinate vesicular fragments. Overall fragments are aphyric, with only rare quartz microphenocrysts observed in larger fragments (~ 30 cm) or liberated in the matrix. This matrix-supported sequence is fairly homogeneous, with sorting and grading absent. The matrix is composed of relict, ash-sized angular to subangular shards or by an alteration mosaic of fine-grained quartz, albite and/or carbonate. Former glass shard morphology ranges from angular and elongate, to slightly curvilinear, blocky and equant, along with rare cuspate shards. However, because of the alteration, the shape of shards less than about 10 µm in size cannot be resolved. Most shards appear intact, with moderate degrees of flattening (Figure 5A). Larger glass fragments (4 mm) are characterized by spherulites or concentric fractures (perlitic cracks; Figure 5B). Spherulites comprise axiolitic or plumose forms that originate from point sources. Groups of axiolitic spherulites are organized into spectacular 'feather-like' patterns.

Vesicular and gray, flow-banded to massive fragments range from lapilli- to breccia-sized. Vesicular fragments are subordinate, making up < 2% of the facies. These fragments are typically lapilli-sized and recognized by intense epidote to chlorite alteration, which gives them a yellowish-green hue on the outcrop. Larger pumice fragments are subspherical to oval in shape, whereas smaller fragments form elongated shreds with irregular serrated terminations, some having cuspate outlines due to fracturing along vesicle walls. Vesicularity ranges from 10-20% (visual estimate), represented by < 1 mm quartz amygdules. Vesicle deformation is moderate, whereby former vesicles are subspherical to elongate in form, but no collapsed vesicles were recognized (i.e., fiamme), which may be an artifact of alteration. Gray, phenocryst-poor, flow-banded clasts are found in a wide range of sizes, from 2 mm up to a maximum of 40 cm, throughout the sequence. Alteration prevents a reliable estimate of their abundance, but it is probably up to 30%. Vesicularity is low, as evident by 5% quartz amygdules. Clast groundmass is typically brown or green in thin section, depending on the degree of alteration to carbonate, chlorite, and/or epidote. The characteristic flow bands are defined by linear arrays of spherulites similar to those observed in the underlying lobate facies. This, together with its similar alteration and vesicularity, suggests that these clasts are probably dacitic in composition as determined by *Jones* [1992]. Fragment morphology is varied. Some are very irregular; others are subrounded that are outlined by bands of spherulites, or subangular tabular to equant forms.

The columnar joints (Figure 5C) range from 0.2 to 2.0 m-wide (~10m long) and are restricted to the thicker sections of the sequence, being oriented perpendicularly to the lower and upper surfaces of the unit. Also noteworthy is a linear, rootless, 40 cm-long fines depleted zone, consisting of white altered fragments, that cuts the stratigraphy at right angles and terminates at the contact with the overlying imbricated unit (Figures 4A and 4C). We interpret this to be a gas-escape structure.

2.4.2 Imbricated upper division. This is also a lapilli tuff breccia, but is coarser grained and shows conspicuous clast imbrication (Figure 5D). Its thickness is variable, but seems to thicken to the east to a maximum of about 1.5 m. The upper contact with the basalt is a sharp contact over the entire exposed length. Fragmental material varies from breccia to tuff, but the latter is minor, making up less than 10% by volume (visual estimate). Imbrication is defined by breccia-to lapilli-sized clasts concentrated along the lower contact with the massive lower division, with clasts oriented at a near constant angle of 30° to the underlying contact with the massive division (Figure 4A and 5D). The size of

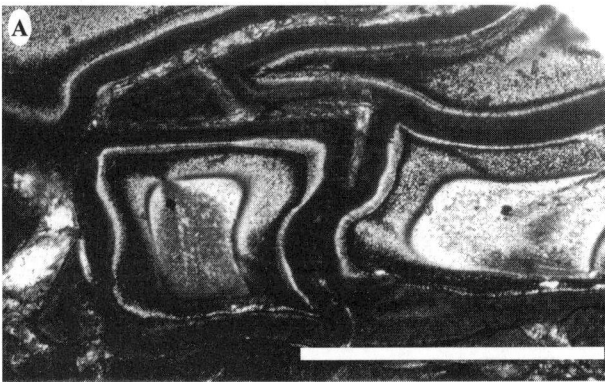 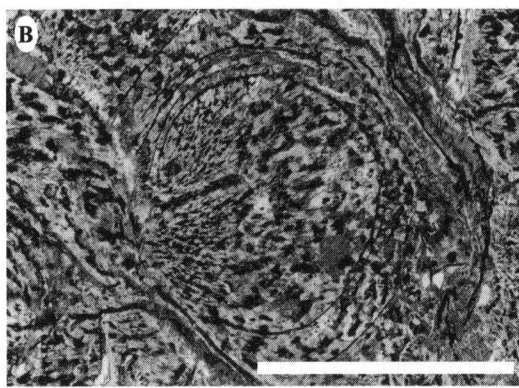

Figure 5. A) Agglomerated former glass shards showing signs of deformation due to weak to moderate compaction of the massive volcaniclastic facies. Alteration defines chilled margins. Scale bars are 0.25 mm; B) Perlitic fractures within former glassy fragments. Scale bars are 0.25 mm; C) Photograph documenting columnar joints. Photo is looking south toward stratigraphic top (hammer for scale). These occur within the massive unit and terminate at the contact with the overlying imbricated facies; D) Photograph documenting an example of the imbricated clasts (outlined by dashed lines). The compass lies along the contact between the laminated and massive facies (solid line). Typically the angle of imbrication with the contact is 35°. Stratigraphic top is south (arrow) and the direction of sediment transport was to the east (image left).

imbricated clasts diminishes within the exposures over a distance of ~500m with the largest clasts being located in the west. Imbricated clasts are elongated and are composed of gray, phenocryst-poor fragments that are massive to flow-banded, analogous to those hosted in the lower massive division and lobate facies. These clasts are also observed in blocky and fluidal forms. Fluidal forms are defined by apparent plastic deformation forming various shapes, including 'C'-shapes similar to those described by *Mueller and White* [1992]. Such deformation suggests that the fragments were still hot and became contorted during transport [*Mueller and White*, 1992]. Other fragments include minor tabular to subspherical vesicular clasts and relict glass particles.

Stratigraphically above the imbricated zone, discontinuous lenses of lapillistone occur which are composed of subangular to subrounded, phenocryst-poor fragments that have up to 5% quartz amygdules.

2.5 Basaltic Lava Flow

A massive basaltic lava flow overlies the volcaniclastic lithofacies and completes the observed stratigraphy in the study area (Figures 2 and 3). The contact between the basalt and the volcaniclastic deposit is sharp, but typically outlined by flattened, smooth surfaced, ellipsoidal forms (Figure 4D), a mm-thick aphyric band, and/or perpendicular fractures that propagate from the contact into the basalt. Flattened forms appear to be lava bodies, and may represent a micro-version of pseudopillows described by *Yamagishi* [1991]. There are minor amounts of vesicles within the lava bodies at the macroscopic scale.

3. MODEL FOR THE VOLCANICLASTIC DEPOSIT

The initial idea of a submarine environment is supported by the local geology (Figure 2). Pillowed basalts and chert horizons throughout the region attest to subaqueous deposition. As well, in situ brecciation forming jigsaw patterns observed in the dacite facies (Figure 4A) indicate quench fragmentation during lava-water interaction [e.g. page 54, *Cas and Wright*, 1987].

The massive lower and upper imbricated volcaniclastic units are considered to form a single depositional sequence based on their contact relationship, and identical fragment compositions. Both lithofacies contain flow-banded and massive gray, aphyric fragments, also suggesting they originated from identical magmas. The abundance of vitrophyric shards and vesicular lapilli demonstrates that fragments were produced by volcanic processes. We propose that both units were deposited from high particle concentration currents, though the nature of the currents' interstitial fluids was likely different.

On the outcrop scale, individual polygonal joint-bounded columns are continuous over 10 m in length, are as much as 1 m in diameter (minimum; visual estimate), appear to intersect at triple points with juxtaposed columns (Figure 5C), are restricted to the lower massive unit, and are oriented perpendicular to lower and upper contacts (Figure 5C); this suggests that they are columnar joints formed by thermal contraction of a completely hot deposit. Columnar jointing in ignimbrite deposits (i.e. Bishop Tuff) is typically associated with moderately to highly welded sections [page 215, *Fisher and Schmincke*, 1984]; where thick accumulations of hot, juvenile fragments results in viscous deformation (i.e. compaction). In contrast, the Hurd deposit is only weakly welded and relatively thin. However, the massive sequence is dominantly composed of dense, non-vesicular lithic fragments that probably resist viscous deformation, which may also cause it to mechanically behave like a lava, thereby facilitating thermal contraction. The large aspect ratio (l/d) [*DeGraff and Aydin*, 1987] and distinct triple intersections (Y-type) [*Aydin and DeGraff*, 1988; *Pollard and Aydin*, 1988] argue against cold-state contraction of a homogeneous material (i.e. loess or soils) and desiccation cracking of mud, respectively [e.g. *DeGraff & Aydin*, 1987]. Columnar jointing in the Hurd deposit is on the same scale as that developed in the Bishop Tuff (1 m diameter, Figure 8-35; *Fisher and Schmincke*, 1984). Furthermore, their preferred orientation and spacing is very different from inferred cooling fracture trends observed in welded and unwelded tuffs, which also frequently cut flow boundaries, and are interpreted to mimic the original topography [e.g. *Fuller and Sharp*, 1992]. Petrographic support for particle-heat retention includes moderate compaction of former glass shards (Figure 5A), intact perlitic fractures (Figure 5B) in mm-sized shards, spherulitic crystallization, and slight flattening of vesicles. Together these textures suggest that the massive unit was deposited in a hot state en masse. Lacking a description of the rheological and other properties of the material, it is not possible to estimate the temperature of deposition. However, the columnar joints are inferred to result from cooling-contraction following hot deposition and the compacted glass shards indicate a temperature around that of the glass transition temperature and certainly above that of the stability of liquid water. Thus it is reasonable to conclude that the lower unit contained little or no aqueous water and that the supporting fluid in the depositing current was likely a gas. This concept is supported by the presence of the previously described fines depleted feature that we interpret to have been a gas-pipe (Figure 4C). The upper imbricated unit has similar features to the underlying massive facies, such as spherulitic crystallization and an abundance of primary volcanic fragments, along with one distinctive fragment that was molded by plastic deformation. However, all of these are features of individual fragments, which could have been transported to their final depositional site. The lack of fine material in this deposit could be the result of flow-top stripping, whereby fine material is elutriated and removed from the moving flow by turbulent mixing of water. Finally, the ellipsoidal lava bodies along the basal contact of the basalt with the imbricated unit suggest interaction with water. Thus we propose that this imbricated unit was deposited by a water-supported current.

Textural features and alteration patterns of flow-banded and massive, gray, aphyric fragments of the volcaniclastic facies indicate that they originated from the same magmatic source as the underlying massive and lobate dacitic lavas. Moreover, a disturbed contact in the basal part of the lower massive volcaniclastic unit with the lobate facies indicates a possible intrusion of viscous lava, similar to that envisioned for cryptodome formation, in which viscous magma intrudes into unconsolidated material. Therefore, not only are the lobate and volcaniclastic facies related, but they appear to be contemporaneous.

3.1 Organization of the Flow Unit

One of the striking features of the Hurd flow unit is the presence of large imbricated clasts found at the base of the upper stratified flow unit. In addition, we observed that the upper termination of the columnar joints also coincides with the base of the upper stratified flow unit. These observations lead us to postulate that the upper part of an inferred stratified flow was relatively water rich in comparison to the underlying high-concentration part, and that clasts in the upper part cooled quickly. Water would also have reduced the viscosity of the upper flow [e.g. *Stix*, 1991], thereby generating a turbulent flow and producing the observed sedimentary stratification. Although the volcaniclastic unit can be divided into massive and stratified subunits, we interpret the two as having been deposited from a single event based on the continuity in composition and clast type. The presence of large imbricated clasts demonstrates that the upper flow was highly energetic and suggests that it had a large horizontal component of velocity. The experimental work of *Postma et al.* [1988] demonstrated that the sudden release of high-density (35–40 vol.%) suspensions on steep slopes (>25°) results in the material separating into a lower, highly concentrated inertia-flow layer, and an upper faster moving turbulent-flow layer having associated turbulent suspension clouds. The separation between the two flow-regimes was defined by a sharp interface, or physical discontinuity formed by a pronounced change in flow density and viscosity. The large clasts settled through the turbulent upper part of the flow, but the upper-flow's greater velocity produced imbrication of clasts at the contact as they glided along the interface with the lower inertially driven part of the flow. *Postma et al.* [1988] made an analogy to water skiing wherein the force provided by the tow-boat keeps the skis gliding along an interface between two fluids having vastly different density and viscosity. In like manner, we conclude that the overall volcaniclastic flow that deposited both units was set in motion on a steep slope, and that the fast flow of the upper water-rich turbulent part of the flow imbricated the clasts along the interface between the two depositional units.

3.2 Style of Eruption

Whilst the volcanic vent is not identified for the Hurd subaqueous primary pyroclastic deposits, bounding lithofacies and regional geology suggest a completely submarine eruption, likely below wave-base (>200 m). The low clast vesicularity, combined with the thin massive lithofacies, suggests a small volume eruption.

We envisage that the eruptive sequence is related to the initial construction of a small submarine dacitic dome characterized by lobate structures and flow top and in-situ breccia. Growth of the dome of dacitic magma increased internal shearing, which evolved from initial viscous deformation (e.g. as indicated by flow-banding) to final brittle failure. *Alidibirov and Dingwell* [2000] suggest that such non-Newtonian behavior is possible under a constant applied stress (e.g. hydrostatic pressure) because the deformation rate of the melt increases due to a reduction in the 'transient friction coefficient' of the melt with increased shearing. Therefore continued growth of this dacitic dome resulted in collapse (brittle or gravitational). Collapse would have rapidly decompressed magma within the dome thereby triggering magma fragmentation and ejection of fragments [*Alidibirov and Dingwell*, 2000]. A laterally directed eruption is favored for production of the Hurd deposits, as this orientation restricts mixing with ambient water [i.e. *Kokelaar and Busby*, 1992] as compared to a vertically directed eruption column. Therefore we envision a gravitational dome collapse similar to those documented for some of the subaerial dome eruptions at Unzen, Japan [*Sato*, 1992] and Soufriere Hills, Montserrat. Collapse of the massive, solidified dome margin forms the oversized tabular breccia fragments that are ejected and mixed with water forming a high-concentration sedimentary gravity flow. Behind this margin, decompression causes the formation of a 'fragmentation wave front' [e.g. *Alidibirov and Dingwell*, 2000], as gas expands to ambient pressure, which both fragments and accelerates the disintegrating magma. The sequence of events at Hurd generated a pyroclastic flow that was protected from ambient water by the combination of a steam envelope and initial formation of the overriding high-concentration flow, both of which acted as insulating layers. Upon deposition, the high temperature massive deposit experienced only weak compaction due to low load pressures from the thin overlying sequence. Columnar joints developed perpendicular to its cooling surface and propagated through the entire massive pyroclastic deposit, terminating at the base of the laminated facies.

4. DISCUSSION AND CONCLUSIONS

Field and petrographic evidence demonstrates that the volcaniclastic material was emplaced hot and under subaqueous conditions. There is abundant evidence, at a regional scale, that Kinojevis rocks were dominated by submarine effusion. Locally there exist pillows and chert-horizons within the section. The jig-saw fit breccias and clasts having quenched margins are also good evidence of subaqueous deposition. We acknowledge that recognizable welded textures and columnar joints in the massive unit are not ubiquitous, but rather limited to the thickest sections of the deposit. The lack of observed fiamme and only weak to moderate degree of glass shard deformation suggests moderate degrees of welding controlled by load pressure. Perlite fractures and plastically deformed flow-banded fragments indicate that the clasts themselves were deposited hot, whereas in situ deformation/agglomeration of former glass shards (Figure 5A) and columnar joints (Figure 5C) indicates that the deposit as a whole was emplaced hot. The supporting medium is difficult to ascertain, but evidence of at least one probable gas segregation pipe indicates that some gas was present.

Textural features and contact relationships between the various lithofacies outline some important temporal and spatial relationships. Alteration patterns and the similar fragment populations suggest they are genetically related. Furthermore, dacitic lobes appear to have initially intruded into the base of the unit while it was still unconsolidated, with the lobes subsequently slightly flattened during welding (Figure 4B). This not only indicates contemporaneous emplacement, but a proximal location to the dome itself, which is considered to be located to the west based on closure directions of lava lobes and the orientation of imbricated clasts in the volcaniclastic deposits. We interpret the volcaniclastic units to represent a single eruptive event, based on similar fragment populations that suggest a continuum of events originating from the same source. Finally, the complete sequence was covered by penecontemporaneously emplaced basalt that protected the underlying flows from subsequent erosion, as well as providing extra load pressure to facilitate welding in an otherwise thin deposit. This timing is based on observed micro-pillows along the lower contact with the imbricated unit. Water originating from the unconsolidated imbricated facies probably formed these structures at the base of an otherwise 'sealed' basalt.

The interpreted setting, proximal to a subaqueous dacitic dome with a small volume pyroclastic deposit, is analogous to block and ash deposits surrounding most subaerial felsic domes (e.g. Novarupta, Alaska, Plate 19.1, *McPhie et al.* 1993). We envision a similar gravitational collapse mechanism to those witnessed from Soufriere Hills, Montserrat. The problem is that most documented block and ash eruptions are not associated with explosive decompression and are dominantly composed of angular breccia-sized fragments of former dome material with only rare cases of welding (page 111, *Cas and Wright*, 1987). However, in this case the abundance of juvenile fragments (vesicular and former glass shards) point to fragmentation of fluidal magma within the dome, which occurred after the solidified dome margin failed, resulting in decompression of a slightly vesicular magma. As for most dome-collapse flows, this flow was dominated by clasts with temperatures high enough to form steam upon contact with water, thereby effectively partially insulating part of the flow from its submarine environment. Furthermore, the overlying basalt may have soon afterwards isolated the entire system from water, and also provided the additional weight necessary to create the unique situation that welded this small volume flow.

All these details suggest that these two volcaniclastic lithofacies represent a single eruptive event originating from the gravitational collapse of a submarine dacitic dome. As the outer margin fragmented it readily mixed with water to form a coarse-grained, high concentration sedimentary gravity flow. The collapse resulted in localized decompression that disintegrated hot fluidal magma, thereby generating a hot, gas-fragment mixture, which then produced a primary subaqueous pyroclastic flow. The overall eruptive mechanism may be a subaqueous equivalent to a subaerial block and ash flow.

Acknowledgments. This work stems from a recently completed honors project by the second author. Previous work in this field area by M. Jones is noted, particularly for regional geology and geochemistry. Jamie Peets is thanked for having brought the imbricated clasts to our attention. We thank Bill Arnott for sharing his insight into processes of sediment flow. C. R. Scott and A. D. Fowler would like to thank the American Geological Union for funding received to attend the Chapmen Conference in Dunedin, New Zealand, where the ideas for this manuscript were first presented to the volcanological community. We are particularly grateful to reviewers S. Allen and C. Breitkreuz, particularly to editor J. D. L. White, for incisive and thoughtful reviews, which significantly improved the content and clarity of this manuscript.

REFERENCES

Alidibirov, M. and D. B., Dingwell, Three fragmentation mechanisms for highly viscous magma under rapid decompression, *J. Volcanol. Geotherm. Res.*, 100, 413-421, 2000.

Allen, S.R., Reconstruction of a major caldera-forming eruption from pyroclastic deposit characteristics: Kos Plateau Tuff, eastern Aegean Sea, *J. Volcanol. Geotherm. Res.*, 105, 141-162, 2001.

Aydin, A and J.M. DeGraff, Evolution of polygonal fracture patterns in lava flows, *Science*, 239, 471-476, 1988.

Ayer, J., Y. Amelin, F. Corfu, S. Kamo, J. Ketchum, K. Kwok, and N. Trowell, Evolution of the southern Abitibi greenstone belt based on U-Pb geochronology: autochthonous volcanic construction followed by plutonism, regional deformation and sedimentation, *Precamb. Res.*, *115,* 63-95, 2002.

Batiza, R. and J.D.L. White, Submarine lavas and hyaloclastite. *Encyclopedia of Volcanoes*, Academic Press, 361-382, 2000.

Busby-Spera, C.J., Large volume rhyolitic ash flow eruptions and submarine caldera collapse in the lower Mesozoic Sierra Nevada,California, *J. Geophys. Res.*, *89,* 8417-8427, 1984.

Carey, S.N. and H. Sigurdsson, The Roseau Ash: deep-sea tephra deposits from a major eruption on Dominica, Lesser Antilles arc, *J. Volcanol. Geotherm. Res.*, *7,* 67-86, 1980.

Cas, R.A.F. and J.V. Wright, *Volcanic Succession: Modern and Ancient*, Allen and Unwin, 528p, 1987.

Cas, R.A.F. and J.V. Wright, Subaqueous pyroclastic flows and ignimbrites: an assessement, *Bull Volcanol.*, *53,* 357-380, 1991.

DeGraff, J.M. and A. Aydin, Surface morphology of columnar joints and its significance to mechanics and direction of joint growth, *Geol. Soc. Am. Bull.*, *99,* 605-617, 1987.

Fisher, R.V., Proposed classification of volcaniclastic sediments and rocks, *Geol. Soc. Amer. Bull.*, *72,* 1409-1414, 1961.

Fisher, R.V., Rocks composed of volcanic fragments and their classification, *Earth. Sci. Rev.*, *1,* 287-298, 1966.

Fisher, R.V., Flow transformations in sediment gravity flows. *Geology*, *11,* 273-274, 1983.

Fiske, R.S., and T. Matsuda, Submarine equivalents of ash flows in the Tokiwa Formation, Japan, *Am. J. Sci.*, *262,* 76-106, 1964.

Fritz, W.J., and C.J. Stillman, A subaqueous welded tuff from the Ordovician of County Waterford, Ireland, *J. Volcanol. Geotherm. Res.*, *70,* 91-106, 1996.

Fuller, C.M., and J.M. Sharp, Permeability and fracture patterns in extrusive volcanic rocks: implications from the welded Santana Tuff, Trans-Peco Texas, *Geol. Soc. Am. Bull.*, *104,* 1485-1496, 1992.

Howells, M.F., D.G Campbell, and A.J. Reedman, Isolated pods of subaqueous welded ash-flow:a distal facies of the Capel Curig Volcanic Formation (Ordovician), North Wales, *Geol. Mag.*, *122,* 175-180, 1985.

Jackson, S.L., J.A. Fyon, and F. Corfu, Review of Archean supracrustal assemblages of the Southern Abitibi greenstone belt in Ontario,Canada:products of microplate interaction within a large-scale plate- tectonic setting, *Precam. Res.*, *65,* 183-205, 1994.

Jones, M. I., *Variolitic basalts: Relations to Archean Epigenetic Gold Deposits in the Abitibi Greenstone belt*, Unpub. M.Sc. thesis, University of Ottawa, 320p., 1992.

Kokelaar, P., and C. Busby, Subaqueous explosive eruption and welding of pyroclastic deposits, *Science*, *257,* 196-201, 1992.

Kokelaar, P., and S. Königer, Marine emplacement of welded ignimbrite:the Ordovician Pitts Head Tuff, North Wales, *J. Geol. Soc. Lond.*, *157,* 517-536, 2000.

LaCroix, A., *La Montagne Pelée et ses eruptions*, Messon, Paris, 1904.

Legros, F. and T.H. Druitt, On the emplacement of ignimbrite in shallow-marine environments, *J. Volcanol. Geotherm. Res.*, *95,* 9-22, 2000.

Mandeville, C.W., S. Carey, and H. Sigurdsson, Sedimentology of the Krakatau 1883 submarine pyroclastic deposits, *Bull. Volcanol.*, *57,* 512-529, 1996.

McArthur, A.N., R.A.F. Cas, and G.J. Orton, Distribution and significance of crystalline, perlite and vesicular textures in the Ordovician Garth Tuff (Wales), *Bull. Volcanol.*, *60,* 260-285, 1998.

McPhie, J., M. Doyle, and R. Allen, *Volcanic textures: a guide to the interpretation of textures in volcanic rocks*, University of Tasmania, Centre for Ore Deposit and Exploration Studies, 198p, 1993.

Mueller, W., and J.D.L. White, Felsic fire-fountaining beneath Archean Seas: pyroclastic deposits of the 2730 Ma Hunter Mine Group, Québec, Canada, *J. Volcanol. Geotherm. Res.*, *54,* 117-134, 1992.

Pollard, D.D. and A Aydin, Progress in understanding jointing over the past century, *Geol. Soc. Am. Bull.*, *100,* 1181-1204, 1988.

Postma, G., W. Nemec, and K.L. Kleinspehn, Large floating clasts in turbidites: a mechanism for their emplacement, *Sed. Geol.*, *58,* 47-61, 1988.

Sato, H., S. Fujii, and S. Nakada, Crumbling of dacite dome lava and generation of pyroclastic flows at Unzen Volcano, *Nature*, *360,* 664-666, 1992.

Stix, J., Subaqueous, intermediate to silicic-composition explosive volcanism: a review, *Earth-Sci. Rev.*, *31,* 21-53, 1991.

White, J.D.L., Subaqueous eruption-fed density currents and their deposits, *Precamb. Res.*, *101,* 87-109, 2000.

White, M.J., and J. McPhie, A submarine welded ignimbrite–crystal-rich sandstone facies association in the Cambrian Tyndall Group, western Tasmania, Australia, *J. Volcanol. Geotherm. Res.*, *76,* 277-295, 1997.

Whitham, A.G., The behaviour of subaerially produced pyroclastic flows in a subaqueous environment:evidence from the Roseau eruption, Dominica, West Indies, *Mar. Geol.*, *86,* 27-40, 1989.

Yamagishi, H., Morphological features of Miocene submarine coherent lavas from the Green Tuff basins: examples from basaltic and andesitic rocks from the Shimokita Peninsula, northern Japan, *Bull. Volcanol.*, *53,* 173-181, 1991.

Young, S., S. Sparks, R. Roberston, L. Lynch, and W. Aspinall, Eruption of Soufriere Hills Volcano in Montserrat continues, *EOS*, *78,* 402-409, 1997.

C.R. Scott, Geological Sciences, Cal State Univ. Northridge, 18111 Nordoff St., Northridge, CA 91330-8266

D. Richard, Earth Sciences, University of Southern California, 3651 Trousvale, Science Hall, Rm 170, Los Angeles, CA 90089-0740 (pommelle@hotmail.com)

A.D. Fowler, Department of Earth Sciences and Ottawa Carleton Geoscience Centre, 140 Louis Pasteur, Ottawa, ON, Canada, K1N 6N5 (afowler@uottawa.ca)

Deep Marine Pumice from the Woodlark and Manus Basins, Papua New Guinea

Raymond A. Binns

CSIRO Exploration and Mining, North Ryde, Australia
and
Geology Department, Australian National University, Canberra, Australia

Two new deep marine occurrences of pumice from Papua New Guinea are located in extensional rift settings rather than island arcs. Peralkaline rhyolite pumice occurs at water depths of 700 to 950 m on the wall of a rift ~80 km west of the propagating Woodlark seafloor spreading axis, within a submarine extension of the dormant D'Entrecasteaux volcanic province. Outcrops up to 100 m long surveyed by manned submersible possess rind zones with polygonal fracturing and local ropy surfaces, and are interpreted as coherent lava flows with no evidence of explosive fragmentation. In the eastern Manus Basin a dredge haul of rhyodacite pumice with exceptionally high vesicularity, from a possible outcrop at the 2,100 m-deep foot of a major extensional fault scarp, resembles exotic pumice fragments recovered from ooze elsewhere in the basin. Their low-K geochemistry conforms to no likely source among subaerial or submarine volcanic centers in the region. A mafic xenolith within this pumice shows features consistent with derivation of the pumice by fractional crystallization within a lower crustal magma chamber followed by tectonic decompression, vesiculation, and eruption to the deep seafloor. If the presence of outcropping pumice at the fault scarp location is confirmed by future photography or submersible observation, a new style of deep marine pumice occurrence will become established that needs to be taken into consideration when interpreting ancient sequences.

INTRODUCTION

Pumice is often perceived as a product only of subaerial or relatively shallow submarine eruptions [e.g., *Doyle and McPhie*, 2000]. Discoveries arising when manned submersible geoscientific dives commenced at continental margin and island arc settings in the late 1980's [*Yuasa*, 1992, 1995; and *Binns*, 1993] might have dispelled this concept some time ago had they been more widely publicized. More recently, two examples have been described of deep marine pyroclastic eruptions that generated abundant silicic pumice in water depths from 500 to 2000 m, the Healy Seamount caldera, Kermadec arc north of New Zealand [*Wright and Gamble*, *1999*; and *Wright et al.*, 2003] and the Myojin Knoll caldera, frontal Izu-Bonin arc south of Japan [*Fiske et al.*, 2001]. Equivalent occurrences have been noted in the southern Mariana arc [*Bloomer et al.*, 2001]. Further examples of such sites may be anticipated in other island arc settings, involving eruption of volatile-rich silicic melts at water depths that approach or exceed the ≤1 km depth limit previously proposed for submarine fragmentation of magma froths [*McBirney*, 1963; and *Cas and Wright*, 1987].

Two occurrences are described below of deep marine pumice encountered during the PACLARK-PACMANUS program of research cruises in Papua New Guinea waters. Neither lies within an island arc; rather, they are at sites of extensional rifting in contrasting tectonic settings (Figure 1). At the first, near the western end of the Woodlark Basin ahead of a propagating seafloor spreading axis, submersible observations have confirmed the presence of large outcrops of peralkaline rhyolite pumice. The second, a dredged sample of rhyodacite pumice at the foot of a major fault scarp bounding a pull-apart rift zone in the eastern Manus back-arc basin, is more contentious but warrants description because it appears, in view of arguments to be provided, to have been derived from an outcrop exposed at the adjacent fault scarp.

PERALKALINE RHYOLITE PUMICE NEAR DOBU SEAMOUNT, WESTERN WOODLARK BASIN

Numerous pumice samples up to 80 cm in size were recovered by dredging during the 1988 PACLARK-III cruise of RV *Franklin* north and west of Dobu Seamount, a small, submarine basaltic andesite volcano [*Binns*, 1993; and *Dril et al.*, 1997] 13 km east of Dobu Island in Dawson Strait, D'Entrecasteaux Islands (Figures 1 and 2). Camera tows in the same area recorded scattered large blocks up to several meters across of a pale, highly vesicular rock that appeared to be the source of this material. Although some images were suggestive of outcrop, the pumice blocks were at first regarded as exotic products derived from subaerial peralkaline rhyolite (comendite) volcanoes on the nearby D'Entrecasteaux Islands [*Fisher*, 1957; *Morgan*, 1966; *Smith*, 1976; *Smith and Johnson*, 1981; and *Stolz et al.*, 1993]. However, a submersible dive during the 1990 SUPACLARK expedition of RV *Akademik Mstislav Keldysh* clearly established the presence of large pumice outcrops, interpreted as submarine lava flows, from many of which scalloped blocks like those photographed had broken loose to become dispersed down slope.

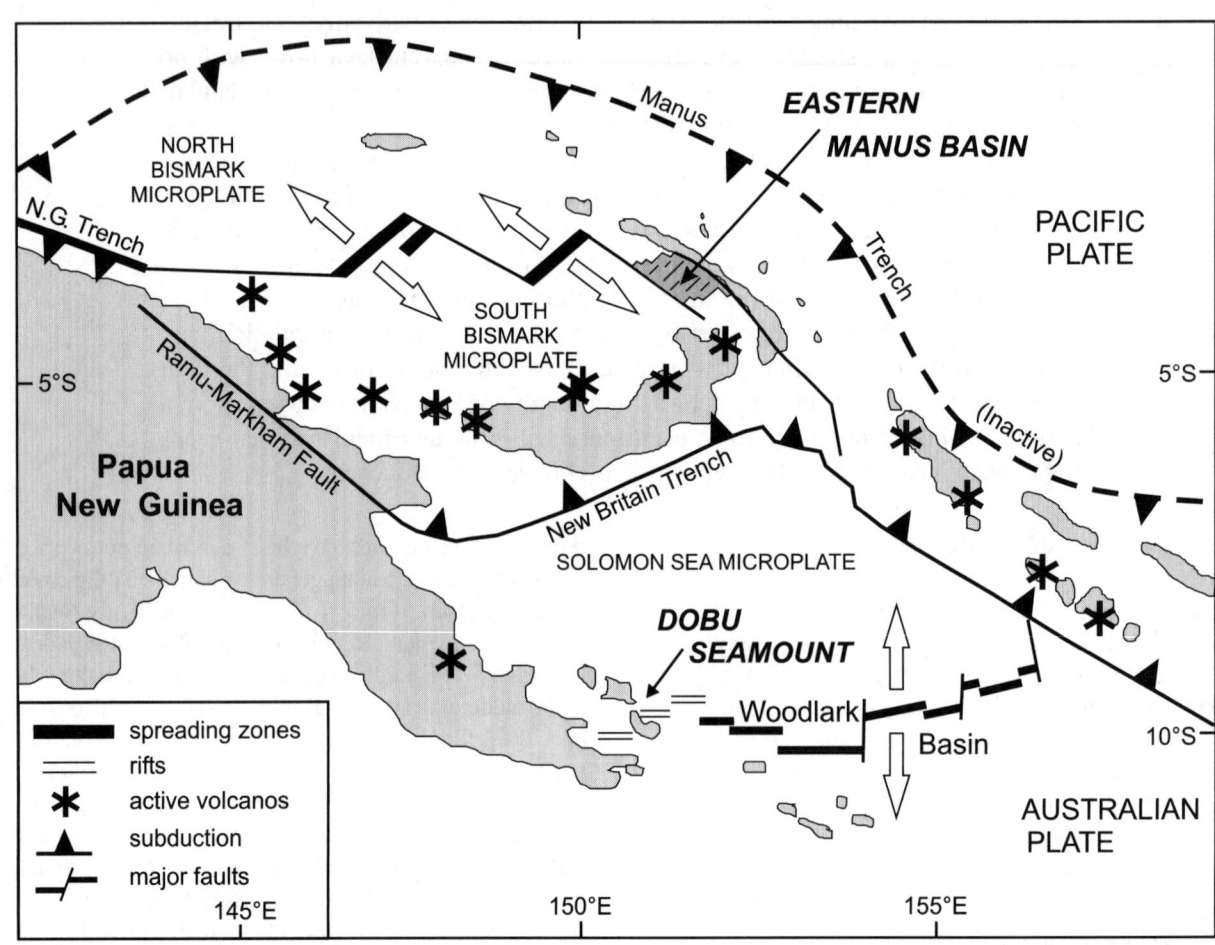

Figure 1. Regional tectonic setting of deep marine pumice occurrences near Dobu Seamount, Woodlark Basin, and in the Eastern Manus Basin, Papua New Guinea.

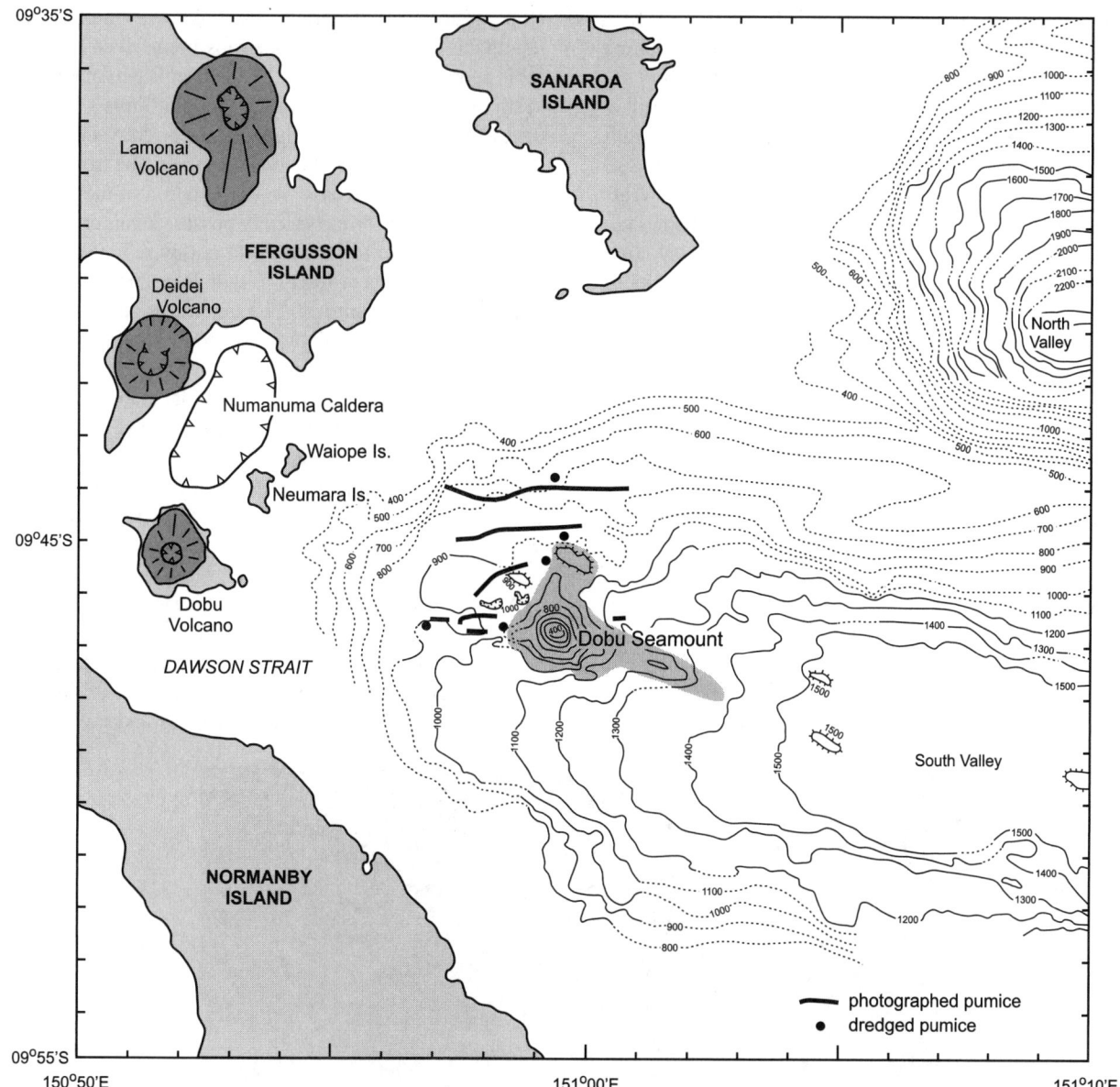

Figure 2. The westernmost Woodlark Basin, near the D'Entrecasteaux Islands, showing the principal seafloor sites where peralkaline rhyolite pumice has been dredged and photographed. Lamonai, Deidei and Dobu Island volcanoes are dormant cones composed of peralkaline rhyolite pyroclastic deposits and obsidian. The shaded area around Dobu Seamount is occupied by younger basaltic andesites and andesites. South Valley and North Valley are rifts formed by crustal extension ahead of the Woodlark seafloor spreading axis. Solid isobaths are from multibeam echosounding (HMAS *Cook*, PACLARK-III cruise, 1988), and dashed isobaths are interpolated from narrow-beam echosounding traverses (RV *Franklin*, PACLARK-I-IV cruises, 1986–1991).

The known distribution of this submarine pumice is shown on Figure 2. It occurs at water depths between 950 and 800 m on a narrow ridge almost 200 m high, about 3 km west of Dobu Seamount, and spreads across a series of low ridges and depressions to as high as 700 meters below sea level (mbsl) on the step-faulted northern wall of a 1600 m-deep graben (Figure 2). The location lies at the sloping western head of South Valley, one of several narrow rifts in continental crust [*Benes et al.*, 1994, 1997] that occur in front of the westward propagating Woodlark seafloor spreading axis of the Solomon Sea [*Taylor et al.*, 1995; *Mutter et al*, 1996; and *Goodliffe et al.*, 1997]. Crustal extension commenced in the Woodlark Basin some 6 Myr ago, prior to the continuing episode of seafloor spreading initiated some time before 3.5 Ma in the east [*Weissel et al.,* 1982]. The most westerly occurrence of mid ocean ridge-type basalt at the present tip of the Woodlark spreading axis lies 78 km east of Dobu Seamount. Applying an average 15 cm/1000yr sedimentation rate over the past 20,000 years [*Barash and Kuptsov,* 1997] to its maximum 360 m sediment fill [*Benes et al.,* 1994], rifting of South Valley commenced around 2–2.5 Myr ago. Basement in the vicinity of the submarine pumice outcrops is believed to consist of Cretaceous to early Tertiary metamorphic and volcanic rocks as exposed on adjacent Fergusson and Normanby Islands [*Benes et al.,* 1994].

Submersible Observations

During Dive M2177 of the SUPACLARK cruise, the *Mir-2* submersible conducted a circuit of the ridge located west of Dobu Seamount, covering a zone 2.2 to 3.4 km west of the seamount summit before returning to the latter feature [*Binns,* 1990]. Numerous continuous or semi-continuous outcrops of pumice were observed on the ridge, many more than 10 m across and several ranging between 30 and 100 m in traversed length. They protrude above level or gently sloping surfaces underlain by unknown thicknesses of hemipelagic ooze, but the outcrops themselves have negligible sediment cover, partly as a consequence of erosion by seafloor currents. Many outcrops are platforms a few meters or less high, but others include spines or domes rising 10 to 20 meters above ooze. Where determined, the trends of such outcrops parallel the east-west orientation of the ridge.

The flatter platforms typically possess an outer platy layer or rind, a few cm to 15 cm thick, cut by reticulate or locally hexagonal joints or fractures that do not penetrate into the deeper interior (Figure 3). The outermost surfaces of some rind zones have a broadly crenulated or ropy structure, with cm-scale amplitude and spacing.

Taller outcrops have deep, irregular subvertical joints, and along these large scalloped blocks have locally fallen aside or broken free. Platy surfaces with the rind-like polygonal fracturing or ropy exterior occur on the sides as well as the tops of some large outcrops. At one site, several meters of exposed interior beneath such a rind showed crude columnar jointing.

A laminated appearance of joint surfaces on outcrops and scalloped surfaces of fallen blocks is caused by subparallel sets of planes populated by large, flattened or elongate vesicles. Regrettably, no observations were made of the relative orientations of these laminar structures and the platy rind

Figure 3. Submersible view (2 to 3 m wide) of an outcrop of pumiceous peralkaline rhyolite, on the ridge west of Dobu Seamount. The outcrop extends beyond low cliffs in the background for 50 to 70 m and reaches 10 m in height. It is partly covered by hemipelagic ooze. The platy outer rind, 10 to 15 cm thick, shows reticulate joints that do not penetrate into the interior. Some such surfaces have a ropy structure.

zones. No evidence was seen on the ridge of talus banks or finer tephra deposits suggestive of explosive eruption.

Petrography and Geochemistry

Dredged pumice samples from the north and west of Dobu Seamount are pale gray or buff in color, and contain scarce but conspicuous phenocrysts of anorthoclase up to 5 mm long. Mostly the vesicles are not stretched, but they display marked size variation. Vesicles larger than a few mm have formed by coalescence of smaller bubbles, judging from linings with shreds and protuberances of formerly fluid glass. This is especially apparent for "giant" vesicles 1-3 cm across, lying in vaguely defined shear planes, where stretching of vesicles is more evident. The vesicularity of a typical specimen (106424, Table 1) is 66 vol.% as calculated [cf. *Klug and Cashman*, 1994] using its bulk specific gravity (0.809 ±0.005, from the weight of a sawn and measured rectangular block carefully leached free of sea salt and dried overnight at 50°C) and that of a finely ground powder of similarly leached fragments (2.364 ±0.007 measured by helium pycnometry after further drying at 50°C). This rises to 69% (bulk SG 0.74) in a subsample from near the trails of larger vesicles.

In thin section, vesicles that appear to represent single bubbles range from 20 to 500 µm in diameter, all sizes of vesicle being packed tightly together (Figure 4A). Larger vesicles in this size range have somewhat irregular outlines, imposed by adjacent smaller neighbors. Bubble walls are less than 20 µm thick, and cusps or bridges at bubble junctions are around 50 µm across or less. The fabric is generally isotropic, but there are zones where variable elongation and flattening of vesicles reflects ductile flow. An extreme development of the latter fabric occurs in a sample of platy rind with a ropy surface, collected under submersible observation (Figure 4B). A second sample from just under the rind shows less pronounced flattening and stretching.

Phenocrysts in different samples range in volumetric proportion, by visual estimate in thin sections, from 0.3% to 1% of the pumice. They include predominant anorthoclase laths ($Ab_{65-69}Or_{28-33}An_{1-5}$), minor ferroaugite prisms ($Ac_6En_{26}Fs_{29}Wo_{39}$), and rarer Ti-rich magnetite (19.6 wt% TiO_2), ilmenite (49.9% TiO_2) and manganiferous fayalitic olivine ($Fo_{24}Fa_{72}Tp_4$). The remainder of most samples is entirely very pale brown to colorless glass, but in some dredged samples this contains a proportion of tiny feldspar laths within bubble walls and cusps. In zones of ductile flow, larger anorthoclase laths are often broken and drawn apart.

The bulk chemical compositions of pumices from five dredge hauls near Dobu Seamount (Figure 2) are effective-

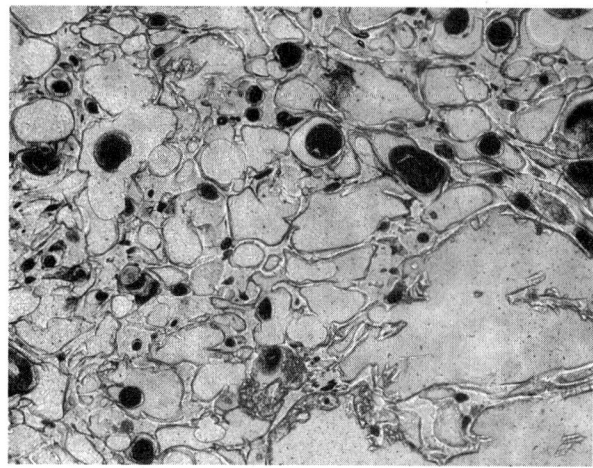

Figure 4. Photomicrographs of pumice from near Dobu Seamount. A: Typical fabric of isotropic pumice showing variable vesicle size and evidence of bubble coalescence (lower right). Dark patches are air bubbles in the epoxy impregnation, filled with grinding medium. B: Stretched fabric (predominantly planar and parallel to outer surface) of a sample of platy rind zone. Circular objects are air bubbles in cloudy epoxy. Widths of fields, 1.3 mm.

ly indistinguishable. Sample 106524 (Table 1) is typical, and geochemically very similar to subaerial comendite pumices and obsidians from nearby Fergusson, Dobu and Sanaroa Islands in the D'Entrecasteaux peralkaline province [*Smith*, 1976; *Smith and Johnson*, 1981; and *Morgan*, 1966]. All share a high peralkalinity index (mol ratio [Na2O+K2O]/Al2O3 averages 1.14 for the submarine pumices), low Ba and Sr contents, elevated Rb, Zr, Nb and Y contents, and high contents of rare earth elements (REE) characterized by light REE enrichment and marked negative Eu anomalies. One probable difference is the higher water

Table 1. Compositions of deep marine pumices from the Woodlark and Manus Basins, Papua New Guinea.

No.		106524	132223	132223G
SiO_2	wt%	68.94	68.11	72.71
TiO_2		0.26	0.57	0.48
Al_2O_3		13.91	14.32	13.27
$Fe_2O_3(t)$		2.98	4.63	2.95
MnO		0.09	0.14	0.15
MgO		0.26	1.58	0.81
CaO		0.38	4.39	2.94
Na_2O		6.48	4.13	4.45
K_2O		4.79	1.13	1.38
P_2O_5		0.03	0.14	0.15
LOI		1.53	1.02	1.21
Total		99.64	100.16	100.46
S	ppm	340	159	253
V		3.7	89	39
Cr		1.0	9	6
Ni		1.5	4	7
Ga		31	17	16
Rb		143	17	20
Sr		5.4	265	222
Y		70	28	32
Zr		923	90	113
Nb		27	<1	<1
Cs		5.7	0.59	0.72
Ba		57	234	283
La		63	10.2	11.9
Ce		135	23.1	26.7
Pr		15.6	3.75	4.27
Nd		56	14.3	16.2
Sm		11.0	3.97	4.45
Eu		0.61	1.32	1.36
Gd		12.1	4.53	5.25
Tb		2.12	1.01	1.13
Dy		13.7	5.38	6.09
Ho		2.98	1.44	1.65
Er		9.2	3.66	4.07
Tm		1.38	0.68	0.73
Yb		9.9	3.79	4.51
Lu		1.57	0.72	0.83
Pb		35	7.4	7.1
Th		23.4	1.24	1.52
U		6.2	0.84	1.00

106524: Dredge D34, 9°46.7'S 150°58.5'E, 860 mbsl, hauled east. 132223: Dredge MD50, 3°22.9'S 151°12.8'E, 2,103 mbsl hauled northwest. 132223G: Glass fraction separated from 132223.
Major elements by XRF (M. Hart, CSIRO); Ni, Ga, Rb, Cs, REE, U and Th by ICPMS and remaining traces by ICPAES (L.Dotter, CSIRO). LOI, loss on ignition at 1050°C.

content of the submarine pumice, as indicated by elevated loss on ignition. Since samples were carefully leached free of seawater (with repeated testing of the leachate by $AgNO_3$) and dried overnight at 50°C, this is evidently contained primarily within the glass component, which shows no sign of alteration or perlitic hydration. Higher pressures under submarine eruption conditions would favor such retention.

Discussion

The pumice outcrops near Dobu Seamount clearly represent a submarine extension of the D'Entrecasteaux peralkaline igneous province. Their characteristics, especially of the rind zones, are those expected for submarine extrusion of moderately viscous, volatile-rich and frothing magma, which elsewhere in the region has erupted explosively under subaerial conditions. The thickness of these lava flows is unknown, since observations of their interiors are restricted to a few meters. Dredged pumice samples with feldspar microlites in their glass suggest that deeper within the flows there may be more-pronounced crystallization.

The age of the submarine lavas relative to late Pleistocene to Recent subaerial activity in the D'Entrecasteaux peralkaline province [*Smith*, 1976] is not established. Sediment cores at ooze pockets in their vicinity failed to penetrate more than a few cm, indicating a youthful character possibly comparable with the Dobu Island, Lamonai and Deidei volcanic edifices, whose dormancy is suggested by physiography and their association with subaerial and shallow submarine hotspring activity [*Fisher*, 1957; *Davies*, 1973; and *Itikarai and de Saint-Ours*, 1990]. Topographic relationships and the presence of a remelted pumice xenolith in basaltic andesite from the submarine Dobu Seamount edifice indicate the latter to be younger than the submarine pumice lavas.

Seafloor mapping conducted to date does not delineate the eruptive source of the submarine pumice flows, in particular whether it is local to the observed outcrops or higher on the flanks of the subaerial sector of the D'Entrecasteaux province. An alternate but less likely source is the submerged Numanuma Caldera (Figure 2), whose presence was predicted [*Lowenstein*, 1982] from the structure of pyroclastic deposits (older than the Dobu Island, Deidei and Namonai volcanoes) on low-lying areas of southeastern Fergusson Island and on Waiope and Neumara Islands (Figure 2). Orthogonal echosounder traverses [*Binns and Wheller*, 1991] confirmed the presence of this 2x4 km feature, elongate north-northeast, which has steep 100–150 m high submarine walls on the eastern, western and southern sides and a relatively flat floor descending gently to a maximum measured depth of 168m toward its southern end. The northern wall is represented by cliffs on the shore of Numanuma Bay, Fergusson Island, where comendites are reported [*Smith*, 1976].

The external structure of submarine pumice flows near Dobu Seamount resembles that described for Miocene subaqueous rhyolite from Shimane Peninsula, Japan [*Kano et al.*, 1991], except surficial fragmentation is lacking in the former. Other analogues include the submarine-source eruptions inferred for pumiceous volcaniclastic deposits at Shinjima Island, Japan [*Kano et al.*, 1996] and at Yali, Greece [*Allen and McPhie*, 2000]. At both these sites, meter-scale blocks of pumice, considered to be sourced from the carapaces of submarine extrusions, have surficial polygonal jointing attributed to thermal contraction. Marginal polygonal jointing and internal columnar jointing ascribed to cooling in nearshore, subaqueous rhyolite lava lobes at Yakumo, Japan [*Yamagishi and Goto*, 1992] also resemble features described above from the pumice outcrops near Dobu Seamount.

RHYODACITE PUMICE AT THE WESTERN END OF THE EASTERN MANUS BASIN RIFT ZONE

Isolated unabraded fragments of white, silicic, pumice up to 20 cm in size and characterized by exceptionally low density were recovered, immersed in hemipelagic ooze, during numerous dredging operations of the PACMANUS program throughout the eastern Manus Basin in the Bismarck Sea of Papua New Guinea. Though clearly exotic, the source of these was a puzzle. In 1996 a dredge (MD-50) aimed at the foot (2,100 mbsl) of a major extensional fault scarp at the western end of the Manus Basin, where seismic profiling suggested an exposure of basement (Figure 5), recovered around 20 kg of freshly-broken pumice fragments up to 30 cm across, together with large talus slabs of Pliocene mudstone and sandstone [*Parr and Binns*, 1997]. The light-duty equipment used rarely breaks rock and it seemed likely to the operators that this dredge hit a large immovable body of pumice, possibly an outcrop, rather than a loose block incorporated along with sedimentary rocks within talus. The pumice appears far too fresh to have been derived from the Pliocene sequence. This inferred *in situ* nature of the pumice near the foot of the MD-50 scarp has not yet been confirmed by bottom-tow photography or submersible observation.

The rapidly opening Manus Basin [*Martinez and Taylor*, 1996] occupies a backarc position relative to the New Britain—Bismarck volcanic arc, between the active New Britain Trench and the inactive Manus Trench (Figure 1). The eastern Manus Basin descends to 2,740 mbsl via a set of northeast trending, low angle, normal faults that create graben and half-graben partly filled with Pleistocene to Recent sediments overlying more deformed Miocene to Pliocene sequences. Basement is thought to comprise Eocene-Oligocene arc volcanic and sedimentary rocks formed during an earlier phase of subduction at the Manus Trench. Formation of the Manus Basin commenced at about 3.5 Ma [*Taylor*, 1979].

An east-west belt of young, highstanding submarine neovolcanic edifices, the Eastern Manus Volcanic Zone [*Binns and Scott*, 1993; and *Binns et al.*, 2002], cuts across the eastern half of the eastern Manus rift zone. Individual edifices range from picritic basalt to rhyodacite in composition. Analyzed lavas are comparable geochemically and isotopically with the subaerial New Britain arc to the south [*Woodhead and Johnson*, 1993; *Kamenetsky et al.*, 2001], and are conspicuously vesicular [*Waters et al.*, 1996]. The MD-50 fault scarp is the most prominent of a set that defines the northwestern wall of the eastern Manus rift zone (Figure 5). It is removed by 45 km to the west from the nearest edifices of the Eastern Manus Volcanic Zone, except for a single 500-m high cone of basaltic andesite 17 km east of the dredge site.

Petrography and Geochemistry

The MD-50 pumice is massive and white to very pale buff in color. Some fragments have a surface thinly stained black by Mn oxide, but no rind structures like those near Dobu Seamount were recovered. Vesicularity, calculated for sample 132223 (Table 1) as described above from bulk (0.256 ±0.002) and powder (2.438 ±0.006) specific gravities, is very high at 89 vol.%.

Vesicles range from 10 µm to 5 mm in size, with a preponderance in the 50–150 µm range wherein many vesicles approach polyhedral shapes (Figure 6A). The smallest vesicles are equant, while larger vesicles are less regular in outline and commonly slightly flattened. Most bubble walls are very thin (~ 5 mm or less).

Visually estimated phenocryst and rarer glomerocryst abundance in thin sections is around 1% by volume. Euhedral plagioclase laths 0.5 to 1 mm long predominate. These range in composition from An_{63} to An_{50} and show slight normal or normal-oscillatory zoning. Occasional laths have separate, more calcic cores (An_{74-80}). Rare rounded xenocrysts of anorthite (An_{92-93}) also occur. Pyroxene prisms (0.2–0.5 mm) and clusters are subordinate by comparison. Many of these have small pieces or selvages

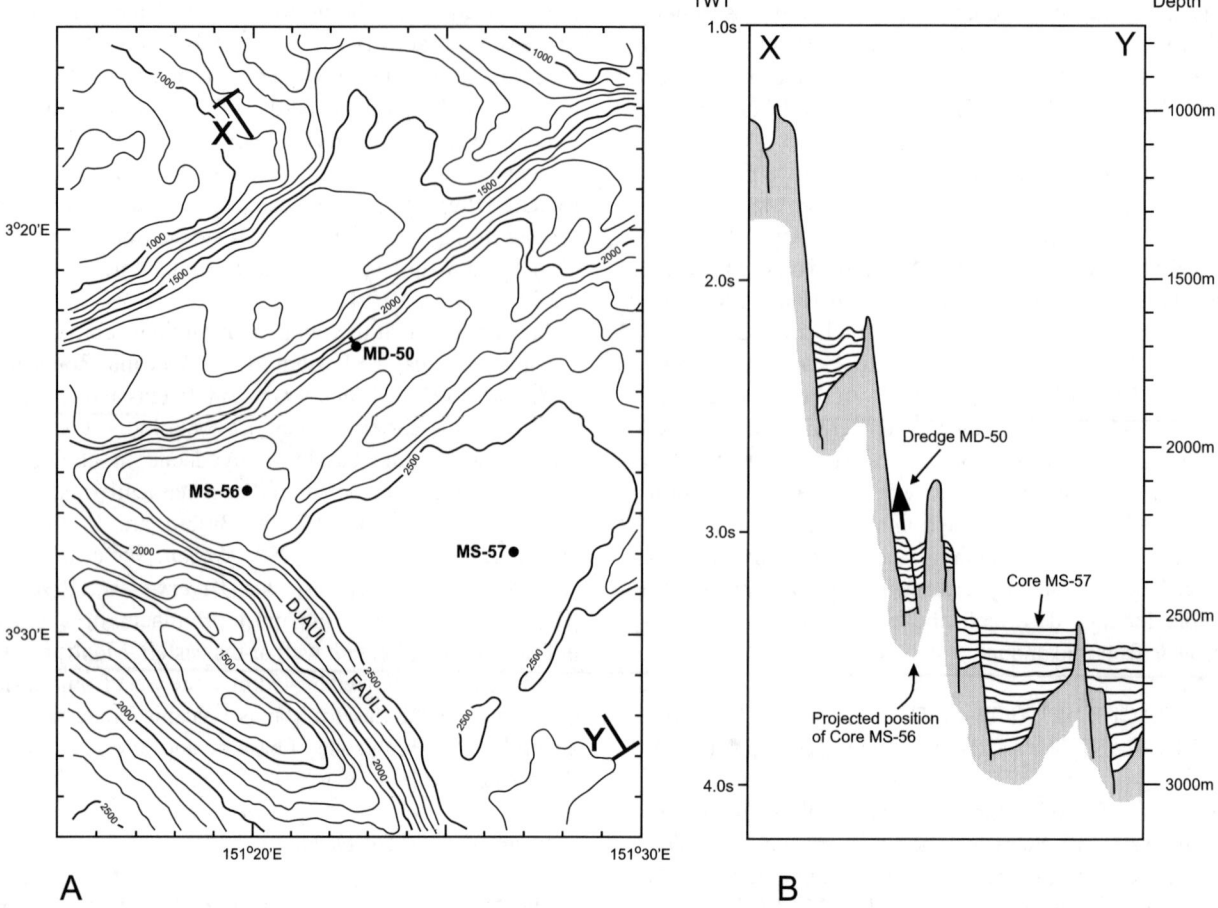

Figure 5. A: Bathymetry of the western end of the Eastern Manus Basin, where apparently outcropping rhyodacite pumice was dredged (MD-50) at the foot of a northeast-trending extensional fault scarp. The southeast-trending Djaul Fault is a major transform structure of the Manus Basin. MS-56 and MS-57 are sediment cores that penetrated hemipelagic ooze lacking any debris from the inferred pumice eruption. Chart derived from an IFREMER digital compilation of multibeam echosounding from various cruises in the Manus Basin (courtesy of J. -M. Auzende). B: Interpreted seismic profile along line X-Y of Fig. 5A, showing extensional faults and half-graben of mildly deformed, presumed Pleistocene-Recent sediments overlying more highly deformed Pliocene sediments (shaded). *Moana Wave* MW8517 Line 10, 1985, courtesy of B. Taylor and K.A.W. Crook.

attached to their surface of material comparable with the groundmass of a xenolith described below, suggesting they also are xenocrysts. Clinopyroxenes varying little in composition from $En_{41}Fs_{15}Wo_{44}$ with low Al_2O_3 (1.3–3.2 wt%), Na_2O (0.28%) and TiO_2 (0.4%) contents are more common than orthopyroxene ($En_{66}Fs_{31}Wo_3$). Titaniferous magnetite with 8.5–8.9 wt% TiO_2 forms minor subhedra 0.1–0.2 mm across, some containing tiny globules of Fe monosulfide.

Table 1 lists the bulk composition for the MD-50 pumice, together with that of a glass-rich concentrate prepared from its powder by heavy liquid and electromagnetic removal of most phenocrysts. The bulk compositions of apparently exotic pumices from other dredge hauls in the eastern Manus Basin are comparable and define trends on Harker diagrams (Figure 7) that can only partly be ascribed to differing phenocryst abundances. The glass composition (see also Table 2 and Figure 8) indicates that low K and elevated Ca are characteristics of the former melt component in this mildly fractionated series. Although the geochemistry of MD-50 pumice, including the very low Nb and the REE abundances (with light REE enrichment and no Eu anomaly), is broadly similar to that of fractionated rhyodacites in the Eastern Manus Volcanic Zone [*Binns and Scott*, 1993; and *Kamenetsky et al.*, 2001) and also those in the subaerial New Britain arc [*Smith and Johnson*, 1981], subtle but important differences will be outlined below.

against its host. Microxenoliths a few mm or less across of similar material are also scattered through the adjacent pumice. Phenocrysts in order of decreasing abundance in the main xenolith include euhedral plagioclase laths 0.2 to 2 mm long, stubby euhedral clinopyroxene prisms with similar dimensions, subhedral and corroded olivines 0.5 to 3.5 mm long lacking reaction coronas, subhedral titaniferous magnetite grains 0.1 to 0.5 mm across, and rare orthopyroxene prisms 0.5 mm long. Melt inclusions (5–50 μm) of pale brown glass with tiny fluid bubbles are present in plagioclase, pyroxene and olivine phenocrysts.

The groundmass is composed of roughly equal proportions of small plagioclase laths (~20 μm long) and clinopyroxene granules (~10 μm), plus 30–50% of very pale brown interstitial glass. Minute titaniferous magnetite euhedra (5 μm, some with "telegraph pole" habit) are less common, and there are rare olivine granules (~10 μm). Also, the groundmass has quite abundant small vesicles (5–10%; 10–50 μm) ranging from sub-spherical to quite angular in shape.

Individual phenocrysts in the xenolith are largely unzoned, except for thin margins (2–5 μm) with compositions similar to the equivalent groundmass phases. Olivine phenocrysts fall into two composition populations, subhedral crystals varying little from Fo_{73} with negligible Ni, and presumably earlier corroded crystals close to Fo_{84} with 1400 ppm Ni. Narrow rims on the latter, and the groundmass olivines, range from Fo_{69} to Fo_{71}. Plagioclase phenocrysts are exceptionally calcic, An_{87} to An_{94}, while their narrow rims and the groundmass laths are An_{71} to An_{75}. Clinopyroxene phenocrysts show little compositional variation from an average $En_{43}Fs_{13}Wo_{44}$ with 3.0% Al_2O_3, 0.18% Na_2O and 0.37% TiO_2. Their narrow rims, and the groundmass clinopyroxenes, are $En_{36}Fs_{24}Wo_{40}$ with 5% Al_2O_3, 0.22% Na_2O and 1.1% TiO_2. A single analyzed orthopyroxene phenocryst is $En_{68}Fs_{29}Wo_3$. Magnetite phenocrysts are less titaniferous than groundmass magnetites (5–6% and 8% TiO_2 respectively) and also richer in minor Mg and Al, and in trace Cr.

A remarkable feature of the xenolith is the presence of scattered, sharply bounded, ellipsoidal to irregular ovoid cavities 0.2 to 2 mm across, not visibly linked to the host pumice or associated with fractures, that are partly filled by one or more globules of very pale brown vesiculated glass lacking microlites (Figure 6B). The typically 100-200 μm sized vesicles *within* these globules constitute 50% to 60% of their volume. Commonly the vesicles are stretched, with radial dispositions. Examined at high magnification, the globule glass appears continuous with the interstitial glass of the xenolith groundmass. Analytical scanning electron

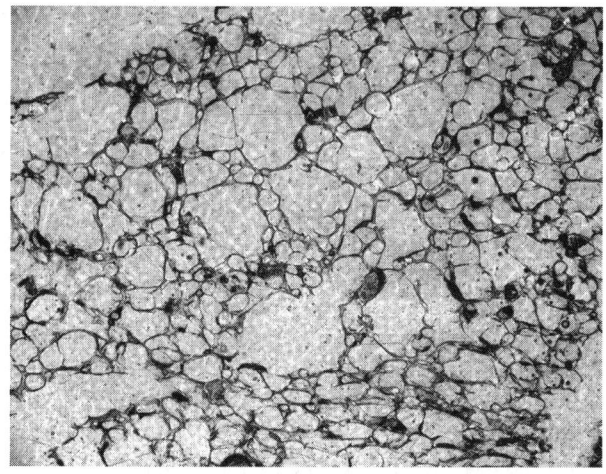

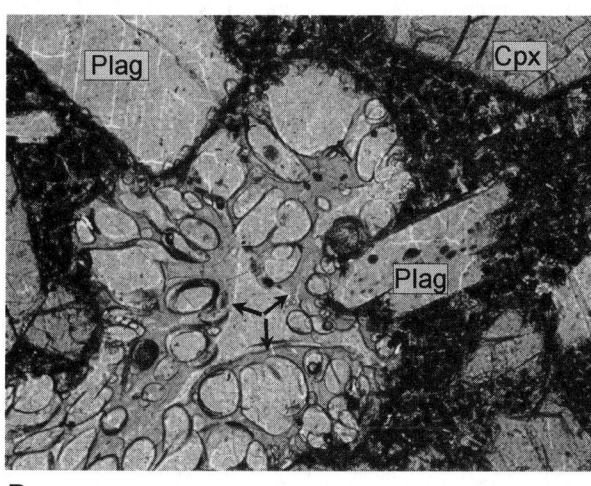

Figure 6. Photomicrographs of MD-50 pumice and its xenolith. A: Partial development of polyhedral foam structure in the pumice. Dark patches are polishing imperfections. B: Ovoid, early-formed vesicle in the porphyritic basaltic andesite xenolith. Arrows in the center of the image point to the boundaries (more obvious in color) of three globules of pale brown vesiculated glass that almost fill the cavity. Later vesicles in the globules display a crude radial elongation (more evident in other examples). Glass within the globules is continuous with interstitial glass in the xenolith groundmass, which is clouded by tiny plagioclase laths, pyroxene needles and magnetite granules. Euhedral clinopyroxene (Cpx) and plagioclase (Plag) phenocrysts occur within the xenolith. Width of fields; 1.3mm.

Xenolith in MD-50 Pumice

A dark gray porphyritic xenolith 5 cm across found in one MD-50 pumice fragment is angular, with sharp boundaries

Table 2. Average microprobe compositions of glass in MD-50 pumice and in cavities (early-formed vesicles) within its contained xenolith.

	Host	Cavity A	Cavity B	Cavity C
n	5	5	4	4
SiO_2 wt%	70.92	68.10	70.92	68.95
TiO_2	0.46	0.59	0.48	0.53
Al_2O_3	12.43	13.09	12.26	12.93
FeO(t)	2.05	4.55	2.86	4.07
MnO	0.10	0.06	0.12	0.15
MgO	0.51	1.06	0.67	0.96
CaO	2.19	4.02	2.87	3.49
Na_2O	4.08	3.61	3.97	3.81
K_2O	1.37	1.14	1.28	1.23
Total	94.12	96.23	95.42	96.12

n: number of points averaged
Analyses by the author on a JEOL 6400 SEM with EDS analyzer at the Electron Microscopy Unit, Australian National University. Beam rastered over a 10x10 mm area to minimize Na loss; beam current 1.0 nA, excitation potential 15 kV; mineral standards.

microscope compositions of globule glasses from three different ellipsoidal cavities are more mafic than the host pumice glass (Table 2), and on Harker diagrams they form a coherent fractionation series with the latter (Figure 8). There is clear but minor variability within globules and between the analyzed ellipsoids. Low analytical totals despite careful calibration and effort to avoid Na loss indicate that the globule glasses are appreciably hydrous (~4%). The more felsic glass in ellipsoid A approaches the composition of that in the host pumice, while the more mafic glasses of ellipsoid A are similar in composition to the xenolith groundmass glass for which, however, publishable analyses were not obtained using a conventional electron microprobe as a consequence of Na loss. Similarly-imperfect microanalyses of melt inclusions within phenocrysts of the xenolith indicate a low-Si basaltic andesite parent melt (~ 57% SiO_2), with ~0.30% K_2O and ~9.3% CaO. On Harker diagrams for these and other elements (except spurious Na), this parent composition is also consistent with fractionation trends for the MD-50 and related pumices, provided olivine has been extracted at the more mafic stages.

Source of MD-50 Pumice

Judging from the 15.5 cm/1000 yr sedimentation rates (over the last 10,000 yr) in the eastern Manus Basin [*Barash and Kuptsov*, 1997], the pumice fragments recovered from sloppy ooze in many dredge hauls are unlikely to be older than a few thousand years. Possible exotic sources warranting consideration for these and the MD-50 pumice, in view of their known generation of pumice rafts, include eruptions at Rabaul related to the geochemical Main Series [*Wood et al.*, 1995] and at Tuluman Island [*Johnson and Smith*, 1974; and *Johnson et al.*, 1978], respectively 140 km southeast and 440 km west-northwest of the MD-50 site. Both are ruled out, however, by their distinctly more potassic character (Figure 7). The older, partly pumiceous Raluan Ignimbrite at Rabaul and its possible source [*Wallace et al.*, 2002], the nearby offshore Tavui (or Raluan) Caldera [*Tiffin et al.*, 1990], have a similar low-K nature to MD-50 pumice but are more siliceous and contain quartz and hornblende phenocrysts [*Wood et al.*, 1995]. Of other potential subaerial sources among recently active New Britain arc volcanoes, only the Witori-Pago and Lolobau centers (170–250 km south of MD-50) have erupted silicic products with low-K characteristics similar to the MD-50 pumice, but neither appears to have recently discharged other than fine ash into the ocean [*Fisher*, 1957; *Blake and*

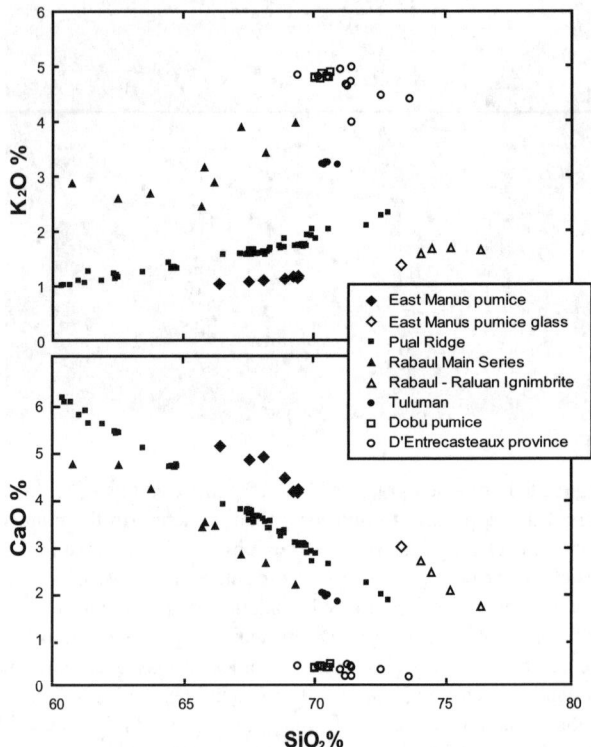

Figure 7. Selected Harker diagrams to demonstrate that the low-K MD-50 pumice and its exotic equivalents from other dredge hauls do not conform geochemically to potential submarine sources in the eastern Manus Basin or to subaerial eruptions in the region. Data from references cited in text, and CSIRO analyses (Pual Ridge). Analyses have been normalized to volatile-free 100% totals in these plots. Dobu submarine pumice and peralkaline rhyolites from the D'Entrecasteaux province are also plotted to demonstrate their limited compositional range.

Ewart, 1974; Blake, 1976; Smith and Johnson, 1981; Torrence et al., 2000; and Global Volcanism Program, 2003]. Little is known about a 12,000 year-old caldera eruption at Lolobau, probably too old to be a realistic source.

Submarine volcanic edifices in the eastern Manus Basin also differ geochemically from the MD-50 and related pumices. Figure 7 illustrates fractionation trends for Pual Ridge, the least potassic edifice in the now well explored Eastern Manus Volcanic Zone, yet which is distinguishably more potassic and less calcic than the MD-50 pumice. Only mafic volcanic rocks have been recovered from the central Manus Basin [Martinez and Taylor, 1996; and Sinton et al., 2003]. Numerous volcanoes, some with relatively shallow calderas, that have erupted silicic glass and pumice have been reported at the western spreading ridge of the Manus Basin, 430 km west of the MD-50 site [Auzende et al., 2000] but geochemical data are not available.

Given an inability to identify a source in the region, the possibility that the MD-50 pumice is not exotic but instead derives from an extrusive outcrop at the foot of the major extensional fault scarp deserves consideration. This, of course, remains unsubstantiated until the location receives more detailed examination by towed camera or submersible, yet the implications would be considerable.

Eruptive Mechanism Suggested by the Xenolith and the Location of MD-50

Glass and mineral compositions discussed above suggest that the host pumice and the glass globules present in the ellipsoidal cavities of the MD-50 xenolith are consanguineous (Figure 8), and that this, rather than an accidental origin, applies also to the mafic xenolith and its hydrous, basaltic andesite parent melt. Accepting such an interpretation, the xenolith provides evidence in microcosm of a process whereby pumice could be erupted at the foot of a major extensional fault scarp through fractional crystallization in a subseafloor chamber, followed by decompression, exsolution of volatiles, and major volume expansion of pockets of silicic residual melt.

The ellipsoidal cavities are interpreted as early-formed vesicles within the part-crystallized basaltic andesite melt now represented by the xenolith, into which progressively fractionating residual melt became squeezed to form the glass globules. An external origin seems convincingly ruled out for these globules which are entirely isolated from the host pumice except at the xenolith margin, where indeed one, evidently still molten, globule appears in thin section to have extruded out into and become mingled with the adjacent host pumice. Similar mm-scale droplets of residual liquid are in fact not uncommon at the edges of large, early-formed vesicles in the interior, partly crystallized portions of glassy submarine andesite and dacite lavas from the Eastern Manus Volcanic Zone, where their formation during cooling clearly requires an appropriate interplay between fluidity of the residual liquid and rigidity of the vesicle walls and enclosing lava body.

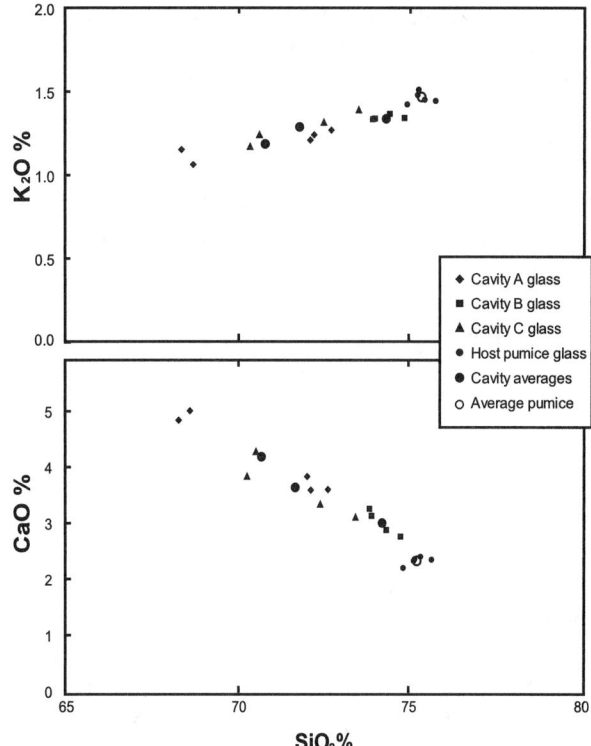

Figure 8. Selected Harker diagrams to illustrate the fractionation relationship of glasses (analyses normalized to volatile-free 100% totals) from globules within early-formed vesicles of a mafic xenolith (Figure 6B) with glass from the host pumice.

The exact timing of globule formation relative to crystallization of microlites and exsolution of tiny vesicles in the xenolith groundmass is not clear, although the latter features are likely to be later and perhaps contemporaneous with vesiculation of both the globules and the host pumice. The habits of vesicles within the globules and those within the host pumice are comparable. This is evident especially for the globule at the xenolith margin that appears to have extruded into the host (the two being differentiated by their relative proportions of vesicles), suggesting that the vesicles in the globules and the host pumice formed simultaneously as a consequence of reduction in pressure. Unloading as a result of sudden movement on a listric extensional fault like that where the MD-50 pumice was recovered is proposed as the likely rapid decompression process. It would lead to pronounced expansion within the magma chamber, and potentially to eruption toward the seafloor.

Pursuing these speculations further on the basis that the MD-50 pumice was indeed recovered from an *in situ* outcrop, and recalling its exceptionally high proportion of vesicles (89%), an estimate is possible for the depth of the source magma chamber inferred above. Assuming the frothy silicic melt behaved as an equilibrium closed system *en route* to the seafloor, at a temperature of 850°C and with its vesicles filled by purely aqueous fluid, and applying a 24 cm^3/g specific volume of H_2O at 210 bars (2100 m eruption depth) and 850°C [*Kennedy and Holser*, 1966] to a recombination of vesicle fluid and glass, the unvesiculated silicic melt at its source contained around 10% dissolved H_2O. Extrapolating solubility relationships from lower pressures [*Holloway and Blank*, 1994; and *Sparks et al.*, 1994], a silicic melt of this kind would be water-saturated at 5–6 kb pressure, or around 15 km deep in sub-seafloor crust. This estimated depth would decrease if considerations were given to vesicle expansion during cooling after eruption, and to the influence of CO_2 [cf. *Yang and Scott*, 1996; and *Kamenetsky et al.*, 2001]. From seismic refraction studies [*Wiebenga*, 1973] and isostatic gravity modeling [*Martinez and Taylor*, 1996], the estimate is consistent with the differentiating magma chamber being located close to the base of the crust, or perhaps near the decollement zone inferred, but not yet seismically imaged, under the Eastern Manus Basin.

The preceding discussion proposes the feasibility of eruption of silicic pumice at great water depths in a province undergoing both tectonic extension and magmatism, such as the eastern Manus Basin. If future seafloor mapping at the MD-50 site or equivalent fault scarps were to confirm the presence of outcrops, then this constitutes a new style of submarine pumice occurrence that needs to be considered when interpreting ancient marine sequences.

Tectonic decompression associated with extensional faulting may also have occurred during subaerial and submarine peralkaline volcanism in the D'Entrecasteaux province within the Woodlark Basin rift zone. For the submarine pumices here, similar calculations to those used above indicate a rather shallower maximum depth limit of 0.5 to 2 km below the seabed for a potential source magma chamber, depending on what H_2O content (0–2%) is assumed for the present glass component.

FRAGMENTATION

The pumice eruptions at Healy Seamount [*Wright and Gamble, 1999*; and *Wright et al.*, 2003] and Myojin Knoll [*Fiske et al.*, 2001] were clearly explosive, and their discovery requires revision of previously proposed depth limits for submarine fragmentation of silicic froths. The pumice outcrops described here near Dobu Seamount in the western Woodlark Basin, by contrast, are interpreted as coherent lava flows, with locally ropy surfaces and characteristic polygonal fracturing of rind zones. Pyroclastic pumiceous deposits were not observed in their vicinity during a submersible dive. However, it warrants mention that a sediment core taken 22 km east of Dobu Seamount on the floor of South Valley (Figure 2) intersected an upwards-fining bed of sand-sized pumice particles and anorthoclase grains at 102–118 cm below mudline. From the sedimentation rates measured nearby [*Barash and Kuptsov*, 1997] its age is around 7000 years, and the layer is more likely derived from an explosive eruption of Numanuma Caldera (Figure 2) than from the almost certainly younger submarine pumice lavas near Dobu Seamount.

Elevated pressures (80–95 bars if Dobu pumice eruption sites are close to the observed outcrops) would explain this lack of fragmentation. Besides their direct effects in terms of fragmentation models applied to subaerial plinian and ignimbrite eruptions [e.g., *Sparks et al.*, 1994; *Papale*, 1999; *Zhang*, 1999], higher pressures during submarine eruption will facilitate retention of more H_2O in the frothing melt phase with consequent reduction of viscosity and vesiculation volume. Contrasted physical properties of peralkaline versus island arc calcic or calk-alkaline melts might also influence limiting depths for fragmentation.

Lacking visual observations at this time, the question of fragmentation at the inferred MD-50 outcrop site in the Manus Basin cannot be effectively assessed. To test the possibility, sediment cores were taken in two adjacent sediment basins (Figure 5), penetrating 17 cm (MS-57) and 70 cm (MS-56) through sequences of thin turbidite beds without intersecting pumiceous fragments or "ash" layers. The exotic pumice blocks recovered from ooze in other dredge hauls certainly indicate that some, coarse, fragmentation occurred at their primary sources which, however, need not be the MD-50 fault scarp since many equivalent sites exist in the basin. These two facts in combination perhaps add circumstantial weight to the contrary argument that the MD-50 occurrence is also exotic, emphasizing the uncertainty associated with the discussions offered above.

Polyhedral ("ripened") foam structures and extreme vesicularities like those of the MD-50 pumice are reportedly unknown among subaerial silicic pumices [*Cashman and Mangan*, 1994], although they are common in high temperature, low viscosity basaltic froths, for example from fire fountains. Again with the caveat that the *in situ* nature of MD-50 pumice has yet to be confirmed, reduced viscosity consequent upon the higher temperatures and H_2O contents permitted by eruption at high pressure on the deep ocean floor could promote development of such characteristics in silicic pumice.

CONCLUSIONS

Two further examples have been presented here of siliceous pumice occurrences in deep marine environments. One from the Woodlark Basin is a confirmed case of outcrop, and of effusive rather than explosive volcanism. The second, in the Eastern Manus Basin characterized by very hydrous magmatism, is only an inferred outcrop, dredged from about 2100 m depth at the foot of a major extensional fault scarp. Feasible mechanisms exist for pumice eruption at such a site, involving decompression in deep-seated, differentiating magma chambers at sites of contemporary magmatism and tectonic extension. The potential implications of this proposed new style of pumice source for interpretation of ancient sequences warrant high priority being assigned to deep-tow photographic surveys or submersible dives at the site or equivalents elsewhere. Confirmation of pumice outcrops in such a deep water setting will mean that future attributions of subaerial or shallow marine origins to pumice occurrences in ancient sequences will need support from other evidence, such as that employed [*Allen and McPhie*, 2000] for the Yali pumice deposits.

Anecdotally, silicic pumice fragments are not uncommonly recovered from dredge hauls in other continental margin and even mid-ocean environments, but they are generally assigned an exotic origin without further consideration. An example where controversy exists is the Central Indian Basin occurrences, variously attributed to remote eruptions such as Krakatau [*Mudholkar and Fujii*, 1995] or to local submarine activity [*Iyer and Sudhakar*, 1993]. Despite clear evidence that floating and suspended pumice can be widely dispersed [e.g., *Coombs and Landis*, 1966; *Worthington et al.*, 1999; and *Ward and Little*, 2000], tectonic unloading and decompression of subjacent differentiating magma chambers even at mid-ocean extensional faults also warrants consideration for such occurrences.

Acknowledgments. I thank Professor Alexander Lisitzin of the Russian Academy of Sciences for the privilege of diving in *Mir-2* at Dobu Seamount, and scientific colleagues plus the ship's officers and crew for assistance during RV *Franklin* operations in the Woodlark and Manus basins, for which the Australian ORV National Facility Steering Committee granted ship time. Access to the Electron Microscopy Unit at Australian National University is gratefully acknowledged. Reviews by Tim Worthington and James White greatly improved the paper.

REFERENCES

Allen, S. R. and McPhie, J., Water-settling and resedimentation of submarine rhyolitic pumice at Yali, eastern Aegean, Greece, *J. Volcanol. Geothrm. Res.*, 95, 285-307, 2000.

Auzende, J.-M., Ishibashi, J.-I., Beaudoin, Y., Charlou, J.-L., Delteil, J., Donval, J.-P., Fouquet, Y., Ildefonse, B., Kimura, H., Nishi, Y., Radford-Knoery, J. and Ru¾llan, E., Extensive magmatic and hydrothermal activity documented in Manus Basin, *Eos Trans. AGU*, 81, 449, 453, 2000.

Barash, M. S. and Kuptsov, Late Quaternary palaeoceanography of the western Woodlark Basin (Solomon Sea) and Manus Basin (Bismarck Sea), Papua New Guinea, from planktic foraminifera and radiocarbon dating, *Marine Geol.*, 142, 171-187, 1997.

Benes, V., Scott, S. D. and Binns, R. A., Tectonics of rift propagation into a continental margin: Western Woodlark Basin, Papua New Guinea, *J. Geophys. Res.*, 99, 4439-4455, 1994.

Benes, V., Bocharova, N., Popov, E., Scott, S. D. and Zonenshain, L., Geophysical and morpho-tectonic study of the transition between seafloor spreading and continental rifting, western Woodlark Basin, Papua New Guinea, *Marine Geol.*, 142, 85-98, 1997.

Binns, R. A., Report on SUPACLARK Dive 3 on Mir-2 submersible: Dobu Seamount, Woodlark Basin, PNG, *CSIRO Division of Exploration Geoscience, Report 180R*, 36 pp., 1990.

Binns, R. A., Submarine felsic volcanism in the eastern Manus and western Woodlark basins, Papua New Guinea, in *Ancient Volcanism and Modern Analogues*, edited by M. Duggan and J. Knutson, Abstracts, IAVCEI General Assembly, September 1993, Canberra Australia, p. 10, 1993.

Binns, R. A., Barriga, F. J. A. S., Miller, D. J. *et al.*, Anatomy of an active felsic-hosted hydrothermal system, eastern Manus Basin, *Proc. ODP, Init. Rep., 193 [CD-ROM]*, Ocean Drilling Program, Texas A&M University, College Station, Texas, 2002.

Binns, R. A. and Scott, S. D., Actively forming polymetallic sulfide deposits associated with felsic volcanic rocks in the eastern Manus backarc basin, Papua New Guinea, *Econ. Geol.*, 88, 2226-2236, 1993.

Binns, R. A. and Wheller, G. E., Report on the PACLARK-V/PACMANUS-I cruise, RV Franklin, Woodlark and Manus basins, Papua New Guinea, *CSIRO Division of Exploration Geoscience, Report 263R*, 107 pp., 1991.

Blake, D. H., Pumiceous pyroclastic deposits of Witori volcano, New Britain, Papua New Guinea, in *Volcanism in Australasia*, edited by R. W. Johnson, pp. 191-200, Elsevier, New York, 1976. Blake, D. H. and Ewart, A, Petrography and geochemistry of the Cape Hoskins volcanoes, New Britain, Papua New Guinea, *J. Geol. Soc. Aust.*, 21, 319-331, 1974.

Bloomer, S. H., Stern, R. J. and COOK17 Shipboard Party., Mantle inputs to the subduction factory: Detailed studies of the southern Mariana seamount province, *Eos Trans. AGU*, 82(47), Fall Meet. Suppl., F1201-1202, 2001.

Cas, R. A. F. and Wright, J. V., *Volcanic Successions Modern and Ancient*, Allen and Unwin, London, 528 pp., 1987.

Cashman, K. V. and Mangan, M. T., Physical aspects of magmatic degassing II. Constraints on vesiculation processes from textural studies of eruptive products, in *Volatiles in Magmas*, edited by M. R. Carroll and J. R. Holloway, *Reviews in Mineralogy* (Mineralogical Society of America), 30, 447-448, 1994.

Coombs, D.S., and C.A. Landis, Pumice from the South Sandwich eruption of March 1962 reaches New Zealand, *Nature*, 209, 289-290, 1966.

Davies, H. L., Fergusson Island, Papua New Guinea – 1:250,000 geological series, *Explanatory Notes SC/56-5*, Bur. Miner. Resour., Geol., and Geophys., Canberra, Australia, 1973.

Doyle. M. G. and McPhie, J., Facies architecture of a silicic intrusion-dominated volcanic center at Highway-Reward, Queensland, Australia, *J. Volcanol. Geotherm. Res.*, 99, 79-96. 2000.

Dril, S. I., Kusmin, M. I., Tsipukova, S. S. and Zonenshain, L. P., Geochemistry of basalts from the western Woodlark, Lau and Manus basins: implications for their petrogenesis and source rock compositions, *Marine Geol.*, 142, 57-83, 1997.

Fisher, N.H., *Catalogue of the active volcanoes of the world including solatara fields, Part V, Melanesia*, 105 pp., International Volcanological Association, Naples, 1957.

Fiske, R. S., Naka, J., Iizasa, K., Yuasa, K. and Klaus, A., Submarine silicic caldera at the front of the Izu-Bonin arc, Japan: Voluminous seafloor eruptions of rhyolite pumice, *Geol. Soc. Amer. Bull.*, 113, 813-824, 2001.

Global Volcanism Program, Volcanoes of offshore New Guinea and the Admiralty Islands, *www.volcano.si.edu/gvp* (accessed 19/3/03).

Goodliffe, A. M., Taylor, B., Martinez, F., Hey, R., Maeda, K. and Ohno, K., Synchronous reorientation of the Woodlark basin spreading center, *Earth Planet. Sci. Lettr.*, 146, 233-242, 1997.

Holloway, J. R. and Blank, J. G., Application of experimental results to C-O-H species in natural melts, in *Volatiles in Magmas*, edited by M. R. Carroll and J. R. Holloway, *Reviews in Mineralogy* (Mineralogical Society of America), 30, 187-230, 1994.

Itikarai, I. and de Saint-Ours, P., Esa'Ala volcanoes, *Bull. Global Volc. Network (Smithsonian Institution)*, 15, 4-5. 1990.

Iyer, S, D, and Sudhakar, M., Coexistence of pumice and manganese nodule fields; evidence for submarine silicic volcanism in the Central Indian Basin, *Deep Sea Res.*, 40, 1123-1129, 1993.

Johnson, R. W. and Smith, I. E., Volcanoes and rocks of St. Andrew Strait, Papua New Guinea, *J. Geol. Soc. Aust.*, 21, 333-352, 1974.

Johnson, R. W., Smith, I. E. and Taylor, S. R., Hot-spot volcanism in St Andrew Strait, Papua New Guinea: Geochemistry of a Quaternary bimodal rock suite, *BMR J. Aust. Geol. Geophys.*, 3, 55-69, 1978.

Kamenetsky, V. S., Binns, R. A., Gemmell, J. B., Crawford, A. J., Mernagh, T. P., Maas, R. and Steele, D., Parental basaltic melts and fluids in eastern Manus backarc Basin: implications for hydrothermal mineralisation, *Earth Planet. Sci. Lettr.*, 184, 685-702, 2001.

Kano, K., Takeushi, K., Yamamoto, T. and Hoshizumi, H., Subaqueous rhyolite block lava in the Miocene Ushikiri Formation, Shimane Peninsula, SW Japan, *J. Volcanol. Geotherm. Res.*, 46, 241-253, 1991.

Kano, K., Yamamoto, T. and Ono, K., Subaqueous eruption and emplacement of the Shinjima Pumice, Shinjima (Moeshima) Island, Kagoshima Bay, SW Japan, *J. Volcanol. Geotherm. Res.*, 71, 187-206, 1996.

Kennedy, G. C. and Holser, W. T., Pressure-volume-temperature and phase relations of water and carbon dioxide, in *Handbook of Physical Constants* (edited by S. P. Clark Jr.), Memoir 97 (Revised Edition), Geological Society of America, pp. 371-383, 1966.

Klug, C. and Cashman, K. V., Vesiculation of May 18, 1980, Mount St. Helens magma, *Geology*, 22, 468-472, 1994.

Lowenstein, P. L., Report on the occurrence of a new thermal spring at Neumara Island in the D'Entrecasteaux group, Milne Bay Province, *PNG Geol. Surv. Tech. Note, 17-82*, 5 pp., 1982.

Martinez, F. and Taylor, B., Backarc spreading, rifting, and microplate rotation between transform faults in the Manus basin, *Marine Geophys. Res.*, 18, 203-224, 1996.

McBirney, A. R., Factors governing the nature of submarine volcanism, *Bull. Volcanol.*, 26, 455-469, 1963.

Morgan, W. R., A note on the petrology of some lava types from east New Guinea, *J. Geol. Soc. Aust.*, 13, 583-591, 1966.

Mudholkar, A. and Fujii, T., Fresh pumice from the Central Indian Basin; a Krakatau 1883 signature, *Marine Geol.*, 125, 143-151, 1995.

Mutter, J. C., Mutter, C. Z. and Fang, J., Analogies to oceanic behaviour in the continental breakup of the western Woodlark basin, *Nature*, 380, 333-336, 1996.

Nairn, I. A., McKee, C. O., Talai, B. and Wood, C. P., Geology and eruptive history of the Rabaul Caldera area, *J. Volcanol. Geotherm. Res.*, 69, 255-284, 1995.

Parr, J. M. and Binns, R. A., Report on the PACMANUS-III cruise, RV Franklin, eastern Manus Basin, Papua New Guinea, *CSIRO Division of Exploration and Mining Report 345R*, 179 pp., 1997.

Papale, P., Strain-induced magma fragmentation in explosive eruptions, *Nature*, 397, 425-428, 1999.

Sinton, J. M., Ford, L. L., Chappell, B. and McCulloch, M. T., Magma genesis and mantle heterogeneity in the Manus back-arc basin, Papua New Guinea, *J. Petrol.*, 44, 159-195, 2003.

Sparks, R. S. J., Barclay, J., Jaupart, C., Mader, H. M. and Phillips, J. C., Physical aspects of magmatic degassing I. Experimental and theoretical constraints on vesiculation, in *Volatiles in Magmas*, edited by M. R. Carroll and J. R. Holloway, *Reviews in Mineralogy* (Mineralogical Society of America), 30, 413-445, 1994.

Smith, I. E. M., Peralkaline rhyolites from the D'Entrecasteaux Islands, Papua New Guinea, in *Volcanism in Australasia*, edited by R. W. Johnson, pp. 275-265, Elsevier, New York, 1976.

Smith, I. E. M. and Johnson, R. W., Contrasting rhyolite suites in the Late Cenozoic of Papua New Guinea, *J. Geophys. Res.*, 86, 10257-10272, 1981.

Stolz, A. J., Davies, G. R., Crawford, A. J. and Smith, I. E. M., Sr, Nd and Pb isotopic compositions of calc-alkaline and peralkaline silicic volcanics from the D'Entrecasteaux Islands, Papua New Guinea, and their tectonic significance, *Miner. Petrol., 47*, 103-126, 1993.

Taylor, B., Bismarck Sea: Evolution of a back-arc basin, *Geology, 7*, 171-174, 1979.

Taylor, B., Goodliffe, A., Martinez, F. and Hey, R., Continental rifting and initial sea-floor spreading in the Woodlark basin, *Nature, 374*, 534-537, 1995.

Tiffin, D. L., Taylor, B. D., Tufar, W. and Itikarai, I., A Seabeam and sampling survey of newly discovered Tavui Caldera near Rabaul, Papua New Guinea, *SOPAC Cruise Report 132*, South Pacific Applied Geoscience Commission, Fiji, 18 pp., 1990.

Torrence, R., Pavlides, C., Jackson, P. and Webb, J., Volcanic disasters and cultural discontinuities in Holocene time, in West New Britain, Papua New Guinea, in *The Archaeology of Geological Catastrophes*, edited by W. G. McGuire, D. R. Griffiths, D. R. Hancock and I. S. Stewart, pp. 225-244, Geol. Soc. London, Spec. Publ. *171*, 2000.

Wallace, P., Arculus, R. and Eggins, S., A further installment in the Rabaul story: the plot thickens, *Geol. Soc. Aust. Abstracts, 67*, 253, 2002.

Ward, W. T. and Little, I. P., Sea-rafted pumice on the Australian east coast; Numerical classification and stratigraphy, *Aust. J. Earth. Sci., 47*, 95-109, 2000.

Waters, J. C., Binns, R. A. and Naka, J., Morphology of submarine felsic rocks on Pual Ridge, Eastern Manus Basin, Papua New Guinea, *EOS Trans. Amer. Geophys. Union, 77 (WPGM Suppl.)*, W120, 1996.

Weissel, J. K., Taylor, B. and Karner, G. D., The opening of the Woodlark Basin, subduction of the Woodlark spreading system, and the evolution of northern Melanesia since mid-Pliocene time, *Tectonophys., 87*, 253-277, 1982.

Wiebenga, W. A., Crustal structure of the New Britain-New Ireland region, in *The Western Pacific: Island arcs, Marginal Seas, Geochemistry*, edited by P. J. Coleman, Uni. West Aust. Press, 163-177, 1973.

Wood, C. P., Nairn, I. A., McKee, C. O. and Talai, B., Petrology of the Rabaul Caldera area, Papua New Guinea, Woodhead, J. D. and Johnson, R. W., Isotopic and trace-element profiles across the New Britain island arc, Papua New Guinea, *Contr. Min. Petrol/, 113*, 479-491, 1993.

Worthington, T. J., Gregory, M. R. and Bondarenko, V., The Denham caldera on Raoul volcano: dacitic volcanism in the Tonga-Kermadec arc, *J. Volcanol. Geother. Res., 90*, 29-48, 1999.

Wright, I. C. and Gamble, J. A, Southern Kermadec submarine caldera arc volcanoes (SW Pacific): caldera formation by effusive and pyroclastic eruption, *Mar. Geol., 161*, 207-227, 1999.

Wright, I. C., Gamble, J. A. and Shane, P. A. R., Submarine silicic volcanism of the Healy caldera, southern Kermadec arc (SW Pacific): 1 – volcanology and eruption mechanisms, *Bull. Volcanol., 65*, 15-29, 2003.

Yamagishi, H. and Goto, Y., Cooling joints of subaqueous rhyolite lavas at Kuroiwa, Yakumo, southern Hokkaido, Japan, *Bull. Volcanol. Soc. Japan, 37*, 205-207, 1992.

Yang, K. and Scott, S. D., Possible contribution of a metal-rich magmatic fluid to a sea-floor hydrothermal system, *Nature, 383*, 420-423, 1996.

Yuasa, M., Submarine pumice volcano in the Izu-Ogasawar (Bonin) arc; a submersible study of the Myojin Knoll, *Abstracts, 29th International Geological Congress, 29*, 496, 1992.

Yuasa, M., Myojin Knoll, Izu-Ogasawara arc: submersible study of submarine pumice volcano, *Bull. Volcanol. Soc.* Japan, 40, 277-284, 1995 (in Japanese with English abstract).

Zhang, Y., A criterion for the fragmentation of bubbly magma based on brittle failure theory, *Nature, 402*, 648-650, 1999.

R.A. Binns, CSIRO Exploration and Mining, PO Box 136, North Ryde, NSW, 1670, Australia, Ray.Binns@csiro.au

Morphology, Distribution, and Estimated Eruption Volumes for Intracaldera Tuffs Associated With Volcanic-Hosted Massive Sulfide Deposits in the Archean Sturgeon Lake Caldera Complex, Northwestern Ontario

George J. Hudak[1,2], Ronald L. Morton[2], James M. Franklin[3], and Dean M. Peterson[2,4]

The Archean Sturgeon Lake Caldera Complex (SLCC) comprises a well-preserved, north-facing homoclinal sequence of greenschist facies metamorphosed intrusive, volcanic, and sedimentary strata. This piecemeal caldera complex is at least 25 km in strike length and contains nearly 3000 meters of dominantly subaqueously deposited intracaldera fill. Episodes of subaerial and subaqueous explosive felsic volcanism produced rhyodacitic to rhyolitic tuffs and lapilli tuffs. Progressing stratigraphically upward, the most voluminous are: a) the High Level Lake Tuff (~16km^3); b) the Mattabi Tuff (~27km^3); and c) the Middle L Tuff (~7km^3). The subaerially erupted, subaerially and locally subaqueously deposited High Level Lake Tuff comprises an 80–300 meter-thick unit composed of basal, poorly sorted, massive to normal graded, quartz-phyric, locally spherulitic tuffs and lapilli tuffs (30–150m thick) that are overlain by thin-bedded tuffs (<1–5m thick). The subaqueously erupted and deposited Mattabi Tuff contains up to thirteen individual flow units, each comprising two distinct depositional facies: a) lower, quartz-phyric, poorly sorted, ungraded, massive tuffs and lapilli tuffs (20–250 meters thick); and b) upper, laminated to medium bedded, typically normal graded tuffs (1–13 meters thick). The subaqueously erupted and deposited Middle L Tuff is also characterized by two distinct lithofacies: a) lower graded, quartz- and, rarely, potassium feldspar-phyric tuffs and lapilli tuffs (5–120m thick); and b) overlying, well-sorted, laminated to thickly bedded, normal graded tuffs (<1–5m thick). These three voluminous tuff deposits host all known volcanic-hosted massive sulfide (VHMS) ore bodies in the SLCC. At Sturgeon Lake, VHMS ore deposition appears to be favored by processes associated with the generation of voluminous subaqueous explosive eruptions.

1. INTRODUCTION

Volcanologists realize that it is essential to study ancient volcanic sequences in order to understand the genetic processes and developmental dynamics of modern volcanic systems [*Lafrance et al.*, 2000; *Mueller et al.*, 2000]. Archean volcanic rocks, especially those that have experienced only low-grade metamorphism and minor structural deformation, often contain exceptionally well-preserved textures and stratigraphic sequences. Studies of ancient volcanic sequences have shed significant light on subaqueous eruptive and depositional processes that are impossible, or extremely difficult, to study in modern systems [*Stix*, 1991; *Kokelaar and Busby*, 1992; *Lipman*, 1997; *Mueller et al.*, 2000]. They have also been indispensable for increasing our understanding of the interrelationships between subaqueous

---

[1]Department of Geology, University of Wisconsin Oshkosh, Oshkosh, Wisconsin

[2]Department of Geosciences, University of Minnesota – Duluth, Duluth, Minnesota

[3]Franklin Geosciences Ltd., Nepean, Ontario, Canada

[4]Economic Geology Group, Natural Resources Research Institute, Duluth, Minnesota

Explosive Subaqueous Volcanism
Geophysical Monograph 140
Copyright 2003 by the American Geophysical Union
10.1029/140GM23

volcanic activity and synvolcanic ore-forming processes [*Allen et al.*, 1996a, 1996b; *Cas*, 1992; *Gibson et al.*, 1999].

The Sturgeon Lake region is located 230-250 km northwest of Thunder Bay Ontario within the Archean Wabigoon volcano-sedimentary greenstone belt (Figure 1). This area has been the focus of several field-oriented volcanological studies since the mid-1980s [*Groves*, 1984; *Groves et al.*, 1988; *Hudak*, 1989, 1996; *Jongewaard*, 1989; *Morton et al.*, 1991, 1999; *Walker*, 1993], and has been interpreted to contain a piecemeal, Archean, subaqueous caldera complex known as the Sturgeon Lake Caldera Complex (SLCC) [*Morton et al.*, 1989, 1991, 1999]. Field, petrographic, and geochemical studies have enabled vertical and lateral facies transitions within the intracaldera strata to be evaluated. While all intracaldera tuffs contain sub-economic volcanic-hosted massive sulfide (VHMS) mineralization, only the three most voluminous intracaldera tuffs host VHMS ore.

This relationship suggests a possible link between the scale of subaqueous explosive volcanism, and the formation of VHMS ore deposits in the SLCC.

The purpose of this paper is to: a) briefly describe the geological setting and stratigraphic development of the SLCC; b) provide descriptions of, and discuss the eruptive and depositional processes associated with, the three most voluminous intracaldera felsic tuff units; and c) evaluate the apparent genetic relationships between the scale of subaqueous explosive volcanism and the formation of VHMS within this ancient caldera complex.

2. REGIONAL GEOLOGY

Detailed and reconnaissance mapping indicate the SLCC is at least 25 kilometers in strike length [*Morton et al.*, 1991] and contains up to 3000 meters of intracaldera fill.

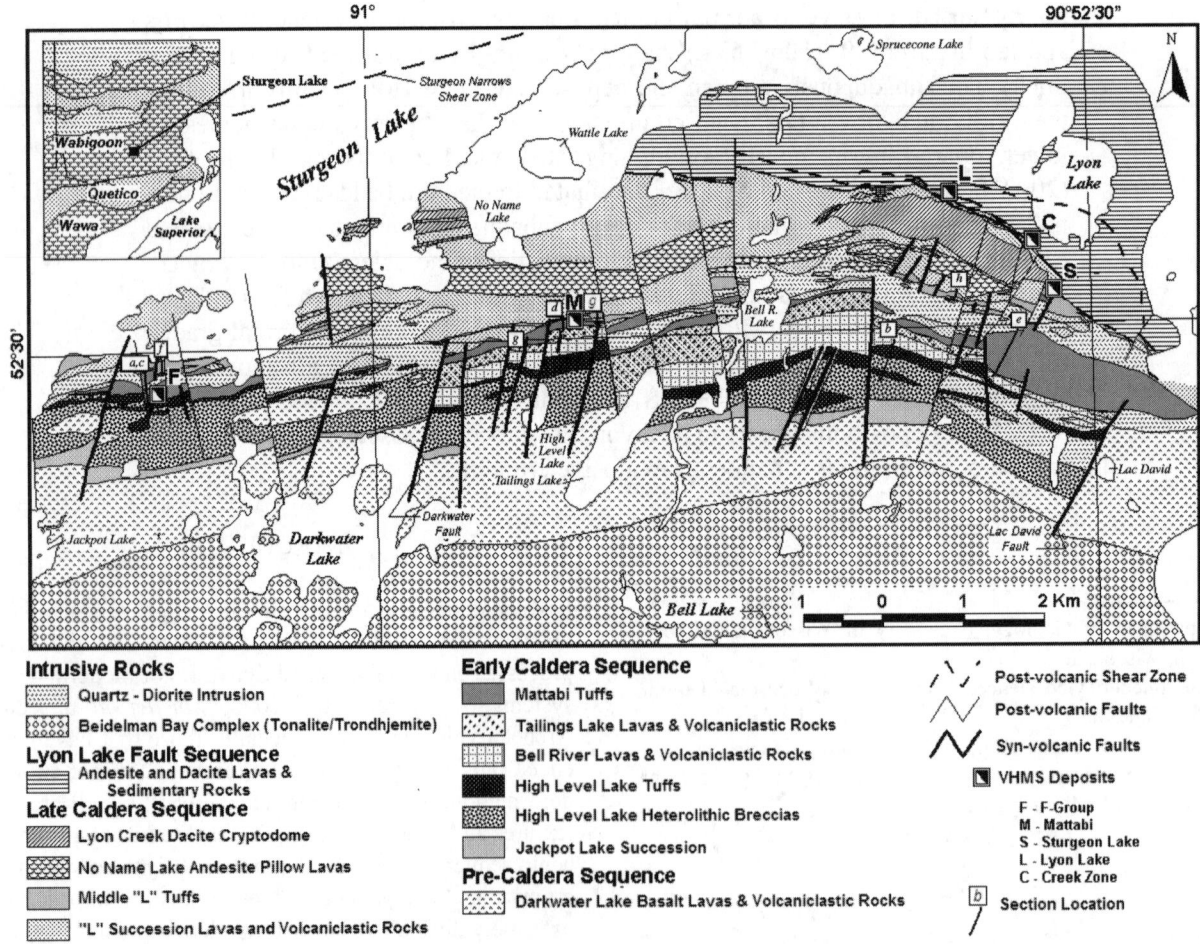

Figure 1. Location map (inset) and geological map of the south Sturgeon Lake region. Strata have been classified into Pre-caldera, Early-Caldera, Late-Caldera and Lyon Lake Fault Sequences after *Hudak* [1996]. Modified after *Morton et al.* [1999] and *Galley et al.* [2000].

This includes six past-producing VHMS orebodies that yielded 18.7 million tons of ore grading 1.06% Cu, 8.5% Zn, 0.91% Pb, and 119.7 grams/ton Ag between 1972 and 1991 [*Franklin*, 1996]. The eastern caldera wall has been interpreted by *Morton et al.* [1991, 1999] to be the Lac David Fault; the western margin of the caldera has been truncated by the northeast-trending Sturgeon Narrows Shear Zone and thus remains poorly constrained (Figure 1). All rocks in the SLCC have been metamorphosed (greenschist and locally amphibolite facies); however, the prefix "meta-" has been omitted in subsequent descriptions of the strata for brevity.

Thirteen supracrustal stratigraphic successions have been grouped into four stratigraphic sequences based on their temporal relationships to caldera development, and their stratigraphic position relative to the Mattabi VHMS orebody [*Hudak*, 1996]. Progressing stratigraphically up-section, these are the Pre-caldera Sequence, the Early caldera Sequence, the Late-caldera Sequence, and the Lyon Lake Fault Sequence. A major shear zone [*Koopman*, 1993] or thrust fault [*Hudak*, 1996] marks the contacts between the Pre-, Early, and Late-caldera Sequences with the overlying Lyon Lake Fault Sequence (Figure 1).

The Pre-caldera Sequence comprises a 200-2100m thick succession of subaerial and shallow subaqueous basalt lava flows, scoria-rich volcaniclastic deposits, and minor associated rhyolite lava flows [*Groves et al.*, 1988; *Morton et al.*, 1991]. Pillow lavas and hyaloclastite are conspicuously absent in this sequence, except in the easternmost regions of the south Sturgeon Lake area [*Jongewaard*, 1989]. *Groves et al.* [1988] have interpreted the volcanic environment as a subaerial to shallow subaqueous shield volcano with local fields of scoria cones and tuff cones.

The Early caldera Sequence contains a 650-1300m thick succession of volcanic and volcaniclastic strata. Up-section, these include: a) subaerial ash fall tuff deposits (Jackpot Lake Succession); b) interstratified heterolithic breccias and subaerial and subaqueously deposited quartz-phyric lapilli tuff and tuff deposits (High Level Lake Succession); c) subaerial felsic lava flows, mafic-intermediate lapilli tuffs and volcaniclastic deposits (Bell River Succession); d) interstratified subaqueous heterolithic breccias, volcanic sandstones and mudstones, and dacitic to andesitic lava flows and tuffs (Tailings Lake Succession); and e) subaqueous, massive to locally well-bedded, quartz-phyric lapilli tuff and tuff deposits (Mattabi Succession). *Hudak* [1996] interpreted this stratigraphic sequence as characteristic of the early stages of caldera development [c.f. *Smith and Bailey*, 1968; *Busby-Spera*, 1984; *Lipman*, 1976, 1997]. The distribution of voluminous felsic volcaniclastic units and associated heterolithic breccias suggests that initial caldera collapse occurred during the deposition of the High Level Lake Succession. A synvolcanic basin bounded by the Darkwater and Lac David Faults (Figure 1) may represent a nested caldera produced simultaneously with the eruption of the Mattabi Tuffs.

The Late-caldera Sequence is composed of a 500-1500m thick succession of quartz- and quartz-feldspar-phyric tuff and lapilli tuff deposits, volcaniclastic sedimentary rocks, andesitic to dacitic lava flows, lava domes (locally cryptodomes), and banded iron formations [*Koopman*, 1993; *Hudak*, 1996]. *Hudak* [1996] interpreted this sequence as characteristic of a maturing, late stage caldera complex [c.f. *Smith and Bailey*, 1968; *Campbell et al.*, 1987; *Chesner and Rose*, 1991: *Cronan et al., 1995*; *Lipman*, 1997]. The most voluminous explosive eruptions in the Late-caldera Sequence produced the Middle L Tuffs. These tuff and lapilli tuff deposits host the VHMS ore deposits in the Late-caldera Sequence (Table 1).

Two extensive synvolcanic intrusive complexes, the Pike Lake Gabbroic Complex and the composite tonalitic to trondhjemitic Beidelman Bay Intrusive Complex (BBIC), intruded the Pre-caldera Sequence at relatively shallow levels in the crust [*Galley et al.*, 2000]. The similar REE geochemical signatures [*Campbell et al.*, 1982; *Lesher et al.*, 1986; *Galley et al.*, 2000] and coeval U-Pb zircon age dates for the BBIC (2733.8 ± 1.4my) and Late-caldera Sequence volcanic rocks (2735.5 ±1.5my) [*Davis and Trowell*, 1982; *Davis et al.*, 1985] suggest that the upper parts of the intracaldera fill were, in part, derived from eruptions of the BBIC.

3. METHODS

The geology within the SLCC has been determined by means of reconnaissance and detailed (1:100) outcrop mapping, relogging of >300,000 m of surface and underground diamond drill core, petrographic analysis of >2100 thin sections, and evaluation of nearly 1900 major and trace element geochemical analyses. Geographic information system analysis has been completed on all strata in the south Sturgeon Lake area, and has enabled precise determination of unit areas. The estimated volumes of the various tuff units have been calculated as cylinders utilizing the dip-corrected average thickness of the unit, and a radius equal to one-half the strike length of the unit [c.f. *Kokelaar and Busby*, 1992].

4. DESCRIPTIONS OF INTRACALDERA TUFFS AND MINERALIZATION

4.1. High Level Lake Tuffs

The oldest intracaldera tuffs associated with the SLCC are the High Level Lake Tuffs (Figure 1). These rhyodacitic

Table 1. Summary of the physical, depositional, and eruptive characteristics of the High Level Lake, Mattabi, and Middle L intracaldera tuffs.

Tuff Unit	High Level Lake	Mattabi	Middle L
Unit Thickness Estimated	80–300 meters	15–650 meters	15–150 meters
Volume (km^3)	16	27	7
Lithofacies Present	– Poorly sorted, ungraded, very thick-bedded (30-150m), matrix-supported, spherulitic quartz-phyric tuffs and lapilli tuffs – Poorly-sorted, faintly normal graded, very thick-bedded, matrix-supported quartz-phyric tuffs and lapilli tuffs up to 25 m thick – Well-sorted, typically normal graded, laminated to thin-bedded (<0.01–0.3m), aphyric to quartz-phyric tuffs (up to 10 m thick) that immediately overlie or are replaced by VHMS mineralization	– Poorly-sorted, non-graded, very thick-bedded (10–155m), matrix-supported quartz-phyric tuffs and lapilli tuffs that locally contain spherulites and lithophysae (MLT) – Poorly-sorted, ungraded to normal graded quartz-phyric tuffs and lapilli tuffs ranging from 6-48 m thick (MLT) – Well-sorted, typically normal graded, but locally inverse graded, laminated to medium-bedded (<0.01–0.3m) aphyric to quartz-phyric tuff (up to 13 m thick) that is locally replaced by VHMS mineralization (MBT) – Massive to laminated, aphyric to sparsely quartz-phyric tuffs from 20–150 m thick (MMT)	– Ungraded to normal graded, massive to very thick-bedded (3–10m), matrix-supported quartz- ± k-spar-phyric tuffs and lapilli tuffs (5–120 m thick) locally interstratified with semi-massive and massive VHMS deposits (MLLT) – Well-sorted, typically normal graded, laminated to very thick-bedded (<0.1–5m) aphyric tuff and quartz-phyric tuff (MLBT) – Matrix-supported, inverse graded volcanic breccia containing 30–60% subangular to angular, 1–35cm diameter jigsaw puzzle-fit, poorly-vesiculated quartz-phyric rhyolite lava fragments, pumice lapilli, and <1-2% VHMS clasts in area 2km east of Bell River Lake.
Bounding Facies Stratigraphically Below Unit	Gradational contact over 10–15 meters with poorly-sorted, unbedded matrix-supported polymict breccias (eastern caldera); gradational contact over 5–10 meters with poorly-sorted, matrix-supported, inverse- to normal-graded, thin- to very thick-bedded polymict breccias, sandstones and mudstones (western caldera) which lack wave-generated sedimentary structures.	Normal to inverse graded, subaqueously deposited, matrix-supported volcanic breccias, volcanic sandstones, and mudstones (Mattabi and Sturgeon Lake areas); VHMS lenses and exhalites within subaqueously deposited tuff and lapilli tuff (F-Group area). Wave-generated sedimentary structures are absent.	Ungraded to normal graded, polymict, commonly crystal-rich volcanic breccias, volcanic sandstones, and graphite-bearing mudstones which lack wave-generated sedimentary structures.
Bounding Facies Stratigraphically Above Unit	Subaerial rhyodacite lavas (eastern caldera); subaqueously deposited tuffs, lapilli tuffs, and VHMS (western caldera) which lack wave-generated sedimentary structures. Locally interstratified with subaqueous breccias, sandstones, and mudstones in western caldera.	Non-graded to normal graded water-settled tuffs and lapilli tuffs interstratified with VHMS lenses and subaqueous lava flows. Wave generated sedimentary structures are absent.	Ungraded to normal graded, polymict, commonly crystal-rich volcanic breccias, volcanic sandstones, and graphite-bearing mudstones which lack wave-generated sedimentary structures
Evidence for Subaqueous Eruption	– Lacking in eastern caldera – Not well constrained in western caldera due to intense hydrothermal alteration (poor preservation of original vitric clasts), b) a poorly-defined western margin for the caldera complex (due to post-volcanic structural deformation associated with the Sturgeon Narrows Shear Zone); and c) a lack of sufficient outcrop or diamond drilling in the region – Deposit characteristics are consistent with subaqueous deposition from either subaerial or subaqueous eruptions	– Lack of fines-depletion in massive units – Lack of inversely graded pumice suggests initially hot pumice (sinkers) – Heat retention features indicate exclusion of water from basal tuffs and lapilli tuffs near Mattabi VHMS – Angular (water-shattered) shards are confined primarily to bedded tuff deposits overlying massive units (Mattabi and Sturgeon Lake areas) – Ponding of chemically stratified Mattabi Tuffs against Lac David and Darkwater Lake Faults suggests intracaldera eruption	– Subaqueous volcanic breccia 2km east of Bell River Lake is compositionally similar to, and grades up into, overlying Middle L Tuffs – Vertical fining-upward sequences are consistent with deposition from collapse of subaqueous eruption column – Local erosive contacts represent deposition from eruption-fed turbidity currents – Ponding in eastern and western margins of SLCC suggests intracaldera eruption
VHMS Ore Deposits Present	F-Group	Mattabi B, C, and D ore lenses	Mattabi A-lens and Sturgeon Lake (Lyon Lake, Creek Zone, and Sub-Creek Zone) deposits
Interpretation	Subaerially erupted, subaerially deposited pyroclastic flows and airfall ash deposits (eastern caldera); syn- to post-eruptive high- to low-density mass flows and turbidity currents and water-settled ash (western caldera) deposited below wave base (>200m).	Subaqueously erupted and subaqueously deposited pyroclastic flows, eruption-fed high particle concentration flows, and syn- to post-eruptive low particle concentration aqueous density currents and water-settled ash deposited at water depths between 200m and 1000m.	Subaqueously erupted and subaqueously deposited eruption-fed high to low particle concentration aqueous density currents and syn- to post-eruptive water-settled ash deposited at water depths between 200m and 1000m.

– Volcaniclastic classifications based on *Fisher* [1961, 1966]

to rhyolitic [*Hudak*, 1989; *Jongewaard*, 1989; *Walker*, 1993], massive quartz-phyric tuff and lapilli tuff (Figure 2a) and bedded tuff deposits occur as lenses (up to 1.5 km in length and up to 100 m thick) within heterolithic breccias (High Level Lake breccias, HLBX), and as an 80–300m thick unit that overlies the breccias and can be traced along strike for 22 km. The contacts between the HLBX and overlying High Level Lake Tuffs are gradational (Table 1); they are marked by an upward increase in the size and percentage of quartz crystals, and a marked decrease in the number of lithic clasts (< 5%) within the HLBX deposits [*Hudak*, 1989; *Morton et al.*, 1991].

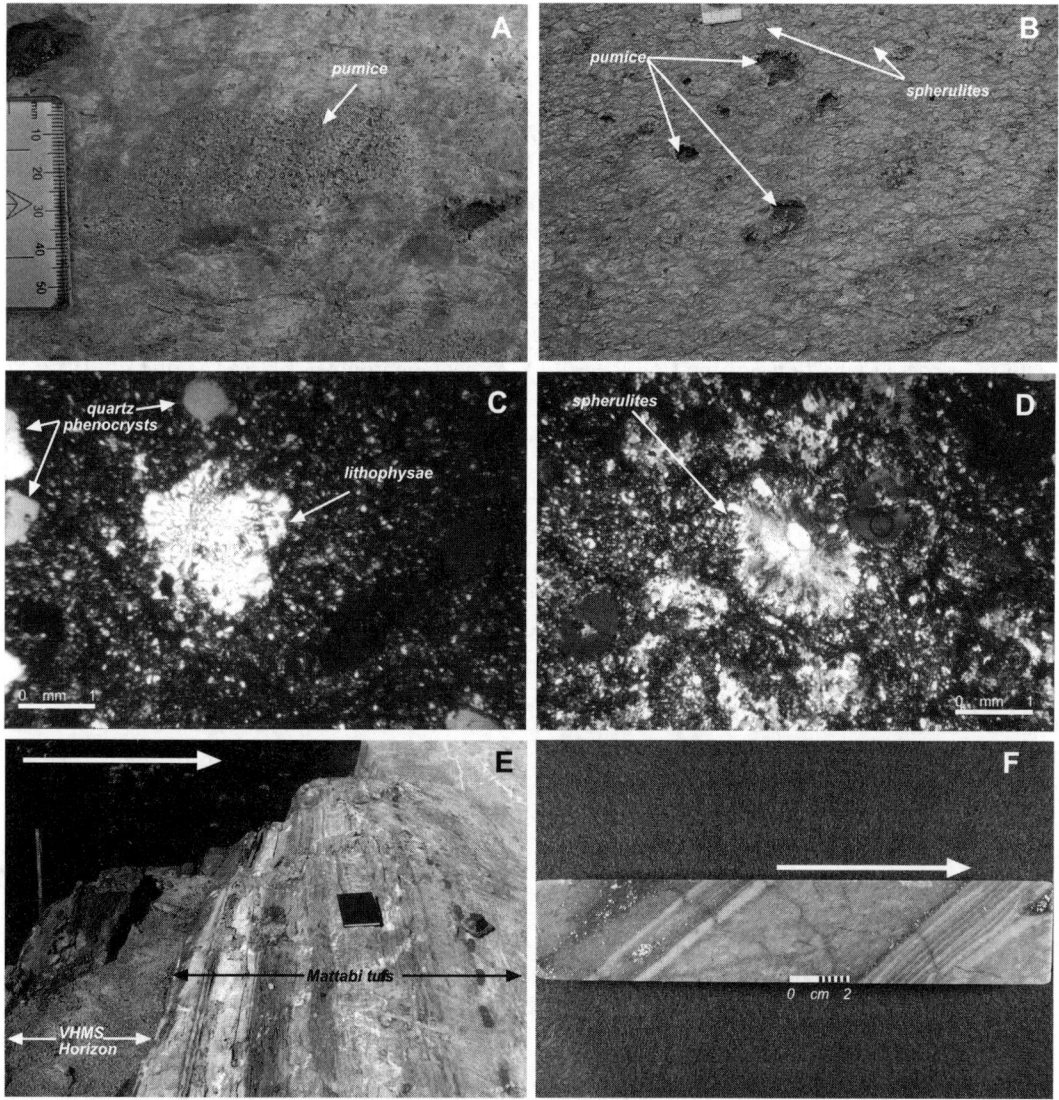

Figure 2. Photographs showing selected features of the High Level Lake and Mattabi tuffs. (a) massive quartz-phyric lapilli tuff from the High Level Lake Succession. (b) amoeboid and angular pumice lapilli within a massive basal deposit (MLT) of the Mattabi Tuffs. (c) photomicrograph illustrating lithophysae within massive basal Mattabi Tuff deposits (MLT) near the Mattabi VHMS ore body. Field of view 8mm by 5.5 mm. (d) photomicrograph of spherulite with the basal Mattabi Tuff deposits. Field of view 8mm by 5.5 mm. (e) laminated to medium-bedded aphyric to quartz-phyric Mattabi Tuff (MBT; right side of photo) in the hanging wall to a small, sphalerite-rich VHMS deposit (left side of photo) which replaces massive quartz-phyric Mattabi Tuff (MLT) approximate 500 meters north of the F-Group VHMS ore body. Large arrow at upper left of photo indicates topping direction. The field book used for scale is 20cm by 13 cm. (f) drill core sample of laminated to thin-bedded Mattabi Tuff (MBT) deposits. Bedding planes locally contain pyrite and sphalerite mineralization. Scale is 2cm long. Large arrow indicates topping direction.

The High Level Lake Tuffs exhibit marked lateral facies changes across the width of the SLCC (Figures 1 and 3). From the Lac David Fault westward to the Darkwater Fault, *Jongewaard* [1989] and *Walker* [1993] found that these deposits comprise basal sections of massive, spherulitic (25–65%), quartz-phyric (1–3%, ≤ 1mm) tuff which is locally overlain by massive quartz-phyric (2–3%, 1mm) tuff that contains up to 20% pumice lapilli, as well as lenses of lapilli- to block-sized angular siliceous lithic fragments (5–10%). Accessory fragments of trondhjemite and granodiorite (up to 1%) are locally present. South of the Sturgeon Lake Mine, the High Level Lake Tuffs are interstratified with, and overlie, HLBX deposits which are similar in morphology to mesobreccias and megabreccias described in subaerial felsic caldera complexes by *Lipman* [1976, 1997]. Neither the HLBX nor the tuffs occur east of the Lac David fault. *Morton et al.* [1989, 1991] interpreted this structure as the eastern ring fracture for the SLCC. The tuffs are overlain by subaerial lava flows of the Bell River Succession.

Beneath the Mattabi Mine, the High Level Lake Tuffs overlie the HLBX, and, in turn, are overlain by massive mafic-intermediate lapilli tuffs of the Bell River Succession. To date, only sub-economic zinc-silver-lead stringer mineralization, similar to that which occurs in subaerial epithermal systems [*White and Hedenquist*, 1990, 1995], has been recognized in the High Level Lake Tuffs in the eastern part of the SLCC.

In the western part of the SLCC, near the F-Group VHMS deposit (Figure 1), the High Level Lake Tuffs are interstratified with, and directly overlie, the HLBX (Figure 3; Table 1). The HLBX in this region are different than those in the eastern part of the caldera, being composed of massive, very thick-bedded, matrix-supported breccias overlain by thin- to very thick-bedded, matrix-supported, normal and locally inverse graded breccias, sandstones, and mudstones. The contact between the HLBX and an interstratified lens of High Level Lake Tuff is gradational over 5–10m. The interstratified unit of High Level Lake Tuff is composed of two horizons (Figure 3): a) a basal unit of massive, faintly normal graded, aphyric to quartz-phyric (1–3%, <1–1.5mm) tuff which is overlain by thin- to medium-bedded tuff containing inverse graded pumice lapilli; and b) an overlying, 15m thick unit of massive, crudely normal graded quartz-phyric tuff. High Level Lake Tuff deposits which overlie the HLBX deposits are bedded, and comprise basal, massive, quartz-phyric (1–10%, locally up to 35%, ≤1mm) tuff and lapilli tuff horizons (up to 25m thick) that are overlain by 1-10m of laminated to very thin plane-parallel bedded, ungraded to normal graded, tuffs. Rare, euhedral quartz crystals are locally present in the tuffs; more commonly, however, quartz crystals are broken into sliver-like shards with smooth conchoidal fracture surfaces.

The F-Group VHMS ore body (Figure 1) occurs at the contact between basal, massive quartz-phyric tuffs and lapilli tuffs, and overlying laminated to bedded tuffs. The ore body occurs as small, separated lenses, each of which contains a chalcopyrite-pyrite-galena footwall stringer zone that passes upward into basal copper-rich ore, a central core of zinc-silver-rich ore, and a pyrite-rich cap [*Mumin*, 1988]. Sulfide mineralization that occurs along bedding planes, and the local presence of phenocrysts and clasts within the ore, suggest a subseafloor replacement origin for the F-Group VHMS. Alteration mineral assemblages adjacent to the VHMS mineralization are aluminous (andalusite, kyanite, pyrophyllite) and consistent with a pre-metamorphic acid-leached alteration zone [*Hudak*, 1989]. The morphology, alteration mineralogy, and ore mineralogy are similar to shallow subaqueous epithermal-like VHMS in modern and ancient environments [*Osterberg et al.*, 1987; *Sillitoe et al.*, 1996; *Gibson et al.*, 1999; *Hannington et al.*, 1999].

4.2. Mattabi Tuffs

The Mattabi Tuffs represent the most voluminous eruptive event within the SLCC (Table 1), forming rhyodacitic to rhyolitic volcaniclastic deposits that can be traced along strike for at least 20 kilometers (Figure 1). The Mattabi Tuffs comprise two morphologically distinct units: a) a lower, bedded unit (MTQ) composed of alternating deposits of massive quartz-phyric tuff and lapilli tuff (MLT) and plane-parallel laminated to medium-bedded aphyric to quartz-phyric tuff (MBT); and b) an upper, massive to bedded tuff (MMT) [*Morton et al.*, 1991].

The MTQ ranges from 100-650m thick within a synvolcanic basin bounded by the Darkwater and Lac David faults (Figure 1). MTQ deposits directly overlie subaqueously deposited volcanic breccias, volcanic sandstones and mudstones (Tailings Lake Succession) within this basin (Figure 3) [*Jongewaard*, 1989; *Walker*, 1993]. Here, the MLT deposits (Figure 2b) comprise poorly sorted, ungraded, matrix-supported quartz-phyric (5-35%, <1–2mm) tuffs and lapilli tuffs that range from 10–155 meters in thickness. Lapilli and bomb-sized pumice are amoeboid to subrounded, well-vesiculated (40–70%), and exhibit no sorting or density stratification within individual MLT units. Petrographic analyses have identified spherulites and, locally, lithophysae within these deposits (Figures 2c and 2d). The MLT are overlain by individual 1–13m thick horizons of plane-parallel laminated to medium-bedded, typically normal graded (but locally inverse graded) aphyric to sparsely quartz-phyric tuff deposits (MBT; Figures 2e and

2f). Spherulites and lithophysae have not been recognized in the MBT deposits.

Up to 13 individual flow units [*Fisher and Schminke*, 1984, p. 195] composed of MLT and overlying MBT deposits have been distinguished by mapping and lithogeochemical analyses in this basin (Figure 3) [*Morton et al.*, 1991; *Bezenek*, 1992]. Locally, flow units are separated by <1–3m of intercalated medium-bedded to very thick-bedded heterolithic breccias. Individual MLT deposits are chemically stratified, with high Zr (550–1025 ppm), Y (75–120 ppm) and Nb (30–80 ppm) at and/or near their bases that gradually decrease upwards toward the MBT deposits, which have low Zr (80–230 ppm), Y (24–45 ppm) and Nb (1–4 ppm) contents; TiO_2 and Ba appear to have opposite trends [*Morton et al.*, 1991]. This zoning pattern holds true regardless of the size or percentage of quartz crystals found in the MLT, and regardless of the thickness of the beds. The Mattabi B, C and D VHMS ore bodies occur at breaks between MLT deposits (hanging wall) and MBT deposits (footwall) within the MTQ unit.

East of the Darkwater Fault, MTQ deposits are locally overlain by massive to bedded tuffs that vary from 20–150m in thickness (MMT; Table 1). The MMT contains a basal unit of massive, aphyric tuff, and locally, quartz-phyric tuff that contains up to 2% conchoidal, sliver-like quartz crystal shards (<1mm in length) enclosed by a re-crystallized ash matrix made up of very fine-grained quartz. This massive unit is overlain by 5–35m of laminated to thin-bedded tuff that is compositionally and chemically similar to the underlying massive tuff deposit. The MMT contains no known massive sulfide mineralization, and lacks the distinctive geochemical zoning exhibited by the MTQ.

The MTQ deposits thin rapidly west of the Darkwater Fault (Figure 1), ranging from 20-80m thick between the Mattabi and F-Group deposits, and varying from 125–200m thick within a synvolcanic basin in the vicinity of the F-Group VHMS deposit [*Hudak*, 1989; *Morton et al.*, 1999]. West of the Darkwater Fault, the MTQ deposits immediately overlie the High Level Lake tuffs and HLBX deposits. The MTQ are characterized by poorly sorted, massive to normal graded quartz-phyric (1–10% <1–2mm) pumiceous (3–30%) tuffs and lapilli tuffs (6–48m thick) that are overlain by well-sorted, typically normal graded, planar laminated to thin-bedded tuff and crystal tuff horizons (<1–5m thick; Figure 3). Lithophysae and spherulites are absent from these deposits. As well, they lack the geochemical cyclicity of the MTQ in the eastern two-thirds of the SLCC.

Although a small, sub-economic VHMS deposit occurs at the contact between the uppermost MLT and overlying bedded MBT deposits (Figure 2e) north of the F-Group ore body, all of the known mineralization within the MTQ deposits west of the Darkwater Fault is composed of very localized Cu-Zn-Ag stringers and thin (<0.1m thick) sulfide replacement deposits along bedding planes. The MMT deposits have not been identified by mapping or drill core logging west of the Darkwater Fault.

4.3 Middle L Tuffs

The Middle L Succession is up to 150 meters thick, and contains rhyodacitic to rhyolitic aphyric to quartz ± potassium feldspar-phyric tuffs and lapilli tuffs that can be traced for at least 15 km across the SLCC (Figures 1 and 3; Table 1). These deposits are thickest immediately west of the Sturgeon Lake Mine, and in the vicinity of the F-Group VHMS deposit; they thin dramatically to <70m north of the Mattabi ore body [*Walker*, 1993]. The Middle L Tuffs host all known VHMS ore bodies in the Late-caldera Sequence (Table 1). The Lyon Lake, Creek Zone, and Sub-Creek Zone VHMS ore bodies were originally parts of the Sturgeon Lake ore body that have been moved into their present positions by post-volcanic faulting [*Hudak*, 1996; *Morton et al.*, 1999].

The morphology of the Middle L Tuffs is generally similar across the caldera complex (Figure 3) and consists of two lithofacies: a) ungraded to normal graded, massive to very thick-bedded quartz- ± potassium feldspar-phyric tuffs and lapilli tuffs (MLLT) that range from 3–60 m thick (Figure 4a); and b) overlying aphyric to quartz-phyric, normal graded, planar laminated to thick-bedded tuff deposits (MLBT) that range from 0.1–15m thick (Figures 4b and 4c). Within the MLLT, quartz crystals (<1-6mm, 1–18%) vary from euhedral and locally embayed, to conchoidally fractured and sliver-shaped; they also exhibit a pronounced normal grading. Angular to oval pumice lapilli (up to 35%) are up to 2cm in length, and locally exhibit inverse size grading. The MLBT deposits are typically aphyric, and vary from planar laminated to medium bedded, although rare thick-bedded horizons have been mapped by *Hudak* [1996] in the eastern one-third of the SLCC. Small (<1mm) quartz crystals and crystal shards typically exhibit normal grading within the MLBT beds (Figure 4c). Contacts between the MLBT and overlying MLLT deposits are locally wavy and scoured (Figure 4b), whereas contacts between MLLT and overlying MLBT deposits are gradational. Although similar in appearance, the MLLT and MLT deposits can be distinguished petrographically by the coarse, inequigranular matrix which characterizes the MLLT deposits. Middle L deposits in the western part of the SLCC are characterized by a higher proportion of massive, quartz-phyric tuffs and lapilli tuffs relative to thin-bedded tuffs and crystal tuffs when compared with equivalent deposits in the central part of the SLCC (Figure 3).

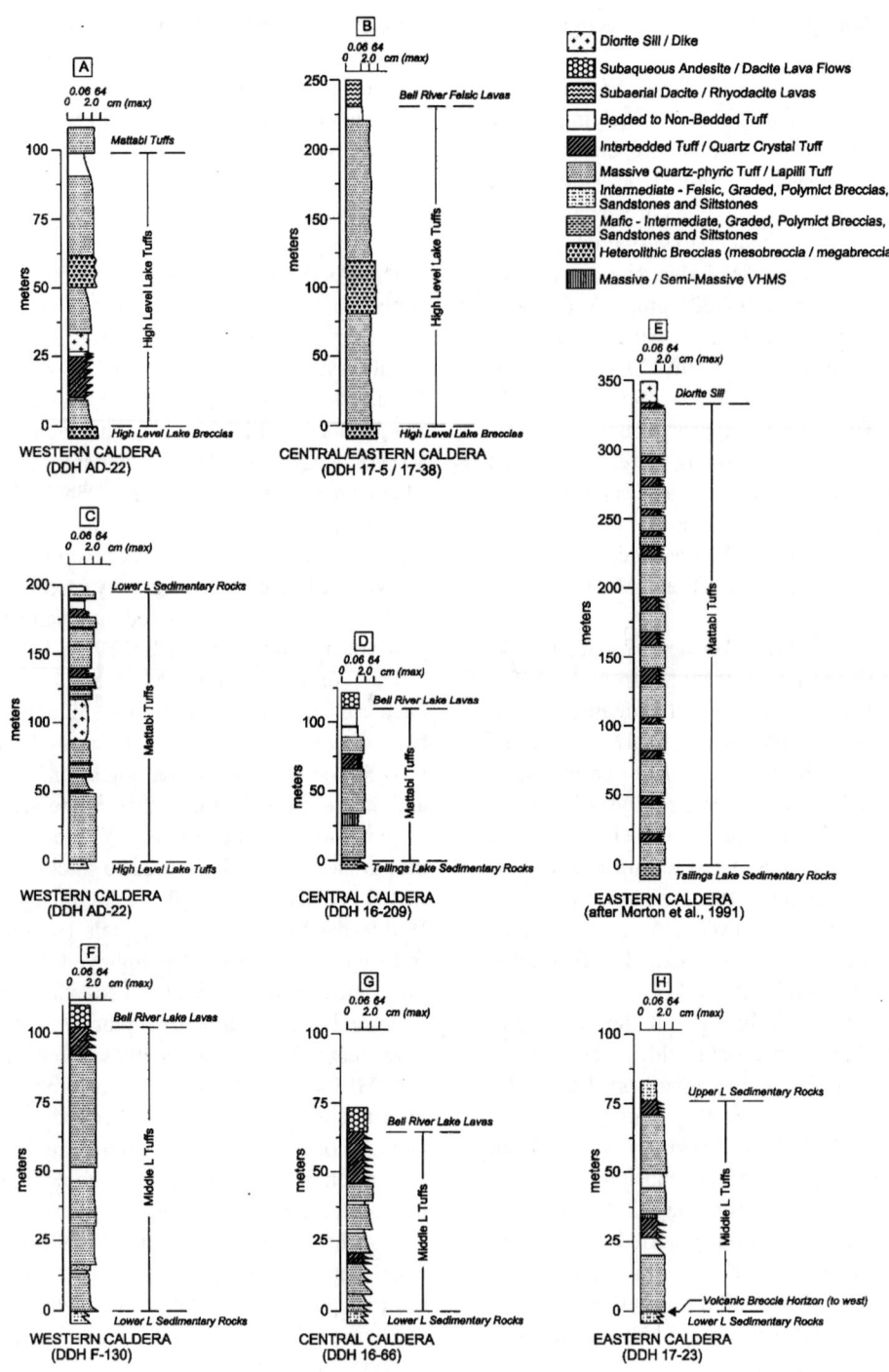

Figure 3. Detailed sections of the High Level Lake, Mattabi, and Middle L tuffs in the eastern, central, and western parts of the Sturgeon Lake Caldera Complex. Locations of sections shown on Figure 1.

A third lithofacies, volcanic breccia, occurs at or near the base of the Middle L Sequence approximately 2km east of Bell River Lake (Figure 1). This breccia is up to 15m thick, and has a strike length of less than 500m [*Hudak*, 1996]. The matrix-supported breccia contains 30–60% subangular to angular, 1–35cm diameter, crudely inverse graded, jigsaw puzzle-fit, poorly vesiculated, quartz-phyric rhyolite lava fragments, as well as <1–2% massive sulfide lapilli (Figure 4d). Carbonate-altered, locally spherulitic pumice lapilli up to 5cm in diameter comprise up to 10% of the unit. The matrix of the breccia is composed of a mixture of recrystallized ash, quartz crystal fragments and shards, and angular rhyolite fragments up to 2mm in diameter. The volcanic breccia has a gradational contact with the overlying MLLT and MLBT deposits.

5. DISCUSSION AND CONCLUSIONS

5.1. Interpretation of the High Level Lake Tuffs

The interstratification of, and the gradational contacts between, the High Level Lake Tuffs and the HLBX indicate that these units formed simultaneously. The heterolithic nature of the HLBX, the presence of only precaldera succession clasts in the HLBX, the variation in clast size from centimeters to several hundreds of meters in the HLBX, and the abutment of the HLBX against the Lac David fault are consistent with this unit representing mesobreccias and megabreccias [*Lipman*, 1976, 1997] that formed during the formation of the SLCC. The High Level Lake Tuffs, therefore, are interpreted to represent deposits from the pyroclas-

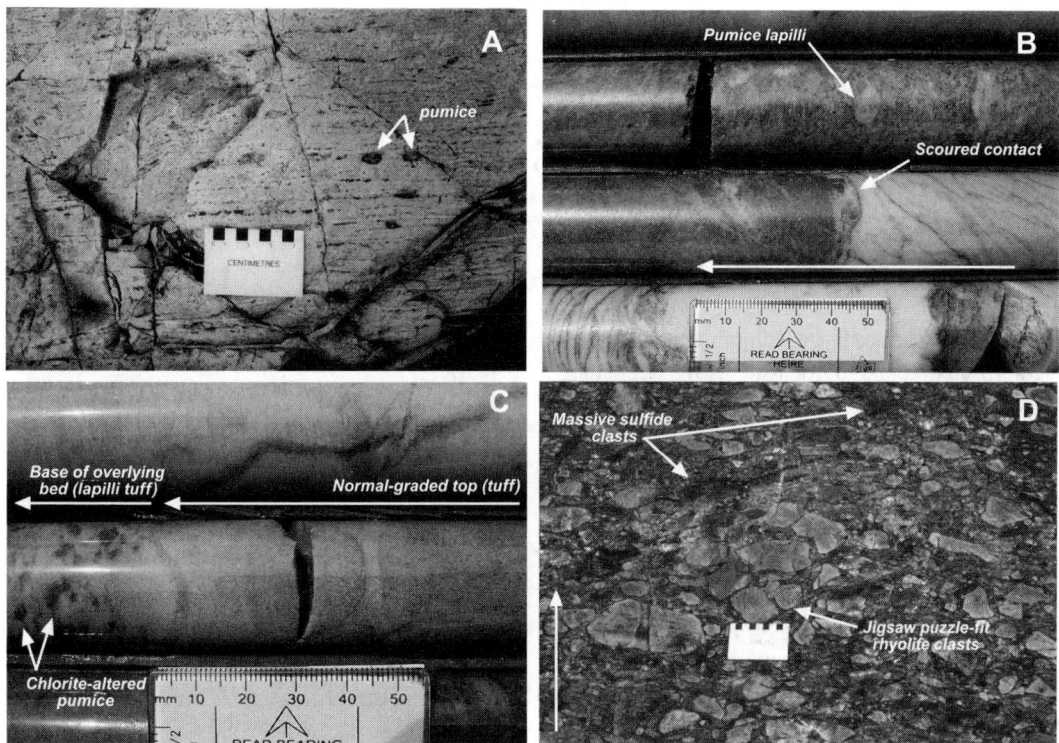

Figure 4. Photographs of various physical features of the Middle L Tuffs. (a) Outcrop appearance of massive Middle L Succession tuff (MLLT) immediately west of the Sturgeon Lake VHMS deposit. Dark markings on scale are equal to 1cm. (b) Drill core appearance of scoured contact between overlying MLLT deposits (top and left side of photo) and underlying MLBT deposits (bottom and right side of photo). Large arrow indicates topping direction. Scale measured in millimeters. (c) Normal graded ash top of MLBT bed. Arrows indicate topping direction. Scale measured in millimeters. (d) Volcanic breccia at base of Middle L Succession in the east-central part of the SLCC. Note jigsaw puzzle-fit fragments of rhyolite lava, as well as massive sulfide clasts. Large white arrow in photograph indicates topping direction. Dark markings on scale are equal to 1cm.

tic eruptions associated with initial collapse of the SLCC. These tuffs and lapilli tuffs are interpreted to have been erupted subaerially, and deposited both subaerially (central and eastern parts of the caldera) and subaqueously (F-Group region of the western part of the SLCC). We envision this eruption to be an Archean analog of modern subaerial eruptions that have created subaqueous caldera complexes at places like Santorini [*Druitt and Francaviglia*, 1992; *Perissoratis*, 1995], Krakatau [*Sigurdsson et al.*, 1991; *Self*, 1992; *Mandeville et al.*, 1996], and Kuwae [*Monzier et al.*, 1994; *Robin et al.*, 1994].

Several types of evidence support the interpretation that the High Level Lake Tuffs were erupted and deposited subaerially across the eastern two-thirds of the SLCC (Table 1). First, bounding facies for the High Level Lake tuffs are subaerial. The immediately underlying (and locally intercalated) HLBX was deposited on top of subaerial mafic lava flows and cinder cone deposits [*Groves et al.*, 1988], and contains no sedimentary units or textures indicative of deposition in a subaqueous environment. The immediately overlying strata comprise subaerial lava flows and mafic to intermediate lapilli tuffs of the Bell River Succession. Second, the High Level Lake Tuffs have morphological and compositional characteristics similar to subaerially deposited pyroclastic flow deposits and associated airfall tuff deposits. The thick, massive basal units (HLLT) containing both juvenile (dense vitric fragments and well-vesiculated pumice) and accessory fragments (trondhjemite and granodiorite) are interpreted to be pyroclastic flow deposits, whereas the overlying tuff is interpreted as airfall ash deposits. Third, hydrothermal mineralization within the High Level Lake Tuffs in this part of the SLCC is consistent with subaerial epithermal stringer zinc-silver (lead) veins in both modern and ancient settings [*White and Hedenquist*, 1990; 1995; *Larson and Hutchinson*, 1993; *Taner*, 2002]; no VHMS deposits have been found in the High Level Lake Tuffs in the eastern two-thirds of the SLCC. In addition, comparison with other caldera-associated ignimbrites (Figure 5a) suggests that considerable volumes of High Level Lake Tuff may have been deposited outside the SLCC from subaerial volcanic plumes.

Subaqueous deposition of the High Level Lake Tuffs in the western part of the caldera is supported by bounding facies, the depositional characteristics of the High Level Lake Tuffs, and the presence of VHMS mineralization (Table 1). First, HLBX deposits underlying, intercalated with, and locally, gradationally in contact with the High Level Lake Tuffs are interpreted to be deposits from a series of subaqueous cohesive mass flows and associated high- to low-density turbidity currents as described by *Bouma*

[1962] and *Lowe* [1982]. The HLBX appear to have been deposited from subaqueous mass flows and turbidity currents that formed when masses of Pre-caldera strata collapsed into a subaqueous portion of the SLCC that was created during the High Level Lake eruption. The overlying Mattabi deposits have facies characteristics consistent with deposition from high and low concentration, eruption-fed

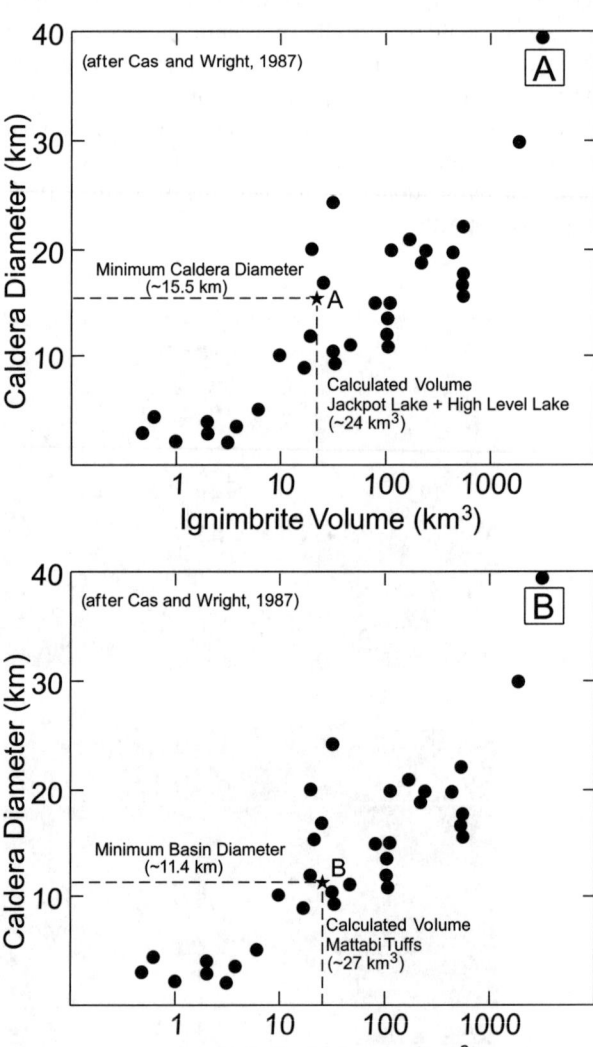

Figure 5. Comparison of estimated eruption volume and caldera diameter for modern subaerial and subaqueous caldera-associated eruptions (after Cas and Wright, 1987) and the caldera-forming eruptions associated with the Sturgeon Lake Caldera Complex. The combined estimated volumes for the Jackpot Lake and High Level Lake eruptions (associated with initial SLCC formation) are shown on Figure A, whereas the estimated eruption volume of the Mattabi eruption and the diameter of the interpreted nested caldera between the Darkwater and Lac David faults are shown in Figure B.

aqueous density currents (discussed below), and are the host for small, non-economic VHMS deposits. Second, the High Level Lake Tuffs in the F-Group area are characterized by vaguely normal graded basal lapilli tuffs (which are crystal enriched relative to the lapilli tuffs in the eastern two-thirds of the SLCC) that are overlain by non-graded to normal graded, plane parallel laminated tuff deposits. These deposits are interpreted to be the products of deposition from low- to high-concentration submarine mass flows and subsequent suspension sedimentation. That these deposits represent primary deposition of pyroclastic material, rather than remobilized unconsolidated tephra, is supported by the homogenous composition of the lapilli, the dominance of vitric grains over lithic grains [*Cousineau and Mueller*, 1993], crystal enrichments which are generally not excessive [2–4 times, rarely 10 times], and the presence of caldera-associated breccia that underlies and is intercalated with the deposits. Third, the F-Group VHMS ore body, which is hosted in the High Level Lake Tuffs, formed subaqueously where sufficient water depth to prevent extensive boiling of metaliferous hydrothermal fluid (at least several hundreds of meters) existed [cf. *Herzig and Hannington*, 1995]. The confinement of the VHMS mineralization (western SLCC) and epithermal-style stringer mineralization (eastern SLCC) to the High Level Lake Tuffs (and not the overlying Bell River lavas), suggests a direct connection between physical, chemical, and temporal processes in an evolving magma chamber, volcanic eruptions, mineralizing fluids in subaerial and subaqueous environments, and water depth.

Northeast-trending rhyolite dikes 1km southeast of Bell River Lake (Figure 1) may indicate proximity to vents associated with the eruption of the High Level Lake Tuffs. Eruptive vents in the western part of the SLCC have not been positively identified because: a) intense hydrothermal alteration has destroyed original textures in the tuffs, precluding detailed facies analysis; b) the western margin for the caldera complex is poorly defined (due to post-volcanic structural deformation associated with the Sturgeon Lake Shear Zone); and c) sufficient outcrop or diamond drilling for volcanological analysis is not available. Within the western part of the SLCC, the lowermost High Level Lake Tuffs may represent initially subaerial pyroclastic flows which entered and mixed with water to become high- to low-concentration turbidity currents. The increased abundance of quartz crystals in the High Level Lake Tuffs relative to the eastern two-thirds of the SLCC, the high proportion of bedded tuff to massive tuff and lapilli tuff, and the presence of inversely graded pumice are consistent with the distal facies of a subaereal pyroclastic flow which was deposited subaqueously [*Cole and DeCelles*, 1991]. During caldera collapse, subaerial vents may have evolved into subaqueous vents along the caldera ring fracture. The uppermost High Level Lake Tuff units may have been deposited from subaqueous eruption-fed high- to low-concentration aqueous density currents and suspension fallout as discussed by *Smellie* [2000] and *White* [2000].

5.2. Interpretation of the Mattabi Tuffs

The Mattabi Tuffs represent the most voluminous eruptive event associated with the SLCC (Table 1). This eruptive event apparently formed a nested caldera between the Darkwater and Lac David faults (Figure 1). Modern analogues may be the Myojin Knoll caldera of the Izu-Bonin arc [*Fiske et al.*, 2001], or the Healy Caldera of the southern Kermadec arc [*Wright and Gamble*, 1999].

Within this volcanic edifice, the Mattabi Tuffs are examples of deposits which appear to be subaqueous pyroclastic flows, but cannot be unambiguously interpreted as such due to a lack of heat retention features (e.g. welding) or reliable paleomagnetic signatures [*Kokelaar and Busby*, 1992; *White*, 2000]. This interpretation is supported by: a) the similar morphology, composition, and geochemical cyclicity of the bedded quartz crystal units across the caldera, which appears to preclude significant mixing with water; b) the thick, massive, poorly sorted nature of the basal beds which are overlain by bedded and laminated tuffs; c) the presence of abundant, well-vesiculated, commonly amoeboid-shaped pumice in the MLT deposits; d) the lack of fines-depletion in the MLT deposits; e) the local presence of spherulites and lithophysae where the tuffs have escaped intense syn- and post-depositional hydrothermal alteration; f) immediate deposition upon subaqueously deposited sedimentary rocks; and g) the occurrence of VHMS and quartz-pyrite exhalative deposits throughout the unit and, locally, subaqueous debris flow deposits separating flow units. As well, comparisons of our volume calculations with other caldera-forming ignimbrites suggest that nearly all the Mattabi tuffs were deposited within the SLCC (Figure 5b). Although a vent or vents for the Mattabi eruptions have not been identified, they likely occur along the margins of, or within, the nested caldera between the Darkwater and Lac David Faults. Intense subaqueous hydrothermal alteration [*Holk et al.*, in press] in this area suggests proximity to vent-sites [*Kokelaar and Busby*, 1992; *Gibson et al.*, 1999].

Deposits of the Mattabi Tuffs in the vicinity of the F-Group VHMS ore body are interpreted to be outflow sheets of the MTQ deposits east of the Darkwater Fault. These outflow sheets were deposited by high- and low-concentration

Group II [*White*, 2000] eruption-fed aqueous density currents and water-settled ash. This is supported by a) the dramatic thinning of the Mattabi tuffs east of the Darkwater fault; b) the lack of distinctive geochemical zonation within individual flow units in this part of the SLCC; c) the normal graded lapilli tuff and tuff deposits; d) the abundance of shattered crystals in the deposits; e) the homogenous composition of the vitric-dominated lapilli; and f) the presence of small VHMS deposits.

The subaqueous explosive eruptions that formed the Mattabi ash flow tuffs are envisioned to have produced high concentration, laterally directed, subaqueous pyroclastic flows similar to those described by *Kokelaar and Busby*, [1992] and *White* [2000]. Hydrostatic pressure above the subaqueous vent was sufficient to suppress the eruption column, leading to laterally directed "boiling-over" or "frothing" of magma from fissures. The fine vitric material produced during the eruption allowed little interaction of the tephra with water, except along the margins of the flow (flow top stripping), in the eastern two-thirds of the SLCC. Sustained, high concentration, high density particulate flows produced the massive basal beds. Processes that formed the overlying plane-parallel laminated to medium-bedded, typically normal graded tuff deposits include; a) deposition of materials stripped from flow tops by dilute aqueous turbidity currents; b) suspension sedimentation from material elutriated from the flows by flow stripping; and c) ingress of water into the vent during the waning stages of the eruption which produced low-concentration, eruption-fed density currents.

The explosive eruptions associated with the formation of the Mattabi tuffs were episodic, giving rise to the formation of numerous alternating massive quartz-phyric tuff and lapilli tuff units. The time period between eruptions was sufficient to allow VHMS deposits to form and, as the geochemical evidence suggests [*Morton et al.*, 1991], to enable compositional restratification of the source magma chamber. The episodic explosive eruptions were likely a manifestation of the balance between volatile content of the evolving magma and hydrostatic pressure as constrained by water depth [*Morton et al.*, 1991]. Eruptions with high volatile contents may have been relatively long-lived and gave rise to the formation of thick basal beds; those with lower volatile contents were shorter lived and produced thinner basal beds. As volatile contents decreased the nature of the eruption changed, and magmatic effusion was insufficient to produce high concentration flows. Eventually, as volatile contents decreased, the hydrostatic pressure eventually exceeded the volatile pressure of the magma and effectively terminated the explosive eruption.

The massive MMT deposits in the upper part of the Mattabi Succession are interpreted to represent continuous deposition from eruptions in a shallow subaqueous environment, one where hydrostatic pressure had little effect on the eruptive process. The bedded units overlying the massive basal beds were deposited by syn- to post-eruptive, low particle concentration aqueous density currents as well as water-settled ash. In morphology, the MMT deposits are similar to tuffs described by *Busby-Spera* [1984] from the Mineral King roof pendant, and by *Morton and Nebel* [1983, 1984] from the Wawa area of northwestern Ontario.

All known VHMS mineralization in the Mattabi Succession occurs as replacement bodies within the MTQ at contacts between bedded tuff in the footwall and massive tuff in the hanging wall. This spatial and temporal relationship suggests a correlation between processes in the magma chamber, the evolution of hydrothermal fluids, and VHMS ore formation.

5.3 Interpretation of the Middle L Tuffs

The Middle L Tuffs, like the Mattabi Tuffs, are interpreted to have been erupted and deposited in a subaqueous environment (Table 1). A possible subaqueous vent for these deposits may be located in proximity to the volcanic breccia deposits which occur at or near the base of the Middle L Succession. The association of jigsaw puzzle-fit lava fragments, massive sulfide clasts, and pumice suggests that this volcanic breccia formed by the dissection of an active subaqueous lava dome which had associated VHMS mineralization. The gradational contact between the volcanic breccia and the overlying Middle L Tuffs suggests that these two units are genetically related. Ingress of water into the subaqueous dome during dissection may have led to the explosive fragmentation of magma which produced the Middle L Tuffs, particularly in the eastern part of the SLCC. The Sturgeon Lake VHMS deposit occurs less than 1km east of this dome, and reflects a region of anomalous thermal activity which is consistent with a vent-proximal setting. The relatively high proportion of MLLT deposits relative to MLBT deposits in the western part of the SLCC may also indicate the presence of subaqueous vents in the western SLCC. To date, however, specific vents have not been identified. The thick Middle L deposits in the eastern and western parts of the SLCC, and the thin nature of the deposits in the central part of the SLCC, suggests a palaeo-topographic high in the vicinity of the Mattabi VHMS deposit at the time of Middle L volcanism.

Water interacted with the Middle L tephra during both its eruption and deposition. The MLLT generally exhibit a pronounced grading of quartz crystals and pumice lapilli. The grading in the basal units, combined with the relatively coarse matrix of these tuffs, suggests that significant amounts of fine tephra may have been elutriated from a subaqueous eruption column and/or eruption-fed mass flows. Facies characteristics suggest deposition from Group II [*White*, 2000] aqueous density currents. Effusion rates may have been insufficient to produce the volumes of extremely fine-grained tephra needed to prevent mixing and entrainment of water in the erupting tephra [*Kokelaar and Busby*, 1992; White, 2000].

5.4 Relationship Between SLCC Subaqueous Eruptions and VHMS Mineralization

The presence of VHMS deposits at specific horizons which correspond to voluminous pyroclastic eruptions suggests a direct link between subaqueous explosive volcanism and VHMS genesis in the SLCC. In modern caldera systems, such large eruptions tend to be dominantly driven by magmatic processes [e.g. *Self*, 1992; *Worthington et al.*, 1999], although inputs from hydrovolcanic eruptions have been documented in many subaerial to subaqueous caldera-forming events (*Druitt and Francaviglia*, 1992; *Robin et al.*, 1994; *Nairn et al.*, 1995). Subaqueous explosive eruptions of felsic magma are generally believed to only be possible under relatively shallow (< 1000m) sub-aqueous conditions [*Cas*, 1992; *Gibson et al.*, 1999] due to the limiting effects of hydrostatic pressure. At Sturgeon Lake, evidence suggests that the subaqueous explosive eruptions were essentially throttled by hydrostatic pressure [Morton et al., 1991]. This being the case, the felsic explosive eruptions that occurred may well have terminated while the subvolcanic magma chamber still contained more volatiles than would be present if the same magma had erupted subaerially (assuming shallow intrusions emplaced under similar 1–3 km lithostatic loads). Thus, immediately following caldera-forming subaqueous eruptions, five important components would be present which, together, would promote a link between large explosive eruptions in the subaqueous environment and the formation of VHMS deposits: a) newly developed, high permeability syn-volcanic faults zones; b) an abundance of hot, possibly metal-rich magmatic hydrothermal fluid (the additional amount of fluid would be the difference between the volatiles that shut off similar subaerial and subaqueous eruptions); c) permeable, chemically reactive glassy volcaniclastic strata that can be easily replaced by sub-seafloor VHMS deposition; d) an abundance of seawater; and e) assuming that the upper parts of a magma chamber can be breached by caldera formation, deep circulating convecting hydrothermal fluids could have access to magma or hot, recently cooled intrusive rocks. This would promote the development of high temperature, metalliferous evolved hydrothermal fluids.

Our present understanding of the SLCC hydrothermal system indicates that the ore compositions, alteration mineral assemblages, and oxygen isotope evidence [*Holk et al.*, in press] are consistent with the formation of the VHMS deposits from a sub-volcanic, intrusion-driven, seawater-dominated convective hydrothermal system [*Hudak*, 1989, 1996; *Jongewaard*, 1989; *Walker*, 1993]. The role of magmatic fluids in the genesis of the VHMS deposits in the SLCC remains poorly understood. It is well established, however, that caldera-forming eruptions represent tremendous magmatic degassing events [*Christiansen*, 2001]. Several studies from possible modern analogues to Sturgeon Lake have documented magmatic inputs into shallow water VHMS systems [*Yang and Scott*, 1996, 2002; *Gamo et al.*, 1997; *Marani et al.*, 1997; *Hannington et al.*, 1999]. Thus, the five components described above appear to be the key factors which relate VHMS mineralization to subaqueous explosive volcanism.

In summary, the Archean SLCC contains deposits characterized by various degrees of water-tephra interaction, and mineralization in an environment evolving from initially subaerial and subaqueous to entirely subaqueous. The spatial and temporal relationships between the production of voluminous, explosive, caldera-forming subaerial and subaqueous eruptions, and the subsequent formation of subaerial and shallow water epithermal-like stringer mineralization and deeper water VHMS deposits, suggests a link between explosive felsic volcanism, evolution of near-surface magma chambers and their associated hydrothermal systems, and the formation of VHMS ore deposits.

Acknowledgments. Thanks to Mattabi Mines Limited, Noranda Mining and Exploration, Ltd., Minnova Inc. (INMET), and Rio Algom for their financial support and access to their properties, diamond drill core, and geological data. Jamie Walker and Peter Jongewaard providing us with their unpublished data. Financial support was provided through the Canada-Ontario Mineral Development Agreement (MDA) and from the Faculty Development Program at the University of Wisconsin Oshkosh. A previous version of this paper benefited greatly from critical reviews by J. Smellie and W. Mueller.

REFERENCES

Allen, R. L., Lundström, I., Ripa, M., Seimeonov, A., and Christofferson, H., Facies analysis of a 1.9Ga, continental margin, back-arc, felsic caldera province with diverse Zn-Pb-Ag-

(Cu-Au) sulfide and Fe oxide deposits, Bergslagen region, Sweden, *Econ. Geol. 91*, 979-1008, 1996a.

Allen, R. L., Weihed, P., and Svenson, S.Å, Setting of Zn-Cu-Au-Ag-massive sulfide ores in the evolution and facies architecture of a 1.9 Ga marine volcanic arc, Skellefte District, Sweden, *Econ. Geol. 91*, 1022-1053, 1996b.

Bezenek, S. R., *Relationship Between the Mattabi Successions Geochemically Zoned Lower Crystal-rich Ash Beds and the Mattabi Massive Sulfide Deposit*, unpublished B. S. Honors Thesis, University of Minnesota – Duluth, 1992.

Bouma, A. H., *Sedimentology of some flysch deposits*, Elsevier, 1962.

Busby-Spera, C. J., Large volume rhyolitic ash flow eruptions and submarine caldera collapse in the lower Mesozoic Sierra Nevada, California. *J. Geophys. Res. 89*, 8417-8427, 1984.

Campbell I. H., Franklin, J. M., Gorton, M. P., Hart, T. R., and Scott, S. D., The role of synvolcanic sills in the generation of massive sulfide deposits, *Econ. Geol. 76*, 2248-2253, 1982.

Campbell, S. D. G., Reedman, A. J., Howells, M. F., Mann, A. C., The emplacement of geochemically distinct groups of rhyolites during the evolution of the Lower Rhyolitic Tuff Formation caldera (Ordivician), North Wales, U. K., *Geol. Mag. 124*, 501-511, 1987.

Cas, R. A. F., Submarine volcanism: eruption styles, products and relevance to understanding the host rock successions to volcanic hosted massive sulfide deposits, *Econ. Geol. 87*, 511-541, 1992.

Cas, R. A., F., Wright, J. V., *Volcanic Successions: Modern and Ancient*, Allen and Unwin, London, 1987.

Chesner, C. A., Rose, W. I., Stratigraphy of the Toba tuffs and the evolution of the Toba caldera complex, Sumatra, Indonesia, *Bull. Volcanol. 53*, 343-356, 1991.

Christiansen, R. L., The Quaternary and Pliocene Yellowstone plateau volcanic field of Wyoming, Idaho, and Montana, *U. S. Geol. Surv. Prof. Paper 729G*, 2001.

Cole, R. B., DeCelles, P. G., Subaerial to submarine transitions in early Miocene pyroclastic flow deposits, southern San Joaquin basin, California, *Geol. Soc. Amer. Bull 103*, 221-235, 1991.

Cousineau, P. A., and Mueller, W., Subaqueous pyroclastic volcanism: From subaqueous pyroclastic flows to volcanogenic epiclastic deposits, *Explor. Mining Geol. 2*, 395-396, 1993.

Cronan, D. S., Varnavas, S., Perissoratis, C., Hydrothermal sedimentation in the caldera of Santorini, Hellenic volcanic arc, *Terra Nova 7*, 289-293, 1995.

Davis, D. W., and Trowell, N. F., U-Pb zircon ages from the eastern Savant Lake-Crow Lake metavolcanic-metasedimentary belt, northwestern Ontario, *Can. J. Earth Sci. 19*, 868-877, 1982.

Davis, D. W., Krogh, T. E., Hinzer, J., and Nakamura, E., Zircon dating of polycyclic volcanic at Sturgeon Lake and implications for base metal mineralization, *Econ. Geol. 80*, 1942-1952, 1985.

Druitt, T. H., and Francaviglia, V., Caldera formation on Santorini and the physiography of the islands in the late Bronze Age, *Bull. Volc. 54*, 484-493, 1992.

Fisher, R. V., Proposed classification of volcaniclastic sediments and rocks, *Geol. Soc. Am. Bull. 72*, 1409-1414, 1961.

Fisher, R. V., Rocks composed of volcanic fragments, *Earth Sci. Rev. 1*, 287-298, 1966.

Fisher, R. V., and Schminke, H.-U., *Pyroclastic Rocks*, Springer-Verlag, New York, 1984.

Fiske, R. S., Naka, J., Iizasa, K., Yuasa, M., and Klaus, A., Submarine silicic caldera at the front of the Izu-Bonin Arc, Japan: voluminous seafloor eruptions of rhyolite pumice, *Geol. Soc. Amer. Bull. 113*, 813-824, 2001.

Franklin, J. M., Volcanic-associated massive sulfide base metals, in *Geology of Canadian Mineral Deposit Types*, edited by O. R. Eckstrand, W. D. Sinclair, and R. I. Thorpe, 158-183, 1996.

Galley, A., van Breemen, O., and Franklin, J. M., The relationship between intrusion-hosted Cu-Mo mineralization and the VMS deposits of the Archean Sturgeon Lake mining camp, northwestern Ontario, *Econ. Geol. 95*, 1543-1550, 2000.

Gamo, T., Okamura, K., Charlou, J-L., Urabe, T., Auzende, J-M., Ishibashi, J., Shitashima, K., Chiba, H., and Shipboard Scientific Party of the ManusFlux Cruise, Acidic and sulfate-rich hydrothermal fluids from the Manus back-arc basin, Papua New Guinea, *Geology 25*, 139-142, 1997.

Gibson, H. L., Morton, R. L., and Hudak, G. J., Submarine volcanic processes, deposits and environments favourable for the location of volcanic-associated massive sulphide deposits, *Rev. Econ. Geol. 8*, 13-51, 1999.

Groves, D. A., *Stratigraphy and alteration of the footwall volcanic rocks beneath the Archean Mattabi massive sulfide deposit, Sturgeon Lake, northwestern Ontario*, unpublished M. Sc. Thesis, University of Minnesota – Duluth, 1984.

Groves, D. A., Morton, R. L., and Franklin, J. M., Physical volcanology of the footwall rocks near the Mattabi massive sulfide deposit, Sturgeon Lake, Ontario, *Can. J. Earth Sci. 25*, 280-291, 1988.

Hannington, M. D., Poulsen, K. H., Thompson, J. F. H., and Sillitoe, R. H., Volcanogenic gold in the massive sulfide environment, *Rev. Econ. Geol. 8*, 325-356, 1999.

Herzig, P. M. and Hannington, M. D., Polymetallic massive sulfides at the modern seafloor – a review, *Ore Geol. Rev. 10*, 95-115, 1995.

Holk, G. J., Taylor, B. E., and Galley, A. G., Oxygen isotope mapping of the Archean Sturgeon Lake Caldera Complex, northwestern Ontario, Canada, *Min. Deposita*, in press.

Hudak, G. J., *The physical volcanology and hydrothermal alteration associated with the F-Group Archean massive sulfide deposit, Sturgeon Lake, northwestern Ontario*, unpublished M. Sc. Thesis, University of Minnesota – Duluth, 1989.

Hudak, G. J., *The physical volcanology and hydrothermal alteration associated with late caldera volcanic and volcaniclastic rocks and volcanogenic massive sulfide deposits in the Sturgeon Lake region of northwestern Ontario*, unpublished. Ph. D. Dissertation, University of Minnesota, 1996.

Jongewaard, P. K., *Physical volcanology and hydrothermal alteration of the footwall rocks to the Archean Sturgeon Lake massive sulfide deposit*, unpublished. M. Sc. Thesis, University of Minnesota – Duluth, 1989.

Kokelaar, P. and Busby, C., Subaqueous explosive eruptions and welding of pyroclastic deposits, *Science 257*, 196-201, 1992.

Koopman, E. R., *Stratigraphy, structural geology, and stratigraphic controls of ore distribution of the Lyon Lake massive sulphide deposit, Sturgeon Lake, Ontario*, unpublished. M. Sc. Thesis, Carleton University, 1993.

Lafrance, B., Mueller, W. U., Daigneault, R., and Dupras, N., Evolution of a submerged composite arc volcano: volcanology and geochemistry of the Normétal volcanic complex, Abitibi greenstone belt, Québec, Canada. *Precamb. Res.* 101, 277-311, 2000.

Larson, J. E., and Hutchinson, R. W., The Selbaie Zn-Cu-Ag deposits, Quebec, Canada: an example of evolution from subaqueous to subaerial volcanism and mineralization in an Archean caldera environment, *Econ. Geol.* 88, 1460-1482, 1993.

Lesher, C. M., Goodwin, A. M., Campbell, I. H., and Gorton, M. P., Trace-element geochemistry of ore-associated and barren felsic metavolcanic rocks in the Superior Province, Canada, *Can. J. Earth Sci.* 23, 222-237, 1986.

Lipman, P. W., Caldera collapse breccias in the western San Juan Mountains, Colorado, *Geol. Soc. Am. Bull.* 87, 1397-1410, 1976.

Lipman, P. W., Subsidence of ash-flow calderas: relation to caldera size and magma chamber geometry, *Bull. Volcanol.* 59, 198-218, 1997.

Lowe, D. R., Sediment gravity flows: II. Depositional models with special reference to the deposits of high density turbidity currents, *Jr. Sed. Pet.* 52, 279-297, 1982.

Mandeville, C. W., Carey, S., and Sigurdsson, H., Sedimentology of the Krakatau 1883 submarine pyroclastic deposits, *Bull. Volc.* 57, 512-529, 1996.

Marani, M. P., Gamberi, F., and Savelli, C., Shallow water polymetallic sulfide deposits in the Aeolian island arc, *Geology 25*, 815-818, 1997.

Monzier, M., Robin, C., and Eissen, J.-P., Kuwae (~1425 A.D.): the forgotten caldera, *J. Volc. Geotherm. Res.* 59, 207-218, 1994.

Morton, R. L., Hudak, G. J., and Franklin, J. M., Geology, south Sturgeon Lake area, Ontario. *Geol. Surv. Canada Open File Rpt. 3642*, 1999.

Morton, R. L., Hudak, G. J., and Franklin, J. M., The Mattabi ash flow tuff and its relationship to the Mattabi massive sulfide deposit, in *New Mexico Bur. Mines Min. Res. Bull. 131*, 196, 1989.

Morton, R. L., Nebel, M., Physical character of Archean felsic volcanism in the vicinity of the Helen iron mine, Wawa, Ontario, Canada, *Precamb. Res.* 20, 39-62, 1983.

Morton, R. L., Nebel, M., Hydrothermal alteration of felsic volcanic rocks at the Helen siderite deposit, Wawa, Ontario, *Econ. Geol.* 79, 1319-1333, 1984.

Morton, R. L., Walker, J. S., Hudak, G. J., and Franklin, J. M., The early development of an Archean submarine caldera complex with emphasis on the Mattabi ash flow tuff and its relationship to the Mattabi massive sulfide deposit, *Econ. Geol.* 86, 1002-1011, 1991.

Mueller, W., Chown, E. H., and Thurston, P. C., Processes in physical volcanology and volcaniclastic sedimentation: modern and ancient, *Precamb. Res.* 101, 81-85, 2000.

Mumin, A. H., *Tectonic and structural controls on massive sulfide deposition in the south Sturgeon Lake volcanic pile, northwestern Ontario, and hydrothermally altered rocks associated with the Lyon Lake massive sulfide ore deposits, Sturgeon Lake, northwestern Ontario*, unpublished M. Sc. Thesis, University of Toronto, 1988.

Nairn, I. A., McKee, C. O., Talai, B., and Wood, C. P., Geology and eruptive history of the Rabaul Caldera Area, Papua New Guinea, *J. Volc. Geotherm. Res.* 69, 255-284, 1995.

Osterberg, S. A., Morton, R. L., and Franklin, J. M., Hydrothermal alteration and physical volcanology of the Headway Coulee massive sulfide occurrence, Onaman area, northwestern Ontario, *Econ. Geol.* 82, 1505-1520, 1987.

Perissoratis, C., 1995, The Santorini volcanic complex and its relation to the stratigraphy and structure of the Aegean arc, Greece, *Marine Geol.* 128, 37-58, 1995.

Robin, C., Monzier, M., and Eissen, J.-P., Formation of the mid-fifteenth century Kuwae caldera (Vanuatu) by initial hydroclastic and subsequent ignimbrite eruption, *Bull. Volc.* 56, 170-183, 1994.

Self, S., Krakatau revisited: the course of events and interpretation of the 1883 eruption, *GeoJournal 28*, 109-121, 1992.

Sigurdsson, H., Carey, S., and Mandeville, C., Krakatau, submarine pyroclastic flows of the 1883 eruption of Krakatau volcano, *Nat. Geog. Res. & Expl.* 7, 310-327, 1991.

Sillitoe, R. H., Hannington, M. D., and Thompson, J. F. H., High sulfidation deposits in the volcanogenic massive sulfide environment, *Econ. Geol.* 91, 204-212, 1996.

Smellie, J. L., Subglacial Eruptions, in Sigurdsson, H. (ed.), *Encyclopedia of Volcanoes*, Academic Press, London, 403-418, 2000.

Smith, R. L., and Bailey, R.A., Resurgent Cauldrons, in Coats, R. R., Hay, R. L., and Anderson, C. A. (eds.), *Studies in Volcanology (Howell Williams Volume), G. S. A. Memoir 116*, 153-210, 1968.

Stix, J., Subaqueous, intermediate to silicic-composition explosive volcanism: a review, *Earth Sci. Rev.* 31, 21-53, 1991.

Taner, M. F., The geology of the volcanic-associated polymetallic (Zn, Cu, Ag, Au) Selbaie deposits, Abitibi, Quebec, Canada, *Expl. Min. Geol.* 9, 189-214, 2002.

Walker, J. S., *Physical volcanology and hydrothermal alteration of the footwall rocks to the Archean Mattabi massive sulfide deposit, northwestern Ontario*, unpublished. M. Sc. Thesis, University of Minnesota – Duluth, 1993.

White, J. D. L., Subaqueous eruption-fed density currents and their deposits, *Precamb. Res.* 101, 87-109, 2000.

White, N. C. and Hedenquist, J. W., Epithermal environments and styles of mineralization: variations and their causes, and guidelines for exploration, *J. Geochem. Expl.* 36, 445-474, 1990.

White, N. C. and Hedenquist, J. W., Epithermal gold deposits: styles, characteristics and exploration, *SEG Newsletter 23*, 1-13, 1995.

Worthington, T. J., Gregory, M. R., and Bondarenko, V., The Denham Caldera on Raoul Volcano: dacitic volcanism in the Tonga-Kermadec arc, *J. Volc. Geotherm. Res. 90*, 29-48, 1999.

Wright, I. C., and Gamble, J. A., Southern Kermadec submarine caldera arc volcanoes (SW Pacific): caldera formation by effusive and pyroclastic eruption, *Mar. Geol. 161*, 207-227, 1999.

Yang, K., and Scott, S. D., Possible contribution of a metal-rich magmatic fluid to a sea-floor hydrothermal system, *Nature 383*, 420-423, 1996.

Yang, K., and Scott, S. D., Magmatic degassing of volatiles and ore metals into a hydrothermal system on the modern sea floor of the Eastern Manus back-arc basin, Western Pacific, *Econ. Geol.* 97, 1079-1100, 2002.

George J. Hudak, Geology Department, University of Wisconsin Oshkosh, 800 Algoma Blvd., Oshkosh, WI 54904.

Ronald L. Morton, Department of Geosciences, University of Minnesota – Duluth, 10 University Drive, Duluth, MN 55812.

James M. Franklin, Franklin Geosciences Ltd., 24 Commanche Drive, Nepean, ON, Canada K2E 6E9.

Dean M. Peterson, Economic Geology Group, Natural Resources Research Institute, 5013 Miller Trunk Highway, Duluth, MN 55811.

Analysis of VHMS-Hosting Ignimbrites Erupted at Bathyal Water Depths (Ordovican Bald Mountain Sequence, Northern Maine)

Lowell G. Kessel and Cathy J. Busby

Department of Geological Sciences, University of California at Santa Barbara

Granulometric analysis of ignimbrites that enclose the Ordovician Bald Mountain volcanic-hosted massive sulfide (VHMS) deposit in northern Maine indicates that fine-ash poor (<10%) pumiceous volcanic deposits may result from deep marine explosive silicic volcanic eruptions. The VHMS is enclosed in a 1 km thick section of ignimbrite (footwall and hangingwall ignimbrites) that fill nested calderas formed at >1.45 km water depth. The footwall ignimbrite is a poorly sorted and nonstratified mixture of pumice lapilli and medium to coarse ash, in units tens of meters to 150 m thick separated by basalt lava flows. It is dominantly nonwelded and has blocky pumices indicating that hydroclastic fragmentation followed magmatic fragmentation. The footwall ignimbrite is interpreted as the product of a steady, high-discharge, volatile-rich phreatomagmatic explosive eruption. Like the footwall ignimbrite, the hangingwall ignimbrite is a fine-ash poor mixture of pumice and medium to coarse ash, but it occurs as thinner nonstratified units 1–150 m thick separated by well-sorted and stratified lapilli tuff and tuff. The hangingwall ignimbrite contains abundant compacted pumice that may record welding, and lacks blocky pumices or other evidence of hydroclastic fragmentation entirely. The hangingwall ignimbrite is interpreted as the product of an unsteady eruption, where lower volatile content (relative to the footwall ignimbrite) produced a lower eruption column, leading to minimal interaction with ambient water in a purely magmatic eruption. We speculate that fine-ash-poor ignimbrites may be typical of very deepwater (bathyal marine) explosive eruptions, where hydrostatic pressures may be sufficient to suppress magmatic vesiculation and reduce the violence of the eruptions, thereby dramatically decreasing the production of bubble wall shards.

INTRODUCTION

More than seventy percent of the Earth's surface is below sea level, and seafloor exploration has revealed that volcanic eruptions are more common under the sea than on land [*Fisher et al.,* 1997]. Yet, the effects of ambient water and hydrostatic pressure on silicic volcanic eruptions in deep marine settings are not clearly understood. This is because they have not been directly observed, and observations of the deposits of recent deep-water eruptions are incomplete. For this reason, most of our knowledge of deep-water silicic volcanic processes comes from inferences based on ancient successions [*Fisher and Schmincke,* 1984; *Francis,* 1993; *Kokelaar and Busby,* 1992]. A major short-

Explosive Subaqueous Volcanism
Geophysical Monograph 140
Copyright 2003 by the American Geophysical Union
10.1029/140GM24

coming of previous studies of ancient successions, however, is that paleo water-depths of eruption are far more poorly constrained than they are at modern volcanoes [*Cas and Wright*, 1991; *Cashman and Fiske*, 1991; *Lentz et al.* 1998; *Mandeville et al.*, 1994; *Schneider et al.*, 1992; *Self and Sparks*, 1978; *Sparks et al.*, 1980; *Wright and Mutti*, 1981].

Our study of an ancient volcanic succession is unique because we have strong constraints on the paleo-water depth of eruption of two large-volume silicic ignimbrites. These are intracaldera ignimbrites that form the footwall and hangingwall of the Ordovician Bald Mountain volcanic hosted massive sulfide (VHMS) in northern Maine (Figure 1A and 1B) [*Busby et al., in press; Foley, in press; Foose et al., in press; Slack et al., in press*]. This VHMS is one of the best preserved such deposits in the world, and the textural and grain size characteristics of the enclosing ignimbrites

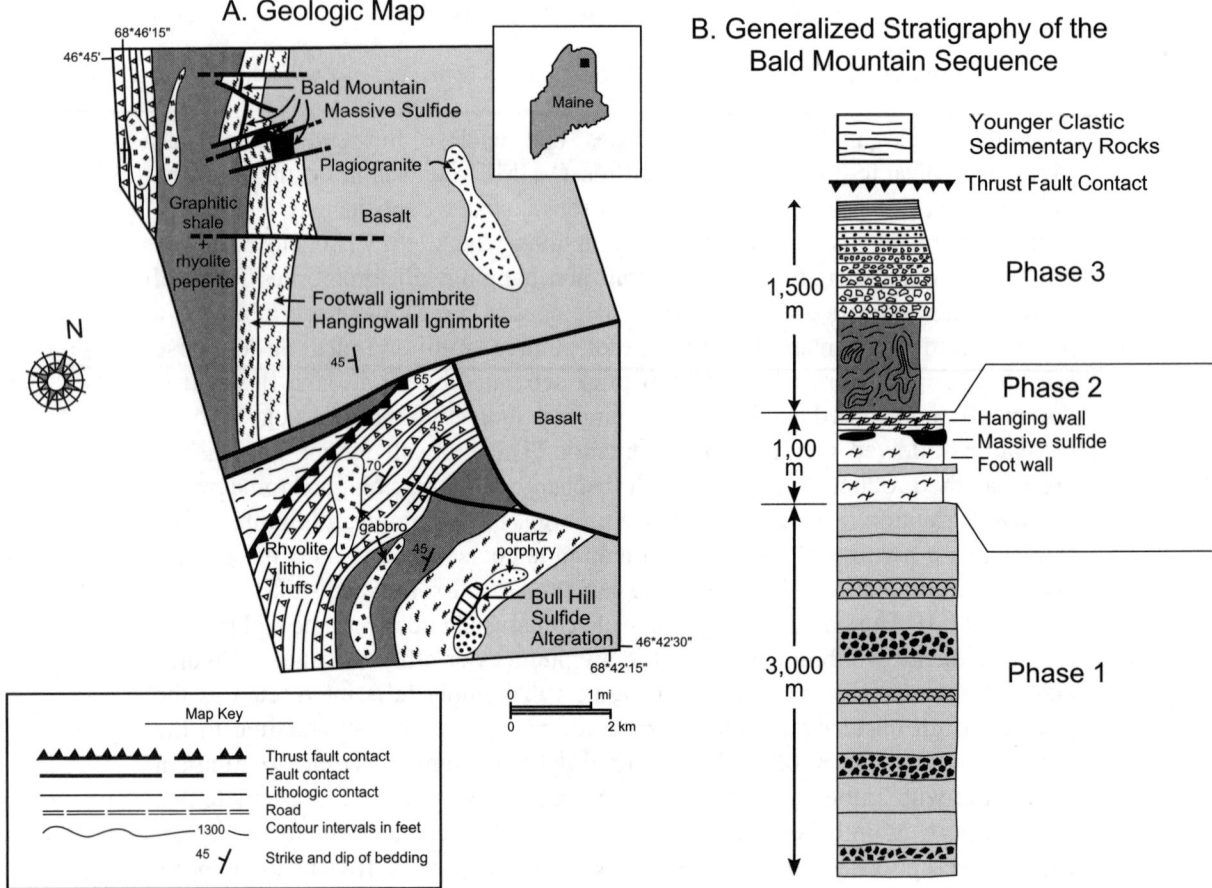

Figure 1. A) Generalized geologic map of the Late Ordovician Bald Mountain sequence in north eastern Maine (with location inset) [*Busby et al.*, in press]. For full description of tectonic and geologic setting, see *Foose et al.* [in press]. Strata dip about 45° to the west, and faults trending at high angles to the section formed as syndepositional normal faults, and were reactivated as thrust faults in Devonian time. The metamorphic grade is low (prehnite-pumpellyite facies), cleavage is generally lacking, and primary volcanic and sedimentary microtextures are very well preserved. B) Generalized stratigraphy of the Bald Mountain area [*Busby et al.*, in press]. The volcanic evolution of the terrane is divided into three main phases, all of which show evidence for ponding within a graben: Phase 1 is characterized by outpouring and ponding of basaltic lavas and breccia-hyaloclastite in a primitive arc-rift basin. Phase 2, the focus of this paper, is dominated by explosive deepwater ignimbrite eruptions and associated exhalative hydrothermal activity. The footwall and hangingwall ignimbrites were deposited in nested calderas, and the Bald Mountain massive sulfide formed in a small synvolcanic graben that was the vent for the footwall ignimbrite. Phase 3 records deposition of carbonaceous mudstone and distal tuffs, as well as graded megabeds generated by collapse of rhyolite lava domes; these in turn were disrupted by peperitic rhyolite intrusions.

are also remarkably well preserved. In this paper, we document the textural and granulometric characteristics of deepwater-erupted ignimbrites at Bald Mountain to address the following questions: (1) What are the effects of hydrostatic pressure and ambient water on deepwater silicic explosive eruptions and how is this manifested in grain morphology and grain size distribution?; and (2) Can granulometric analysis provide a methodology to distinguish subaerial (or shallow marine) from deepwater-erupted ignimbrites?

EVIDENCE FOR BATHYAL MARINE ERUPTION AND DEPOSITION OF THE FOOTWALL AND HANGING-WALL IGNIMBRITES AT BALD MOUNTAIN

Sedimentological features of the entire 5 km thick section above and below the massive sulfide and enclosing ignimbrites in the Bald Mountain sequence (*Busby et al., in press*), indicate deposition in deep water (Figures 1A and 1B). The section forms a conformable sequence that lacks any wave-generated sedimentary structures indicative of shallow marine environments. There are no in situ shallow marine fauna, nor even any resedimented shallow marine fauna. Furthermore, there is no evidence for subaerial exposure or erosion, or fluvial sedimentation. By analogy with modern marine settings, these observations suggest that the entire section was deposited below wave base, in water depths certainly greater than ~60 m and possibly greater than ~150 to 200 m [*Draper, 1967; Butnam et al., 1979*]. Additional evidence is that the carbonaceous mudstones that occur throughout the section typically, but not exclusively, form in deepwater environments >200 m deep [*Pickering et al., 1989*]. Other features throughout the Bald Mountain sequence that are consistent with (but not restricted to) deepwater paleoenvironments include: pillow lava and interstratified hyaloclastite; deposition of material coarser than mud grade from gravity currents (mostly turbidity currents); and presence of laminated fine-grained tuffs interpreted to represent subaqueous fallout [*Busby et al., in press*]. Geochemical data on the footwall and hangingwall ignimbrites at Bald Mountain show a primitive island-arc signature [*Schulz and Ayuso, in press*]. This is consistent with the inferred deepwater environment of deposition, since younger primitive arcs lie in deep water.

The best constraints on paleowater depths of deposition for the footwall and hangingwall ignimbrites at Bald Mountain come from the VHMS they enclose. The VHMS was produced by construction and fragmentation of chimneys by black and white smokers, followed by hydrothermal replacement [*Slack et. al., in press*]. Foley [*in press*] used fluid inclusion analysis of chimney fragments from the VHMS to determine hydrostatic pressure, and arrived at a paleo-water depth estimate of greater than 1.45 km. This estimate also constrains the paleowater depth of eruption of the footwall ignimbrite, because the massive sulfide grew in the vent for the footwall ignimbrite; this happened immediately after eruption of the footwall ignimbrite, since no sedimentary rocks or lava flows intervene [*Busby et al., in press*]. The vent for the footwall ignimbrite lies within a small (370m x 275 m) but deep (215 m) synvolcanic graben, with footwall ignimbrite agglutinated onto its walls in vertical welding fabrics [*Busby et al., in press*]. In addition to venting the "sheet-forming" ignimbrite, it also vented gas-poor magma in silicic fire fountains, to build a "cone" or proximal ejecta lobe of rhyolite hyalotuff that lies within the ignimbrite sheet (*Busby et al., [in press]*; see core CL-55 in Figure 2A). We have not located a vent for the hangingwall ignimbrite, but the footwall and hangingwall ignimbrites are interpreted to be the fill of two nested calderas that collapsed in rapid succession (probably much less than a half million years; *Busby et al., [in press]*; see Figures 2A and 2B). We therefore infer that the hangingwall ignimbrite was erupted in the same water depths as the underlying VHMS deposit (1.45 km).

TERMINOLOGY

The volcaniclastic terminology used in this paper largely follows that of *Fisher and Schmincke* [1984] and *Heiken and Wohletz* [1992]. The term "pyroclastic flow" refers to a highly-concentrated pyroclastic density current composed entirely of freshly-erupted pyroclastic debris (e.g. glass shards, crystals, pumice and rock fragments); *sensu-strictu* the fluid phase is hot gases, but since the temperature of emplacement can rarely be determined in ancient settings, we use the term *sensu-lato*, i.e., regardless of emplacement temperature. Pyroclastic flow deposits are predominantly massive (i.e. nonstratified) and poorly sorted [*Freundt et al., 2000*]. Pyroclastic flows may be classified according to the degree of vesiculation shown by essential fragments [*Heiken and Wohletz, 1992; Wright and Mutti, 1981*]. The term "ignimbrite" is used for pumiceous, ash-rich pyroclastic flow deposits, which may or may not be welded; some other workers use the term "ash-flow tuff" for these instead [*Freundt et al., 2000*].

Subaqueous pyroclastic flow deposits are the subaqueous equivalent of subaerial pyroclastic flow deposits [*Fiske, 1963; Fiske and Matsuda, 1964; Gibson et al., 2000*]. Several workers have proposed that the term "subaqueous pyroclastic flow" be applied only to hot, gas-supported density currents [*Cas and Wright* 1987, 1991; *Stix, 1991;*

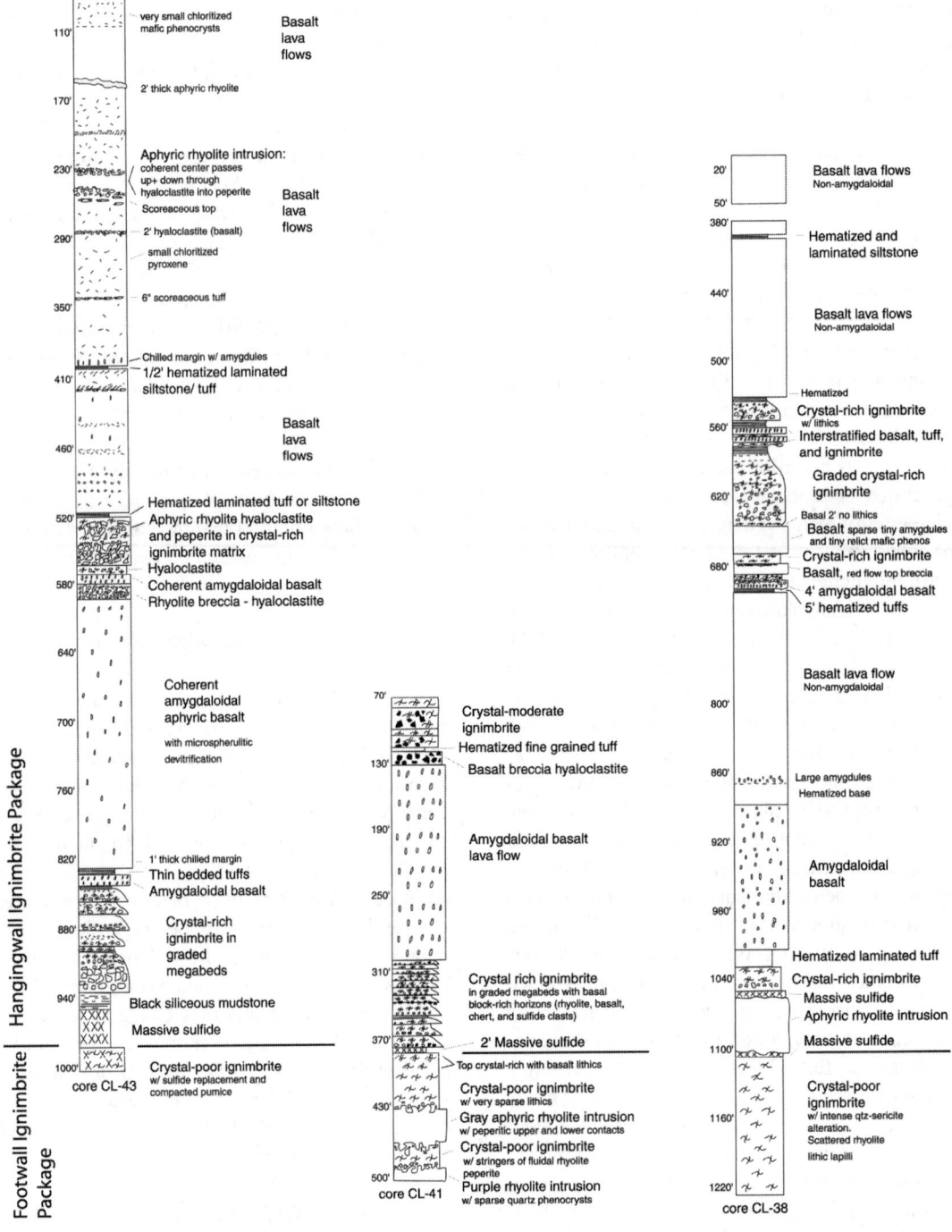

Figure 2. A) Stratigraphic correlation diagram of selected cores through the footwall and hangingwall ignimbrites in the Bald Mountain caldera [positions of cores given in *Busby et al.*, in press]. The footwall ignimbrite occurs in massive very thick units, with no internal stratification or grading, separated by basalt lavas. These are interpreted as intracaldera deposits [*Busby et al.*, in press]. The hangingwall ignimbrite occurs as thinner flow units, including massive nonstratified nongraded pumiceous pyroclastic flow units 1-60m thick, or graded megabeds 4 to 33 m thick. These are interpreted as accidentally ponded outflow from the Bull Hill caldera [*Busby et al.*, in press].

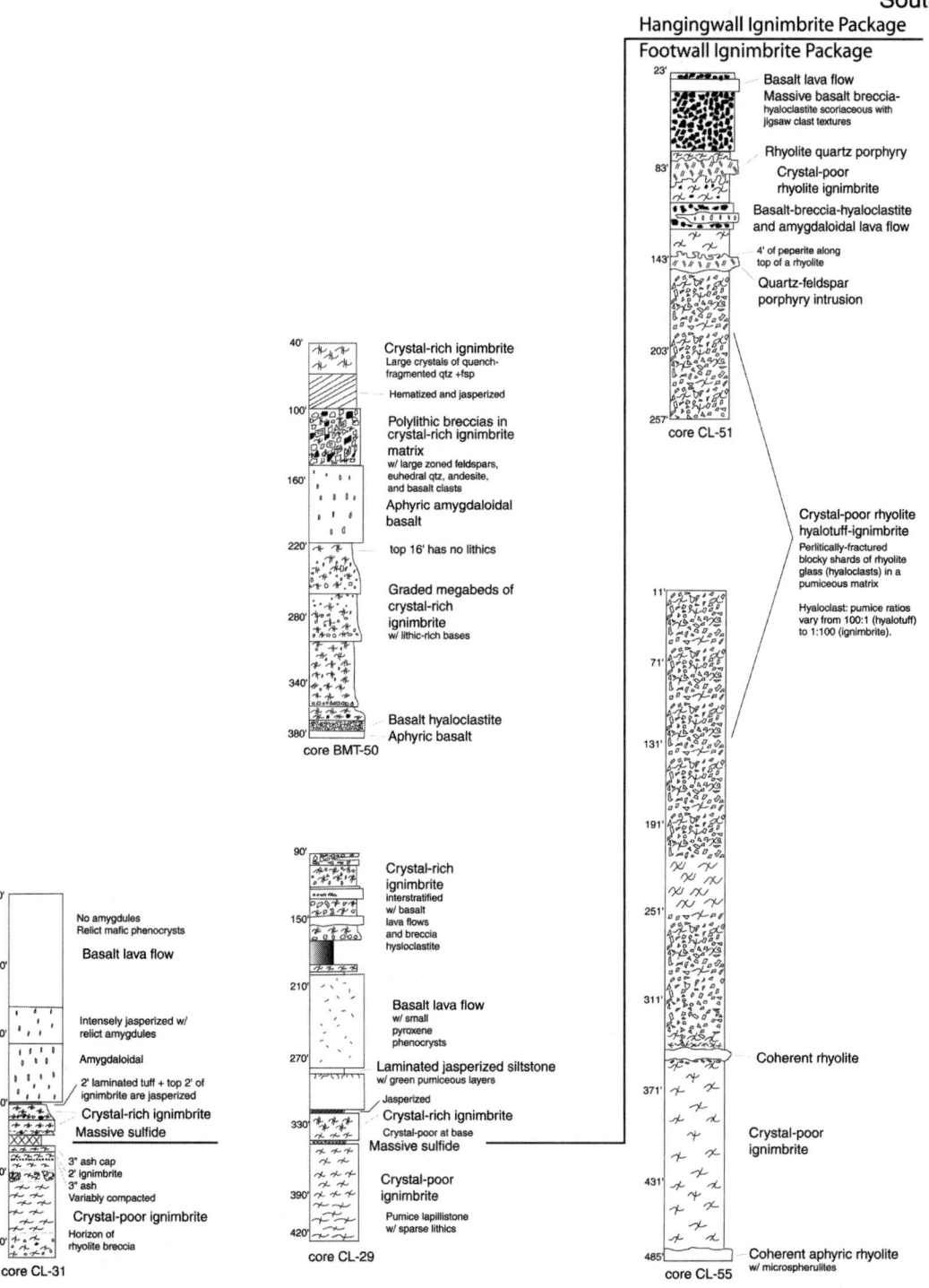

Figure 2. A) *(continued)* Stratigraphic correlation diagram of selected cores through the footwall and hangingwall ignimbrites in the Bald Mountain caldera [positions of cores given in *Busby et al.*, in press]. The footwall ignimbrite occurs in massive very thick units, with no internal stratification or grading, separated by basalt lavas. These are interpreted as intracaldera deposits [*Busby et al.*, in press]. The hangingwall ignimbrite occurs as thinner flow units, including massive nonstratified nongraded pumiceous pyroclastic flow units 1–60m thick, or graded megabeds 4 to 33 m thick. These are interpreted as accidentally ponded outflow from the Bull Hill caldera [*Busby et al.*, in press].

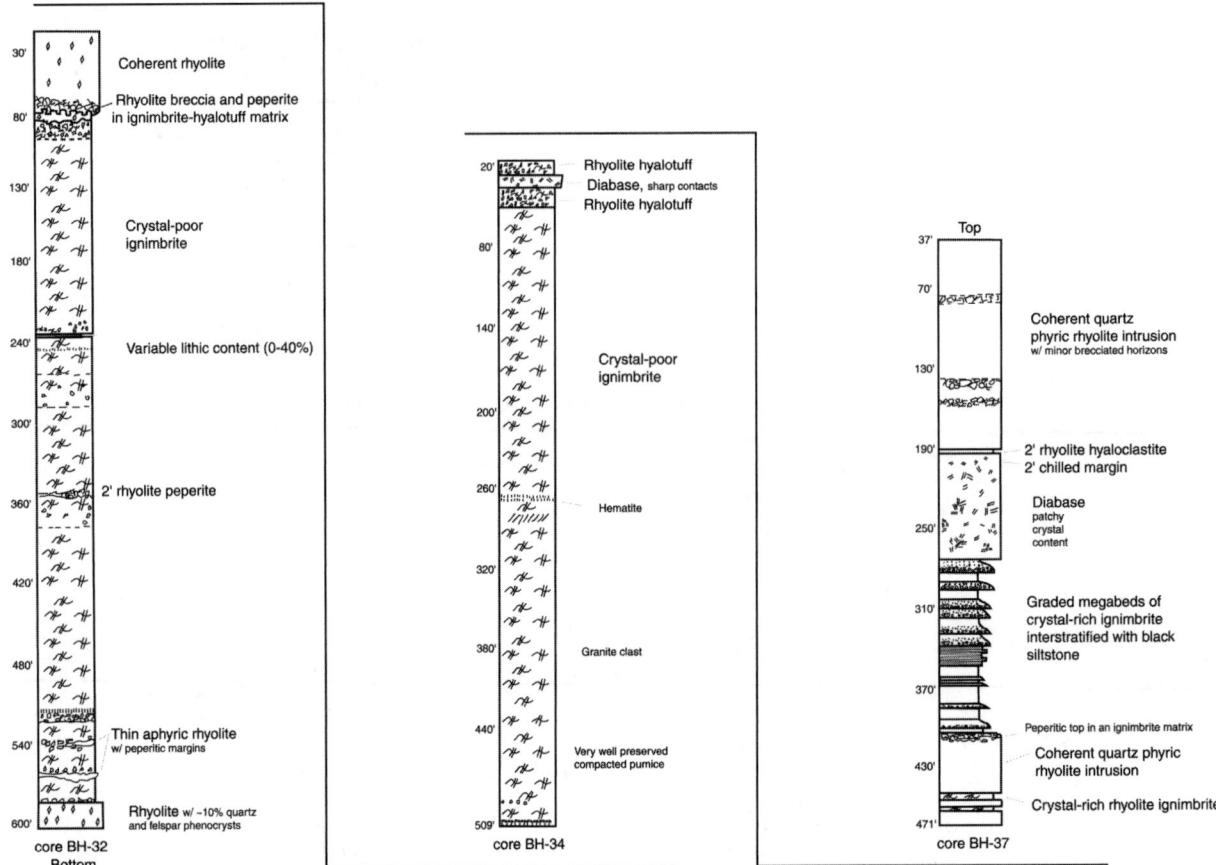

Figure 2. B) Composite vertical section through part of the hangingwall ignimbrite in the Bull Hill caldera (positions of cores given in *Busby et al.*, [in press]). As in the Bald Mountain caldera, both massive (i.e. nonstratified, nongraded) megabeds and graded megabeds are present, but the massive megabeds are more common. This is because the hangingwall ignimbrite is an intracaldera ignimbrite here, and it is an accidentally-ponded outflow ignimbrite in the Bald Mountain caldera [*Busby et al.*, in press].

McPhie et al., 1993]; subaqueous water-supported pyroclastic flows were termed "volcaniclastic mass flows". This distinction cannot be applied to most ancient pyroclastic flow deposits, where one can rarely prove whether the flow was above or below 100°C at deposition [*Busby-Spera*, 1986; *Gibson et al*, 2000; *White*, 2000]. Furthermore, this nomenclature does not attempt to separate eruptive-fed density currents from those stemming from much later resedimentation [*Kano, 1996; White*, 2000]. We agree with *Gibson et al.* [2000] and *White* [2000] that the nomenclature of *Cas and Wright* [1991] masks the origin of subaqueous pyroclastic flows as the direct product of an eruption. Most volcanic eruptions on Earth take place under water, and it is important to name their products accordingly where they are recognized. A deposit that is predominantly massive (i.e. lacking stratification or traction structures), poorly graded, poorly sorted, very thick, and composed entirely of fresh pyroclastic debris of uniform composition is highly likely to be the direct result of an eruption (i.e., it is a primary deposit), whether or not one can determine temperature of emplacement. In this paper, we use the term "ignimbrite" for the deposits of pumiceous pyroclastic flow deposits, regardless of their environment of eruption and deposition (subaerial or subaqueous).

Blocky pumice clasts and shards may form by hydroclastic fragmentation of pyroclasts contemporaneous with gas exsolution and vesiculation [*Fisher and Schmincke*, 1984; *Heiken and Wohletz*, 1992; *Busby et al.*, in press]. In this case, hydroclastic fragmentation is the result of hot gases within large pumice rapidly contracting upon contact with cold external water, resulting in collapsed and brittle fragmented pumice clasts. Blocky pumice is similar to the blocky scoria described by *Doucet et al.*, [1994] and phreatomagmatic tephra described by *Morissey et al.*, [2000].

The components (i.e., glass shards, pumice and crystals) of dominantly massive pyroclastic flow deposits may also be present in more stratified or cross-stratified, highly sorted deposits that may show grading or partial Bouma sequences. These are referred to here as tuff turbidites, and were probably transported by high density or low-density turbidity currents with a high percentage of interstitial water. Finally, subaqueous fallout tuffs are accumulations of pyroclastic material that settled through water, wherein sorting and stratification are greatly accentuated by this settling process [*Fisher*, 1964].

METHODS

We examined 380 thin section samples of the Bald Mountain ignimbrites from drill core and outcrop, and selected twenty-five well-preserved, representative samples of the footwall and hangingwall ignimbrites for textural analysis, including: (1) grain morphology and (2) granulometric properties. The thin sections were cut orthogonal to bedding, and stained for K-feldspar and calcic feldspar. Five hundred points were counted for each sample, using an electronic stage-advancing unit with a Zeiss polarizing microscope. We used the Gazzi–Dickinson method, where crystals are counted as such, whether or not they are contained within lithic clasts.

Grain size distribution analysis requires the measuring of the long axis of free crystals, lithic fragments, pumice clasts, glass shards, and other grains, and statistically analyzing the variation in grain size distributions using parameters such as mean, median, standard deviation, kurtosis, and skewness [*Sheridan*, 1971; 1987]. We measured the apparent long axis of each grain on the thin section samples, using a micrometer eyepiece on a polarizing microscope; analyses are therefore expressed as volume percentages, rather than weight percentages [*Lewis and McConchie*, 1994; *Wright and Mutti*, 1981]. The grains are seen in two dimensions, so the apparent long axis measurement is a minimum estimate. The grain size data were tabulated and converted from microscope measurements to the standard Phi (ϕ) scale [*Inman*, 1952; *Sheridan*, 1971]. Cumulative volume percentages were calculated and the Inman parameters determined [*Inman*, 1952].

We compiled published granulometric data from subaerial and subaqueous pyroclastic flows and tuff turbidites (summarized in Table 1) for comparison with our granulometric data (Table 2).

TEXTURAL AND GRANULOMETRIC ANALYSIS OF DEEP-WATER ERUPTED IGNIMBRITES

Representative cores through the footwall and hangingwall ignimbrites are illustrated and described in Figures 2A and 2B. The footwall ignimbrite is crystal poor, with less than a few percent quartz and plagioclase feldspar, whereas the hangingwall ignimbrite has the same minerals, but generally in much higher percentages. The footwall and hangingwall ignimbrite are also distinguishable on geochemical grounds [*Schulz and Ayuso, in press*]. The modal compositions of phenocrysts in both ignimbrites are >95% quartz and calcic plagioclase (Table 2). The low percentage of K-feldspars (<2%) is consistent with a whole-rock geochemistry of low K_2O content, except where hydrothermally altered with minor sericite [*Schulz and Ayuso, in press*].

The footwall ignimbrite is a poorly sorted and wholly nonstratified mixture of blocky pumice lapilli and medium to coarse ash (Figures 3A through 3C), referred to here as pumice lapillistone. It occurs in massive flow units tens of meters to 150m thick separated by basalt lava flows and intruded by rhyolites (Figure 2A). It lacks stratification or grading, and it contains no ash interbeds (Figure 2A).

Like the footwall ignimbrite, the hangingwall ignimbrite is a fine-ash poor mixture of pumice and lesser medium to coarse ash (pumice lapillistone), but it occurs as thinner flow units 1–150 m thick separated by well-sorted and stratified lapilli tuff and tuff (Figure 2A). Flow units of the hangingwall ignimbrites occur as (1) massive nonstratified nongraded pumiceous pyroclastic flow units 1–150 m thick (Figure 2B), and lesser (2) graded megabeds 4 to 33 m thick (Figure 2A). The graded megabeds pass upward from volcanic lithic breccia (with interstitial pumice) or tuff breccia (pumice-rich matrix-supported) through lithic lapilli tuff through pumice lapilli tuff, to coarse-grained crystal tuff, and medium- to fine-grained crystal vitric tuff. These graded megabeds are massive, with laminations restricted to the uppermost several inches. The hangingwall ignimbrite lacks blocky pumice (Figures 4A through 4C).

5.1. Microtextural Features

The footwall ignimbrite is pumice-rich (50–81 volume %) and contains approximately 1–20% lithic fragments (Figures 4A through 4C and Table 2). Some crystals have jigsaw puzzle textures. Hematite replacement on the surface of many pumice clasts outlines their blocky geometry (Figures 3B and 3C). The angular block-like pumice, with blunt terminations, is abundant throughout the footwall ignimbrite. The matrix is fine ash poor, with all clasts measuring greater than 0.8 mm. Bubble-wall shards are generally absent and blocky-rigid pumices are closely packed (Figure 3C).

The hanging wall ignimbrite differs from the footwall ignimbrite in its lack of blocky rigid pumices (compare Figures 3 and 4). Only the graded tops of a few hangingwall ignimbrite flow units have blocky pumices. Instead, the

Table 1. Summary descriptions of published granulometric data shown in Figures 5 and 6.

Unit name	Age	Location	Inferred depositional environment	Deposit characteristics or Sedimentary structures	Inferred depositional processes
Subaerial Erupted Ignimbrites					
Bishop Tuff (Ignimbrites) (Sheridan, 1971)	640k	Long Valley Caldera Bishop, California USA	Subaerial	Non-welded to welded Ign. composed of pumice, shards, lithic fragments, and crystals	Subaerial ignimbrite flows
Vulsini Ignimbrites (Sparks, 1976)	Quaternary	Vulsini volcano 100 km north of Rome, Italy	Subaerial	Non-welded ignimbrites composed of pumice, shards, lithic fragments, and crystals	Subaerial ignimbrite flows
Krakatau Ignimbrites (Mandeville et al., 1994)	year 1883	Krakatau Volcano Indonesia	Subaerial to subaqueous	Non-welded to welded Ign. composed of pumice, shards, and crystals	Subaqueous pyroclastic flow with no mixing of the flow with water. Total retention of fine ash and coarse material.
Roseau Ignimbrites * weakly fines-poor (Whitham, 1989)	30k	Roseau Valley Dominica, West Indies	Subaerial	Massive, pumice rich (up to 58%), and poorly sorted	Subaerial ignimbrite flows
Subaqueous Erupted Pumiceous Mass Flow					
Shinjima Pumice (Kano et al., 1996)	6-11k	Shinjima Island, Kagoshima Bay, SW Japan	Subaqueous	Fines-poor pumice lapilli tuff, diffusely stratified with upward coarsening pumice	Turbulent mixing of eruption column with water Water logged mass flows
Subaqueous Tuff Turbidites					
Roseau subaqueous deposits (Whitham, 1989)	30k	Roseau Valley Dominica, West Indies	Subaqueous	Silty and fine grained tuffs with rounded fragments and greater lithic content	High and low concentration sediment gravity flows I.e., turbidity currents
Dali Ash (Wright and Mutti, 1981)	Tertiary	Rhodes, Greece	Subaqueous	Massive lower unit overlain by bedded upper unit. No evidence for heat retention Rich in fine ash size material	High and low concentration sediment gravity flows I.e., turbidity currents
Joetsu Ash (Kano et al. 1996)	Pleistocene	Japan	Subaqueous	Massive to stratified beds that are fine ash poor	Tuff turbidite and subaqueous volcaniclastic mass flow

Table 2. A) Granulometric data from Bald Mountain with irresolvable groundmass counted as cement.

Ignimbrites with the irresolvable groundmass counted as cement

Unit	sample	Phi 95	Phi 84	Phi 75	Phi 50	Phi 25	Phi 16	Phi 5	Md	GMPhi	SD Phi	IGSD Phi	IGSk	GK	Facies
Hanging Wall	BH-39-554	3.54	1.81	1.13	0.22	-0.52	-1.00	-2.57	0.22	0.34	1.40	1.63	0.11	1.52	xlRchlg
Hanging Wall	CL-30-291	3.13	2.32	1.81	0.88	0.13	-0.26	-0.78	0.88	0.98	1.29	1.24	0.13	0.95	xlRchlg
Hanging Wall	CL-30-696	2.54	1.67	1.13	0.22	-0.85	-1.85	-4.32	0.22	0.35	1.76	1.67	-0.09	1.42	xlRchlg
Hanging Wall	CL-32-234	1.81	1.22	0.60	-0.33	-1.43	-1.85	-2.72	-0.33	-0.18	1.54	1.35	0.06	0.91	xlRchlg
Hanging Wall	CL-38-536	2.06	1.25	1.06	-0.03	-1.26	-2.17	-2.79	-0.03	-0.01	1.71	1.36	-0.06	0.86	xlRchlg
Hanging Wall	CL-41-102	2.13	0.81	-0.03	-1.72	-4.09	-5.64	-5.64	-1.72	-1.67	3.23	2.40	0.01	0.79	xlRchlg
Hanging Wall	CL-41-298	3.54	2.32	1.81	0.81	-0.52	-2.96	-4.81	0.81	0.87	2.64	1.97	-0.14	1.47	xlRchlg
Hanging Wall	CL-41-81	2.54	1.96	1.32	0.49	-0.33	-0.52	-1.04	0.49	0.64	1.24	1.16	0.17	0.89	xlRchlg
Hanging Wall	CL-43-518	4.13	2.81	2.54	1.43	0.61	0.18	-0.57	1.43	1.47	1.32	1.37	0.10	0.99	xlRchlg
Hanging Wall	CL-43-528	2.81	1.96	1.54	0.74	0.04	-0.19	-0.75	0.74	0.83	1.08	1.08	0.15	0.97	xlRchlg
Hanging Wall	CL-46-318	2.47	1.06	0.47	-1.07	-2.30	-2.66	-3.12	-1.07	-0.77	1.86	1.69	0.27	0.83	xlRchlg
Hanging Wall	CL-46-476	3.06	1.74	1.06	0.25	-0.75	-2.09	-2.43	0.25	-0.03	1.91	1.79	-0.10	1.25	xlRchlg
Hanging Wall	CL-46-528	1.25	0.27	-0.19	-1.07	-1.96	-2.58	-2.96	-1.07	-1.13	1.43	1.35	0.02	0.98	xlRchlg
Hanging Wall	BMT-18-155	0.47	-0.64	-1.03	-1.72	-2.59	-2.91	-3.59	-1.72	-1.65	1.13	1.10	0.10	1.07	xlpoorlg
Hanging Wall	CL-49-372	1.74	1.06	0.60	-0.58	-1.67	-2.17	-2.72	-0.58	-0.40	1.62	1.36	0.12	0.81	xlpoorlg
Hanging Wall	BH-32-68B	3.13	2.13	1.54	0.22	-1.04	-1.72	-3.43	0.22	0.21	1.92	1.96	-0.06	1.04	xlpoorlg
Foot Wall	CL-51-220	2.54	0.15	-0.85	-1.85	-2.72	-3.17	-3.59	-1.85	-1.62	1.66	1.76	0.32	1.34	xlpoorlg
Foot Wall	CL-35-442.5	1.25	0.15	-0.85	-1.96	-2.66	-2.66	-2.79	-1.96	-1.49	1.41	1.31	0.55	0.91	xlpoorlg
Foot Wall	CL-35-455	2.32	0.74	0.15	-1.28	-2.58	-3.02	-3.43	-1.28	-1.19	1.88	1.81	0.16	0.86	xlpoorlg

Tuff turbidites

Unit	sample	Phi 95	Phi 84	Phi 75	Phi 50	Phi 25	Phi 16	Phi 5	Md	GMPhi	SD Phi	IGSD Phi	IGSk	GK	Facies
Foot wall	BH-39-325	3.06	1.74	1.25	0.15	-1.11	-1.61	-2.37	0.15	0.09	1.68	1.66	0.01	0.94	xlLtf
Foot wall	BH-39-239	2.81	2.13	1.67	0.96	0.22	-0.19	-0.78	0.96	0.97	1.16	1.12	0.02	1.01	xlLtf
Hanging wall	BH-39-151	3.54	3.13	2.81	2.13	1.32	1.04	0.32	2.13	2.10	1.04	1.01	-0.08	0.89	fgxltf
Hanging wall	Cl-49-136	3.06	2.06	1.47	0.47	-0.47	-0.90	-1.59	0.47	0.55	1.48	1.44	0.09	0.98	xlLtf
Hanging Wall	BMT-50-41	3.54	2.32	1.67	0.49	-0.46	-0.70	-1.53	0.49	0.70	1.51	1.53	0.21	0.98	xlLtf
PIII / BH	BMT-12-216	2.81	1.67	1.22	0.22	-0.78	-1.00	-1.75	0.22	0.30	1.34	1.36	0.11	0.93	xlLtf

Key to granulometric parameters
Md = Median grain size
GMPhi = Graphic mean
SD Phi = Standard deviation or non-inclusive graphic standard deviation
GSD Phi = Inclusive graphic standard deviation
IGSk = Inclusive graphic skewness
Gk = Graphic kurtosis

Key to facies symbols
xlpoorlg = crystal poor ignimbrite
xlmodlg = crystal moderate ignimbrite
xlRchlg = crystal rich ignimbrite

Table 2. B) Granulometric data from Bald Mountain with with irresolvable groundmass counted as very fine ash.

Ignimbrites with irresolvable groundmass counted as very fine ash

Unit	sample	Phi 95	Phi 84	Phi 75	Phi 50	Phi 25	Phi 16	Phi 5	Md	GMPhi	SD Phi	IGSD Phi	IGSk	GK	Facies
Hanging Wall	BH-39-554	6.08	5.07	5.07	0.61	-0.36	-0.57	-2.57	0.61	1.70	2.82	2.72	0.42	0.65	xlRchlg
Hanging Wall	CL-30-291	6.10	5.07	3.54	1.32	0.37	-0.04	-0.70	1.32	2.12	2.56	2.31	0.44	0.88	xlRchlg
Hanging Wall	CL-30-696	3.13	1.81	1.22	0.22	-0.83	-1.85	-4.32	0.22	0.06	1.83	2.04	-0.18	1.49	xlRchlg
Hanging Wall	CL-32-234	2.81	1.25	0.74	-0.26	-1.43	-1.80	-2.72	-0.26	-0.27	1.53	1.60	0.05	1.04	xlRchlg
Hanging Wall	CL-38-536	5.06	1.74	1.25	0.06	-1.11	-2.17	-2.79	0.06	-0.12	1.95	2.17	0.07	1.36	xlRchlg
Hanging Wall	CL-41-102	5.80	3.06	1.25	-0.94	-3.12	-5.64	-5.64	-0.94	-1.18	4.35	3.91	0.05	1.07	xlRchlg
Hanging Wall	CL-41-298	4.06	4.13	2.81	1.22	-0.12	-1.10	-4.81	1.22	1.42	2.61	2.65	-0.12	1.24	xlRchlg
Hanging Wall	CL-41-81	6.10	5.13	3.54	1.04	0.00	-0.36	-0.89	1.04	1.94	2.75	2.43	0.47	0.81	xlRchlg
Hanging Wall	CL-43-518	6.13	3.54	2.81	1.67	0.74	0.22	-0.52	1.67	1.81	1.66	1.84	0.24	1.32	xlRchlg
Hanging Wall	CL-43-528	5.13	2.81	1.96	0.88	0.18	-0.19	-0.73	0.88	1.17	1.50	1.64	0.37	1.35	xlRchlg
Hanging Wall	CL-46-318	6.08	4.06	2.06	-2.63	-2.09	-2.66	-3.07	-2.63	-0.41	3.36	3.06	0.95	0.90	xlRchlg
Hanging Wall	CL-46-476	6.10	5.13	2.47	0.60	-0.40	-1.47	-2.43	0.60	1.78	3.30	2.68	0.46	1.22	xlRchlg
Hanging Wall	CL-46-528	5.90	4.06	1.06	-0.78	-1.72	-2.43	-2.96	-0.78	0.28	3.24	2.97	0.50	1.31	xlRchlg
Hanging Wall	BMT-18-155	4.32	0.47	-0.59	-1.61	2.59	-2.91	-3.59	-1.61	-1.35	1.69	2.04	0.37	-1.02	xlpoorlg
Hanging Wall	BH-32-68B	6.00	4.97	2.81	0.96	-0.78	-1.43	-3.43	0.96	1.50	3.20	3.03	0.16	1.08	xlpoorlg
Hanging Wall	CL-49-372	2.06	1.25	0.74	-0.53	-1.64	-2.17	-2.72	-0.53	-0.48	1.71	1.58	0.06	0.82	xlpoorlg
Foot Wall	CL-51-220	6.00	5.07	4.99	-0.89	-2.43	-3.12	-3.43	-0.89	0.35	4.09	3.48	0.46	0.47	xlpoorlg
Foot Wall	CL-35-442.5	5.06	1.74	-3.91	-0.99	-2.26	-2.66	-2.72	-0.99	-0.64	2.20	2.28	0.40	-1.93	xlpoorlg
Foot Wall	CL-35-455	6.00	1.54	0.47	-1.26	-2.43	-3.02	-3.43	-1.26	-0.91	2.28	2.57	0.39	1.33	xlpoorlg

Key to granulometric parameters
Md = Median grain size
GMPhi = Graphic mean
SD Phi = Standard deviation or non-inclusive graphic standard deviation
GSD Phi = Inclusive graphic standard deviation
IGSk = Inclusive graphic skewness
Gk = Graphic kurtosis

Key to facies symbols
xlpoorlg = crystal poor ignimbrite
xlmodlg = crystal moderate ignimbrite
xlRchlg = crystal rich ignimbrite

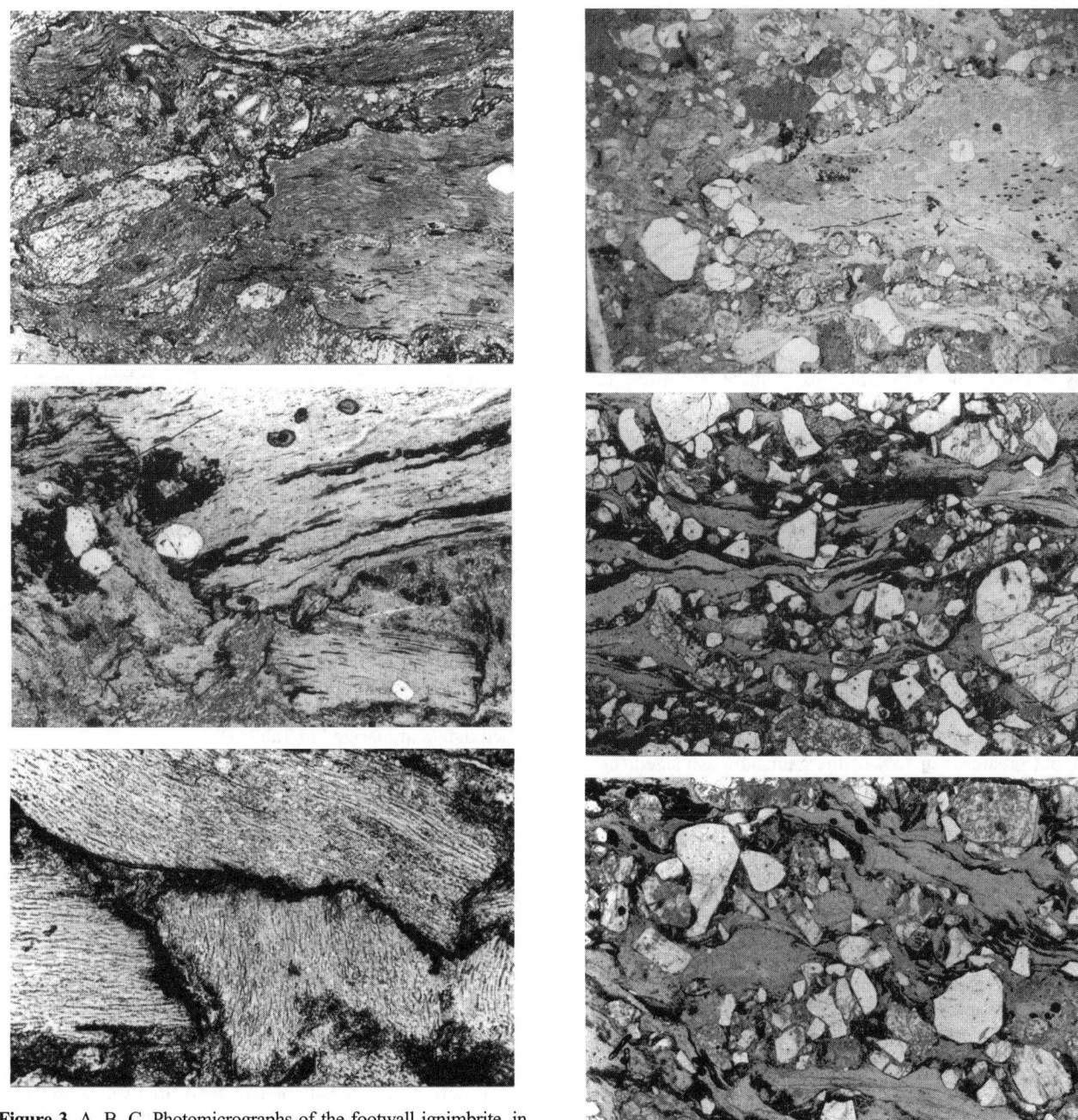

Figure 3. A, B, C. Photomicrographs of the footwall ignimbrite, in thin sections cut perpindicular to bedding, arranged from lowest to highest magnification. These photomicrographs all show the typical pumice-rich, fine-ash-poor texture of the footwall ignimbrite, referred to as pumice lapillistone. Hematitic alteration outlines pumice well. Note rigid, relatively unflattened nature of the pumice, possibly due to retardation of magmatic vesiculation by hydrostatic pressure. Their blocky-angular shapes are suggestive of hydroclastic fragmentation. Field of view for A = 7 mm (sample BMT18-155'), for B = 4.5 mm (sample CL35-455') and for C = 0.9 mm (sample CL35-455'). Sample numbers refer to cores (and footage from top of core) presented in stratigraphic correlation diagrams of *Busby et al.* [in press].

Figure 4. Photomicrographs of the hangingwall ignimbrite, in thin sections cut perpindicular to bedding. Sample numbers refer to cores (and footage from top of core) presented in stratigraphic correlation diagrams of *Busby et al.* [in press]. Photo A shows moderate compaction of pumices and relict bubble-wall shards (field of view 7 mm, sample CL41-298'). Photo B shows strong compaction of pumices (field of view 7 mm, sample CL30-291'). Photo C shows closeup of strongly compacted pumices (field of view 4.5 mm, sample CL30-291').

hanging wall ignimbrites are dominated by pumice compaction textures. The pumices are plastically deformed around phenocrysts and lithic fragments, and show wispy terminations (Figures 4B and 4C). The compacted pumice appears to have collapsed vesicles, although the identification of clearly collapsed vesicles at high magnification is difficult to verify as a result of the hydrothermal and diagenetic alteration. Two possible mechanisms for pumice compaction include primary welding and diagenetic compaction. We prefer, but cannot prove, a primary welding interpretation because: (1) the degree of diagenesis required to generate apparent eutaxitic textures should completely replace the matrix and obscure the compacted shards and wispy terminations of the shredded pumice lapilli and (2) blocky (quenched) pumices in the underlying footwall ignimbrite show no compaction at all, suggesting that diagenetic compaction was not important in either ignimbrite.

5.2. Granulometry of Bald Mountain Ignimbrites and Comparison with Published Examples of Subaerial and Subaqueous Pyroclastic Flows and Tuff Turbidites

Granulometric parameters of Bald Mountain ignimbrites and tuff turbidites (i.e. median and Inman sorting coefficient) are compared with previously published data from other pyroclastic deposits in Table 1 and Figures 5 and 6.

The footwall ignimbrite has 0–2% very-fine ash (<1/16 mm), but greater than 75% of the grains are ash sized (i.e., 1/16–2 mm) and the rest are lapilli sized (Table 2). Coarse lapilli to block-sized pumices are also generally absent from the cores and outcrops [*Busby et al., in press*]. The footwall ignimbrite samples contain 7–24% groundmass that is irresolvable by petrographic microscope, due to devitrification, chloritization, diagenetic alteration, and/or cementation of the particles; these are common processes in volcaniclastic rocks [*Surdam and Boles*, 1979; *Schneider et al.*, 1992; *WoldeGabriel et al.*, 1996].

The hanging wall ignimbrite has a generally high phenocryst content and therefore, like other crystal-rich ignimbrites, has grain size distributions strongly controlled by typical phenocryst sizes (0.1–2.0 mm; Table 2) [*Sheridan et al.*, 1987]. Like the footwall ignimbrites, the hanging wall ignimbrite samples have less than 10% very fine ash. The hanging wall ignimbrites have predominantly coarse ash to fine lapilli sized grains, with <5 % coarse fraction greater than Phi −3 (Figure 6). The hanging wall ignimbrite has 2.8–24% irresolvable groundmass.

We use two end-member models to estimate the original percentages of very fine ash (Table 2). In the first model, we assume the irresolvable groundmass was produced by cementation (e.g., authigenic feldspar or zeolite) and not by replacement of fine ash fragments. The resultant grain size distribution, shown in Figure 5A, is well sorted and fines poor. In the second model, all of the irresolvable groundmass is counted as fine ash content, ranging from 7–28%. The resultant grain size distribution, shown in Figure 5B, is less sorted compared to model one but nonetheless remains in the "fines-depleted" pyroclastic flow field outlined by *Walker* [1983]. We prefer end-member model one because the best-preserved footwall ignimbrite samples (approximately 33%), with only 7.2% irresolvable groundmass, and the best-preserved hangingwall ignimbrite samples (approximately 25%), with only 2–3% irresolvable groundmass, contain less than 2% very fine ash. This suggests that the other samples lacked very fine ash as well.

A newly-devised granulometric plot of fine ash *vs.* coarse ash (Figure 6) may prove useful in future attempts to distinguish pyroclastic flow deposits from fines-poor equivalents or from tuff turbidites. This plot clearly illustrates the fine-ash-poor character of the Bald Mountain ignimbrite deposits relative to poorly-sorted subaerial or shallow marine pyroclastic flow deposits reported in the literature (sources of published data In Table 1). Our new plot also shows that well stratified, graded beds of pyroclastic material, referred to here as tuff turbidites, lie in a separate field (using our data from Bald Mountain tuff turbidites and published data summarized in Table 1).

POSSIBLE ORIGINS OF FINE-ASH POOR PYROCLASTIC DEPOSITS

The origin of fine-ash-poor pyroclastic deposits is briefly discussed as a prologue to presentation of our eruptive models for the hangingwall and footwall ignimbrites. There are several ways that fine-ash-poor deposits may form.

In subaerial settings, deposits composed entirely of pumice (with little ash) are commonly interpreted as Plinian fall deposits, but these are far better stratified and much thinner than ignimbrites [*Cioni et al.*, 2000]. *Walker* [1971; 1983] defined a fines-depleted field for subaerial ignimbrites (Figures 5A and 5B); this field included volumetrically small features such as gas elutriation pipes and "layer 1" deposits, as well as volumetrically significant "fines-depleted ignimbrite", inferred to result from deflation during flow.

Facies models for explosive eruptions from relatively shallow subaqueous vents (a few hundred meters or less) commonly attribute a fines-poor matrix to elutriation by mixing with water during eruption and/or transport, but all of those density current deposits are well-stratified, with at

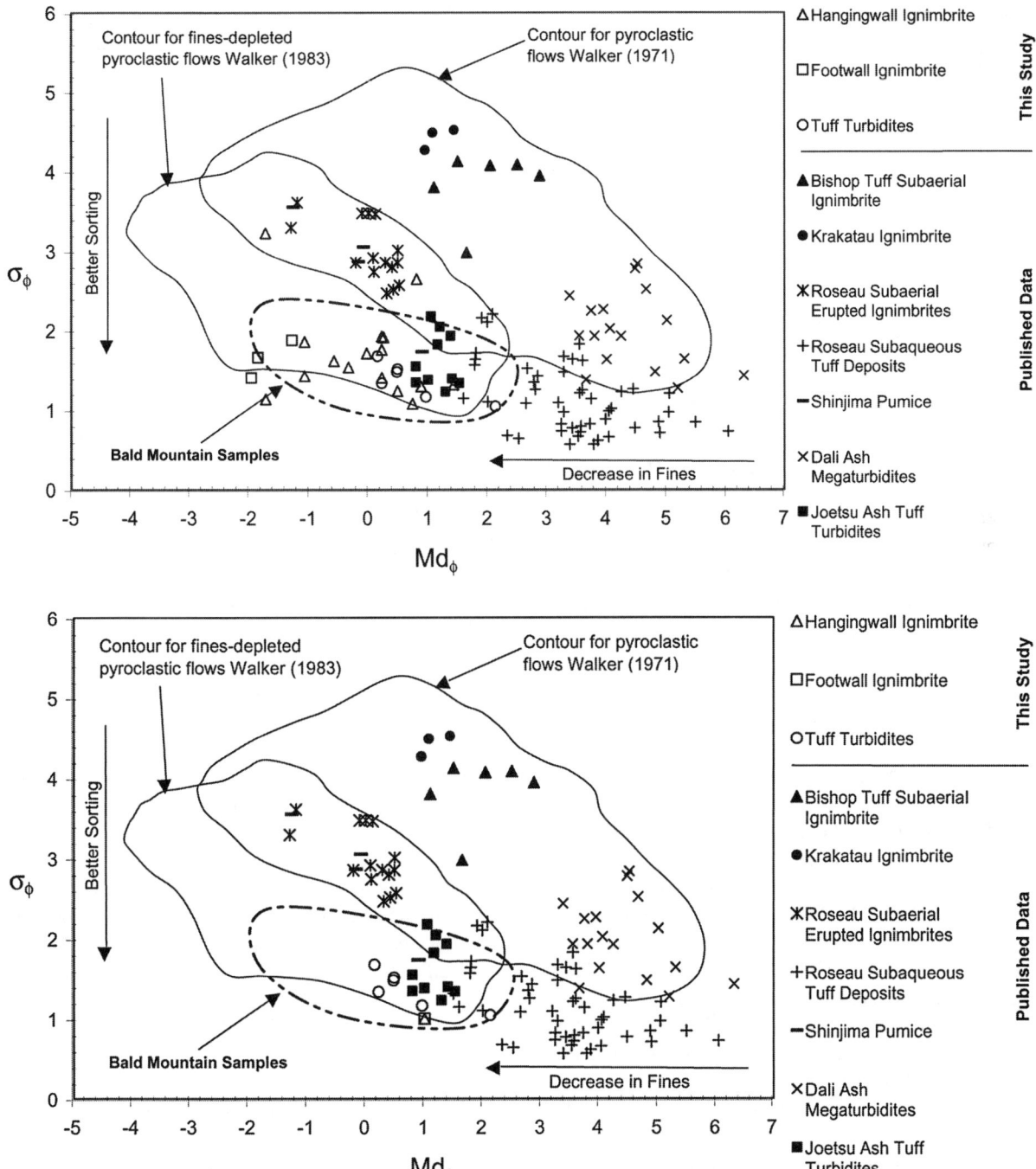

Figure 5. A) Plot of median diameter (Md) vs. Inman sorting coefficient (standard deviation {sigma} for the footwall ignimbrite and hangingwall ignimbrite of this study, compared with published data described in Table 1. Bald Mountain sequence samples with irresolvable groundmass counted as cement. B) Plot of median diameter (Md) vs. Inman sorting coefficient (standard deviation {sigma} for the footwall ignimbrite and hangingwall ignimbrite of this study, compared with published data described in Table 1. Bald Mountain sequence samples with irresolvable groundmass counted as fine ash.

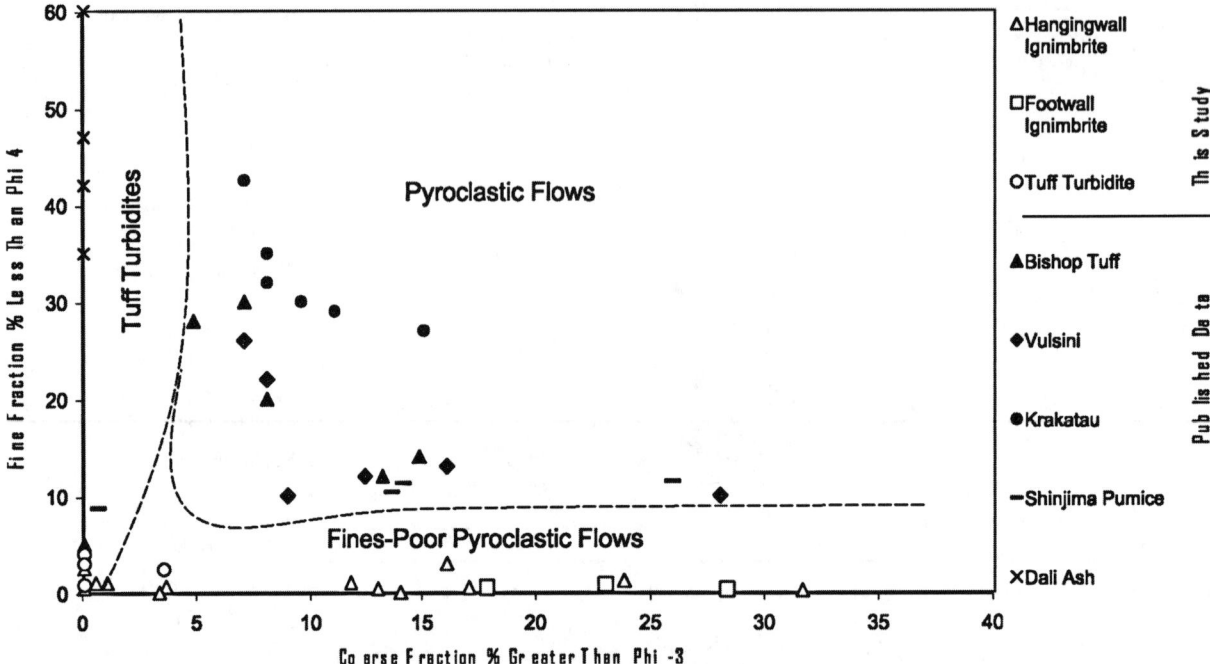

Figure 6. Plot of median diameter (Md) vs. Inman sorting coefficient (standard deviation {sigma}) for the footwall ignimbrite and hangingwall ignimbrite of this study, compared with published data described in Table 1.

least some beds that show sorting of all size fractions [*Kano et al., 1996; Allen and Cas, 1998; Allen and McPhie, 2000; White, 2000; Allen, 2001*]. In this model, much of the fine ash component is deposited later as fine ash fallout above the flow deposit or down current from it [*Cashman and Fiske, 1991; Kano et al., 1996; Smith and Smith, 1985*]. For example, the fines-depleted Shinjima Pumice was erupted during the formation of the Wakamiko caldera several thousand years ago at about 100–140m below sea level [*Kano et al., 1996*]. It contains parallel to wavy stratification throughout, and occurs in beds 1–10m thick with sharp bases and inversely graded tops [*Kano et al., 1996*]. Similarly, the >150 m thick Quaternary Yali pumice breccia, which was deposited by a combination of water settling and mass-flow resedimentation, is stratified in beds 0.1–2 m thick [*Allen and McPhie, 2000*]. Similar well-sorted, well-stratified water-lain tephra result from lacustrine and subglacial eruptions [summarized by *White, 2000*].

Recent surveys of modern deepwater volcanoes have shown that pumiceous deposits are erupted from silicic calderas in water depths up to 1.4–1.6 km [*Wright and Gamble, 1999; Fiske et al, 2001*] where VHMS deposits are forming today [*Iizasa et al., 1999; Glasby et al., 2000; Yuasa and Kano, in press*]. However, the nature of the pyroclastic deposits in these modern deepwater calderas is not yet well known. Studies of ancient deposits suggest that deepwater explosive eruptions have hydrostatically suppressed columns that may feed pyroclastic flows with high particle concentrations, which in some cases retain enough heat to weld [*Kokelaar and Busby, 1992; White, 2000*]. We use observations from Bald Mountain and from other ancient deepwater ignimbrites to speculate that fine-ash-poor ignimbrites may be typical of very deepwater (bathyal marine) explosive eruptions. In these cases, hydrostatic pressure may be sufficient to suppress magmatic vesiculation and reduce the violence of the eruptions, thereby dramatically decreasing the production of bubble wall shards.

ERUPTION OF THE FOOTWALL AND HANGINGWALL IGNIMBRITES AT >1.45 KM WATER DEPTH

The abundance of pumice in both the footwall and hangingwall ignimbrites indicates that the eruptions were driven by magmatic explosivity; this suggests the magma had high initial gas/melt ratios [*Sparks, 1978; Wohletz, 1986*], which should have generated ash as well as pumice. The pumices of the footwall ignimbrite have high vesicularity and exhibit primarily hydroclastic textures; therefore, we infer that the footwall ignimbrite was produced by phreatomagmatic eruptions of volatile-rich magma (Figure 7). The eruption column was apparently high enough to allow sufficient mixing with water to generate the hydroclastic textures, as well

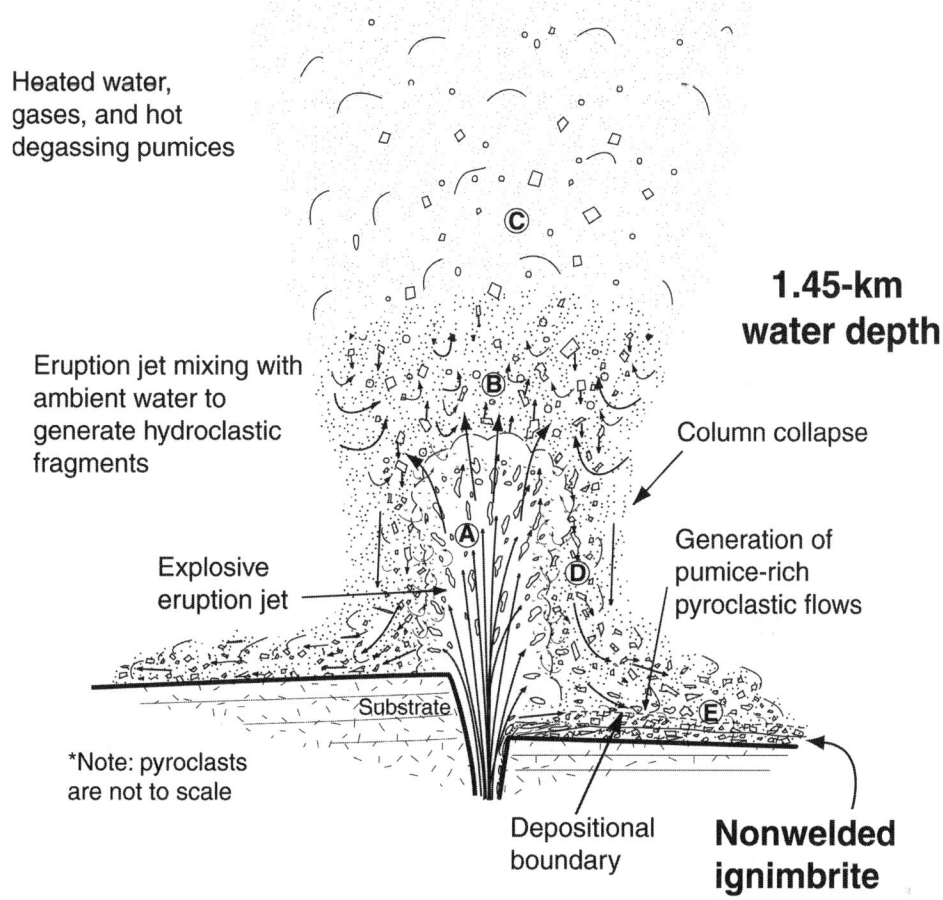

Figure 7. Eruption model for explosive deepwater eruption of the footwall ignimbrite, involving both magmatic and hydroclastic fragmentation. Steady, high-discharge eruption results in a massive deposit. Gas-rich nature of magma results in relatively high eruption column (despite great hydrostatic pressure), leading to mixing with ambient water and quench fragmentation of pumices. **A.** Eruption of hot gases and long tube pumice lapilli and blocks. **B.** Mixing between eruption plume and ambient water, and generation of steam. Rising pumice hydroclastically fragment due to rapid chilling with the water and form blocky pumice fragments. **C.** Gases (e.g., S, CO_2, CO, F) and steam rise to the surface leaving behind some water logged pumice. Some pumice continue to rise buoyantly to the surface and/or drift away down current. **D.** Gaseous vapors condense leading to collapse of dense pyroclastic mass. Column collapse results in the generation of pumice-rich pyroclastic flows i.e. ignimbrites (E).

as cooling the pyroclasts sufficiently to keep them from welding upon deposition. Nonetheless, the total lack of stratification, grading or sorting (of material coarser than fine ash) suggests that the pyroclastic flows had high particle concentrations, with a relatively low amount of admixed water.

The pumices of the hangingwall ignimbrite have high vesicularity but lack hydroclastic textures. Instead, they exhibit irregular grain morphology commonly with wispy terminations and perhaps relict welding texture. Therefore, we infer that the hangingwall ignimbrite was produced by explosive eruptions of magma with a lower volatile content than that of the footwall ignimbrite, resulting in a lower eruption column and minimal interaction with ambient water (Figure 8). The lack of mixing with water may have been aided by development of a steam cupola or carapace around the eruption column and pyroclastic flows similar to those proposed by *Kokelaar and Busby*, [1992], *Mueller et al.,* [1994], *Schneider et al.,* [1992], and *Sparks et al.,* [1980]. The graded tops present above some of the flow units may have resulted from episodic collapse of the steam cupola and/or disruption of the steam carapace at the magma-water interface. This episodic collapse/disruption was probably the result of unsteady or pulsating discharge,

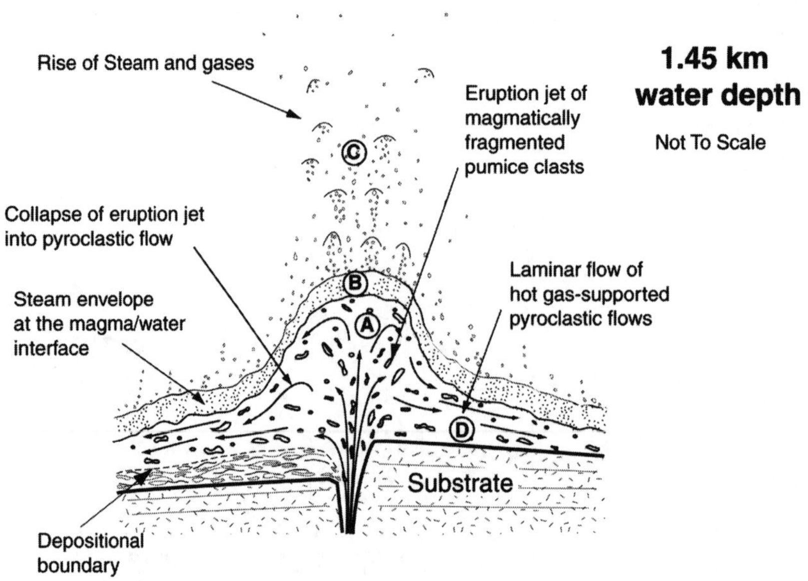

Figure 8. Eruption model for deepwater eruption of the hangingwall ignimbrite, involving magmatic but no hydroclastic fragmentation. Unsteady, relatively low-discharge-rate eruption results in mulitple flow units, some capped by graded megabeds. Relatively gas-poor nature of magma results in low eruption column, leading to minimal interaction with ambient water. This results in a lack of quench-fragmented pumices, and possible retention of heat to form (?) welding fabrics. A) Eruption of hot gases, crystals and pumice lapilli. Fine ash is absent due to hydrostatic suppression of magmatic vesiculation and fragmentation. B) Low eruption column develops steam envelope, by the rapid conversion of water to steam. The steam envelope provides a barrier from which mixture of the eruption jet and pyroclastic flow with ambient water is minimized, preventing quench fragmentation of pumices. Elutriation of fines may occur here due to degassing and loss at the magma-water interface. C) Relatively little ash or pumice rises with steam to the surface. D) Collapse of eruption jet into pyroclastic flows. Some of the flow units contain normally graded tops, suggest mixing at the magma/water interface, or collapse of the steam cupola into secondary pyroclastic flows during pulsating (unsteady) eruptions.

which produced more flow units than are present in the footwall ignimbrite.

Our model for the eruption of the hangingwall ignimbrite suggests that there may be another mechanism for production of fines-poor ignimbrite besides elutriation by mixing with water during eruption and/or transport. We speculate that that mechanism may be hydrostatic suppression of magmatic fragmentation.

COMPARISON OF THE BALD MOUNTAIN FOOTWALL AND HANGINGWALL IGNIMBRITES WITH OTHER PUMICE LAPILLISTONE DEPOSITS

The footwall ignimbrite at Bald Mountain is texturally and sedimentologically similar to the footwall at the Rosebury mine in Tasmania (field and thin section observations of CJB). The footwall at the Rosebury mine has been described as a "very thick, mass-flow emplaced pumice breccia", and referred to as "ignimbrite-like" by *Allen and Cas* [1990]. We would describe the Rosebury footwall as a pumice lapillistone, and not a pumice breccia, because it is not made of block-sized clasts; it lacks them almost entirely. Like the footwall ignimbrite at Bald Mountain, it consists of lapilli- to coarse ash-sized pumice lapilli, is crystal poor, and lacks fine ash, and it is hundreds of meters thick, with no internal stratification or grading. By analogy with Bald Mountain, we suggest it is a true ignimbrite, erupted at bathyal water depths; unfortunately, there are no direct controls on the paleo-water depths of eruption of the Rosebury footwall ignimbrite. We speculate, however, that fines-poor ignimbrites may be typical of bathyal marine ignimbrite eruptions associated with massive sulfides.

The textural and sedimentological characteristics of the footwall ignimbrite at Bald Mountain are also similar to the Tamarack tuff of Devonian age in the Northern Sierra terrane, California (field and thin section observations of CJB). Like the footwall ignimbrite at Bald Mountain, the Tamarack tuff lies in a deepwater primitive arc succession, is a plagioclase rhyolite, and is a very thick (60–200 m) massive nongraded deposit of closely packed lapilli-sized blocky-rigid pumices (pumice lapillistone). Also like the footwall ignimbrite, the Tamarack tuff contains horizons of silicic fire fountain deposits, preserved as local spatter accumulations, as well as intrusions of silicic globular peperite and peperite pseudo-pillows. The vent for the Tamarack tuff, or any massive sulfides that may have been associated with it, are apparently not exposed in the cross-sectional, 30 km long view afforded by present–day outcrop (CJB unpublished mapping).

CONCLUSIONS

The footwall and hangingwall ignimbrites to the Ordovician Bald Mountain VHMS deposit in northern Maine were erupted at bathyal water depths (>1.45 km) within nested calderas. The Bald Mountain VHMS formed in the vent for the footwall ignimbrite immediately after it erupted, and is overlain by the hangingwall ignimbrite. Both ignimbrites are poor in fine ash, but lack the stratification and grading typical of pyroclastic flows elutriated of their fines by water. We propose that hydrostatic pressure at a water depth of >1.45 km suppressed the fragmentation processes and violence of the eruptions sufficiently to reduce production of fine ash.

The footwall ignimbrite is a fine-ash-poor pumice lapillistone with hydroclastic fragmentation textures that forms a nonstratified, nonwelded deposit. The gas-rich nature of the magma resulted in a relatively high eruption column, despite great hydrostatic pressure. This prompted sufficient mixing with ambient water to cause quench fragmentation of the pumices, without diluting the ejecta enough to generate low-density flows that produce sorted, stratified deposits. We infer the footwall ignimbrite to be dominantly the product of a steady, large-volume, volatile-rich explosive phreatomagmatic eruption that produced high-concentration pyroclastic flows in very deep water.

The hanging wall ignimbrite is a fine-ash-poor deposit of compacted pumices that occur in massive flow units 1-150 m thick, in places interstratified with graded and massive to laminated tuff turbidites of the same composition. Unsteady or pulsating eruptions resulted in deposition of multiple flow units, some capped by graded megabeds. The slightly more gas-poor nature of the magma (relative to the footwall ignimbrite) resulted in a lower eruption column, leading to minimal interaction with ambient water. This resulted in a lack of quench-fragmented pumices, and possible retention of heat to form welding(?) fabrics.

In summary, our model for deepwater eruption of the hangingwall ignimbrite involves magmatic but no hydroclastic fragmentation. We speculate that the fines-poor character of the hangingwall ignimbrite resulted from suppression of explosivity at great water depths, rather than elutriation by water, and that this may have also contributed to the fine-ash-poor nature of the footwall ignimbrite (in addition to elutriation by water).

Acknowledgments. This paper is the result of LGK's master's thesis under CJB, and we gratefully acknowledge committee members R.V. Fisher and William Wise for their input into the thesis research and writing. CJB would like to further honor the memory of R.V. Fisher by saying that he had the single greatest positive impact on her career of any other scientist. We all love and miss him. We also gratefully acknowledge discussions with K. Kano, J. McPhie, and R.S. Fiske. We also thank K. Kano, and W. Mueller for helpful reviews of an earlier version of this manuscript.

CJB is grateful to Project Chief John Slack for inviting her to join the USGS Bald Mountain project, and for the support the project provided to her and the senior author. We are also grateful to the other members of the Bald Mountain project team for their cooperation and support: K. Schulz, M. Foose, R.A. Ayuso and N.K. Foley. We also wish to thank the USGS summer students who assisted us in the field and the core store, R.M. Gentry and C.S. Gallarano. Financial support was also provided to LGK by a Geological Society of America research grant, and by the department of Geological Sciences at the University of California, Santa Barbara.

We thank Black Hawk Mining Inc. of Toronto for unrestricted access to surface properties, drill cores and exploration reports. Our work also benefited from maps and cross sections prepared by Superior Mining Company, Chevron Resources, and Bolide Resources, Inc.

REFERENCES

Allen, S.R., Reconstruction of a major caldera-forming eruption from pyroclstic flow characteristics: Kos Palteau Tuff, eastrn Aegean Sea. *J. Volcanol. and Geoth. Res.,* 105, 141-162, 2001.

Allen, R.L., and R.A.F. Cas, The Rosebery controversy: distinguishing prospective submarine ignimbrite-like units from true subaerial ignimbrites in the Rosebery-Hercules ZnCuPb massive sulfide district, Tasmania, *Geol. Soc. Aust. Abs ,* 25, 31-32, 1990.

Allen, S.R. and Cas, R.A.F., Rhyolitic fallout and pyroclastic density current deposits from a phreatomagmatic eruption in the eastern Aegean Sae, Greece, *J. Volcanol. and Geoth. Res.,* 105, 219-251, 1998.

Allen, S.R. and J. McPhie, Water-settling and resedimentation of submarine rhyolitic pumice at Yali, eastern Aegean, Greece, *J. Volcanol. Geoth. Res.,* 95, 285-307, 2000.

Busby, C.J., L.G. Kessel, K. Schulz, M. Foose, and J. Slack, Deepwater Silicic Explosive Volcanic Setting of the Ordovician Bald Mountain Massive Sulfide Deposit, Northern Maine, *Econ. Geol. Mono.,*11, in press.

Busby-Spera, C. J., Depositional features of rhyolitic and andesitic volcaniclastic rocks of the Mineral King submarine caldera complex, Sierra Nevada, California, *J. Volcanol. Geoth. Res.,* 27, 43-76, 1986.

Butman, B., Noble, M.; Folger, D. W., Long-term observations of bottom current and bottom sediment movement on the Mid-Atlantic continental shelf, *J. Geophys. Res.*, *84*, 1187-1205, 1979.

Cas, R.A.F., and J. V. Wright, *Volcanic successions: Modern and ancient*, 528 pp., Allen and Unwin Ltd, London, 1987.

Cas, R.A.F., and J. V. Wright, Subaqueous pyroclastic flows and ignimbrites: an assessment, *Bull. Volcanol.*, *53*, 357-380. 1991.

Cashman, K.V., and R. S. Fiske, Fallout of pyroclastic debris from submarine volcanic eruptions, *Science*, *253*, 275-280, 1991.

Cioni, R, P. Marianelli, R. Santacroce, and A.. Sebrana, Plinian and subplinian eruptions: in, H. Sigurdsson, (ed), *Encyclopedia of Volcanoes*: Academic Press, San Diego, p. 477 – 494, 2000.

Doucet, P., W. Mueller, and F. Chartrand, Archean, deep-marine, volcanic eruptive products associated with the Coniagas massive sulfide deposit, Quebec, Canada, *Can. J. of Earth Sci.*, 31, 1569-1584, 1994.

Draper, L., Wave activity at the sea bed around northwestern Europe. *Marine Geol.*, *5*, 133, 1967.

Fisher, RV, Maximum size, median diameter, and sorting of tephra: *J. Geophys. Res.* 69: 341-355, 1964.

Fisher, R.V., G. Heiken, and J.B. Hulen, *Volcanoes: Crucibles of change*. Princeton University Press, New Jersey, 1997.

Fisher, R.V., and H. U. Schmincke, *Pyroclastic rocks*, 472 pp., Springer-Verlag Berlin Heidelberg, New York, Tokyo, 1984.

Fiske, RS, Subaqueous pyroclastic flows in the Ohanapecosh Formation, Washington. *Geol. Soc. Am. Bull.*, *74*, 391-406, 1963.

Fiske, RS and T. Matsuda, Submarine equivalents of ash flows in the Tokiwa formation, Japan. *Am. J. Sci.*, *262*, 76-106, 1964.

Fiske, R.S., J. Naka, K. Iizasa, and M. Yuasa, Submarine silicic caldera at the front of the Izu-Bonin arc, Japan: Voluminous seafloor eruptions of pumice. *Geol. Soc. America Bull.*, *113*, 813-824, 2001.

Foley, N.K., Thermal and chemical evolution of ore fluids and massive sulfide mineralization at Bald Mountain, Maine, *Econ. Geol. Mono.*,11, in press.

Foose, M.F., J.F. Slack, C.J. Busby, K.J. Schulz, and M.V. Scully, Geologic and structural setting of the Bald Mountain volcanogenic massive sulfide deposit, northern Maine - Cu-Zn-Au-Ag mineralization in a synvolcanic sea floor graben, *Econ. Geol. Mono.*, 11, in press.

Francis, P., *Volcanoes: A planetary perspective*, Clarendon Press, Oxford, 1993.

Freundt, A., C.J.N. Wilson, and S.N. Carey, Ignimbrites and block-an-ash flow deposits: in, H. Sigurdsson, (ed), *Encyclopedia of Volcanoes*: Academic Press, San Diego, p. 581- 600, 2000.

Gibson, H.L., R. L. Morton, and G. J. Hudak, Submarine volcanic processes, deposits, and environments favorable for the location of volcanic-associated massive sulfide deposits, in Volcanic-associated massive sulfide deposits: processes and examples in modern and ancient settings, in Reviews in Economic Geology, edited by T. Barrie, and Hannington, M. D., pp. 13-48, *Soc. of Econ. Geol.*, 2000.

Glasby, G.P., K. Iizasa, M. Yusa, and A. Usui, Submarine hydrothermal mineralization on the Izu-Bonin arc, south of Japan: An overview. *Marine Georesources and Geotechnology*, *18*, 141-176, 2000.

Heiken, G., and K. Wohletz, *Volcanic Ash*, 246 pp., University of California Press, Berkeley, Los Angeles, 1992.

Inman, D.L., Measures of describing the size distribution of sediments, *J. Sed. Petrol.*, *22*, 125-145, 1952.

Iizasa, K., Fiske, R.S., Ishizuka, O., Yuasa, M., Hshimoto, J., Naka, J., Horii, Y., Fujiwara, Y., Imai, A., and Koyama, S., A Kuroko-type polymetallic sulfide deposit in a submarine silicic caldera, *Science*, *283*, 975-977, 1999.

Kano, K., A Miocene coarse volcaniclastic mass-flow deposit in the Shimane Peninsula, SW Japan: product of a deep submarine eruption? *Bull. Volcanol.*, *58*, 131-143, 1996.

Kano, K., T. Yamamoto, and K. Ono, Subaqueous eruption and emplacement of the Shinjima Pumice, Shinjima (Moeshima) Island, Kagoshima Bay, SW Japan, *J. Volcanol. Geotherm. Res.*, *71*, 187-206, 1996.

Kokelaar, P., and C. Busby, Subaqueous explosive eruption and welding of pyroclastic deposits. *Science*, *257*, 196-200, 1992.

Lentz, D.R., J. A. Walker, and McCutheon, S. R., Pyroclastic volcanism and volcanogenic massive sulfide deposit genesis: Resolving the depth dilemma, New Nouveau Brunswick. Abstracts, 1998: 23rd Annual review of activities. IC 98-3, 35, 1998.

Lewis, W.L., and D. McConchie, *Analytical Sedimentology*, 197 pp., Chapman and Hall, New York, NY, 1994.

Mandeville, C.W., S. Carey, H. Sigurdsson, and J. King, Paleomagnetic evidence for high-temperature emplacement of the 1883 subaqueous pyroclastic flows from Krakatau Volcano, Indonesia, *J. Volcanol. Geotherm. Res.*, *99* (B5), 9487-9504, 1994.

McPhie, J., M. Doyle, R. Allen, *Volcanic Textures: a Guide to the Interpretation of Textures in Volcanic Rocks*, Tasmania: Pongrats, 1993.

Morissey, M.Z., B., K. Wohletz, and R. Buettner, Phreatomagmatic Fragmentation, in *Encyclopedia of Volcanoes*, edited by H. Sigurdsson, pp. 431-446, Academic Press, San Diego, CA, 2000.

Mueller, W., P. Doucet, and F. Chartrand, Archean, deep-marine, volcanic eruptive products associated with the Coniagas massive sulfide deposit, Quebec, Canada, *Can. J. Earth Sci.*, 31, 1569-1584, 1994.

Pickering K.T., R.N. Hiscott, and F.J. Hein, *Deep Marine Environments*: Unwin, Hyman, London, 416 pp, 1989.

Schneider, J.L., C. Fourquin, and J. C. Paichler, Two examples of subaqueously welded ash-flow tuffs: the Visean of southern Vosges (France) and the Upper Cretaceous of northern Anatolia (Turkey), *J. Volcanol. Geotherm. Res.*, *49*, 365-383, 1992.

Schulz, K.J., and R. A. Ayuso, Volcanic geochemistry and paleotectonic setting of the Bald Mountain massive sulfide deposit, northern Maine, *Econ. Geol. Mono.*,11, in press.

Self, S., and R. S. J. Sparks, Characteristics of widespread pyroclastic deposits formed by the interaction of silicic magma and water, *Bull. Volcanol.*, *41 (3)*, 196-212, 1978.

Sheridan, M.F., Particle-size characteristics of pyroclastic tuffs, *J. Geophys. Res., 76 (23)*, 5627-5634, 1971.

Sheridan, M.F., K. H. Wohletz, and J. Dehn, Discrimination of grain-size subpopulations in pyroclastic deposits, *Geology*, *15*, 367-390, 1987.

Slack, J.F., M. P. Foose, M. J. K. Flohr, M. V. Scully, and H. E. Belkin, Exhalative and sub-seafloor replacement processes in the formation of the Bald Mountain massive sulfide deposit, northern Maine, *Econ. Geol. Mono.*,*11*, in press.

Smith, G.A., and R. D. Smith, Specific gravity characteristics of recent volcaniclastic sediment: Implications for sorting and grain size analysis, *J. Geol.*, *93*, 619-622, 1985.

Sparks, R.S.J., Grain size variations in ignimbrites and implications for the transport of pyroclastic flows, *Sedimentology*, *23*, 147-188, 1976.

Sparks, R.S.J., The dynamics of bubble formation and growth in magmas: a review and analysis. *J. Volcanol. Geotherm. Res.*, *3*, 1-37, 1978.

Sparks, R.S.J., H. Sigurdson, and S. N. Carey, The entrance of pyroclastic flows into the sea, II. Theoretical consideration on subaqueous emplacement and welding, *J. Volcanol. Geotherm. Res.*, *7*, 97-105, 1980,

Stix, J, Subaqeous, intermediate to silicic-composition explosive volcanism: a review. *Earth Sci Rev.*, *31*, 21-35, 1991.

Surdam, R.C., and J. R. Boles, Digenesis of volcanic sandstones, *Soc. of Econ. Paleontol. Min., Spec. Pub.*, *26*, 227-242, 1979.

Walker, G.P.L., Grain-size characteristics of pyroclastic deposits, *J. Geol.*, *79*, 696-714, 1971.

Walker, G.P.L., Ignimbrite types and Ignimbrite problems, *J. Volcanol. Geotherm. Res.*, *17*, 65-88, 1983.

Walker, G.P.L., J.V. Wright, B.J. Clough, and B. Booth, Pyroclastic geology of the rhyolitic volcano of La Primavera, Mexico. *Geology Rundsch*, *70*, 1100-1118, 1981.

White, J.D.L., Subaqueous eruption-fed density currents and their deposits: *Precambrian Res.*, *101*, 87-109, 2000.

Whitham, A.G., The behavior of subaerially produced pyroclastic flows in a subaqueous environment: Evidence from the Roseau eruption, Dominica, West Indies, *Marine Geology*, *86*, 27-40, 1989.

Wohletz, K.H. Explosive magma-water interactions: thermodynamics, explosion mechanisms, and field studies, *Bull. Volcanol.*, *8*, 245-264, 1986.

WoldeGabriel, G., D. E. Broxton, and F. M. Byers Jr., Mineralogy and temporal relations of coexisting authigenic minerals in altered silicic tuffs and their utility as potential low-temperature dateable minerals, *J. Volcanol. Geotherm. Res.*,, *71*, 155-165, 1996.

Wright, I.C., and J.A. Gamble, Southern Kermadec submarine caldera arc volcanoes (SW Pacific): caldera formation by effusive and explosive eruption, *Marine Geology*, *161*, 207-227, 1999.

Wright, J.V., and E. Mutti, The Dali Ash, Island of Rhodes, Greece: a problem in interpreting submarine volcanogenic sediments, *Bull. Volcanol.*, *44*, 153-166, 1981.

Yuasa, M., and Kano, K., Submarine silicic calderas on the northern Shichito-Iwojima Ridge, Izu-Ogasawara (Bonin) Arc, western Pacific, this volume.

Lowell Kessel and Cathy Busby, Department of Geological Sciences, University of California, Santa Barbara,CA 93106.